TABLE A-4   Values of *t* for Given Probability Levels

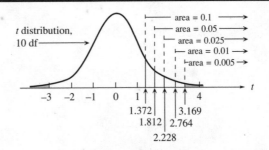

Degrees of Freedom ν	α, Area in the Right Tail Under the Curve				
	0.1	**0.05**	**0.025**	**0.01**	**0.005**
	Critical Values of *t*				
	$t_{\nu,0.1}$	$t_{\nu,0.05}$	$t_{\nu,0.025}$	$t_{\nu,0.01}$	$t_{\nu,0.005}$
1	3.078	6.314	12.706	31.821	63.657
2	1.886	2.920	4.303	6.965	9.925
3	1.638	2.353	3.182	4.541	5.841
4	1.533	2.132	2.776	3.747	4.604
5	1.476	2.015	2.571	3.365	4.032
6	1.440	1.943	2.447	3.143	3.707
7	1.415	1.895	2.365	2.998	3.499
8	1.397	1.860	2.306	2.896	3.355
9	1.383	1.833	2.262	2.821	3.250
10	1.372	1.812	2.228	2.764	3.169
11	1.363	1.796	2.201	2.718	3.106
12	1.356	1.782	2.179	2.681	3.055
13	1.350	1.771	2.160	2.650	3.012
14	1.345	1.761	2.145	2.624	2.977
15	1.341	1.753	2.131	2.602	2.947
16	1.337	1.746	2.120	2.583	2.921
17	1.333	1.740	2.110	2.567	2.898
18	1.330	1.734	2.101	2.552	2.878
19	1.328	1.729	2.093	2.539	2.861
20	1.325	1.725	2.086	2.528	2.845
21	1.323	1.721	2.080	2.518	2.831
22	1.321	1.717	2.074	2.508	2.819
23	1.319	1.714	2.069	2.500	2.807
24	1.318	1.711	2.064	2.492	2.797
25	1.316	1.708	2.060	2.485	2.787
26	1.315	1.706	2.056	2.479	2.779
27	1.314	1.703	2.052	2.473	2.771
28	1.313	1.701	2.048	2.467	2.763
29	1.311	1.699	2.045	2.462	2.756
30	1.310	1.697	2.042	2.457	2.750
$\vdots$	$\vdots$	$\vdots$	$\vdots$	$\vdots$	$\vdots$
∞	1.282	1.645	1.960	2.326	2.576

Table A-4 is taken from Table III of Fisher and Yates: *Statistical Tables for Biological, Agricultural and Medical Research*, published by Longman Group Ltd., London (previously published by Oliver & Boyd, Edinburgh), and by permission of the authors and publishers.

STATISTICS

in a World
of Applications

STATISTICS

in a World
of Applications

· · · · · · ·

Fourth Edition

Ramakant Khazanie
Humboldt State University

▟▙ HarperCollins*CollegePublishers*

To the memory of my grandparents, my father, and my brother, Suresh

Sponsoring Editor: Kevin Connors
Developmental Editor: Kathy Richmond
Project Editor: Cathy Wacaser
Design Administrator: Jess Schaal
Cover and Text Design: Kay Fulton
Cover Photo: Gay Bumgarner/Tony Stone Images
Photo Researcher: Diane Peterson-Blanas
Production Administrator: Randee Wire
Project Coordination: Elm Street Publishing Services, Inc.
Compositor: Interactive Composition Corporation
Printer and Binder: R. R. Donnelley & Sons Company
Cover Printer: Phoenix Color Corporation

Statistics in a World of Applications, Fourth Edition

Copyright © 1996 by HarperCollins College Publishers

HarperCollins® and ⬛® are registered trademarks of HarperCollins Publishers Inc.

Library of Congress Cataloging-in-Publication Data
Khazanie, Ramakant.
    Statistics in a world of applications / Ramakant Khazanie.—4th
ed.
      p.   cm.
    Rev. ed. of: Elementary statistics in a world of applications. 3rd
ed. c1990.
    Includes index.
    ISBN 0-673-46768-6
    1. Statistics. I. Khazanie, Ramakant.   Elementary statistics in
a world of applications.   II. Title.
QA276.12.K5   1996
519.5—dc20   95-23434

95 96 97 98 9 8 7 6 5 4 3 2 1

Brief Contents

Contents

Preface

Statistical methodology has become an important component of scientific reasoning. It is now integral to such fields as engineering, education, agriculture, business, biology, medicine, fishery management, forestry, geology, communication, psychology, and ecology. Therefore, more and more academic disciplines require a course in introductory statistics. *Statistics in a World of Applications,* fourth edition, acquaints students with the fundamentals of statistics and provides, by touching on diverse areas of interest, a basis for studying specialized methods within their fields.

This textbook covers topics in elementary statistics that, in recent years, have been offered at the freshman–sophomore level in various colleges and universities. It appeals primarily to readers' intuition and employs mathematics at the most elementary level to stimulate interest among students. A background in high-school algebra will suffice.

In a first course such as this, students can hardly be expected to master the vast subject of statistics. The goal of this book, therefore, is to introduce students to some aspects of statistical methodology. Students who acquire a sound understanding of the topics developed here can realize the importance of statistical reasoning in work and in life, discern the basic statistical assumptions underlying a given situation, and choose an appropriate test if the problem falls within the framework of their introduction to the subject. But more than anything, I hope that students will be able to interpret the results of statistical investigation and decipher the daily barrage of statistical information.

APPROACH

The overriding concern in this book is to present material clearly so that students can read it on their own with minimal difficulty. Each concept is introduced through an example, often drawn from a real-life situation, followed by numerous detailed examples. Some of these examples are based on hypothetical data. Exercises at the end of each section give students ample opportunities to test their skills and to clarify understanding of the subject matter. These exercises serve an additional purpose—to show how pervasively statistical applications are used in everyday life.

Chapters 3, 4, 5, and 6 involve probability concepts which students have traditionally found difficult to master. Special care has been taken to keep the presentation accessible and lucid.

Chapter 7 (Estimation—Single Population), Chapter 8 (Estimation—Two Populations), and Chapter 9 (Tests of Hypotheses) are by far the most important chapters. A concerted effort is made to convey all the basic concepts by appealing to students' intuition.

MINITAB* Labs, occurring at the end of each chapter, include line-by-line statements for input. The labs are useful for learning how to perform statistical computations. But, more importantly, they allow students to explore chapter concepts further using technology.

The following features of the book are meant to make it reader friendly:

1. Definitions are highlighted.
2. Wherever appropriate, a step-by-step approach is provided for performing statistical computations.
3. Important formulas are boxed.
4. Key words and key formulas are listed at the end of each chapter.
5. The properties of important statistical distributions are enumerated.
6. Table summaries are provided at the end of Chapters 7, 8, and 9 to give a programmed approach to setting confidence intervals and carrying out tests of hypotheses.

CHANGES IN THE FOURTH EDITION

The fourth edition preserves the basic format of the previous editions and builds on their style, clarity, and readability. The following significant changes and additions characterize this edition.

1. In Section 1-3, *Hoodwinking with Statistics* shows the student how statistics can be manipulated to support different points of view.
2. The material in Chapters 4 and 5 of the third edition has been reorganized by placing discrete variables in Chapter 4 and continuous variables in Chapter 5.
3. In response to suggestions of some current users of the text, Section 4-3 on *hypergeometric distribution* is introduced in Chapter 4. Due to time constraints, it may not be possible to cover this section; hence, it is listed as optional.

---

* MINITAB is a registered trademark of Minitab, Inc., 3081 Enterprise Drive, State College, Pennsylvania 16801; telephone: 814/238-3280; telex: 881612; fax: 814/238-4383. The author wishes to thank Minitab, Inc. for providing the software.

4. In Section 5-2, the student is introduced to the *normal probability plot* for determining whether a given data set warrants the assumption of a normal distribution.

5. Section 6-3 discusses the importance of the *design of experiments,* with some preliminary material in the first chapter.

6. The material on *Criteria for Good Estimators* in Section 7-1 of the third edition is omitted now because the topic is rarely covered in an introductory course.

7. *Newspaper clippings* are placed throughout the text to add real-world relevance to the material.

8. There are *three large, real-data sets.* These data sets, found as appendices, record real data from (1) the sales of homes in Humboldt County in California, (2) SAT scores, and (3) a diabetes study at a medical school.

9. *Concept and discussion questions* are provided at the end of each chapter to promote better understanding of statistical concepts and to stimulate critical thinking.

10. *Graphics calculator exercises* have been added to the ends of many chapters and provide opportunities for a student to delve deeper into relevant material, show graphical results, or perform repetitive tasks so that he or she can focus on concepts.

COURSE ORGANIZATION

The contents in this book can be divided into four parts. The first part, consisting of Chapters 1 and 2, gives an account of descriptive statistics and is intended to familiarize the student with field data summarization. The second part, Chapters 3 to 6, covers elementary probability notions. In this part the goal is to build the tools essential to understanding the concepts of statistical inference. The third part, Chapters 7 to 9, deals with the inferential aspects of statistics and is by far the most important. In the fourth part, consisting of the remaining chapters, the order of importance is as follows: Chapters 11, 10, 13, 12. This will vary, however, from instructor to instructor.

The topics treated in this book should prove adequate for an introductory one-semester course meeting three hours a week or a one-quarter course meeting four hours a week. In my teaching of a one-semester course, I lecture three times a week, with two hours more a week set aside for MINITAB Labs. I invariably cover the first nine chapters. There is enough optional material from which to fashion a two-quarter course meeting three hours a week. The chapters on linear regression and linear correlation, goodness of fit, analysis of variance, and nonparametric methods are included to provide the flexibility needed in structuring such an offering.

THE SUPPLEMENT PACKAGE TO ACCOMPANY THIS TEXT

For the Instructor

The *Instructor's Complete Solution Manual* contains complete, worked-out solutions to all of the exercises in the text.

The *HarperCollins Test Generator/Editor with QuizMaster* is available in IBM (both DOS and Windows applications) and Macintosh versions, and is fully networkable. The Test Generator enables instructors to select questions by objective, section, or chapter, or to use a ready-made test for each chapter. The Editor enables instructors to edit any preexisting data or to create their own questions easily. The software is algorithm-driven, allowing the instructor to regenerate constants while maintaining problem type, providing a very large number of test or quiz items in multiple-choice and/or open-response formats for one or more test forms. The system features printed graphics and accurate mathematics symbols. **QuizMaster** enables instructors to create tests and quizzes using the Test Generator/Editor and save them to disk so students can take the test or quiz on a stand-alone computer or network. QuizMaster then grades the test or quiz and allows the instructor to create reports on individual students or entire classes.

StatExplorer offers students software tools to explore the fundamentals of statistics. This program allows a wide range of statistical representations including graphs, center and spreads, and transformations. StatExplorer can be used by instructors for presentation purposes, too.

For use with the text, a *Large Data-Set Disk* is available so that the real data in appendices can be accessed easily for classroom use. Please contact your local HarperCollins representative for more information.

For the Student

The *Student's Solution Manual* includes complete, worked-out solutions to every odd-numbered exercise in the text, with the exception of Concept and Discussion Questions where more than one answer may be appropriate. To order, use ISBN 0-673-46770-8.

ACKNOWLEDGMENTS

In the course of developing this edition, I have used many valuable suggestions made by my colleagues, the reviewers, and the users of the previous three editions. I express my thanks and appreciation to the following reviewers who have contributed to this fourth edition:

Glenn Aguiar, *City College of San Francisco*

Joyce E. Anderson, *Salem State College*

Richard Anderson-Sprecher, *University of Wyoming*

Patricia M. Buchanan, *Pennsylvania State University*

James A. Condor, *Manatee Community College*

Elwyn H. Davis, *Pittsburgh State University*

Martha L. Di Fazio, *Salem State College*

Gary Donica, *Florence Darlington Technical College*

Paul J. Fairbanks, *Bridgewater State College*

James E. Gehrmann, *California State University–Sacramento*

Pat Gilbert, *Diablo Valley College*

Bernard Harris, *University of Wisconsin–Madison*

Subramanyam Kasala, *The University of North Carolina at Wilmington*

John Khoury, *Brevard Community College*

Mary Ann Koehler, *Cuyahoga Community College*

Ronald LaBorde, *Marian College of Fond du Lac*

Edward Leinse, *Oliver Harvey College*

David Meredith, *San Francisco State University*

Robert F. Mooney, *Salem State College*

Ronnie L. Morgan, *West Chester University*

Joseph M. Moser, *San Diego State University*

Marnie Pearson, *Foothill College*

Calvin Schmall, *Solano Community College*

George Van Zwalenberg, *Calvin College*

Many thanks go to Dr. Martha Di Fazio of Salem State College for the writing of the graphics calculator exercises found in the text and the graphics calculator appendices.

I express my thanks to Professor Amode Sen for many helpful discussions over a wide range of topics. Also, my thanks are due to Professor Virgil Anderson of Purdue University who inculcated in me a taste and liking for statistical applications. I extend my appreciation to my colleague, Professor Yoon Kim, for sharing his computer expertise; to another colleague, Professor Maureen Reiner, for her useful suggestions; to Professor Arthur Dull of Diablo Valley College for his keen interest in the book; and to Ronald LaBorde of Marian College of Fond du Lac for his contributions to this edtion. Finally, I thank my family for enduring patiently during the course of this project. My daughters, Amita and Uma, deserve a special mention for their enthusiastic help during different phases of the work. I cannot thank them enough.

R. K.

Organization of Data

USA SNAPSHOTS®

A look at statistics that shape your finances

What type of reward do employees want?

One-store gift certificate 2%

Other 10%

Cash or all-purpose gift certificates 58%

Travel award 8%

Extra vacation 22%

▲ Source: Graphic "What Type of Reward Do Employees Want?" by Suzy Parker, *USA Today*. Data from American Express/David Michaelson & Associates, Survey of 700 Employed Adults. Reprinted by permission of *USA Today*, December 1, 1993, Section B, p. 1.

1

INTRODUCTION

What is statistics? Allusions to statistics are abundant, and a penchant to quote statistics is almost universal. Strangely enough, however, skepticism about statistics is widespread. Much of this reaction stems from misleading conclusions, which often result either from poor planning of an experiment or because some pertinent information is withheld. In this context the following observation is appropriate.

> The secret language of statistics, so appealing in a fact-minded culture, is employed to sensationalize, inflate, confuse and oversimplify. Statistical methods and statistical terms are necessary in reporting the mass data of social and economic trends, business conditions, "opinion" polls, and census. But without writers who use the words with honesty and understanding and readers who know what they mean, the result can only be semantic nonsense.*

Statistics is a science that deals with the methods of collecting, organizing, and summarizing data in such a way that valid conclusions can be drawn from them. It provides an intelligent and objective approach to acquiring

**National Health Expenditures:
1960 to 1981**

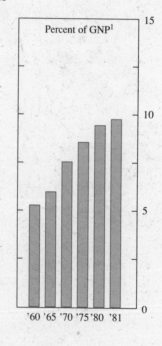

Billions of dollars

- –·– Health expenditures, total
- ···· Private expenditures
- —— Public expenditures

Percent of GNP[1]

**Land Owned
by the Federal Government: 1979**

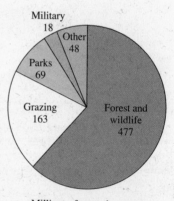

Military 18
Other 48
Parks 69
Grazing 163
Forest and wildlife 477

Millions of acres, by usage

[1]GNP: Gross National Product

Source: Charts prepared by U.S. Bureau of the Census.

Figure 1-1 Statistical Information Presented through Charts

* From Darrell Huff, *How to Lie With Statistics* (New York: W. W. Norton & Co.), p. 8.

TABLE 1-1 Statistical Information Presented as a Table
Self-Employed Persons in the U.S., by Industry
(in thousands)

	1989	1990	1991	1992
Self-employed persons[1]	10,041	10,132	10,368	10,453
Agriculture, forestry, and fisheries	1,434	1,437	1,479	1,415
Farms	1,085	1,058	1,082	1,061
Agricultural services, forestry, and fisheries	349	379	397	354
Mining	25	24	23	23
Construction	1,433	1,473	1,457	1,470
Manufacturing	409	432	421	398
Durable goods	246	258	253	247
Nondurable goods	163	174	168	151
Transportation and public utilities	323	304	313	335
Wholesale trade	349	334	350	349
Retail trade	1,548	1,539	1,544	1,439
Finance, insurance, and real estate	621	634	618	630
Services	3,899	3,955	4,163	4,394

[1] Consists of active proprietors or partners who devote a majority of working hours to their unincorporated businesses. Source: *Survey of Current Business*, August 1993.

knowledge from the data about the world around us. Statistical investigations and analyses of data fall into two broad categories—*descriptive statistics* and *inductive statistics.*

Descriptive statistics deals with processing data without attempting to draw any inferences from them. It refers to the presentation of data in the form of tables, as in Table 1-1, and graphs, as in Figure 1-1, and to the description of some characteristics of the data, such as averages, which we will consider in Chapter 2. To most people it is this notion that is conveyed by the word *statistics:* a compilation of a huge mass of data and charts dealing with taxes, accidents, epidemics, farm outputs, incomes, university enrollments, brands of television in use, sports, and so on.

The following report, released by the Associated Press on April 3, 1993, provides an example of descriptive statistics. The report gives findings but does not presume to forecast. In other words, it does not predict what is in store for years to come.

Disasters Costly in '92

ZURICH, Switzerland (AP)— Natural and man-made disasters caused a record $27.1 billion in damage to insured property and killed more than 24,000 people worldwide last year, a report said Friday

The Swiss Reinsurance Company said damage costs were 87 percent higher than in 1991, mainly because

Source: "Disasters Costly in '92" from the Associated Press, April 3, 1993. Reprinted by permission of the Associated Press.

of hurricane devastation in the United States.

It said Hurricane Andrew caused $15.5 billion in insured damage in August in south Florida and Hurricane Iniki, which hit Hawaii in September, cost insurance companies $1.6 billion.

The report said 13,284 people died last year in natural disasters such as earthquakes, floods and storms. December's earthquake in Indonesia claimed the highest toll—a reported 2,484 dead.

The death toll in 1991 included 140,000 people killed in cyclones in Bangladesh.

The report said 11,140 people were killed in "major" man-made accidents such as fires, explosions and traffic accidents. The tally also included 1,769 people killed in 30 plane accidents.

The figures include accidents or disasters that killed at least 20 people and caused more than $11 million in damage. The report does not include casualties or damage from wars or famines.

Inductive statistics, also called **inferential statistics,** is a scientific discipline concerned with developing and using mathematical tools to make forecasts and inferences. In this role—devising procedures for carrying out analyses of data—statistics caters to the needs of research scientists in such diverse disciplines as industrial engineering, biological sciences, social sciences, marketing research, and economics, to mention just a few. Basic to the development and understanding of inductive statistics are the concepts of probability theory.

An illustration of inferential statistics would be a situation where an educated guess is made, as in the following article which appeared on January 27, 1995. The article claims that *most* Americans favor continued support for public television when actually only 1005 individuals were interviewed.

Poll Backs PBS Aid

WASHINGTON (AP)—

As part of a lobbying campaign to protect millions of dollars in federal funding, public TV stations released a poll today that found most Americans favor continued taxpayer support for public television.

The poll said 49 percent of those interviewed supported a funding increase, 35 percent would like to maintain funding at current levels, and 13 percent favored reduced funding or no funds at all. Three percent had no opinion.

The telephone poll of 1,005 people was conducted Jan. 5–8 by Princeton, N.J.-based Opinion Research Corp. for the Public Broadcasting Service. The survey, which did not cover public radio, had a margin of error of plus or minus 3 percentage points.

Source: "Poll Backs PBS Aid" from the Associated Press, January 27, 1995. Reprinted by permission of the Associated Press.

This chapter and the next concentrate specifically on certain aspects of descriptive statistics. Subsequent chapters are concerned mainly with inductive statistics, with concepts developed in Chapters 3 through 6 and then, in the remainder of the book, applied to decision-making procedures.

1-1 POPULATION, SAMPLE, AND VARIABLES

POPULATION AND SAMPLE

One of the goals of a statistical investigation is to explore the characteristics of a large group of items on the basis of a few. Sometimes, for physical, economical, or some other reason, it is almost impossible to examine each item in a group under study. In such a situation often the recourse is to examine a small subcollection of items from this group and to use whatever information it provides to infer about the larger group. A valid approach to acquiring information about the group begins with the collection of pertinent data.

For example, if in a primary election we are interested in finding the disposition of the voters toward a certain candidate, it would not be feasible in terms of time and money to interview every potential voter. Instead we might interview only a few of them, the selection being carried out in such a way as to provide a good cross section. The data collected might be the response of the interviewees, *yes* or *no,* to the question "Are you in favour of the candidate?" Since we are interviewing only a few and drawing conclusions about the entire set of voters, our inferences are subject to an element of uncertainty, and so the statements should be couched in the language of probability.

A set of data on all the conceivable members of a group that form the target of our study constitute a **population,** or a **universe.**

The word *population* as used today in statistics is in a much wider sense than its usual meaning and has evolved from its early usage in statistics related to census populations. For us, the term *population* will mean a set of data consisting of all the hypothetically possible observations pertinent to the inquiry. For example, if we were interested in investigating the verbal test performance in 1994, the SAT verbal test scores of all the students who took the exam in 1994 would constitute a population of verbal test scores. As another example, suppose we are interested in the durability of the tires of a certain brand manufactured by a company. Then the mileage data that one could possibly collect for all the tires of the brand produced in the past and to be produced in the future would make the relevant population.

An item or an object on which a measurement or an observation can be recorded is called a **sampling unit.**

Thus, a population is a collection of data based upon the observation of all conceivable sampling units. Loosely, the term population is often used in referring to all of the conceivable sampling units themselves. A sampling unit could be fish in the lake, a hospital patient, an ear of corn in the field, a plot of land, a television set, or a tire, depending on the investigation. The totality of sampling units is determined by the scope of the statistical investigation. A clear understanding of this is important because it provides a framework for the data-collecting process. It is also important in determining the group to which the conclusions of a statistical analysis apply.

For instance, suppose the highway department wants to make a study of accidents taking place on a certain stretch of a highway. Before data collection can start, answer an important question: "What is the scope of the investigation?" Is the department interested in accidents occurring during a certain holiday period, such as Labor Day weekends? In that case all the accidents on all such holiday periods will be the totality of sampling units, and the data collection will be carried out on such units. If the interest is in accidents occurring during weekdays, then the population will be the set of data on all the accidents on all weekdays. It may be that the interest is in accidents during peak hours on weekdays, necessitating a different framework for data collection.

EXAMPLE 1 Suppose an ornithologist is interested in investigating migration patterns of birds in the Northern Hemisphere. Then all the birds in the Northern Hemisphere will represent the sampling units of interest. This choice imposes restriction, for it does not include birds that are native to Australia and do not migrate to the Northern Hemisphere.   ●

EXAMPLE 2 Every ten years the U.S. Bureau of the Census conducts a survey of the entire population of the United States accounting for every person regarding sex, age, and other characteristics. The last such survey was carried out in 1990. Only the federal government, with its vast resources, can undertake a task of this magnitude. In this case the data collected from the entire population of the United States is the population in the statistical sense.   ●

> A set of data values actually collected on some of the sampling units from the population is called a **sample.**

The terms *population* and *sample* are relative. A collection that constitutes a population in one context may well be a sample in another context. For instance, if we wish to learn how people in Peoria, Illinois, feel about a certain national issue, then the data on all the residents of Peoria would constitute the relevant population. Assume, however, that Peoria represents a cross section of the U.S. population. If we use the response from these residents to understand the feelings about the issue among all the U.S. residents, then the data on the residents of Peoria would represent a sample.

Statistical theory provides a logical process for using samples to arrive at conclusions regarding the population from which the sample is selected. That is, based on this theory, we can draw inferences from a subcollection to the entire collection, taking into account the uncertainty that exists because we have no way of knowing exactly the outcomes for the entire group.

EXAMPLE 3 A fisheries researcher is interested in the study of Dungeness crab (Cancer Magister Dane) along the Pacific Coast of North America. It would be inconceivable and impossible to investigate every crab individually. The only way to make any kind of educated guess about their behavior would be by examining a small subcollection, that is, by collecting a sample. ●

EXAMPLE 4 Suppose a machine has produced 10,000 electric bulbs and we are interested in getting some idea about the life of the bulbs, that is, how long a bulb will last in general. It would not be practical to test all the bulbs because the bulbs that are tested will never reach the market. So, we might choose 50 of these bulbs to test. Our interest is in learning about the 10,000 bulbs and we investigate 50. The length of time that each of the 10,000 bulbs could last is the population, and the data on the life of the 50 bulbs is a sample. ●

In Examples 3 and 4 above actual populations do exist. When they are surveyed by selecting samples from respective populations, we learn about their characteristics. Consider, on the other hand, the following example where no actual population exists.

EXAMPLE 5 A dietitian wants to study the effect of a diet on reducing weight. A group of individuals will be put on a diet regimen for a certain length of time and their weight loss will be measured at the end of the period. In this case the values for the entire population do not actually exist. However, there is a conceptual collection of population values, namely, the values for all the individuals who will be, or could be, given the diet. The goal of the study is, presumably, to assess the effect of the diet on this potential collection of individuals. A properly planned method for collecting data will shed light on this inquiry. The data on the group of individuals who are actually included in the study constitute a sample. ●

A survey of an existing population carried out by adopting a sampling procedure will tell us only about the objects that are observed and, using appropriate statistical methods, can throw light on the characteristics of the underlying population. This is often referred to as **observational study.** The observational approach, however, is not suitable if we wish to establish causality. An example would be hospital records that give data on consumption of alcohol during pregnancy and the health of the child born. Studies indicate that a high proportion of abnormal children are born to mothers who consume alcohol during pregnancy. But this fact alone provides no convincing reason to believe that alcohol is the cause of abnormalities. It is quite possible that there are other competing factors such as smoking and depression, which are among the factors closely associated with drinking alcohol, that are responsible for the consequences.

For establishing causality one has to plan an experiment in such a way that the factor whose effect is to be assessed is manipulated appropriately by devising a suitable design for collecting the data. As an illustration, suppose you wish to find out if chewing gum affects the health of dental gums. You might consider a simple experiment with two similar groups of individuals with healthy gums. Treat both groups to the same kind of food during the experimental period, but have the members of only one of the groups chew gum. If the study reveals a significant difference in the condition of gum tissue between the two groups, then there would be a reasonable basis to attribute it to the practice of chewing gum. This kind of study is called an **experimental study.** No populations actually exist. There are conceptual populations formed by the experimenter and composed of those units which could be their members. The data of the experiment serve as samples from these conceptual populations under study. The method for collecting data by devising a planned experiment falls in the realm of **designs of experiments,** which will be studied later in Chapter 6.

Reasons for sampling   There are a number of reasons for resorting to sampling, including those listed below.

1 Since a sample contains considerably fewer units than the population, the cost of collecting data and analyzing them is much lower for a sample.

2 A sample can provide relevant information about the population in relatively less time than is possible by taking a complete census. This is a very important consideration when the information is needed quickly. This would certainly be true, for example, when a political candidate wishes to know the disposition of the voters just prior to an election.

3 The nature of the investigation may be such that the process of acquiring information from each unit results in the unit's destruction, as in the preceding Example 4.

4 The entire population may not be accessible at the time it is needed for testing. This happens typically in manufacturing processes. For example, if a tire manufacturer wants to know about the quality of tires likely to be produced by a new plant, the only recourse is to examine a sample of tires in the initial production lot. Also, consider the preceding Example 3.

5 The population may be infinite in size.

6 There are situations, such as testing a new drug, where great risks are involved in considering the entire population.

7 Since we investigate only a fraction of the total population in a sample, the administrative problems involved in the collection of data, in processing, and in the supervision of field work are more manageable. For a given cost, a sample can elicit more detailed information through the use of qualified personnel who receive intensive training.

Professional pollsters such as Gallup, Louis Harris, and Roper who survey public opinion, various advertising agencies interested in consumer preferences, and research scientists invariably use samples to make projections regarding population characteristics. The pollsters often base their results on interviews usually conducted by telephone, of about 1000 individuals. To give a nationwide picture of monthly unemployment, the U.S. Bureau of Labor Statistics regularly puts out reports based on interviews of 60,000 households. The unemployment information that is published monthly refers to the statistical population consisting of all U.S. residents who are civilians aged 16 years and over.

VARIABLES

Statistical data or information that we gather is obtained by interviewing people, by inspecting items, and in many other ways. The characteristic that is being studied is called a *variable*.

> A **variable** is a characteristic observed on sample units and that can vary from unit to unit.

Heights of people, grades on a high school test, the time it takes for the bus to arrive at the bus stop, profession of an individual, and hair color are examples of variables. There are two kinds of variables—*qualitative* and *quantitative*.

> A **qualitative variable** can be identified simply by noting its presence. It describes observations as belonging to one of a set of categories. Hence a qualitative variable is also called a **categorical variable.**

The color of an object is an example of a qualitative variable. In the same way, the outcome of tossing a coin is an example of a qualitative variable. The outcome is either heads or tails and has no numerical value. Although a qualitative variable has no numerical value, it is possible to assign numerical values to a qualitative variable by giving values to each quality. For example, we could assign the value of 1 to heads and the value of 0 to tails and in this way quantify the qualitative information.

> A **quantitative variable** consists of values measured on a numerical scale. Data collected for a quantitative variable are often referred to as *metric data*.

The height of an individual when expressed in feet or inches is a quantitative variable. Diameters of bolts produced by a machine (in inches), waiting time at a bus stop (in minutes), the price of a stock (in dollars), the annual income of a family (in dollars), the number of bacteria in a culture, and the volume of sales in a day (in dollars) are other examples of quantitative variables.

A familiar example of a variable given qualitatively as well as quantitatively is a score on an examination. If the score is given numerically as 80 points, 65 points, and so on, then the variable is quantitative; but if the score is given in letter grades A, B, C, D, F, then the same variable becomes qualitative.

When, for instance, the grade points 4, 3, 2, 1, 0 are associated with the grades *A*, *B*, *C*, *D*, *F*, respectively, we use what is called the *nominal scale*.

Here, a specific number is associated with each category of the variable. In this context, however, it should not be construed that grade *A*, for example, is four times as much as grade *D*.

As another example, the height of an individual measured in feet is a quantitative variable. The same variable recorded as tall, medium, or short is a qualitative variable. If we were to assign numbers 3, 2, 1 for *tall, medium, short,* respectively, we would be using the nominal scale.

We can classify quantitative variables further as continuous or discrete.

If a quantitative variable can assume any numerical value over an interval or intervals, it is called a **continuous variable.**

The height of an adult person is a continuous variable because it can be *any* value in the interval, say, 3 feet to 8 feet.

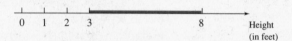

Weight and time are also examples of continuous variables.

A **discrete variable** is one whose possible values consist of breaks between successive values. That is, the possible values of a discrete variable can be counted. It can have a finite number of possible values, or as many values as there are integers.

The number of patients treated in an emergency ward of a hospital during a day is an example of a discrete variable. The only values that the variable can take are integral values such as 0, 1, 2, and so forth. For instance, we would not report that there were 8.5 patients treated on a certain day.

Other examples of discrete variables are the number of bacteria in a culture, the number of defective bulbs, the number of telphone calls received during a one-hour period, the number of unfilled orders at the end of a day, the size of a shirt collar, the income tax paid by a wage earner, and the postage due on a first-class letter.

The classification of variables can be displayed schematically as follows.

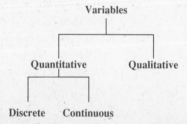

EXAMPLE 6

The following is an excerpt from an article that appeared in *Time* magazine in the economy and business section.

> The "little guy" is back in the stock market in a big way and, as it turns out, is not so likely to be a guy. Women have rushed to buy stocks during the past two years, more than men. They constitute 57 percent of all new shareholders.

The information in the statement "Women have rushed to buy stocks during the past two years, more than men" is qualitative in nature where the variable of interest is the sex of the shareholder—male or female.

Presumably the study is conducted to describe the new shareholders during the period mentioned. Naturally, then, "all new shareholders during the two-year period" would constitute the population. ●

EXAMPLE 7

The following table based on data from the U.S. International Trade Commission, *Synthetic Organic Chemicals,* annual, is reproduced from the Statistical Abstract of the United States 1982–83.

There are several variables involved in describing the data in Table 1-2.

TABLE 1-2 Synthetic Organic Pesticides—Production and Sales

Item	Unit	1960	1965	1970	1975	1976	1977	1978	1979	1980
Production, total	Mil. lb	648	877	1,034	1,603	1,364	1,388	1,416	1,429	1,468
Herbicides	Mil. lb	102	263	404	788	656	674	664	657	806
Insecticides	Mil. lb	366	490	490	660	566	570	605	617	506
Fungicides	Mil. lb	179	124	140	155	142	143	147	155	156
Production, value[1]	Mil. dol	307	577	1,058	2,900	2,880	3,116	3,342	3,685	4,269
Sales, total	Mil. lb	570	764	881	1,317	1,193	1,263	1,300	1,369	1,406
Sales value[1]	Mil. dol	262	497	870	2,359	2,410	2,808	3,041	3,631	4,078

[1] Manufacturer's unit value multiplied by production.

Source: U.S. Dept. of Agriculture, Agriculture Stabilization and Conservation Service, *The Pesticide Review,* 1979. Based on data from the U.S. International Trade Commission, *Synthetic Organic Chemicals,* annual.

One of the variables is the type of pesticide. It is a qualitative variable with three qualities, namely, herbicides, insecticides, and fungicides. Another variable is the number of pounds of each type of pesticide, which is a quantitative variable. In theory we can record weight to any degree of accuracy, so this is a continuous variable. In practice, however, the weights must have been recorded to the nearest pound. Two other variables are production value (in dollars) and sales value (in dollars). They both are quantitative variables and are discrete. ●

PARAMETERS, STATISTICS

A numerical value computed from a given data set is called a *quantitative measure.*

Any quantitative measure that describes a characteristic of the population is called a **parameter.**

In general a population will have several parameters. These are the constants that are peculiar to that population. If all the population data were available, in principle we could compute all its parameters. Usually such information is not available, and, consequently, the population parameters are by and large unknown to the investigator. It is often the goal of a statistical inquiry to obtain information about them.

A quantitative measure that describes a characteristic of the sample is called a **statistic.** It is a numerical measure computed from the sample data.

The value of a statistic will, more than likely, differ from sample to sample because each sample will have a different set of values. In practice, the investigator will have the sample data available after conducting a survey or carrying out an experiment; he or she will use the statistics computed from this sample to gain information about the corresponding population parameters.

For example, the proportion of Democrats among all the voters in the United States is a parameter. If we interview 1000 people among all the voters in the United States, then the proportion of Democrats in this sample is a statistic.

As another example, suppose we are interested in the heights of the people in the United States at a given time. Then the height of the tallest person at that instant in the United States is a parameter. The height of the tallest person in a sample of 500 people, for example, is a statistic.

Data collected in an investigation and not organized systematically are called **raw data.** The arrangement of these data in an ascending or descending order of magnitude is called an **array.** (Without electronic sorting capabilities, arranging such data could be very tedious, especially with a large number of observations.) The difference between the largest and the smallest value is called the **range.** For example, suppose the prices of a stock during a five-day trading period were $10, $8, $7, $9, and $12. Since $7 was the lowest price and $12 was the highest price, the range of prices is $12 − $7, or $5.

Table 1-3 records the heights (in inches) of 8 students. Column 1 presents the raw data and Column 2 illustrates the arrangement in an array. The largest value is 73 and the smallest value is 65. Thus, the range is 73 − 65, or 8 inches.

A range computed from the sample data gives the *sample range;* it is a statistic. A range computed from the entire population gives the *population range;* consequently, it is a parameter.

Samples sometimes contain just a few observations that differ abnormally from the majority of the observed values in the data set. Such wild values are referred to as **outliers.** They can result from obvious external causes such as faulty experimentation with inadequate controls, wrong

TABLE 1-3
Heights Presented as Raw Data and as an Array

Raw Data	Array
66	65
68	66
72	66
65	68
66	68
73	69
68	72
69	73

recording of data, and so on. It is also quite possible that they arise in a natural way, representing a new phenomenon that deserves further study. One way to deal with data that include maverick observations is to employ statistical methods that are not too sensitive to such extreme values. The range, for instance, is very sensitive to extreme values and would not give a realistic picture of the spread in the data if there is an outlier.

Section 1-1 Exercises

1. Which of the following variables are qualitative and which are quantitative?
   (a) The life of a light bulb (in hours)
   (b) The scent of a flower
   (c) A person's nationality
   (d) A monthly telephone bill (in dollars)
   (e) The salary of an individual (in dollars)
   (f) A person's sex
   (g) The tensile strength of an alloy
   (h) The verdict of a jury
   (i) A dividend (in dollars) paid by a company
   (j) The type of security (whether common stock, preferred stock, or bond) owned by an individual
   (k) The religion of an individual
   (l) The concentration (mg) of volatile fatty acids in blood

2. Give three other examples each of qualitative and quantitative variables.

3. Indicate which of the following quantitative variables are continuous and which are discrete.
   (a) The life of a light bulb
   (b) The number of children born in a family
   (c) The height of a person
   (d) The number of registered Republicans in some given year
   (e) The salary of an individual
   (f) The speed of a car
   (g) The breaking strength of a certain type of cable
   (h) The length of time one has to wait for a bus to arrive
   (i) The amount of water in a reservoir
   (j) The number of redwood trees on an acre of land
   (k) The prize (in dollars) won in a lottery
   (l) The daily dietary intake (μg) of selenium in wheat products
   (m) The amount bid for a painting in an auction
   (n) A person's pulse rate

4. Give three other examples each of discrete and continuous variables.

5. An operator at a plant that manufactures bicycle tires is interested in the durability of these tires. Describe what constitutes the population in this case. Describe a suitable sample that the operator might draw.

6. An opinion survey is conducted to find out the attitude of adult Americans toward legalizing marijuana.
   (a) Describe the population.
   (b) Describe a suitable sample that you might draw.

7. When a coin was tossed twenty times, it showed heads twelve times.
  (a) Describe the population.
  (b) Describe the sample.

8. The following is an excerpt from the August 2, 1993, issue of *Time* magazine.

> Gunshots now cause 1 of every 4 deaths among American teenagers, according to the National Center for Health Statistics. Bullets killed nearly 4,200 teenagers in 1990, the most recent year for which figures are available, up from 2,500 in 1985. An estimated 100,000 students carry a gun to school, according to the National Educational Association. In a survey released last week, pollster Louis Harris found alarming evidence of a gun culture among the 2,508 students he polled in 96 schools across the U.S. Fifteen percent of students in the sixth through 12th grades said they have carried a handgun in the past 30 days, 11% said they have been shot at, and 59% said they know where to get a gun if they need one.

  (a) Name two quantitative variables and one qualitative variable in the context of the investigation.
  (b) "*Gunshots now cause 1 of every 4 deaths among American teenagers . . . .*" Indicate the population of interest.
  (c) Is the value 1/4, implied by "*one out of every 4,*" a value of a parameter or the value of a statistic?
  (d) "*An estimated 100,000 students carry a gun to school . . . .*" Does the information provide descriptive statistics or inference statistics?
  (e) "*Louis Harris found alarming evidence of a gun culture among the 2,508 students he polled . . . .*" Does the figure 2,508 represent the number in a sample or in a population?
  (f) "*Fifteen percent of students in the sixth through 12th grades said they have carried a handgun . . . .*" Is the value 0.15, which is implied by "*fifteen percent,*" a value of a statistic or the value of a parameter?

In Exercises 9–13 identify the variables in the clippings and classify them.

9.

Quake Centered off Shore

5.3 Temblor Strikes 170 Miles off Eureka

EUREKA—
An earthquake struck about 170 miles offshore of Eureka at 11:45 A.M. Saturday in an area known for its frequent seismic activity, a spokeswoman for the U.S. Geological Survey said.

The USGS in Menlo Park registered the tremor at 4.8 on the Richter scale and characterized it as "mild," spokeswoman Pat Jorgenson said. Nobody reported feeling it to the USGS, she said.

10.

Econowatch

WASHINGTON (AP)—
Sales of new homes fell 3.1 percent in August, the government said today, puzzling economists who were looking for an increase because of low mortgage rates.

But Americans' personal income rebounded strongly in August following rare back-to-back declines the previous two months, a second government report said.

Incomes jumped 1.3 percent last month to a seasonally adjusted annual rate of $5.43 trillion.

Meanwhile, sales of new homes fell in every region of the country except the Northeast. Nationally, they totaled 616,000 at a seasonally adjusted annual rate, the Commerce Department said.

It was the fifth drop this year and put the sales level at the lowest in five months.

11.

Rats Subjected to Anemia Tests

CAPE CANAVERAL, Fla. (AP)—
Columbia's astronauts stuck tiny catheters into the tails of five white rats Saturday to draw blood and inject iron and iodine, part of a space anemia study.

It was the first time since the 14-day medical research mission began Monday that the crew handled some of the 48 rodents aboard the shuttle.

The rats, 2- to 3-month-old males weighing less than a pound each, are being subjected to some of the same tests as the astronauts.

Saturday's rat tests focused on anemia, one of the many side effects of space travel.

Source: "EconoWatch" from the Associated Press, September 30, 1993. Reprinted by permission of the Associated Press.

Source: "Rats Subjected to Anemia Tests" from the Associated Press, October 24, 1993. Reprinted by permission of the Associated Press.

12.

Leisure Time Down, Work Up, Study Says

Americans work more hours each week than they did in 1973, women have less time to relax than men do and Hispanics have the least amount of leisure time of all, according to a survey commissioned by Philip Morris Companies Inc.

The survey, conducted by Louis Harris and Associates Inc., shows that the amount of time spent each week at work and commuting increased to 46.8 hours in 1987, up 15 percent from 40.6 hours in 1973.

As a result, leisure hours have dropped 37 percent, to 16.6 hours a week last year from 26.2 hours in 1973, according to those surveyed. Relaxation time has dropped 8 percent from 18.1 hours in 1984.

Hispanics have 13 hours of leisure time a week; blacks, 15 hours; and whites, 17 hours, according to the survey of more than 1,500 households questioned last year.

The survey is the fifth such poll by Harris and the third straight commissioned by Philip Morris.

13.

Small Companies

Between 1988 and 1991 companies employing fewer than 500 people added *all* of the net new jobs to the U.S. economy. Five out of six American workers earn paychecks from companies with fewer than 1000 employees and two thirds of that group work for companies that employ fewer than 100 people.

14. The Commerce Department reported the following figures for monthly big ticket factory orders (billions of dollars) from October 1987 to September 1988. The figures are rounded off.

111.5	111.5	115.0	113.5	114.0	115.5
117.5	115.5	125.5	116.0	122.5	117.5

Arrange the data in an ascending array. Compute the range.

15. The following figures represent weekly lumber production (million board feet) during fourteen weeks in the twelve-state western region in the United States.

390	406	446	420	370	328	410
320	368	392	280	323	382	290

Calculate the range.

---

16. The following figures refer to the amount (in dollars) spent on groceries during twenty weeks.

41.10	36.20	32.78	27.63	13.41
29.80	12.85	8.31	34.81	22.67
28.50	52.68	37.38	18.69	64.85
18.60	10.87	23.76	17.68	15.87

Find the following.
(a) The highest amount spent during a week
(b) The lowest amount spent during a week
(c) The range of spending
(d) The number of weeks during which less than $30 were spent

17. The following data give the ages of twenty volunteers working in a hospital.

30	28	32	43	62	39	48	41	70	62
28	54	46	49	56	31	52	71	58	53

Find the following.
(a) The range of age
(b) How many volunteers are over the age of 45

18. Refer to the real estate data in Appendix IV, Table B-1. Suppose you are interested in sale prices of houses sold in the South Bay and North Bay areas in Humboldt County in California from 1989 to 1994. What is your target population? If you want to learn about characteristics of houses sold in Humboldt County since 1980, state whether the data in Appendix Table B-1 is a population or a sample. Describe the variables in the table and classify them.

1-2 FREQUENCY DISTRIBUTIONS

This section considers the organization of data in tabular form for the purpose of giving *frequency distributions*. A statistical table is useful because it summarizes and often readily identifies some of the striking features of the data. Frequency distributions can be prepared for both qualitative and quantitative data.

QUALITATIVE DATA

When describing qualitative data, to avoid ambiguities the categories are formulated in such a way that any item falls in precisely one category. The **frequency of a category** is defined as the number of items falling in that category. The frequency of a category answers the question, "How *frequently* does the category occur in the listing?" The frequency table is prepared by listing all the distinct categories and the corresponding frequencies as in Example 1.

EXAMPLE 1

Hospital records of 36 peptic ulcer patients provided the following information about their blood types *A, B, O, AB*.

O	A	B	O	A	A	A	O	O
O	A	O	A	B	O	O	O	AB
B	A	A	O	O	A	A	O	AB
O	A	A	B	A	O	A	O	O

Prepare a frequency table based on the blood types of the peptic ulcer patients.

SOLUTION   There are four distinct categories, namely, the four blood types *A, B, O, AB*, which we list in the first column of Table 1-4. In Column 2 we make tally marks (/) next to each category if the category is found as we scan through the information. When the tallies are counted, we get the frequencies. These are presented in Column 3. To ease counting, the tallies are grouped in blocks of five.

TABLE 1-4   Frequency Distribution of Blood Types

Blood Type	Tally	Frequency
O	∕∕∕ ∕∕∕ ∕∕∕ /	16
A	∕∕∕ ∕∕∕ ////	14
B	////	4
AB	//	2
Total		36

If we wish, we can compute the relative frequencies and percentage frequencies as explained in the following discussion.   ●

QUANTITATIVE DATA

For quantitative data the distributions may be ungrouped in the sense that the table provides distinct values of the variable and the corresponding frequencies. Or, it may be grouped in that the data are broken into distinct classes and the table gives the classes and the corresponding frequencies.

UNGROUPED DATA

As we saw in Section 1-1, an array is an arrangement of data in an ascending or descending order of magnitude. In forming an array, any value is repeated as many times as it appears. The number of times a value appears in the listing is referred to as its **frequency.** Thus, if the value 20.5 appears three times, we will list it that many times and call 3 the frequency of the value 20.5.

When the data are arranged in tabular form according to frequencies, the table is called a *frequency table.* The arrangement itself is called a **frequency distribution.**

The **relative frequency** of any observation is obtained by dividing the actual frequency of the observation by the total frequency, that is, by the sum of all the frequencies. If the relative frequencies are multiplied by 100 and thus expressed as percentages, the resulting values are called **percentage frequencies.**

For example, suppose a value occurs 9 times out of a total of 60 observations. Then the relative frequency of the value is 9/60, or 0.15, and its percentage frequency is 0.15 × 100, or 15 percent.

An advantage of expressing frequencies as relative frequencies is that frequency distributions of two sets of data can be compared.

EXAMPLE 2

The following data were obtained when a die was rolled thirty times.

1	2	4	2	2	6	3	5	6	3
3	1	3	1	3	4	5	3	5	3
5	1	6	3	1	2	4	2	4	4

Construct a frequency table.

SOLUTION   There are six observed values, namely, 1, 2, 3, 4, 5, 6, and so we get the frequency distribution given in Table 1-5.

TABLE 1-5   Frequency Table Giving the Number of Times Each Face on the Die Appears in Thirty Throws

Number on the Die	Tally	Frequency	Relative Frequency	Percentage Frequency
1	/N/	5	5/30 = 0.1667	16.67
2	/N/	5	5/30 = 0.1667	16.67
3	/N/ ///	8	8/30 = 0.2666	26.66
4	/N/	5	5/30 = 0.1667	16.67
5	////	4	4/30 = 0.1333	13.33
6	///	3	3/30 = 0.1000	10.00
Sum		30	1.0	100.00

The face that shows up is recorded in Column 2 by a tally mark. The number of tallies associated with the face, when counted, gives its frequency.

The relative frequency column is obtained by dividing each frequency by 30, the total frequency. The percentage frequency column is then obtained by multiplying the corresponding relative frequencies by 100. Observe that the sum in the relative frequency column is 1 and in the percentage frequency column is 100. ●

GROUPED DATA, CLASSES OF EQUAL WIDTHS

When there is a huge mass of data and when too many of the observed values are *distinct* values, it is inconvenient and almost impossible or meaningless to give a frequency table based on individual values, as we did in Example 2. In such a situation, it is better to divide the entire range of values and group the data into classes. For example, if we are interested in the distribution of ages of people, we could form the classes 0–19, 20–39, 40–59, 60–79, and 80–99. A class such as 40–59 represents all the people with ages between 40 and 59 years, inclusive. When the data are arranged in this way, they are called *grouped data*. The number of individuals in a class is called the frequency of the class or, simply, *class frequency*.

As an example, consider the figures recorded in Table 1-6, which represent the exam scores of eighty students. Observe that the data are quantitative.

TABLE 1-6   Scores Earned by Eighty Students on an Exam

62	73	85	42	68	54	38	27	32	63
68	69	75	59	52	58	36	85	88	72
52	52	63	68	29	73	29	76	29	57
46	43	28	32	9	66	72	68	42	76
38	38	39	28	19	12	78	72	92	82
72	33	92	69	28	39	85	59	68	52
85	59	76	80	72	74	54	48	29	36
10	82	58	88	68	58	46	37	29	35

The instructor wants to evaluate the students' performance. Are the students performing poorly? Are there many good students? From a look at the mass of data, not much can be said regarding their performance. Arranging the data as we did in Table 1-5, listing the distinct values with corresponding frequencies, would result in a table with 39 rows of distinct values. Such a distribution tells us little more than the raw data itself. The problem would be even more unwieldy if there were, say, a thousand students. Forming classes and condensing the data will allow us to shed light on the instructor's questions.

Formation of classes   We should be careful that, in the process of forming classes, pertinent information is not lost by taking too few classes; nor should there be so many classes that the data are not condensed enough to make the arrangement into classes worthwhile. There are different rules* to determine the number of classes, but depending upon the bulk of data, we use our own judgment and, as a general guide, choose anywhere from six to twenty classes.

---

* One such rule is called **Sturgess's Rule**, which gives k, the number of classes, when there are n measurements in the data set, by the formula $k = 1 + 3.31 \log(n)$. Based on this formula, the following is a handy table giving k for different values of n.

n	9	17	33	66	133	266	537	1073
k	4	5	6	7	8	9	10	11

We will divide the data given in Table 1-6 into six classes. If we scan the data, we find that the range of scores is from a low of 9 points to a high of 92 points. The difference between the highest score and the lowest score is 83 points, which is the range that we have to cover. Since we have agreed to form six classes, we have 83/6, or so to say, about 13.8 points to be covered by each class. The size of the class is given by rounding this figure *up*. Thus, the width of each class could be taken as 14 points. We round this up a little higher to 15 points in order to provide convenient intervals. You must realize, however, that you can go only so far in rounding up and still have the desired number of classes. With a class width of 15 points, we can form the following six classes.

5–19,   20–34,   35–49,   50–64,   65–79,   80–94

The first class is taken as 5–19 because the left ends 5, 20, 35, 50, 65, and 80 of the resulting classes are convenient to work with. You could have taken the first class as 7–21, for example, if you wished. However, whichever starting class you select, be sure that the classes obtained cover the entire range of values.

Once the classes are formed, we scan the data systematically and match the values with the classes in which they fall. There is no overlap between classes and, consequently, any observation belongs to only one class. We identify the values by tally marks. The number of values that fall in a class represents the frequency of the class and is called the **class frequency.** Table 1-7 gives the classes, the corresponding frequencies, the relative frequencies (obtained by dividing the class frequencies by 80), and the percentage frequencies. It is a good practice to include these four columns to describe a standard frequency distribution.

TABLE 1-7   Distribution of Test Scores Given in Table 1-6

Class	Tally	Frequency	Relative Frequency	Percentage Frequency
5–19	////	4	0.0500	5.00
20–34	/// /// //	12	0.1500	15.00
35–49	/// /// ///	15	0.1875	18.75
50–64	/// /// /// /	16	0.2000	20.00
65–79	/// /// /// /// //	22	0.2750	27.50
80–94	/// /// /	11	0.1375	13.75
Sum		80	1.0000	100.00

An overall picture emerges from the frequency distribution shown in Table 1-7. For example, we see that the class 65–79 had the maximum number of students, namely, 22; and that 22 + 11 (or 33) students scored over 64 points.

Steps to follow in forming a frequency distribution   The following steps are suggested for preparing a frequency distribution from raw data:

1. *Range.* Scan the raw data and find the smallest and the largest values. Their difference gives the range.
2. *Number of classes.* Decide on a suitable number of classes, depending upon the bulk of data and the information that the table is supposed to bring out.
3. *Class width.* Divide the range by the number of classes you have decided to have. This will give the approximate class width.

$$\text{approximate class width} = \frac{\text{largest value} - \text{smallest value}}{\text{number of classes}}$$

   Round this figure up to a convenient value to obtain the class width; then form the classes.
4. *Classes.* For the first class pick a starting point (lower class limit) that is slightly less than the lowest value in the data set. To get the starting point for the second class, add the class width to the starting point of the first class, and so on. Be sure that the classes completely cover the entire range of values.
5. *Frequency.* Find the number of observations in each class by using tally marks.

EXAMPLE 3   The following data give the amounts (in dollars) spent on fast-food meals by a young couple during forty weeks.

32	22	19	18	43	42	40	43	18	21
31	26	22	25	47	40	26	32	22	34
28	35	47	26	35	38	35	28	19	38
35	38	36	25	22	45	48	26	34	41

Construct a frequency distribution using seven classes.

SOLUTION   The smallest value is 18 and the largest value is 48. Therefore, the range is $48 - 18$, or 30.

Since we are asked to form seven classes, the approximate length of a class interval is 30/7, or 4.29. Rounding up to 5, we take the length of a class interval as 5.

A convenient value to start the first class is at 15. Thus the first class would be 15–19. Other classes, each 5 units long, are 20–24, 25–29, . . . , 45–49. The resulting frequency distribution is given in Table 1-8.   ●

GROUPED DATA, CLASSES OF UNEQUAL WIDTHS

It is recommended that as far as possible, all the classes be of the same width. However, this may not always be possible. The class lengths will often be dictated by the nature of the data and by the particular aspect of the distribution that we wish to emphasize.

TABLE 1-8 Frequency Distribution of the Amount Spent on Fast-Food Meals during Forty Weeks

Class	Tally	Frequency	Relative Frequency	Percentage Frequency
15–19	////	4	0.100	10.0
20–24	/\/\/	5	0.125	12.5
25–29	/\/\/ ///	8	0.200	20.0
30–34	/\/\/	5	0.125	12.5
35–39	/\/\/ ///	8	0.200	20.0
40–44	/\/\/ /	6	0.150	15.0
45–49	////	4	0.100	10.0
Sum		40	1.000	100.0

Look at the data given in Table 1-9, which lists the prices of forty stocks on a given day. The lowest price is $10 and the highest price is $600. However, a quick glance shows that a majority of the prices are in the range of $10 to $70. If we try to cover the entire range in equal intervals of 10 units each, then we will need about sixty class intervals. Too many! But if we decide to make ten equal classes of $60 each, we will put most of the data in one class interval, namely, 1–60. Such concentration of data in one class will condense the data to such an extent that we will sacrifice relevant information. A reasonable set of classes would be 10.0–19.9, 20.0–29.9, 30.0–39.9, 40.0–49.9, 50.0–59.9, 60.0–69.9, and 70.0–600.0.

TABLE 1-9 Prices of Forty Stocks on a Given Day (in dollars)

12.5	36.0	30.6	30.9	30.0	55.8	39.6	48.6
18.5	42.0	26.5	28.5	150.0	200.0	45.0	10.0
28.0	48.0	22.0	22.0	260.0	28.0	120.5	68.2
39.0	51.3	17.6	272.0	85.0	62.0	185.0	70.0
28.7	55.7	600.0	556.0	20.6	538.0	48.0	70.0

CLASS INTERVALS, CLASS MARKS, AND CLASS BOUNDARIES

In Table 1-7 we presented a frequency distribution of the scores earned by eighty students. The blocks such as 5–19, 20–34, and so on are called **class intervals.** The lower ends of the class intervals are called *lower limits,* and their upper ends are called *upper limits.* Thus, 5, 20, 35, 50, 65, and 80 constitute the lower limits and 19, 34, 49, 64, 79, and 94 the upper limits. A class that does not have either an upper limit or a lower limit is called an *open-ended class.* The **class size,** also called class width, is equal to the difference between two consecutive upper limits around that class. For example, the two consecutive upper limits around the class 20–34 are 19 and 34. Thus, the size of the class 20–34 is 34 − 19, or 15.

The **class mark** is defined as the midpoint of a class interval. It is computed by adding the lower and upper class limits of a class interval and then dividing the sum by 2.

$$\text{Class mark} = \frac{(\text{lower limit of the class} + \text{upper limit of the class})}{2}$$

The midpoints of the class intervals 5–19 and 20–34 are, respectively,

$$\frac{5 + 19}{2} = 12 \quad \text{and} \quad \frac{20 + 34}{2} = 27.$$

An open-ended class has no class mark because it will not have either a lower limit or an upper limit.

Because the class mark is the midpoint of the class, it serves as a representative value of the class. This is especially realistic if the observations in the class are evenly distributed over the class.

A point that represents the halfway, or separation, point between successive classes is called a **class boundary.**

So a class boundary is a point where the lower class ends and the higher class begins. In general, it is obtained by adding the upper limit of the lower class and the lower limit of the upper class and then dividing the sum by 2, and it is reported to one more decimal place than the original data.

The halfway point between the two class intervals 5–19 and 20–34 is 19.5. Note that 19.5 is equal to (19 + 20)/2. For the class intervals 20–34 and 35–49, the halfway point is 34.5.

Table 1-10 provides a summary with reference to the frequency distribution of test scores given in Table 1-7. Note that the upper boundary of one class is the lower boundary of the next.

TABLE 1-10   Class Limits, Class Boundaries, and Class Marks for Frequency Distribution Presented in Table 1-7

Class	Lower Limit	Upper Limit	Lower Boundary	Upper Boundary	Class Mark
5–19	5	19	4.5	19.5	12
20–34	20	34	19.5	34.5	27
35–49	35	49	34.5	49.5	42
50–64	50	64	49.5	64.5	57
65–79	65	79	64.5	79.5	72
80–94	80	94	79.5	94.5	87

EXAMPLE 4   The following frequency distribution gives the amount of time (in minutes) that a doctor spent with twenty-eight patients.

Class	Frequency
2–6	5
7–11	7
12–16	10
17–27	6

Find the following.

(a)   The upper and lower limits of all the classes

(b)   The lengths of all the class intervals

(c)   The class marks of the classes

(d)   The boundaries of the classes

SOLUTION

(a)   For example, the lower end of the class interval 7–11 is 7, and so 7 is a lower limit. The upper end of 7–11 is 11, and therefore 11 is an upper limit. Other lower and upper limits are obtained similarly and are as given in Columns 2 and 3 of Table 1-11.

(b)   The length of the class interval 7–11 is $11 - 6$ or 5, the difference between two consecutive upper limits. The lengths of other class intervals are given in Column 4 of Table 1-11.

(c)   The class mark of the class 7–11 is obtained by dividing the sum of the lower and upper class limits by 2. Thus, the class mark of 7–11 is $(7 + 11)/2$, or 9. Other class marks are given in Column 5 of Table 1-11.

(d)   The lower boundary of the class 7–11 is the halfway point between the classes 2–6 and 7–11. Therefore, the lower class boundary of the class 7–11 is $(6 + 7)/2$, or 6.5. Notice also that 6.5 is the upper boundary of the class 2–6. We complete Columns 6 and 7 following this procedure.

TABLE 1-11   Class Limits, Class Marks, and Class Boundaries

Class	CLASS LIMITS		Length of Class Interval	Class Mark	CLASS BOUNDARIES	
	Lower	Upper			Lower	Upper
2–6	2	6	5	4	1.5	6.5
7–11	7	11	5	9	6.5	11.5
12–16	12	16	5	14	11.5	16.5
17–27	17	27	11	22	16.5	27.5

Section 1-2 Exercises

1. A quality control engineer is interested in determining whether a machine is properly adjusted to dispense 1 pound (16 ounces) of sugar. The following data refer to the net weight (in ounces) packed in thirty one-pound bags after the machine was adjusted.

15.6	15.9	16.2	16.0	15.6	16.2
15.9	16.0	15.6	15.6	16.0	16.2
15.6	15.9	16.2	15.6	16.2	15.8
16.0	15.8	15.9	16.2	15.8	15.8
16.2	16.0	15.9	16.2	16.2	16.0

Arrange the data in a frequency table.
Hint: Notice that there are only a few distinct values.

2. Prepare a frequency table having the classes 2.2–2.3, 2.4–2.5, 2.6–2.7, 2.8–2.9, 3.0–3.1, 3.2–3.3, 3.4–3.5, and 3.6–3.7 for the following set of data.

3.0	3.2	2.9	3.4	3.3	3.1	3.3	3.5	3.3	2.9
3.2	3.4	3.4	3.0	3.2	3.6	3.4	3.6	3.5	2.2
3.1	3.2	2.9	3.6	3.1	3.0	2.3	2.6	2.7	3.7
2.5	3.3	3.5	2.4	2.6	3.1	2.7	3.6	2.7	3.0

3. Arrange the following data into a frequency table using the classes 5.00–9.99, 10.00–14.99, 15.00–19.99, 20.00–24.99, 25.00–29.99, and 30.00–34.99.

21.10	16.20	12.78	7.63	13.41	16.48	6.89
9.80	12.85	8.31	14.81	22.67	10.45	8.88
28.50	32.68	17.38	18.69	14.85	11.78	9.99
18.60	10.87	23.76	17.68	15.87	29.99	8.76

4. The weekly earnings of workers in a factory vary from a low of $278.35 to a high of $431.50. If it is agreed to have eight classes with equal class intervals, give suitable classes for grouping the data into a frequency table.

5. The following are the hourly wages (in dollars) of thirty factory workers.

9.60	11.50	8.85	12.20	8.75	9.30	10.10	9.90
9.25	9.10	11.35	11.20	8.90	9.60	9.75	10.25
9.80	10.65	10.15	9.75	11.10	10.15	10.85	9.70
9.35	10.60	10.15	10.60	10.45	11.20		

Construct a frequency distribution by arranging the data into a suitable number of classes.

6. The following figures give the telephone bills (in dollars) of forty residents of Metroville.

15.80	23.05	17.72	44.18	33.38	23.20	68.50	43.47
34.05	16.10	18.10	29.65	52.25	27.28	68.90	57.12
46.04	27.00	36.07	19.16	17.78	48.19	16.89	26.55
25.28	37.13	18.50	38.59	50.25	40.51	37.51	47.97
33.26	16.81	31.72	65.49	58.64	21.30	34.21	20.40

(a) What percent of the residents paid over $24?
(b) What percent of the residents paid less than $36 but more than $24?
(c) Arrange the data into a frequency distribution.

7. Fifty students were interviewed regarding the number of hours per week that they spend on their studies. The responses were grouped in a frequency table as follows.

Number of Hours	Number of Students
0–9	3
10–19	7
20–29	17
30–39	12
40–49	10
50 and over	1

Answer the following questions, where possible. Find how many students studied:
(a) fewer than 20 hours
(b) fewer than 22 hours
(c) fewer than 29 hours
(d) more than 20 hours
(e) more than 22 hours
(f) more than 29 hours.

8. The following frequency table gives the distribution of the amounts (in dollars) of charitable contributions made during a year by eighty families.

Amount Contributed	Number of Families
0.00–49.99	5
50.00–99.99	14
100.00–149.99	19
150.00–199.99	17
200.00–249.99	12
250.00–299.99	10
300 and over	3

Where possible, find how many families contributed:
(a) less than \$200
(b) more than \$200
(c) \$150 or more
(d) less than \$250
(e) \$50 or more but less than \$200.

9. The number of quarts of milk consumed during one week by a certain number of families is presented in a frequency table. If the classes are given as 6–10, 11–15, 16–20, 21–25, 26–30, and 31–35, find the following:
(a) The class marks
(b) The class boundaries
(c) The sizes of the class intervals

10. A civil engineer analyzed the breaking strength (in pounds) of one hundred cables. The following table gives the distribution of the findings.

Breaking Strength	Number of Cables
1400–1499	8
1500–1599	20
1600–1699	12
1700–1799	35
1800–1899	18
1900–1999	7

Find the following.
(a) The class marks
(b) The class boundaries
(c) The class boundary below which there are 75 percent of the cables
(d) The class boundary above which there are 60 percent of the cables

11. Find the relative frequency and percentage frequency of each class in Exercise 10.

12. Mac, a meat cutter, recorded the amount of ground chuck that he sold on a particular day. He sold a total of ninety packages and the following is the distribution of the weight (in ounces) of the packages.

Weight	Number of Packages
12.1–16.0	8
16.1–20.0	14
20.1–28.0	20
28.1–32.0	23
32.1–40.0	18
40.1–48.0	7

Find the following:
(a) The class marks
(b) The class boundaries

13. Find the relative frequency and percentage frequency of each class in Exercise 12.

14. The following figures pertain to the distribution of the voltage of one hundred batteries.

Voltage	Number of Batteries
8.75–8.80	12
8.81–8.88	20
8.89–9.00	28
9.01–9.06	26
9.07–9.15	10
9.16–9.30	4

Find the following.
(a) The width of each class
(b) The class marks
(c) The class boundaries

15. The distribution of the earnings of a free-lance photographer during sixty months is as follows.

Earnings (in Dollars)	Number of Months
900–1099	5
1100–1299	9
1300–1499	12
1500–1899	10
1900–2299	11
2300–2999	4
3000–3899	9

(a) Why is it desirable to arrange the frequency distribution with unequal classes?
(b) Find the class marks.
(c) Find the class boundaries.
(d) Find the length of the longest class.
(e) Find the class with the largest frequency.

16. The class marks in a frequency distribution of the tips (in dollars) earned by a waiter during a week are as follows.

    84.5   104.5   124.5   144.5   164.5

Assuming equal class sizes, find the following.
(a) The size of each class interval
(b) The class boundaries
(c) Each class

17. The class marks in a frequency distribution of the lives of a number of light bulbs (in hours) are as follows.

    400   425   450   475   500   525

Assuming that the class intervals are equal, find the following.
(a) The size of each class interval
(b) The class boundaries
(c) Each class

1-3   GRAPHS AND CHARTS

As previously discussed, data presented in tabular form can reveal some features which we would not be able to discern simply by looking at raw data. It is often even more helpful to present this information in a *graphical* form so that it can make a stronger visual impact. Pictorial presentation has the advantage that even a novice without any technical expertise can assimilate the information by looking at a chart or graph.

We can make a point pictorially in several ways if we are imaginative enough. Charts such as Figures 1-2, 1-3, and 1-4 are probably familiar to most people through their reading of newspapers and magazines.

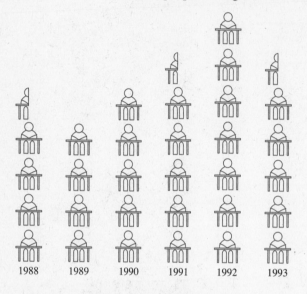

Each symbol = 10,000 students

Figure 1-2 Number of Students Enrolled in the Tippegawa School District during 1988–93

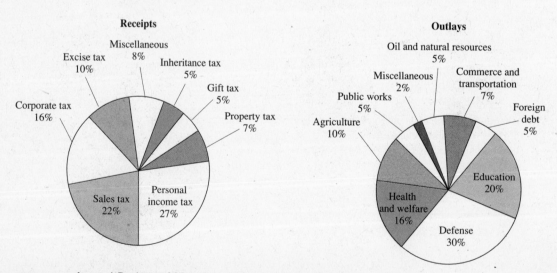

Figure 1-3 Annual Budget of Mondavia during the Fiscal Year 1992–93

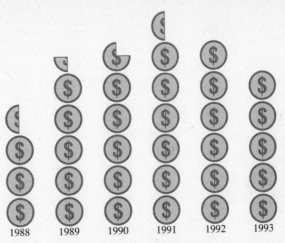

1988   1989   1990   1991   1992   1993

Each symbol = $1,000,000

Figure 1-4 The Annual Budget of
Tippegawa School District during 1988–1993

THE PIE CHART

The diagram in Figure 1-3 is called a **pie chart** for the obvious reason that
it resembles a pie cut into wedge-shaped slices. Pie charts are useful for
presenting information regarding qualitative variables when we are interested
in showing what percentage of the total is accounted for by each category.

Suppose a category has a relative frequency 0.05 or, as a percentage
frequency, 5 percent. Since the total number of degrees in a circle is 360 and
5 percent of 360 is 360(0.05), that is, 18 degrees, we allot a central angle of
18 degrees to this category. The pie chart in Figure 1-5 is constructed for the
data in Table 1-12, which gives the distribution of responses of 150 patients
regarding the relief provided when a pain-killing drug was administered to
them.

Figure 1-5 Pie Chart for Data
in Table 1-12

TABLE 1-12   Response Regarding the Relief Provided by a
Pain-Killing Drug

Response	Frequency	Relative Frequency	Central Angle (in Degrees)
Excellent	30	0.20	(0.20)(360) = 72
Satisfactory	66	0.44	(0.44)(360) = 158.4
Fair	36	0.24	(0.24)(360) = 86.4
Poor	18	0.12	(0.12)(360) = 43.2
Sum	150	1.00	360

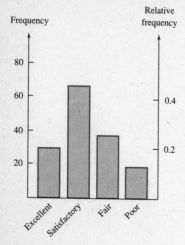

Frequency   Relative frequency

Figure 1-6 Bar Graph for Data in Table 1-12

The data in Table 1-12 could also be presented as a **bar graph** as in Figure 1-6, where vertical bars are erected for each category. The height of each bar reflects the frequency of the corresponding category.

For frequency distributions describing quantitative data, graphical arrangements include the histogram and the ogive.

THE HISTOGRAM

A **histogram** is a graphic presentation of a frequency distribution of a quantitative variable where vertical rectangles are erected on the horizontal axis for each class in the frequency distribution. The base of a rectangle for a class extends from its lower boundary to its upper boundary. So, you might say that a histogram is a bar graph for quantitative data.

A histogram is called a *frequency histogram* when frequencies are plotted along the vertical axis. Instead of frequencies we might plot relative frequencies along the vertical axis to obtain a *relative frequency histogram*. The two histograms can be drawn to look identical by choosing scales along the vertical axis appropriately.

EXAMPLE 1

The first two columns of Table 1-13 present the distribution of temperature increase (in degrees Celsius) of a coolant used in a compressor chamber recorded on fifty occasions.

Draw a frequency histogram.

TABLE 1-13   Distribution of Temperature Increase

Temperature Increase	Frequency f	Relative Frequency $f/50$	Class Boundary
			0.95
1.0–1.9	6	0.12	
			1.95
2.0–2.9	8	0.16	
			2.95
3.0–3.9	14	0.28	
			3.95
4.0–4.9	16	0.32	
			4.95
5.0–5.9	6	0.12	
			5.95

SOLUTION   For drawing a frequency histogram, we represent temperature increase, the underlying variable, along the horizontal axis and label it as such. The boundaries computed in Column 4 of Table 1-13 are then marked along this axis adopting a scale so that all the classes are properly accommodated. Frequencies are marked along the vertical axis with a scale so that the top of the scale is at or slightly above the maximum frequency in the fre-

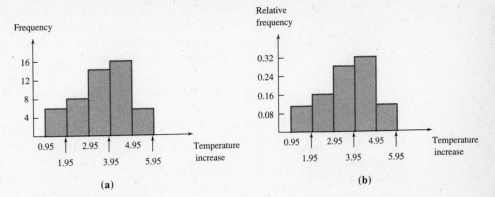

Figure 1-7 Histograms

a. Frequency histogram for temperature increase   b. Relative frequency histogram for temperature increase

quency distribution table. In our example the maximum frequency is 16 and so 16 marks the top of the scale. Once this is done, we erect a rectangle of height 6 over the first interval 0.95–1.95, a rectangle of height 8 over the interval 1.95–2.95, and so on obtaining the histogram in Figure 1-7(a).

In Figure 1-7(b) we obtain exactly the same graph by plotting relative frequencies along the vertical axis. Of course, along the vertical axis we have used a scale that is fifty times that in the first graph. Both the graphs can be presented together as shown in Figure 1-8.

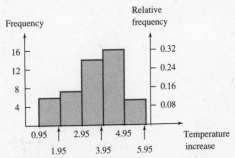

Figure 1-8 The Frequency Histogram and Relative Frequency Histogram Presented Together

In the preceding definition of a histogram a tacit assumption is that all the classes are of the same width. If this is not the case, then we should first divide the frequency of each class by the corresponding class width to obtain the **frequency densities** of the classes.

$$\text{frequency density of a class} = \frac{\text{frequency of the class}}{\text{the class size}}$$

We now plot frequency densities along the vertical axis to complete the histogram.

EXAMPLE 2   The first two columns of Table 1-14 below show the distribution of test scores of seventy-seven students. Plot the histogram.

TABLE 1-14   Distribution of Test Scores

Class	Frequency f	Class Size	Boundary Point	Frequency Density (f/Class Size)
			4.5	
5–20	8	16		8/16 = 0.50
			20.5	
21–40	12	20		12/20 = 0.60
			40.5	
41–55	15	15		15/15 = 1.00
			55.5	
56–87	40	32		40/32 = 1.25
			87.5	
88–95	2	8		2/8 = 0.25
			95.5	

SOLUTION   Since the class sizes are unequal, in Column 5 we compute the frequency densities. Column 4 gives the boundary points.

We now use a suitable scale to locate the class boundaries on the horizontal axis. Next, with the interval between successive class boundaries representing the base, we use a suitable scale and erect rectangles with their heights to correspond to the frequency densities of the respective classes. Thus, we obtain the histogram in Figure 1-9.

Graphical presentation of data in the form of a histogram will be helpful later in our study of a probability distribution, which is the theoretical counterpart of a frequency distribution.

If the histogram of a distribution is such that when the graph is folded along a vertical axis the two halves coincide (as in Figure 1-10, where the axis is indicated by the dotted line), then the distribution is a **symmetric distribution.** Otherwise it is an **asymmetric distribution.**

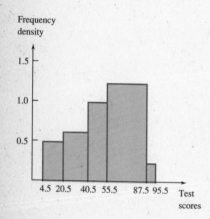

Figure 1-9 Histogram of Test Scores in Table 1-14

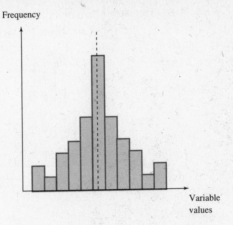

Figure 1-10 Distribution Symmetric about the Dotted Line

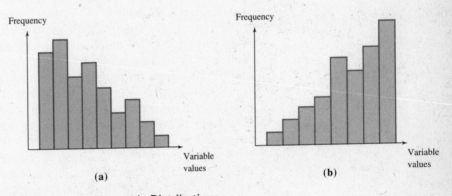

Figure 1-11 Asymmetric Distributions

For an asymmetric distribution, if there is a long tail on the right-hand side, as in Figure 1-11(a), the distribution is said to be *skewed to the right*, and if there is a long tail to the left, as in Figure 1-11(b), it is said to be *skewed to the left*.

The histogram given on the next page for the systolic blood pressure data (SYSBP) in Appendix IV, Table B-3, was obtained using MINITAB software. Notice the almost symmetric nature of the graph.

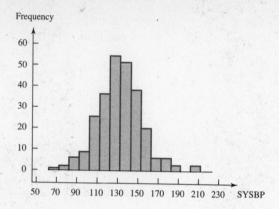

STEM-AND-LEAF DIAGRAM

A histogram is one of the descriptive methods for describing the data. In recent years several new descriptive methods have been developed allowing the investigator to take a closer look at the raw data before proceeding with the traditional data analysis for drawing inferences. The ideas were pioneered by Professor John Tukey* and comprise what is called **exploratory data analysis,** or **EDA.** The emphasis is strictly on exploring the data. The EDA tools that provide new insight into the data are easy-to-use graphical methods and involve simple arithmetic computations. The scrutiny of the data may reveal some patterns which might form a basis for future investigation. It may bring to light some observations that are out of line with the rest of the data, prompting the investigator to check the original data, the method of collecting them, and so on. Professor Tukey characterized EDA as detective work—numerical detective work (counting or graphical detective work) prior to the evaluation of the evidence by the jury and the judge. He likened the analysis where inferences about the population are made to the evaluation of the evidence and the court verdict in a trial. We shall now examine the *stem-and-leaf diagram,* which is one of the EDA tools for looking more effectively at the data. Another EDA tool called the *boxplot* will be studied in Chapter 2.

A **stem-and-leaf diagram** gives the actual data in a display that resembles a histogram. First we must write each number in two parts. The first part of a number serves as a stem; the part that follows it is a leaf.

It is recommended that each number be partitioned such that there are between 5 and 20 stems. If this results in the leaves having more than one digit, then, of the digits in the leaves, delete those after the first. As a general rule, a leaf consists of one digit. The stems are then displayed in a vertical column with the smallest stem at the top of the column and the largest stem at the bottom. The leaves associated with a stem are recorded along the corresponding horizontal row.

---

* The interested reader is referred to *Exploratory Data Analysis* (EDA) by John Tukey, 1977, Addison-Wesley Publishers.

As variations of this basic stem-and-leaf diagram with one line per stem, we can also have diagrams with two or five lines per stem. Also, we can arrange in order the leaves belonging to each stem so that 0s are placed before 1s which, in turn, are placed before 2s, and so on.

We illustrate the basic approach in the following two examples, without ordering the leaves in the stems.

EXAMPLE 3

The following figures represent the number of rooms occupied in a motel on thirty days. Draw two stem-and-leaf diagrams, with one line per stem and two lines per stem, respectively.

20	14	21	29	43	17	15	26	8	14
39	23	16	46	28	11	26	35	26	28
30	22	23	7	32	19	22	18	27	9

SOLUTION   We consider each number as a two-digit number. The first digit in each case constitutes a stem; the second, a leaf. For instance, 2 is the stem and 0 is the leaf for the first number, 20; 1 is the stem and 4 is the leaf for the second number, 14; and so on. Altogether, there are five stems, namely 0, 1, 2, 3, and 4, since our data range from 07 to 46.

For the first diagram, these stems are written to the left of a vertical line and the corresponding leaves to the right as in Figure 1-12(a). From the diagram, reading horizontally along stem 3, for instance, we see that 39, 35, 30, and 32 rooms were occupied on four days.

In the second illustration we construct a stem-and-leaf diagram with two lines per stem. For example, for stem 1 we will have two lines, with the values from 10 to 14 entered on one line and the values from 15 to 19 entered on a separate line. The display is shown in Figure 1-12(b).   ●

EXAMPLE 4

Consider the following information on per-share annual dividends offered by twenty-four corporations listed on the New York Stock Exchange.

1.25	0.80	1.52	1.44	1.56	1.48
1.40	1.48	0.96	1.56	1.28	1.32
1.36	1.32	1.20	1.28	0.96	1.12
1.24	0.84	0.92	1.00	1.10	1.24

Plot a stem-and-leaf diagram.

SOLUTION   Because of the nature of the data, we will take the stem for each number to be the first part of the number, which includes the first decimal place. For example, for the first number 1.25, the part 1.2 will constitute a stem and the part that follows, 5, will form a leaf. Scanning the data systematically we obtain the diagram in Figure 1-13.   ●

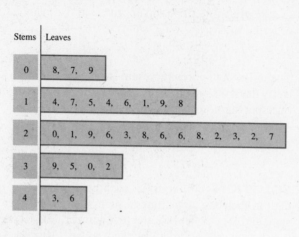

(a)

Figure 1-12a Stem-and-Leaf Diagram—One Line Per Stem

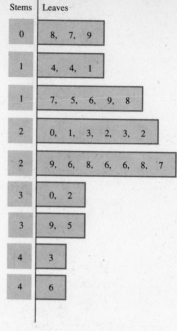

(b)

Figure 1-12b Stem-and-Leaf Diagram—Two Lines Per Stem

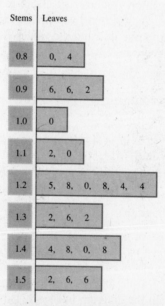

Figure 1-13 Stem-and-Leaf Diagram for the Dividend Offered

THE OGIVE

Before we can define and sketch an *ogive* (pronounced ō–jīv), we should explain what is meant by a *cumulative frequency distribution*. As the name suggests, it represents accumulated frequencies. For any class boundary, a **cumulative frequency distribution** gives the total number of observations that have a value "less than" that boundary. Thus, whereas the frequency distribution gives the number of values falling in an interval, the cumulative frequency distribution gives the number of values that fall below specified values.

EXAMPLE 5

The data given in Table 1-15 refer to the net weight of sugar (in ounces) contained in 185 one-pound bags. Determine the cumulative frequency distribution.

SOLUTION   First, we find the class boundaries and form the "less than" column, Column 1 in Table 1-16. To obtain the cumulative frequency distribution, we have to determine how many observations are less than any given boundary point. For example, how many bags weigh less than 15.05 ounces? There are none. Therefore, the cumulative frequency of bags weighing less than 15.05 ounces is 0. The cumulative frequency up to and including the class interval 15.7–15.9, that is, the cumulative frequency of bags weighing less than 15.95 ounces, is 5 + 10 + 45, or 60. Continuing, we get Table 1-16, which gives the cumulative frequency distribution.

TABLE 1-15

Weight Class	Number of Bags
15.1–15.3	5
15.4–15.6	10
15.7–15.9	45
16.0–16.2	65
16.3–16.5	46
16.6–16.8	14

TABLE 1-16   The "Less Than" Cumulative Frequency Distribution

Weight	Cumulative Frequency (Number of Bags)			
less than 15.05	0			
less than 15.35	5			
less than 15.65	15			
less than 15.95	60			
less than 16.25	125			
less than 16.55	171	=	125 +	46
less than 16.85	185		↑	↑
			Previous cumulative frequency	Frequency of the class

An **ogive** is a line graph obtained by representing the upper class boundaries along the horizontal axis and the corresponding cumulative frequencies along the vertical axis. It is also referred to as a *cumulative frequency polygon.* If relative cumulative frequencies, which are obtained by dividing cumulative frequencies by the total frequency, are plotted along the vertical axis we get what is called the **relative frequency ogive.**

For the distribution of sugar weights in Example 5 we obtained the cumulative frequency distribution in Table 1-16. The ogive for this distribution is now plotted in Figure 1-14.

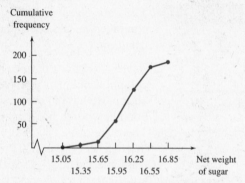

Figure 1-14 Ogive of the Distribution of Net Weight of Sugar

The cumulative frequency distribution and ogive are useful in many situations. Cumulative frequency tables are used, for instance, in insurance mortality tables giving the relative frequency of individuals living up to a certain age or, equivalently, the relative frequency of those living past a given age group. The ogive can be used in computing statistical quantities called *percentiles,* which will be discussed in the next chapter.

HOODWINKING WITH GRAPHS

Graphical methods provide a good complement to other statistical analyses of the data. Graphs and charts are greatly appreciated in board meetings and seminars, and serve to liven up discussions because most people are comfortable with visual messages. However, one should exercise caution in reading and interpreting graphs. Some graphs and charts are misleading because of pure and simple ignorance on the part of the individual preparing them. Often, a misleading and distorted picture may emerge in spite of honesty and correctness of technique employed in preparation. Because the investigator wishes to promote a certain point of view, he or she chooses to emphasize certain aspects of the true nature of the data and to deemphasize others. We consider a few situations on the following pages.

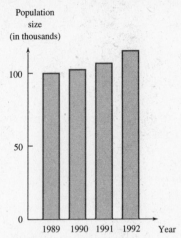

Figure 1-15 Population Growth in Tippegawa County during 1989–1992

The bar graph in Figure 1-15 shows how the population (in thousands) in Tippegawa county has changed during 1989–1992. This graph gives the basic picture and shows the steady increase in the county population.

Some variations of the above graph that you might encounter are shown in Figure 1-16. There is nothing unethical about these graphs in that there is no falsification of data. But you should guard against first impressions conveyed. In Figure 1-16(a) the height of the bar for 1992 is over one and a half times that for 1989. An unsuspecting reader could be misled into drawing wrong conclusions by just looking at the relative sizes of the bars. You should pay attention to numbers and note that the scale on the vertical axis starts at 80 and not at 0 as in Figure 1-15.

In Figure 1-16 (b) the vertical scale starts at 0 as in Figure 1-15 but there is a scale break along the vertical axis which results in shortening the bars. Again, everything is above board. However, the possibility of being misled remains.

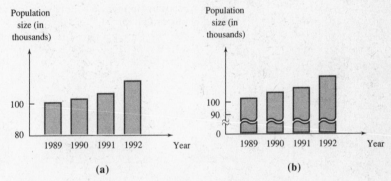

(a)   (b)

Figure 1-16 Variations of Bar Graph in Figure 1-15

A shrewd and widely exploited technique to emphasize or deemphasize some features of the data is to manipulate scales along the two coordinate axes. Consider the graphs in Figure 1-17, which are drawn to bring out the pattern of population growth in Tippegawa County. Figure 1-17(a) deemphasizes the growth pattern whereas Figure 1-17(b) exaggerates it.

Distortion and abuse often occur when pictographs are used to show amounts such as expenditure or production as plane or solid objects. In this approach, if the amount is multiplied by a factor, that same factor is mistakenly applied to all the dimensions of the object. The fallacy is that when area is used to describe the amount, the area gets multiplied by the *square* of the factor. When a solid oject is used, the volume is multiplied by the *cube* of the factor. An example is illustrated in Figure 1-18, where the diagrams are supposed to show that the revenue of Tippegawa County in 1992 was double that in 1982. The figure fails to do that. In fact, Figure 1-18(a) conveys the message that the revenue was quadrupled and Figure 1-18(b) shows it to be eight times as much.

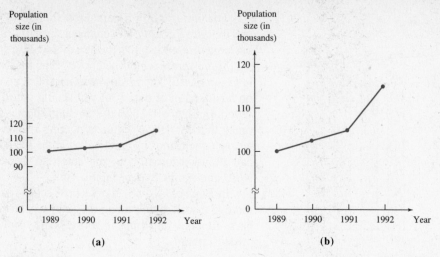

Figure 1-17

Growth pattern in Tippegawa county with the vertical scale compressed in (a) and elongated in (b)

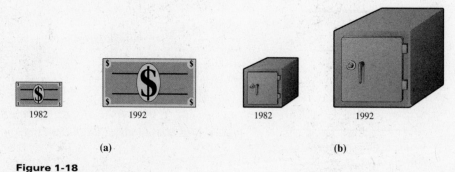

Figure 1-18

Pictographs supposed to show that revenue doubled. However, the impression when bills are used is that it quadrupled and when safes are used is that it was eight times as much as in 1982.

The figure on page 43, which appeared in *Time* magazine on March 29, 1993, also misleads the reader. The diagram purports to provide a comparative picture of increase in temporary employment versus increase in all employment. Yet there is a conspicuous flaw in the way the diagram is drawn. The height of the person shown is taken accurately as 12.5, since 250 = (12.5)(20). No problem with that. However, comparing the blue area for increase in temporary employment with the gray area for all employment yields a distorted picture altogether. The blue area is drawn such that it is not just 12.5 times the gray area, but at least 20 times.

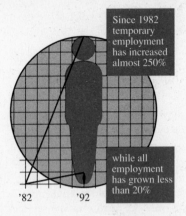

Since 1982 temporary employment has increased almost 250%

while all employment has grown less than 20%

'82   '92

The lesson is that, as with anything in life, one should not go by first impressions. Study the graph carefully and pay attention to numbers. In the final analysis, depend upon specific numbers.

Section 1-3 Exercises

1. Complete the following table for the categorical data.

Category	Frequency	Relative Frequency	Central Angle (in Degrees)
Category I	85	0.2742	(0.2742)(360) = 98.7
Category II	156	_____	_____
Category III	69	_____	_____
Sum	310	1.000	360

2. In a genetic experiment, when crosses were made involving a variety of pink-flower plants, it was found that among 800 flowers collected there were 224 red, 380 pink, and 196 white. Construct a pie chart showing these results.

3. The following figures give the distribution of land in a certain county.

Forest land	Farm land	Urban	Other land
30%	40%	10%	20%

Describe the distribution with a pie chart.

4. Of the 200 patients in a hospital, 60 were of blood type O, 75 of type A, 45 of type B, and 20 of type AB. Construct a pie chart to depict the distribution.

5. At the end of a training program, 400 workers in a plant are judged by a panel of examiners as *excellent, very good, good,* or *fair.* The following data give the distribution.

Excellent	Very good	Good	Fair
80	140	120	60

Represent the figures by a pie chart.

6. The following are the results of a poll which asked the question, "To what degree are you satisfied with the outcome of the 1992 U.S. presidential election?"

A great deal	21%
Somewhat	45%
Not at all	30%
No opinion	4%

   Draw a pie chart to show the response. What other graphical presentation could you give to describe the response?

7. The *Los Angeles Times* polled 1,294 adults statewide in California by telephone from March 20 to 22, 1993. Telephone numbers were chosen from a list of all exchanges in the state. Random-digit dialing techniques were used to ensure that both listed and unlisted numbers could be contacted. One of the questions posed was: "How would you rate the quality of the services provided by your local public schools?" The response among parents is summarized below as percentages.

Excellent	15%
Adequate	42%
Inadequate	22%
Very poor	15%
Don't know	6%

   Draw a pie chart and a bar graph to reflect the response.

8. The data in the following table refer to the number of foreign students enrolled, by region, in institutions of higher education in the United States and outlying areas in 1988–1989. (Source: *Digest of Education Statistics,* 1990, National Center for Education Statistics, U.S. Department of Education)

Region	Number of Students
Europe	42,770
Latin America	45,030
Middle East	40,200
South and East Asia	191,430
Africa	26,730

   Use a suitable graphical method to describe the data.

9. The two bar graphs on page 45, which give the number of births and deaths in Humboldt County from 1987 to 1993, are based on the data provided by the Department of Vital Statistics, Humboldt County Health Department. Round the numbers for each year to the nearest fifty and prepare a bar graph for the net gain (excess of births over deaths) during 1987–1993. Interpret the bar graph obtained.

10. The bar graph on page 45, which is based on a report of the National Safety Council, gives data for total accidental deaths by cause in 1990.
    (a) Interpret the bar graph.
    (b) Which cause is responsible for the greatest number of accidental deaths?
    (c) Prepare a relative frequency table by cause.
    (d) Convert the bar graph into a pie chart

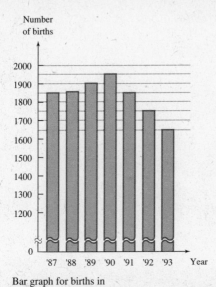

Bar graph for births in
Humboldt County, 1987–1993

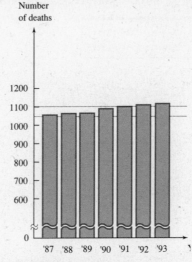

Bar graph for deaths in
Humboldt County, 1987–1993

Total Accidental Deaths by Cause

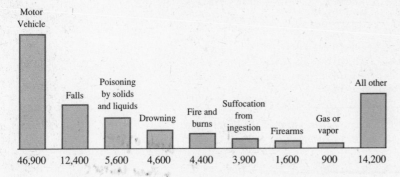

11. Draw the histogram for the following frequency distribution of the lives of 400
light bulbs.

Life of Bulb (in Hours)	Number of Bulbs
600–699	85
700–799	77
800–899	124
900–999	78
1000–1099	36

Source: "Total Accidental Deaths by Cause" by the National Safety Council. Data for all
accident statistics from the National Safety Council, *Accident Facts*, 1990 Edition.

12. Draw the histogram for the following distribution of the heights of fifty students.

Height (in Inches)	Number of Students
60–62	4
63–65	12
66–68	14
69–71	9
72–74	11

13. The following measurements were obtained for the leaf lengths (in centimeters) of thirty rhododendron leaves.

10.5	12.5	8.5	7.0	11.5	7.0	11.0	11.0
13.5	15.5	7.2	8.3	12.5	12.0	8.2	6.5
13.5	9.5	12.8	13.2	8.7	9.5	11.7	12.2
7.9	10.3	7.6	14.7	9.1	11.7		

(a) Choose a suitable number of classes and obtain a frequency distribution.
(b) Construct a histogram representing the data.

14. Obtain the cumulative frequency distribution from the following distribution of the number of days absent during a year among eighty employees in a factory.

draw ojive

Number of Days Absent	Number of Employees
0–5	6
6–10	19
11–15	23
16–20	14
21–25	18

15. The following data give the cumulative frequency distribution of the weights (in milligrams) of the adrenal glands of 120 mice. (*Journal of the Science of Food and Agriculture* 35: 1984, pp. 41–46)

draw ojive
+ then convert

Weight (in Milligrams)	Cumulative Frequency
less than 2.45	0
less than 2.95	6
less than 3.45	26
less than 3.95	49
less than 4.45	74
less than 4.95	102
less than 5.45	120

Find how many glands weighed:
(a) less than 3.95 milligrams
(b) not less than 4.45 milligrams
(c) 3.45 milligrams or more
(d) less than 4.45 milligrams but 3.45 milligrams or more.

16. Draw the ogive of the distribution of the lives of 400 light bulbs given in the following table.

Life of Bulb (in Hours)	Number of Bulbs
600–699	85
700–799	77
800–899	124
900–999	78
1000–1099	36

17. Draw the ogive of the distribution of incomes presented below of 1000 families in an economically depressed area.

Income (in Dollars)	Number of Families
less than 8,000	58
8,000–10,999	120
11,000–13,999	278
14,000–16,999	384
17,000–19,999	121
20,000–22,999	39

18. Sketch the ogive of the distribution of the weights of ninety packages of ground chuck given in the following table.

Weight (in Ounces)	Number of Packages
12.1–16.0	8
16.1–20.0	14
20.1–24.0	20
24.1–28.0	23
28.1–32.0	18
32.1–36.0	7

19. The highway police department conducted a survey and clocked the speeds of 200 cars on a highway. The following distribution was obtained.

Speed (in Miles Per Hour)	Number of Cars
48–50	12
51–53	32
54–56	50
57–59	85
60–62	15
63–65	6

(a) Obtain the cumulative frequency distribution.

(b) Draw the histogram of the distribution.

(c) Draw the ogive of the frequency distribution.

20. The starch content (mg/100 g fresh tissue) of ripe pepino fruits was determined in thirty-five fruits by enzymic hydrolysis with amynoglucose. The following data were obtained.

43.3	43.7	67.2	38.8	45.0	49.3	50.6
42.8	57.0	28.8	60.6	37.2	49.9	52.8
83.5	50.5	47.1	21.9	52.5	77.2	58.8
57.5	66.4	89.3	49.5	44.5	59.5	48.4
57.0	40.1	31.8	73.1	52.0	41.8	82.3

(a) Construct a frequency table having eight equal classes with the initial class starting at 16.

(b) Sketch the histogram.

(c) Plot the ogive.

21. Solanidine glycosides are natural constituents of potatoes and in high amounts can cause toxicity. Eighty fresh potato tubers were analyzed and the amount of solanidine (mg/kg) determined. The data are presented in the following frequency table.

SOLANIDINE	
(mg/kg)	**Frequency**
20–29	10
30–39	24
40–49	18
50–59	12
60–69	8
70–79	5
80–89	3

(a) Find the class marks.

(b) Compute the relative frequencies and percent relative frequencies of the classes.

(c) Plot the histogram.

(d) Sketch the ogive.

22. Ninety determinations were made of the amount (mg N/1) of ammonia nitrogen in the water in a bay. The data are presented in the following frequency table.

Ammonia Nitrogen	Frequency
0.35–0.39	10
0.40–0.44	6
0.45–0.49	18
0.50–0.54	24
0.55–0.59	8
0.60–0.64	11
0.65–0.69	8
0.70–0.74	5

(a) Find the class marks.
(b) Plot the histogram.
(c) Sketch the ogive.

23. Plot the stem-and-leaf diagram for the following cholesterol readings (mg/deci-liter of blood) of twenty patients.

| 185 | 230 | 195 | 186 | 240 | 190 | 238 | 254 | 225 | 237 |
| 210 | 224 | 214 | 197 | 203 | 233 | 198 | 215 | 216 | 205 |

24. The following figures give the percentage recovery of chemically reactive lysine in the presence of glucose by a modified trinitrobenzene sulphonic acid procedure.

57	80	83	66	65	78	70
66	56	67	72	80	63	75
65	74	71	72	71	70	74
70	73	71	61			

Plot a stem-and-leaf display.

25. Accumulation of sulfides in bottom deposits are among the main ecologic factors influencing the chemistry and the biochemical conditions in bodies of water. The primary source of hydrogen sulfide and other sulfides is bacterial reduction of sulfates to hydrogen sulfide.

The following data contain the concentration of acid-soluble sulfides (mg/liter fresh ooze) in the bottom sediments of a lagoon. Thirty-two samples were analyzed.

998	938	971	949	1013	982	941	979
1000	1021	1002	995	956	976	1047	986
1010	1015	967	987	990	974	965	998
954	974	984	988	980	1018	985	984

Construct a stem-and-leaf display.

26. The *Digest of Education Statistics,* 1990, published the following distribution of Scholastic Aptitude Test scores in mathematics for female students for 1988–1989. The figures are based on 566,994 females who took the test and give the percentage of students with a specified score or higher.

Scores	Percent of Female Students
200 or higher	100.00
250 or higher	98.87
300 or higher	92.75
350 or higher	81.42
400 or higher	66.77
450 or higher	51.27
500 or higher	35.37
550 or higher	22.15
600 or higher	12.11
650 or higher	5.90
700 or higher	2.03
750 or higher	0.39

(a) Prepare a *less than* table giving percent students scoring less than 200, less than 250 and so on.

(b) How many females scored 400 or higher?

(c) How many students scored less than 450 points?

(d) How many females scored at least 450 but less than 650?

(e) Draw a graph to show the information in the table.

27. As consumers are becoming more aware of the dangers associated with high-fat and high-cholesterol diets, they are switching to low-fat foods. The following pie chart describing the percent of shoppers who choose low-fat cereals, low-fat versions of dairy foods, or low-fat meats appeared in *USA Today*, October 1993, and is based on a national consumer survey conducted by HealthFocus, Inc., Emmaus, Pa.

(a) Do the percentages given in the pie chart add up to what you would expect?

(b) What percentage of respondents reported that they always or usually choose low-fat cereals, dairy foods, and meats?

(c) Present the information contained in the pie chart as a bar graph.

(d) Suppose there were 850 respondents in the survey. Determine how many would be from each category.

**Percent of Shoppers Who Choose Low-fat Cereals,
Low-fat Versions of Dairy Foods, or Low-fat Meats**

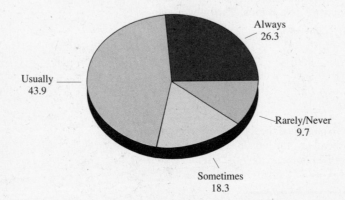

28. The following article on motorcycle fatalities which appeared in the *San Francisco Chronicle* shows how statistical figures are conveniently used and abused by contesting parties to promote their points of view. Provide your own critical analysis, taking into account the arguments presented by the backers and foes of the helmet law.

Source for figure in Exercise 27: Graphic "Percent of Shoppers Who Choose Low-fat Cereals, Low-fat Versions of Dairy Foods, or Low-fat Meats" from *USA Today*, October 1993. Data from 1992 Health Focus® Survey. Base: Health Active Shoppers (90% of food shoppers). Reprinted by permission of The Society for the Advancement of Education.

Motorcycle Fatalities Decline

Backers Cite Helmet Law, but foes Point to Fewer Riders, Accidents

By T. Christian Miller
Chronicle Staff Writer

Deaths and injuries from motorcycle accidents have dropped dramatically since California's helmet law went into effect January 1, but opponents of the measure say the figures are misleading.

Motorcyle fatalities for the first six months of this year fell 43 percent statewide compared to the same period last year, from 227 to 130 deaths, according to California Highway Patrol figures.

The number of motorcycle injuries dropped from 3,443 in the first three months last year to 2,369 for the same period this year, a 32 percent decline.

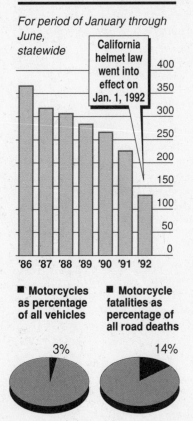

MOTORCYCLE DEATHS

For period of January through June, statewide

California helmet law went into effect on Jan. 1, 1992

'86 '87 '88 '89 '90 '91 '92

■ Motorcycles as percentage of all vehicles

■ Motorcycle fatalities as percentage of all road deaths

3%   14%

Source: "Motorcycle Fatalities Decline" by T. Christian Miller from the *San Francisco Chronicle,* July 28, 1992; Graphic "Motorcycle Deaths" by the *Chronicle*. Graphic's data from the California Highway Patrol. Copyright© San Francisco Chronicle. Reprinted by permission.

"It appears the enactment of this bill has reduced deaths and saved society millions of dollars," said Bob Terry, an aide to Assemblyman Dick Floyd, D-Carson and sponsor of the helmet bill.

But safety experts and opponents of the law said the new figures do not tell the whole story.

Although no hard numbers are available, many motorcycle groups estimate that ridership is down 40 to 50 percent since passage of the law. Coupled with a 41 percent decrease in the number of accidents this year, that means there are more deaths for fewer riders and fewer accidents than last year, they say.

"If you're in an accident now, you're more likely to die," said Paul Lax of the American Brotherhood Aimed Toward Education. "For those of us who ride, we're at greater risk."

Although the statistics dispute that claim—there were 2.3 deaths per 100 accidents for the first three months of this year compared to 2.4 last year —Lax insisted that the new law has made riding more dangerous.

"Motorcycle riding is more deadly since the helmet law," he said. "That's the simple fact."

Lax said the helmets obstruct peripheral vision and hearing, give a false sense of security to riders and increase the chance of a neck injury due to their weight.

Other motorcycle advocates said it is too early to look at the effects of the statistics.

"I think the CHP is jumping to conclusions," said Wes Yeargin of the Modified Motorcycle Association. "It's going to take a year or two, but statistics will show that helmets are not saving lives."

For many, there is another issue at stake that goes beyond whatever the numbers might eventually say.

"The statistics are bogus as hell," said Jim Graessler, a Harley-Davidson rider for 23 years. "We feel we should choose whether we want to wear one of the damn things or not."

Safety experts deemed the news an early vindication of the law, but said more study was needed.

"The preliminary, tentative results are encouraging, but they are not the final word," said Jess Kraus, director of the Southern California Injury Prevention Research Center. "It's a very complicated issue."

Kraus, who is doing a two-year study of the law's effects, said other issues that needed exploration included finding out whether the deaths and injuries were attributable to head injuries and whether fake helmets were worn by those who died.

CHAPTER **1** Summary

✔ *CHECKLIST: KEY TERMS AND EXPRESSIONS*

❑ descriptive statistics, page 3
❑ inductive, or inferential statistics, page 4
❑ population, or universe, page 5
❑ sampling unit, page 5
❑ sample, page 6
❑ observational study, page 7
❑ experimental study, page 8
❑ designs of experiments, page 8
❑ variable, page 9
❑ qualitative variable, page 9
❑ categorical variable, page 9
❑ quantitative variable, page 9
❑ continuous variable, page 10
❑ discrete variable, page 10

❑ parameter, page 12
❑ statistic, page 12
❑ raw data, page 12
❑ array, page 12
❑ range, page 12
❑ outlier, page 12
❑ frequency of a category, page 17
❑ frequency, page 18
❑ frequency distribution, page 18
❑ relative frequency, page 19
❑ percentage frequency, page 19
❑ class frequency, page 21
❑ class intervals, page 23
❑ class size, page 23
❑ class mark, page 24

❑ class boundary, page 24
❑ pie chart, page 31
❑ bar graph, page 32
❑ histogram, page 32
❑ frequency density, page 34
❑ symmetric distribution, page 34
❑ asymmetric distribution, page 34
❑ exploratory data analysis (EDA), page 36
❑ stem-and-leaf diagram, page 36
❑ cumulative frequency distribution, page 39
❑ ogive, page 40
❑ relative frequency ogive, page 40

KEY FORMULAS

class mark

$$\text{class mark} = \frac{\text{lower limit of the class} + \text{upper limit of the class}}{2}$$

frequency density of a class

$$\text{frequency density of a class} = \frac{\text{frequency of the class}}{\text{the class size}}$$

CHAPTER **1** Concept and Discussion Questions

1. The main purpose of a statistical investigation is to study the data collected in a sample. True or false?

2. In 1992 there was much debate in the halls of Congress about how to raise revenues without increasing federal income tax. Two items considered for tax increases were gasoline and cigarettes. Describe your approach to determine the approximate additional revenue that might be collected if the tax on cigarettes was increased by a quarter per pack. Take into account all the possible factors and implications.

3. What purpose do graphical methods of presenting data serve?

4. A college administrator had the following information on newly admitted freshman students: age in years, gender, numerical SAT scores, family income recorded as low-middle-high, and the ethnic designation. Suppose the administrator wants to present the data graphically using pie chart, bar graph, or histogram. Show what graph(s) would be appropriate for each of the variables mentioned.

5. The table below, which appeared in the *U.S. News & World Report*, November 21, 1994, gives percentages of families in different income brackets receiving Medicare benefits.

   (a) Does the table describe the frequency distribution of a variable? Explain.

   (b) Suppose there are two million families in the income bracket $75,000–$99,999. Determine the number of families in this income bracket that do not receive Medicare benefits.

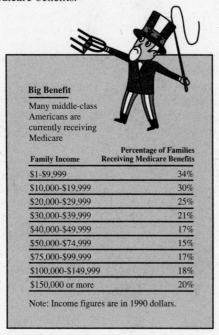

Big Benefit

Many middle-class Americans are currently receiving Medicare

Family Income	Percentage of Families Receiving Medicare Benefits
$1-$9,999	34%
$10,000-$19,999	30%
$20,000-$29,999	25%
$30,000-$39,999	21%
$40,000-$49,999	17%
$50,000-$74,999	15%
$75,000-$99,999	17%
$100,000-$149,999	18%
$150,000 or more	20%

Note: Income figures are in 1990 dollars.

6. A frequency distribution gives the number of values falling in an interval. What does the cumulative frequency distribution give?

7. When a social worker investigated a sample of inner-city children in a certain city, it was found that 33 percent of them were growing up in a single parent home. The social worker published the findings, stating that 33 percent of the inner-city children in the United States were growing up in a single parent home. Comment on the conclusion.

8. Describe the kind of information a histogram provides. What meaning does the area under a histogram carry?

9. Explain why one should not have too few or too many classes in preparing a frequency histogram.

10. Suppose you are given data in bulk. Explain the various aspects you would consider for summarizing it.

Source: Graphic "Big Benefit" from *U.S. News & World Report,* November 21, 1994. Basic data from the Congressional Budget Office. Copyright © 1994 by U.S. News & World Report. Reprinted by permission.

CHAPTER **1** **Review Exercises**

1. Define the following terms.
   (a) population; sample
   (b) parameter; statistic
   (c) continuous variable; discrete variable
   (d) qualitative variable; quantitative variable
   (e) frequency; relative frequency; percentage frequency
   (f) class mark; class boundary
   (g) cumulative frequency

2. Classify the following variables as to whether they are qualitative or quantitative. If the variable is quantitative state whether it is continuous or discrete.
   (a) Number of tickets issued by a policeman
   (b) Size of a shirt specified large, medium, small
   (c) Number of letters on a page
   (d) Height of a redwood tree (in feet)
   (e) Depth of an ocean basin (in feet)
   (f) Number of eggs in a bird nest
   (g) Number of units a student carries in a semester
   (h) Time it takes to solve a certain problem (in minutes)
   (i) Number of organizations in the United Nations
   (j) Price of a house (in dollars)
   (k) Score on the Scholastic Aptitude Test
   (l) Color of the eye of a drosophila fly
   (m) Attitude toward an election proposition (possible responses: favor, do not favor, indifferent)
   (n) Number of patients served by a clinic
   (o) Amount of monosaccharide (mg/100g) in white bread
   (p) The price of a stock traded on New York Stock Exchange
   (q) Shoe size of a person
   (r) Marital status of an individual
   (s) Concentration of citric acid (mg) in the edible pericarp of a fruit

3. The pie charts on the following page describe the percentages of white and black households in the following income groups: $<\$10,000$, $\$10,000-24,999$, $\$25,000-49,999$, $>\$50,000$. Suppose there were 1200 black households and 1600 white households in the survey. Prepare a pie chart for the distribution in the four income groups when the 2800 households in the survey are considered together.

White Households in Income Groups

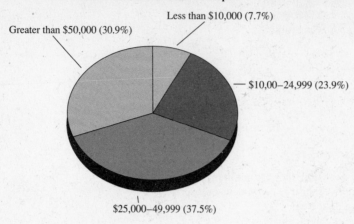

Greater than $50,000 (30.9%)

Less than $10,000 (7.7%)

$10,00–24,999 (23.9%)

$25,000–49,999 (37.5%)

Black Households in Income Groups

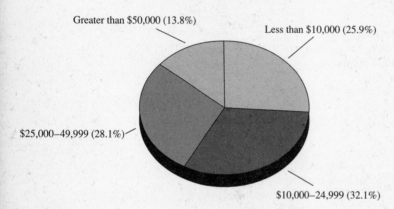

Greater than $50,000 (13.8%)

Less than $10,000 (25.9%)

$25,000–49,999 (28.1%)

$10,000–24,999 (32.1%)

4. A laboratory experiment to measure the tensile strength (in tons per square inch) of thirty specimens of an alloy yielded the following readings.

2.58	2.65	2.40	2.46	2.44	2.31	2.30	2.71
2.38	2.46	2.73	2.43	2.39	2.59	2.68	2.73
2.64	2.52	2.80	2.64	2.71	2.73	2.43	2.59
2.48	2.46	2.54	2.68	2.66	2.60		

(a) Arrange the data in a frequency distribution having the classes: 2.30–2.38, 2.39–2.47, 2.48–2.56, 2.57–2.65, 2.66–2.74, 2.75–2.83.

(b) Obtain the class marks and class boundaries.

(c) Draw the histogram of the distribution.

Source for charts in Exercise 3: *The Struggle for Democracy*, C. Neal Tate, HarperCollins Publishers

(d) Convert the distribution obtained in Part a into a cumulative frequency distribution.

(e) Draw the ogive of the frequency distribution.

5. Plot a stem-and-leaf diagram for the data in Exercise 4.

6. Thirty pigs were each fed a diet containing 1.8 percent linoleic acid (percent of dry feed) from 20 kilograms to 35 kilograms of their weight followed by 1.4 per- cent linoleic acid from 35 kilograms to slaughter at 85 kilograms live weight. The following figures relate to the fat (percent of wet weight) in the tissue composition of the backfat.

78.4	71.5	81.9	84.2	76.9	76.6	82.8
70.3	65.0	81.4	86.7	74.0	69.8	88.5
79.5	79.1	72.9	72.3	76.5	74.4	83.4
75.7	81.0	73.5	84.2	87.7	76.4	84.9
78.1	80.6					

(a) Prepare a frequency distribution with five equal classes, with the first class starting at 65.0.

(b) Plot the histogram.

(c) Sketch the ogive.

7. The following figures give the distribution of sales (in dollars) made by a mail-order service during the pre-Christmas period.

Sales (in Dollars)	Number of Orders
0–199	120
200–399	85
400–599	74
600–799	69
800–999	6

(a) Obtain the cumulative frequency distribution.

(b) Draw the histogram of the distribution.

(c) Draw the ogive of the frequency distribution.

8. The fishing lines that anglers use sometimes break during fishing and are dis-carded on the beach, creating a hazard for beach walkers, birds, and animals. A survey was conducted at a fishing beach and the lengths (cm) of the discarded fishing lines measured. The following table gives the frequency distribution of the lengths.

Length	Frequency
1–20	35
21–40	70
41–60	40
61–80	56
81–100	22
101–120	8
121–140	19
141–160	10

(a) Find the class marks and class boundaries.

(b) Sketch the histogram.

(c) Plot the ogive.

9. Thirty-five milk-yielding Ayrshire cows were fed a diet of silage that was supplemented with a protein concentrate. The following figures refer to the milk fat concentration (g/kg) for each cow.

30	28	27	24	42	35	32
41	20	36	25	30	34	35
26	33	31	34	35	35	33
34	30	34	38	42	37	33
40	21	33	30	46	37	36

Plot a stem-and-leaf display with two lines per stem.

10. The following data give the concentration of a poison (parts per million) found in the brains of twenty-five birds that were found dead near an industrial zone.

8.2	7.3	14.2	7.6	9.8
8.9	9.6	14.5	10.6	7.5
14.4	12.3	11.5	8.8	11.6
11.9	8.8	14.8	12.2	10.8
12.0	12.5	13.0	14.9	8.4

Plot a stem-and-leaf display.

11. The following table shows the concentration of starch (g/100 g) in pepper (Piper nigrum L.) when peppercorns were collected and analyzed twenty-six weeks after pollination. Thirty samples were picked for analysis.

30.2	21.4	27.2	34.1	41.0	28.3	24.6
32.5	32.4	27.0	33.5	23.8	28.2	28.8
37.9	42.0	35.2	31.3	24.3	22.9	26.9
42.7	32.8	36.6	33.8	27.3	23.0	33.3
30.6	33.7					

(a) Prepare a frequency distribution consisting of eight classes with the intial class starting at 20.0.

(b) Sketch the histogram.

(c) In the above data disregard the digit that comes after the decimal point. For example, read 30.2 as 30, 42.7 as 42, and so on. Present a stem-and-leaf display for the new converted data.

12. The following table records the distribution of the amount of magnesium (mg per liter) found in sixty samples of drinking water.

Amount of Magnesium (mg per Liter)	Frequency
6.1–7.0	6
7.1–8.0	8
8.1–9.0	14
9.1–10.0	16
10.1–11.0	10
11.1–12.0	6

(a) Obtain the cumulative frequency distribution.

(b) Draw the histogram of the distribution.

(c) Draw the ogive.

13. In a survey conducted by the Alaska Cooperative Extension Service, the following question was asked: "What is the most serious environmental problem facing the world today?" When 1900 people responded, 20% said pollution, 17% said overpopulation, 14% said deforestation, and the other 49% said such things as global warming, reduced water quality, waste, loss of habitats and animal species, etc.

(a) Draw a pie chart depicting the distribution in the four categories.

(b) Prepare a frequency table by categories.

The following exercise is suited for MINITAB.

14. Lead arsenate is an insecticide commonly used to control light brown apple moths and other pests in vineyards. To investigate the effect of this insecticide on table wines, fifty bottles of white wine from different wineries in a certain region where the insecticide was used were sampled, and the amount of lead (mg/l) was determined by carbon rod atomic absorption spectrophotometry. The following data present the amount of lead in each bottle.

0.44	0.50	0.45	0.42	0.46	0.34
0.44	0.34	0.39	0.41	0.43	0.37
0.47	0.37	0.36	0.45	0.47	0.34
0.39	0.37	0.30	0.40	0.41	0.45
0.45	0.36	0.31	0.39	0.35	0.43
0.43	0.37	0.44	0.38	0.42	0.38
0.46	0.38	0.40	0.42	0.40	0.38
0.47	0.47	0.41	0.44	0.47	0.43
0.46	0.34				

(a) Arrange the data as an ascending array.

(b) Plot stem-and-leaf displays for the above data. First obtain a display with an increment from one stem to the next of 0.02 and then a display with an increment of 0.05.

(c) Plot a histogram where the intial class mark is at 0.30 and the size of each class in 0.03.

WORKING WITH LARGE DATA SETS

Project: Refer to the SAT scores in Appendix IV, Table B-2.

(a) Name the variables involved.

(b) Describe what type of graphic presentations would be suitable for each variable.

(c) Give graphic presentations for the two variables 'sex' and 'age'.

(d) If you are familiar with MINITAB, use the software to graphically highlight the features of the variables 'math scores', 'verbal scores', and 'high school GPA'. Comment on such aspects as skewness, etc.

GRAPHING CALCULATOR INVESTIGATIONS

Data released by the Office of the Secretary of Education in Massachusetts profiled various aspects of school districts within the state. A sampling of the data for 33 school districts is given in the table below. For Exercises 1–6, please refer to this data table. For instructions on how to use your graphics calculator, please refer to the Graphics Calculator Appendix at the back of this text.

1. Enter the data for total 1992 per pupil spending in List L_1 of the TI-82. Next, use the Sort option to arrange these data in ascending order. Reexamine the new sorted list L_1 and note the minimum and maximum values. Create a histogram for these data using the STAT PLOT option. Decide on a suitable viewing window for your histogram using the minimum and maximum values of the list for the range of x values.

2. Use the TRACE button to determine the class boundaries of the histogram you created for 1992 per pupil spending. Note the class boundaries and corresponding frequencies. Use this information to create a frequency table for these data.

3. Enter the data for the per pupil spending on Sports in list L_2 and the per pupil spending on Textbooks in list L_3. Sort both these lists in ascending order and use the new sorted lists to establish a back-to-back stem leaf graph for these data. What is your "visual" impression about the comparative spending in these two areas, based on your graph?

4. Use the sports data in list L_2 to create a histogram for per pupil spending on Sports. Do the same for per pupil spending on Textbooks in list L_3.

5. Enter the data for the percent of children who watch 4 or more hours of TV per day in the 8th grade in list L_4 and in the 12th grade in list L_5. Create a back-to-back stem leaf graph contrasting 8th and 12th grade TV viewing. Comment on similarities and differences that you can observe from visually inspecting this graph. Next, produce two separate histograms for the 8th and 12th grade data.

6. From the variables that remain in the list (for example, drop-out rate, pupil-to-staff ratio, teacher's average salary, etc.) select one and summarize it both graphically in terms of a histogram and tabularly in terms of a frequency table. Present your results to the class.

CITY/TOWN	Total 1992 per Pupil Spending	PER PUPIL SPENDING ON		PCT. OF STUDENTS PERFORMING AT OR ABOVE GRADE LEVEL*				% STUDENTS WITH LESS THAN 1 HR. HOMEWORK		% STUDENTS WHO WATCH 4 OR MORE HRS. OF TV DAILY**		Student Dropout rate%	Early Childhood Program?	Pupils–to–Staff Ratio	Average Teacher Salary
		Sports	Txtbks	MATH 8th	12th	SCIENCE 8th	12th	8th	12th	8th	12th				
Arlington	$5,914	$43	$16	30	28	32	23	57	37	38	13	0.08	No	13.4	$39,856
Belmont	$5,635	$78	$25	45	43	47	43	72	22	35	10	1.1	Yes	14.4	$43,809
Beverly	$4,828	$39	$39	31	28	36	34	63	30	43	8	2.8	NA	14.6	$37,976
Braintree	$4,958	$53	$25	25	38	34	39	49	25	32	8	1.6	NA	14.6	$41,920
Boston	$6,273	$27	$40	12	18	13	14	33	21	29	23	10.7	Yes	13.6	$42,903
Brookline	$6,691	$42	$31	40	40	40	38	77	14	56	10	0.3	NA	11.7	$34,746
Cambridge	$8,586	$34	$37	23	21	22	22	36	20	33	21	2.4	NA	11.7	$42,484
Chelsea	$3,997	$9	$10	.5	14	8	7	38	28	10	35	13.3	Yes	17.1	$34,981
Cohasset	$4,859	$91	$31	31	27	46	40	81	38	29	0	0.3	NA	14.0	$37,490
Dedharn	$5,788	$93	$27	28	31	30	36	58	35	44	17	1.7	NA	13.0	$37,339
Everett	$4,071	$42	$45	19	21	20	22	41	41	19	27	3.3	Yes	19.6	$40,678
Gloucester	$4,683	$36	$17	23	31	26	37	18	38	67	18	4.1	Yes	14.8	$34,124
Hingham	$4,876	$79	$46	32	36	44	30	42	18	72	15	0.2	NA	16.3	$46,137
Haul	$4,611	$45	-	22	22	18	12	7	35	52	19	2.8	Yes	16.0	$40,193
Lexington	$6,511	$91	$36	46	46	56	52	55	22	73	1	0.1	NA	13.9	$47,843
Lynn	$4,129	$20	$20	10	18	18	20	24	39	59	19	7.8	NA	16.4	$33,817
Lynnfield	$5,547	$146	$33	32	39	41	37	52	29	77	10	0.9	NA	15.2	$40,076
Malden	$4,508	$36	$30	13	22	18	25	19	36	68	20	6.6	NA	17.6	$41,060
Marblehead	$5,474	$63	$24	39	34	45	47	49	22	87	2	0.3	Yes	13.7	$38,689
Medford	$5,739	$33	$21	19	19	28	22	29	39	72	13	3.0	NA	13.6	$40,230
Melrose	$4,832	$77	$23	21	28	36	33	45	37	75	12	0.7	NA	15.7	$41,691
Milton	$4,795	$57	$25	26	30	33	35	43	28	70	6	0.5	Yes	14.6	$34,766
Nahant	$4,874	$2	$40	-	-	-	-	-	-	-	-	-	-	15.2	$37,439
Needham	$5,482	$65	$31	29	41	49	40	79	19	60	12		Yes	16.9	$42,663
Newton	$6,260	$50	$13	39	45	48	47	79	19	52	8	0.6	No	17.8	$55,880
Peabody	$4,859	$60	$50	25	22	29	24	51	41	24	12	3.7	NA	16.0	$36,529
Quincy	$5,443	$55	$37	25	29	32	32	51	38	24	10	5.2	NA	14.5	$38,946
Revere	$4,507	$47	$27	19	24	23	23	50	37	16	18	6.5	Yes	18.2	$42,560
Salem	$4,899	$38	$42	20	27	24	27	38	35	12	16	6.1	NA	14.0	$34,948
Saugus	$5,037	$30	$24	26	22	29	24	66	43	23	15	0	NA	14.2	$35,668
Somerville	$5,947	$22	$9	19	20	21	19	29	46	25	11	5.4	NA	14.3	$39,971
Stoneham	$5,336	$71	$15	22	32	35	34	61	40	35	4	1.1	NA	14.2	$34,904
Samnscott	$4,896	$76	$70	31	32	35	42	80	27	29	19	0.5	NA	15.6	$38,573
Wakefield	$5,079	$62	$42	17	28	36	36	61	35	34	12	2.4	NA	14.4	$35,869
Waltham	$6,114	$71	$46	-	-	-	-	56	34	25	13	3.3	NA	13.9	$38,447
Watertown	$6,430	$98	$32	-	-	-	-	54	30	80	13	1.6	NA	13.1	$41,088
Weymouth	$4,342	$6	$33	19	27	25	29	61	41	15	13	2.4	NA	15.3	$38,340
Winchester	$5,506	$80	$19	-	-	-	-	90	24	64	6	0.7	Yes	15.7	$46,678
Woburn	$5,225	$71	$39	12	26	27	27	48	49	16	8	1.6	NA	12.7	$36,559

Source: "How Schools Districts Stack Up" from the *Boston Globe*, June 6, 1993. Data from the Office of the Secretary of Education. Reprinted by permission of the *Boston Globe*.

* NOTE: Proficiency scores in math and science are based on Massachusetts Educational Assessment Program (MEAP) testing scores. The state uses as "grade level" proficiency a MEAP level of III or IV.   ** NOTE: Both daily homework and TV watching figures are based on what students taking the MEAP tests reported in a questionnaire.

The Massachusetts Dept. of Employment and Training released a summary of the unemployment rates for selected communities north of Boston. The data for October 1994 and November 1994 are given in the table.

7. Enter the unemployment data for October 1994 in one list and November 1994 in another list on the TI-82. Sort both data sets in ascending order and present a back-to-back stem leaf graph contrasting the unemployment rates for October and November of 1994. Discuss what this graph indicates about possible similarities and differences. Next, create two separate histograms: one for the October data and the other for the November data.

Unemployment Rates

City/Town	Oct '94	Nov. '94
Amesbury	4.2	4.7
Baxford	4.8	4.2
Beverly	4.4	3.9
Chelsea	7.6	6.7
Danvers	4.2	3.3
Essex	3.8	4.3
Everett	6.2	5.6
Georgetown	5.7	5.0
Gloucester	6.9	7.2
Groveland	5.5	4.7
Hamilton	3.3	3.3
Haverhill	7.5	6.3
Lpswich	4.0	3.8
Lynn	6.8	5.6
Lynnfield	4.7	3.4
Malden	6.1	5.0
Manchester	3.2	3.5
Marblehead	3.8	3.5
Melrose	4.5	3.9
Merrimac	6.3	4.6
Nahant	6.0	4.0
Newbury	5.2	4.3
Newburyport	5.6	4.6
Peabody	5.2	4.5
Revere	7.3	6.3
Rockport	4.6	6.7
Rowley	3.5	3.7

Unemployment Rates (continued)

City/Town	Oct '94	Nov. '94
Salem	5.5	4.8
Salisbury	6.2	6.3
Sangus	6.1	4.9
Swarpcott	4.9	4.3
Wenharn	1.7	1.9
Winthrop	6.8	5.4
West Newbury	4.6	3.5

Source: "Unemployment Rates" from the *Boston Globe*, December 25, 1994. Data from the Mass. Dept. of Employment & Training: Economic Research & Analysis Dept. Reprinted by permission of the *Boston Globe*.

Health officials and consumer groups are concerned about possibly unnecessary cesarean births. Data released by the Massachusetts Department of Public Health in January 1989 compared the percent of births by cesarean in the years 1986 and 1987. Data for 57 Massachusetts hospitals are given in the table:

8. Enter the data for the percent by cesarean in 1986 in one list and the 1987 data in another list. Sort both lists in ascending order and use this to create a back-to-back stem leaf graph for the data. Discuss what this graph suggests about possible differences in the percent of births by cesarean in 1986 and 1987. Then create two separate histograms, displaying the 1986 data on one histogram and the 1987 data on another.

Massachusetts Cesarean Rates: Going Up

	NUMBER OF BIRTHS		% BY CESAREAN	
	1986	**1987**	**1986**	**1987**
Addison Gilbert	291	308	29.9%	29.2%
Anna Jacques	793	860	17.9%	21.5%
Bay State Med. Cntr.	5,637	5,864	26.5%	25.8%
Berkshire Med. Cntr.	1,412	1,428	23.5%	23.1%
Beverly	1,288	1,257	21.1%	22.1%
Bon Secours	1,209	1,158	26.0%	29.8%
Boston City	1,679	1,726	17.9%	21.8%
Brigham and Women's	9,564	9,996	27.7%	26.5%
Brockton	1,222	1,201	17.9%	19.9%
Burbank	935	1,031	18.0%	15.4%

(continued)

Massachusetts Cesarean Rates: Going Up (continued)

	NUMBER OF BIRTHS		% BY CESAREAN	
	1986	1987	1986	1987
Cambridge City	490	526	17.6%	18.7%
Cape Cod	1,167	1,891	25.0%	28.5%
Charlton Memorial	1,915	1,891	21.2%	22.5%
Cooley Dickenson	1,104	1,060	22.8%	24.9%
Emerson	1,873	1,870	19.2%	21.8%
Fairview	116	75	33.6%	30.7%
Falmouth	519	533	27.4%	29.3%
Framingham Union	2,108	2,140	24.8%	25.5%
Franklin Cy. Public	864	926	20.3%	19.4%
Goddard Memorial	2,238	2,321	21.1%	24.3%
Harrington Memorial	548	571	15.7%	18.4%
Haverhill Municipal	596	751	26.5%	20.2%
Henry Heywood Memorial	205	335	16.1%	23.9%
Hunt Memorial	690	745	19.7%	21.1%
Jordan	792	819	23.7%	21.4%
Lawrence General	1,659	1,793	18.3%	21.6%
Leominster	1,141	1,072	19.8%	21.9%
Leonard Morse	658	749	24.7%	26.2%
Lowell General	2,184	2,386	23.0%	22.8%
Lynn	923	908	27.2%	24.4%
Malden	2,093	2,170	22.6%	25.9%
Martha's Vineyard	150	176	32.0%	28.0%
Mary Lane	246	241	17.5%	25.7%
Melrose-Wakefield	1,340	1,244	29.7%	30.6%
Milford-Whittinsville	428	460	33.2%	33.0%
Morton	659	649	29.9%	27.3%
Mount Auburn	1,032	1,149	24.3%	26.0%
Nantucket Cottage	74	81	14.9%	19.8%
New England Memorial	864	704	23.5%	30.5%
Newton-Wellesley	2,534	2,828	20.2%	20.1%
North Adams Reg.	367	390	19.9%	17.9%
Norwood	983	931	24.2%	25.2%

(continued)

Massachusetts Cesarean Rates: Going Up (continued)

	NUMBER OF BIRTHS		% BY CESAREAN	
	1986	**1987**	**1986**	**1987**
Providence	1,276	1,405	17.2%	18.5%
Quincy City	779	866	28.8%	31.9%
Salem	1,387	1,502	26.1%	29.8%
South Shore	1,934	1,941	29.2%	26.6%
St. Elizabeth's	1,117	1,186	29.0%	26.9%
St. Joseph's	313	296	28.4%	35.8%
St. Luke's	2,021	2,053	23.8%	26.9%
St. Margaret's	3,564	3,381	28.5%	32.0%
St. Vincent's	1,790	1,955	16.8%	14.8%
Sturdy Memorial	988	988	26.8%	26.0%
Tobey	284	351	19.0%	19.4%
Waltham	784	699	26.7%	27.9%
Winchester	1,411	1,606	24.2%	27.5%
Worcester Hahnemann	1,075	1,124	26.1%	23.7%
Worecester Memorial	3,145	3,260	25.4%	27.9%
State total	83,722	86,233	23.7%	24.6%

Source: "Massachusetts Cesarean Rates" from the *Boston Globe,* January 27, 1989. Data from the Massachusetts Department of Public Health. Reprinted by permission of the *Boston Globe.*

9. Consider the following data concerning the presidents of the United States.

Presidents of the United States

Name	Age at Inauguration	Age at Death
1. George Washington	57	67
2. John Adams	61	90
3. Thomas Jefferson	57	83
4. James Madison	57	85
5. James Monroe	58	73
6. John Quincy Adams	57	80
7. Andrew Jackson	61	78
8. Martin Van Buren	54	79
9. William Henry Harrison	68	68
10. John Tyler	51	71
11. James Polk	49	53
12. Zachary Taylor	64	65
13. Millard Fillmore	50	74
14. Franklin Pierce	48	64
15. James Buchanan	65	77

(continued)

Presdents of the United States (continued)

Name	Age at Inauguration	Age at Death
16. Abraham Lincoln	52	56
17. Andrew Johnson	56	66
18. Ulysses S. Grant	46	63
19. Rutherford Hayes	54	70
20. James Garfield	49	49
21. Chester Arthur	51	57
22. Grover Cleveland	47	71
23. Benjamin Harrison	55	67
24. Grover Cleveland	55	71
25. William McKinley	54	58
26. Theodore Roosevelt	42	60
27. William Howard Taft	51	72
28. Woodrow Wilson	56	67
29. Warren Harding	55	57
30. Calvin Coolidge	51	60
31. Herbert Hoover	54	90
32. Franklin D. Roosevelt	51	63
33. Harry S. Truman	60	88
34. Dwight D. Eisenhower	62	78
35. John F. Kennedy	43	46
36. Lyndon B. Johnson	55	64
37. Richard Nixon	56	
38. Gerald Ford	61	
39. Jimmy Carter	52	
40. Ronald Reagan	69	
41. George Bush	64	
42. Bill Clinton		

Source: "Presidents of the U.S." from *The World Almanac and Book of Facts, 1993*. Reprinted with permission from *The World Almanac and Book of Facts 1993*. Copyright © 1992. All rights reserved. The World Almanac is an imprint of Funk & Wagnalls Corporation.

(a) Form two groups of presidents. The first will be those inaugurated before 1870 (George Washington to Ulysses Grant) and the second will include more recent presidents (Rutherford Hayes to Bill Clinton). Enter the age of death in two separate lists in the TI-82. Sort both lists and create back-to-back stem leaf plots for the age of death of "early" vs. "recent" presidents. What do you observe ?

(b) Do the same for the age of inauguration for "early" vs. "recent" presidents. What do you observe?

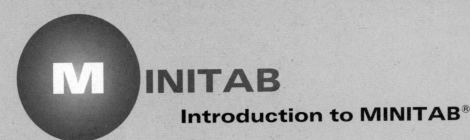

MINITAB

Introduction to MINITAB®

In statistics we must make numerical computations. Fortunately, the computer can assume much of the drudgery of this task and relieve the user of the fear of computational errors. In this capacity, however, the computer is only a tool. Like any tool, we must learn how to use it.

MINITAB* provides an excellent opportunity to simulate theoretical populations, thereby giving relevance to statistical theory. This aspect will be emphasized wherever suitable and appropriate.

MINITAB is a computer program designed as an easy to learn and use software package that does most statistical computations you will encounter in your course. To use the MINITAB program, you must first learn to operate your computer system. One interacts with a computer at a terminal. Each computer system differs slightly, and the procedure necessary to get on and off a given computer terminal can be easily learned by consulting the individuals in charge of the computer facility at your institution.

The following steps must be executed each time you use the MINITAB software package.

1. *Access your computer system:* Check with your computer system for logging-on procedure.

2. *Access MINITAB:* This may involve typing something like FIND MINITAB or just the word MINITAB. Check with your installation. Once you have accessed MINITAB you will see the MINITAB prompt, which, in the latest version of MINITAB, is MTB >. MINITAB is now your servant and awaits your command.

3. *Use MINITAB.*

4. *Exit MINITAB:* When you are through using MINITAB, type the command STOP to leave the program. You are now in your computer's operating system.

5. *Exit your computer system:* Check with your computer system for logging-off procedure.

---

* MINITAB is the registered trademark of Minitab Inc.

MINITAB allows the user to enter data as uppercase letters, lowercase letters, or a combination of both. For consistency, we will use uppercase letters.

As an overview, MINITAB's worksheet has two components: a **spreadsheet** and a **pad.** The spreadsheet is a large electronic "sheet of paper" arranged in 1,000 columns and is, initially, empty. The columns are generically named C1, C2, C3, . . . , C1000 and each can hold up to 1,000 entries.

The pad is an electronic "pad of paper" containing 1,000 "pages" called *constant boxes.*

These constant boxes are generically called K1, K2, K3, . . . , K1000. Only a single number can be placed in a constant box (in contrast to a column which can hold up to 1,000 entries). To start with, the first 998 constant boxes are empty, K999 contains the natural number e ($= 2.71828 \ldots$), and K1000 contains the circumference ratio π ($= 3.14159 \ldots$). Hence, the intial state of MINITAB's worksheet looks as shown below.

```
                        MINITAB Worksheet
           Spreadsheet
           C1   C2    ...     C1000              Pad
        +----+----+----+----+----+        +------------+------+
 Row 1  |    |    |    |    |    |        |            | K1   |
        +----+----+----+----+----+        +------------+------+
     2  |    |    |    |    |    |        |            | K2   |
        +----+----+----+----+----+        +------------+------+
     3  |    |    |    |    |    |        |            | K3   |
        +----+----+----+----+----+        +------------+------+
   ...  |    |    |    |    |    |        |            | ...  |
        +----+----+----+----+----+        +------------+------+
   999  |    |    |    |    |    |        | 2.71828... | K999 |
        +----+----+----+----+----+        +------------+------+
  1000  |    |    |    |    |    |        | 3.14159... | K1000|
        +----+----+----+----+----+        +------------+------+
```

The latest MINITAB release contains a dictionary of about 200 mnemonic English-language commands to which the system responds. These commands operate on the data stored in the worksheet. For MINITAB analysis the data are entered into columns. A command always starts with a command word followed by a list of arguments. Only the first four characters of a command name are needed, together with the numerical arguments, such as column numbers pertinent to the command, and so on. All other text in a command is simply to document the command for the reader's benefit. Be aware that this extra text cannot contain any numbers other than the numerical arguments.

Note: All lines of instruction must end by hitting the RETURN key. The terminal will not send your instruction to the computer until you hit RETURN. Thus, if you want the computer to do something, and the screen is just staring at you with a blinking "cursor" (a little blinking box or dash), then you have probably not hit RETURN.

---

* The treatment of MINITAB in this textbook is for interactive mode only. MINITAB Inc. has developed a windows version for PCs. This version allows you to type commands interactively or to use menus and dialogue to access the analysis.

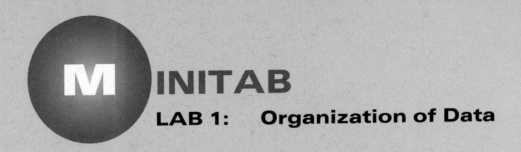

MINITAB

LAB 1:   Organization of Data

I n this lab we consider MINITAB fundamentals and use the features of MINITAB to enter data and produce graphical displays. First log on and access MINITAB.

1 Type in each of the following lines (those in CAPITAL LETTERS) as directed and observe the results. Most MINITAB commands are self-explanatory English-language phrases. Each command must be typed after the prompt and on a separate line.

```
MTB >NOTE ENTER DATA INTO COLUMN C1

MTB >SET DATA INTO COLUMN C1

DATA >2  4  7  5

DATA >1  4,

DATA >3,1,  8

DATA >END OF DATA

MTB >PRINT C1
```

The NOTE command is useful for providing explanatory notes and does not cause the computer to do anything. It is useful for future reference as a record of your operation of the program. When NOTE is the first word on a line, MINITAB overlooks the text after it on the line and does not process it. The symbol # can also be used in place of NOTE for the purpose of annotation. Unlike NOTE, it can be typed anywhere on a line, the text on the line after that being ignored.

Upon giving the SET command above, the prompt changes from MTB > to DATA >, indicating that you should insert the data. When all the data are entered with the SET command, type END OF DATA, signaling that data entry is completed. Notice that the prompt changes back to MTB >.

The SET command above has put nine numbers—2, 4, 7, 5, 1, 4, 3, 1, 8, in this order—into a single column C1. (*Note:* You should not type SET DATA INTO COLUMN 1. Columns are to be specified as C1, C2, . . . and so on.) Data are entered across lines with any number of entries on a line. If we choose to, we could enter the nine values on nine different lines.

Both the blank and the comma act as data separators. Notice that the number of spaces between two numbers is not restricted. However, it is advisable to maintain a consistent pattern as you enter data. It will improve the readability of the data and also save you aggravation in other computer courses that you might take. Since a comma acts as a data separator, it cannot appear within a number. For example, 2,385 will not be considered as "two thousand three hundred and eighty five," but as the two numbers 2 and 385.

2   Now type in the following lines.

```
MTB >SET DATA INTO C1

DATA >2.5  4.8  7.9  8.4  9.6  7.3  6.8  7.2

DATA>3.8 7.3  6.0  4.5  4.8  4.5  2.9  5.3

DATA>3.7 5.2  4.8  3.3  8.6  6.7  5.2  8.9

DATA>END OF DATA

MTB >PRINT C1
```

As a result of this new SET command, you will find that the old data in Column C1 are erased and the new data are entered in that column. Therefore, before you reuse a column make certain that you do not need the old data in that column. If you need to retain the data in an old column, enter the new data in a different column.

3   The command SORT arranges the data given in a column into an ascending array. Since the sorted data are again a column of values, we must specify in which column to place them. For example, the following instructions will sort the data in Column C1, place the sorted data in C10, and print the data in both columns.

```
MTB >SORT C1 PUT ARRAY IN C10

MTB >PRINT C1 C10
```

4   The commands LET, DELETE, and INSERT are useful for editing data. You can change a particular entry in a column with the command LET. In the following lines the LET command replaces the number in Row 6 of Column C1, which is at present 7.3, with the new number 3.8.

```
MTB >LET C1(6) = 3.8

MTB >PRINT C1
```

The LET command is particularly useful when you wish to correct data that have been entered incorrectly. The command can also be used to carry out arithmetic operations, as we will see in the next chapter.

If inadvertently you enter data in a column, you can omit it with the command DELETE. For example, to delete the 3rd, 6th, and 9th entries from Column C1, we would give the command

```
DELETE ROWS 3, 6, 9 FROM C1
```

If the rows are consecutive rows, say Rows 2 to 6, we can abbreviate by using a colon as follows

```
DELETE ROWS 2:6 FROM COLUMN C1
```

If you leave out a set of data from a single column, the command INSERT can be used to enter the omitted values. Suppose you omitted 120, 130, 140 between rows 3 and 4 in column C1. You can enter them as follows and terminate with the command END.

```
MTB >INSERT BETWEEN ROWS 3 AND 4 OF COLUMN C1

DATA>120, 130, 140

DATA>END
```

5 Plotting a histogram is simple with the command HISTOGRAM.

```
MTB >NOTE HISTOGRAM OF DATA IN A COLUMN

MTB >HISTOGRAM OF DATA IN COLUMN C1
```

As an incidental observation, since only the four characters of a command are used by the computer, we could have given an abbreviated command HIST C1 in place of HISTOGRAM OF DATA IN COLUMN C1. In what follows the instructions will be spelled out in detail.

MINITAB chooses its own intervals and constructs the histogram. This is referred to as the default version. It is possible to control MINITAB's display of the histogram by specifying the first midpoint, that is, class mark and the interval width. For example, we could obtain a histogram with the first midpoint at 2.30 and interval width 1.20 with a set of subcommands as follows.

```
MTB >HISTOGRAM OF C1;

SUBC >INCREMENT = 1.20;

SUBC >START = 2.30.
```

The semicolon (;) at the end of the line for the command HISTOGRAM changes the prompt to SUBC >. We now provide the subcommands. Each subcommand line except the last one must end with a semicolon. The subcommand sequence is terminated by placing a period (.) at the end of the last subcommand line. At this point the prompt reverts to MTB >. (As we progress we will come across other commands that have subcommands allowing control over the way the command works. Be sure to place the semicolons and the period appropriately when giving subcommands under a command.)

The HISTOGRAM command, in effect, also gives us the frequency distribution by giving all the class marks (with the heading "midpoints") and the class frequencies (with the heading "counts").

6 MINITAB also has a built-in command to give a stem-and-leaf display. The following example illustrates the procedure.

```
MTB >NOTE STEM-AND-LEAF DISPLAY

MTB >NOTE DATA IN EXAMPLE 3 IN TEXT PAGE (37)

MTB >SET DATA IN C5

DATA>20  14  21  29  43  17  15  26   8  14

DATA >39 .23  16  46  28  11  26  35  26  28

DATA>30  22  23   7  32  19  22  18  27   9

DATA>END OF DATA

MTB >PRINT C5

MTB >STEM-AND-LEAF OF C5
```

The command STEM-AND-LEAF prints a stem-and-leaf plot for data in column C5. The number in parentheses in the first column of the print-out, (6), is the frequency of the middle interval. The numbers in that column preceding (6) represent the cumulative frequencies starting at the lower end of the scale, while the numbers that follow (6) are the cumulative frequencies starting at the upper end. The numbers in the second column represent the stems.

Notice that the display on the monitor is similar to that in Figure 1-12(b) in Section 1-3, except for reordering within each stem. This is because MINITAB has listed each stem on two lines. (In general, MINITAB allows for one, two, or five lines per stem.) To get a display similar to that in Figure 1-12(a) with one stem per line, we would require an increment of 10 from one stem to the next. We specify this under the subcommand INCREMENT as follows and alter the display. (Be sure to place the semicolon and period as shown.)

```
MTB >STEM-AND-LEAF C5;

SUBC >INCREMENT = 10.
```

The display on the screen now resembles the one in Figure 1-12(a), except for reordering. To get five lines per stem, specify INCREMENT = 2 . . Try this.

7 The data also can be entered into columns by using the command READ. With the READ command the data are entered one row at a time, whereas with the SET command they are entered one column at a time.

```
MTB >NOTE ENTER DATA WITH READ COMMAND

MTB >READ DATA INTO COLUMNS C1, C3, C5

DATA >0.93  7.60  8.25
```

```
DATA >0.55  3.95   4.31

DATA >0.87  6.54   6.52

DATA >0.62  4.27   5.15

DATA >0.72  5.85   7.31

DATA >0.79  5.41   6.30

DATA >END OF DATA

MTB  >PRINT C1 C3 C5
```

With the READ command we enter six numbers in each of the columns C1, C3, and C5. It is important to realize that, when we use the READ command, the data for each column are typed one value on each line. This is in contrast to the SET command, which allows any number of values to be typed on a line. In place of the READ command we could, of course, use three separate SET commands, SET C1, SET C3, and SET C5, to enter the above data into columns C1, C3, and C5. (We have to type END OF DATA with the READ command as we do with the SET command.)

Since only one observation per line can be entered with the READ command, it is not as efficient as the SET command if we want to enter data in just one column. For example, to enter data in Column C8, SET C8 is much faster than using READ C8 because in the latter case we have to hit the RETURN key every time we enter a value on a line. In summary, of the two ways to enter data, use the SET command for entering data in a single column and use the READ command for entering data in more than one column.

When data are read in consecutive columns, say C5, C6, C7, C8, one can use a dash and type the command READ C5–C8.

Finally, to be able to enter the data in different columns with a single READ command, each column *must* contain the same number of values.

8 It is often useful to give headings to columns, and the NAME command accomplishes this. Each column name can be up to eight characters long and should be enclosed between single quotation marks with the condition that neither the first nor the last character may be a blank.

```
MTB >NOTE NAME THE COLUMNS

MTB >NAME C1 'REST'

MTB >NAME C3 '5 MIN', C5 '10 MIN'

MTB >NOTE COLUMNS WILL BE PRINTED WITH HEADINGS

MTB >PRINT C1 C3 C5
```

The print command PRINT C1, C3, C5 could also have been given as PRINT 'REST' C3 C5 or as PRINT 'REST' C3 '10 MIN' or any other similar variation. Try these variations.

9   As the final topic of this lab we consider the cumulative frequencies and the ogive.

```
MTB >NOTE OBTAIN CUMULATIVE FREQUENCIES AND PLOT
        OGIVE

MTB >NOTE DATA EXAMPLE 5 FROM TEXT PAGE (39)

MTB >NOTE COLUMN C1 CONTAINS BOUNDARY VALUES

MTB >NOTE COLUMN C2 CONTAINS FREQUENCIES

MTB >READ C1 C2
DATA>15.05     0
DATA>15.35     5
DATA>15.65    10
DATA>15.95    45
DATA>16.25    65
DATA>16.55    46
DATA>16.85    14
DATA>END OF DATA
MTB >PARSUMS C2 PUT IN C5
MTB >NAME C1 'BDRY VAL', C5 'CUM FREQ'
MTB >PRINT C1 C2 C5
MTB >PLOT C5 C1
```

The PARSUMS command calculates the cumulative frequencies of the frequencies given in Column C2 and places them in Column C5. The PLOT command then sketches the ogive. It is important that the column for which the values are plotted along the y-axis (the vertical axis) be mentioned first in the PLOT command.

10   You can save your data in case you wish to use it in the future. Assign a suitable name to the data file, such as LAB1 or MYDATA, and use the command SAVE with the filename enclosed between single quotes. Thus, you could save your data, giving it the filename LAB1, as follows.

```
MTB>SAVE 'LAB1'
```

In the saved file MINITAB will save all the columns of data and the constants. You can retrieve any spreadsheet file by typing RETRIEVE along with the name of the file, enclosed in single quotes, as follows.

```
MTB>RETRIEVE 'LAB1'

MTB>INFORMATION
```

After retrieving the data file type the command INFORMATION and verify that the retrieved file has the correct data.

11 You have learned several MINITAB commands in this lab. If you wish to learn more about these commands, MINITAB has the command HELP, which will allow you to learn more about them. For instance, try the following.

```
MTB >HELP SET

MTB >HELP HISTOGRAM

MTB >HELP PARSUMS
```

To see what commands MINITAB has, type the line

```
MTB >HELP COMMANDS
```

To obtain a virtual course in MINITAB, type this line.

```
MTB >HELP OVERVIEW
```

Assignment Work out Exercise 14 in the review section (page 59).

12 You have completed this session of the lab and now want to quit. Type STOP followed by the system command to log off. That's it!

```
MTB>STOP
```

Log off.

Commands you have learned in this session:

NOTE	SET	DELETE
INSERT	SORT	PRINT
NAME	SAVE	RETRIEVE
HISTOGRAM	STEM-AND-LEAF	PARSUMS
HELP	LET	

Commands useful for statistical work.

```
HISTOGRAM ;                    STEM-AND-LEAF ;

  INCREMENT ;                     INCREMENT .

  START .
```

Numerical Description of Quantitative Data

USA SNAPSHOTS®

A look at statistics that shape the nation

Weight of federal taxes

Federal per capita tax bills increased 6.8% from fiscal year 1993 to 1994. The federal tax bite:

Conn. $7,105

N.J. $6,680

Alaska $5,839

USA $4,701

Miss. $2,975

Utah $3,350

W.Va. $3,353

Lowest

Average

Highest

▲ Source: Graphic "Weight of Federal Taxes" by Cindy Hall and Raul Febles, *USA Today*. Data from Tax Foundation. Reprinted by permission of *USA Today*, July 7, 1994, Section A, p. 1.

INTRODUCTION

In the preceding chapter we considered useful methods of summarizing and presenting data to bring out an overall picture of the data. We now discuss methods of summarizing data by computing values from the observed data. These computations come under two broad headings—*averages,* also called *measures of central tendency,* and *measures of dispersion.*

To the "average" reader the word *average* is familiar from its frequent everyday use. For example, we talk about a batting average in baseball, an average family, the average income of a person, the average annual rainfall, a grade point average, and so on. But most people's definition of an "average family" may only be a vague notion of a "typical" or "representative" family, whatever that means.

In statistical terminology an **average** is a value that typically and effectively represents a given set of data. Since an extreme value, such as the smallest or the largest, is not the most typical observation, a better choice for a representative value should be a number on a numerical scale somewhere at the center of the distribution of the data. For this reason an average is referred to as a **measure of central tendency** or a *measure of location* of the distribution.

The commonly used measures of central tendency are (1) the arithmetic mean, (2) the median, and (3) the mode.

A measure of central tendency describes only one of the important characteristics of a distribution. To understand the distribution well, we must also know the extent of **variability.** The variability of the data is also referred to as the *spread, dispersion,* or *scatter* of the data. Incomes of people with a certain level of education will exhibit some spread, as will heights of people, life lengths of electric bulbs, scores of students on a test, the amount of coffee packed in jars, the number of accidents on a day, and so on.

The commonly used measures of variability are (1) the range, (2) the mean deviation, and (3) the standard deviation.

2-1 MEASURES OF CENTRAL TENDENCY

THE ARITHMETIC MEAN

The most important and widely used measure of central tendency for quantitative data is the *arithmetic mean,* also called the *mean.* In everyday usage, however, this term is erroneously regarded as synonymous with the term *average.* Actually, the mean is one of the averages; the median and the mode are two other averages. We talk about the average income of workers, the average daily intake of calories, the grade point average, and so on. The term *average* is ambiguous and leaves room for convenient misinterpretation of the data.

THE ARITHMETIC MEAN

For any set of measurements, the **arithmetic mean,** or, simply, the **mean,** is computed by adding all the values in the data and then dividing this total by the number of values in the data set. Identifying the observed variable as x,

$$\text{mean} = \frac{\text{sum of the values of } x}{\text{number of values in the data}}$$

Thus, if the quarterly premiums for car insurance quoted by five companies are (in dollars) 238, 325, 274, 258, and 309, then the mean of the quoted premiums is given by

$$\text{mean} = \frac{238 + 325 + 274 + 258 + 309}{5}$$
$$= 280.80 \text{ dollars.}$$

The mean 280.80 is the balancing point of the data as shown in Figure 2-1.

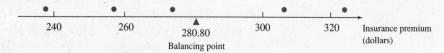

Figure 2-1 The Mean as the Balancing Point

Because the given data constitute a sample, the mean is referred to as the **sample mean.** If the variable values are denoted by x, the symbol customarily used for the sample mean is $\bar{x}$. (Read this as "ex bar".) Suppose there are n values in the data set, written as $x_1, x_2, \ldots, x_n$, where x_1 stands for the first value, x_2 for the second value, and so on. Then, the formula for the mean can be written compactly as

$$\bar{x} = \textbf{mean} = \frac{\sum\limits_{i=1}^{n} x_i}{n}$$

where $\sum\limits_{i=1}^{n} x_i$ is a shorthand expression for the sum of the x_i values for i ranging from 1 to n. That is,

$$\sum_{i=1}^{n} x_i = x_1 + x_2 + \cdots + x_n$$

The symbol Σ, the uppercase Greek sigma, is called sigma notation, or **summation notation.**

Whenever all the x_i values under consideration in a given context are involved in a sum, we shall dispense with the formal notation and simply write $\Sigma\, x_i$ in place of the explicit notation $\overset{n}{\underset{i=1}{\Sigma}}\, x_i$. Just realize that there is a sum involved and that the terms of the sum are given as the index i ranges from 1 to n.* For example,

$$\sum (x_i - 2) = (x_1 - 2) + (x_2 - 2) + \cdots + (x_n - 2)$$

$$\sum f_i x_i = f_1 x_1 + f_2 x_2 + \cdots + f_n x_n$$

and so on. Thus, we shall write

$$\bar{x} = \text{mean} = \frac{\sum x_i}{n}$$

Weighted Mean

Occasionally, it may turn out that the values in a data set carry different relative importance or *weights*. Such would be the case, for example, if you wanted to compute the semester mean in a course, based on two hourly exams and a final exam when the exams have different relative importance. Suppose the second exam counts twice as much as the first exam and the final exam counts three times as much as the first exam. If the scores on the first and the second exam are, respectively, 78 and 72, and the score on the final exam is 86, then the weighted mean is given as

$$\text{weighted mean} = \frac{78 + 2(72) + 3(86)}{1 + 2 + 3} = 80$$

Once again, weighted mean is the balancing point of the data, as shown in Figure 2-2.

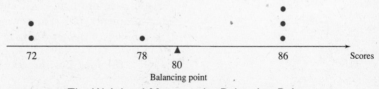

| 72 | 78 | ▲ 80
Balancing point | 86 | Scores |

Figure 2-2 The Weighted Mean as the Balancing Point

---

* The interested reader will find additional material on summation notation in Appendix I.

> ### WEIGHTED MEAN
>
> If the values $x_1, x_2, \ldots, x_n$ in a data set have different weights and the value x_i has weight w_i, then a general formula for computing the **weighted mean** is
>
> $$\text{weighted mean} = \frac{\sum w_i x_i}{\sum w_i}$$

In the formula, each value x_i is multiplied by its weight w_i. The sum of the resulting values then is divided by the sum of the weights.

A word about rounding   As a rule, in any set of computations, rounding should never by carried out in the intermediate steps, but only when presenting the final answer. Other than this, the number of significant figures that should appear in the final answer depends upon the problem involved. Keeping many decimal places in the final answer may not contribute much by way of information. On the contrary, it might only serve in giving an unnecessary stamp of precision.

As a rule, decimals from 0 to 4 are eliminated by rounding down. Thus, a decimal number such as 7.53 is rounded down to 7.5. Decimals from 5 to 9 are eliminated by rounding up. For example, 3.837 is rounded up to 3.84.

In presenting the mean we will have one significant place more than in the original data. If the mean is to be used in subsequent formulas, then as many significant figures as possible should be retained.

Properties of the Mean

The mean has the following important properties.

1. Suppose there are two sets of data and the mean for the first set, with m measurements, is $\bar{x}$ and that for the second set, with n measurements, is $\bar{y}$. Then, weighting the mean of each data set by the number of observations in it,

$$\text{the mean of the combined data} = \frac{m\bar{x} + n\bar{y}}{m + n}.$$

   In the special case when the two data sets have the same number of measurements, the mean of the combined data is $(\bar{x} + \bar{y})/2$.

2. If a fixed value d is added to each of the observations in the data, then

$$\text{the mean of the new data} = d + \text{\textit{mean of the old data}}.$$

   For example, suppose the mean monthly salary of the empolyees working for a company is 1600 dollars. If each empolyee gets a monthly raise of 100 dollars, then the mean monthly salary after the raise will be $100 + 1600$, or 1700 dollars.

3. If each observation in the data is multiplied by a fixed constant c, then

the mean of the new data

= c *times* the mean of the old data.

For example, suppose the height of students is measured in feet and the mean height is found to be 4.8 feet. If it is necessary to express the mean height in inches, then the mean height is 12 (4.8), or 57.6 inches. (This saves us the trouble of converting the height of each student in the original data set into inches and then finding the mean.)

THE MEDIAN

The *median* is another commonly used measure of central tendency. One often hears about median annual salary, median age of a group of individuals, median family income, and so on. The **median** conveys the notion of the middle value, dividing the distribution into two halves when the values are arranged in increasing or decreasing order.

When there is an *odd* number of observations in the data, the median of a set of values is defined as the middle value when the observations are arranged in an array in order of magnitude from the smallest to the largest.

For example, suppose we want to find the median amount spent on groceries when seven customers spent the following amounts (in dollars).

25   28   20   4   47   15   52

Arranging the data in an ascending order of magnitude, we get the following.

Customer number	1	2	3	4	5	6	7
Dollars spent	4	15	20	25	28	47	52

Bottom 3 values   Median   Top 3 values

Three of the customers spent less than $25 and three spent more than $25. Therefore, the median amount spent on groceries is $25.

If the number of observations is *even*, then the median is defined as the mean of the two middle values. As an example, suppose we want to find the median amount when eight customers spent the following amounts (in dollars)

25   28   20   4   47   15   52   62

Arranging the data in an ascending order of magnitude, we get

Customer number	1	2	3	4	5	6	7	8
Dollars spent	4	15	20	25	28	47	52	62

$$\text{Median} = \frac{25 + 28}{2} = 26.5$$

Since there is an even number of observations, the median is the mean of the two middle values, 25 and 28. Therefore, the median is $(25 + 28)/2$, or $26.50.

We can now summarize as follows.

THE MEDIAN

Suppose the observations are arranged in an array in order of magnitude from the smallest to the largest.

If the number of observations is *odd*, then the median is the middle value. If the number is *even*, then the median is the mean of the two middle values.

Remark   With proper interpretation one can define the median of n observations as the $(n + 1)/2$th value in the ascending array. The proper interpretation is called for when n is even. For instance, when $n = 8$, we get $(n + 1)/2 = (8 + 1)/2 = 4.5$. In this case, we will interpret the median as the mean of the fourth and fifth observations, the two ordinal numbers flanking 4.5. *Note:* The median is *not* $(n + 1)/2$. It is the $(n + 1)/2$th largest observation.

EXAMPLE 1   Find the median of 14, 15, 14, 14, and 23.

SOLUTION   The data can be arranged in an ascending order of magnitude as 14, 14, 14, 15, and 23. There are five observations—an odd number. The median is the third value, since $(5 + 1)/2 = 3$. The median is 14.   ●

EXAMPLE 2   The number of school years completed by ten adults was given as follows.

| 6 | 5 | 15 | 6 | 14 | 0 | 18 | 6 | 11 | 10 |

Find the median school years completed.

SOLUTION   We first arrange the data in an ascending order of magnitude as shown in the following array.

Adult number	1	2	3	4	5	6	7	8	9	10
Years completed	0	5	6	6	6	10	11	14	15	18

$$\text{Median} = \frac{6 + 10}{2} = 8 \text{ years}$$

There are ten adults—an even number. Since $(10 + 1)/2 = 5.5$, the median is equal to the mean of the fifth and sixth values, which are 6 and 10, respectively. Their mean gives the median as $(6 + 10)/2$, or 8 years.   ●

While we are discussing the median, it is appropriate to introduce some other measures of location as well. These are *percentiles* and *quartiles*.

PERCENTILES

The sample *p*th **percentile** is a value such that, after the data have been ordered from the smallest to the largest, at least *p* percent of the observations are at or below the value and at least $(100 - p)$ percent are at or above this value.

For example, the 35th percentile of a distribution is a value such that at least 35 percent of the observations are less than or equal to it and at least 65 percent of the observations are larger than or equal to it.

SIDEBAR

*I*t was reported in *Parade* magazine, June 20, 1993, that, with his $200,000 presidential salary, President Clinton still is in the top 1% of American earners. This would suggest that, if the earnings were ranked from low to high, then the figure 200,000 is higher than the 99th percentile. Thus, at least 99 percent of the earners earn less than the president of the United States.

The 25th percentile is called the **first quartile** (or lower quartile) usually denoted by Q_1 or Q_L; the 75th percentile is called the **third quartile** (or upper quartile) usually denoted by Q_3 or Q_U. The median, which is the 50th percentile, is also called the **second quartile.**

The basic notion in defining the quartiles is that they split the entire collection of observations, when arraged in ascending order, into four, essentially equal, contiguous quarters, each containing 25 percent of the values. Of course, the interpretation of this concept becomes intriguing when, for example, we have to divide twenty-two observations into four equal parts. The method for locating quartiles differs depending upon the interpretation. But no matter which interpretation is used, all give answers that are rather close.

As mentioned in the remark on page 82, in order to determine where the median is located we first compute $(n + 1)(1/2)$, that is, $(n + 1)(2/4)$, and interpret the location of the median accordingly. With the same principle in mind, to locate Q_1 and Q_3 we first compute $(n + 1)(1/4)$ and $(n + 1)(3/4)$, respectively, and interpret these numbers to find Q_1 and Q_3. This is shown in Figure 2-3. We will see how the procedure works in the following example.

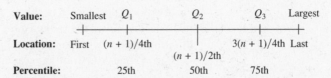

Figure 2-3 Locations of Quartiles

EXAMPLE 3   The following figures represent weekly lumber production (million board feet) during fourteen weeks in the twelve-state western region in the United States.

390	406	446	420	370	328	410
320	368	392	280	325	382	290

Compute Q_1 and Q_3.

SOLUTION   First we arrange the data in an ascending order as follows.

Rank	Value
1	280
2	290
3	320
4	325
5	328
6	368
7	370
8	382
9	390
10	392
11	406
12	410
13	420
14	446

Location of Q_1: Because $n = 14$, we see the following.

$$(n + 1)\frac{1}{4} = (14 + 1)\frac{1}{4} = 3 + \frac{3}{4}$$

We interpret this to mean that the first quartile is between the third and fourth observations, three-fourths of the way above the third observation as shown in Table 2-1. As can be seen from the ranked observations, the third observation is 320. Because the difference between the third and fourth observations is $(325 - 320) = 5$, three-fourths of the way above is $(3/4)5 = 3.75$. Thus,

$$Q_1 = 320 + 3.75 = 323.75.$$

Location of Q_3: Here

$$(n + 1)\frac{3}{4} = (14 + 1)\frac{3}{4} = 11 + \frac{1}{4}.$$

This tells us that the third quartile is between the eleventh and the twelfth observations, one-fourth of the way above the eleventh observation as shown in Table 2-1. From the ranked data the eleventh observation is 406. Since the difference between the eleventh and the twelfth observations is $(410 - 406) = 4$, one-fourth of the way above is $(1/4)4 = 1$. Therefore

$$Q_3 = 406 + 1 = 407.$$

Incidentally, the median is the mean of the seventh and the eighth observations, and is equal to $(370 + 382)/2 = 376$. In summary,

$$Q_1 = 323.8, \ Q_3 = 407.0, \text{ and median} = 376.0$$

TABLE 2–1   Location of Quartiles

Rank	Value	
1	280	
2	290	
3	320	←——3rd observation
		} ← Q_1 located in this area
4	325	←—— 4th observation
5	328	
6	368	
7	370	←——7th observation
		} ← Median located in this area
8	382	←—— 8th observation
9	390	
10	392	
11	406	←——11th observation
		} ← Q_3 located in this area
12	410	←—— 12th observation
13	420	
14	446	

Computation of the quartiles can be very tedious even for a moderate sample size, especially because the values need to be arranged into an array. But the method itself is fairly straightforward.

The **interquartile range** for a set of data, abbreviated **IQR,** gives an interval that covers, approximately, the middle 50 percent of the observations.

INTERQUARTILE RANGE

An **interquartile range, IQR,** of a set of data is the distance between the upper and lower quartiles and is defined as follows.

$$IQR = Q_3 - Q_1$$

For example, the interquartile range for the weekly lumber production data in Example 3 is $407.0 - 323.8 = 83.2$. An interval of length 83.2 contains the middle 50 percent of the values.

BOXPLOTS

A graphic display that brings into focus the middle fifty percent of the observations is called a **boxplot** or a **box-and-whiskers diagram.** It is a useful tool of exploratory data analysis and describes features of the data such as their center of location, spread, departure from symmetry, and length of tails.

One could use the quartiles Q_1 and Q_3 to consider the middle 50 percent of the data, and some authors do indeed describe a boxplot using these values. But because there are different interpretations of the quartiles, the locations of the lower 25 percent and the upper 25 percent are determined by computing what are called the *hinges*. The **lower hinge** is the median of the lower half of the data and the **upper hinge** is the median of the upper half.

If n, the total number of observations in the data set, is even, then the lower half will contain the lower $n/2$ values and the upper half, the upper $n/2$ values.

If n is odd, the lower half of the data will consist of the lower $(n + 1)/2$ values, that is, all the lower values up to and including the median of the original data set. The upper half of the data will contain the upper $(n + 1)/2$ values, that is, all the values greater than or equal to this median. So the median is in both the lower half and the upper half.

We now show the approach by describing how to display a boxplot for the data in the preceding Example 3.

EXAMPLE 4   Draw a boxplot for the following data.

| 390 | 406 | 446 | 420 | 370 | 328 | 410 |
| 320 | 368 | 392 | 280 | 325 | 382 | 290 |

SOLUTION   Follow the steps listed below.

1. First, we arrange the data in ascending order of magnitude and compute the median for the entire data set. We have already done this in Example 3 and have computed the median as 376.
2. Next, we calculate the lower and the upper hinges. They are, respectively, the medians of the bottom half of the data and the upper half of the data.

The lower hinge   From the data arranged in Example 3, the lower half data consist of the following values.

| 280 | 290 | 320 | 325 | 328 | 368 | 370 |

The median of these values is 325. Thus, the lower hinge is 325. *Note:* This value is close to $Q_1 = 323.8$.

The upper hinge   The upper half of the data is as follows.

| 382 | 390 | 392 | 406 | 410 | 420 | 446 |

Since the median of this data is 406, the upper hinge is 406. *Note:* This value is close to $Q_3 = 407$.

3. Finally, we sketch the boxplot as follows. First, we draw a horizontal axis and above this axis locate the smallest value in the data set, the lower hinge, the median, the upper hinge, and the largest value in the data set.

    A boxplot consists of a box drawn in such a way that its left edge is at the lower hinge and the right edge is at the upper hinge.

Thus, the box encloses essentially the middle 50 percent of the values. Also, a vertical line is drawn inside the box at the median. This marker, which corresponds to the center of location, brings out the symmetry, or the lack of it, based on the departure of the median line from the center of the box. If the median is near the center of the box, then there is a strong indication that the middle 50 percent of the data is symmetrically distributed.

To indicate the extent of the spread of the data, lines are drawn connecting the left edge of the box to the smallest value and the right edge of the box to the largest value as shown in Figure 2-4.

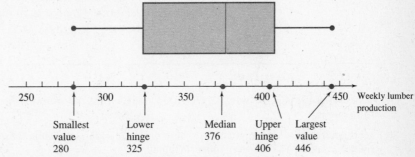

Figure 2-4 Boxplot for Weekly Lumber Production

Boxplots are useful for comparing distributions of different sets of data by displaying the plots next to one another.

THE MODE

The word *mode* in French means "fashionable," and in the context of a frequency distribution, mode means the most common value.

The **mode** is defined as the value that occurs most frequently in the data, that is, it is the value that has the highest frequency.

Of all the measures of central tendency, the mode is the easiest to obtain once the data are arranged in a frequency table. We find it simply by inspection.

It may happen that there are some values in the data that occur equally often, but more frequently than the rest of the values. Each of these values is a mode. For example, suppose the data consist of the following.

1.0   1.1   1.2   1.2   1.2   1.4   1.4   1.4

The values 1.2 and 1.4 occur equally often—three times—which is more often than the other observations, 1.0 and 1.1. Our data have two modes, 1.2 and 1.4. Therefore, the data are called *bimodal*. In general, if the data have more than one mode, they are said to be *multimodal*.

| EXAMPLE 5 | Discuss the nature of the mode for the following sets of data. |

(a)   2, 3, 7, 7, 7, 8, 9, 9

(b)   2, 3, 7, 7, 7, 7, 8, 9, 9, 9, 9

SOLUTION

(a)   Since 7 is the value that occurs most often (three times), 7 is the mode. Also, 7 is the only mode.

(b)   This set of values has two modes, 7 and 9. They each occur four times. Thus, the data are bimodal.   ●

When the data are grouped in a frequency distribution, the class with the highest frequency is called the *modal class*. The class mark of the modal class is then called the mode. Of course, in this context we must require that all the classes be of the same length.

Section 2-1 Exercises

1. The scores of five students were 52, 68, 76, 55, and 84. Find the arithmetic mean of the scores.

2. Eight children in a kindergarten class were observed separately while they as-sembled a puzzle. The observed times (in minutes) for them were 15, 8, 13, 9, 16, 13, 18, 10. Compute the arithmetic mean of the time to assemble the puzzle.

3. The following temperatures (in Fahrenheit) were recorded in a city on twelve days in winter: 18, 5, 22, −9, 5, 4, −10, 8, 18, −18, 0, −8. Compute the mean temperature.

4. Ten tobacco farmers reported that they had received the following prices (in dollars) per pound of tobacco: 1.64, 1.53, 1.49, 1.60, 1.42, 1.55, 1.68, 1.40, 1.38, 1.58. Compute the mean price per pound.

5. In a class of ten students, the lowest score on a test was 40 points and the highest score was 80 points. State which of the following statements are true.
   (a) The mean score is (80 + 40)/2, or 60.
   (b) The mean score is greater than 40.
   (c) The mean score is less than 80.

6. The mean speed of six race drivers was found to be 120 miles per hour. It was realized, however, that the speed of one of the drivers was not recorded and that the speeds of the other five drivers were 140, 110, 120, 130, and 110 miles per hour. Find the missing speed.

7. During the first ten weeks of 1988, the mean weekly receipts of a city transit company were $840,000. In the next ten weeks, the mean weekly receipts were $860,000. Find the mean weekly receipts for the first twenty weeks.

8. If the mean weekly receipts of a transit company for the first ten weeks were $840,000 and for the next twenty weeks were $860,000, find the mean weekly receipts during the first thirty weeks.

9. The mean weight of ten items in one box was 30 pounds, of fifteen items in a second box was 60 pounds, and of twenty items in a third box was 20 pounds. Find the mean weight of all the items in the three boxes.

10. The mean score of a student on the first three tests is 81 points. How many total points must the student score on the next two tests if the overall mean score has to be 85 points?

11. (a) Suppose the mean of the first twenty observations is 29.8. What would the next observation have to be for the mean of the first twenty-one observations to be 30?

    (b) Suppose the mean of the first 1000 observations is 29.8. What would the next observation have to be for the mean of the first 1001 observations to be 30?

    If you compare your answers in Parts a and b above, what conclusion can you draw?

12. Using the formula for the mean, find the mean of the data in Part a. Use this answer to find the means of the data in Parts b and c.
    (a) 24, 32, 28, 26, 25
    (b) 74, 82, 78, 76, 75
    (c) 10,124; 10,132; 10,128; 10,126; 10,125
    Hint: The data in Part b are obtained by adding 50 to each observation in Part a; and the data in Part c are obtained by adding 10,100 to each observation in Part a.

13. Using the formula for the mean, find the mean of the data in Part a. Use this answer to find the means of the data in Parts b and c.
    (a) 8, 6, 9, 5
    (b) 24, 18, 27, 15
    (c) 56, 42, 63, 35
    Hint: The data in Part b are obtained by multiplying by 3 each observation in Part a; and the data in Part c are obtained by multiplying by 7 each observation in Part a.

14. A worker made ten flags. The mean length of cloth per flag was found to be 2.5 yards. Express the mean length in inches.

15. Ace Garden Nursery employs nine student helpers. Suppose each student is paid $5.50 an hour and the number of hours they work per week during the school year are as follows:

    5   7   8   7   5   9   5   9   11

    (a) Find the mean number of hours a week that a student works at the nursery during the school year.
    (b) Suppose the owner of the nursery wants each student to work an additional four hours per week during the summer. Use the answer in Part a to compute the mean amount that will be paid per student during a week in summer.
    (c) Find the amount that each student will be paid during a week in summer. Using these figures, find the mean amount paid to a student per week during summer. Compare this answer with that in Part b and comment.

16. A quality control engineer was interested in whether a machine was properly adjusted to dispense 1 pound of sugar in each bag. Thirty bags filled on the machine were weighed. The data given below refer to the net weight of sugar (in ounces) packed in each bag.

15.6	15.9	16.2	16.0	15.6	16.2
15.9	16.0	15.6	15.6	16.0	16.2
15.6	15.9	16.2	15.6	16.2	15.8
16.0	15.8	15.9	16.2	15.8	15.8
16.2	16.0	15.9	16.2	16.2	16.0

Find the mean weight per bag.

17. The number of guests in a hotel during nine weeks were as follows.

455	150	327	169	321	2568	398	1352	387

Find the median number of guests.

18. The following are the speeds (in miles per hour) attained by eight race drivers.

128	140	126	137	148	130	142	133

Find the median speed. Interpret the value by describing what it tells you regarding the distribution of the data.

19. The following figures refer to the diameters (in inches) of six bolts produced by a machine.

2.0	2.3	2.1	2.0	2.1	2.1

Find the median diameter.

20. Uma, a seventh grader at Sunny Brae Middle School, participated in the Jump Rope for Heart program. She had 28 sponsors and they pledged the following amounts:

5.00	3.50	3.50	7.50	10.00	5.00	7.50
10.00	5.00	7.50	7.50	5.00	5.00	3.50
7.50	3.50	3.50	3.50	3.50	3.50	7.50
7.50	7.50	5.00	5.00	7.50	7.50	7.50

Find the median amount pledged.

21. There are thirty companies which are considered in the computation of the Dow Jones Industrial Average, an index which serves as a barometer of the economic activity of the U.S. market. The earnings per share and the dividends paid by these companies in 1993 are presented below.

	1993*	
	Earnings	**Dividend**
Allied-signal	2.31	0.58
Aluminium Co. of America	0.79	1.60
American Express	2.30	1.00
AT&T	3.15	1.38
Bethlehem Steel	−0.18	0.00
Boeing	3.66	1.00
Caterpillar Inc.	3.36	0.30

	1993*	
	Earnings	Dividend
Chevron	2.80	1.75
Coca-Cola	1.68	0.68
Disney	1.81	0.24
Dupont	2.46	1.76
Eastman Kodak	2.56	2.00
Exxon	4.21	2.88
General Electric	3.03	1.26
General Motors	2.13	0.80
Goodyear	3.23	0.58
IBM	0.00	1.00
Int'l Paper	2.54	1.68
McDonald's	1.46	0.21
Merck	2.33	1.03
Minnesota Mining	2.91	1.66
Morgan (J.P)	8.48	2.48
Philip Morris	4.06	2.60
Procter & Gamble	3.04	1.17
Sears, Roebuck	4.56	1.60
Texaco	3.98	3.20
Union Carbide	1.00	0.75
United Technologies	3.30	1.80
Westinghouse	0.76	0.40
Woolworth	−3.76	1.15

*Source: *From the Value Line* investment survey, December 23, 1994. Copyright © 1994 by Value Line Publishing, Inc. Reprinted by permission. All rights reserved.

(a) Compute the mean and the median earnings per share of the companies.

(b) Compute the mean and the median of the annual dividend paid in 1993 by the companies.

22. During the three-hour period from 9 A.M. to 12 noon on a certain day, a bank teller observed that customers arrived at the following hours.

9:05	9:35	9:50	10:15	10:30	10:40
10:55	11:15	11:28	11:40	11:58	

If the time between any two successive arrivals is called *interarrival time,* find the following.

(a) The mean interarrival time   (b) The median interarrival time

23. The following data give the saponin content (mg/100 g) of unprocessed samples of eight high-yielding varieties of chick-peas (*Cicer arietinum*).

4138	3952	3980	3967
3992	4197	4015	4057

Calculate the following.

(a) The mean saponin content

(b) The median saponin content

24. The following figures give the β-lactoglobulin, as a percent of total nitrogen, in
twelve samples of cow's milk.

9.6	9.7	10.0	8.8	9.5	9.4
9.2	9.8	9.5	8.5	10.1	9.3

Calculate the following.
(a) The mean β-lactoglobulin
(b) The median β-lactoglobulin

25. In its survey of what people earn, *Parade* magazine, June 20, 1993, reported the
following: "In 1992, the median wage for workers was $23,140, (which is) 3.4%
higher than in 1991." Interpret the information contained in this statement.

26. To test the braking system on a new model of a car, a consumer agency measured
the stopping distances (in feet) with eighteen cars driven at 40 miles per hour by
drivers of proven ability. The data are recorded below.

114	122	113	114	96	132
112	99	119	107	136	103
115	98	131	128	104	134

Construct the stem-and-leaf diagram and use it to locate the median.

27. The sample sizes n are as follows. In each case compute $(n + 1)(1/4)$ and give
the ranks of the observations between which Q_1 lies. Also, tell how much of the
way above the lower-ranked observation it lies.
(a) $n = 28$ (b) $n = 33$ (c) $n = 42$ (d) $n = 55$
(e) $n = 48$ (f) $n = 89$ (g) $n = 70$ (h) $n = 39$

28. The sample sizes n are as follows. In each case compute $(n + 1)(3/4)$ and give
the ranks of the observations between which Q_3 lies. Also, tell how much of the
way above the lower-ranked observation it lies.
(a) $n = 28$ (b) $n = 33$ (c) $n = 42$ (d) $n = 55$
(e) $n = 48$ (f) $n = 89$ (g) $n = 70$ (h) $n = 39$

29. Sections of bacon each weighing 100 grams and consisting of both back fat and
lean muscle tissue were homogenized in a food processor and then chemically
treated. The copper content was then assayed by submitting the samples for
copper analysis by flame atomic absorption spectroscopy. The following data
give the copper content of the bacon ($\mu g/g$).

0.55	0.46	0.72	0.32	0.45	0.47
0.51	0.39	0.36	0.62	0.71	

Calculate the quartiles Q_1, Q_2, and Q_3, and interpret them.

30. The starch content (mg/100 g fresh tissue) of ripe pepino fruits was determined
in thirty fruits by enzymic hydrolysis with amynoglucose.

43.3	43.7	67.2	38.8	45.0	49.3	50.6
42.8	57.0	28.8	60.6	37.2	49.9	52.8
83.5	50.5	47.1	21.9	52.5	77.2	58.8
57.5	66.4	89.3	49.5	44.5	59.5	48.4
57.0	40.1					

(a) Compute the following.
   i. The mean starch content
   ii. The median starch content
   iii. The quartiles Q_1 and Q_3

(b) Prepare a histogram and locate the mean, median, Q_1, and Q_3 along the horizontal axis

(c) Sketch the boxplot.

31. The following data give the biomass of the zooplankton (g/m^3) in Korchevatsky Bay in Russia during 1975–1979. (From "The contributions of zooplankton to the food of fish grown in tanks and ponds near power plants" by S.A. Krazan and S.A. Fil.)

2.44	0.52	3.30	0.60	1.20
0.80	0.26	1.29	0.48	2.00
0.47	0.80	0.80	0.26	0.60
0.15	0.41	0.14	0.30	0.17

(a) Compute the following.
   i. The mean biomass amount
   ii. The median biomass amount
   iii. The quartiles Q_1 and Q_3

(b) Sketch the boxplot.

(c) Comment on the symmetry of the data.

32. Because of their nutritional importance, an analysis was conducted to find the basic organic constituents of the tissue of the edible freshwater fish *Mystus vittatus*. Adult fish were freshly collected and biochemical estimations of the total carbohydrate content were made using Anthrone reagent. The following data give the amount of total carbohydrates in the liver tissues of the fish.

35.7	34.9	35.3	35.7	35.4	36.0	35.9	37.0
33.4	37.5	35.8	37.6	33.9	37.3	36.7	

(a) Find the following.
   i. The mean carbohydrate level
   ii. The median carbohydrate level
   iii. The quartiles Q_1 and Q_3

(b) Sketch the boxplot.

Find the mode(s) for the sets of data in Exercises 33–39.

33. 14, 15, 16, 14, 14, 15, 16, 10

34. 14, 14, 13, 12, 15, 15

35. 14, 17, 18, 27, 26, 17

36. 4.2, 3.5, 3.9, 3.9, 3.8, 3.9, 4.2, 4.2

37. −3.1, −4.0, −5.3, −6.8, −4.0, −6.8

38. 2.8, 2.8, 3.1, 4.5, 3.1, 2.8, 3.1

39. −4, −5, −7, −7, −5, −7, −4, −4, −5, −2, 0

40. On a fishing trip Harry caught ten trout. The following figures refer to their sizes (in inches).

10.2	11.5	10.6	10.6	12.2
11.5	12.1	11.5	10.2	10.2

Find the mode.

41. A hardware store has in stock nails of the following different sizes (in inches).

0.5   1.0   1.5   2.0   2.5
3.0   3.5   4.0   4.5   5.0

If 10 percent of the nails are of size 2.0, 15 percent are of size 3.0, and 53 percent are of size 3.5, can you find the mode? Justify your answer.

42. The following table gives the distribution of child occupants, ages 1–5, killed in motor vehicle crashes from 1976–1980. (*Am. J. Public Health* 76 (1986): 31–34.)

Age (Years)	Number of Children
1	698
2	747
3	594
4	538
5	513

Find the modal age of children killed.

2-2   COMPARISON OF MEAN, MEDIAN, AND MODE

As a brief summary of the definitions, the *mean* is the arithmetic mean, the *median* is the middle value when the observations are arranged in ascending or descending order, and the *mode* is the most frequent value.

If the data are displayed as a relative frequency histogram, then the median is the point on the *x*-axis such that half of the area under the histogram is to either side of the median. The mean, $\bar{x}$, is the balancing point of the histogram. The mode has the tallest rectangle erected above it.

If the distribution of the data is symmetric with a single mode, then the mean, median, and mode will coincide, as in Figure 2-5(a). However, if the distribution is skewed, the three measures will differ from each other. If the data are skewed to the right, then the mean will be greater than the median as in Figure 2-5(b). On the other hand, if the data are skewed to the left then the opposite will be true, as shown in Figure 2-5(c).

Of the three measures of central tendency that we have discussed, the arithmetic mean is by far the most commonly used measure and the mode is the least used. The mode is not even well-defined in many data sets since it many not be unique. We list below the advantages and disadvantages of these measures.

1   The mean can always be computed when all the measurements are available, and every observation is involved in its computation. Computation of the median or the mode does not take into account individual values in the data. If the mode exists, it is the easiest measure to find.

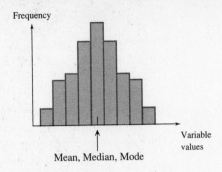

(a)

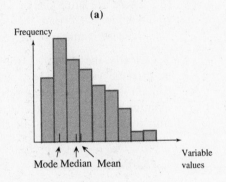

(b)

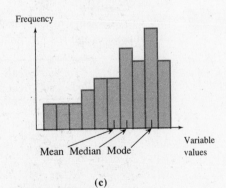

(c)

Figure 2-5
a. Symmetric distribution; mean, median, mode coincide
b. Distribution skewed to the right
c. Distribution skewed to the left

2   From the standpoint of inductive statistics, the mean is more tractable mathematically than either the median or the mode and has many desirable properties for sound tests.

3   The mean is sensitive to extreme values, which are not representative of the rest of the data; that is, it is strongly affected by an extremely large value or an extremely small value especially with small sample sizes. The critique of the EconoWatch article at the end of this section illustrates that the sensitivity of the mean to the outliers is reduced when the sample size is very large. This sensitivity does not occur with either the median or the mode, which are positional averages. For example, the median of 100, 200, and 300 is 200, and the median of 100, 200, and 30,000 is also 200. Similarly, the mode of the set of values 10, 15, 15, 18, and 16 is 15; and that of −2, 15, 15, 2000, and 30,000 is also 15.

4   If we are given means for several sets of data, together with the number in each set, we can find the mean for the combined data without having to compute it from the original data. This is not possible for the median and mode.

5   There are situations in which the mean is to be preferred over the median, even when there are extreme values present. This would be the case if one was interested in the sum or the total.

Clearly, the main disadvantage of the mean is that it is affected strongly by extreme values, that is, outliers. It gives a distorted picture if the data contain just a few observations that are disproportionately higher (or lower) than the rest of the data. The median or the mode is to be preferred in such situations. For example, consider the annual incomes of the following five families.

$18,000 $20,800 $29,720 $22,760 $195,000.

The mean, equal to $57,256, gives a distorted picture of the incomes, whereas the median, which is $22,760, gives a more realistic picture.

EXAMPLE   Two supermarkets in California, Food Baron and Food Market, gave the following sales figures for meat sold (in pounds) on seven days.

Food Baron (Pounds of Meat Sold)	Food Market (Pounds of Meat Sold)
1700	1000
1600	1100
1400	1600
1300	4200
1600	1000
1300	1300
1600	1000

Compare the performances of the supermarkets by considering the measures of central tendency.

SOLUTION   First we obtain the measures of central tendency as follows.

	Food Baron	Food Market
Mean	1500	1600
Median	1600	1100
Mode	1600	1000

If we used the mean as the criterion for deciding which supermarket fared better, we would certainly have to conclude that Food Market is ahead in its performance. However, a closer look at the sales figures conveys an entirely different picture. The performance of Food Baron is very consistent, having consistently sold around 1500 pounds. On the other hand, the performance of Food Market is more erratic, and the one figure that has inflated the mean of the Food Market data is 4200 pounds. The measures of spread of data will be the topic of discussion in the next section.

If we look at the median amounts sold, we get a more realistic picture. On 50 percent of the days, Food Baron sold at least 1600 pounds. The corresponding figure for Food Market is only 1100 pounds.

Finally, considering the mode, we find that Food Baron sold 1600 pounds most often, namely, three times. In contrast, the amount sold most often by Food Market is 1000 pounds.

In terms of consistent performance of amount of meat sold, the Food Baron fared better.   ●

SIDEBAR

*I*n the Associated Press Econowatch article, published in the *Times-Standard of Eureka*, October 11, 1993, the *average* prices for gasoline are reported based on a survey of 10,000 gasoline pumps, a rather large sample indeed.

What specific average is the report giving? Is it the mean, or the median, or the mode? The average price most probably refers to the mean in this case since the figure is reported to four places of decimals.

EconoWatch

LOS ANGELES (AP)—

Gasoline prices rose more than 4 cents a gallon over the past two weeks because of a new federal tax at the pumps, an industry analyst said Sunday.

The average price of self-serve and full-serve gas, incluling all grades and taxes, was $1.1754 per gallon on Sept. 24, according to the Oct. 8 Lundberg Survey of 10,000 gasoline stations nationwide.

The increase of 4.33 cents over the previous survey was mainly due to the new 4.3-cent-a-gallon federal gas tax that went into effect Oct. 1, said analyst Trilby Lundberg.

At self-service pumps, the average price for regular unleaded gasoline was $1.1062 per gallon.

Source: EconoWatch from the Associated Press, October 11, 1993. Reprinted by permission of Associated Press.

Since you are aware of the distinction between measures, in the future when you come across any reference to average, you should show the curiosity and determine, if possible, what specific measure the average represents. Otherwise, the reported figure could be misleading, and should be taken with a pinch of salt. This confusion is often conveniently exploited by contesting antagonists in a dispute. For example, in salary contract negotiations the management will often quote the average that happens to be the largest as a measure of their generosity and the labor union will quote the one that happens to be the lowest to boost their demand for higher wages.

According to the article, the average price of self-serve and full-serve gas, including all grades and taxes, was $1.1754 per gallon on Sept 24. The sum of all the prices at the 10,000 gas pumps will, therefore, be $1.1754 \times 10,000 = 11,754$ dollars. Suppose, for the sake of discussion, there exists a gas station which charges an exorbitant price of $20.00 per gallon and you want to include it in the survey. By how much would the mean change? The mean based on the 10,001 pumps will be

$$\frac{11754 + 20}{10001} = 1.1773 \text{ dollars}$$

The change in the mean price is $1.1773 - 1.1754 = 0.0019$, not appreciable at all. This goes to show that if the mean is based on a very large sample, it is not all that susceptible to the influence of outliers.

Section 2-2 Exercises

1. We know that if the distribution is symmetric with a single mode, then the mean, the median, and the mode coincide. Is it true that if the mean, the median, and the mode coincide, then the distribution must be symmetric?

2. What measure of central tendency would you use if you were seeking the most common, or typical, value?

3. What measure of central tendency would you use if an exceptionally high or low value was affecting the arithmetic mean?

4. Which of the following statements are true?
   (a) The mean, the median, and the mode represent frequencies of values.
   (b) The mean, the median, and the mode represent values of the variable.
   (c) Only the median and the mode represent frequencies.

5. Find the mean and the median when all the observations in the data are identical, say, equal to 20.

6. The following figures give the medical expenses (in dollars) incurred during a year by ten families.

   820   430   390   10,760   550   1950   476   340   25,570   330

Find the mean medical expenses and the median medical expenses. Which measure do you think is more appropriate as a measure of central tendency?

7. The following figures represent the closing prices of a stock (in dollars) on eight trading days: 12, 11, 15, 12, 13, 18, 48, and 91. Find the mean closing price, the median closing price, and the modal closing price. Which measure of central tendency do you think is least satisfactory?

8. In a frog race with ten frogs, three frogs strayed away. For the remaining frogs that completed the race, the following times (in seconds) were recorded; 185, 200, 140, 145, 245, 580, and 600. Choose an appropriate measure of central tendency for the time to complete the race by the ten frogs and find its value.

9. A psychologist assigned an identical task to each of eight students in special education classes. The following figures give the time (in minutes) that each child took to complete the task: 28, 36, 38, 50, 52, 60, 140, and 168. What is the most appropriate measure of central tendency? What is its value?

10. Suppose you are a retailer of sportswear and have data available on sizes of jackets sold during this fiscal year. If you are interseted in replenishing your stocks for the next year, what measure of central tendency would be of greatest use to you?

11. Suppose the mean annual salary this year of ten faculty members in the statistics department of a university is $49,500 and the median annual salary is $47,800. If the highest paid faculty member is the only one who gets a raise of $3000 next year, what is the new mean annual salary? What is the new median annual salary?

12. The net income (in millions of dollars) of seven companies in the first quarter of 1993 was reported as follows. (*Note:* Negative net income should be interpreted to mean that the company incurred a loss.)

Company	Net Income
Merck	614
Bankamerica	484
Sears	435
United Technologies	64
Delta	-134
IBM	-285
AT&T	-5,644

Compute a measure of central tendency that would be most appropriate for this type of data. Justify your choice.

13. The following chart, which appeared in *Time* magazine, November 15, 1993, shows the current population of families on welfare grouped according to the total number of years they are likely to receive welfare payments, whether or not those years are consecutive.

Limiting Welfare: Two Years and Out

Last week the Clinton administration permitted Wisconsin to impose a two-year limit for welfare recipients, and a two-year limit is often mentioned as part of President Clinton's own plans to reform welfare. The majority of first-time welfare recipients spend less than two years in the program, but 45% of first-time recipients eventually go back on welfare. In the chart, the current population of families on welfare is broken down according to the total number of years they are likely to receive welfare payments, whether those years are consecutive or not.

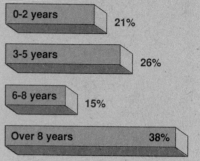

0-2 years 21%

3-5 years 26%

6-8 years 15%

Over 8 years 38%

(a) Which measure of central tendency would you use if you were interested in computing the average number of years that families are likely to receive welfare payments, whether or not those years are consecutive?

(b) Give as narrow an interval as possible indicating where the measure lies.

14. Provide a critical analysis of the following editorial review, with particular reference to the highlighted text.

With Only $1,000 in the Bank, No Wonder We're Feeling Glum

The economists keep wondering why, in this third—or is it fourth?—year of "economic recovery," so many Americans still think we're in a recession. That's easy: They're broke. Flat broke. More broke than they've been in a half-century, and getting broker by the minute.

Just what that means can be discovered in the ironically titled "The Wealth of U.S. Families in 1991 and 1993" a recently completed piece of financial excavation by Chevy Chase, Md.-based Capital Research Associates. "Wealth," that outfit's work makes clear from its analysis of approximately 38,000 families, is a wildly generous description: Median net financial assets in 1991 were only $1,500, and by 1993 only $1,000. Adjusted for inflation, that amounts to a 37% plunge. Throw in the value of net equity in vehicles and businesses owned and the 1993 median climbs to $6,550, a 25.2% drop from 1991. Moreover, while younger families predictably have fewer assets than older, more established ones, there's cold comfort in those figures, too: The 1993 median for families headed by someone 45 to 54, for example, was a mere $2,600.

Not only do those figures represent a "woeful low," as characterized by John L. Steffens of Merrill Lynch, which commissioned the analysis, but the low has been getting lower just as the nation's economic pilots have been fretting that their beast is soaring too high. **What's going on? Apparently just more of that familiar refrain** about the rich getting richer. As the Capital Research Associates study also makes clear, if with some qualifiers, *mean* family wealth jumped smartly during the same two-year span that *median* wealth swooned. In other words, while at least half of all Americans were watching their already shriveled nest eggs diminish even more, there is at least some statistical suggestion that a small minority was growing very much better off.

In this recovery, many more boats seem to have been swamped than lifted.

2-3 MEASURES OF DISPERSION

As useful as the measures of central location are in providing some understanding about the data in a distribution, total reliance on the information conveyed by the mean, the median, and the mode can be misleading, as we now illustrate.

Suppose a test is given to two sections of a class, Section A and Section B. Table 2-2 gives the scores (in points) recorded in the two sections.

We can verify that both sections have the same mean score (60), the same median score (60), and the same mode (60). Solely on the basis of this information, however, it would be wrong to infer that the two sets of data are similar. In Section A, the group of students is homogeneous, all of them scoring in the vicinity of 60 points. In Section B, on the other hand, the performance is very erratic, from a low of 30 points to a high of 90 points. If we choose a student from Section A, we can assume that the student's score will be close to 60 points. We cannot say the same about a student chosen from Section B.

Often, a revealing picture emerges by considering histograms of the distributions, as in Figure 2-6. The two distributions give the tensile strength of cables manufactured by two processes. In both cases, we have the same mean, median, and mode. However, there is smaller variability in the data for the histogram in Figure 2-6(a) than for the histogram in Figure 2-6(b)

Variability of values in data collected is a very common phenomenon, and its importance should be acknowledged. For instance, a company manufacturing electric bulbs will be interested not only in the average life of the bulbs, but also in how consistent the performance of the bulbs is. The manufacturer interested in marketing bulbs with a mean life of 1000 hours will not

TABLE 2-2 Test Scores in Sections A and B

Section A Scores	Section B Scores
56	30
58	35
60	60
60	60
60	60
62	85
64	90

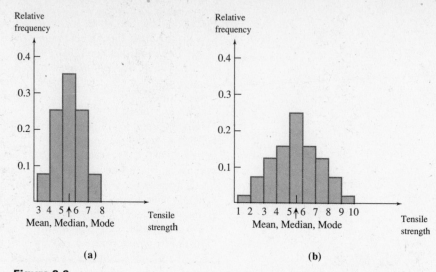

(a) **(b)**

Figure 2-6

Two distributions having the same mean, median, and mode, with the spread of data in (b) more than in (a)

be satisfied even if this goal is realized if, in fact, there is a very high ratio of bulbs that only last for up to 200 hours. In other words, the success of the bulbs will depend upon the variability of life; the smaller the variability the better.

THE RANGE

The simplest measure of dispersion is the **range.** As we saw in Section 1-1, the range is defined as the difference between the largest and smallest value in the data.

$$\text{range} = \text{largest value} - \text{smallest value}$$

A familiar example where range is used as the measure of variability is a weather forecaster's daily temperature report given as, for example, a low of 40 degrees to a high of 70 degrees. The range is 70 − 40, or 30, degrees.

If we take the highest and the lowest observations, the range is easy to compute. However, it ignores the intermediate values and provides no information at all about the set of observations between the highest and lowest values. Consider, for example, the following two sets of data.

    3   7   9   11   16   and   3   12   12   12   16

Each set has the same range, but the two distributions are markedly different. Also, by describing data solely in terms of extreme values, the presence of an unusually extreme value will inflate the range out of proportion. The range, therefore, is the least satisfactory of all measures of dispersion that we consider in this chapter.

MEAN DEVIATION

The variability of the data depends upon the extent to which individual observations are spread about a measure of central tendency. If the values are widely scattered, the variability will be large. On the other hand, if the values are compactly distributed about this measure, the variability will be small. Because the arithmetic mean is by far the most important measure of central tendency, the deviations of the individual observations are usually taken as deviations from the mean $\bar{x}$.

For any value x_i in the data set of n values, $x_i - \bar{x}$ is its deviation from the mean. The n values will give rise to n deviations from the mean. Some of these deviations will be positive, others negative. Their sum, however, is *always* zero, irrespective of the scatter of the data. This is a mathematical fact. Thus we always have

$$\sum (x_i - \bar{x}) = 0$$

Because the sum of the deviations is always zero, $\sum (x_i - \bar{x})$ is of no value to us as a measure of variability. But notice that if the mean is 50, then the value 42 is as much removed from the mean as is the value 58, and the two will contribute to the same extent toward the variability. This allows us to overcome the difficulty posed by the fact that $\sum (x_i - \bar{x}) = 0$, by giving weight only to the magnitude of each deviation and considering all the deviations positive. These values are called *absolute deviations* from the mean, and are denoted as $|x_i - \bar{x}|$. The arithmetic mean of the absolute deviations is customarily called the **mean deviation,** though, perhaps, calling it *mean absolute deviation* would seem more appropriate. It is found by dividing the sum of the absolute deviations by the total number of observations.

MEAN DEVIATION

If $x_1, x_2, \ldots, x_n$ are n values in the data set, then

$$\text{mean deviation} = \frac{\sum |x_i - \bar{x}|}{n}$$

In contrast with the range, the mean deviation takes into account the magnitude of all the observations in the data.

EXAMPLE 1 A secretary typed six letters. The times (in minutes) spent on these letters were 6, 8, 13, 10, 8, and 9. Find the mean deviation.

SOLUTION   We see that the mean is

$$\bar{x} = \frac{6 + 8 + 13 + 10 + 8 + 9}{6} = 9.$$

Therefore,

$$\text{mean deviation} = \frac{|6 - 9| + |8 - 9| + |13 - 9| + |10 - 9| + |8 - 9| + |9 - 9|}{6}$$

$$= \frac{3 + 1 + 4 + 1 + 1 + 0}{6} = \frac{5}{3}.$$

VARIANCE AND THE STANDARD DEVIATION

The use of absolute values causes considerable problems in treating mean deviation mathematically. For this reason, mean deviation is not suitable, and we will seldom refer to it again in the remainder of the book. The preferred measures of spread or dispersion of data are **sample variance** and **sample standard deviation**. The sample variance is computed from the deviations $(x_i - \bar{x})$, by squaring them and dividing the sum of these squares by $n - 1$. The deviations are squared because squaring produces positive numbers for both positive and negative deviations. The sample standard deviation is defined as the *positive* square root of the variance. The sample variance is denoted by s^2 and the sample standard deviation by s where s is positive.

SAMPLE VARIANCE AND STANDARD DEVIATION

If a sample consists of n numerical measurements $x_1, x_2, \ldots, x_n$, then the **sample variance** s^2 is defined by

$$s^2 = \frac{\sum (x_i - \bar{x})^2}{(n - 1)}$$

and the **sample standard deviation** s by

$$s = \sqrt{\frac{\sum (x_i - \bar{x})^2}{(n - 1)}}$$

If you analyze the formula for s, the following steps are involved in its computation.

1. Compute the mean.
2. Compute the deviations from the mean. (As a computational check, the sum of these values should be 0.)
3. Square the deviations.
4. Add the squared deviations.

5. Divide the sum of the squared deviations by $n - 1$. In this step we have computed s^2 or the sample variance.

6. Find the *positive* square root. This step yields the standard deviation.

If the observations in the data are too widely spread out, then quite a few of them will be far removed from the mean. For these observations $(x_i - \overline{x})^2$ will be large and, as a consequence, $\Sigma (x_i - \overline{x})^2$ will be large. This will result in making both s^2 and s large. Thus if the data are too scattered, the variance and the standard deviation both will be large.

Rationale for Dividing by $n - 1$

For finding s^2, the sum of squares of deviations from $\overline{x}$ is divided by $n - 1$ instead of by n. The rationale is as follows. Suppose we are told to pick *any* three numbers with the condition that their sum is to be 18. Are we really free to pick three numbers? Of course not. We have freedom to pick only two (that is, $3 - 1$) numbers since the third number is determined automatically by the constraint. About the same sort of thing happens when finding the sample variance. We are told that we have n observations from which to compute the sample variance, but first we must compute the sample mean using these observations since it is involved in the formula. It turns out that we have, in effect, $n - 1$ observations with which to compute the sample variance. The number $n - 1$ is called the *degrees of freedom* (abbreviated *df*) associated with the sample variance.

EXAMPLE 2

The price (in dollars) of a certain commodity on eight trading sessions was as follows.

| 38 | 11 | 8 | 60 | 52 | 68 | 32 | 19 |

Find the variance and standard deviation.

SOLUTION The computations are presented in Table 2-3. Following the steps listed on page 104 we proceed as follows.

1. In Column 1 we first find the mean $\overline{x} = 288/8$, or 36.

2. Then in Column 2 we compute deviations of the prices from the mean price; that is, we compute $x_i - 36$. As a check, notice that the sum of the entries in this column is zero.

3. The third column presents squares of deviations from the mean.

4. Next, we compute the sum of the squares in Column 3. We get

$$\sum (x_i - \overline{x})^2 = 3574$$

5. In this step we obtain the variance. Since $n = 8$, the variance (in dollars^2) is given as

$$s^2 = \frac{\sum (x_i - \overline{x})^2}{n - 1} = \frac{3574}{8 - 1} = 510.57$$

TABLE 2-3 Computation of the Variance of the Price of a Commodity

Price x	Deviation $x - 36$		Square of Deviation $(x - 36)^2$	
38	2		4	
11	−25		625	
8	−28		784	
60	24	Step 2	576	Step 3
52	16		256	
68	32		1,024	
32	−4		16	
19	−17		289	

$\sum x_i = 288 \qquad \sum (x_i - \bar{x}) = 0 \qquad \sum (x_i - \bar{x})^2 = 3574 \quad \leftarrow$Step 4

Step 1→   mean $\bar{x} = 36$

6. Finally, in this step we find the positive square root of the variance, getting the standard deviation. The standard deviation is expressed in dollars and is given as

$$s = \sqrt{510.5714} = 22.6 \text{ dollars}$$

If you examine Column 3 of squared deviations, the largest squared deviation is 1,024 and is contributed by the data value 68. The other major contributors are 60 and 52 at the upper end, and 8 and 11 at the lower end. All these values are considerably farther from the mean price 36 than the rest of the values.

●

A Word about Rounding

Retain two decimal places more than in the original data in presenting the sample variance and one decimal place more than in the original data in presenting the standard deviation. Be careful not to find the standard deviation from the rounded variance.

Shortcut Formula

Through some algebraic simplifications, the preceding formula for s leads to the following shortcut formula for the standard deviation and it is convenient for computational purposes.

SHORTCUT FORMULA FOR THE STANDARD DEVIATION

$$s = \sqrt{\frac{1}{(n-1)}\left[\sum x_i^2 - \frac{\left(\sum x_i\right)^2}{n}\right]}$$

In the shortcut formula, you need to compute $\Sigma\, x_i^2$ and $\Sigma\, x_i$, which is a lot easier than computing the deviations and their squares. In words, the shortcut formula is

$$s = \sqrt{\frac{1}{(n-1)}\left[\left(\begin{array}{c}\text{sum of squares}\\ \text{of sample values}\end{array}\right) - \frac{(\text{sum of sample values})^2}{n}\right]}$$

EXAMPLE 3

Compute s for the data in Example 2 using the shortcut formula.

SOLUTION   We have determined that $n = 8$ and $\Sigma\, x_i = 288$. The sum of the squares of the values in the data set is

$$\Sigma\, x_i^2 = (38)^2 + (11)^2 + (8)^2 + (60)^2 + (52)^2 + (68)^2$$
$$+ (32)^2 + (19)^2$$
$$= 13{,}942$$

Substituting in the shortcut formula, we then get

$$s = \sqrt{\frac{1}{(8-1)}\left[13{,}942 - \frac{(288)^2}{8}\right]} = 22.6$$

This is the same answer we obtained in Example 2.   ●

Use of Pocket Calculators

The statistical function s is built into many scientific pocket calculators and can be obtained simultaneously with the mean $\bar{x}$. Instead of using the definitional formula on page 104, the calculator uses the shortcut formula for computing s. It accumulates n, $\Sigma\, x_i$, and $\Sigma\, x_i^2$, in three special registers as the data are entered. The calculator uses the accumulated entries in the registers n and $\Sigma\, x$ to compute the mean when the key for the mean is pressed. It uses the accumulated entries in all the three registers and the shortcut formula to give s when the key for s is pressed. A few authors choose to define the standard deviation by dividing the sum of the squares of the deviations by n instead of by $n-1$. Some calculators, therefore, have two keys, labeled s_n and s_{n-1}, or σ_n and σ_{n-1}. Since we divide by $n-1$ in this text, you should use the key s_{n-1} or σ_{n-1} for s.

You will be performing many statistical computations and should own a versatile statistical calculator to save you the drudgery of tedious computations. Also, familiarize yourself thoroughly with the calculator's accompanying manual.

A word about the units to be employed in expressing the variance and the standard deviation   Because the variance is based on the sum of squared differences, it is expressed in squared units. The standard deviation is the square root of the variance. Therefore, it is expressed in terms of the units that were employed in measuring the data.

For example, if the data are given in feet, then the units used for expressing the variance are feet squared, abbreviated $(ft)^2$, and those for the standard deviation are feet. In the same way, if the data are expressed in minutes, then the units for expressing the variance are minutes squared, or $(min)^2$, and for the standard deviation the units are minutes.

PROPERTIES OF THE STANDARD DEVIATION

The standard deviation of data has the following two important properties:

1. If a fixed value d is added to each of the observations in the data, then the standard deviation is unchanged, that is,

$$\left(\begin{array}{c} \text{the standard deviation} \\ \text{of the new data} \end{array}\right) = \left(\begin{array}{c} \text{the standard deviation} \\ \text{of the original data} \end{array}\right)$$

    For example, if the salary of each employee is raised by $100, then the standard deviation of salaries before the raise is the same as the standard deviation after the raise.

2. If each observation in the data is multiplied by a fixed constant c, then

$$\left(\begin{array}{c} \text{the standard deviation} \\ \text{of the new data} \end{array}\right) = c \text{ times} \left(\begin{array}{c} \text{the standard deviation} \\ \text{of the original data} \end{array}\right)$$

    For example, suppose the height of students in a classroom is measured in feet and the standard deviation is 0.8 feet. If the height is measured in inches, then the standard deviation will be (12)(0.8), or 9.6 inches.

SIGNIFICANCE OF THE STANDARD DEVIATION

When a researcher analyzes a bulk of numerical data, the two measures that will often be quoted are the mean and the standard deviation. There is a special relationship between these two statistics in communicating considerable information about the extent of the spread of the data. This relationship is explained in the theorem developed by the great Russian mathematician P. L. Chebyshev (1821–1894) and is named after him.

If k is a positive number, an interval which is k standard deviations from the mean is defined as the interval $(\bar{x} - ks, \bar{x} + ks)$ as shown in Figure 2-7. Thus, for example, if $\bar{x} = 16.8$, $s = 1.02$, then an interval which is two standard deviations from the mean is $(16.8 - 2(1.02), 16.8 + 2(1.02))$, that is, $(14.76, 18.84)$

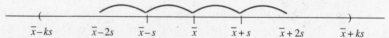

Figure 2-7 An Interval That Is k Standard Deviations from the Mean

For any positive number k, Chebyshev's rule states the minimum percentage of observations that will fall in the interval $(\overline{x} - ks, \overline{x} + ks)$, that is, within k standard deviations of the mean. This percentage is given by $\left(1 - \dfrac{1}{k^2}\right) \cdot 100$. In the following presentation we interpret the rule for certain values of k.

CHEBYSHEV'S RULE

For any data set, irrespective of its frequency distribution (whether it be symmetric, skewed to the right, skewed to the left, or any other variation) the following is true.

1. *At least* 0 percent of the observations will fall within 1 standard deviation of the mean, that is, in the interval $(\overline{x} - s, \overline{x} + s)$. This assertion is quite obvious.
2. *At least* 75 percent of the observations will fall within 2 standard deviations of the mean, that is, in the interval $(\overline{x} - 2s, \overline{x} + 2s)$.
3. *At least* 89 percent of the observations will fall within 3 standard deviations of the mean, that is, in the interval $(\overline{x} - 3s, \overline{x} + 3s)$.

Chebyshev's rule gives very conservative estimates, as it would have to, since it covers the entire gamut of frequency distributions. If additional information is available about the frequency distribution of the data, then more substantial statements can be made about how many observations will fall within 1, 2, and 3 standard deviations of the mean. For data having an almost symmetric, sort of mound or bell-shaped distribution, where the measurements are concentrated near the center and the histogram is tapering off in either direction from the center, the following Empirical rule is available. It provides a guideline for the *approximate* number of observations. (The basis for the Empirical rule is explained later on in Chapter 5 on page 306.)

THE EMPIRICAL RULE

For a data set with a symmetric frequency distribution as described above, the following hold true.

1. *Approximately* 68.3 percent of the observations will fall in the interval $(\overline{x} - s, \overline{x} + s)$.
2. *Approximately* 95.4 percent of the observations will fall in the interval $(\overline{x} - 2s, \overline{x} + 2s)$.
3. *Approximately* 99.7 percent, that is, practically all the observations will fall in the interval $(\overline{x} - 3s, \overline{x} + 3s)$.

As an illustration, consider the data presented by the dot diagram in Figure 2-8, where each value is denoted by a dot. There are three measurements each equal to 2, one measurement equal to 3, and so on, and altogether there are thirty-six measurements.

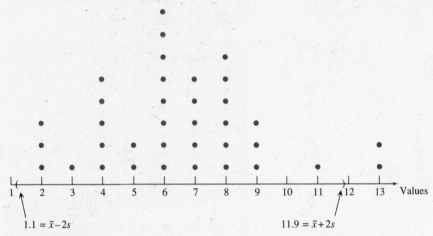

Figure 2-8 Dot Plot.
There are thirty-four values in (1.1, 11.9).

It can be verified that $\bar{x} = 6.5$ and $s = 2.7$. Thus, the interval $(\bar{x} - 2s, \bar{x} + 2s)$ is given by (1.1, 11.9) and the interval $(\bar{x} - 3s, \bar{x} + 3s)$ by (−1.6, 14.6). From the dot diagram it can be seen that there are thirty-four measurements in the interval (1.1, 11.9), that is, 94.4 percent. In the interval (−1.6, 14.6) there are all thirty-six observations, that is, 100 percent. In the following table we present the actual percentages and those predicted by the two rules.

Interval	Chebyshev (at least) %	Empirical (Approx.) %	Actual %
(1.1, 11.9)	75	95.4	94.4
(−1.6, 14.6)	89	99.7	100

Notice that the actual percentages are definitely in keeping with the predictions of Chebyshev's rule. Also, the actual percentages are not too far off from those given by the Empirical rule, even though, from the dot diagram, we could not quite say that the distribution was symmetrical and bell-shaped.

EXAMPLE 4   A psychologist claims to have a new instructional method for improving the performance of students. The psychologist tries the instructional method with eighty students and reports the summary of the results as follows: $\bar{x} = 120$ points and $s = 22$ points. What does this information tell us about the student scores?

SOLUTION   Of course, we have no idea about the individual scores. But by using Chebyshev's rule we are in a position to make the following statements.

(a)   Because $\bar{x} = 120$ and $s = 22$, we have the following.

$$\bar{x} - 2s = 120 - 2(22) = 76$$
$$\bar{x} + 2s = 120 + 2(22) = 164$$

We can infer that at least 75 percent of the eighty students, that is, at least sixty students, had scores between 76 and 164 points.

(b)   For three standard deviations from the mean we have the following.

$$\bar{x} - 3s = 120 - 3(22) = 54$$
$$\bar{x} + 3s = 120 + 3(22) = 186$$

We can infer that at least 89 percent of the students, that is, at least seventy-one students, had scores between 54 and 186 points.

If the psychologist adds to the report that the frequency distribution was almost mound shaped, then we could improve our estimates using the Empirical rule as follows.

Approximately 95.4 percent, that is, approximately seventy-six students, had scores between 76 and 164 points. Also, practically every student in the study had a score between 54 and 186 points.   ●

Section 2-3 Exercises

1. The scores of six students on a test are as follows.

   75   82   56   69   87   43

Arrange the data in an ascending order of magnitude and find the range of scores.

2. The amount of time spent on each of ten telephone calls was recorded (in minutes) as follows.

   25   15   8   2   5
   12   34   3   12   3

Find the range of the time spent on these calls.

3. The altitudes (in feet) of six locations (above or below sea level) are as follows.

   1200 above   140 above   100 below
   10,500 above   400 below   800 above

Find the range of heights.

4. The minimum daily temperature readings (°F) of a ski resort on eight days were as follows.

    22   −18   0   −8   16   30   5   −20

    Find the temperature range for these days.

5. If the range of the price of a stock during a week was $15.75, and if the lowest price was $82.50, find the highest price during the week.

6. For the set of values 8, 6, 12, 10, and 14, find the following.

    (a) $\sum (x_i - \bar{x})$ **(b)** $\sum |x_i - \bar{x}|$ **(c)** The mean deviation

7. The following figures refer to the times (in minutes) that eight customers spent from the time they entered a store to the time they left it.

    6   3   2   10   2   5   4   6

    Find the following.
    (a) The arithmetic mean $\bar{x}$ of the time the customers spent in the store
    (b) The absolute deviations
    (c) The mean deviation

In Exercises 8–11, compute mean deviations for the given sets of observations.

8. 14, −4, 10, 6, 10, −15, and 7

9. 1.8, 2.8, 2.3, 4.5, and 1.1

10. 6, −7, −5, −4, and 5

11. −8, −7, −6, −5, −10, and −12

12. Seven buses are expected to arrive at the bus depot at 4 A.M., 9 A.M., 10 A.M., 11 A.M., 12 noon, 1 P.M., and 2 P.M. They arrive at 8:04 A.M., 9:10 A.M., 9:52 A.M., 10:56 A.M., 11:55 A.M., 1:10 P.M., and 2:00 P.M., respectively. Find the mean deviation of the amount of time by which a bus is late. (If a bus arrives ahead of schedule, for example, by 4 minutes, then consider the late arrival time is negative, or −4.)

13. The following figures give the price per pound of ground chuck (in dollars) at a supermarket during six weeks.

    2.98   2.89   2.80   2.88   2.84   2.95

    Find the number of weeks when the price is within one mean deviation of the mean price.

14. For the set of values 18, 19, 16, 12, 7, 10, 23, find the following.
    (a) The arithmetic mean $\bar{x}$
    (b) The deviations from the mean
    (c) The squared deviations
    (d) The sample variance
    (e) The sample standard deviation

In each of Exercises 15–17 compute the variance and the standard deviation for the given sets of observations.

15. 14, −4, 10, 6, 10, −15, and 7

16. 1.8, 2.8, 2.3, 4.5, and 1.1

17. 6, −7, −5, −4, and 5

18. What is the standard deviation of a set of scores when all the observations are identical?

19. What does it mean when the variance of a set of observed values is zero?

20. If the variable x is measured in ounces, what units should be used in expressing the variance and the standard deviation?

21. The following readings were obtained for the tensile strength (in tons per square inch) of six specimens of an alloy.

2.58   2.65   2.40   2.46   2.44   2.41

Find the mean tensile strength and the standard deviation of the tensile strengths.

22. The heights of four fifth-grade students are 58, 56, 60, and 62 inches, and those of five sixth-grade students are 66, 59, 63, 61, and 66 inches. Which grade has the higher standard deviation?

23. A club has ten members whose weights are $x_1, x_2, \ldots, x_{10}$. Find the mean weight and the variance of the weights if $\Sigma \, x_i = 1500$ and $\Sigma \, x_i^2 = 325,000$.

24. Five samples of coal, each of which weighs 2.00 grams, lose the following amounts of moisture (in grams) after air drying in an oven.

0.158   0.160   0.156   0.153   0.163

Find the variance of the amount of moisture lost.

25. In a chemistry experiment seven students obtained the following percentages of chlorine in a sample of pure sodium chloride.

60.65   60.68   60.70   60.60
60.64   60.65   60.70

Find the standard deviation of the percentage of chlorine.

26. Using the formula for the variance, find the variance of the data in Part a. From this answer find the variance of the data in Parts b and c.
(a) 7, 8, 6, 8, and 6
(b) 27, 28, 26, 28, and 26
(c) 107, 108, 106, 108, and 106

Hint: The observations in Part b are obtained by adding 20 to each observation in Part a; and those in Part c are obtained by adding 100 to each observation in Part a.

27. After grading a test, Professor Adams announced to the class that the mean score was 68 points with a standard deviation of 12 points. However, a student brought to his attention that one of the questions on the test was incorrect and impossible to solve. As a result, Professor Adams agreed to add 10 points to each student's score. Find the mean, the variance, and the standard deviation of the new set of scores.

28. The mean salary of the employees in a certain company is $22,300, with a standard deviation of $2500. The company announces a flat raise of $1000 for each employee during the following year. Find the mean and the standard deviation of the new salaries.

*Stem + leaf
+ Table like
P HO*

29. The following data are the chlorophyll content (mg/g) of thirty wheat leaves.

2.6	2.9	3.2	3.0	2.6
3.2	3.2	3.0	2.9	3.0
2.6	2.6	3.0	3.2	2.9
3.2	2.6	2.9	3.2	2.6
3.2	2.8	3.2	3.0	3.0
2.8	2.9	3.2	2.8	2.8

(a) Find the variance and the standard deviation of the chlorophyll content.

(b) How many observations are within two standard deviations of the mean?

30. To show the impact of the increase in federal tax effective October 1, 1993, Mobil Corporation published the following figures for major markets where Mobil does most of its business. The data give federal, state, and local taxes, in cents, that go into the price of gas at the pump.

Metropolitan Area	Federal Tax	State Excise Tax	Other State and Local Taxes	Total
Chicago	18.63	19.00	21.66	59.29
Long Island, NY	18.63	8.00	26.49	53.12
New Haven, CT	18.63	29.00	5.28	52.91
Buffalo, NY	18.63	8.00	25.79	52.42
Albany, NY	18.63	8.00	25.79	52.42
Los Angeles	18.63	17.00	11.84	47.47
Providence, RI	18.63	28.00	0	46.63
Tampa	18.63	4.00	21.17	43.80
Miami	18.63	4.00	20.17	42.80
Baltimore	18.63	23.50	0	42.13
Philadelphia	18.63	12.00	10.35	40.98
Boston	18.63	21.00	.50	40.13
Fairfax, VA	18.63	17.50	3.13	39.26
Detroit	18.63	15.00	5.61	39.24
Dallas	18.63	20.00	.59	39.22
Phoenix	18.63	18.00	1.00	37.63
Newark, NJ	18.63	10.50	4.04	33.17
St. Louis	18.63	13.00	0	31.63

Gasoline Taxes in Cents per Gallon in Some Key Locations:

Average of these areas: 44.13

Source: "Pump Price and Tax Collectors-VI" by Mobil Corporation. Copyright © 1993 Mobil Corporation. Reprinted by permission of Mobil Corporation.

You can verify that the average of 44.13 given at the bottom of the table refers to the mean and that it is computed from the column giving the total taxes. Compute the standard deviation of the total taxes. Without using the actual data any more, what would the mean tax and the standard deviation of the taxes be in the listed areas if the federal tax of 18.63 cents was eliminated?

31. The following data for calcium and oxalic acid contents (in mg/g) of alfalfa meal are reproduced from the article *Oxalic Acid Content of Alfalfa Hays and Its Influence on the Availability of Calcium, Phosphorus, and Magnesium to Ponies*, by H. F. Hintz et al., from the *Journal of Animal Science*, vol. 58, no. 4, 1984. Reprinted by permission of the American Society of Animal Science.

Location	Calcium	Oxalate
Alabama	16.4	3.6
Alabama	15.1	3.7
Ontario, Can.	19.6	3.3
Ontario, Can.	12.3	2.4
Ontario, Can.	14.0	2.8
Colorado	17.2	3.3
Illinois	13.8	4.1
Iowa	15.6	4.3
Iowa	15.2	3.2
Nebr. or Kan.	13.0	3.7
Nebr. or Kan.	16.2	3.7
Nebr. or Kan.	15.1	4.2
Nebr. or Kan.	12.8	6.6
Minnesota	16.3	2.4
Michigan	16.0	2.7
Nebraska	16.8	3.3
Nebraska	14.2	3.8
Ohio	15.7	2.0
Ohio	17.3	2.3
Ohio	13.5	3.5
Texas	12.0	4.7
Washington	14.7	5.1

(a) Prepare stem-and-lead diagrams for calcium content and oxalic acid content.
(b) Use the stem-and-leaf diagrams to compute the median contents of the two variables.
(c) Construct histograms for the two variables.
(d) Compute the mean and the standard deviation of the two variables.
(e) For each variable locate the mean along the horizontal axis of the histogram and mark the interval that is two standard deviations from the mean.
(f) Compute (*calcium/oxalate*) ratios. Then compute the mean and the standard deviation of the resulting ratios.

32. Benzo(a)pyrine is one of the polycyclic hydrocarbons released into the marine environment by the oil pollution of the seas. The amount of benzo(a)pyrine (μg/kg) in the tissues of mussels, a mollusk, was determined in a coastal region by the spectral luminescence method. Ten samples were analyzed from the unpolluted section of the region and eight samples from the oil-polluted section.

BENZO(a)PYRINE (μ/kg)

Unpolluted	Polluted
0.36	382.1
0.39	292.6
0.35	313.9
0.38	373.8
0.37	322.8
0.37	337.1
0.35	381.2
0.39	351.3
0.38	
0.37	

Calculate the following.

(a) The mean benzo(a)pyrine in the unpolluted section and also in the oil-polluted section

(b) The standard deviation of the benzo(a)pyrine content in each section of the coastal region

33. Three fish-rearing ponds were fertilized with fodder yeast and mineral fertilizer containing superphosphate and ammonium nitrate. Five determinations of dissolved oxygen concentration (mg/l) were made in two of the ponds and four determinations were made in the third pond. The data are shown in the following table.

DISSOLVED OXYGEN (mg/l)		
Pond 1	Pond 2	Pond 3
5.5	5.3	5.8
5.8	5.2	5.9
5.6	5.4	5.7
5.5	5.1	5.6
5.4	5.3	

In Chapter 12 we will investigate such data in more detail. For now, answer the following questions:

(a) Find the mean dissolved oxygen concentration in each of the three ponds.

(b) Find the overall mean dissolved oxygen concentration in the three ponds taken together. (When different sets of data are regarded as one set, the data are said to be *pooled*.)

(c) Find the mean of the three means you computed in Part a. (*Careful!* Consider Property 1 on page 80.) Is your answer the same as in Part b?

(d) Find the standard deviation of the dissolved oxygen concentration in each of the three ponds.

(e) Find the overall standard deviation of the dissolved oxygen concentration in the three ponds taken together.

34. *Coridothyme capitatus* is a wild spice plant widely distributed in Israel and on the West Bank of the Jordan River. An area of approximately 0.1 hectare was randomly chosen for sample collection. A sample of plant material was gathered by taking one branch of each *C. capitatus* plant growing in the selected area. Essential oil was distilled from each sample separately. The thymol (carvacrol isomer) and carvacrol contents in essential oils of *C. capitatus* from different sites are presented in the following table. [Alexander Fleisher et al., Chemovarieties capitatus I. Rchb. growing in Israel; *J. Sc. Food Agric.* 35 (1984): 495–499.]

Site Location	Thymol Concentration in Essential Oil (%)	Carvacrol Concentration in Essential Oil (%)
Rosh-Hanikra, beach, Western Galilee	49.3	9.8
Haifa, Mt. Carmel	63.6	5.2
Hacarmel beach, coastal plain	55.0	5.1
Kfar Galim, coastal plain	41.7	24.2

Site Location	Thymol Concentration in Essential Oil (%)	Carvacrol Concentration in Essential Oil (%)
Atlit, coastal plain	42.0	22.9
Habonim beach, coastal plain	13.7	55.7
Moshav Habonim, coastal plain Sample A	45.2	10.9
Moshav Habonim, coastal plain Sample B	19.2	46.3
Dor beach, coastal plain	6.7	59.4
Zichron Jaacob, Mt. Carmel	47.5	20.3
Hertzelia beach, coastal plain	48.9	9.5
Umtzafa forest, West Bank	52.8	7.4
Nes-Ziona, inner plain	56.8	5.3
Nablus, West Bank	47.6	19.8
Bet-El, West Bank	51.1	6.6
Bet-Lechem, West Bank	51.8	10.8
Gush-Atzion, West Bank	16.0	44.3
Kibbutz Nativ Halamed Hey, West Bank	43.6	7.7
Beer-Sheva-Arad road, Negev	53.8	5.8
Kibbutz Saad, Negev	61.2	5.0

Compute the following.
(a) The mean thymol and carvacrol concentrations
(b) The standard deviations of thymol and carvacrol concentrations

35. Mr. Chan owns a retail store in a small town. When he checked his computer spreadsheet at the end of the calendar year 1994, there were the following balances (in dollars) in 33 accounts receivable.

52	74	50	57	62	101	57	99	68	114	48
76	84	52	63	97	76	73	64	78	85	95
67	61	82	90	51	75	88	65	25	160	72

(a) Compute the mean balance and the median balance.
(b) Prepare a histogram and locate the mean and the median.
(c) Compute the standard deviation and determine the number of observations in the interval $(\bar{x} - 2s, \bar{x} + 2s)$. Exhibit this interval on the histogram.

36. Fifteen 5- to 6-week-old male rats each weighing approximately 100 grams were fed a test diet containing casein treated with a methanol (HCHO) solution containing 40 grams HCHO/liter. The rats were housed for 14 days in metabolism cages and were fed 10 grams fresh weight of diet per day. During the final 7-day period a total collection of feces and urine was carried out and the samples were analyzed for total N (nitrogen). The following data give figures for N digestibility of the rats.

0.91	0.91	0.94	0.95	0.93
0.93	0.92	0.92	0.91	0.92
0.94	0.92	0.92	0.93	0.92

Determine the following:
(a) The mean nitrogen digestibility of the rats
(b) The median nitrogen digestibility from the stem-and-leaf diagram
(c) The standard deviation of the nitrogen digestibility
(d) The interquartile range. (Provide an interpretation, as well).

37. The following figures give the cost (in cents) for a quart of a particular brand of oil at eight automotive supply centers.

    80   85   95   88   92   85   92   95

    Find the number of places where the price is within one standard deviation of the mean.

38. When ten 5-gram samples of an alloy were analyzed for silver content by the method of electrodeposition, the following amounts of silver (in grams) were found.

    | 1.273 | 1.276 | 1.265 | 1.276 | 1.274 |
    | 1.277 | 1.273 | 1.275 | 1.285 | 1.278 |

    (a) Find the mean and the standard deviation of the amount of silver.
    (b) Determine the interval which represents two standard deviations from the mean.
    (c) How many samples have amounts of silver in the interval obtained as the solution to Part b?

39. The following figures give the percentage recovery of chemically reactive lysine in the presence of glucose by a modified trinitrobenzene sulphonic acid procedure.

    | 57.2 | 79.7 | 82.7 | 66.3 | 64.7 |
    | 66.1 | 55.6 | 67.3 | 72.1 | 80.5 |
    | 64.8 | 74.0 | 71.2 | 71.5 | 70.7 |
    | 69.9 | 73.3 | 70.9 | 61.4 | 78.4 |
    | 70.2 | 62.8 | 74.7 | 70.0 | 73.9 |

    (a) Compute the mean percentage recovery.
    (b) Compute the variance of the percentage recovery.
    (c) Determine the interval $(\bar{x} - 2s, \bar{x} + 2s)$ and compute the percentage of readings that fall in it.
    (d) Determine the interval $(\bar{x} - 3s, \bar{x} + 3s)$ and compute the percentage of readings that fall in it.
    (e) Are the data described satisfactorily by the Empirical rule?

40. The following data contain figures for the concentration of acid-soluble sulfides (mg/l fresh ooze) in the bottom sediments of a lagoon. Twenty-five samples were analyzed.

    | 998 | 938 | 971 | 949 | 1013 |
    | 1000 | 1021 | 1002 | 995 | 956 |
    | 1010 | 1015 | 967 | 987 | 990 |
    | 954 | 974 | 984 | 988 | 982 |
    | 941 | 976 | 1047 | 974 | 965 |

(a) Calculate the following.
   i. The mean concentration and the median concentration of sulfides
   ii. The standard deviation
(b) Find the number of observations in these intervals.
   i. $(\bar{x} - 2s, \bar{x} + 2s)$
   ii. $(\bar{x} - 3s, \bar{x} + 3s)$
(c) Are the data described satisfactorily by the Empirical rule?

41. Thirty milk-yielding Ayrshire cows were fed a diet of silage supplemented with a protein concentrate. The figures below refer to the milk fat concentration (g/kg) for each cow.

30.5	28.7	27.3	24.1	42.3
41.5	20.5	36.3	25.5	30.9
26.3	33.6	31.9	34.7	35.0
34.5	30.6	34.0	38.2	42.5
40.3	21.7	35.7	32.7	34.6
35.1	35.9	33.3	37.4	33.3

Calculate the following.
(a) The mean milk fat concentration
(b) The standard deviation of the milk fat concentration
(c) The number of cows with milk yield within two standard deviations of the mean
(d) The interquartile range (Interpret this quantity in terms of the variable under consideration.)
(e) The number of cows with milk yield falling in the interquartile range

42. For the data in Exercise 41 prepare a histogram with 6 classes and discuss its symmetry. Locate the mean and the median on the horizontal axis along the base of the histogram. Comment on the closeness of the mean and the median considering the symmetry of the histogram or the lack of it.

43. Sixty dry cell batteries were tested. It was found that the mean life was 96 hours with a standard deviation of 12 hours. What does this tell you about the lives of the batteries? (*Hint:* Consider Chebyshev's rule and see Example 4 on page 111.)

44. When dopamine levels (nmoles/g) were measured in the brains of 36 rats, the mean amount was 6.18 nmoles/g and the standard deviation was 0.68 nmoles/g. What does this information tell you about the dopamine levels of the 36 rats? (Use the hint for Exercise 43, above.)

2-4   MEAN AND STANDARD DEVIATION FOR FREQUENCY DISTRIBUTIONS

The basic formula for computing the sample mean and the sample standard deviation have been given in the preceding sections. Now, if the data are given in a frequency table, the formulas can be expressed in the following form.

SAMPLE MEAN AND STANDARD DEVIATION FOR A FREQUENCY DISTRIBUTION

If the data consisting of n observations are given in a frequency table with the frequency of the value x_i equal to f_i, then the sample mean $\bar{x}$ and the standard deviation s are given by

$$\bar{x} = \frac{\sum f_i x_i}{n}$$

$$s = \sqrt{\frac{\sum f_i (x_i - \bar{x})^2}{n - 1}}$$

where $\sum f_i = n$.

Notice that in the formula for the mean, the frequencies of the values are simply their weights.

A more convenient version of the formula for computing the sample standard deviation is the following.

$$s = \sqrt{\frac{1}{(n-1)}\left[\sum f_i x_i^2 - \frac{\left(\sum f_i x_i\right)^2}{n}\right]}$$

EXAMPLE 1   The data in Table 2-4 give the distribution of the diameters (in inches) of twenty tubes manufactured by a machine. Find the following.

(a)   The mean diameter

(b)   The standard deviation

SOLUTION   The necessary computations are presented in Table 2-5.

(a)   The sums in Columns 2 and 3 yield

$$n = \sum f_i = 20 \quad \text{and} \quad \sum f_i x_i = 50.0$$

respectively. Therefore, the mean (in inches) is

$$\bar{x} = \frac{\sum f_i x_i}{n} = \frac{50.0}{20} = 2.5.$$

(b)   From Column 5 of Table 2-5 we have $\sum f_i x_i^2 = 127.62$. Substituting in the formula

$$s = \sqrt{\frac{1}{(n-1)}\left[\sum f_i x_i^2 - \frac{\left(\sum f_i x_i\right)^2}{n}\right]}$$

TABLE 2-4 Diameters of Tubes Manufactured by a Machine

Diameter (in Inches)	Frequency
2.0	2
2.2	4
2.3	6
2.8	3
3.0	5

TABLE 2-5   Computations for Finding the Mean and the Variance

Diameter x	Frequency f	fx	x^2	fx^2
2.0	2	4.0	4.00	8.00
2.2	4	8.8	4.84	19.36
2.3	6	13.8	5.29	31.74
2.8	3	8.4	7.84	23.52
3.0	5	15.0	9.00	45.00
Sum	20	50.0		127.62

we get the sample standard deviation as

$$s = \sqrt{\frac{1}{(20-1)}\left[127.62 - \frac{(50)^2}{20}\right]}$$

$$= \sqrt{0.138}$$

$$= 0.37 \text{ inches}$$

In the next example, in which we compute the mean and the standard deviation for grouped data, the secret is to find the class marks first and then to treat them as though they are the observed values.

EXAMPLE 2   **(Grouped data)** In Columns 1 and 2 of Table 2-6 scores of eighty students are grouped into a frequency distribution.

TABLE 2-6   Computations to Find the Mean and the Standard Deviation of Student Scores

Class	Frequency f	Class Mark x	fx	fx^2
5–19	4	12	48	576
20–34	12	27	324	8,748
35–49	15	42	630	26,460
50–64	16	57	912	51,984
65–79	22	72	1,584	114,048
80–94	11	87	957	83,259
Sum	80		4,455	285,075

Compute the following.

(a)   The mean

(b)   The standard deviation

SOLUTION   The class marks are computed in Column 3 of Table 2-6. Other computations are presented in Columns 4 and 5 of the table.

(a) From Columns 2 and 4, $n = 80$ and $\Sigma f_i x_i = 4{,}455$. Therefore, the mean (in points) is

$$\bar{x} = \frac{4455}{80} = 55.7.$$

(b) From Column 5 of Table 2-6 we have $\Sigma f_i x_i^2 = 285{,}075$. Therefore, substituting in the formula, the standard deviation is

$$s = \sqrt{\frac{1}{(80-1)}\left[285{,}075 - \frac{(4455)^2}{80}\right]}$$

$$= \sqrt{468.1922}$$

$$= 21.6 \text{ points}$$

Section 2-4 Exercises

1. Among forty people, fifteen weighed 150 pounds each, twenty weighed 170 pounds each, and five weighed 200 pounds each.
   (a) Find the mean weight.   **(b)** Find the standard deviation.

2. A company employs eighty junior executives, twenty-six senior executives, and eight vice-presidents. During one year the mean salary of junior executives was $58,000, that of senior executives was $78,000, and that of vice-presidents was $108,000. Compute the mean salary of the 114 executives.

3. The following distribution gives the top speeds (in miles per hour) at which thirty racers were clocked in an auto race.

Top Speed	Number of Racers
145	9
150	8
160	11
170	2

Find the mean, the variance, and the standard deviation of the distribution of the top speeds.

4. Find the variance and the standard deviation of the following distribution of earnings of 100 graduate students during a year.

Earnings (in Dollars)	Number of Graduate Students
5,550	13
7,550	26
9,550	36
11,550	25

5. Sixty-eight samples of milk were examined for the percentage of milk fat. The following information was obtained.

Percentage Milk Fat	Number of Samples
3.50	2
3.55	7
3.60	8
3.62	6
3.65	12
3.68	8
3.70	15
3.75	10

Compute the mean percentage milk fat and the standard deviation.

6. A motel owner has a motel with twenty-five rooms. The following data show the room occupancy records during 1994.

Number of Rooms Occupied	Number of Days
8	2
12	6
16	39
18	48
20	58
21	56
22	36
23	30
24	40
25	50

Calculate the mean number of rooms occupied during 1994 and the standard deviation.

7. In the preceding Exercise 6, if the charge per room is $35 per night, what are the mean receipts per night?

8. The following data are the net weights of sugar (in ounces) packed in thirty one-pound bags.

15.6	15.9	16.2	16.0	15.6
15.9	16.0	15.6	15.6	16.0
15.6	15.9	16.2	15.6	16.2
16.0	15.8	15.9	16.2	15.8
15.8	16.2	16.2	16.0	16.2
15.9	16.2	15.8	16.2	16.0

Find the following.
(a) The mean amount per bag
(b) The variance and the standard deviation
(c) Determine the interval that is two standard deviations around the mean.

How many values in the data set lie in this interval? Is this in keeping with Chebyshev's rule?

Note: Arrange in a frequency table first.

9. Complete the following table and compute the variance of the data given.

Class	Frequency f_i	Class Mark x_i	$f_i x_i$	$f_i x_i^2$
2–6	23	—	—	—
7–11	14	—	—	—
12–16	34	14	476	6664
17–21	15	—	—	—
22–26	14	—	—	—
Sum	—	—	—	—

10. For the distribution of life (in hours) of 400 cathode tubes, given in the following table, find the variance and the standard deviation of the life of the tubes.

Life of Tube (in Hours)	Number of Tubes
600–699	85
700–799	77
800–899	124
900–999	78
1000–1099	36

11. The following table gives the distribution of the breaking strength (in pounds) of sixty cables.

Breaking Strength	Number of Cables
140–149	8
150–159	10
160–169	18
170–179	12
180–189	8
190–199	4

Find the mean and the standard deviation of the breaking strength of the cables.

12. Fifty candidates entering an astronaut training program were given a psychological profile test. The following table gives the distribution of their scores.

Score (in Points)	Number of Candidates
60–79	8
80–99	16
100–119	12
120–139	8
140–159	6

(a) Find the mean score and the standard deviation of the scores.
(b) Prepare a histogram, locate the mean along the horizontal axis, and mark an interval that is 1.5 standard deviations around the mean.
(c) Considering the data in the table, at least how many candidates have scores that fall in the interval? Give the maximum answer you can.
(d) Does this agree with Chebyshev's rule?

13. Gary has thirty books on a shelf. When he grouped them according to the number of pages each has, he obtained the following distribution.

Number of Pages	Number of Books
100–199	3
200–299	6
300–399	8
400–499	8
500–599	5

(a) Compute the mean number of pages and the standard deviation.
(b) From the data, at least how many books have total page counts that are within 1.2 standard deviations of the mean? Give the maximum answer you can.

14. The following table gives the distribution of the amount of rainfall (in inches) in December in Tippegawa County for a period of thirty years.

Rainfall in December (in Inches)	Number of Years
2.8–4.7	3
4.8–6.7	6
6.8–8.7	8
8.8–10.7	8
10.8–12.7	5

Find the following.
(a) The mean amount of rainfall in December
(b) The standard deviation of the amount of rainfall

15. The office of the Dean of Graduate Studies has compiled the following information regarding the ages of twenty-six students who will be awarded their Ph.D.s.

Age (in Years)	Number of Students
20–24	1
25–29	9
30–34	8
35–39	5
40–44	2
45–49	1

Find the following.
(a) The mean age of the students awarded Ph.D.s
(b) The standard deviation of the age

16. A university has a faculty of 330 members. The following table shows the distribution of the number of years, to the nearest year, that they have served on the faculty.

Number of Years	Number of Faculty
1–5	20
6–10	55
11–15	76
16–20	84
21–25	60
26–30	35

(a) Sketch the histogram.
(b) Find the mean number of years in service.
(c) Locate the mean on the *x*-axis.

17. During the breeding season male frogs vocalize to attract females. In a study published in the *Journal of Herpetology,* Mac F. Given used playbacks to simulate a vocal intrusion on a male's calling site in order to explore how male frogs interact acoustically. Among other aspects of the investigation, time intervals between the end of a stimulus call and the beginning of a natural toad call were recorded. The following histogram was obtained for the distribution of the lengths of the time intervals. Treat the class marks in the histogram as observed interval lengths, find the frequency of each class mark, and use this information to compute the mean length of the time intervals between the end of a stimulus call and the beginning of a natural toad call. Also, find the standard deviation.

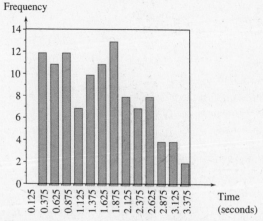

Histogram of time intervals between the end of a stimulus call and the beginning of a natural frog call.

Source: Graphic from the *Journal of Herpetology,* vol. 27, no. 4, pp. 447–452. Reprinted by permission.

CHAPTER 2 Summary

✓ *CHECKLIST: KEY TERMS AND EXPRESSIONS*

❑ average, page 77
❑ measures of central tendency, page 77
❑ variability, page 77
❑ arithmetic mean, page 78
❑ mean, page 78
❑ sample mean, page 78
❑ summation notation, page 78
❑ weighted mean, page 80
❑ median, page 81
❑ percentile, page 83
❑ first quartile, page 83
❑ third quartile, page 83
❑ second quartile, page 83

❑ interquartile range, page 85
❑ boxplot, page 85
❑ box-and-whiskers diagram, page 85
❑ lower hinge, page 86
❑ upper hinge, page 86
❑ mode, page 87
❑ range, page 102
❑ mean deviation, page 103
❑ sample variance, page 104
❑ sample standard deviation, page 104
❑ Chebyshev's rule, page 109
❑ Empirical rule, page 109

KEY FORMULAS

mean (ungrouped)

$$\text{mean} = \frac{\sum x_i}{n}$$

$$\text{weighted mean} = \frac{\sum w_i x_i}{\sum w_i}, \text{ where } w_i \text{ is the weight of } x_i$$

mean (grouped)

$$\text{mean} = \frac{\sum f_i x_i}{n}, \text{ where } x_i \text{ is the class mark and } f_i \text{ is its frequency}$$

range

$$\text{range} = \text{largest value} - \text{smallest value}$$

mean deviation

$$\text{mean deviation} = \frac{\sum |x_i - \bar{x}|}{n}$$

sample standard deviation (ungrouped)

$$s = \sqrt{\frac{\sum (x_i - \bar{x})^2}{n-1}} = \sqrt{\frac{1}{(n-1)}\left[\sum x_i^2 - \frac{(\sum x_i)^2}{n}\right]}$$

sample standard deviation (ungrouped frequency distribution)

$$s = \sqrt{\frac{\sum f_i(x_i - \bar{x})^2}{n-1}}, \text{ where } f_i \text{ is the frequency of } x_i \text{ and } \sum f_i = n$$

$$= \sqrt{\frac{1}{(n-1)}\left[\sum f_i x_i^2 - \frac{(\sum f_i x_i)^2}{n}\right]}, \text{ (shortcut formula)}$$

sample standard deviation (grouped)

$$s = \sqrt{\frac{1}{(n-1)}\left[\sum f_i x_i^2 - \frac{\left(\sum f_i x_i\right)^2}{n}\right]}, \text{ where } x_i \text{ is the class mark and } f_i \text{ is its frequency}$$

sample variance

$$\text{sample variance} = s^2$$

CHAPTER 2 Concept and Discussion Questions

1. Suppose just one observation in the data is changed. Which measure of central tendency is bound to be affected?

2. Do the median and the mode represent frequencies, or do they refer to the value of the variable?

3. In what way do the deviations from the mean serve as a check on your computations?

4. Suppose you are given the values of $\bar{x}$ and s. Does this give you a complete picture of the underlying distribution? In what way does this information help in knowing the distribution?

5. Describe an actual situation in which the median would be a better measure of location than the mean.

6. Ingrid's score on the SAT math test was in the 85th percentile. Interpret the statement.

7. Suppose you have collected some numerical field data. What numerical measures would you consider to summarize the data? Describe also any graphical methods you know that would highlight the spread of the data.

CHAPTER 2 Review Exercises

1. Describe the three measures of central tendency you have learned and compare their advantages and disadvantages.

2. On a trading day on the New York Stock Exchange, the mean gain based on 30 industrial stocks was 35 cents per share. On the same day, the mean loss based on 20 transportation stocks was 56 cents per share. Determine the mean gain or loss per share when all the 50 stocks are considered together.

3. Give a set of data for which the mean, the median, and the mode coincide.

4. The following figures were reported for annual suspended sediment (in cubic yards per square mile) from the Casper Creek watershed in Northern California

during the period 1963–1967. (Raymond Rice et al. of the Pacific Southwest Forest and Range Experiment Station)

Hydrologic Year	Suspended Sediment
1963	101.15
1964	94.00
1965	612.54
1966	744.04
1967	138.24

Find the following.
(a) The mean annual suspended sediment during the period
(b) The median annual sediment during the period

5. A secretary typed twelve letters. The times (in minutes) spent on these letters were as follows.

    6   8   13   10   8   9   10   6   8   8   6   10

Find the median amount of time spent.

6. A shipment of twenty boxes containing a certain type of electrode was received. Upon inspection it was found that three boxes contained no defectives, ten boxes contained two defectives each, four boxes contained three defectives each, and the remaining boxes contained four defectives each. Find the following.
(a) The mean number of defectives in a box
(b) The median number of defectives in a box
(c) The modal number of defectives in a box

7. The following data refer to the weight gain (in pounds) of thirteen women during pregnancy.

    51.9   28.5   36.4   42.9   35.6   48.2   30.6
    41.5   38.8   40.5   29.5   34.7   52.5

(a) What is the mean weight gain in the sample?
(b) What is the median weight gain in the sample?

8. The following figures represent *Daphnia magna* biomass as percent of the total zooplankton biomass in samples obtained from eight fish-rearing ponds.

    34.2   23.5   28.2   19.6   38.2   25.9   22.7   25.1

Determine the following.
(a) The mean percent biomass of *Daphnia magna*
(b) The median percent biomass
(c) The variance of the percent biomass
(d) The standard deviation of the percent biomass

9. Ten hens, approximately 64 weeks of age, were slaughtered and the pH values of breast meat were measured about 24 hours after slaughter. The data on pH values are as follows.

    5.9   6.1   6.0   5.9   6.0   6.1   5.9   6.1   6.0   6.3

Determine the following.

(a) The mean of the pH value in breast meat

(b) The variance of the pH value in breast meat and the standard deviation

(c) The number of values within two standard deviations of the mean

10. Oyster mushrooms were cultivated in peat and, after harvesting, the moisture content was determined by oven drying and the total fat content was determined gravimetrically after ether extraction. (*Journal of the Science of Food and Agriculture*, 1986, 37, pp. 833–838.)

Moisture (%)	Fat (%)
90.3	2.3
89.6	2.2
90.2	2.2
89.6	2.3
90.7	2.0
91.2	2.5
90.4	1.8
91.1	1.9
90.1	2.2
89.0	2.0

Calculate the following.

(a) The mean moisture content and the mean fat content

(b) The median of the moisture content and that of the fat content

(c) The variance of the moisture content and that of the fat content

(d) The standard deviation of the moisture content and that of the fat content

11. In recent years, members of the public have become increasingly aware of the importance of fiber in their diets. The following data represent the amount of dietary fiber (g/100 g) in ten samples of white bread sold in the market.

    4.8   4.7   4.7   5.0   4.5   5.1   4.7   4.8   4.6   4.9

Determine the following.

(a) The mean dietary fiber

(b) The median dietary fiber

(c) The standard deviation of the dietary fiber

12. The following data are the number of rooms occupied in a motel on Interstate 101 over a period of 40 days.

20	14	21	29	43	8	11	14
17	15	26	10	14	34	14	31
39	23	14	30	28	13	7	18
11	26	35	19	24	30	27	14
29	10	18	15	22	24	20	23

(a) Compute the mean and the standard deviation of the number of rooms occupied.

(b) Prepare a stem-and-leaf diagram and use it to locate the median.

13. The following table gives the distribution of the amount of gasoline (in gallons) sold at a gas station on a certain day.

Amount Sold	Number of Customers
0.0–3.9	17
4.0–7.9	70
8.0–11.9	92
12.0–15.9	63
16.0–19.9	58

Find the following.
(a) The mean amount of gasoline sold per customer
(b) The modal class and the mode

14. Over a 40-day period, Tom has obtained a frequency distribution of the number of minutes that he had to wait for the bus to arrive, which is shown in the following table.

Extent of Wait (in Minutes)	Frequency f
0–4	11
5–9	14
10–14	8
15–19	4
20–24	3
Sum	40

Compute the following.
(a) The mean waiting time
(b) The variance of the waiting time and the standard deviation

15. Cholestyramine is a powerful drug that lowers cholesterol. Suppose six patients, who had abnormally high levels of cholesterol, were kept on a low-fat diet and were treated for a certain period with the drug. The following figures give, for each individual, the cholesterol level (in mg/dl of blood) before and after the treatment.

	CHOLESTEROL (mg/dl)	
Patient	Before	After
Bob	280	224
Judith	290	240
Raphael	250	194
George	260	216
Laura	285	242
Maria	310	262

(a) Compute percent reduction of cholesterol for each individual.
Hint: For example, in Bob's case the percent reduction is $100(280 - 224)/280$.
(b) Compute the mean percent reduction.

16. Define the concept of variability in experimental data and compare the different measures of it.

17. Suppose the data consist of resistance of resistors measured in ohms. Give the units in which s^2 and s should be expressed.

18. The U.S. Department of Agriculture reported the following figures (in billions of dollars) for net cash farm income during the period 1977–1981.

Year	Net Cash Farm Income
1977	17.4
1978	25.9
1979	27.4
1980	21.9
1981	19.0

Calculate the mean deviation of the annual net cash farm income during the given period and the standard deviation.

19. Of twenty dental patients, three had one cavity, six had three cavities, five had four cavities, three had five cavities, and three had eight cavities. Find the variance and the standard deviation of the number of cavities a patient has.

20. The distribution of the earnings of a free-lance photographer during 40 months is as follows.

Earnings (in Dollars)	Number of Months
900–1299	5
1300–1699	9
1700–2099	17
2100–2499	3
2500–2899	6

(a) Prepare a histogram of the distribution and locate the mean along the horizontal axis.

(b) Compute the variance and the standard deviation of the earnings.

21. The following figures show the yield of tobacco per harvested acre for sixteen states during 1985.

State	Yield per Acre (100 Pounds)	State	Yield per Acre (100 Pounds)
Colorado	16.6	North Carolina	22.2
Florida	26.8	Ohio	21.4
Georgia	22.8	Pennsylvania	19.0
Indiana	22.4	South Carolina	23.0
Kentucky	23.0	Tennessee	20.6
Maryland	13.3	Virginia	21.0
Massachusetts	15.9	West Virginia	18.8
Missouri	21.8	Wisconsin	21.9

Compute the mean, the range, and the standard deviation of the data.

22. Twelve sheep were fed a special diet consisting of dried grass and barley for three months. At the end of the period the plasma insulin concentration (μU/ml) was determined for each sheep. The data are as follows.

20.5	26.2	18.8	21.2
22.1	21.6	25.4	26.8
22.9	22.8	30.8	28.4

Calculate the following.
(a) The mean
(b) The median
(c) The quartiles Q_1 and Q_3
(d) The interquartile range. What does the interquartile range explain about the distribution of the plasma insulin concentration of the twelve sheep?

23. Ten 17-week-old turkeys were slaughtered and the total haem pigments (mg/g tissue) were extracted with phosphate buffer (pH 6.8). The following data give the total haem pigments in breast and thigh meats of the ten birds. (*Journal of the Science of Food and Agriculture*, 1986, 37, pp. 1236–1240.)

HAEM PIGMENT (mg/g)	
Breast Meat	**Thigh Meat**
0.57	2.11
0.51	2.21
0.60	2.20
0.57	2.19
0.56	2.00
0.58	2.22
0.65	2.13
0.68	2.10
0.57	2.17
0.59	1.92

Compute the following.
(a) The mean haem pigment in breast meat
(b) The mean haem pigment in thigh meat
(c) The median haem pigment in breast meat
(d) The median haem pigment in thigh meat
(e) The quartiles Q_1 and Q_3 for haem pigment in breast meat
(f) The quartiles Q_1 and Q_3 for haem pigment in thigh meat
(g) The interquartile range for haem pigment in breast meat
(h) The interquartile range for haem pigment in thigh meat

24. For assigning grades to his students, Professor Alvarez adopts a procedure that is shown in the following figure.

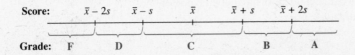

For example, any student getting a score greater than $\bar{x} + s$ but less than or equal to $\bar{x} + 2s$ will earn a letter grade of B. In a statistics course that Professor Alvarez taught, the following scores were recorded for twenty students.

75	71	88	99	80
77	76	70	74	68
49	61	87	75	88
72	74	69	77	59

Assign grades to each of the students.

25. In Exercise 24, according to Chebyshev's rule, at least how many students should get a grade of D, C, or B? Is your assignment of grades in agreement with this?

26. Rainwater was collected in water collectors at thirty different sites near an industrial basin and the amount of acidity (pH level) was measured. The following stem-and-leaf diagram shows the figures (ranging from 2.6 to 6.3).

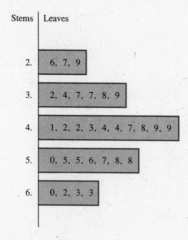

Stems	Leaves
2.	6, 7, 9
3.	2, 4, 7, 7, 8, 9
4.	1, 2, 2, 3, 4, 4, 7, 8, 9, 9
5.	0, 5, 5, 6, 7, 8, 8
6.	0, 2, 3, 3

Find the following.
(a) The mean acidity
(b) The median acidity
(c) The standard deviation of acidity

27. The following number of issues were traded on the New York Stock Exchange during seven trading sessions.

1994	2008	2005	1998
1992	2003	2007	

Find the mean and the standard deviation of the number of issues traded adopting the following procedure: subtract 2000 from each number; find the mean, variance, and standard deviation for the new data obtained; then, apply the properties of the mean and the standard deviation.

28. According to the journal *Agricultural Statistics 1987*, published by the U.S. Department of Agriculture, the taxes (in dollars) levied by each state on farm real estate per $100 of full value during 1985 were as follows.

AL....	0.15	IN.....	·76	NE....	1.36	SC....	.28
AK....	.49	IA.....	.81	NV....	.24	SD....	1.02
AZ....	.48	KS.....	.73	NH....	.91	TN....	.38
AR....	.31	KY.....	.24	NJ.....	.73	TX....	.28
CA....	.42	LA.....	.18	NM....	.12	UT....	.31
CO....	.43	ME.....	.88	NY.....	2.32	VT....	1.04
CT....	.77	MD....	.41	NC.....	.40	VA....	.42
DE....	.13	MA.....	.97	ND.....	.63	WA....	.44
FL....	.46	MI....	2.28	OH.....	.76	WV....	.17
GA....	.51	MN.....	.63	OK....	.33	WI....	1.92
HI....	.35	MS....	.26	OR....	.62	WY....	.33
ID....	.40	MO....	.40	PA....	.83		
IL.....	.98	MT....	.53	RI..:..	1.36		

(a) Compute the mean and the standard deviation of the data.
(b) Compute the intervals $(\bar{x} - 2s, \bar{x} + 2s)$ and $(\bar{x} - 3s, \bar{x} + 3s)$.
(c) Compute the percentages of measurements falling in the intervals in Part b.
(d) Are the percentages in Part c in keeping with the statements of Chebyshev's rule?
(e) Are the data described satisfactorily by the Empirical rule?

29. To investigate the effect of the insecticide lead arsenate on table wines, thirty-five bottles of white wine from different wineries in a certain region where the insecticide was used were sampled and the amount of lead (mg/l) was determined by carbon rod atomic absorption spectrophotometry. The findings were as follows.

0.44	0.50	0.45	0.42	0.46	0.34	0.47
0.44	0.34	0.39	0.41	0.43	0.37	0.41
0.47	0.37	0.36	0.45	0.47	0.34	0.44
0.39	0.37	0.30	0.40	0.41	0.45	0.47
0.45	0.36	0.31	0.39	0.35	0.43	0.43

(a) Display the data graphically by means of a histogram.
(b) Compute the mean and the median. Then locate them along the *x*-axis of the histogram.
(c) Compute the range and the standard deviation of the amount of lead.
(d) Find the percentages of the observations falling in the intervals $(\bar{x} - 2s, \bar{x} + 2s)$ and $(\bar{x} - 3s, \bar{x} + 3s)$. Are these percentages close to those given by the Empirical rule?

Exercises 30 and 31 below are suited for MINITAB.

30. The chlorophyll *a* content (mg/m³) and suspended organic matter (SOM) (g/m³) were determined in the northern and northwestern zones of Lake Onega, including Petrozavodsk and Kondopoga inlets in the Soviet Union. (*Hydrobiological*

Journal 1982, 18, pp. 91–94.) The chlorophyll contents were determined at twenty-one stations, while the concentration of SOM was determined at nineteen stations. The data are as follows.

Chlorophyll (mg/m^3)	SOM (g/m^3)
1.73	3.2
1.80	3.2
1.44	3.8
2.60	3.8
1.58	2.7
1.08	3.2
0.62	2.7
2.86	3.9
0.65	1.8
0.67	2.0
0.80	1.9
0.71	2.5
0.86	5.4
1.21	4.6
1.51	4.2
1.50	4.1
1.33	4.0
1.11	3.0
1.09	1.7
1.56	
3.83	

Compute the following.
(a) The mean chlorophyll *a* content
(b) The median chlorophyll *a* content
(c) The mean SOM content
(d) The median SOM content
(e) The standard deviation of the chlorophyll *a* content
(f) The variance of the SOM content
(g) The quartiles for chlorophyll *a* content and for SOM content
(h) The interquartile ranges for chlorophyll *a* content and for SOM content

31. The following information gives the price to earnings (P/E) ratios of 30 industrial stocks and 26 transportation stocks as reported in the September 1993 *Stock Guide* booklet, published by Standard & Poor's. Reprinted by permission of Standard & Poor's, a division of McGraw-Hill Inc.

Industrial Stocks

Stock	P/E	Stock	P/E
AlliedSgnl	16	Goodyear	13
Alcoa	54	Hershey	15
AmExpress	13	IntPaper	26
AmT&T	20	McDonalds	18
BethSteel	18	Merck	17

Industrial Stocks (continued)

Stock	P/E	Stock	P/E
Boeing	11	MinnMnMf	16
Caterpillar	27	MorganJP	10
Chevron	20	PhilipMor	10
CocaCola	25	PolaroidCorp	25
Disney	22	Sears	10
Dupont	17	Texaco	16
EKodak	22	UnCarbide	20
Exxon	16	UtdTech	17
GenElec	19	Westnghse	15
GenMotor	21	Woolworth	14

Transportation Stocks

Stock	P/E	Stock	P/E
AMR	42	FedlExp	30
AirbrnFrt	21	GlobalCarrier	30
AlaskaAir	53	GreyhoundLines	9
AmPresdnt	12	IllinoisCentral	14
ArnoIndus	18	NorflkSo	16
BuildersTransp	50	OTRExp	13
BurlNthn	14	RoadwSvc	18
CSX	16	RyderSys	18
CannonExp	15	SowestAir	37
CanadianPac	32	SwiftTransport	20
CarolFrght	37	TNTFrghtWays	24
Conrail	15	UnPacific	17
Confrght	20	XTRA	21

Analyze and compare the data for the two types of stocks using methods learned in this chapter and Chapter 1. Among other things, consider side-by-side display of boxplots and computations of measures of central tendency and measures of dispersion.

WORKING WITH LARGE DATA SETS

Project: Refer to the medical data from the diabetes study in Appendix IV, Table B-3. Using MINITAB, provide descriptive statistics for FBSGLU measurements. Draw a histogram and locate these statistics on the graph. Determine the fraction of measurements falling in the intervals $(\bar{x} - s, \bar{x} + s)$, $(\bar{x} - 2s, \bar{x} + 2s)$, and $(\bar{x} - 3s, \bar{x} + 3s)$. Also, obtain a box plot and discuss the symmetry or the lack of it. Do the same for the other variables HTCM, WTKG, SYSBP, DIASBP, and CHOLES.

Present a written report on your findings.

 GRAPHING CALCULATOR INVESTIGATIONS

Refer to the school district data given in the calculator exercises for Chapter 1.

1. If your data for sports and textbook expenditures still is stored in two separate lists on your TI-82, then proceed to create boxplot displays of these two data sets on the same graph using the STAT PLOT option. If you have cleared these data sets from your calculator then reenter them and proceed to create the boxplot displays. Comment on similarities and differences in the two boxplots. Do they indicate comparable spending for the majority of cities and towns? Do they suggest that expenditures are higher for one group? What about variability in expenditures? Does it seem to be about the same for sports and textbook expenditures?

2. Select the CALC option from the STAT menu to determine the sample mean, standard deviation, minimum, maximum, Q1, median, and Q3 for both sports expenditures and textbook expenditures. Focus on the summary statistics for the sports data. Does a comparison of the mean and median lead you to believe the distribution of sports expenditures is approximately symmetrical? How does this compare with your visual impression when you created the histogram for the sports data? Now focus on the textbook expenditures. Comment on the symmetry or lack thereof by comparing the sample mean and median and by inspecting the histogram.

3. If your data for the percent of children who watch 4 or more hours of TV per day still is stored in lists on your TI-82, then proceed to create boxplot displays of these data sets on the same graph using the STAT PLOT key. If your data are no longer stored in your calculator, then reenter them in two separate lists. Comment on the similarities and differences in these box plots. Does it appear that one grade level has a higher percent who watch 4 or more hours of TV per day? Which group, and why? What about the variability in the data? How can you compare the width of the boxes to make a statement about similarities or differences in the variability of each group? Which group, if either, is more variable with respect to the percent that watch more than 4 hours of TV per day?

4. Select the CALC option from the STAT menu to determine the sample mean, standard deviation, minimum, maximum, Q1, median, and Q3 for the percent who watch more than 4 hours of TV per day for both 8th and 12th graders. Compare the sample means and medians, as well as the histograms, and comment on the symmetry or skewness of each data set.

Refer to the unemployment data given in the calculator exercises for Chapter 1.

5. Display boxplots for the October 1994 and November 1994 unemployment data on the same graph. Determine the sample means, standard deviations, and quartiles for both groups. Comment on possible similarities and differences between the October and November unemployment rates. Finally, comment on the symmetry or skewness of these data sets.

Refer to the presidential data given in the calculator exercises for Chapter 1.

6. (a) Display boxplots for the age at inauguration for the "early" vs "recent" presidents on the same graph. Determine the sample means, standard deviations, and quartiles for both groups. What do you observe about the age at inauguration for these two groups?

   (b) Do the same for the age at death for "early" vs "recent" presidents.

MINITAB

LAB 2: Numerical Description of Quantitative Data

Our main goal in this lab is to see how MINITAB can be used to compute the three measures of central tendency (mean, median, and mode) and the three measures of dispersion (range, mean deviation, and standard deviation) for a given set of data. We will also learn more about the LET command, which is a powerful tool for performing arithmetic computations.

The first order of business is to log on the computer and access MINITAB. Then, as in Lab 1, type in the lines in capital letters.

1 In Lab 1, you saw that the LET command can be used to correct numbers entered in a column. It also allows us to perform the standard arithmetic operations: addition, subtraction, multiplication, and division. The symbols used are + for addition, − for subtraction, * for multiplication, and / for division. There is also the symbol ** for raising to a power. The appropriate power is designated by the number that comes after the two stars.

```
MTB >NOTE ARITHMETIC WITH COLUMNS USING LET COMMAND
MTB >READ DATA INTO C1, C2
DATA>2.4    7.9
DATA>1.8    8.3
DATA>3.0    9.6
DATA>2.5    8.6
DATA>END
MTB >PRINT C1 C2
MTB >LET C10 = C1 + C2
MTB >LET C11 = C1 − C2
MTB >LET C12 = C1 * C2
MTB >LET C13 = C1/C2
MTB >LET C14 = C1 ** 2
MTB >LET C15 = C1/100
```

```
MTB >NAME C10 'SUM' C11 'DIFF' C12 'PRODUCT' C13
     'QUOTIENT'

MTB >NAME C14 'SQUARE'

MTB >PRINT C10-C15
```

Consider the command LET C10 = C1 + C2. The instruction is to add numbers in Column C1 to the corresponding numbers in Column C2. The result is a column of numbers and, when the answer is a column of numbers, we have to tell the computer the column in which we want the answer placed. We have placed the answer in Column C10. Similar comments apply to the other LET commands. It is important to note that, unlike other commands, no extra text is permitted in a LET command.

2 The following instructions show how to store constants and perform arithmetic with them.

```
MTB >NOTE HOW TO STORE CONSTANTS AND PERFORM
     ARITHMETIC

MTB >LET K1 = 3.7

MTB >LET K5 = 2

MTB >LET K10 = K1 + 3 * K5 + 8

MTB >LET K11 = K5 * K5

MTB >LET K12 = K5 ** 3

MTB >LET K13 = K5 ** 0.5

MTB >PRINT K1 K5 K10-K13
```

First the constants 3.7 and 2 are stored as K1 and K5 and then the results of the arithmetic operations are stored as constants K10, K11, K12, and K13. Thus, for example, K10 = 3.7 + 3(2) + 8 and the constant in K13 is the square root of K5, that is, $\sqrt{2}$.

3 The SUM command in MINITAB does just what it says. It computes the sum of the entries in a column. There are two ways to compute the sum, as described in the following instructions.

```
MTB >NOTE SUM OF DATA IN A COLUMN

MTB >SET DATA IN C10

DATA>3.1 2.4 4.5 2.8 2.8 3.3 3.5 3.6 2.8

DATA>END OF DATA

MTB >SUM C10 PUT IN K6

MTB >NOTE ALTERNATE WAY TO FIND SUM USING LET

MTB >LET K20 = SUM (C10)

MTB >PRINT K20
```

The first method in the preceding instructions, SUM C10 PUT IN K6, displays the answer on the screen and also stores it in K6 for later use. The other method using the LET command stores the answer in a preassigned location, which in our case is K20, but we have to give the additional command PRINT asking for its display on the monitor.

Simply typing the command SUM C10 will display the answer. But if you want to use the sum later in subsequent computations, then you need to tell the computer to store it in some location.

Note: When you compute a numerical value with the LET command using data in a column, the column must be enclosed in parentheses. For example, see LET K20 = SUM(C10) in the preceding instructions.

4   With the built-in commands MEAN and MEDIAN we can compute the mean and the median for data contained in a column as follows.

```
MTB >NOTE MEASURES OF CENTRAL TENDENCY

MTB >MEAN OF DATA IN C10 PUT IN K1

MTB >MEDIAN OF DATA IN C10 PUT IN K2

MTB >HISTOGRAM C10
```

We could have also computed the mean and the median with the LET commands

```
LET K1 = MEAN(C10)

LET K2 = MEDIAN(C10)
```

and then asked for the printout of these constants with the command PRINT.

There is no command that gives the mode directly. However, it can be obtained easily by considering the histogram, and picking the value having the most stars.

5   The standard deviation of data in a column can be computed and stored as a constant in two ways as follows.

```
MTB >NOTE STANDARD DEVIATION OF DATA IN A COLUMN

MTB >STDEV OF C10 PUT IN K6

MTB >NOTE ALTERNATE WAY TO COMPUTE STANDARD
        DEVIATION

MTB >LET K6 = STDEV(C10)

MTB >PRINT K6
```

The STDEV command directly prints out the standard deviation of the data in a column. As you now know, if it is computed using the LET command, then a PRINT command must be given to display the answer.

There are no predefined commands in MINITAB to compute the range and the mean deviation. They can be computed by using the LET command effectively, as shown by the following.

6   We can compute the range in two ways as follows.

```
MTB >NOTE RANGE OF DATA IN C10

MTB >MAXIMUM C10 PUT IN K1

MTB >MINIMUM C10 PUT IN K2

MTB >NOTE K3 BELOW GIVES THE RANGE

MTB >LET K3 = K1 − K2

MTB >PRINT K3

MTB >NOTE ALTERNATE WAY TO FIND RANGE

MTB >LET K4 = MAXIMUM(C10) − MINIMUM(C10)

MTB >PRINT K4
```

Basically, in both methods, we obtain the maximum and the minimum with the commands MAXIMUM and MINIMUM, and use the LET command to find the difference which gives the range.

7   The following short program computes the mean deviation.

```
MTB >NOTE MEAN DEVIATION OF DATA IN C10

MTB >LET C11 = C10 − MEAN(C10)

MTB >LET C12 = ABSO(C11)

MTB >MEAN C12
```

As can be seen, the deviations from the mean are placed in Column C11. The absolute values of these deviations, indicated by ABSO(C11), are entered in Column C12. The mean of this column then yields the mean deviation.

8   A quick summary of many of the useful descriptive statistics based on a column of data is provided by the DESCRIBE command.

```
MTB >DESCRIBE C10
```

DESCRIBE gives the number of observations (N), mean, median, standard deviation (STDEV), minimum (MIN), maximum (MAX), first quartile (Q1), and third quartile (Q3). It also provides trimmed mean (TRMEAN), and standard error of the mean (SEMEAN), which will be considered in Chapter 6. The trimmed mean is obtained by trimming the smallest 5 percent and the largest 5 percent of the values.

MINITAB will also display a boxplot with the single command BOXPLOTS. We will use this command to obtain a boxplot for the data in Example 4 from Section 2-1, after storing the data in Column C5. Simultaneously, we also ask for a quick summary of the data with DESCRIBE.

```
MTB >NOTE BOXPLOT FOR DATA IN EXAMPLE 4 SECTION 2-1

MTB >SET DATA IN C5

DATA>390  406  446  420  370  328  410

DATA>320  368  392  280  325  382  290

DATA>END OF DATA

MTB >PRINT C5

MTB >DESCRIBE C5

MTB >BOXPLOT FOR DATA IN C5
```

Compare the values of the quartiles with those obtained in Example 3 in Section 2-1 and the display of the boxplot with that in Figure 2-4.

9 MINITAB provides no direct command for computing the mean when a frequency distribution is given. But it is fairly simple to compute it with the repeated use of the LET command as follows.

```
MTB >NOTE MEAN WHEN FREQUENCY DISTRIBUTION IS GIVEN

MTB >NOTE DATA FROM EXAMPLE 1 ON PAGE 120 IN TEXT

MTB >READ DATA IN COLUMNS C1 C2

DATA>2.0  2

DATA>2.2  4

DATA>2.3  6

DATA>2.8  3

DATA>3.0  5

DATA>END

MTB >NOTE WILL PLACE 'VALUE TIMES FREQUENCY' IN C5

MTB >LET C5 = C1 * C2

MTB >NAME C1 'DIAM' C2 'FREQ' C5 'PROD'

MTB >PRINT C1 C2 C5

MTB >LET K1 = SUM(C5)/SUM(C2)

MTB >NOTE K1 IS THE MEAN

MTB >PRINT K1
```

The standard deviation can be computed with the following program.

```
MTB >NOTE COLUMN OF SQUARED DEVIATIONS FROM THE
     MEAN
MTB >LET C10 = (C1 − K1)**2
MTB >NOTE COLUMN OF FREQUENCY TIMES SQUARED
     DEVIATION
MTB >LET C11 = C2*C10
MTB >NOTE K4 GIVES VARIANCE
MTB >LET K4 = SUM(C11)/(SUM(C2) − 1)
MTB >NOTE K5 GIVES STANDARD DEVIATION
MTB >LET K5 = K4 ** 0.5
MTB >PRINT K4 K5
```

Compare these answers with those in the example in the text.

Assignment   Work out Exercises 30 and 31 in Chapter 2 Review Exercises.

10   Having completed the MINITAB session, type STOP.

```
MTB >STOP
```

Now log off.

Commands you have learned in this session

```
LET        SUM      MEAN
MEDIAN     STDEV    MAXIMUM
MINIMUM    ABSO     DESCRIBE
BOXPLOT
```

Commands useful for statistical work

```
MEAN       MEDIAN    STDEV
DESCRIBE   BOXPLOT
```

CAREER *PROFILE*

Name	Judith C. Schafer
Occupation	Manager, Business Planning and Economic Statistics
Employer	American Petroleum Institute, Washington, DC

I'd like to be able to say that I became interested in statistics because math was always my favorite subject in school, math was something that I've always felt comfortable with, and because I've always known how useful math can be.

The truth is, I developed a classic case of "math anxiety" early on. As a result, math was my least favorite subject, and dealing with it made me very uncomfortable (to the point that I usually felt "clinically" stupid). I never really believed that I was going to need Algebra II/Trig or Analytical Geometry in the "real world."

It wasn't until I took an introductory statistics course in college that I realized how useful this stuff can be. Statistics, it seemed to me, was math with a purpose. It could be used to solve all kinds of practical problems across a wide variety of disciplines, ranging from psychology to engineering. In fact, I became interested enough in statistics that I forgot I was afraid of math, and I discovered that I actually had an aptitude for the subject. This led me to pursue a master's degree in operations research and management science from George Washington University's School of Engineering and Applied Science.

At the American Petroleum Institute, where I manage a small group of analysts, we rely on statistics a lot. My staff and I provide information on the supply and demand of U.S. petroleum to various groups, including Congress, the news media, and consumers. This involves designing and conducting surveys to collect data, and developing methodologies using statistical analysis to estimate data.

Many of the techniques that we employ are relatively simple. For example, we typically begin a project by looking at scatter plots of the data, measures of central tendency (mean, range), and measures of variability (standard deviation). This gives us a great deal of specific information about the data, which we attempt to describe in a linear regression model. Then, using R-squares and p-values, along with plots of the residuals, we examine and fine-tune the model to ensure that our assumptions about the data are being met and that the proper variables are being used.

This is how we estimate how much crude oil, motor gasoline and heating fuel is being produced and stored in the United States. These data are used by other organizations and petroleum industry analysts to conduct economic and financial assessments, develop forecasts, and examine trends.

API is also increasing its use of statistics in "bench marking" studies that measure the efficiency and effectiveness of petroleum industry operations in a wide range of areas, from accounting to risk management insurance.

If this sounds intimidating, take the word of a former math anxiety victim—it really isn't. Statistics is a powerful real-world tool *that anyone can use.* It helps you do your job better and faster. And the subject isn't difficult to learn if you're willing to spend some time studying it.

My advice would be to try statistics. You just might like it a lot.

Probability

3

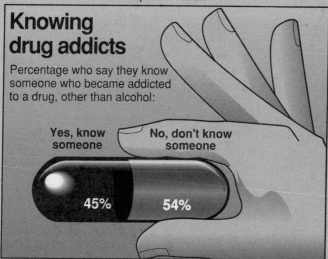

INTRODUCTION

....................................➤

I n every walk of life we make statements that are probabilistic and that carry overtones of chance. For example, we might talk about the probability that a bus will arrive on time, or that a child to be born will be a son, or that the stock market will go up, and so on. What is the characteristic feature in all the phenomena above? It is that they are not deterministic. Past information, no matter how voluminous, will not allow us to formulate a rule to determine precisely what will happen in the future. The theory of probability involves the study of this type of phenomena, called *random phenomena*.

Probability theory has a central place in the theory and applications of statistics since we are analyzing and interpreting data that involve an element of chance or uncertainty. The reliability of our conclusions is supported by accompanying probability statements.

Probability theory originated in the middle of the seventeenth century. It probably began with Blaise Pascal (1623–1662). His interest and that of his contemporaries, Pierre Fermat among them, was occasioned by problems in gambling initially brought to their attention by a French nobleman, Antoine Gombauld, known as the Chevalier de Méré.

As interest in the natural sciences proliferated, so did demands for new laws of probability. Among prominent early contributors were the nineteenth-century mathematicians Laplace, De Moivre, Gauss, and Poisson. Today, probability theory finds applications in diverse disciplines such as biology, economics, operations research, and astronomy. A biologist might be interested in the distribution of bacteria in a culture, an economist in economic forecasts, a production engineer in the inventory of a particular commodity, an astronomer in the distribution of stars in different galaxies, and so on. Considerable interest is also shown in this field as a mathematical discipline in its own right. The name of the brilliant Russian mathematician A. N. Kolmogorov (1903–87) should be mentioned as a pioneer in this area.

3-1   THE EMPIRICAL CONCEPT OF PROBABILITY

A statement such as "There is a 90 percent chance that the bus will arrive on time" is not uncommon. When we consider a particular bus, either it arrives on time or it does not. Why does "90 percent chance" figure in our statement? This statement presumably is based on a person's experience over a long period of time. If we were to keep track of all the buses arriving at the bus stop to find the ratio of the buses that arrived on time to the total number of buses that stopped at the stop, then that ratio would be approximately 90 percent. The empirical, or experimental, idea behind this statement is the kind of rationale that is, in general, at the root of most probabilistic statements.

As another illustration of the *empirical concept of probability*, consider an experiment involving the tossing of a normal coin. A person tossing the coin 1000 times might obtain 462 heads and 538 tails, thus getting heads 46.2

percent of the times. In general, if we were to toss the coin a large number of times, invariably we would find the ratio of the number of heads to the total number of tosses to be *approximately* 50 percent. The larger the number of tosses, the closer the approximation will be to 50 percent. This approximation (close to $\frac{1}{2}$) that is inherent in the situation lends support to the widely accepted statement that the chance of getting heads on a coin is 1 out of 2.

In our discussion regarding the coin, the emphasis should not be on the fact that the ratio approaches $\frac{1}{2}$, but that it stabilizes. In practice, this stable value is the measure of probability. For example, suppose we toss a coin repeatedly and, as in the preceding illustration, compute the ratio of the number of heads to the total number of tosses. If, after tossing the coin a large number of times, the ratio stabilizes around $\frac{1}{3}$, then we would say with a high degree of confidence that the coin is a biased coin, or not fair, and the chance of getting heads on this particular coin is 1 out of 3.

EMPIRICAL CONCEPT OF PROBABILITY

The **empirical concept of probability** is that of "relative frequency," the ratio of the total number of occurrences of a situation to the total number of times the experiment is repeated.

When the number of trials is large, the relative frequency provides a satisfactory measure of the probability associated with a situation of interest.

The intuitively important notion of probability as "long-range relative frequency" in the case of repeatable experiments was put on a firm footing by Jacob Bernoulli (1654–1705), who made many outstanding contributions to probability.

In interpreting a probability statement, we think in terms of repeated performance of the experiment, as in the following illustrations.

a.   The probability of drawing a spade from a deck of cards is 0.25. Empirically this means that if we were to pick a card from the deck repeatedly, every time returning the card to the deck and shuffling the deck properly, approximately 25 percent of the times we will get a spade. If the experiment is repeated a large number of times, the approximation will be very close to 0.25.

b.   The probability of precipitation today is 0.80. The weatherman has calculated from past records that when the weather conditions are such as they are today, it has rained approximately 80 percent of the time.

c.   A patient is told by the doctor that the probability of the patient's recovery is 95 percent. Should the patient derive comfort from the doctor's statement? What the doctor's statement means presumably is that with similar patients, about 95 percent have been cured. If the doctor has been in practice for many years and has had many such patients, the figure 95 percent would be reliable as a measure of the doctor's success. Of course, the particular patient can take it for what it is worth as far as his or her own recovery is concerned.

Quite often, however, it may not be possible to assign probabilities on empirical grounds because a repeated long series of identical trials may not be feasible. For example, empirically how can one expect to assign a value to the probability that a given horse running in the Kentucky Derby will win the race? In such situations where repeatability is not always possible probabilities are sometimes assigned more on the basis of the subjective judgment of an individual. Such probabilities are called **subjective probabilities.** The assignment may simply reflect an expert opinion of the experimenter or just an educated guess.

Section 3-1 Exercises

Give empirical interpretations of the probability statements in Exercises 1–10. Philosophize if you have to.

1. For a person who enters a certain contest, the probability of winning a prize is 0.12.

2. The probability that a child born in the ghetto will be successfully employed when an adult is 16 percent.

3. The probability that the stock market will go up on Monday is 0.62.

4. The probability of striking oil when a well is drilled in a certain region is 0.28.

5. In the United States, the probability that a person will catch cold during winter is $\frac{3}{4}$.

6. The probability that the accused will be acquitted is 80 percent.

7. The probability that Senator Thornapple will vote for the farm bill is very slim.

8. The probability that transistors of a certain brand will last beyond 500 hours is 0.7.

9. The probability that the airline pilots will go on strike is 0.42.

10. The probability that Candidate X will win in the mayoral primary election is 0.58.

11. Answer true or false:
    (a) A random phenomenon means chaotic and haphazard behavior.
    (b) A random phenomenon exhibits regularity or pattern over time in a series of repetitions.

3-2   SAMPLE SPACE, EVENTS, AND THE MEANING OF PROBABILITY

Because notions of set theory in mathematics are at the heart of probability theory, we begin by introducing some related terms.

> A **set** is a gathering of objects that we choose to isolate because they share some common characteristic.

The intuitively familiar notion of a set is that it is a collection of objects, requiring only that we can determine unambiguously whether any given object is a member of the collection. For example, the set of beautiful flowers is not a well-defined set because whether or not a flower is beautiful is a subjective judgment. On the other hand, the set of all the words in *Webster's New World Dictionary*, 1990 Edition, that begin with the letter *d* is a well-defined set. Even though we may not know all such words, we can decide whether a given combination of letters starting with letter *d* is indeed a word in the dictionary.

When a complete list of the members, or elements, of a set is given, it is customary to write the list within braces, separated by commas. For example,

{apple, peach, pear, banana}

is a set listing four kinds of fruit.

Since we are interested only in which objects are in the set, there is no reason the members should be written in any particular order. Thus, the sets

{apple, peach, pear, banana}   and   {banana, peach, pear, apple}

represent the same collection and consequently the same set. Sets are commonly denoted using uppercase letters. For example, we might use the letter *F* (for fruit) to denote the preceding set. Thus,

$$F = \{apple, peach, pear, banana\}.$$

Any subcollection of members from a set *A* is called a **subset** of *A*.

Thus, if *A* = {Tom, Dick, Harry, Jane} and *B* = {Dick, Jane}, then *B* is a subset of *A*.

Sets and relations between them may be illustrated conveniently by geometric configurations. In such diagrams, called **Venn diagrams,** named for the English mathematician John Venn (1834–1923), sets are identified as regions and the members of sets as points in those regions. Venn diagrams are especially useful for visualizing the relationship between two or more sets.

The Venn diagrams in Figure 3-1 display realtionships between two sets *A* and *B*. Part a shows that *B* is a subset of *A*; Part b shows that *A* and *B* have no members in common; and Part c shows that *A* and *B* have some members in common indicated by the shaded portion.

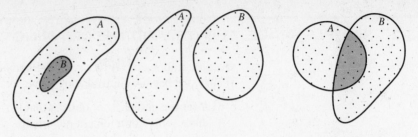

(a) *B* is a subset of *A*.   **(b)** *A* and *B* have no members in common.   **(c)** The shaded portion shows members in common to *A* and *B*.

Figure 3-1 Venn Diagrams Displaying Relationships Between Two Sets *A* and *B*

SAMPLE SPACE AND EVENTS

Now we turn our attention back to probability. If asked the chance of getting a head when a coin is tossed, how many of us would not say that it is one chance in two? Put mathematically, the chance is $\frac{1}{2}$. The reasoning would be that there are two possibilities on a toss, a head or a tail, and there is only one head. If we rolled a die, the chance of getting an even number would be $\frac{1}{2}$. Once again, our argument would be that there are six possible outcomes, namely, 1, 2, 3, 4, 5, 6, of which 2, 4, and 6 are even—three outcomes out of six.

Our intuitive notion lays the groundwork for our formal definitions. The tossing of a coin or the rolling of a die constitutes an *experiment*. In probability and statistics the term **experiment** is used in a very wide sense and refers to any procedure that yields a collection of outcomes. In this sense, "a baby being born" is an experiment whose possible outcomes are: boy, girl. As we saw, the knowledge of all the possible outcomes when a coin is tossed, or a die is rolled, is important. This is always the case for determining probabilities.

The set whose elements are all possible outcomes of an experiment is called the **sample space.** The elements of the sample space are called **sample points.**

Every possible outcome of the experiment is listed in the sample space. When the experiment is actually performed, it will result in exactly one outcome, that is, one sample point, of this set. The sample space, which will be denoted by the letter *S*, provides the basis for our discussion; we work within the framework of this set. Usually the Venn diagram for the sample space is shown as a rectangle as in Figure 3-2.

A sample space is called a **finite sample space** if it has a finite number of outcomes in it. If there are *N* outcomes in *S*, we can identify them as e_1, $e_2, \ldots, e_N$. This would permit us to write the sample space *S* as

$$S = \{e_1, e_2, \ldots, e_N\}.$$

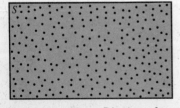

Figure 3-2 Venn Diagram for the Sample Space

As an example, consider a conventional die in the form of a cube with one to six dots on its faces. We can think of tossing such a die as an experiment. There are six possible outcomes.

Therefore,

Here $N = 6$, and e_1 would then stand for the outcome ⚀, e_2 for ⚁, and so on.

If the number of outcomes in the sample space is not finite, then it is said to be an **infinite sample space.** For an example of an infinite sample space, suppose when Jill throws a dart at a circular dart board, we are interested in the point where she hits the board. The sample space is infinite because there is an infinite number of points on the board where the dart could conceivably hit. In what follows, the discussion is restricted to finite sample spaces.

A particular situation of interest, such as "an even number of dots appear when a die is tossed," is called an *event. An event is said to have occurred if **one** of the outcomes that make up the event takes place.* Thus, the event "an even number of dots appear when a die is tossed" will occur if either two, four, *or* six dots show up. (We certainly do not expect two, four, *and* six dots to show up simultaneously on one toss; they cannot.) In other words, this particular event will be realized if *any one* of the outcomes in the set

⚁, ⚃, ⚅ takes place. Thus, any verbal description of a situation

can be characterized by a subset of outcomes of the sample space. We, therefore, define an event as follows.

An **event** is a subset of the sample space. A **simple event,** or *elementary event,* is one that contains only one outcome; that is, it contains one sample point.

An event E with reference to the sample space S can be expressed by means of a Venn diagram as in Figure 3-3.

Given any sample space with N outcomes $e_1, e_2, \ldots, e_N$, there are exactly N simple events $\{e_1\}, \{e_2\}, \ldots, \{e_N\}$. Any event E can be represented as comprised of simple events of outcomes in E. For example, if $E = \{e_2, e_5, e_8\}$, then E is made up of the three simple events $\{e_2\}, \{e_5\}, \{e_8\}$.

An **impossible event** is one that has no outcomes in it and, consequently, cannot occur. On the other hand, a **sure event** is one that has all the outcomes of the sample space in it and will, therefore, definitely occur when the experiment is performed. Thus, the sample space constitutes a sure event.

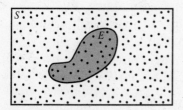

Figure 3-3 Venn Diagram Displaying Event E as a Subset of S

For example, suppose we make a list of the names of all the presidents of the United States before 1992, and choose one name from the list. The event "the name picked is that of a female president" is an impossible event; the events "the name picked is that of a male president" and "the name is that of a president born after 1400 A.D." are sure events.

EXAMPLE 1

Mr. Shapiro owns three computer stocks, IBM, COMPAQ, and Apple, and two oil stocks, XON and TX. He calls his broker, Laura, and asks her to sell one stock of her choosing. "Which stock might the broker sell?" is the question that Mr. Shapiro has in mind.

SOLUTION   Just one possibility is that the broker will sell COMPAQ. All the possibilities that he contemplates comprise the sample space S, which can be written as

$$S = \{IBM, COMPAQ, Apple, XON, TX\}.$$

There are five simple events:

$$\{IBM\}, \{COMPAQ\}, \{Apple\}, \{XON\}, \{TX\}.$$

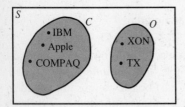

Figure 3-4 Venn Diagram Displaying the Sample Space S and the Events C and O

In the following we describe some events by giving their verbal descriptions as well as their set listings. These events are also shown in Figure 3-4.

C: The broker sells a computer stock = {IBM, COMPAQ, Apple}

O: The broker sells an oil stock = {XON, TX}

Incidentally, the two events C and O make up the entire sample space S. ●

EXAMPLE 2

As a slight variation of Example 1, suppose Mr. Shapiro asks the broker to sell two stocks of her choice.

(a)   Give the appropriate sample space.

(b)   List as sets the outcomes in the following events.
  i. C: Both stocks are computer stocks.
  ii. O: No computer stock is sold; that is, both stocks are oil stocks.
  iii. B: The broker sells one computer stock and one oil stock.

SOLUTION
(a)   It is conceivable that the broker sells IBM and COMPAQ. We represent this outcome as [IBM, COMPAQ]. There are nine other possibilities. The entire set S of ten outcomes is given by

$$S = \{[IBM, COMPAQ], [IBM, Apple],$$
$$[IBM, XON], [IBM, TX], [COMPAQ, Apple],$$
$$[COMPAQ, XON], [COMPAQ, TX],$$
$$[Apple, XON], [Apple TX], [TX, XON]\}$$

Each *pair* is a sample point.

(b) i. "Both stocks are computer stocks" describes those outcomes of S where both entries are computer stocks. So,

$$C = \{[\text{IBM}, \text{COMPAQ}], [\text{IBM}, \text{Apple}], [\text{COMPAQ}, \text{Apple}]\}$$

ii. Here we consider those outcomes of S in which there is no computer entry. There is only one such outcome, [TX, XON]. Hence,

$$O = \{[\text{TX}, \text{XON}]\}$$

O is a simple event since it has only one outcome.

iii. In this case the outcomes comprising the event will clearly have one computer entry and one oil entry. Therefore,

$$B = \{[\text{IBM}, \text{XON}], [\text{IBM}, \text{TX}], [\text{COMPAQ}, \text{XON}], [\text{COMPAQ}, \text{TX}], [\text{Apple}, \text{XON}], [\text{Apple}, \text{TX}]\}$$

Note that the three events C, O, and B together account for the entire sample space. ●

EXAMPLE 3

Vincent and Joel play a game where they simultaneously exhibit their right hands with one, two, three, or four fingers extended.

(a) Write all the outcomes in the sample space.

(b) List the outcomes in the following events.
   i. M: Vincent and Joel extend the same number of fingers.
   ii. F: Vincent and Joel together extend four fingers.
   iii. E: Vincent shows an even number of fingers.

SOLUTION

(a) We give the outcomes as *ordered pairs* where the first component represents the number of fingers on Vincent's hand and the second component, those on Joel's hand. The sample space can be given as

$$S = \{(1, 1), (1, 2), (1, 3), (1, 4), (2, 1), (2, 2), (2, 3),$$
$$(2, 4), (3, 1), (3, 2), (3, 3), (3, 4), (4, 1), (4, 2),$$
$$(4, 3), (4, 4)\}$$

and has sixteen outcomes.

(b) The events are now easily given as subsets of S as follows.
   i. M: Vincent and Joel extend the same number of fingers $= \{(1, 1), (2, 2), (3, 3), (4, 4)\}$.
   ii. F: Vincent and Joel together extend four fingers $= \{(1, 3), (2, 2), (3, 1)\}$. The sum of the components of each sample point is 4.
   iii. E: Vincent shows an even number of fingers $= \{(2, 1), (2, 2), (2, 3), (2, 4), (4, 1), (4, 2), (4, 3), (4, 4)\}$. Notice that the first component of each of the outcomes in this event is an even number, either a 2 or a 4. ●

THE THEORETICAL DEFINITION OF PROBABILITY

Now that we have discussed the important concepts of sample space and events, we can define the probability of an event. As in the preceding section, the discussion is restricted to sample spaces having a finite number of sample points. Let us assume that the sample space S has N outcomes $e_1, e_2, \ldots, e_N$, so that there are N simple events $\{e_1\}, \{e_2\}, \ldots, \{e_N\}$.

CONDITIONS ON THE SAMPLE POINT PROBABILITIES

The **probability of a simple event** $\{e\}$ is a number denoted by $P(\{e\})$ and satisfies the following conditions.

1. $P(\{e\})$ is always between zero and one; that is, $0 \leq P(\{e\}) \leq 1$.
2. The sum of the probabilities of all the simple events is 1; that is,

$$P(\{e_1\}) + P(\{e_2\}) + \cdots + P(\{e_N\}) = 1.$$

The requirements for the sample point probabilities in the context of our finite sample space are based on Kolmogorov's axioms, which apply to a much wider class of sample spaces. Having defined the requirements on probabilities assigned to events having single outcomes, we now define the probability of a general event containing a collection of outcomes.

PROBABILITY OF AN EVENT

The **probability of an event A,** denoted by $P(A)$, is defined as the sum of the probabilities assigned to the simple events that comprise the event A. The impossible event has probability 0 and the sure event S has probability 1.

Thus, if $A = \{e_2, e_4, e_7\}$, then $P(A) = P(\{e_2\}) + P(\{e_4\}) + P(\{e_7\})$.

S I D E B A R   *Improbable, but Not Impossible*

The range of probability values is from 0 to 1, both values inclusive.

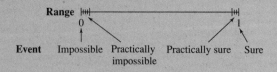

We have defined an impossible event as one with no outcomes in it. So, it will not occur when the experiment is performed and has probability 0. A sure event has all the outcomes and will certainly occur. It has probability 1.

Intuition and empirical evidence collected bear out that an event which has probability very close to zero is *practically impossible,* in that we may assume, for all practical purposes, that it will not occur on any *given* performance of the experiment. On the other hand, an event that has a probability close to one is *practically sure* to occur. Whether a particular value such as 0.001, which is at the lower end of the spectrum of the range of probability values, is low enough to categorize the event as practically impossible is a moot matter. The same can be said in classifying an event as practically certain when given a probability value close to 1, such as 0.995. An understanding of these subtleties will be important when we consider the inferential aspects of statistics. Only practical considerations can provide guidance in this matter. If there is a chance of 0.001 of a nuclear meltdown in a nuclear power plant, we would consider the risk too high and would not dismiss the event as practically impossible. However, if we are told that the probability of erroneously saying that a diet is effective in reducing weight is 0.001, we would not give a second thought to the consequences and would decide that such an erroneous decision is practically impossible. Thus, in the final analysis, whether a particular phenomenon is to be considered as practically impossible or practically certain will depend on the seriousness of the consequences. For example, with the safety of the traveling public as the highest priority, the FAA classifies *probable* to mean a value of 0.00001 or greater, and *improbable* to mean a value less than 0.00001.

EXAMPLE 4   A loaded die is rolled.

(a) Show that the following assignment of probabilities would be acceptable.

$$P(\{1\}) = \frac{1}{3} \qquad P(\{4\}) = \frac{1}{12}$$

$$P(\{2\}) = \frac{1}{4} \qquad P(\{5\}) = \frac{1}{6}$$

$$P(\{3\}) = \frac{1}{12} \qquad P(\{6\}) = \frac{1}{12}$$

(b) Find the probabilities of the following events.
  i. The number is a multiple of 3.
  ii. The number is even.
  iii. The number is even or a multiple of 3.

SOLUTION

(a)   We see that the probabilities of all the simple events are between 0 and 1. Also,

$$P(\{1\}) + P(\{2\}) + P(\{3\}) + P(\{4\}) + P(\{5\}) + P(\{6\})$$

$$= \frac{1}{3} + \frac{1}{4} + \frac{1}{12} + \frac{1}{12} + \frac{1}{6} + \frac{1}{12}$$

$$= 1.$$

Therefore, the given assignment of probabilities would be acceptable.

(b)   i. The probability that the number is a multiple of 3 is

$$P(\{3, 6\}) = P(\{3\}) + P(\{6\}) = \frac{1}{12} + \frac{1}{12} = \frac{1}{6}.$$

Empirically, this means that if we roll this particular die many times, approximately $\frac{1}{6}$th of the times either a 3 or a 6 will appear.

ii. The probability that the number is even is

$$P(\{2, 4, 6\}) = P(\{2\}) + P(\{4\}) + P(\{6\})$$

$$= \frac{1}{4} + \frac{1}{12} + \frac{1}{12} = \frac{5}{12}.$$

iii. The probability that the number is even or a multiple of 3 is

$$P(\{2, 3, 4, 6\}) = P(\{2\}) + P(\{3\}) + P(\{4\}) + P(\{6\})$$

$$= \frac{1}{4} + \frac{1}{12} + \frac{1}{12} + \frac{1}{12} = \frac{1}{2}.$$

The probability is $\frac{1}{2}$. An interpretation of this probability statement is that in a large series of rolls of this die, approximately, fifty percent of the times an even number or a multiple of 3 will occur.   ●

EXAMPLE 5   The candy company Mars, Inc., claims that the color ratios in a bag of plain M&Ms are as shown in the display. Suppose we pick one M&M from a huge pile after mixing it thoroughly. Find the probability that the color of the M&M picked

(a)   has the second letter r

(b)   has the last letter n

(c)   has its first letter coming after the letter j in the alphabet.

SOLUTION   We are interested in the color, so the relevant sample space can be taken to be

$$S = \{\text{brown, red, yellow, orange, green, tan}\}$$

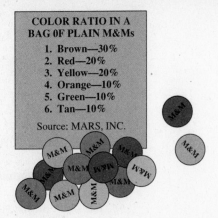

**COLOR RATIO IN A
BAG 0F PLAIN M&Ms**
  1. **Brown—30%**
  2. **Red—20%**
  3. **Yellow—20%**
  4. **Orange—10%**
  5. **Green—10%**
  6. **Tan—10%**

Source: MARS, INC.

Based on the claim of the company, the following assignment of probabilities to the simple events is appropriate.

$$P(\{brown\}) = 0.3 \qquad P(\{red\}) = 0.2$$
$$P(\{yellow\}) = 0.2 \qquad P(\{orange\}) = 0.1$$
$$P(\{green\}) = 0.1 \qquad P(\{tan\}) = 0.1$$

Notice that all the probabilities add up to 1.

(a)   Let R: the second letter in the color is r. Then,

$$R = \{brown, orange, green\} \quad and$$
$$P(R) = 0.3 + 0.1 + 0.1 = 0.5$$

(b)   Let N: the last letter in the color is n. Then,

$$N = \{brown, green, tan\} \quad and$$
$$P(N) = 0.3 + 0.1 + 0.1 = 0.5$$

(c)   Let F: the first letter of the color comes after the letter j. Then,

$$F = \{red, yellow, orange, tan\} \quad and$$
$$P(F) = 0.2 + 0.2 + 0.1 + 0.1 = 0.6 \qquad \bullet$$

EQUALLY LIKELY OUTCOMES

We now consider an important special case, which arises when all the outcomes are assigned the same probability. In this case the outcomes are said to be **equally likely.**

Suppose there are N outcomes in the sample space, all equally likely. Since one of the requirements is that the sum of the probabilities of the simple events should be 1, it is easy to see that each outcome will have a probability equal to $1/N$. Consequently, if an event A contains r outcomes, then based on the definition of probability of an event, we get $P(A)$ as

$$P(A) = \underbrace{\frac{1}{N} + \frac{1}{N} + \cdots + \frac{1}{N}}_{r \text{ times}} = \frac{r}{N} = \frac{\text{number of outcomes in } A}{\text{number of outcomes in } S}$$

This leads to the following result for the probability of an event when there are a finite number of equally likely outcomes in the sample space.

PROBABILITY IN THE EQUALLY LIKELY CASE

If the sample space contains a finite number of outcomes, all equally likely, then

$$P(A) = \frac{\text{number of outcomes in } A}{\text{number of outcomes in } S}.$$

The outcomes in A are often referred to as the outcomes favorable to A. Letting $n(E)$ be the number of outcomes in an event E, we can write

$$P(A) = \frac{n(A)}{n(S)},$$

for any event A when the outcomes are equally likely.

In dealing with problems of probability, we will often use statements such as "An unbiased coin is tossed," "A fair die is rolled," "An object is picked at random," and so on. Such expressions are meant to suggest that the outcomes of the experiment are equally likely.

EXAMPLE 6   Suppose a fair die is rolled. Find the probabilities of the following.

(a) E: An even number of dots show up.

(b) T: The number of dots showing up is a multiple of 3.

(c) B: The number of dots showing up is even or a multiple of 3.

SOLUTION   The sets describing these events are as follows.

Because the die is a fair die, all six outcomes in the sample space are equally likely, each with a probability of $\frac{1}{6}$. There are three outcomes in E, two in T, and four in B, giving the following probabilities for the three events.

(a) $P(E) = \dfrac{3}{6} = \dfrac{1}{2}$ **(b)** $P(T) = \dfrac{2}{6} = \dfrac{1}{3}$ **(c)** $P(B) = \dfrac{4}{6} = \dfrac{2}{3}$

As an experimental interpretation, in a long series of rolls with a fair die, an even number will show up approximately 50 percent of the times, a multiple of 3 will show up approximately 33.3 percent of the times, and an even number or a multiple of 3 will show up approximately 66.7 percent of the times.   ●

EXAMPLE 7   In Example 2, suppose the broker sells two stocks chosen at random. Find the probabilities of the following events.

(a) C: Both stocks are computer stocks.

(b) O: No computer stock is sold.

(c) B: One computer stock and one oil stock are sold.

SOLUTION   We have already discussed the sample space S and the events C, O, and B in Example 2. Counting the outcomes in each of these events, we get the following totals.

$$n(S) = 10 \qquad n(C) = 3 \qquad n(O) = 1 \qquad n(B) = 6$$

Since the two stocks are chosen at random, making the outcomes equally likely, we have the following probabilities.

$$P(C) = \frac{3}{10} \qquad P(O) = \frac{1}{10} \qquad P(B) = \frac{6}{10} \qquad ●$$

EXAMPLE 8   With reference to Example 3, find the probabilities of the following events.

(a) M: Vincent and Joel extend the same number of fingers.

(b) F: Vincent and Joel together extend four fingers.

(c) E: Vincent shows an even number of fingers.

SOLUTION   As we have seen, there are four outcomes in the event M, three in the event F, and eight in the event E. Also, there are sixteen outcomes in S, which we assume to be equally likely. Therefore,

(a) $P(M) = \dfrac{4}{16} = \dfrac{1}{4}$ **(b)** $P(F) = \dfrac{3}{16}$ **(c)** $P(E) = \dfrac{8}{16} = \dfrac{1}{2}$

Thus, if Vincent and Joel were to play the game a large number of times, approximately 25 percent of the times they will extend the same number of fingers, approximately 19 percent of the times they will together extend four fingers, and approximately 50 percent of the times Vincent will show an even number of fingers.   ●

Section 3-2 Exercises

1. Three children are born in a family. On any birth the child could be a son or a daughter. Using s to represent a son and d to represent a daughter, give the sample space S.

2. A committee of two is selected from a group consisting of five people: Juan, Dick, Mary, Paul, and Jane. Write the following events as sets, listing the outcomes they contain.
   (a) The sample space
   (b) Both members are males
   (c) Exactly one member is a male

3. A box contains seven tubes, three of which are defective. Two tubes are selected at random.
   (a) List six of the outcomes in the sample space.
   (b) List six of the outcomes in the event specifying one defective and one nondefective tube.

   Hint: In this exercise, identify the seven tubes $G_1, G_2, G_3, G_4, D_1, D_2, D_3$, where G_1, G_2, G_3, G_4 represent the good tubes and D_1, D_2, D_3 represent the defective tubes. Now list the outcomes in a suitable way.

4. Let a penny, a nickel, and a die be tossed. If you are observing the outcome on the penny, the nickel, and the die in that order, list all the outcomes in the following sets.
   (a) The sample space
   (b) A: Three dots appear on the die
   (c) B: An odd number of dots appear on the die
   (d) C: Heads appear on the two coins
   (e) D: A head appears on the penny and a tail on the nickel

5. Suppose the sample space $S = \{e_1, e_2, e_3\}$. Indicate the cases where we have an acceptable assignment of probabilities to the simple events.
   (a) $P(\{e_1\}) = \dfrac{1}{3}, \quad P(\{e_2\}) = \dfrac{1}{2}, \quad P(\{e_3\}) = \dfrac{1}{3}$
   (b) $P(\{e_1\}) = \dfrac{1}{3}, \quad P(\{e_2\}) = -\dfrac{1}{3}, \quad P(\{e_3\}) = 1$
   (c) $P(\{e_1\}) = \dfrac{1}{5}, \quad P(\{e_2\}) = \dfrac{3}{5}, \quad P(\{e_3\}) = \dfrac{1}{5}$
   (d) $P(\{e_1\}) = \dfrac{1}{2}, \quad P(\{e_2\}) = 0, \quad P(\{e_3\}) = \dfrac{1}{2}$

6. Suppose your instructor has assigned three problems for homework and has announced in class that only one problem will be graded. Your friend who has taken the same course from the instructor tells you from his experience that the probability of Problem 1 being graded is 0.3, that of Problem 2 being graded is 0.6, that of Problem 3 being graded is 0.4. Is this a valid assignment of probabilities?

7. Suppose your friend hands you three floppy disks and tells you that exactly one of them contains a data file that you should be working with. She also tells you that the probability that Disk 1 contains the file is 0.3, that Disk 2 contains it is 0.6, and that Disk 3 contains it is 0.1. Is this a valid assignment of probabilities?

8. A letter of the English alphabet is chosen at random. Find the probability of the event that the letter selected
   (a) is a vowel
   (b) is a consonant
   (c) follows the letter p
   (d) follows the letter p and is a vowel
   (e) follows the letter p or is a vowel.
   In each case provide an empirical interpretation.

9. An urn contains ten tubes of which four are known to be defective. If one tube is picked at random, find the probability that it is:
   (a) defective
   (b) nondefective.

10. Four hundred people attending a party are each given a number, 1 to 400. One number is called at random. Find the probability that the number called:
    (a) is 123
    (b) has the same three digits
    (c) ends in 9.
    In each case, explain what the probabilities mean empirically.

11. A fair die is rolled once.
    (a) List all the outcomes in the event "an even number or a number less than 3 shows up."
    (b) What is the probability that an even number or a number less than 3 shows up?

12. Bertha has the following four books on a shelf: history, calculus, statistics, and English. If she picks two books at random, list all the outcomes in the following events.
    (a) The sample space
    (b) She picks the statistics book as one of the two books.

13. A die is rolled twice. List the outcomes in each of the following events and then find the probability of each event.
    (a) The sample space
    (b) The same number is on the two rolls.
    (c) The total score is 7.
    (d) The total score is even.
    (e) The total score is less than 6.
    (f) The outcome on the first toss is greater than 4.
    (g) Each toss results in the same even number.

14. A letter is chosen at random from the letters of the English alphabet. Find the probability of each of the following events.
    (a) The letter is j.
    (b) The letter is one of the letters in the word *board*.
    (c) The letter is not in the word *board*.
    (d) The letter is between d and m, both inclusive.
    (e) The letter is in the word *card* or the word *board*.
    (f) The letter is in both the words *card* and *board*.
    (g) The letter is in the word *board* but not in the word *card*.

15. Suppose probabilities are assigned to the simple events $S = \{e_1, e_2, e_3, e_4\}$ as follows.

$$P(\{e_1\}) = 3P(\{e_2\}), \; P(\{e_3\}) = P(\{e_1\}), \; P(\{e_4\}) = 4P(\{e_2\})$$

   (a) Find the probabilities assigned to the simple events.
   (b) Find $P(A)$, where $A = \{e_1, e_3, e_4\}$.

16. A person decides to take a vacation and has three destinations in mind: Las Vegas, the Bahamas, and Hawaii. She is twice as likely to go to Hawaii as to Las Vegas, and three times as likely to go to Las Vegas as to the Bahamas. Find the probability that she vacations in Las Vegas.

17. The probabilities that a student will receive an A, B, C, D, F, or incomplete in a course are, respectively, 0.05, 0.15, 0.20, 0.25, 0.30, and x. Find the probability that the student will receive:
   (a) an incomplete   (b) at most a C or an incomplete   (c) a B or a C.

(handwritten:)
a) $P(\text{incomplete}) = 0.05$
b) $P(\leq C) = 0.75$
c) $P(B \text{ or } C) = 0.35$

18. Suppose a coin is tossed three times yielding the sample space $S = \{HHH, HHT, HTH, THH, HTT, TTH, THT, TTT\}$, where HTH means heads on the first toss, tails on the second toss, and heads on the third toss. Assume the assignment of probabilities to the simple events as follows.

$$P(\{HHH\}) = 0.064, \quad P(\{HHT\}) = 0.096, \quad P(\{HTH\}) = 0.096,$$
$$P(\{THH\}) = 0.096, \quad P(\{HTT\}) = 0.144, \quad P(\{TTH\}) = 0.144,$$
$$P(\{THT\}) = 0.144, \quad P(\{TTT\}) = 0.216$$

   Find the probability of:
   (a) heads on the first toss   (b) heads on the second toss
   (c) exactly one head   (d) at least one head.

19. Suppose there are two video arcades, V_1 and V_2, in your town and three video arcades, V_3, V_4, and V_5, in the neighboring town. You pick one video arcade at random to visit over the weekend.
   (a) Write down a suitable sample space.
   (b) Find the probability that you pick:
       i. Arcade V_3
       ii. an arcade in the neighboring town.

20. Consider the video arcades described in Exercise 19. Suppose you decide to visit the arcades on two occasions and on each occasion pick an arcade at random.
   (a) Write down a suitable sample space.
   (b) Find the probability that you go to:
       i. Arcade V_3 both times
       ii. the same arcade both times
       iii. a different arcade each time.

21. In the TV game show "The Price Is Right" a contestant is shown a television set, a refrigerator, and a sofa set. She is given three price tags of $825, $650, and $750 and asked to match the price tags correctly with the items to win all the items. The contestant will win all three items if all the price tags are matched correctly, but will not win any prize if there is a single mismatch. Suppose the person matches the tags at random.
   (a) Write down the sample space.

(b) Find the probability that the contestant will:
  i. win the three items by matching all the price tags correctly
  ii. match only one price tag correctly
  iii. match exactly two price tags correctly
  iv. match no price tag correctly
  v. not win any of the items.

22. Fifteen students are enrolled in a beginning statistics course. Their class level, major, and gender are as shown.

Student	Class Level	Major	Gender
1	senior	nursing	male
2	freshman	sociology	female
3	sophomore	sociology	male
4	junior	nursing	female
5	senior	nursing	female
6	junior	sociology	male
7	junior	psychology	female
8	senior	sociology	female
9	senior	nursing	female
10	junior	psychology	female
11	sophomore	nursing	male
12	freshman	sociology	female
13	junior	nursing	female
14	sophomore	sociology	male
15	senior	nursing	male

A student is selected at random.
(a) How many outcomes are there in the sample space? 15
(b) Find the probability that the student is
  i. a junior 5/15
  ii. a nursing major
  iii. a female
  iv. a junior, a nursing major, and a female 2/15
  v. a junior or a nursing major or a female
  vi. not a senior
(c) Interpret the probabilities in Part b.

23. Suppose the following table categorizes the U.S. population based on ideology. (The figures are based on the Gallup poll, March 1993.)

Conservative	24%
Libertarian	19%
Populist	27%
Liberal	20%
Undesignated	10%

Suppose a U.S. citizen is picked at random and you are interested in the ideology of the individual.
(a) List all the elements of the sample space.
(b) Find $P(A)$ if A is the event that the citizen picked is a conservative or is undesignated.

   (c) Find $P(B)$ if B is the event that the citizen is neither a populist nor a liberal.

   (d) Find $P(C)$ if the event C represents an individual with an ideology that starts with the letter L.

24. The following information gives the annual dividend (in dollars) offered by eight companies during 1992 and 1993.

Company	1992	1993
A	1.28	1.36
B	3.25	2.80
C	0.80	1.10
D	1.60	1.60
E	2.20	1.80
F	1.80	0.00
G	4.5	2.80
H	0.00	0.80

    Suppose a company is selected at random. Determine the probability of picking a company that
    (a) reduced its dividend
    (b) changed its dividend to $2.80
    (c) increased its dividend by more than 10 cents.

25. Suppose the classification of the population on the basis of the community and attitude toward laws covering the sale of firearms is as given below.

	ATTITUDE TO THE LAWS		
Community	More Strict	Kept as Are	No Opinion
Urban	30%	12%	3%
Suburban	19%	6%	1%
Rural	22%	5%	2%

    An individual is picked at random and we are interested in his or her community and attitude to the laws.
    (a) List the elements of the sample space.
    (b) What is the probability that the individual
       i. believes in more strict laws
       ii. is from a rural community
       iii. is suburban and believes that the laws should be kept as they are?

26. In the game of American roulette, the wheel has 38 slots of which 18 are labeled black, 18 red, and 2 green. The wheel is balanced in such a way that when it is spun the ball is equally likely to land in any slot. Suppose you are interested only in the color where the ball lands when the wheel is spun. Describe an appropriate sample space and assign probabilities to the simple events.

3-3   SOME RESULTS OF ELEMENTARY PROBABILITY

This section presents some results that follow from our definition of probability.

> **THE ADDITION LAW**
>
> If A and B are two events, then the probability that A or B (or both)* occur is equal to the sum of their probabilities minus the probability of their simultaneous occurrence. That is,
>
> $$P(A \text{ or } B) = P(A) + P(B) - P(A \text{ and } B)$$

To justify the result consider the Venn diagrams in Figure 3-5. The term $P(A \text{ or } B)$ on the left side of the formula $P(A \text{ or } B) = P(A) + P(B) - P(A \text{ and } B)$ stands for the probability assigned to the sample points in the shaded portion in Figure 3-5(a).

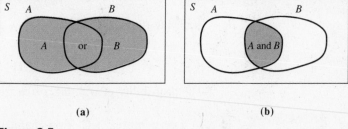

(a)   (b)

Figure 3-5
a. The shaded region describes the event A or B
b. The shaded region describes the event A and B

As a first step in justifying the right side of the equation, we might add $P(A)$, the probability assigned to the sample points in A, to $P(B)$, the probability assigned to the sample points in B, to get $P(A) + P(B)$. This is certainly not the same as $P(A \text{ or } B)$ because the sample points in the overlap "A and B" are included twice; once with A and once with B. The overlap is shown in Figure 3-5(b). Therefore, to compensate for counting these points twice, we must subtract $P(A \text{ and } B)$ from $P(A) + P(B)$ to obtain $P(A \text{ or } B)$.

The formula given above involves four quantities: $P(A \text{ or } B)$, $P(A \text{ and } B)$, $P(A)$, and $P(B)$. Given any three quantities, you can find the fourth.

---

* In set theory the event "A or B" is denoted as $A \cup B$ and is called the *union* of events A and B. Also, the event "A and B" is denoted as $A \cap B$ and is called the *intersection* of the events.

EXAMPLE 1   On a TV quiz show a contestant is asked to pick an integer at random from the first 100 consecutive positive integers, that is, the integers 1 through 100. If the number picked is divisible by 12 or 9, the contestant will win a free trip to the Bahamas. What is the probability that the contestant will win the trip?

SOLUTION   Since 1 integer is picked at random from the 100 integers, the sample space consists of 100 equally likely outcomes and is given by $S =$ $\{1, 2, 3, \ldots, 100\}$. Let the two events be defined as follows.

   A: The number is divisible by 12.
   B: The number is divisible by 9.

Then:

$$A = \{12, 24, 36, 48, 60, 72, 84, 96\}$$
$$B = \{9, 18, 27, 36, 45, 54, 63, 72, 81, 90, 99\}.$$

Now the event "A and B" consists of the set of integers that are divisible by 12 *and* 9. This is the set of integers divisible by 36 and, therefore, consists of the set $\{36, 72\}$. Thus we have the following probabilities.

$$P(A) = \frac{8}{100} \qquad P(B) = \frac{11}{100} \qquad P(A \text{ and } B) = \frac{2}{100}$$

Applying the addition law, we now get

$$P(A \text{ or } B) = \frac{8}{100} + \frac{11}{100} - \frac{2}{100} = \frac{17}{100}.$$

The probability of winning a free trip is 0.17.   ●

EXAMPLE 2   In a certain area, television channels 4 and 7 are affiliated with the same national network. The probability that Channel 4 carries a particular sports program is 0.5, that Channel 7 carries it is 0.7, and that they both carry it is 0.3. What is the probability that José will be able to watch the program on either of the channels?

SOLUTION   Let the events be defined as follows.

   A: Channel 4 carries the program.
   B: Channel 7 carries the program.

We wish to find the probability that Channel 4 carries the program *or* channel 7 carries the program, that is, $P(A \text{ or } B)$. Since $P(A) = 0.5$, $P(B) = 0.7$, and $P(A \text{ and } B) = 0.3$, from the addition law we get

$$P(A \text{ or } B) = 0.5 + 0.7 - 0.3 = 0.9.$$

Thus, the probability that José will be able to watch the program is 0.9. ●

EXAMPLE 3 The town of Metroville has two ambulance services: the city service and a citizens' service. In an emergency, the probability that the city service responds is 0.6, the probability that the citizens' service responds is 0.8, and the probability that either of the services responds is 0.9. Find the probability that both services will respond to an emergency.

SOLUTION If A represents the event that the city service responds, and B the event that the citizens' service responds, then we are interested in finding $P(A \text{ and } B)$. We are given the following information.

$$P(A) = 0.6 \qquad P(B) = 0.8 \qquad P(A \text{ or } B) = 0.9$$

By the addition law, we have

$$P(A \text{ or } B) = P(A) + P(B) - P(A \text{ and } B).$$

Therefore, transposing appropriately, we have

$$P(A \text{ and } B) = P(A) + P(B) - P(A \text{ or } B)$$
$$= 0.6 + 0.8 - 0.9$$
$$= 0.5.$$

In the long run, about 50 percent of the times both services will respond to an emergency. ●

MUTUALLY EXCLUSIVE EVENTS

We now consider a special case of the addition law. Suppose we draw a card from a standard deck. Consider the two events A and B, where A stands for getting a spade and B for getting a red card. We know that if a card is a spade it cannot be red and, conversely, if it is a red card then it cannot be a spade. Two events of this type are said to be mutually exclusive and are defined as follows.

> **MUTUALLY EXCLUSIVE EVENTS**
>
> Two events A and B are **mutually exclusive events** if they do not have any outcome in common and, consequently, cannot occur simultaneously. It follows, therefore, that
>
> $P(A \text{ and } B) = 0$ for mutually exclusive events A and B.

By the addition law of probability, we then have the following consequence.

> If A and B are mutually exclusive events, then
>
> $$P(A \text{ or } B) = P(A) + P(B).$$

This result can be extended to any finite number of mutually exclusive events, that is, events that cannot occur together in pairs. For instance, if A, B, and C are three events, then

$$P(A \text{ or } B \text{ or } C) = P(A) + P(B) + P(C)$$

if A, B, and C are mutually exclusive. Here note that A and B cannot occur simultaneously. Similarly, A and C cannot occur together, nor can B and C.

EXAMPLE 4

A box contains three red, four green, and five white balls. One ball is picked at random. What is the probability that it will be red or white?

SOLUTION   Let us write the events as follows:

    R: A red ball is picked. W: A white ball is picked.

The events R and W are mutually exclusive, because if a ball is red it cannot be white and vice versa. Therefore, $P(R \text{ and } W) = 0$ and

$$P(R \text{ or } W) = P(R) + P(W)$$
$$= \frac{3}{12} + \frac{5}{12} = \frac{8}{12} = \frac{2}{3}. \qquad \bullet$$

EXAMPLE 5

How well is Jane prepared for her test? Suppose there is a probability of 0.1 that she will get an A, a probability of 0.4 that she will get a B, and a probability of 0.3 that she will get a C. What is the probability that she will get at least a C?

SOLUTION   Jane will get at least a C if she gets an A *or* she gets a B *or* she gets a C. Now the events "getting an A," "getting a B," and "getting a C" are mutually exclusive. One cannot get two grades on one test at the same time. Therefore,

$$P(\text{Jane gets at least a C})$$
$$= P(\text{Jane gets an A}) + P(\text{Jane gets a B})$$
$$+ P(\text{Jane gets a C})$$
$$= 0.1 + 0.4 + 0.3 = 0.8. \qquad \bullet$$

EXAMPLE 6

Consider a family with four children. A survey indicates that it is reasonable to believe that the probabilities of having 0, 1, 2, 3, or 4 sons in such a family are, respectively, 0.1, 0.3, 0.35, 0.2, and 0.05. Find the probability that in a family with four children there will be:

(a)   at least two sons   **(b)**   at most two sons.

SOLUTION

(a) There will be at least two sons if there are two sons *or* three sons *or* four sons. (Here, when we say two sons, for example, we mean *exactly* two sons. In what follows we will adopt this usage of the phrase in similar contexts.) The three events are mutually exclusive since, for example, a family that has exactly three sons cannot, at the same time, have exactly two sons or exactly four sons. Therefore,

$$P(\text{at least two sons})$$
$$= P \text{ (two sons)} + P \text{ (three sons)} + P \text{ (four sons)}$$
$$= 0.35 + 0.2 + 0.05$$
$$= 0.6.$$

(b) Arguing as in Part a, we have

$$P(\text{at most two sons})$$
$$= P \text{ (zero sons)} + P \text{ (one son)} + P \text{ (two sons)}$$
$$= 0.1 + 0.3 + 0.35$$
$$= 0.75.$$ ●

EXAMPLE 7

From her long experience, a car salesperson has found that the probability that a customer will buy a sports car is 0.1, that the customer will buy a hatchback is 0.2, and that the customer will buy a truck is 0.15. What is the probability that a prospective customer will buy one of the given types of vehicles?

SOLUTION Consider the three events "buying a sports car," "buying a hatchback," and "buying a truck." It is, of course, possible that a customer will buy both a sports car *and* a hatchback. In that case, certainly, the events are not mutually exclusive. Assume that our car salesperson knows that this is not the case, that is, that a prospective customer is looking to purchase a single vehicle and that the three events are indeed mutually exclusive. Therefore,

$$P \begin{pmatrix} \text{customer buys a sports car} \\ or \text{ a hatchback} \\ or \text{ a truck} \end{pmatrix} = 0.1 + 0.2 + 0.15$$
$$= 0.45.$$ ●

Example 7 illustrates that events may not be taken for granted to be mutually exclusive. Always satisfy yourself that you have not made this assumption blindly. The case where 3 or more events are not mutually exclusive is beyond the scope of our study.

LAW OF THE COMPLEMENT

If we consider an event A, the event "*nonoccurrence of A*" is called the **complement of A.** It contains precisely those outcomes that are not in A as shown in Figure 3-6. For example, the complement of the event that a student will pass a course is the event that the student will fail. As another example, if A stands for "at least three successful launches of a missile in five attempts," then the event "at most two successful launches in five attempts" is its complement. We shall, therefore, write the complement of A as *not A*. In the literature it is written variously as A', $\overline{A}$, A^C, and so on.

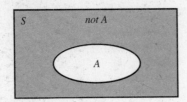

Figure 3-6
Complement of A contains
outcomes of S that are outside A

From the Venn diagram in Figure 3-6 it can be seen that

$$P(A) + P(not\ A) = 1$$

Therefore, $P(not\ A) = 1 - P(A)$, and we have the following **law of the complement.**

> **LAW OF THE COMPLEMENT**
>
> The probability that an event A will not occur is equal to 1 minus the probability that it will occur. That is, for any event A
>
> $$P(not\ A) = 1 - P(A)$$

As a few examples of the applications of this law: If the probability that the stock goes up is 0.6, then the probability that it will not go up is $1 - 0.6 = 0.4$; if the probability of precipitation is 0.8, then the probability of no precipitation is $1 - 0.8 = 0.2$; if the probability of at least three sons in a family of five children is 0.4, then the probability of at most two sons is $1 - 0.4 = 0.6$.

EXAMPLE 8   A person owns five stocks. The probabilities that exactly 0, 1, 2, 3, 4, or 5 of the stocks go up in price on a given day are, respectively, 0.1, 0.2, 0.3, 0.22, 0.1, and 0.08. Find the probability that:

(a) at least one stock goes up   **(b)** at most four stocks go up.

SOLUTION   Let us write $A_0, A_1, A_2, A_3, A_4,$ and A_5, respectively, for the six events that exactly 0, 1, 2, 3, 4, or 5 of the stocks go up in price. Note that the six events are mutually exclusive.

(a) P (at least one stock goes up)

$$= P(A_1 \text{ or } A_2 \text{ or } A_3 \text{ or } A_4 \text{ or } A_5)$$
$$= P(A_1) + P(A_2) + P(A_3) + P(A_4) + P(A_5)$$
$$= 0.2 + 0.3 + 0.22 + 0.1 + 0.08$$
$$= 0.9$$

Alternatively, notice that the event "at least one stock goes up" negates the event "no stock goes up." Therefore, by the law of complement,

$$P \text{ (at least one stock goes up)} = P \text{ (}not\text{ (0 stocks go up))}$$
$$= 1 - P \text{ (0 stocks go up)}$$
$$= 1 - P(A_0)$$
$$= 1 - 0.1 = 0.9.$$

The latter approach is much more efficient, especially in a case where, for example, there were twenty stocks involved instead of five.

(b) P (at most four stocks go up)

$$= P(A_0 \text{ or } A_1 \text{ or } A_2 \text{ or } A_3 \text{ or } A_4)$$
$$= P(A_0) + P(A_1) + P(A_2) + P(A_3) + P(A_4)$$
$$= 0.1 + 0.2 + 0.3 + 0.22 + 0.1$$
$$= 0.92$$

An alternate method is to note that the event of interest negates the event that all five stocks go up. Therefore,

$$P \text{ (at most four stocks go up)} = P \text{ (}not\text{ (five stocks go up))}$$
$$= 1 - P \text{ (five stocks go up)}$$
$$= 1 - P(A_5)$$
$$= 1 - 0.08 = 0.92. \quad\bullet$$

EXAMPLE 9   Two employees on a university campus, Tom from Plant Operations and Becky from Public Safety, are assigned to check that a certain building on the campus is locked over the weekend. Suppose that on any weekend the probability that Tom checks is 0.96, that Becky checks is 0.98, and that they both check is 0.95. What is the probability that neither Tom nor Becky check on a weekend?

SOLUTION   Let us write the events as follows.

T: Tom checks.
B: Becky checks.

The event "neither Tom nor Becky checks" can be considered as the event "it is *not* the case that either Tom *or* Becky checks." Thus, the event that we are interested in is "not (T or B)." Now

$$
\begin{aligned}
P\,(\text{not }(T \text{ or } B)) &= 1 - P\,(T \text{ or } B) \\
&= 1 - [P\,(T) + P\,(B) - P\,(T \text{ and } B)] \\
&= 1 - [0.96 + 0.98 - 0.95] \\
&= 0.01.
\end{aligned}
$$

Over the long haul, on approximately 1 percent of the weekends neither Tom nor Becky will check.   ●

ODDS

The concept of odds is frequently associated with gambling games. The odds in favor of an event A are expressed as the ratio of $P\,(A)$ to $P\,(not\ A)$ as follows

$$
\text{odds in favor of } A = \frac{P\,(A)}{P\,(not\ A)} = \frac{P\,(A)}{1 - P\,(A)}
$$

Thus, for example, since the probability of getting two heads when two coins are tossed is $\frac{1}{4}$,

$$
\text{the odds in favor of getting two heads} = \frac{1/4}{3/4} = \frac{1}{3}
$$

In common parlance this is often stated as follows: *odds in favor of getting two heads,* when tossing two coins, *are 1 to 3.* In general, if one states that the odds in favor of an event are a to b it means that

$$
\frac{P\,(A)}{1 - P\,(A)} = \frac{a}{b}
$$

A little algebra leads to $P\,(A) = a/(a + b)$

In summary,

> "The **odds in favor of an event** A are a to b" means that
>
> $$
> P(A) = \frac{a}{a + b}.
> $$

Thus, for instance, if the weatherperson reports that the probability of snow on a weekend is $\frac{2}{5}$, then the odds in favor of snow are,

$$
\frac{2/5}{3/5} = \frac{2}{3}
$$

That is, the odds in favor of snow are 2 to 3.

If a student feels that her odds of being selected to play on the basketball team are 5 to 2, then the probability of her being selected is $5/(5 + 2) = \frac{5}{7}$. (Note we have a = 5, b = 2.)

> If the odds in favor of A are a to b, then the **odds against A,** that is, in favor of *not A*, are b to a.

9/19/96

Section 3-3 Exercises

1. If $P(A) = 0.4$, $P(B) = 0.5$, and $P(A \text{ and } B) = 0.25$, find $P(A \text{ or } B)$.

2. If $P(A) = 0.7$, find $P(not\ A)$.

3. If $P(A) = 0.7$, $P(not\ B) = 0.4$, and $P(A \text{ and } B) = 0.5$, find $P(A \text{ or } B)$.

4. If $P(not\ A) = 0.5$, $P(not\ B) = 0.4$, and $P(A \text{ and } B) = 0.3$, find $P(A \text{ or } B)$.

5. If $P(A) = 0.15$, $P(B) = 0.65$, and $P(A \text{ or } B) = 0.72$, find $P(A \text{ and } B)$.

6. If $P(not\ A) = 0.3$, $P(B) = 0.5$, and $P(A \text{ or } B) = 0.9$, find $P(A \text{ and } B)$.

7. If $P(A \text{ and } B) = 0.25$, $P(A \text{ or } B) = 0.85$, and $P(B) = 0.45$, find $P(not\ A)$.

8. Find $P(A)$ if the odds in favor of A are:
   (a) 5 to 9   (b) 10 to 1
   (c) 14 to 7   (d) 1 to 99.

9. Find $P(A)$ if the odds against A are:
   (a) 3 to 4   (b) 2 to 1
   (c) 1 to 100   (d) 8 to 6.

10. Express the following probabilities in terms of odds in favor of A.
    (a) $P(A) = \frac{7}{9}$ (b) $P(A) = \frac{1}{11}$
    (c) $P(A) = \frac{5}{8}$ (d) $P(A) = 0.1$
    (e) $P(A) = 0.3$ (f) $P(A) = 0.8$

11. Suppose the probability that a surgery is successful is $\frac{6}{7}$. What are the odds against the surgery?

12. The odds that a company will declare a dividend at the annual meeting are 3 to 17. What is the probability that the company will not declare a dividend?

13. The odds that a shipment will reach its destination in good condition are 9995 to 5. What is the probability that it will not reach its destination in good condition?

14. A person fires two shots. Suppose you are told that the probability of scoring on the first shot is 0.8 and that of scoring on the second shot is 0.6. Could this be considered a valid assignment of probabilities?

15. The probability that a radioactive substance emits a least one particle during a one-hour period is 0.008. What is the probability that it does not emit any particle during the period?

16. When is the following a true statement?

$$P(A \text{ or } B) = P(A) + P(B)$$

17. Suppose the probability that a person chosen at random is over 6 feet in height or weighs more than 160 pounds is 0.3. What is the probability that the person is neither over 6 feet in height nor weighs more than 160 pounds?

18. Suppose α represents the probability of concluding that a drug is good when in fact it is not good.
    (a) Give the maximum and minimum values that α can have.
    (b) Describe an event for which $1 - \alpha$ is the probability.

19. If A and B are events for which $P(A) = 0.6$, $P(B) = 0.7$, and $P(A \text{ or } B) = 0.9$, find $P(not(A \text{ and } B))$.

20. Is it possible to have two events A and B such that $P(A) = 0.6$, $P(B) = 0.7$, and $P(A \text{ and } B) = 0.2$? Explain.

21. Explain why it is *not* possible to have two events A and B with the following assignments of probabilities.
    (a) $P(A) = 0.6$ and $P(A \text{ and } B) = 0.8$
    (b) $P(A) = 0.7$ and $P(A \text{ or } B) = 0.6$

22. Two fair dice are rolled. Find the probability that the total on the two dice is not equal to 5.

23. The probability that an adult male is a Democrat is 0.6; the probability that he belongs to a labor union is 0.5; and the probability that he is a Democrat or belongs to a labor union is 0.75. Find the probability that a randomly picked adult male is:
    (a) a Democrat and belongs to a labor union
    (b) not a Democrat nor does he belong to a labor union.

24. From past experience, Scotty knows that the probability that his mother will serve ice cream for dessert is 0.5, the probability that she will serve pie is 0.7, and the probability that she will serve ice cream or pie (or both) is 0.9. Find the probability that his mother will serve:
    (a) both ice cream and pie
    (b) neither ice cream nor pie.

25. In a certain club, the probability that a member picked at random is a lawyer is 0.64, the probability that the member is male is 0.75, and the probability that the member is a male lawyer is 0.50. Find the probability that:
    (a) the member is a lawyer or male
    (b) the member is neither a lawyer nor male.

26. The probability that a person goes to a concert on Saturday is $\frac{2}{3}$, and the probability that the person goes to a baseball game on Sunday is $\frac{4}{9}$. If the probability of going to either program is $\frac{7}{9}$, find the following probabilities.
    (a) The person goes to both programs.
    (b) The person goes to neither program.

27. A student is taking two courses, history and English. If the probability that she will pass either of the courses is 0.7, that she will pass both courses is 0.2, and that she will fail in history is 0.6, find the probability that:
    (a) she will pass history
    (b) she will pass English.

28. Past records have led a bank to believe that the probability that a customer opens a savings account is 0.4, the probability that a customer opens a checking account is 0.7, and the probability that a customer opens a savings and a checking account is 0.25.

   (a) What is the probability that a customer will open either a savings account or a checking account?

   (b) What is the probability that a person will open neither a savings account nor a checking account?

   [handwritten: $P(S \text{ or } C) = P(S) + P(C) - P(S \cap C)$
   $X = 0.4 + 0.7 - 0.25$
   $X = 0.85$
   $P(S \text{ nor } C) = 0.15$]

29. The probability that an American tourist traveling to Europe will visit Paris is 0.8, the probability that the tourist will visit Rome is 0.6, and the probability that the tourist will visit either of the cities is 0.9. If a tourist is selected at random from all American tourists traveling to Europe, find the probability that he or she will visit:

   (a) both cities

   (b) neither city.

30. The probabilities that a typist makes 0, 1, 2, 3, 4, 5, or 6 mistakes on a page are, respectively, 0.05, 0.1, 0.3, 0.25, 0.15, 0.1, and 0.05. Find the probability that the typist makes:

   (a) at least one mistake

   (b) at least three mistakes

   (c) at most five mistakes

   (d) an even number of mistakes.

31. The probabilities that an office will receive, in a half-hour period, 0, 1, 2, 3, 4, or 5 calls are 0.05, 0.1, 0.25, 0.3, 0.1, and 0.2, respectively. Find the probability that in a half-hour period:

   (a) more than two calls will be received

   (b) at most four calls will be received

   (c) four or more calls will be received.

32. The proportion of university students who own an automobile is 0.3 and the proportion of university students who live in the dorm is 0.6. If the proportion of university students who either own a car or live in a dorm is 0.8, find the proportion of students who live in a dorm and own a car. (Assume that the proportions are taken from a large student body.)

33. A handicapper feels that the probability that Jolly Prince will win the horse race is 0.37. What are the odds in favor of Jolly Prince winning?

34. Suppose the handicapper estimates that the odds in favor of Esmeralda winning the race are 7:13. What is the probability that Esmeralda will not win the race?

35. Mr. and Mrs. Arnold own two cars, a Toyota and a Honda. The odds that the Toyota will not need repairs in the next six months are 9 to 1, and the odds that the Honda will need repairs during the same period are 1 to 4. Also, the odds that both cars will need repairs during the period are 1 to 24. What are the odds that at least one of the cars will need repairs in the next six months? *Hint:* Convert the odds into probabilities and use the appropriate result.

   [handwritten: $9:1$ $1:4$
   $\frac{9}{1+9} = \frac{9}{10}$ $\frac{1}{4+1} = \frac{1}{5}$]

36. Suppose you have two midterm tests next week, one in statistics and the other in biology. Assume that the odds of your passing the statistics test are 100 to 2, the odds of passing the biology test are 46 to 5, and the odds of passing at least one of the courses are 202 to 2. Compute the probability of your passing both courses.

3-4 COUNTING TECHNIQUES AND APPLICATIONS TO PROBABILITY (OPTIONAL)

COUNTING TECHNIQUES: PERMUTATIONS AND COMBINATIONS

As we have seen, the classical definition of the probability of an event E requires that we know $n\,(E)$, the number of outcomes in E, and $n\,(S)$, the number of outcomes in the sample space S. A straightforward means of finding the number of outcomes would be to list them explicitly. However, if there are too many outcomes, this method is impractical. In such situations we often use certain rules based on counting techniques. The basic rule for counting is the multiplication rule, which we will refer to as the basic counting principle. It can be stated as follows

> **THE BASIC COUNTING PRINCIPLE**
>
> If a certain experiment can be performed in r ways, and *corresponding to each of these ways* another experiment can be performed in k ways, then the combined experiment can be performed in rk ways.

The essential reasoning behind this principle can be illustrated through the following example. Suppose a coin and die are tossed. We could represent the outcomes as ordered pairs, where the first component refers to the experiment involving a coin and the second to the one with a die, as follows.

$$(H, 1) \quad (H, 2) \quad (H, 3) \quad (H, 4) \quad (H, 5) \quad (H, 6)$$
$$(T, 1) \quad (T, 2) \quad (T, 3) \quad (T, 4) \quad (T, 5) \quad (T, 6)$$

The first row has six pairs with H as the first entry. The second row also has six pairs, with T as the first entry. So, there are $6 \cdot 2$ outcomes corresponding to the two rows, each with six outcomes.

The principle can also be illustrated by means of a **tree diagram,** as in Figure 3–7. First we list the r outcomes of one experiment and then, corresponding to each of these, the outcomes of the other experiment. The total number of branches (rk) gives the number of all the combined possibilities.

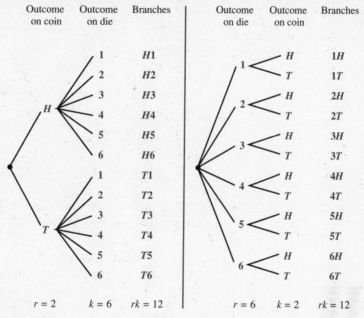

Figure 3-7
Tree diagrams showing twelve branches when a coin and a die are tossed

Thus, applying the basic counting principle, we have the following examples.

(a) The number of possible outcomes when a die is rolled twice is $6 \cdot 6$, or 36.

(b) The number of different ways for a man to get dressed if he has eight different shirts and six different ties is $8 \cdot 6$, or 48.

The basic counting principle can be extended to cases where the combined experiment consists of three or more steps, as in the following.

(c) There are $6 \cdot 6 \cdot 2 \cdot 2$, or 144 possibilities when two dice and two coins are tossed.

(d) The number of different ways for a man to get dressed if he has eight different shirts, six different ties, and five different jackets is $8 \cdot 6 \cdot 5$, or 240.

(e) If a student has a choice of three math courses, two social sciences courses, four physics courses, and three biology courses, and she can take only one course in each discipline, then she can make her schedule in $3 \cdot 2 \cdot 4 \cdot 3$, or 72, different ways.

Next we apply the basic counting rule to find the number of possible ways when some fixed number of items can be picked from a lot containing a given number of items. We will concern ourselves with the special case of sampling *without replacement,* where once an object is picked, it is not returned to the lot. We are thus led to the discussion of *permutations* and *combinations*.

Permutations

Suppose there are four contestants, Ann, Betty, Carl, and Dick, who are competing for three prizes, a first prize, a second prize, and a third prize. In how many ways can they be awarded the prizes so that any person gets at most one prize?

Notice that the outcome where Ann gets the first prize, Betty the second prize, and Dick the third prize is quite different from the outcome where Dick gets the first prize, Betty the second prize, and Ann the third prize, although the same three people are involved in both cases. *Order does make a difference.*

In Table 3-1 we give all the possible ways of awarding the prizes by listing the individuals by their initials *A, B, C, D*. For example, by the outcome *ACB* we mean that Ann gets the first prize, Carl the second prize, and Betty the third prize.

TABLE 3-1   Permutations of
Letters *A, B, C, D*
Taken Three at a
Time

ABC	*ABD*	*ACD*	*BCD*
ACB	*ADB*	*ADC*	*BDC*
BCA	*BAD*	*CAD*	*CBD*
BAC	*BDA*	*CDA*	*CDB*
CAB	*DAB*	*DAC*	*DBC*
CBA	*DBA*	*DCA*	*DCB*

As mentioned earlier, the order in which the letters are written is important. Any particular ordered arrangement is called a **permutation.** In Table 3−1 there are twenty-four permutations altogether. The reason is simple. We are choosing three letters out of four. There are four choices for the first letter. Corresponding to any of these ways, the second letter can be chosen in three ways (remember we are sampling without replacement). Last, for any way of drawing the first two letters, the third letter can be chosen in two ways. Therefore, after applying the basic counting principle, the total number of ways is $4 \cdot 3 \cdot 2$, or 24.

In general, suppose there are n distinct objects and we choose r of these without replacement. We cannot choose more objects than there are in the lot. Therefore r cannot exceed n. Any particular ordered arrangement consisting of r objects is a permutation. We will denote the total number of permutations by the symbol $_nP_r$ and call it *the number of permutations of* n *objects taken* r *at a time.*

> **NUMBER OF PERMUTATIONS**
>
> The number of permutations (or ordered arrangements) of n distinct objects taken r at a time is given by
>
> $$_nP_r = n(n-1)(n-2)\cdots(n-r+1).$$

On the right-hand side of the formula for $_nP_r$, remember that there are r consecutively decreasing factors in the product with the first term as n. Thus, for example, the number of permutations when four objects are picked without replacement from eight distinct objects are

$$_8P_4 = 8 \cdot 7 \cdot 6 \cdot 5 = 1680.$$

Notice that starting with 8 as the first term, we have four factors in the product, corresponding to the four objects picked.

In particular, $_nP_n$, the number of permutations of n objects *taken all together,* is $n(n-1)(n-2)\cdots3 \cdot 2 \cdot 1$. We will denote such a product of consecutive integers by $n!$. This is read **n factorial.** Thus, for example, $4! = 4 \cdot 3 \cdot 2 \cdot 1 = 24$. Also, for mathematical purposes, it is convenient to define $0! = 1$. Using the factorial symbol we can write $_nP_r$ in a compact way as follows.

$$_nP_r = \frac{n!}{(n-r)!}$$

Thus, in summary, we have the following.

> $$_nP_r = n(n-1)\cdots(n-r+1) = \frac{n!}{(n-r)!}$$
>
> $$_nP_n = n(n-1)\cdots3 \cdot 2 \cdot 1 = n!$$
>
> $$0! = 1$$

EXAMPLE 1   How many numbers with three distinct digits are possible using the digits 3, 4, 5, 6, 7, and 8? Write five of these numbers.

SOLUTION   There are six distinct digits and we are picking three. Observe that order is important, because, for example, the numbers 356 and 536 are different even though they have the same digits. Applying the formula, there are

$$_6P_3 = 6 \cdot 5 \cdot 4 = 120$$

distinct numbers. Five of these numbers are 345, 456, 465, 678, and 567. ●

EXAMPLE 2   Five people arrive at the checkout counter at the same time. In how many different ways can these people line up?

SOLUTION   The number of ways of arranging five people in a line is five factorial, that is,

$$5! = 5 \cdot 4 \cdot 3 \cdot 2 \cdot 1 = 120.$$ ●

EXAMPLE 3   (a)   Find the number of ways of arranging the letters in the word *object*.

(b)   Find the number of ways if each arrangement in Part a starts with the letter *j*.

SOLUTION
(a)   There are six distinct letters in the word. Therefore, the number of arrangements is 6!, that is,

$$6! = 6 \cdot 5 \cdot 4 \cdot 3 \cdot 2 \cdot 1 = 720.$$

(b)   Since the first letter is fixed as *j*, the total number of possibilities are found by arranging the remaining five letters in all possible ways. As we know, this can be done in five factorial ways, that is, 120 ways. One of the possibilities is *jtoecb*.   ●

Combinations

As noted, the order in which objects are arranged matters in permutations. However, order does not matter in *combinations*. Consider the following example, in which we invite three volunteers from a group of four students, Ann, Betty, Carl, and Dick. If Ann, Betty, and Carl volunteer, this is in no way different from Carl, Ann, and Betty volunteering. We are interested in who volunteered and not in what order they volunteered. We list the outcome where these three individuals volunteer as *ABC*. Altogether, there are just four possibilities, *ABC*, *ABD*, *ACD*, and *BCD*. Each of these possibilities is called a combination. Thus, there are four combinations.

Turning our attention to the general case, suppose we pick *r* objects from *n* distinct objects without replacement, and *order is not important*. Each such selection is a **combination**. Symbolically, we will denote the number of these unordered arrangements by the symbol $\binom{n}{r}$* and call it *the number of combinations of n objects taken r at a time.*

----

*There are other notations for $\binom{n}{r}$ such as C_r^n, nCr, etc.

NUMBER OF COMBINATIONS

The number of combinations of n things taken r at a time is given by

$$\binom{n}{r} = \frac{n(n-1)(n-2)\cdots(n-r+1)}{r!} = \frac{n!}{r!\,(n-r)!}$$

Note that the numerator in the first expression for $\binom{n}{r}$ is simply equal to $_nP_r$. So, we have

$$\binom{n}{r} = \frac{_nP_r}{r!}$$

The argument for dividing the number of permutations by $r!$ to get the number of combinations is that each combination gives rise to $r!$ possibilities when arranged in all possible ways, thereby giving all the permutations.

EXAMPLE 4   Evaluate the following.

(a) $\binom{8}{4}$ **(b)** $\binom{7}{7}$ **(c)** $\binom{10}{3}$ **(d)** $\binom{4}{0}$

SOLUTION

(a) $\binom{8}{4} = \dfrac{8 \cdot 7 \cdot 6 \cdot 5}{4!} = \dfrac{8 \cdot 7 \cdot 6 \cdot 5}{4 \cdot 3 \cdot 2 \cdot 1} = 70$

(b) $\binom{7}{7} = \dfrac{7!}{7!\,(7-7)!} = \dfrac{7!}{7!\,0!} = 1$

Notice that $0! = 1$, and $7!$ cancels out in the numerator and the denominator of the second fraction.

(c) $\binom{10}{3} = \dfrac{10 \cdot 9 \cdot 8}{3!} = \dfrac{10 \cdot 9 \cdot 8}{3 \cdot 2 \cdot 1} = 120$

(d) $\binom{4}{0} = \dfrac{4!}{0!\,(4-0)!} = \dfrac{4!}{1 \cdot 4!} = 1$ ●

The results of Example 4b and d illustrate what is true, in general, for any positive integer n. We have

$$\binom{n}{n} = 1 \quad \text{and} \quad \binom{n}{0} = 1.$$

Remark A commonly raised question is, "When do we want permutations and when do we want combinations?" To answer this we must first satisfy ourselves that objects are drawn from the lot without replacement. This done, the total number of ways will be $_nP_r$ if the order in which the objects are drawn is important, and $\binom{n}{r}$ if it is not important. Of course, whether order is relevant will depend on the nature of the experiment and the situation of interest. For example, if we are dealing cards from a deck for a game of bridge, the order in which we deal the cards does not matter. What matters is the kinds of cards. Therefore, in such a situation we would be interested in the number of combinations. However, in seating people in a row, the order in which they are seated could be quite important. (People tend to be protocol conscious.) If we are interested in the different seating arrangements, we have a candidate for permutations.

EXAMPLE 5   Find the number of ways in which a hand of five cards can be dealt from a deck of fifty-two cards.

SOLUTION   When cards are dealt from a deck, we are interested in what cards are dealt in a hand and not in the order in which they are dealt. Therefore, we want the number of combinations, which is

$$\binom{52}{5} = \frac{52 \cdot 51 \cdot 50 \cdot 49 \cdot 48}{5 \cdot 4 \cdot 3 \cdot 2 \cdot 1} = 2{,}598{,}960. \qquad \bullet$$

EXAMPLE 6   How many ways are there of choosing a set of three books from a set of eight books?

SOLUTION   Since the order of choosing the books is not important, the answer is

$$\binom{8}{3} = \frac{8 \cdot 7 \cdot 6}{3 \cdot 2 \cdot 1} = 56. \qquad \bullet$$

EXAMPLE 7   A company vice president has to visit four of the twelve subsidiaries that the company owns.

(a)   How many sets of four companies are there from which the vice president can pick one set to visit?

(b)   In how many different ways can she plan her intinerary in visiting four of the twelve subsidiaries?

SOLUTION

(a)   Here, we are simply picking four subsidiaries out of twelve. What matters is which subsidiaries are picked. The order of picking them is not relevant. We have a case where combinations are appropriate. The answer is $\binom{12}{4}$, or 495.

(b) Let us identify the subsidiaries as $s_1, s_2, \ldots, s_{12}$. Suppose the vice president decides to visit the subsidiaries s_1, s_3, s_6, s_7. Then, for example, the route $s_1 \rightarrow s_3 \rightarrow s_6 \rightarrow s_7$ is different from the route $s_3 \rightarrow s_1 \rightarrow s_7 \rightarrow s_6$ though the same four subsidiaries are visited in both cases. Order is relevant and we use permutations. The answer is $_{12}P_4$, or 11,880.   ●

SIDEBAR   *Lottery Numbers*

When choosing six numbers for a lottery ticket, most people will shy away from picking consecutive integers and will pin their hopes on six numbers which are not in sequence. Somehow the feeling prevails that six numbers such as 16, 17, 18, 19, 20, 21 are less likely to form a winning set of numbers than, say, the six numbers such as 7, 18, 24, 31, 42, and 47. This may be because, deep down, there is some vague awareness of the collection of all the outcomes that describe the underlying event, or it may be because of the empirical evidence of lottery results to date. If a lottery involves picking six numbers out of 51, as in the California state lottery, there are $\binom{51}{6} = 18{,}009{,}460$ distinct ways. Is any one way more likely than any other? Absolutely not! Among these myriad possibilities there are 46 sequences, namely, $\{1, 2, 3, 4, 5, 6\}, \ldots, \{46, 47, 48, 49, 50, 51\}$, grossly outnumbered by the ways that do not form sequences. So it is not surprising that our experience shows that the winning tickets are invariably nonsequences. But do not rule out a *particular* sequence in preference to a *particular* nonsequence. One of these days a sequence might just show up.

APPLICATIONS OF COUNTING TO FINDING PROBABILITIES

In the examples that follow, we see how counting techniques can be employed to solve problems in probability.

EXAMPLE 8

Three digits are picked at random from the digits 1 through 9. Find the probability that the digits are consecutive digits.

SOLUTION   We are picking three digits out of nine. The number of ways this can be done is

$$\binom{9}{3} = \frac{9 \cdot 8 \cdot 7}{3 \cdot 2 \cdot 1} = 84.$$

The event of interest consists of the set of outcomes $(1, 2, 3)$, $(2, 3, 4)$, $(3, 4, 5)$, $(4, 5, 6)$, $(5, 6, 7)$, $(6, 7, 8)$, and $(7, 8, 9)$. There are seven outcomes in the event. Therefore, the probability that the digits are consecutive is $\frac{7}{84}$, or $\frac{1}{12}$.   ●

EXAMPLE 9   If the letters of the word *volume* are arranged in all possible ways, find the probability that:

(a)   the word ends in a vowel

(b)   the word starts with a consonant and ends in a vowel.

SOLUTION   The problem clearly states that order is relevant. Since there are six distinct letters in the word *volume,* there are 6!, or 720, possible ways of arranging these letters.

(a)   There are three vowels in the word *volume*. Since the word must end in a vowel, there are three choices for the last letter. This done, there are five choices for the first letter, four for the second, three for the third, two for the fourth, and one for the fifth. By the basic counting rule, we get $3 \cdot 5 \cdot 4 \cdot 3 \cdot 2 \cdot 1$, or 360, ways. Therefore, the probability is $\frac{360}{720}$, or $\frac{1}{2}$.

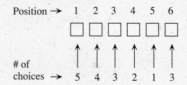

(b)   Since we have three consonants and three vowels, there are three choices for the first letter and three choices for the last. Notice that we now have four choices for the second letter, three for the third, two for the fourth, and one for the fifth. Thus, there are $3 \cdot 3 \cdot 4 \cdot 3 \cdot 2 \cdot 1$, or 216, possible ways. Therefore, the probability is $\frac{216}{720}$, or $\frac{3}{10}$.

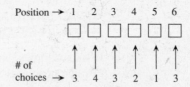

An important type of probability problem can be couched in the following format.

Items in a lot represent a mixture of two or more kinds of items. Suppose we select from these items, without replacement, a sample of a certain number of items. We are interested in finding the number of ways of selecting these items from the lot so that there are so many of each kind.

For example, a person might have nine stocks of which four are utility stocks and five are industrial stocks. If three stocks are selected, we might want to find the number of ways of doing this so that there are two utility stocks and one industrial stock in the selection.

We can represent the situation schematically as follows.

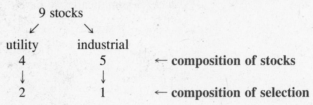

9 stocks

utility	industrial	
4	5	← **composition of stocks**
↓	↓	
2	1	← **composition of selection**

There are $\binom{4}{2}$ ways of selecting two utility stocks out of four. *Corresponding to any one way of doing this,* there are $\binom{5}{1}$ ways of selecting one industrial stock out of five. Applying the basic counting rule, we have $\binom{4}{2}\binom{5}{1}$ ways in which the selection can be made. (Notice that, in effect, we have simply put parentheses around the vertical arrows, omitting the arrows.)

As a further extension of the above example, suppose there are fifteen stocks of which three are utility stocks, seven industrial stocks, and five transportation stocks. If six stocks are selected, then the number of ways in which there are two utility, one industrial, and three transportation stocks is $\binom{3}{2}\binom{7}{1}\binom{5}{3}$, as can be seen from the following scheme.

15 stocks

utility	industrial	transportation	
3	7	5	← **composition of stocks**
↓	↓	↓	
2	1	3	← **composition of selection**

EXAMPLE 10

Mrs. Chavez has eight stocks in her portfolio, of which three are utility and five are industrial. She selects three stocks at random to give to her nephew. Find the probability that the nephew receives:

(a) exactly two utility stocks

(b) only utility stocks

(c) no utility stock.

SOLUTION We are selecting three stocks out of eight and since order is not important, there are

$$\binom{8}{3} = \frac{8 \cdot 7 \cdot 6}{3 \cdot 2 \cdot 1} = 56$$

outcomes in the sample space.

(a) "Exactly two utility stocks" means two utility stocks and one industrial stock. Schematically we have the following situation.

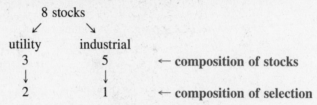

The number of ways of selecting two utility stocks and one industrial stock is

$$\binom{3}{2}\binom{5}{1} = \frac{3 \cdot 2}{2 \cdot 1} \cdot \frac{5}{1} = 15.$$

Consequently, the probability of exactly two utility stocks in the nephew's gift is $\frac{15}{56}$.

(b) If the nephew receives only utility stocks, then he receives three utility stocks and zero industrial. We have the following scheme.

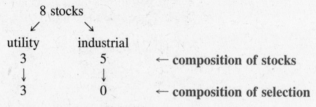

The number of ways is $\binom{3}{3}\binom{5}{0} = 1$. Therefore, the probability of only utility stocks in the gift is $\frac{1}{56}$.

(c) In this case we have the following scheme.

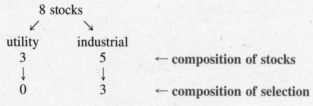

The number of ways is $\binom{3}{0}\binom{5}{3} = 10$ and, consequently, the probability of no utility stock in the gift is $\frac{10}{56}$.

EXAMPLE 11 A geology professor has five silicates, seven pyrites, and eight carbonates in a rock collection. He picks six rocks at random for a student to analyze. Find the probability that the professor picked:

(a) two silicates, one pyrite, and three carbonates

(b) two silicates

(c) two silicates and three pyrites.

SOLUTION Because there are twenty rocks in all, the number of ways of picking six of these is

$$\binom{20}{6} = \frac{20 \cdot 19 \cdot 18 \cdot 17 \cdot 16 \cdot 15}{6 \cdot 5 \cdot 4 \cdot 3 \cdot 2 \cdot 1} = 38{,}760.$$

(a) We have the following schematic representation.

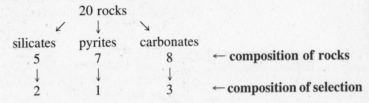

Therefore, the number of ways of picking two silicates, one pyrite, and three carbonates is

$$\binom{5}{2}\binom{7}{1}\binom{8}{3} = \frac{5 \cdot 4}{2 \cdot 1} \cdot \frac{7}{1} \cdot \frac{8 \cdot 7 \cdot 6}{3 \cdot 2 \cdot 1} = 3{,}920.$$

Thus the probability of picking these rocks is $\frac{3920}{38{,}760}$, or approximately 0.101.

(b) We are picking six rocks. If two rocks are silicates, then four rocks will have to be "nonsilicates." For this situation, we have the following scheme.

```
              20 rocks
            ↙        ↘
     silicates      nonsilicates
         5               15          ← composition of rocks
         ↓               ↓
         2               4           ← composition of selection
```

The number of ways is

$$\binom{5}{2}\binom{15}{4} = \frac{5 \cdot 4}{2 \cdot 1} \cdot \frac{15 \cdot 14 \cdot 13 \cdot 12}{4 \cdot 3 \cdot 2 \cdot 1} = 13{,}650.$$

Consequently, the probability of picking two silicate rocks is $\frac{13{,}650}{38{,}760}$, or approximately 0.352.

(c) If among six rocks picked there are two silicates and three pyrites, then there must be one carbonate. We have the following scheme.

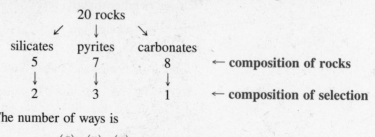

The number of ways is

$$\binom{5}{2}\binom{7}{3}\binom{8}{1} = \frac{5 \cdot 4}{2 \cdot 1} \cdot \frac{7 \cdot 6 \cdot 5}{3 \cdot 2 \cdot 1} \cdot \frac{8}{1} = 2,800.$$

As a result, the probability of picking two silicates and three pyrites is $\frac{2800}{38,760}$, or approximately 0.072. ●

EXAMPLE 12   Eight vice presidents—Mr. Cox, Mr. Evans, Mrs. Fisher, Mr. Vegotsky, Mr. Pritchard, Ms. McKie, Mr. Alvarez, and Mr. Tang—are being considered for four vacancies on the board of directors in a company. If the four candidates are picked at random to fill the vacancies, what is the probability that Mr. Cox and Mrs. Fisher are among the four selected?

SOLUTION   There are $\binom{8}{4}$, or 70, ways of picking four vice presidents

out of eight. Thus the sample space will have seventy possible outcomes. We are interested in the probability that Mr. Cox and Mrs. Fisher are included. The basic scheme can be represented as follows.

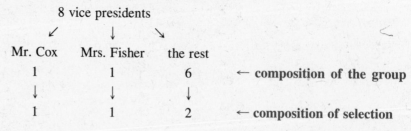

The number of ways is $\binom{1}{1}\binom{1}{1}\binom{6}{2}$, or 15. The probability that Mr. Cox and

Mrs. Fisher are among the four selected to fill the vacancies is $\frac{15}{70}$, or approximately 0.214. ●

> **SIDEBAR**   *Birthday Problem*
>
> **W**hat are the chances of two or more people at a gathering having the same birthday? Surprisingly many people believe that the chances are very small, estimating 1 in a million, or even smaller. The reason for such an answer might be because we have come to expect extremely small odds for winning lotteries and sweepstakes. Actually, the odds are far from that small. First of all, we must realize that the answer will depend on the number of people attending the party. If there are only two people at the party, there is one day of the 365 days in a year on which they could both be born. Thus the chance is 1 in 365. At the other extreme, if there are more than 365 people attending the party, then we can be sure that there will be at least two individuals with the same birthday. It turns out that in a group of twenty three people, the chance for two or more people to have the same birthday is about fifty-fifty. That is, with twenty three people at a party the odds are the same as those of getting heads in the toss of a coin. The odds increase significantly as the number at the party increases. In a gathering of 64 people it is almost certain that at least two individuals will have the same birthday.
>
> The problem is often misunderstood and misstated. In his excellent book, *Innumeracy,* author John Allen Paulos narrates one incident when a guest on Johnny Carson show brought this topic up and bungled it. There were 120 people in the studio audience and Mr. Carson asked how many among them shared his birthday of March 19, a particular birthday. Apparently nobody did. The question should have been "Are there any two people in the audience having a common birthday?" On that score, it is almost certain that he would not have been disappointed.

9/26/96

Section 3-4 Exercises

1. Compute the following.

   (a) $_5P_2$

   (b) $_8P_5$

   (c) $_6P_6$

   (d) $6!$

   (e) $4!\,0!$

   (f) $\dfrac{20!}{18!}$

   (g) $\dbinom{8}{2}$

   (h) $\dbinom{6}{4}$

   (i) $\dbinom{14}{3}$

   (j) $\dbinom{20}{20}$

   (k) $\dbinom{16}{0}$

   (l) $\dbinom{125}{1}$

2. Evaluate $\dbinom{8}{3}$ and $\dbinom{8}{5}$. Provide a rationale justifying why the answer is the same. Generalize by writing a rule that is illustrated by this result.

3. A menu lists two soups, three meat dishes, and five desserts. How many different meals are possible consisting of one soup, one meat dish, and a dessert?

4. A customer who wants to buy an automobile has a choice of three makes, five body styles, and six colors. Find how many choices he has in the selection.

5. A die is rolled four times and a coin is tossed twice. Find how many outcomes there are in the sample space. List three outcomes in the sample space.

6. An individual can be of genotype AA, Aa, or aa at one locus on the chromosome, of genotype BB, Bb, or bb at a second locus, and of genotype CC, Cc, or cc at a third locus. How many possibilities are there for the genotypic composition of an individual with respect to the three loci?

7. A farmer has three concentrations of a nitrogen fertilizer, four concentrations of a phosphate fertilizer, and five concentrations of an organic fertilizer. How many ways are there for the farmer to combine the three types of fertilizers containing one and only one concentration of each type?

8. A hi-fi store has in its inventory six similar speakers, one of which is defective. Find how many ways there are:
   (a) of picking two speakers
   (b) that would not include the defective speaker.

9. Ten people have gathered at a party. Find the number of handshakes when each person shakes hands with everyone else at the party.

10. An IRS agent has 200 income-tax returns on his desk. If he has decided to scrutinize twenty of the returns, find how many sets of twenty returns are possible. (Do not simplify your answer.)

11. A high school wants to buy six personal computers for its laboratory from a local supplier. The supplier has ten personal computers in stock, of which four are foreign-made.
   (a) Find how many ways there are to buy six personal computers from the supplier.
   (b) Find how many ways there are if the high school wants four domestic and two foreign-made computers.

12. A superintendent of education has sixteen schools in her district and wishes to visit four of the schools.
   (a) How many different ways are there for her to pick a group of four schools to visit?
   (b) Suppose the superintendent is also interested in the order in which she will visit the four schools. How many possible routes are there?

13. Sixteen college graduates have applied for six vacancies in a company. How many ways are there in which the company can make six offers?

14. Sixteen college graduates, of whom five are women, have applied for six vacancies in a company. If the company has decided to hire two women, how many ways are there in which the company can make six offers of this type?

15. If there are 100 entries in a contest, find the number of ways in which three different prizes—first, second, and third—can be awarded if no contestant can win more than one prize.

16. A medical researcher has a study group of fifteen patients and wants to assign them to one of three groups for testing the effectiveness of a new drug: current standard drug, new experimental drug, and placebo. In how many ways can the fifteen patients be assigned to these groups so that the groups have, respectively, six, six, and three patients?

17. California State Lottery, LOTTO 6/51, involves selecting six numbers without replacement from 1 to 51. If you buy a ticket and select six numbers at random, what is the probability of your picking the six winning numbers?

18. The Department of Statistics hired three faculty members, none of whom was born during a leap year. What is the probability that:
    (a) none will have January 21 as a birthday
    (b) at least one will have January 21 as a birthday.

19. The ice cream franchise Baskin-Robbins advertises thirty-one different flavors of ice-cream. Suppose you order a triple-scoop cone and tell the vendor to pick the three different flavors at random. Find the probability that you get:
    (a) the three flavors vanilla, chocolate, and butterscotch.
    (b) vanilla as one of the flavors.

20. A bioactive tetrapeptide (a compound of four amino acids linked into a chain) is composed of only amino acids Proline (P) and Valine (V).
    (a) Draw a tree representing all possible tetrapeptides made up of the components proline and valine. List the resulting sample space.
    (b) List the sample points that constitute each of the following events.
        A: The first amino acid is proline.
        B: Exactly two of the amino acids are proline.
        C: The first and the last amino acids are proline.
        D: Proline is the first and last amino acid only.
        E: Valine is found at one or the other end of the chain (inclusive *or*).
    (c) Assuming that each outcome in the sample space is equally likely, find the probability of each event in Part b.

21. In the RNA genetic code, a codon is a three-letter "word" written from a four-letter "alphabet." Each letter represents one of the four ribonucleotides: Uracil (U), Adenine (A), Guanine (G), Cytocine (C). Each codon codes for an amino acid or "stop." The "stop" signals for termination in the construction of a polypeptide chain, the first major step in protein synthesis. For example, ACG codes for threonine.
    (a) How many RNA codons are possible?
    (b) Of the possible RNA codons, sixty-one are known to code for the twenty existing amino acids; the others code for "stop." How many codons code for "stop"?
    (c) A codon codes for valine if and only if it starts with GU. How many synonyms (that is, equivalent codons) are there for valine?
    (d) There are six synonyms for arguinine, six for leucine, four for valine, three for isoleucine, and two for lysine. How many different but equivalent ways can the five-amino-acid chain arguinine-leucine-valine-isoleucine-lysine be coded?

22. Assume that each codon in Exercise 21 is equally likely. What is the probability that a randomly formed codon will:
    (a) begin with G or U, and end with A or G
    (b) begin with U, end with C, and have no repetitions
    (c) begin with U, and end with A or G?

23. The letters of the word *problem* are arranged in all possible ways. If an arrangement is picked at random, find the probability that:
   (a) the two vowels are next to each other
   (b) the arrangement will end with a consonant
   (c) the arrangement will start with *p* and end with *m*.

24. A six-digit number is formed with the digits 1, 2, 3, 4, 5, 7 with no repetitions. Find the probability that:
   (a) the number is even
   (b) the number is divisible by 5, that is, the units digit is 5.

25. Steve, Carol, Barb, and Alan are invited to a party. If they arrive at different times and at random, find the probability that:
   (a) they will arrive in the order Steve, Carol, Barb, and Alan
   (b) Steve will arrive first
   (c) Steve will arrive first and Carol last.

26. If each license plate contains three different nonzero digits followed by three different letters, find the probability that, if a license plate is picked at random, the first digit will be odd and the first letter will be a vowel.

27. In a group there are three lawyers, two professors, and three doctors. If they are seated in a row, find the probability that those of the same profession sit together.

28. From a group of ten lawyers, fifteen accountants, and twelve doctors, a committee of six is selected at random. What is the probability that the committee consists of three lawyers, two accountants, and one doctor?

29. From six married couples, five people are selected at random. Find the probability of selecting two men and three women.

30. A box contains four white balls and five red balls. If three balls are drawn from the box without replacement, find the probability that:
   (a) one ball is white   **(b)** no ball is white
   (c) at least one ball is white   **(d)** at most one ball is white.

31. If a committee of five is selected from a group of eight seniors, six juniors, and four sophomores, find the probability that the committee has:
   (a) two seniors, two juniors, and one sophomore
   (b) no seniors
   (c) at least one senior.

32. There are eight people at a picnic and their ages are as follows: 15, 5, 2, 20, 7, 30, 40, 23. If three people are picked at random, what is the probability that their combined ages will exceed 27 years?

33. A box contains eight balls marked 1, 2, 3, . . . , 8. If four balls are picked at random, find the probability that the balls marked 1 and 5 are among the four selected balls.

34. A box contains eight good transistors, four transistors with minor defects, and three transistors with major defects. Four transistors are picked at random without replacement. Find the probability that:
   (a) no transistor is good; that is, all transistors are defective
   (b) at least one transistor is good
   (c) exactly two transistors are good
   (d) one transistor is good, one has a minor defect, and two have a major defect
   (e) one transistor is good and three have major defects.

3-5 CONDITIONAL PROBABILITY AND INDEPENDENT EVENTS

CONDITIONAL PROBABILITY

If we are interested in the probability of events, given some condition, we are concerned with *conditional probability*. When certain information is available about the outcome of the underlying experiment, we are led to make an appropriate adjustment of the probabilities of the associated events. As an example, consider the city of Burlington, Vermont. Suppose it snows on about 20 percent of the days each year; that is, if we pick a random calendar day, the probability of snow on that day is 0.2. Now suppose the random calendar day is a winter day. With this additional information, we would most likely say that the probability of snow is much higher than 20 percent. Similarly, if we know that the random day is a summer day, we would say that the probability of snow on that day is just about zero. In both situations, information about the day has led us to reassess the probability of the event that it snows on that day.

To put these ideas into more concrete form, consider this example. Suppose a certain sports club has thirty-five members, of whom twenty are adults and fifteen are minors. Among the adults, eight are basketball players, whereas among the minors, five are basketball players. The composition of the club membership is shown in the Venn diagram in Figure 3-8. It can also be presented as a table as follows.

	Adult	Minor	Total
Player	8	5	13
Nonplayer	12	10	22
Total	20	15	35

Suppose a person picked at random is a basketball player. Given this information, what is the probability that the person is an adult?

Let us write the events involved symbolically as follows.

A: The person is an adult.
B: The person is a basketball player.

In probability theory, the notation $P(A \mid B)$ stands for "probability of event A given that event B has occurred" or, for short, "probability of A given B." The vertical line is to be read "*given.*"

Given the condition that the person picked is a basketball player, our concern is with the thirteen basketball players. Of these, eight are adults. (See the Venn diagram in Figure 3-8, or the table.) Therefore, we can say

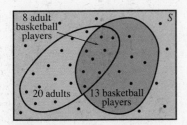

Figure 3-8
Venn Diagram Depicting the Club Membership

$$P(A \mid B) = \frac{8}{13} \quad \left(\text{that is,} \frac{n \,(A \text{ and } B)}{n \,(B)} \right).$$

We can rewrite $P(A|B)$ obtained above to give

$$P(A|B) = \frac{8/35}{13/35} \qquad \left(\text{that is, } \frac{n(A \text{ and } B)/n(S)}{n(B)/n(S)}\right).$$

But

$$\frac{n(A \text{ and } B)/n(s)}{n(B)/n(S)} = \frac{P(A \text{ and } B)}{P(B)}$$

Thus, in the general context we get

$$P(A|B) = \frac{P(A \text{ and } B)}{P(B)}.$$

From the discussion above, we can give the definition of the conditional probability as follows.

CONDITIONAL PROBABILITY

The **conditional probability** of an event A given that an event B has occurred is defined by

$$P(A|B) = \frac{P(A \text{ and } B)}{P(B)}, \text{ provided } P(B) \neq 0.$$

$P(A|B)$ is not defined if $P(B) = 0$.

EXAMPLE 1

An individual is picked at random from a group of fifty-two athletes. Suppose twenty-six of the athletes are female and six of them are swimmers. Also, there are ten swimmers among the males.

(a) Given that the individual picked is a female, find the probability that she is a swimmer.

(b) Given that the individual picked is a swimmer, find the probability that the person is a male.

SOLUTION

(a) Given that the athlete is a female, we are concerned with the twenty-six female athletes. Of these, six are swimmers. Thus, the desired probability is $\frac{6}{26}$, or $\frac{3}{13}$.

Alternatively, we note the following.

$$P(\text{athlete is a female and a swimmer}) = \frac{6}{52}$$

$$P(\text{athlete is a female}) = \frac{26}{52}$$

Therefore, applying the formula,

$$P(\text{athlete is a swimmer} \mid \text{athlete is a female})$$

$$= \frac{P(\text{athlete is a female and a swimmer})}{P(\text{athlete is a female})}$$

$$= \frac{6/52}{26/52} = \frac{3}{13}.$$

(b) There are sixteen swimmers altogether and there are ten athletes who are male and swimmers. Therefore,

$$P(\text{athlete is a swimmer}) = \frac{16}{52} \text{ and}$$

$$P(\text{athlete is a male and a swimmer}) = \frac{10}{52}.$$

Applying the rule for conditional probability,

$$P(\text{athlete is a male} \mid \text{athlete is a swimmer})$$

$$= \frac{P(\text{athlete is a male and a swimmer})}{P(\text{athlete is a swimmer})}$$

$$= \frac{10/52}{16/52} = \frac{5}{8}. \qquad \bullet$$

EXAMPLE 2   It is known that the probability that a bulb will last beyond 100 hours is 0.7 and the probability that it will last beyond 150 hours is 0.28. Given that a bulb lasts beyond 100 hours, determine the probability that it will last beyond 150 hours.

SOLUTION   We will describe the events involved as follows.

L_{100}: the bulb lasts beyond 100 hours

L_{150}: the bulb lasts beyond 150 hours.

We want to find $P(L_{150} \mid L_{100})$. From the definition of conditional probability,

$$P(L_{150} \mid L_{100}) = \frac{P(L_{150} \text{ and } L_{100})}{P(L_{100})}.$$

Now notice that the event "L_{150} and L_{100}" states that a bulb lasts beyond 150 hours and beyond 100 hours. This is simply the event the bulb lasts beyond 150 hours.

$$P(L_{150} \mid L_{100}) = \frac{P(L_{150})}{P(L_{100})} = \frac{0.28}{0.7} = 0.4. \qquad \bullet$$

If we multiply both sides of the formula $P(A \mid B) = P(A \text{ and } B)/P(B)$ by $P(B)$, we get the *general multiplication law of probabilities*.

> **THE GENERAL MULTIPLICATION LAW**
>
> If A and B are two events, then P (A and B) is given by the formula
>
> $$P(A \text{ and } B) = P(B) \cdot P(A \mid B).$$

This law is extremely useful in instances that require finding the probability that two events will occur simultaneously.

EXAMPLE 3   The probability that the stock market goes up on Monday is 0.6. Given that it goes up on Monday, the probability that it goes up on Tuesday is 0.3. Find the probability that the market goes up on both days.

SOLUTION   We are interested in the event that the market goes up on Monday *and* it goes up on Tuesday. Let the events be defined as follows.

  M: market goes up on Monday.

  T: market goes up on Tuesday.

We are given that $P(M) = 0.6$ and $P(T \mid M) = 0.3$, and we want to find $P(T \text{ and } M)$. Applying the general multiplication law of probabilities, we get

$$P(T \text{ and } M) = P(M) \cdot P(T \mid M)$$
$$= (0.6)(0.3) = 0.18. \qquad \bullet$$

EXAMPLE 4   Ted has an instructor who gives a midterm exam and a final exam. The probability that a student passes the midterm exam is 0.6. Given that a student fails the midterm exam, the probability that the student passes the final exam is 0.8. What is the probability that Ted fails the midterm and passes the final?

SOLUTION   We have

$$P(\text{Ted fails the midterm } and \text{ passes the final})$$
$$= P(\text{Ted fails the midterm}) \cdot P(\text{Ted passes the final} \mid \text{Ted fails the midterm})$$
$$= (0.4)(0.8) = 0.32$$

because the probability that Ted fails the midterm is $1 - 0.6$, or 0.4, and the probability of his passing the final, given that he fails the midterm, is 0.8.

$\bullet$

EXAMPLE 5   A cosmetics saleperson makes house calls to different households in a drive to sell a product. The probability that a resident is home is 0.7. Given that a resident is home, the probability that the resident buys the product is 0.3. Find the probability that a resident is home when called and buys the product.

SOLUTION   The probability is equal to the product of the two probabilities P(a resident is home) and P(a resident buys a product | a resident is home) and, consequently, is equal to $(0.7)\,(0.3) = 0.21$. ●

The general multiplication law can be extended to more than two events. For example,

$$P(A \text{ and } B \text{ and } C) = P(A)\,P(B\,|\,A)\,P(C\,|\,A \text{ and } B).$$

INDEPENDENT EVENTS

In some situations the fact that an event B has occurred may not influence the probability of the occurrence (or nonoccurrence) of the event A, so that we have $P(A\,|\,B) = P(A)$. (It turns out that if $P(A\,|\,B) = P(A)$, then $P(B\,|\,A) = P(B)$). As might be expected, such events are called *independent events*.

Suppose A and B are independent events. From the general multiplication law of probabilities, namely $P(A \text{ and } B) = P(B) \cdot P(A\,|\,B)$, it follows then that

$$P(A \text{ and } B) = P(B) \cdot P(A) = P(A) \cdot P(B).$$

INDEPENDENT EVENTS

Two events A and B are said to be **independent events** if the probability of the simultaneous occurrence of A and B is equal to the product of the respective probabilities, that is,

$$P(A \text{ and } B) = P(A) \cdot P(B).$$

Two events that are not independent are said to be **dependent events.**

It is worth noting that if A and B are independent, then so are A and *not B*, or *not A* and B, or *not A* and *not B*. This can be proved, but let us rely on our intuition.

EXAMPLE 6   Past attendance records show that the probability that the chairman of the board attends a meeting is 0.65, that the president of the company attends a meeting is 0.8, and that they both attend a meeting is 0.6. Would you say that the chairman and the president act independently regarding their attendance at the meetings of the board of directors?

SOLUTION   Because $0.6 \neq (0.65)(0.8)$, we see that

$$P(\text{chairman attends a meeting } and \text{ president attends it})$$
$$\neq P(\text{chairman attends a meeting}) \cdot P(\text{president attends a meeting}).$$

Therefore, the two events "the chairman attends a meeting" and "the president attends a meeting" are not independent. ●

EXAMPLE 7

Victor flies from San Francisco to New York via Chicago. He takes Unified Airlines from San Francisco to Chicago, and Beta Airlines from Chicago to New York. The probability that a Unified plane lands safely is 0.9995, and the probability that a Beta plane lands safely is 0.9998. Determine the probability of the following events.

(a) Victor lands safely in Chicago and New York.

(b) Victor lands safely in Chicago, but has a mishap in New York.

SOLUTION   We can assume that independence is inherent in the situation since the safety on one airline should not influence that on the other.

(a)   Since the events are independent,

$$P \text{ (Victor lands safely in Chicago } and \text{ New York)}$$
$$= P \text{ (Victor lands safely in Chicago)}$$
$$\cdot P \text{ (Victor lands safely in New York)}$$
$$= (0.9995)(0.9998) = 0.9993.$$

(b)   Here we want

$$P \text{ (a safe landing in Chicago } and \text{ mishap in New York)}$$
$$= P \text{ (a safe landing in Chicago)} \cdot P(\text{ mishap in New York).}$$

Now the probability of a mishap in New York is $1 - 0.9998 = 0.0002$. Therefore, the probability of the desired event is $(0.9995)(0.0002) = 0.0002$.

●

The multiplication rule for two independent events given above can be extended to the case of three or more independent events. Thus, if A, B, C, and D are *independent events,* we can write

$$P(A \text{ and } B \text{ and } C \text{ and } D) = P(A) \cdot P(B) \cdot P(C) \cdot P(D).$$

Remark   The definition of independence above should be understood in the sense that the occurrence of one event has no bearing on the probability of occurrence of the other event. Students often say, imprecisely, that events are independent if you can multiply probabilities. One can always multiply probabilities just as one can add them. The correct statement is that the probability *of the simultaneous occurrence of a collection of events* can be obtained by multiplying the probabilities of the individual events.

SIDEBAR   **Space Shuttle Challenger Disaster and Independence**

Statistical independence is extremely important in industry for improving reliability of machines whose successful operation depends on proper functioning of component parts. In this context, the following account of what might have led to the space shuttle Challenger disaster in January 1986 is very illuminating. The account is based on the excellent educational video series *Against All Odds: Inside Statistics,* vol. 15, "What is probability?", in the Annenberg / CPB Project Series.

A successful initial launch of the shuttle orbiter depends on proper functioning of six field joints, all of which operate independently. Each of the field joints contains two O-rings—a primary O-ring and a secondary O-ring as a backup in case of the failure of the primary O-ring. The O-rings are meant to contain burning gases within the boosters.

In the original design, the probability of failure of any field joints was 0.023. So the probability of success of any given field joint was 0.977. This reliability rate might seem high enough, but for a successful launch all six field joints have to work correctly, and, due to independence, the probability of success in launching is $(0.977)(0.977)(0.977)(0.977)$ $(0.977)(0.977) = 0.87$. That means the probability of failure in the initial phase of the launch is 0.13, or 13 percent, which cannot be dismissed as negligible, especially when so much is at stake.

In the new design, the probability of proper functioning of the field joints had to be improved to a level much higher than 0.977. This would have to be accomplished by improving the reliability provided by the O-rings. The investigation showed that the two O-rings in the original design did not operate independently. In fact, the O-rings shared a common failure mode called joint rotation. Had they been independent, each one with a small chance of failure, then by the multiplication rule, the probability of failure of both O-rings would have been extremely small. As a result, the probability that at least one O-ring did not fail would have been extremely high. Bear in mind that for the field joint to function properly, only one O-ring has to do its job properly.

In the subsequent design a third O-ring which operates independently of the previous two was added.

Results of the following type involving any finite number of independent events are often useful. Suppose A, B, C, D are four events with the same probability p of occurring. Then

$$P \text{ (all the events occur)} = P \text{ (}A\text{ and }B\text{ and }C\text{ and }D\text{)}$$
$$= P(A)P(B)P(C)P(D)$$
$$= p^4$$

$$P \text{ (none of the events occurs)}$$
$$= P \text{ (}not\ A\text{ and }not\ B\text{ and }not\ C\text{ and }not\ D\text{)}$$
$$= P(not\ A)P(not\ B)P(not\ C)P(not\ D)$$
$$= (1 - p)(1 - p)(1 - p)(1 - p)$$
$$= (1 - p)^4$$

$$P \text{ (at least one event occurs)}$$
$$= P \text{ (}not\text{ (none of the events occurs))}$$
$$= 1 - P \text{ (none of the events occurs)}$$
$$= 1 - (1 - p)^4$$

EXAMPLE 8   A number is picked at random from the digits $1, 2, \ldots, 9,$ and an unbiased coin and a fair die are tossed. Find the probability of picking an odd digit, getting a head on the coin, and getting a multiple of 3 on the die.

SOLUTION   Note the following.

$$P\,(\text{odd digit}) = \frac{5}{9}$$

$$P\,(\text{head}) = \frac{1}{2}$$

$$P\,(\text{multiple of 3 on the die}) = \frac{1}{3}$$

Therefore, since the events are independent,

$$P\,(\text{odd digit } and \text{ a head on the coin}$$
$$and \text{ a multiple of 3 on the die})$$

$$= P\,(\text{odd digit}) \cdot P\,(\text{head on the coin})$$
$$\cdot P\,(\text{a multiple of 3 on the die})$$

$$= \frac{5}{9} \cdot \frac{1}{2} \cdot \frac{1}{3} = \frac{5}{54}.$$ ●

EXAMPLE 9   A message will be transmitted from Terminal A to Terminal B if the relays R_1, $R_2, R_3,$ and R_4 in Figure 3-9 are closed. If the probabilites that these relays are closed are, respectively, 0.5, 0.7, 0.8, and 0.8, find the probability that a message will get through. Assume that the relays function independently.

$$\quad A \qquad\quad R_1 \qquad\quad R_2 \qquad\quad R_3 \qquad\quad R_4 \qquad\quad B$$

Figure 3-9 Message through Four
Independent Relays in Series

SOLUTION   The message will get through if Relay R_1 is closed *and* Relay R_2 is closed *and* Relay R_3 is closed *and* Relay R_4 is closed. Since the relays function independently, it follows that

$$P\,(\text{message gets through}) = P\,(R_1 \text{ is closed}) \cdot P\,(R_2 \text{ is closed})$$
$$\cdot P\,(R_3 \text{ is closed}) \cdot P\,(R_4 \text{ is closed})$$

$$= (0.5)(0.7)(0.8)(0.8)$$
$$= 0.224.$$ ●

Remark   Two events are independent if the occurrence of one does not influence the probability of occurrence of the other. They are mutually exclusive if the two are not compatible in the sense that, if one occurs, the other cannot occur. The concepts of independent events and mutually exclusive events are divergent concepts in the following sense.

If two events with positive probabilities are independent, they cannot be mutually exclusive, and vice versa

This follows simply from the fact that if A and B are mutually exclusive, they are not compatible, and

$$P \ (A \text{ and } B) = 0.$$

But, if they are independent

$$P \ (A \text{ and } B) = P \ (A) \cdot P \ (B) > 0 \quad \text{(because } P \ (A) > 0,$$
$$P \ (B) > 0)$$

and the two cannot be reconciled.

As a concrete example, suppose Mr. Arnold owns a motel. Consider the following two events.

A: Mr. Arnold takes off a weekend.

B: Mr. Arnold's son, Bill, takes off a weekend.

Suppose Mr. Arnold or Bill has to be on duty at the motel on every weekend. Because they cannot both take off the same weekend the two events are obviously mutually exclusive. But what one does depends on what the other does, making the two events dependent events.

On the other hand, it the two decide independently, then it is very likely that they will both take off sometimes on the same weekend and, then, the events are not mutually exclusive.

S I D E B A R   *Shuttle Orbiter Safety*

On May 11, 1990, the Office of Technology Development presented some new concerns for the American space program. The office declared that the NASA shuttle schedule was too ambitious to be safe for astronauts. "If the reliability is 98%, the nation faces a fifty-fifty chance of losing an additional shuttle orbiter in the next 34 flights." The statement assumes that the safety aspects of a flight are independent from flight to flight. This may not be quite correct because each flight provides information that is utilized in fine-tuning the next one.

The probability of not losing a shuttle on a flight is 0.98. Assuming independence of risks, the probability of not losing a shuttle on two flights is $(0.98)^2$. Continuing this argument, the probability of not losing an orbiter in any of the next 34 flights is $(0.98)^{34}$, which is approximately 0.503. Hence the probability of losing at least one shuttle is $1 - 0.503 = 0.497 \approx 0.5$, the figure quoted by the Office of Technology Development.

SIDEBAR   *Law and Probability*

We feel confident in saying that the possibility that an event with extremely low probability will occur on a *specific* trial can be safely excluded. But practically impossible events with small probability, no matter how small, as long as it is not zero, can and do occur. For example, you bump into a former classmate exactly thirty years to the day after graduation, or you hum an old tune, and when you turn on the radio that same song is being played. Some people tend to ascribe such occurrences to divine or mystical forces lurking in a sinister way.

Soldier's Surprise

The letter was addressed "to any soldier" and sent to Saudi Arabia. So when Sgt. Rory Lomas opened it, he was stunned to find it signed by his 10-year-old daughter, Cetericka.

"Its amazing if you think of the odds of him getting his own daughter's letter out of the thousands that were sent," Lomas' wife, Barbara, said Sunday.

The letter from Cetericka, a fifth grader at Windsor Forest Elementary, was one of thousands written by Chatham County children and addressed "to any soldier."

As a matter of fact, in independent repetitions of the underlying experiment a large number of times the probability that a practically impossible event will occur at least once can be appreciable to the point that it cannot be disregarded as negligible. The following case discussed in *Innumeracy* by John Allen Paulos in the section on coincidence and law is quite interesting and pertinent here.

In 1964 in Los Angeles a blond woman with a ponytail who snatched a purse from a woman was seen entering a yellow car driven by a black male sporting a mustache and a beard. Based on this description given by a witness, police tracked down a couple who fit the above description. With the figures in parentheses showing the estimated probability for each characteristic, the prosecution estimated that the probability that a randomly selected couple would have all the characteristics, namely, own a yellow car ($\frac{1}{10}$), man with a mustache ($\frac{1}{4}$), woman with a ponytail ($\frac{1}{10}$), woman with blond hair ($\frac{1}{3}$), black man with beard ($\frac{1}{10}$), interracial couple in a car ($\frac{1}{1000}$), was

$$\frac{1}{10} \cdot \frac{1}{4} \cdot \frac{1}{10} \cdot \frac{1}{3} \cdot \frac{1}{10} \cdot \frac{1}{1000} = \frac{1}{12,000,000}$$

You can see that the characteristics are assumed to be independent in the above computations. The prosecution argued that the estimated prob-

Source: "Soldier's Surprise" from the Associated Press, January 7, 1991. Reprinted by permission of the Associated Press.

ability was so minuscule that the event was practically impossible. The jury went along. The couple was convicted.

When the case was appealed to the Supreme Court of California, however, the conviction was reversed. The following argument was used effectively by the defense.

The probability that a randomly selected couple does not have all the characteristics is

$$1 - \frac{1}{12,000,000} = \frac{11,999,999}{12,000,000}$$

In the city of Los Angeles, with approximately 2,000,000 couples, the probability that none of the couples will have all the characteristics is

$$\left(\frac{11,999,999}{12,000,000}\right)^{2,000,000} \approx 0.846$$

Hence, the probability of *at least one couple* in Los Angeles having all the characteristics is $1 - 0.846 = 0.154$. Based on this figure, the defense successfully argued that since there was at least one couple fitting the description, namely, the convicted couple, the probability of having at least two such couples was approximately 8 percent, not negligible. (We shall develop this last aspect systematically in this section and in Chapter 4.)

SIDEBAR   Law and Probability Revisited

Under the premises considered earlier, in Chapter 4 we shall show that the probability of *at least two couples* in the city of Los Angeles fitting the description of the convicted couple is 0.013. (For the present we shall accept this value.) Earlier we established the probability of at least one couple fitting the description to be 0.154. Therefore,

P(at least two couples fit the description | at least one such couple)

$$= \frac{P(\text{at least one } and \text{ at least two couples fit the description})}{P(\text{at least one couple fits the description})}$$

$$= \frac{P(\text{at least two couples fit the description})}{P(\text{at least one couple fits the description})}$$

$$= \frac{0.013}{0.154} = 0.0844$$

This figure of 8.44 percent probability was presented by the defense in support of the convicted couple.

10/3/96

Section 3-5 Exercises

1. If A and B are two events, find $P(A|B)$ if:
   (a) $P(B) = 0.6$, $P(A \text{ and } B) = 0.2$
   (b) $P(\text{not } B) = 0.7$, $P(A \text{ and } B) = 0.1$
   (c) $P(B) = 0.8$, $P(A \text{ and } B) = 0.5$.

2. If A and B are two events, find $P(A \text{ and } B)$ if:
   (a) $P(A) = 0.5$, $P(B) = 0.7$, and $P(A|B) = 0.4$
   (b) $P(A) = 0.8$, $P(B) = 0.6$, and $P(B|A) = 0.55$
   (c) $P(A) = 0.6$, $P(B) = 0.8$, and $P(A|B) = 0.7$
   (d) $P(\text{not } A) = 0.3$, $P(\text{not } B) = 0.5$, and $P(B|A) = 0.6$.
   Note: Some of the information given is superfluous.

3. In each of the cases in Exercise 2 determine $P(A \text{ or } B)$.

4. In the table below, 400 individuals are classified according to whether they were vaccinated against influenza and whether they were attacked by influenza.

	Vaccinated	Not Vaccinated
Attacked	60	85
Not attacked	190	65

   If a person is chosen at random, find the probability that:
   (a) the person was vaccinated and attacked
   (b) the person was attacked, given the person was vaccinated
   (c) the person was not attacked, given the person was not vaccinated
   (d) the person was vaccinated, given the person was attacked.

5. In a particular geographic region, 60 percent of the people are regular smokers, and evidence indicates that 10 percent of the smokers have lung cancer. Find the percentage of people who are smokers and have lung cancer in this region.

6. A petroleum company exploring for oil has decided to drill two wells, one after the other. The probability of striking oil in the first well is 0.2. Given that the first attempt is successful, the probability of striking oil on the second attempt is 0.8. What is the probability of striking oil in both wells?

7. The probability that an adult male is over 6 feet tall is 0.3. Given that he is over 6 feet tall, the probability that he weighs less than 150 pounds is 0.2. Find the probability that a randomly chosen man will be taller than 6 feet and will weigh less than 150 pounds.

8. The probability that a person contracts a certain type of influenza is 0.2. Past experience has revealed that the probability that a person who gets the influenza does not recover is 0.005. Find the probability that a randomly chosen person:
   (a) contracts the influenza and does not recover
   (b) contracts the influenza and recovers.

9. The probability that it will snow is 0.3. Given that it snows, the probability is 0.4 that the plane will take off on time, and given that it does not snow, the probability is 0.9 that the plane will take off on time. Find the probability that:
   (a) it will snow and the plane will take off on time
   (b) it will not snow and the plane will take off on time
   (c) the plane will take off on time.

10. The probability that a student studies is 0.7. Given that she studies, the probability is 0.8 that she will pass a course. Given that she does not study, the probability is 0.3 that she will pass the course. Find the probability that:
    (a) she will study and pass the course
    (b) she will not study and will pass the course
    (c) she will pass the course.

11. Suppose A and B are independent events. Find $P(A \text{ and } B)$ if:
    (a) $P(A) = 0.5$, $P(B) = 0.4$
    (b) $P(A) = 0.3$, $P(\text{not } B) = 0.6$
    (c) $P(A) = 0.8$, $P(B) = 0.7$
    (d) $P(\text{not } A) = 0.3$, $P(\text{not } B) = 0.6$.

12. In each of the cases given in Exercise 11 determine $P(A \text{ or } B)$.

13. If A and B are independent events with $P(A) = 0.8$ and $P(A \text{ and } B) = 0.24$, find:
    (a) $P(B)$
    (b) $P(A \text{ or } B)$.

14. Suppose A and B are two events with $P(A) = 0.35$ and $P(B) = 0.6$. Find $P(A \text{ or } B)$ if the two events are:
    (a) mutually exclusive
    (b) independent.

15. Suppose A and B are two events for which $P(A) = 0.4$, $P(B) = 0.6$, and $P(A \text{ or } B) = 0.76$.
    (a) Are A and B mutually exclusive?   NO $P(A+B) \neq 0$
    (b) Are A and B independent?   Yes $.76 = .4 + .6 - X$ $X = .24$

16. Orwell wants to buy a special ribbon for his typewriter. The probability that the university bookstore carries it is 0.3 and the probability that Village Stationers carries it is 0.3. Assuming the two stores stock the item independently, find the probability that neither store carries the item.

17. The probability that a plane arrives at the airport before 11:00 AM is 0.7. The probability that a passenger leaving from downtown by bus arrives at the airport by 11:00 AM is 0.6. Find the probability that:
    (a) the plane and the passenger both arrive by 11:00 AM
    (b) the plane arrives by 11:00 AM and the passenger arrives after 11:00 AM.

18. A genetic population consists of a large number of alleles that are of two types, Type A and Type a. It is known that 60 percent of the alleles in the population are of Type A. If two alleles are picked at random, assuming independence, find the probability that:
    (a) both alleles are of Type A
    (b) one allele is of Type A and the other is of Type a
    (c) both alleles are of Type a.

19. The probability that a person picked at random is an octogenarian is 0.008. If two people are picked at random, assuming independence, find the probability that:
    (a) they are both octogenarians
    (b) neither is an octogenarian
    (c) at least one is an octogenarian.

a) $0.008 \times 0.008 = 0.000064$
b) $1 - 0.008 = 0.992$
   $0.992 \times 0.992 = 0.984$
c) $1 - 0.984 = 0.016$

20. It is known from past experience that the probability that a person has cancer is 0.02 and the probability that a person has heart disease is 0.05. If we assume that incidence of cancer and heart disease are independent, find the probability that:
    (a) a person has both ailments
    (b) a person does not have both ailments
    (c) a person does not have either ailment.

21. From the records of the Public Health Division, the percentages of A, B, O, and AB blood types are, respectively, 40 percent, 20 percent, 30 percent, and 10 percent. If two individuals are picked at random from this population, assuming independence, find the probability that:
    (a) they are both of blood type A
    (b) they are both of the same blood type
    (c) neither is of blood type A
    (d) one is of blood type A and the other is of blood type O
    (e) they are both of different blood types.

22. Suppose a penny, a dime, and a nickel with respective probabilities of heads equal to 0.2, 0.3, and 0.4 are tossed simultaneously. Find the probability that there will be:
    (a) no heads
    (b) three heads
    (c) exactly one head.

23. A mouse caught in a maze has to maneuver through three escape hatches in order to escape. The probability that the mouse can maneuver through Hatch 1 is 0.6; given that he maneuvers through Hatch 1, the probability of maneuvering through Hatch 2 is 0.4; finally, given that the mouse maneuvers through Hatches 1 and 2, the probability of maneuvering through Hatch 3 is 0.2. Find the probability that:
    (a) the mouse will make his escape
    (b) the mouse will not be able to escape.

24. The genetic composition of a population is such that if a person is chosen at random, the probabilities that the person is of genotype, *AA*, *Aa*, or *aa* at a locus on a chromosome are, respectively, 0.36, 0.48, and 0.16. The probabilities that the individual is of genotype *BB*, *Bb*, or *bb* at another locus are, respectively, 0.49, 0.42, and 0.09. If the genotypic distributions at the two loci are independent, find the probability that a randomly drawn person is:
    (a) of genotype *AA* at one locus and of genotype *Bb* at the other
    (b) of genotype *Aa* at one locus and *Bb* at the other.

25. The probability that an adult male is shorter than 66 inches is 0.3, that he is between 66 and 70 inches is 0.5, and that he is taller than 70 inches is 0.2. If two males are picked at random in succession, find the probability that:
    (a) the heights of both are between 66 and 70 inches
    (b) the first is shorter than 66 inches and the second is taller than 70 inches
    (c) The first is 70 inches or shorter and the second is 66 inches or taller.

26. The probabilities that an office receives 0, 1, 2, or 3 calls during a half-hour period are, respectively, 0.1, 0.2, 0.4 and 0.3. It is safe to assume that the number of phone calls received during any two nonoverlapping time periods are independent. Find the probability that during the two time periods 11:00–11:30 and 3:00–3:30, together:
(a) no calls will be received
(b) three calls will be received
(c) four calls will be received.

27. A nuclear power plant has a fail-safe mechanism consisting of six protective devices that function independently. The probabilities of failure for the six devices are 0.3, 0.2, 0.2, 0.2, 0.1, and 0.1. If an accident will take place if all the devices fail, find the probability of an accident.

28. On a day in December, the probability that it will snow in Burlington, Vermont, is 0.6. Given that it snows in Burlington, the probability that it will snow in Boston is 0.7. Find the probability that on a random December day it will snow in both cities.

29. On a day in December, the probability that it will snow in Boston is 0.4 and the probability that it will snow in Moscow is 0.7. Assuming independence, find the probability that it will snow:
(a) in both cities
(b) in neither city
(c) only in Moscow
(d) in exactly one city.

30. Suppose current flows through two switches A and B to a radio and back to the battery as shown in the following figure. If the probability that Switch A is closed is 0.8, and given that Switch A is closed the probability that Switch B is closed is 0.7, find the probability that the radio is playing.

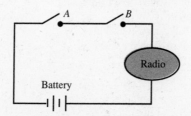

31. Suppose current flows through two switches A and B to a radio and back to the battery as shown in the figure for Exercise 30. If the probability that Switch A is closed is 0.8 and the probability that Switch B is closed is 0.9, assuming that the switches function independently, find the probability that the radio is playing.

32. Suppose current flows through two switches A and B to a radio and back to the battery as shown in the following figure. Suppose the probability that Switch A is closed is 0.8, the probability that Switch B is closed is 0.6, and given that

Switch A is closed, the probability that Switch B is closed is 0.7. Find the probability that the radio is playing.

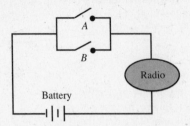

33. Suppose current flows through two switches A and B to a radio and back to the battery as shown in the figure for Exercise 32. If the probability that Switch A is closed is 0.8 and the probability that Switch B is closed is 0.9, and given that the switches function independently, find the probability that the radio is playing.

34. Marvin, a basketball player, has decided to shoot until he makes a basket. Suppose the probability that he makes a basket on any attempt is 0.6 and the results of the attempts are independent. Find the probability that he will need:
    (a) exactly one attempt   (b) exactly two attempts
    (c) exactly three attempts   (d) at least three attempts.

35. Studies done show that the probability of getting AIDS in an unprotected heterosexual encounter is 1 in 500. Assuming that the risks are independent, show that with 346 such encounters there is a fifty-fifty chance of contracting AIDS.

36. Sickle-cell anemia is a genetic blood disease that, based on a survey, is prevalent among approximately 1 in 400 individuals of the Afro-American population. Normal blood cells have two A genes. Carriers of the trait who do not show any symptoms of the disease have one A gene and one S gene which is the culprit. The individuals who show the symptom have two S genes. When two carriers of the trait without the symptoms marry, the chance that an offspring is normal is 1 in 4, the chance of its being a carrier without the symptom is 1 in 2, and the chance of showing the symptoms is 1 in 4. If two children are born to two carriers without the trait, find the probability that
    (a) they are both normal
    (b) only one is normal
    (c) both show the symptoms
    (d) one is a carrier not showing the symptoms and the other shows the symptoms.
        It is well documented that whether one child gets any particular set of genes is independent of another child getting them.

3-6   THE THEOREM OF TOTAL PROBABILITY AND BAYES' FORMULA (OPTIONAL)

Bayes' formula was developed by the Reverend Thomas Bayes (1702–1761), an English Presbyterian minister and mathematician. It is interpreted as a formula for the probability of "causes" or "hypotheses," as in the following illustration. Consider two machines that produce microchips for computers, with Machine 1 producing 8 percent defective chips and Machine 2 producing 6 percent defective chips. Suppose a chip is picked at random from a randomly selected machine. In this context an easily answered question is: If the chip was picked from Machine 2, what is the probability that it is defective? The answer, of course, is 0.06, the fraction of defective chips produced by it. But a harder question to answer is the inverse question: If the chip is defective, what is the probability that it was picked from Machine 2? The answer is provided by Bayes' formula. But before Bayes' formula can be applied, it is important to find the probability that the randomly picked chip is defective. The theorem of total probability is useful in computing this probability. For more insight we proceed with the following example.

EXAMPLE 1

In a used car lot, suppose 50 percent of the cars are manufactured in the United States and 15 percent of these are compact; 30 percent of the cars are manufactured in Europe and 40 percent of these are compact; and, finally, 20 percent are manufactured in Japan and 60 percent of these are compact. If a car is picked at random from the lot:

(a) find the probability that it is a compact

(b) given that the car is a compact, find the probability that it is European.

SOLUTION

(a) A car could be compact in one of the following ways.
(it is a compact *and* it is U.S. manufactured) *or*
(it is a compact *and* it is European) *or*
(it is a compact *and* it is Japanese)

To avoid all this cumbersome writing, it is convenient to introduce the following set notation. Let

$$A = \text{the car is a compact}$$
$$B_1 = \text{the car is U.S. manufactured}$$
$$B_2 = \text{the car is European}$$
$$B_3 = \text{the car is Japanese}$$

Thus, the statement "the car is a compact" can be written

$$A = (A \text{ and } B_1) \text{ or } (A \text{ and } B_2) \text{ or } (A \text{ and } B_3)$$

Notice that the events "A and B_1," "A and B_2," and "A and B_3" are mutually exclusive. Therefore,

$$
\begin{aligned}
P(A) &= P(A \text{ and } B_1) + P(A \text{ and } B_2) + P(A \text{ and } B_3) \\
&= P(A|B_1) \cdot P(B_1) + P(A|B_2) \cdot P(B_2) \\
&\quad + P(A|B_3) \cdot P(B_3) \\
&= (0.15)(0.50) + (0.40)(0.30) + (0.60)(0.20) \\
&= 0.315.
\end{aligned}
$$

●

A tree diagram can very effectively highlight the basic approach by displaying the various possibilities, together with the probabilities, as branches, as in Figure 3-10. After computing the probabilities of the various branches, we add the probabilities of those branches that describe the event of interest.

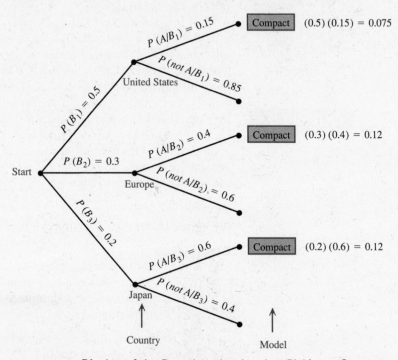

Figure 3-10 Display of the Branches that Lead to Picking a Compact Car from the Lot, and Their Probabilities

(b)   Here we want to find $P(B_2|A)$. Using the formula for conditional probability, this is given by

$$P(B_2|A) = \frac{P(A \text{ and } B_2)}{P(A)}.$$

But we know that $P(A \text{ and } B_2) = P(A|B_2) \cdot P(B_2)$. Thus we have

$$P(B_2|A) = \frac{P(A|B_2) \cdot P(B_2)}{P(A)}.$$

All the quantities on the right-hand side are known. We have $P(A|B_2)$ = 0.40, $P(B_2)$ = 0.30, and from Part a, $P(A)$ = 0.315. Therefore,

$$P(B_2|A) = \frac{(0.40)(0.30)}{0.315} = \frac{8}{21}.$$

In Example 1 we addressed two probability questions. We consider them separately in the following discussion.

We found the probability of an event A, which occurred in conjunction with three mutually exclusive events B_1, B_2, and B_3. Note that events B_1, B_2, B_3 exhaust the entire sample space since, when we pick a car from the lot, it *has* to be either a U.S. car, or a European car, or a Japanese car. Such events are called **exhaustive events.** We showed that if B_1, B_2, and B_3 are mutually exclusive and exhaustive, then

$$P(A) = P(A \mid B_1) \cdot P(B_1) + P(A \mid B_2) \cdot P(B_2) + P(A \mid B_3) \cdot P(B_3).$$

A generalization of this result is called the **theorem of total probability.**

After finding the probability of A, we proceeded to find the conditional probability that the car is European given that it is a compact. In other words, given that the car is a compact, what is the probability that "its being European" was the cause of it? This probability, $P(B_2 \mid A)$, as we saw, is given by

$$P(B_2 \mid A) = \frac{P(A \mid B_2) \cdot P(B_2)}{P(A)}$$

where, of course, $P(A) = P(A \mid B_1) \cdot P(B_1) + P(A \mid B_2) \cdot P(B_2) + P(A \mid B_3) \cdot P(B_3)$. A generalization of this result is called **Bayes' formula.**

These generalizations are as follows.

Suppose $B_1, B_2, \ldots, B_n$ are mutually exclusive and exhaustive events, all having positive probabilities. If A is any event with positive probability, then we have the following results.

THEOREM OF TOTAL PROBABILITY

$$P(A) = P(A \mid B_1) \cdot P(B_1) + \cdots + P(A \mid B_n) \cdot P(B_n)$$

BAYES' FORMULA

$$P(B_i \mid A) = \frac{P(A \mid B_i) \cdot P(B_i)}{P(A \mid B_1) \cdot P(B_1) + \cdots + P(A \mid B_n) \cdot P(B_n)}$$

The probabilities $P(B_1), \ldots, P(B_n)$ are called the *a priori*, or prior, probabilities, and $P(B_1 \mid A)$, $P(B_2 \mid A)$, $\ldots$, $P(B_n \mid A)$ the *a posteriori*, or posterior, probabilities.

EXAMPLE 2

A mouse is inside a room and each of the four walls of the room has a door through which the mouse could attempt to escape. Unluckily for the mouse, there is a trap at each of the doors d_1, d_2, d_3, d_4, and they work with probabilities 0.3, 0.2, 0.3, and 0.5, respectively. If the mouse picks a door at random:

(a) what is the probability that the mouse will escape?

(b) given that the mouse escapes, what is the probability that the mouse chose door d_3 to escape?

SOLUTION

(a) We are interested in the event that the mouse will escape. This could occur by his picking door d_1, *or* by his picking door d_2, *or* by his picking door d_3, *or* by his picking door d_4. So let A represent the event that the mouse escapes and D_1, D_2, D_3, and D_4, respectively, the events that the mouse picks doors d_1, d_2, d_3, or d_4. Applying the theorem of total probability, we can write

$$P(A) = P(A|D_1) \cdot P(D_1) + P(A|D_2) \cdot P(D_2)$$
$$+ P(A|D_3) \cdot P(D_3) + P(A|D_4) \cdot P(D_4).$$

Now, since a door is picked at random, $P(D_1) = P(D_2) = P(D_3) = P(D_4) = 0.25$. Also, the following are easily seen.

$$P(A|D_1) = 1 - 0.3 = 0.7$$
$$P(A|D_2) = 1 - 0.2 = 0.8$$
$$P(A|D_3) = 1 - 0.3 = 0.7$$
$$P(A|D_4) = 1 - 0.5 = 0.5$$

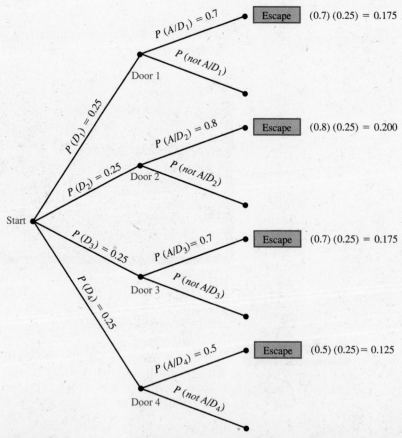

Figure 3-11

Therefore,

$$P(A) = (0.7)(0.25) + (0.8)(0.25) + (0.7)(0.25) \\ + (0.5)(0.25) = 0.675.$$

See the display in the tree diagram in Figure 3-11.

(b) Here we want to find $P(D_3 | A)$, which, applying Bayes' formula, is given by

$$P(D_3 | A) = \frac{P(A | D_3) \cdot P(D_3)}{P(A)}$$

$$= \frac{(0.7)(0.25)}{0.675}$$

$$= 0.259.$$

Section 3-6 Exercises

1. Three boxes contain white, red, and black balls in the numbers given below.

Box	White	Black	Red
1	8	2	6
2	4	5	3
3	2	7	3

A box is picked at random and a ball drawn from it at random.
(a) Find the probability that the ball is red.
(b) Given that the ball is red, what is the probability that it came from Box 2?

2. On a true-false test, the probability that a student knows the answer to a question is 0.7. If she knows the answer, she checks the correct answer; otherwise, she answers the question by flipping a fair coin.
(a) What is the probability that she answers a question correctly?
(b) Given that she answers the question correctly, what is the probability that she knew the answer?

3. The probability that an executive is promoted to a higher position is $\frac{5}{8}$. If he is promoted, he will go on vacation with a probability of $\frac{5}{6}$; however, if he is not promoted, there is a probability of $\frac{1}{3}$ that he will take a vacation.
(a) Find the probability that the executive will go on a vacation.
(b) Given that he does go on a vacation, find the probability that it was due to his having been promoted.

4. Four machines are in operation. Machine *A* produces 5 percent defective items, Machine *B* produces 3 percent, Machine *C* produces 4 percent, and Machine *D* produces 8 percent. A machine is selected at random and an item picked from it.
(a) Find the probability that the item is defective.
(b) Given that the item is defective, find the probability that it came from Machine *B*.

5. A box contains three coins. One of the coins is a two-headed coin, the second is a fair coin, and the third is a biased coin with $P(\text{head}) = 0.2$. Suppose a coin is picked at random and flipped.

   (a) Find the probability that the coin shows heads.

   (b) If the coin shows heads, find the probability that it is the fair coin.

6. A lie detector test is such that when given to an innocent person, the probability of this person being judged guilty is 0.05; on the other hand, when given to a guilty person, the probability of this person being judged innocent is 0.12. Suppose 18 percent of the people in the population are guilty. Given that a person picked at random is judged guilty, what is the probability that he or she is innocent?

7. The probability that a student studies for a test is 0.7. Given that the student studies, the probability of passing the test is 0.8. Given that the student does not study, the probability of passing the test is 0.3. What is the probability that the student will pass the test?

8. Messages are sent on a telegraph by using the signals dot and dash to code them. The probability that a signal is a dash is 0.6. From past experience, it is found that a signal that originates as a dash has a probability of 0.2 of being received as a dot, and a signal that originates as a dot has a probability of 0.1 of being received as a dash.

   (a) What is the probability that a signal is received as a dash?

   (b) Given that a signal is received as a dash, what is the probability that it originated as a dash?

CHAPTER 3 Summary

✓ **CHECKLIST: KEY TERMS AND EXPRESSIONS**

- empirical concept of probability, page 149
- subjective probabilities, page 150
- set, page 151
- subset, page 151
- Venn diagram, page 151
- experiment, page 152
- sample space, page 152
- sample points, page 152
- finite sample space, page 152
- infinite sample space, page 153
- event, page 153
- simple event, page 153
- impossible event, page 153
- sure event, page 153
- probability of a simple event, page 156

- probability of an event, page 156
- equally likely outcomes, page 159
- mutually exclusive events, page 169
- complement of an event, page 172
- law of the complement, page 172
- odds, page 174
- tree diagram*, page 178
- permutation*, page 180
- n factorial, page 181
- combination*, page 182
- conditional probability, page 196
- independent events, page 199
- dependent events, page 199
- exhaustive events*, page 213
- theorem of total probability*, page 213
- Bayes' Formula*, page 213

---

*From optional Sections 3-4 and 3-6

KEY FORMULAS

probability of an event

$P(A)$ = sum of the probabilities assigned to the simple events that comprise event A, always

$$= \frac{\text{number of outcomes in } A}{\text{number of outcomes in } S}, \text{ if the outcomes are finite and equally likely}$$

law of the complement

$P(\text{not } A) = 1 - P(A)$

the addition law (with special cases)

$P(A \text{ or } B) = P(A) + P(B) - P(A \text{ and } B)$, always

$\qquad = P(A) + P(B)$, if A and B are mutually exclusive

$\qquad = P(A) + P(B) - P(A) \cdot P(B)$, if A and B are independent

conditional probability

$$P(A \mid B) = \frac{P(A \text{ and } B)}{P(B)}, \text{ provided } P(B) \neq 0$$

general multiplication law (with special cases)

$P(A \text{ and } B) = P(B) \cdot P(A \mid B)$, always

$\qquad = P(B) \cdot P(A)$, if A and B are independent

n factorial*

$n! = n(n - 1) \cdots 3 \cdot 2 \cdot 1$

$0! = 1$

number of permutations*

$$_n P_r = n(n - 1) \cdots (n - r + 1) = \frac{n!}{(n - r)!}$$

number of combinations*

$$\binom{n}{r} = \frac{n!}{r! \, (n - r)!}$$

theorem of total probability*

$P(A) = P(A \mid B_1) \cdot P(B_1) + \cdots + P(A \mid B_n) \cdot P(B_n)$

Bayes' formula*

$$P(B_i \mid A) = \frac{P(A \mid B_i) \cdot P(B_i)}{P(A \mid B_1) \cdot P(B_1) + \cdots + P(A \mid B_n) \cdot P(B_n)}$$

CHAPTER 3 Concept and Discussion Questions

1. State whether the concept of repeatability of the experiment is at the heart of the empirical notion of probability or subjective probability.

2. Based on the idea of repeated trials, provide an interpretation of the statement $P(\text{The light will be green when a car arrives at a certain intersection}) = 0.4$.

3. A company is manufacturing electric bulbs. How would you estimate the probability that a randomly selected bulb will last beyond 200 hours of life?

---

*From optional Sections 3-4 and 3-6.

4. How would you go about determining the probability that planes of a certain airline land on time?

5. Describe a situation and give two events that are mutually exclusive. Give an intuitive argument to show why the two events you have given cannot be independent.

6. Describe a situation and give two events that are independent. Show why the events cannot be mutually exclusive.

7. Consider the event "an individual is sent to jail when in fact she is not guilty." What is the complement of this event?

CHAPTER 3 Review Exercises

Note: Exercises with an asterisk (*) are solved using counting techniques from Section 3-4.

1. Give an empirical interpretation of the following probability statement: The probability that an electric bulb will last beyond 100 hours is 0.72.

2. What does it mean to say that an event has occurred?

3. Consider three fertilizers each with four levels of concentration. A farmer wishes to form mixes of the fertilizers by combining the fertilizers so that each mix has one and only one concentration level of each fertilizer. How many mixes are possible? Further, suppose the farmer wishes to test each of these mixes on five different soil types. How many different tests will he have to run?

4. Find $P(A \text{ and } B)$ if:
   (a) $P(A) = 0.65$ and $P(B \mid A) = 0.4$
   (b) $P(B) = 0.3$ and $P(A \mid B) = 0.6$
   (c) $P(A) = 0.4$, $P(B) = 0.6$, and the events are independent
   (d) $P(A) = 0.3$, $P(B) = 0.55$, and $P(A \text{ or } B) = 0.7$.

5. Find $P(A \text{ or } B)$ if:
   (a) $P(A) = 0.4$, $P(B) = 0.7$, and $P(A \text{ and } B) = 0.3$
   (b) $P(A) = 0.65$, $P(B) = 0.35$, and $P(B \mid A) = 0.4$
   (c) $P(A) = 0.4$, $P(B) = 0.3$, and $P(A \mid B) = 0.6$
   (d) $P(A) = 0.4$, $P(B) = 0.6$, and the events are independent.

6. If $P(A) = 0.4$, $P(B) = 0.3$, and $P(A \mid B) = 0.6$, find $P(B \mid A)$.

7. Suppose A and B are two mutually exculsive events and the event B is not an impossible event. Find $P(A \mid B)$.

8. In the following cases, state whether the events A and B are independent or mutually exclusive or neither.
   (a) $P(A) = 0.4$, $P(B) = 0.3$, and $P(A \text{ or } B) = 0.56$
   (b) $P(A) = 0.5$, $P(B) = 0.3$, and $P(A \text{ or } B) = 0.8$
   (c) $P(A) = 0.4$, $P(B) = 0.5$, and $P(A \text{ or } B) = 0.7$

9. The probability that the propulsion system of a missile functions properly is 0.90, and the probability that its guidance system functions properly is 0.75. If the two systems operate independently, what is the probability that a launch is successful?

10. Henry is sick in the hospital. The probability that his brother Bill will come to visit him on a certain day is 0.3, and the probability that his coach Ms. Sibley will come to visit him on that day is 0.12. If Bill and Ms. Sibley decide independently, what is the probability that:
(a) at least one of them will pay a visit on that day
(b) neither will pay a visit on that day?

*11. A box contains three 15-watts bulbs, four 30-watt bulbs, two 60-watt bulbs, and one 100-watt bulb. If three bulbs are picked at random, find the probability that there are:
(a) three 15-watt bulbs
(b) one 15-watt bulb, one 60-watt bulb, and one 100-watt bulb
(c) one 15-watt bulb and one 60-watt bulb
(d) no 30-watt bulbs.

12. From past information it is known that Machine *A* produces 10 percent defective tubes and Machine *B* produces 5 percent defective tubes. Two tubes are picked at random, one produced by Machine *A* and the other produced by Machine *B*. Find the probability that:
(a) both tubes are defective
(b) only the tube from Machine *A* is defective
(c) exactly one tube is defective
(d) neither tube is defective.

13. The probability that a driver on a freeway will be speeding is 0.3. The probability that a driver will be speeding and will be detected is 0.1. Find the probability that a driver will be detected, given that the driver is speeding.

14. A radar complex consists of two units operating independently. The probability that one of the units detects an incoming missile is 0.95, and the probability that the other unit detects it is 0.90. Find the probability that:
(a) both will detect a missile
(b) at least one will detect it
(c) neither will detect it.

*15. Suppose the National Aeronautics and Space Administration is interested in recruiting five schoolteachers in the astronaut training program and, by region, the following number of teachers have applied: East 10, West 12, North 6, South 8, and Midwest 5. If the five trainees are picked at random, what is the probability that each region is represented?

16. A high-tech company manufacturing floppy disks has a two-stage inspection plan. Each disk is first subjected to visual inspection and then is checked electronically. The probability that a disk is rejected in the first stage is 0.15. If it is accepted at this stage, the probability of rejecting it under electronic inspection is 0.12. What is the probability that a randomly picked floppy disk will pass the inspection plan?

17. George wants to bet on two horse races at the racetrack. Suppose the probability that he picks a winner in the first race is 0.3 and that he picks a winner in the second race is 0.4. Assuming independence, find the probability that George will pick:
(a) winners in both races   (b) a winner in the first race only
(c) exactly one winner   (d) no winner in any race.

18. As an extension of Exercise 17, suppose George wants to bet on three horse races, and the probabilities that he picks a winner in the first, second, and third race are, respectively, 0.3, 0.4, and 0.6. Assuming independence, find the probability that George will pick:

    (a) winners in all three races

    (b) a winner in the first race only

    (c) exactly one winner

    (d) winners in the first and third race only

    (e) no winner at all.

WORKING WITH LARGE DATA SETS

Project: State lotteries are fairly common these days and, as most of us are aware, the chances of winning are slim. Consider a MINI lottery of your own where three numbers are picked out of the ten numbers 0, 1, 2 . . . 9. List all the distinct "3-number-tickets" that are possible. Check to see if this agrees with the theoretical answer $\binom{10}{3} = 120$. What is the probability of winning the MINI lottery, that is , of picking the combination with all the winning numbers?

According to theory, there are $\binom{3}{2}\binom{7}{1} = 21$ distinct tickets where exactly two numbers will match with the numbers in the winning ticket, $\binom{3}{1}\binom{7}{2} = 63$ tickets with exactly one match, and $\binom{3}{0}\binom{7}{3} = 35$ tickets with no matching numbers. Go through your listing and verify that this is indeed the case if, for example, the winning numbers are 5, 6, and 8. Find the probability for each situation mentioned above. These are the theoretical probabilities.

Next, suppose 1000 people participate in the MINI lottery. Use the following program to simulate this participation.

```
MTB>STORE 'MINILOTO'

STOR>NOECHO

STOR>SAMPLE 3 FROM C1 AND PLACE IN C2

STOR>PRINT C2

STOR>END

MTB>SET C1

DATA>0:9

DATA>END

MTB>EXECUTE 'MINILOTO' 1000 TIMES
```

Now, ask your instructor to pick three numbers which will form the winning numbers. (You could also use the computer for this purpose with the command MTB>EXECUTE 'MINILOTO'). Analyze the simulated data to find the proportion of the ticket holders in the sample who picked all the winning numbers, the proportion of those with exactly two matches with the winning numbers, the proportion of those with just one match, and the proportion of those with no match. Compare with the theoretical answers and prepare a written report.

MINITAB

LAB 3:   Probability

As explained earlier in this chapter, the empirical concept of probability becomes the concept of relative frequency when the basic experiment is repeated a large number of times under identical conditions. The process of performing an experiment a large number of times is a tedious one. The technique of producing data that mimic the actual performance of an experiment is called *simulation*, a task well suited for computers.

Log on and access MINITAB.

1 The following instructions simulate twenty rolls of a fair die.

```
MTB >NOTE SIMULATE 20 ROLLS OF A DIE

MTB >RANDOM 20 DIGITS PUT IN C1;

SUBC>INTEGERS FROM 1 TO 6.

MTB >PRINT C1
```

The main command RANDOM tells how many trials are to be performed and where the results are to be stored. We have decided to store the results in Column C1. The semicolon at the end of the line changes the prompt to SUBC >. With the subcommand INTEGERS we specify the set of integers from which the selection is to be made. Because we want to simulate rolling a die we specify the set 1 to 6. The command is terminated by placing a period at the end of the subcommand line.

Identifying 0 as "tails" and 1 as "heads" we can simulate 10 tosses of a fair coin as follows.

```
MTB >NOTE 10 TOSSES OF A FAIR COIN, 1 = HEADS AND
     0 = TAILS

MTB >RANDOM 10 DIGITS PUT IN C5;

SUBC>INTEGERS FROM 0 TO 1.

MTB >PRINT C5
```

2 Next, we simulate births of 50 infants in a nursery and investigate what happens to the relative frequency of girls as the number of births increases. For this purpose we identify 1 with "girl" and 0 with "boy" and proceed as follows.

```
MTB >NOTE BIRTH OF 50 LIVE INFANTS 1 = GIRL, 0 = BOY

MTB >RANDOM 50 INTEGERS IN C2;

SUBC>INTEGERS FROM 0 TO 1.

MTB >NOTE NUMBER OF INFANTS BORN, RECORDED IN C1

MTB >SET C1

DATA>1:50

DATA>END OF DATA

MTB >NOTE CUMULATIVE NUMBER OF GIRLS, PLACED IN C3

MTB >PARSUM C2 PUT IN C3

MTB >NOTE RELATIVE FREQUENCY OF GIRLS, PLACED IN C4

MTB >LET C4 = C3/C1

MTB >NAME C1 'INFANTS' C2 'GIRL/BOY' C3 'GIRLS' C4
     'REL FREQ'

MTB >PRINT C1-C4

MTB >PLOT C4 VERSUS C1
```

Fifty births are first simulated in Column C2. Next, with the SET command, 50 consecutive integers are generated to correspond to the number of infants born. When the prompt DATA $>$ appears following the SET command we enter 1:50 for data, meaning that we want 50 consecutive integers starting with 1. (If we typed 4:8, for example, we would get the consecutive integers 4, 5, 6, 7, 8.)

The PARSUM command calculates the partial sums, that is, cumulative sums. Finally, the command PLOT gives a plot of the data with C4, the column mentioned first, as the y-axis and C1, the column mentioned next, as the x-axis.

Although there might be large fluctuations in the plot initially, see if the graph levels off with small fluctuations about a horizontal line. Invariably it will.

3 Will at least two people in a group of fifteen have the same birthday? It can be shown that the theoretical probability of this happening is 0.25. We can see how this answer compares with the empirical answer you get by simulating several groups of fifteen people. To carry out the simulation of picking 15 days out of 365 we assume that there are 365 days in a year and number them consecutively from 1 to 365. The following program picks 15 days from 365, at random and with replacement.

```
MTB >STORE 'BIRTHDAY'

STOR>NOECHO
```

```
STOR>NOTE 15 INTEGERS PICKED FROM 1 TO 365

STOR>RANDOM 15 INTEGERS IN C1;

STOR>INTEGERS FROM 1 TO 365.

STOR>SORT DATA IN C1 PUT IN C2

STOR>PRINT C2

STOR>END
```

STORE puts all the MINITAB commands that follow it into a computer file which we have chosen to name "BIRTHDAY." The commands are stored in the file for future execution. Upon receiving the STORE command, the prompt changes to STOR >. The command NOECHO suppresses the printing of any commands except the NOTE commands. The END command signals that the storing of commands is ended. The prompt then reverts back to MTB >.

The commands are stored in a file because we want to simulate the experiment several times. We can execute it 50 times, for instance, with the following instruction.

```
MTB>EXECUTE 'BIRTHDAY' 50 TIMES
```

Each time the program is executed, see whether two consecutive numbers are the same. In that case at least two people have the same birthday. In how many of the fifty groups did at least two have the same birthday? How does this fraction compare with the theoretical answer of 0.25? You stand to get a better approximation by pooling together your results with the rest of the class.

4   How does one pick a sample of a given size without replacement? The command for this purpose is SAMPLE. We must specify the column in which the source values are located and the column in which the sample values are to be placed. For example, the following program will pick 6 lotto numbers from 1 to 51.

```
MTB >NOTE ENTER INTEGERS 1 TO 51 IN COLUMN C1

MTB >SET C1

DATA>1:51

DATA>END

MTB >NOTE PICK THE SAMPLE OF SIX INTEGERS

MTB >SAMPLE 6 FROM C1 AND PLACE IN C2

MTB >PRINT C2
```

Assignment   Repeat the BIRTHDAY simulation for a group of twenty-three people. The theoretical answer in this case is 0.507.

5   Having completed the MINITAB session, type STOP after the prompt.

```
MTB>STOP
```

Now log off.

Commands you have learned in this session:

```
RANDOM;
```

```
INTEGERS.
```

```
PARSUM
```

```
PLOT
```

```
STORE
```

```
EXECUTE
```

```
SAMPLE
```

Discrete Probability Distributions

Fulcrum

INTRODUCTION

I t is not uncommon to hear someone claim to have "developed a system" for a game of chance. That is, given idealized conditions in such a game, this person has built a probability model that incorporates the factors that have a bearing on the particular game, such as, for example, blackjack. Based on this, the individual has developed a strategy that may optimize his or her winnings. In actually trying the system, the person will play the game many times, thus acquiring a mass of data. Using the techniques from Chapters 1 and 2, it would be possible to obtain the frequency distribution, the arithmetic mean, and the variance of the actual winnings. This is, of course, the empirical aspect. The person should also take into account the theoretical counterparts of these while building such a system. The theoretical counterpart of a frequency distribution is called the *probability distribution*, the main focus of this chapter and the next. This important concept is at the very heart of the theoretical study of populations.

4-1   RANDOM VARIABLES—THE DEFINITION

When an experiment is performed, our interest often is not so much in the outcomes of the experiment as in the numerical values assigned to these outcomes. For example, a gambler who will win \$5 if heads shows up and lose \$4 if tails shows up will think in terms of the numbers 5 and −4 rather than in terms of heads and tails. Technically this person is interested in the winnings and is assigning the value 5 to heads and −4 to tails. We refer to "the winnings of the gambler on a toss" as a *random variable*. Its value depends on which outcome occurs when a coin is tossed and, consequently, depends on chance. In general, we give the following definition of a random variable.

> A **random variable** assigns numerical values to the outcomes of a chance experiment. Therefore, mathematically a random variable is a function defined on the outcomes of the sample space.

It is customary to denote random variables by uppercase letters such as X, Y, and Z and their typical values by the corresponding lowercase letters x, y, and z. Thus, suppose we let X represent the 'gambler's winnings on a toss'. Then, instead of making the long-winded statement, "the winnings of the gambler on a toss are \$5," we would simply write $X = 5$. In general, $X = x$ will represent the statement that "the random variable X assumes a value x."

Also, if a and b are any given numbers, then, for example, $a < X \leq b$ will represent the statement that "the random variable X assumes a value greater than a and less than or equal to b." Thus, if X represents the height of a person in inches, then $68 < X \leq 72$ will mean that the height of the person is greater than 68 inches and less than or equal to 72 inches. Similar interpretations for $X \geq b$, $X \leq a$, and so on, are obvious.

EXAMPLE 1   Suppose three statellites are launched into space. Let X denote the number of satellites that go into orbit. Describe the random variable X.

SOLUTION   First we write down the outcomes of the sample space and then the numerical values of X. Writing s if a satellite attains an orbit successfully, and m if it does not, we have the following assignment, where the first row lists the outcomes of the sample space, and the second row lists corresponding numerical values assigned to X.

Sample Point	sss	ssm	sms	mss	mms	msm	smm	mmm
Number of satellites attaining orbit	3	2	2	2	1	1	1	0

The random variable assumes four distinct values: 0, 1, 2, and 3.   ●

EXAMPLE 2   Suppose a bowl contains five beads, of which three are white and two are black, and two beads are picked at random without replacement. If X represents the number of white beads in the sample, describe the random variable X.

SOLUTION   Identifying the white beads as w_1, w_2, and w_3 and the black beads as b_1 and b_2, the outcomes of the sample space and the corresponding values that X assumes are as follows.

Sample Point	w_1w_2	w_1w_3	w_2w_3	w_1b_1	w_1b_2	w_2b_1	w_2b_2	w_3b_1	w_3b_2	b_1b_2
Number of white beads	2	2	2	1	1	1	1	1	1	0

The random variable assigns the values 0, 1, and 2 to the outcomes of the sample space.   ●

EXAMPLE 3   Suppose a person throws a dart at a circular dart board of radius 1 foot. (We assume that the person is bound to hit somewhere on the board.) Let X denote the distance (in feet) by which the center of the board is missed. Describe the random variable X.

SOLUTION   In this case the sample space consists of all the points on the board and X measures the distance (in feet) from the point of impact to the center of the board. Thus X assumes any value x where $0 \leq x < 1$.   ●

Additional examples of random variables are the maximum temperature during a day, the life of an electric bulb, the amount of annual rainfall, the salary that a new college graduate makes, the number of cavities a child has, the number of bacteria in a culture, the number of telephone calls received in an office between 9:00 A.M. and 10:00 A.M., and so on.

A random variable is classified as discrete or continuous, depending upon the range of its values. (The range of a random variable is the set of values it can assume.) As will be recalled, the concept of continuous and discrete variables was introduced in Chapter 1.

For our intuitive grasp, a **discrete random variable** is one that has a definite gap between any two possible values in its range. Formally the number of values that such a random variable assumes can be counted, and there can be as many values as there are positive integers. Thus, we often say that a discrete random variable can assume at most a *countable infinite* number of values.

The salary that a new college graduate earns, the number of bacteria in a culture, the number of cavities that a child has, the price of gold on the exchange on a particular day—these are all examples of discrete random variables.

A **continuous random variable** is one that can assume any value over an interval or intervals.

The maximum daily temperature, the life of an electric bulb, the distance by which a sharpshooter misses a target, the height of a person, and the length of time a person has to wait at a bus stop provide examples of continuous random variables.

We shall study discrete random variables in this chapter and consider continuous variables in the next chapter.

Section 4-1 Exercises

1. A fair die is rolled. Describe the random variable X in each of the following cases.
   (a) A person wins as many dollars as the number that shows up and X represents the person's winnings.
   (b) The same as in Part a, but this time the person pays $1 for the privilege of playing and X represents the net winnings.
   (c) A person wins $4 if an even number shows up and loses $3 if an odd number shows up, and X represents the person's winnings.

In Exercises 2–15, drawing on your past experience, give the range, that is, the set of values that the given random variable assumes. Also, classify the random variables as discrete or continuous.

2. Burning time of an experimental rocket

3. Speed of a car on a freeway

4. Tensile strength of an alloy

5. Typing speed of a typist

6. Number of home runs scored in a baseball game

7. Voltage of a battery

8. Balance of payment during a fiscal year

9. Reaction time of a frog to a given electric stimulus

10. Time taken by a biological cell to divide into two

11. Number of accidents during a random day
12. Number of attempts a pool player has to make to sink a ball
13. Number of stoplights that are green at intersections when a car driver encounters six stoplights on a drive
14. The amount of chlorophyll in a rubber plant leaf
15. The number of votes a candidate gets in a local election with a voting population of 18,080.

4-2   DISCRETE RANDOM VARIABLES

We will consider here only discrete random variables that assume a finite number of values.

> The **probability distribution** of a discrete random variable, assuming a finite number of values, can be described by listing all the values that the random variable can assume, together with the corresponding probabilities. Such a listing is called the **probability function** of the random variable.

To understand a probability function, consider the random variable X described in Example 2 of Section 4-1. The random variable X represents the number of white beads in a sample of two picked at random from a bowl containing five beads, of which three are white and two are black. As we have seen, the random variable assumes values 0, 1, and 2. To find, for example, the probability that both beads picked are white, that is, $P(X = 2)$, we begin with

$$P(X = 2) = P(\{w_1w_2, w_1w_3, w_2w_3\})$$
$$= P(\{w_1w_2\}) + P(\{w_1w_3\}) + P(\{w_2w_3\}).$$

Now, since there are ten outcomes in the sample space, each sample point is assigned a probability of $1/10$. Therefore,

$$P(X = 2) = \frac{3}{10} = 0.3.$$

TABLE 4-1   The Probability Function of the Number of White Beads

Using counting techniques, this result can also be justified by noting that

$$P(X = 2) = \frac{\binom{3}{2}\binom{2}{0}}{\binom{5}{2}} = 0.3.$$

Number of White Beads	Probabilities
0	0.1
1	0.6
2	0.3
Sum	1.0

Similarly, it can be shown that $P(X = 1) = 6/10$ and $P(X = 0) = 1/10$. We can thus write the values that the random variable assumes and their respective probabilities, obtaining the probability function given in Table 4-1. Notice that the total of all the probabilities in Column 2 equals 1, which is a requirement that every probability function must satisfy.

A probability function can be given graphically by representing the values of the random variable along the horizontal axis and erecting ordinates to represent probabilities at these values, as is done in Figure 4-1.

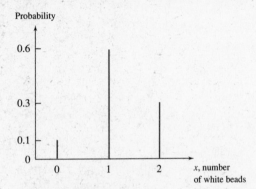

Figure 4-1 Probability Function of the Number of White Beads

The distribution given in Table 4-1 is a theoretical distribution. If, for instance, we actually perform the experiment 200 times—that is, if we repeat the experiment of picking two beads without replacement from the five beads (three white and two black) 200 times—we might get the empirical distribution given in Table 4-2.

TABLE 4-2   The Empirical Distribution When the Experiment Was Performed 200 Times

Number of White Beads	Actual Frequency	Relative Frequency
0	24	0.12
1	126	0.63
2	50	0.25

The agreement between the relative frequencies in Column 3 of Table 4-2 and the corresponding probabilities in Table 4-1 will improve if the experiment is repeated a greater number of times.

In general, suppose a random variable X assumes the values $x_1, x_2, \ldots, x_k$. We can represent the probability that X takes the value x_i by $p(x_i)$. That is,

$$P(X = x_i) = p(x_i)$$

The probability function can then be given in the form of a table, as in Table 4-3. By definition a probability function gives the probabilities with which the values are assumed by the random variable. Therefore,

TABLE 4-3   The Probability Function
of a Random Variable

Value x	Probability $p(x)$
x_1	$p(x_1)$
x_2	$p(x_2)$
$\vdots$	$\vdots$
x_k	$p(x_k)$
Sum	$\sum p(x_i) = 1$

$0 \le p(x_i) \le 1$. Also, a random variable has to assume one of its possible values. Hence, $\sum p(x_i) = 1$. We summarize these results as follows.

PROPERTIES OF A PROBABILITY FUNCTION

1. The probability that a random variable assumes a value x_i is always between 0 and 1. That is,

$$0 \le p(x_i) \le 1.$$

2. The sum of all the probabilities $p(x_i)$ is equal to 1. That is

$$\sum p(x_i) = 1.$$

EXAMPLE 1

The number of telephone calls received in the math office between 12:00 noon and 1:00 P.M. has the probability function given in Table 4-4.

(a) Verify that it is in fact a probability function.

(b) Find the probability that there will be three or more calls.

(c) Find the probability that there will be an even number of calls.

TABLE 4-4   The Probability Function of the Number of Telephone Calls

Number of Calls x	Probability $p(x)$
0	0.05
1	0.20
2	0.25
3	0.20
4	0.10
5	0.15
6	0.05

SOLUTION

(a) Since all the probabilities in Column 2 of Table 4-4 are between 0 and 1, and since their sum is 1, we have a genuine probability function.

(b) Writing X for the random variable "the number of telephone calls," we want $P(X \ge 3)$.

$$P(X \ge 3) = P(X = 3, \text{ or } X = 4, \text{ or } X = 5, \text{ or } X = 6)$$
$$= P(X = 3) + P(X = 4) + P(X = 5)$$
$$+ (X = 6) \text{ (Why?)}$$
$$= p(3) + p(4) + p(5) + p(6)$$
$$= 0.20 + 0.10 + 0.15 + 0.05$$
$$= 0.5$$

The probability statement $P(X \geq 3) = 0.5$ has the following interpretation. If we were to keep records of the calls coming in between 12:00 noon and 1:00 P.M. over many days, then on approximately 50 percent of the days, there will be at least 3 calls.

(c)   Here we want $P(X$ is even$)$, which can be written as

$$P(X \text{ is even}) = P(X = 0, \text{ or } X = 2, \text{ or } X = 4, \text{ or } X = 6)$$
$$= P(X = 0) + P(X = 2) + P(X = 4)$$
$$\quad + P(X = 6)$$
$$= p(0) + p(2) + p(4) + p(6)$$
$$= 0.05 + 0.25 + 0.10 + 0.05$$
$$= 0.45.$$

EXAMPLE 2   Mrs. Crandall has owned five stocks for a number of years. She has concluded during the years she has owned the stocks that the probabilities that exactly 1, 2, 3, 4, or 5 of her stocks go up in price during a trading session are, respectively, 0.18, 0.30, 0.20, 0.15, and 0.08. Let X represent the number of Mrs. Crandall's stocks that go up in price on a trading session.

(a)   Describe the probability function of X.

(b)   Find the probability that at most two stocks go up during a trading session.

SOLUTION

(a)   To describe the probability function we must give the values that the random variable assumes and the corresponding probabilities. We are given the probabilities $p(1) = 0.18, p(2) = 0.30, p(3) = 0.20, p(4) = 0.15$, and $p(5) = 0.08$. But because it is possible that none of the stocks will go up in price, there is a value that X assumes for which no probability is given. That value is 0. The probability $p(0)$ is found using

$$p(0) + p(1) + p(2) + p(3) + p(4) + p(5) = 1.$$

That is,

$$p(0) + 0.18 + 0.30 + 0.20 + 0.15 + 0.08 = 1.$$

Thus,

$$p(0) = 1 - 0.91 = 0.09.$$

The probability function is presented in Table 4-5.

(b)   The probability that at most two stocks go up in price can be written as $P(X \leq 2)$ and can be determined as follows.

$$P(X \leq 2) = P(X = 0, \text{ or } X = 1, \text{ or } X = 2)$$
$$= p(0) + p(1) + p(2)$$
$$= 0.09 + 0.18 + 0.30$$
$$= 0.57$$

TABLE 4-5   The Probability Function of the
Number of Mrs. Crandall's Stocks
Going Up in Price

Number of Stocks Going Up in Price x	Probability $p(x)$	
0	?	←0.09
1	0.18	
2	0.30	
3	0.20	
4	0.15	
5	0.08	
Sum	1.00	

If, over a large number of trading sessions, Mrs. Crandall keeps track of the number of her stocks that go up in each trading session, in approximately 57 percent of the sessions at most two of her stocks will go up.   ●

THE MEAN OF A DISCRETE RANDOM VARIABLE

The mean of the probability distribution of a discrete random variable, or the **mean of a discrete random variable,** is a number obtained by multiplying each possible value of the random variable by the corresponding probability and then adding these terms. Using the Greek letter μ and writing μ_x for the mean of a random variable X, or simply μ if the context is clear, in the case of a discrete random variable we have the following.

THE MEAN OF A DISCRETE RANDOM VARIABLE

The mean μ of a discrete random variable X is defined as

$$\mu = \sum x_i p(x_i)$$

It is common in the literature to refer to the mean μ as the *mathematical expectation* or the *expected value* of the random variable X, and to denote it as $E(X)$. (Read $E(\ )$ as "the expected value of ( )."

The probability distribution of a random variable is a theoretical model for the relative frequency distribution of the underlying population. Therefore, the mean of the random variable is also referred to as the **population mean.**

Motivation for the Definition of μ

What motivates the above definition of the theoretical mean? The clue is to be found in the formula in Section 2-4 for the mean of data given in the form of a frequency distribution. As you will recall, the formula for $\bar{x}$ is $\bar{x} = \dfrac{\sum f_i x_i}{n}$, which can be written as

$$\bar{x} = \sum x_i \frac{f_i}{n}$$

The quantity $\dfrac{f_i}{n}$ in the formula is the proportion of the values that are equal to x_i in the sample data set. When you are dealing with a population, the quantity analogous to the proportion of values equal to x_i in a sample is the proportion of values equal to x_i in the population. This population proportion is just $p(x_i)$, thus leading to the formula $\mu = \sum x_i p(x_i)$ for the theoretical mean.

EXAMPLE 3   Suppose the hourly earnings X of a self-employed landscape gardener are given by the following probability function.

Gardener's hourly earnings x (dollars)	0	6	12	16
Probability $p(x)$	0.3	0.2	0.3	0.2

Find the gardener's mean hourly earnings.

SOLUTION   Applying the formula, we have

$$\mu = 0(0.3) + 6(0.2) + 12(0.3) + 16(0.2) = 8.0$$

The gardener's mean hourly earnings are 8 dollars.   ●

Empirical Interpretation of $E(X)$

In Example 3, what does the number 8.0 mean in practical terms? Suppose we keep track of the gardener's earnings during N hours, and suppose he earns nothing during f_1 hours, \$6 during f_2 hours, \$12 during f_3 hours, and \$16 during f_4 hours (so that $f_1 + f_2 + f_3 + f_4 = N$) Then, using the definition in Chapter 2, the arithmetic mean of his *actual hourly earnings* will be

$$\frac{0 \cdot f_1 + 6 \cdot f_2 + 12 \cdot f_3 + 16 \cdot f_4}{N}$$

$$= 0\left(\frac{f_1}{N}\right) + 6\left(\frac{f_2}{N}\right) + 12\left(\frac{f_3}{N}\right) + 16\left(\frac{f_4}{N}\right).$$

Now the empirical notion of probability tells us that if N is very large, then f_1/N, f_2/N, f_3/N, and f_4/N should be close to the corresponding theoretical probabilities 0.3, 0.2, 0.3, and 0.2. That is,

$$0\left(\frac{f_1}{N}\right) + 6\left(\frac{f_2}{N}\right) + 12\left(\frac{f_3}{N}\right) + 16\left(\frac{f_4}{N}\right)$$

$$\approx 0(0.3) + 6(0.2) + 12(0.3) + 16(0.2)$$

where $\approx$ means is *approximately equal to*. But notice that the right-hand side of this approximation is simply the mean μ. Thus, the implication is that if we keep track of the gardener's earnings over a very long period of time, then the arithmetic mean of his actual earnings would be close to 8.0, the theoretical mean.

In general, the *empirical concept of expectation* entails repeating the experiment a large number of times and taking the arithmetic mean of all the resulting values. This computed mean $\bar{x}$, based on the data collected from many repetitions of the experiment, will be close to the theoretical mean μ.

A formal statement of this result is covered by the following law of large numbers in probability theory.

LAW OF LARGE NUMBERS

It is practically certain that the arithmetic mean of a large number of independent observations of a phenomenon will be close to the true mean of its distribution.

Physical Interpretation of Mean μ

The mean is a numerical measure of the center of a probability distribution. As such, it can be interpreted in a physical sense as the *center of gravity* of a probability distribution. (We saw a similar interpretation of the mean of a set of data in Chapter 2.) This can be seen best if we imagine weights located on a weightless seesaw at the various values that the random variable assumes. If the weights correspond to the probabilities with which these values are assumed, then a fulcrum located precisely at the mean will balance the seesaw. Figure 4-2 illustrates this interpretation with reference to Example 3.

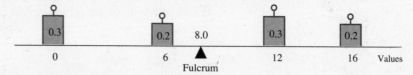

Figure 4-2 Mean, Viewed as the Center of Gravity

EXAMPLE 4

For the probability distribution of the number of telephone calls given in Example 1, find the mean number of telephone calls between 12:00 noon and 1:00 P.M.

SOLUTION   The distribution of the random variable and related computations are presented in Table 4-6. The sum in Column 3 gives the mean number of telephone calls as 2.75.

Interpretation   Suppose data are available regarding the number of telephone calls between 12:00 noon and 1:00 P.M. for many days, say 500 days.

TABLE 4-6   Computations to Find the Mean Number
of Telephone Calls

Number of Calls x	Probability $p(x)$	$x \cdot p(x)$
0	0.05	0.00
1	0.20	0.20
2	0.25	0.50
3	0.20	0.60
4	0.10	0.40
5	0.15	0.75
6	0.05	0.30
Sum	1.00	2.75

Then the mean number of telephone calls per day computed from these data will be approximately 2.75 calls, or the total number of calls will be approximately 500(2.75) = 1375.   ●

EXAMPLE 5

In a raffle the prizes include one first prize of $3000, five second prizes of $1000 each, and twenty third prizes of $100 each. In all, 10,000 tickets are sold at $1.50 each. What are the expected net winnings of a person who buys one ticket?

SOLUTION   We are interested in the expected net winnings. So let X represent the net winnings of the person. The possible values of X are $(3000 - 1.5)$, $(1000 - 1.5)$, $(100 - 1.5)$, and $(0 - 1.5)$; that is, 2998.5, 998.5, 98.5, and −1.5. The respective probabilities are 1/10,000, 5/10,000, 20/10,000, and 9974/10,000. (Why?) The probability function of X is given in Table 4-7. Applying the formula, the mean is given by

$$\mu = (2998.5)\frac{1}{10,000} + (998.5)\frac{5}{10,000}$$
$$+ (98.5)\frac{20}{10,000} + (-1.5)\frac{9974}{10,000}$$
$$= -0.5.$$

TABLE 4-7   Probability Function of
Net Winnings in a Raffle

Net Winnings x	Probability $p(x)$
2998.5	1/10,000
998.5	5/10,000
98.5	20/10,000
−1.5	9974/10,000

The interpretation of this result is that if a person were to participate in many raffles of this type, this person would lose approximately half a dollar per raffle.

THE VARIANCE AND STANDARD DEVIATION OF A DISCRETE RANDOM VARIABLE

The **standard deviation of a probability distribution** provides a numerical measure of the spread of the distribution and is defined as follows along with the companion quantity called the **variance of the distribution.**

VARIANCE AND STANDARD DEVIATION OF A DISCRETE RANDOM VARIABLE

The **variance** of a discrete random variable X is defined by

$$\text{Variance of } X = \sum (x_i - \mu)^2 p(x_i)$$

It is also called the *population variance* referring to the underlying population. Variance of X is often written as $\text{var}(X)$ and is denoted as the square of the lowercase Greek letter σ. Thus, $\text{var}(X) = \sigma^2$ where σ is taken to be positive.

The positive square root σ is called the **standard deviation** of X. So,

$$\sigma = \text{standard deviation of } X = \sqrt{\sum (x_i - \mu)^2 p(x_i)}$$

The standard deviation is often written as $sd(X)$.

In the formula above, the quantity $x_i - \mu$ represents the deviation of the value x_i from the mean μ. The variance is found by multiplying the squared deviation $(x_i - \mu)^2$ by the corresponding probability $p(x_i)$ and then adding the resulting terms. A large value of σ^2 will indicate that extreme values on either side of μ are assigned high probabilities, that is, large deviations from the mean are highly probable. Thus, the magnitude of σ^2 reflects the extent of spread about the mean μ. The larger the variance, the larger the spread.

The procedure for computing the standard deviation is shown in Example 6 and involves the following steps with reference to Table 4-8.

1. Compute the mean μ (sum of the values in Column 3).
2. Compute deviations from the mean (Column 4).
3. Find squares of the deviations (Column 5).
4. Multiply each squared deviation by the corresponding probability (Column 6).

5. Sum all the values obtained in Step 4. This quantity gives the variance (sum of values in Column 6).
6. Find the *positive* square root of the sum in Step 5 to obtain the standard deviation.

EXAMPLE 6

The distribution of the number of raisins in an oatmeal raisin cookie is as follows.

Number of raisins x	0	1	2	3	4	5
Probability p(x)	0.05	0.1	0.2	0.4	0.15	0.1

Find the mean, the variance, and the standard deviation of the number of raisins in a cookie.

SOLUTION The necessary computations are presented in Table 4-8. The sum in Column 3 gives the mean as 2.8, and the sum in Column 6 gives the variance as 1.56. The standard deviation is equal to $\sqrt{1.56}$, or about 1.249.

TABLE 4-8 Computations to Find the Mean μ and the Standard Deviation σ

Number of Raisins	Probability $p(x)$	$x \cdot p(x)$	$(x - \mu)$	$(x - \mu)^2$	$(x - \mu)^2 \cdot p(x)$
0	0.05	0.0	−2.8	7.84	0.392
1	0.10	0.1	−1.8	3.24	0.324
2	0.20	0.4	−0.8	0.64	0.128
3	0.40	1.2	0.2	0.04	0.016
4	0.15	0.6	1.2	1.44	0.216
5	0.10	0.5	2.2	4.84	0.484
Sum	1.0	$\sum x \cdot p(x)$ $= 2.8$			$\sum (x - \mu)^2 \cdot p(x)$ $= 1.56$

It is often possible to simplify the computations involved in finding σ by using the following version of the formula. The formula is also more suited for pocket calculators.

A SHORTCUT FORMULA FOR COMPUTING σ

$$\sigma = \sqrt{\sum x_i^2 p(x_i) - \mu^2}$$

We need only two columns of computations, one column of $x_i p(x_i)$ to find μ and another column of $x_i^2 p(x_i)$ to find $\Sigma \, x_i^2 p(x_i)$.

For example, using this formula, let us compute the variance of the number of raisins in an oatmeal raisin cookie whose distribution is described in Example 6. The necessary computations are shown in Table 4-9.

TABLE 4-9   Computations to Find σ Using the Shortcut Formula

Number of Raisins x	Probability $p(x)$	$x^2 \cdot p(x)$
0	0.05	0
1	0.10	0.1
2	0.20	0.8
3	0.40	3.6
4	0.15	2.4
5	0.10	2.5
Sum	1.0	$\Sigma \, x_i^2 \cdot p(x_i) = 9.4$

In Example 6 we found the mean μ to be 28. Therefore,

$$\sigma = \sqrt{\Sigma \, x_i^2 p(x_i) - \mu^2}$$
$$= \sqrt{9.4 - 2.8^2}$$
$$= \sqrt{1.56} = 1.249$$

This is the same answer that we found in Example 6.

EXAMPLE 7

Suppose X denotes the number that shows up when a fair die is rolled. Find the mean, the variance, and the standard deviation of X.

SOLUTION   The probability function of X and the necessary computations are shown in Table 4-10.

TABLE 4-10   Computations to Find the Mean and the Variance of the Number When a Die Is Rolled

x	$p(x)$	$x \cdot p(x)$	$x^2 \cdot p(x)$
1	1/6	1/6	1/6
2	1/6	2/6	4/6
3	1/6	3/6	9/6
4	1/6	4/6	16/6
5	1/6	5/6	25/6
6	1/6	6/6	36/6
Sum	1	21/6 = 3.5	91/6

From Column 3 of the table we find that the mean $\mu = 3.5$. Because

$$\sum x_i^2 p(x_i) = \frac{91}{6}$$

we get the variance as

$$\sigma^2 = \sum x_i^2 p(x_i) - \mu^2$$

$$= \frac{91}{6} - \left(\frac{21}{6}\right)^2$$

$$= \frac{35}{12} = 2.92.$$

Taking the square root, we get the standard deviation

$$\sigma = \sqrt{35/12} \approx 1.71.$$

Therefore, mean $= 3.5$, variance $= 2.92$, and standard deviation $= 1.71$ ●

There are several important discrete distributions. Two of these, the binomial and the hypergeometric, will be considered in detail in the next two sections. The *Poisson distribution*, another important distribution, is treated in Appendix II.

10/17/96

Section 4-2 Exercises

Note: Exercise with asterisk (*) is solved using counting techniques from Section 3-4.

1. A box contains three defective bolts and three nondefective bolts. Suppose three bolts are picked at random and X represents the number of defective bolts in the sample. Describe a suitable sample space, the value that X assumes at each of the sample points, and the corresponding probabilities.

2. Jimmy tosses an unbiased coin three times in succession. He wins $1 for heads on the first toss, $2 for heads on the second toss, and $3 for heads on the third toss. For each tail on a toss he loses $1.50. (For example, if he gets the sequence *HTH* he will win $1 − $1.50 + $3, or $2.50.) If X represents Jimmy's winnings, find the probability function of X.

*3. A group consists of fifteen Democrats and ten Republicans. A committee is formed by picking six people at random from the group. Let X represent the number of Republicans on the committee.
   (a) What does the event "$X = 2$" mean? Using the formula from the counting techniques Section in Chapter 3, find the probability $P(X = 2)$.
   (b) What does the event "$2 \le X < 5$" mean? Find $P(2 \le X < 5)$.
   (c) Express in terms of X the event "Republicans have a majority." Find the probability of this event.

4. A random variable X has the following probability function.

x	-15	5	10	25	30	40
$p(x)$	0.25	0.15	0.10	0.25	0.15	0.10

Find the following probabilities:

(a) $P(X \geq 15)$ (b) $P(5 \leq X < 28)$
(c) $P(X < 25)$ (d) $P(X \leq 5)$
(e) $P(X \geq 30)$ (f) $P(X \geq 8)$
(g) $P(5 < X < 30)$ (h) $P(5 \leq X \leq 30)$

5. The number of passengers in a car on a freeway has the following probability function.

Number of passengers x	1	2	3	4	5
$p(x)$	0.40	0.30	0.15	0.10	0.05

Find the following:

(a) The expected number of passengers in a car, and provide an interpretation of the value obtained
(b) The variance of the number of passengers in a car, and also the standard deviation.

6. Maria, a substitute teacher, lives 10 miles from Freshwater School, 15 miles from Sweetwater School, and 6 miles from Clearwater School. On any school day the probability that she goes to teach at Freshwater School is 0.4, at Sweetwater School is 0.3, and at Clearwater School is 0.2. Find the expected distance that Maria drives to school on a school day. (Assume there are only three schools in the area.)

7. Bruce pays $2 for the privilege of tossing a coin once. He will win $5 if heads shows up and will lose $3 if tails shows up. If the probability of heads on a toss is 0.6, find Bruce's expected net winnings. In practical terms, how would you intepret the expected winnings?

8. Terry, James, Sandra, and Billie weigh 200, 170, 150, and 120 pounds, respectively. A person in the group is picked at random. Find the following.

(a) The expected weight of the person
(b) The standard deviation of the weight.

9. For an experiment, prepare four chips marked 200, 170, 150, and 120. Pick a chip at random, record the number on it, and return it to the lot. Repeat the process forty times so that you have a set of forty values. Determine the following:

(a) A frequency distribution of the data collected
(b) The arithmetic mean, and compare it with the theoretical answer in Exercise 8a
(c) The standard deviation, and compare it with the theoretical answer in Exercise 8b.

10. A fair die is rolled. If an even number shows up, Dolores will win as many dollars as the number that shows up on the die. If an odd number shows up, she will lose $2.20. Find the following:

(a) The probability function of Dolores's winnings, giving the values of the random variable and the corresponding probabilities
(b) Dolores's expected winnings and the standard deviation of these winnings.

11. Conduct an experiment in which you actually play the game in Exercise 10 sixty times. Prepare a table like the following:

Game Number	Number Showing Up	Winnings	Cumulative Winnings
1	4	4	4
2	5	−2.2	1.8
⋮	⋮	⋮	⋮
60	·	·	·

(a) Find the arithmetic mean and the standard deviation of the winnings from the empirical data. Compare these values with the theoretical answers obtained in Exercise 10.

(b) From Part b of Exercise 10, Dolores's expected winnings on a single roll of die are $0.90. What should she expect her winnings to be for a total of 60 rolls? Compare this answer to the outcome of the experiment.

12. A building contractor pays $250 to bid on a contract. If he gets the contract, the probability of which is 0.2, he will make a net profit of $10,000. On the other hand, if he does not get the contract, the $250 is forfeited. Find the contractor's expected net profit on a bid. Interpret the result.

13. Suppose, in an eight-round match between two boxers, the number of rounds that the match lasts has the following distribution.

Number of rounds	1	2	3	4	5	6	7	8
Probability	0.05	0.15	0.20	0.20	0.15	0.10	0.10	0.05

Find the probability that a match will last:
(a) more than four rounds   (b) less than four rounds
(c) more than two rounds and less than or equal to five rounds.

14. For Exercise 13, find the following:
(a) The expected number of rounds that a match lasts
(b) The standard deviation of the number of rounds.

15. Suppose a random variable X assumes the values 1, 2, and 4. If $P(X = 1) = 0.3$ and the expected value of X is 2.7, find $P(X = 2)$ and $P(X = 4)$.

16. Each day a bakery bakes four fancy cakes at the cost of $8 each, and prices them at $25 each. (Any cake not sold at the end of the day is discarded.) The demand for cakes on any day has the following probability distribution.

Number of cakes sold x	0	1	2	3	4
Probability p(x)	0.1	0.2	0.5	0.1	0.1

(a) If the bakery sells two cakes on a day, find the profit on that day. Find the probability of making this profit.

(b) Continuing as in Part a, find the profits if the number of cakes sold on a day is 0, 1, 3, and 4. Find the corresponding probabilities.

(c) Use the probability function derived in Parts a and b to find the expected profit of the bakery when four cakes are baked on a day.

17. In Exercise 16, find the expected profit if instead of baking four cakes the bakery bakes:

(a) three cakes

(b) two cakes.

In view of the answers in Exercise 16 Part c and in Parts a and b of this exercise, how many cakes should the bakery bake in order to make it most profitable in the long run?

18. The number of traffic accidents in a certain town during a week is a random variable with the following probability function.

Number of accidents x	0	1	2	3
Probability p(x)	0.50	0.30	0.15	0.05

Find the probability that during a week there will be:

(a) no accident

(b) at least one accident

(c) at most one accident.

19. For Exercise 18, compute the following.

(a) The expected number of accidents per week and interpret the figure

(b) The standard deviation of the number of accidents during a week

20. Mr. Robbins wants to invest money in a venture where the probability that he earns $10,000 is 0.38, that he earns $8000 is 0.26, and that he loses $15,000 is 0.36. Describe the probability function of Mr. Robbins's earnings on his investment and find his expected earnings. Interpret the answer.

21. Suppose Mr. Costello plans to go to the horse races on the weekend and wants to bet on Esmeralda to win in the first race and Prince's Folly to win in the second race. The probability of Esmeralda's winning is 0.01, and $75 will be paid on a $1 ticket; the probability of Prince's Folly winning is 0.02, and the amount paid on a $1 ticket is $45. Find the following:

(a) The probability function of Mr. Costello's net winnings at the races on the weekend if he buys a $1 ticket in each race

(b) The expected net winnings

(c) The standard deviation of the net winnings

22. Extending Exercise 21, suppose Mr. Costello plans to go to the races on the weekend and wants to bet on Esmeralda to win in the first race, Prince's Folly to win in the second race, and Devil's Tail to win in the third race. The probability of each horse's winning and the amount paid on a $1 ticket are as follows.

Horse	Probability	Amount Paid
Esmeralda	0.090	10
Prince's Folly	0.020	45
Devil's Tail	0.013	75

If Mr. Costello buys three $1 tickets, one in each race, compute the following.
(a) The probability function of his net winnings
(b) The expected net winnings
(c) The standard deviation of the net winnings

23. Suppose you receive an invitation to enter a national magazine's $2,500,000 sweepstakes. The following table gives the amount that an entrant can win together with the corresponding probability.

Dollars Won x	Probability $p(x)$
20	1/3497
100	1/367081
500	1/1468321
1,000	1/7341601
5,000	1/24472001
10,000	1/73416001
100,000	1/73416001
2,000,000	1/73416001

(a) What is the probability that you will not win any amount?
(b) Find your expected winnings. In the long run would it be worthwhile for you to spend 32 cents for the postage stamp?

24. The distribution given below describes the amount of time (in hours) spent on homework by 9-year-olds during the year 1987–88. The figures are based on the data published by *Digest of Education Statistics* 1990, National Center for Education Statistics, page 114.

Amount of time spent (hrs)	none	0.5	1.5	2.5
Percent of students	33	47	13	7

(a) Find the expected amount of time spent by a 9-year-old during a day in 1987–88.
(b) Find the standard deviation of the amount of time spent.

25. Suppose, based on long-standing experience, an insurance company has determined that, during a year, the probability of a payout of $1,000,000 on a policy is 0.00001, the probability of a payout of $500,000 is 0.00005, the probability of a payout of $100,000 is 0.0005, and the probability of a payout of $8,000 is 0.006, with no payout on the remaining. If the annual premium is $162, is it profitable for the company to issue such policies? If the company sells 15,000 policies, approximately how much profit(loss) does the company expect to make?

26. Suppose you play the game of roulette described in Exercise 26 in Section 3-2 and bet one dollar on the red color. When the wheel is spun, if the ball lands in the red slot you will win one dollar. Otherwise, you forfeit your dollar. Let X be your net winnings on one bet. Describe the random variable and compute your expected winnings and standard deviation of your winnings on one bet. Interpret the value that you get for the expected winnings.

4-3   THE BINOMIAL DISTRIBUTION

THE PROBABILITY DISTRIBUTION

The starting point in discussing a **binomial distribution** is to consider a basic experiment in which the possible outcomes of the experiment can be classified into one of two categories, a **success** or a **failure.** For example, a tossed coin lands heads or tails, a basketball player either makes a free throw or does not make it, a new vaccine is effective or is not effective, a missile is fired successfully or is not fired successfully, and so on. The use of the word *success* does *not* carry the usual connotation of a desirable outcome. The basic experiment, often called a trial, is repeated a given number of times, for instance n times, thus giving n trials, which constitute the **binomial experiment.** In performing the trials, a tacit assumption is that the outcome of one trial does not influence the outcome of any other trial. Thus, we say that the trials are *independent*. This assumption, together with others characterizing a binomial distribution, can be stated as follows.

Characterization of the Binomial Distribution

1   The experiment called the binomial experiment consists of a fixed number n of independent and identically performed trials.

2   The performance of a trial results in an outcome that can be classified either as a *success* or a *failure.*

3   The probability of success is known and remains the same from trial to trial. Furthermore, if P (success) $= p$, then P (failure) $= 1 - p$. It is customary to let $1 - p = q$.

It is common in statistical literature to refer to each trial as a *Bernoulli trial,* in honor of the Swiss mathematician Jacob Bernoulli (1654–1705). The **binomial random variable** X is defined as follows.

$$X = \text{the number of successes in } n \text{ trials}$$

It is easily seen that the possible values of X consist of integers $0, 1, 2, \ldots,$ n. Thus, if X denotes the number of successful missile launches in ten attempts, then X could assume any value $0, 1, \ldots, 10$. Our objective will be to determine the probabilities with which these values are assumed. The values that X assumes together with the corresponding probabilities will give us the distribution of X. Before treating the general case, we will illustrate the approach through the following example.

Suppose a basketball player shoots at the basket four times. If he scores on any shot, we will denote the outcome as S and call it a success; if he misses, we will denote the outcome as F and call it a failure. Let

$X =$ the number of times the player makes the basket in four attempts.

Since the player can make 0 baskets, 1 basket, 2 baskets, 3 baskets, or 4 baskets, the possible values of X are 0, 1, 2, 3, and 4. Assuming that the probability of scoring on any shot is p, we want to find the probabilities with which the values 0, 1, 2, 3, and 4 are assumed. The sample space, which represents the results of four shots, consists of 16 or 2^4, outcomes, and these are listed in Table 4-11 according to the number of baskets made.

TABLE 4-11 Outcomes of the Sample Space according to the Number of Baskets Made in Four Shots

0 Baskets	1 Basket	2 Baskets	3 Baskets	4 Baskets
FFFF	SFFF	SSFF	SSSF	SSSS
	FSFF	SFSF	SSFS	
	FFSF	SFFS	SFSS	
	FFFS	FSSF	FSSS	
		FSFS		
		FFSS		

We will first find the probability of making one basket. From Table 4-11, the possible outcomes in this event are SFFF, FSFF, FFSF, and FFFS given in Column 2. Thus,

$$P(1 \text{ basket}) = P(\{SFFF\}) + P(\{FSFF\})$$
$$+ P(\{FFSF\}) + P(\{FFFS\}).$$

Now, SFFF stands for "score on the first shot" *and* "miss on the second shot" *and* "miss on the third shot" *and* "miss on the fourth shot." Since the trials are independent, we can therefore write

$$P(\{SFFF\}) = P\begin{pmatrix} \text{score on} \\ \text{the first} \\ \text{shot} \end{pmatrix} \cdot P\begin{pmatrix} \text{miss on} \\ \text{the second} \\ \text{shot} \end{pmatrix}$$
$$\cdot P\begin{pmatrix} \text{miss on} \\ \text{the third} \\ \text{shot} \end{pmatrix} \cdot P\begin{pmatrix} \text{miss on} \\ \text{the fourth} \\ \text{shot} \end{pmatrix}$$
$$= pqqq$$
$$= pq^3.$$

(It could be argued that the outcomes of different shots are not independent since the result of one shot could influence the player on the next shot. We shall assume that this is not the case.) As a matter of fact, a similar argument shows that the probabilities $P(\{FSFF\})$, $P(\{FFSF\})$, and $P(\{FFFS\})$ are all equal to pq^3. Therefore,

$$P(1 \text{ basket}) = 4\underbrace{pq^3}.$$

probability of each outcome
number of outcomes

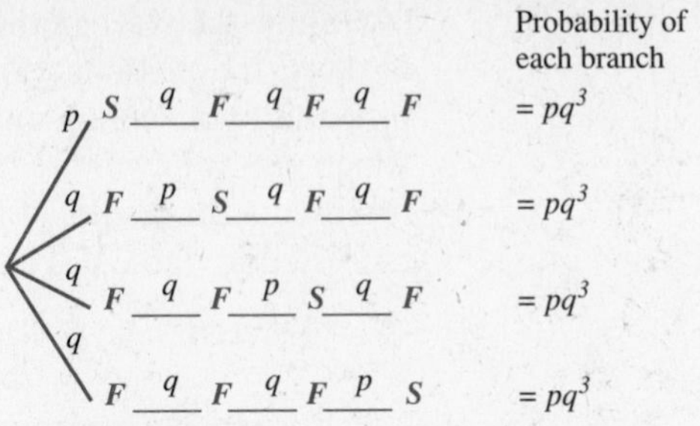

Figure 4-3 The four branches; each results in one success and 3 failures; each has probability pq^3.

TABLE 4-12 The Probability Distribution of the Number of Baskets Made in Four Attempts

Number of Baskets x	Probability of x Baskets
0	q^4
1	$4pq^3$
2	$6p^2q^2$
3	$4p^3q$
4	p^4

The result is also easy to see by considering the four branches, each one of which results in one success and three failures as shown in Figure 4-3 above.

Next, to find the probability of two baskets, we note from Table 4-11 that there are six outcomes describing the event, and each of these outcomes has the same probability, p^2q^2. For example,

$$P(\{SSFF\}) = ppqq = p^2q^2.$$

Thus,

$$P(2 \text{ baskets}) = 6\,p^2q^2.$$

probability of each outcome

number of outcomes

Proceeding in this way, it is now simple to construct Table 4-12, which gives the probability distribution of X.

*Turning to the general case of n trials, the sample space consists of 2^n outcomes. These outcomes are indicated in Table 4-13 according to the number of successes.

We will now find the probability of x successes in n trials, where, of course, x could be any integer 0, 1, 2, . . . , n. One of the ways in which we can have x successes is to have successes on the first x trials and failures on the remaining $n - x$ trials, as

$$SS \ldots SFF \ldots F.$$

x trials $(n - x)$ trials

---

*At this point, the reader who is not familiar with counting techniques can go directly to the paragraph following Example 2, where tables to find binomial probabilities are explained.

TABLE 4-13 Outcomes of the Sample Space When n Trials Are Performed, Arranged according to the Number of Successes

0 Successes	1 Success ...	x Successes ...	$(n-1)$ Successes	n Successes
$FF \ldots F$	$SFF \ldots F$ $FSF \ldots F$	$SS \ldots SF \ldots F$	$FSS \ldots S$ $SFS \ldots S$	$SS \ldots S$
		$SFS \ldots F \ldots SF$		
	$FF \ldots FS$		$SS \ldots SF$	

↑	↑	↑	↑	↑
1 outcome	n outcomes	$\binom{n}{x}$ outcomes in this listing. Each outcome has x successes and $n-x$ failures. Probability of each outcome is $p^x q^{n-x}$	n outcomes	1 outcome

Arguing as in our earlier illustration of the number of baskets made, the probability of this outcome is

$$P(\{SS \ldots SFF \ldots F\}) = \underbrace{pp \ldots p}_{\substack{\uparrow \\ x \text{ times}}}\underbrace{qq \ldots q}_{\substack{\uparrow \\ (n-x) \text{ times}}}$$

$$= p^x q^{n-x}.$$

Now there are $\binom{n}{x}$ outcomes, each having x successes and $n-x$ failures, as this is simply the number of ways of picking x trials out of n for successes to occur on them. Thus, we have the formula

$$P(x \text{ successes}) = \underbrace{\frac{n!}{x!\,(n-x)!}}\ \underbrace{p^x q^{n-x}}.$$

number of outcomes ⟶↑ with x successes and $(n-x)$ failures

↳ probability of any sample point with x successes and $(n-x)$ failures

In summary we have the following result.

BINOMIAL PROBABILITY DISTRIBUTION

In a binomial experiment with a constant probability p of success at each trial, the probability of x successes in n trials is given by

$$P(x \text{ successes}) = \frac{n!}{x! \, (n - x)!} p^x q^{n-x}$$

where x is any integer $0, 1, 2, \ldots, n$, and $q = 1 - p$.

The probability distribution can be presented as in Table 4-14.

TABLE 4-14 The Probability Distribution of the Number of Successes in n Trials

Number of Successes x	Probability
0	q^n
1	npq^{n-1}
⋮	⋮
x	$\dfrac{n!}{x!(n - x)!} p^x q^{n-x}$
⋮	⋮
n	p^n

EXAMPLE 1   The probability that a person who undergoes a kidney operation will recover is 0.6. Find the probability that of the six patients who undergo similar operations:

(a)   none will recover   (b)   all will recover

(c)   half will recover   (d)   at least half will recover.

SOLUTION   Since there are six patients, $n = 6$. If a patient recovers, we call the outcome a success. Thus $p = P$ (success) $= 0.6$ and $q = 1 - 0.6 = 0.4$. We therefore have the following probabilities.

(a) $P(\text{none recover}) = P(0 \text{ successes})$

$= (0.4)^6$ From Table 4-14 $n = 6$,

$= 0.0041$ $x = 0$ and $q = 0.4$.

(b) $P(\text{all recover}) = P(6 \text{ successes})$

$= (0.6)^6$ Use Table 4-14 with $n = 6$,

$= 0.0467$ $x = 6, p = 0.6$.

(c) $P(\text{half recover}) = P(3 \text{ successes})$

$$= \frac{6!}{3! \ 3!} (0.6)^3 (0.4)^3 \qquad \text{Use Table 4-14 with } n = 6, \\ x = 3.$$

$$= 20(0.6)^3 (0.4)^3$$

$$= 0.2765$$

(d) $P(\text{at least half recover}) = P(\text{at least 3 successes})$

Now, "at least 3 successes" means "either 3 successes *or* 4 successes *or* 5 successes *or* 6 successes." Therefore,

$$P(\text{at least half recover}) = P(3 \text{ successes}) + P(4 \text{ successes})$$
$$+ P(5 \text{ successes}) + P(6 \text{ successes})$$

$$= 0.2765 + \frac{6!}{4! \ 2!} (0.6)^4 (0.4)^2 + \frac{6!}{5! \ 1!} (0.6)^5 (0.4)^1$$

$$+ 0.0467 \qquad \text{Use the formula and the results of} \\ \text{Parts c and b.}$$

$$= 0.2765 + 15(0.0207) + 6(0.0311) + 0.0467$$

$$= 0.8203$$

EXAMPLE 2

A test consists of five questions, and to pass the test a student has to answer at least four questions correctly. Each question has three possible answers, of which only one is correct. If a student guesses on each question, what is the probability that the student will pass the test?

SOLUTION Let $p = P$ (correct answer). Because the student guesses on each question, we have $p = \frac{1}{3}$ and, consequently, $q = 1 - p = \frac{2}{3}$. With $n = 5$, we get

$$P(\text{student passes the test}) = P(\text{at least 4 correct answers})$$

$$= P(4 \text{ correct answers}) + P(5 \text{ correct answers})$$

$$= \frac{5!}{4! \ 1!} \left(\frac{1}{3}\right)^4 \left(\frac{2}{3}\right)^1 + \frac{5!}{5! \ 0!} \left(\frac{1}{3}\right)^5 \left(\frac{2}{3}\right)^0$$

$$= 0.0453.$$

The chance is less than 5 in 100 that the student will pass the test.

TABLES TO FIND BINOMIAL PROBABILITIES

It will be clear by now that the computations involving binomial probabilities can become very tedious. Extensive tables giving these probabilities for various values of n and p are available. Each pair of n and p gives a different binomial distribution. For ease of reference, let

$$b(x; n, p) = \text{probability of } x \text{ successes in } n \text{ trials, where} \\ \text{the probability of success on any trial is } p$$

In Appendix III, Table A-1, we give $b(x; n, p)$ for $n = 2, 3, \ldots, 15$ and various values of p. For example, to find $b(5; 12, 0.4)$, we go to the part of Table A-1 where $n = 12$ and look under the column headed by 0.4 corresponding to $x = 5$, obtaining the value 0.227. Thus, the probability of five successes in twelve trials when the probability of success on a trial is 0.4 is equal to 0.227.

SIDEBAR   *Still Another Look at Law and Probability*

With $p = 1/12,000,000$ as the probability that a randomly selected couple in Los Angeles would fit the description of the convicted couple, we now figure out the probability of at least two couples fitting the description. Let

X = number of couples among 2,000,000 fitting the description.

What we want is $P(X \geq 2)$, which is equal to $1 - P(X = 0) - P(X = 1)$. We have already established that $P(X = 0) = 0.846$. Now, using the binomial formula with $n = 2,000,000$ and $p = 1/12,000,000$, we get

$$P(X = 1) = \binom{2,000,000}{1}\left(\frac{1}{12,000,000}\right)^1\left(1 - \frac{1}{12,000,000}\right)^{1,999,999}$$

$$= 0.141$$

Hence, $P(X \geq 2) = 1 - 0.846 - 0.141 = 0.013$. This figure was used in Law and Probability Revisited on page 205.

EXAMPLE 3   Suppose that 80 percent of all families own a VCR. If ten families are interviewed at random, use the table of binomial probabilities (Table A-1) to find the probability that:

(a)   seven families own a VCR

(b)   at least seven families own a VCR

(c)   at most three families own a VCR.

SOLUTION   Let $p = P$(a family owns a VCR). Then $p = 0.8$.

(a)   Because $n = 10$, referring to the binomial table with $n = 10$, $p = 0.8$, and $x = 7$, we get

$$P(7 \text{ families own VCR}) = b(7; 10, 0.8)$$

$$= 0.201.$$

(b) "At least seven families own a VCR" means that either seven families *or* eight families *or* nine families *or* ten families own a VCR. Thus,

$$P\left(\begin{array}{c}\text{at least 7 families}\\ \text{own VCR}\end{array}\right) = b(7; 10, 0.8) + b(8; 10, 0.8)$$
$$+ b(9; 10, 0.8) + b(10; 10, 0.8)$$
$$= 0.201 + 0.302 + 0.268$$
$$+ 0.107$$
$$= 0.878.$$

(c) "At most three families own a VCR" means that either zero families *or* one family *or* two families *or* three families own a VCR. Thus,

$$P\left(\begin{array}{c}\text{at most 3 families}\\ \text{own VCR}\end{array}\right) = b(0; 10, 0.8) + b(1; 10, 0.8)$$
$$+ b(2; 10, 0.8) + b(3; 10, 0.8)$$
$$= 0 + 0 + 0 + 0.001$$
$$= 0.001. \qquad \bullet$$

EXAMPLE 4 A machine has fourteen identical components that function independently. It will stop working if three or more components fail. If the probability that a component fails is equal to 0.1, find the probability that the machine will be working.

SOLUTION The machine will be working if at most two components fail, which means that either zero components fail *or* one component fails *or* two components fail. Since $n = 14$ and $p = 0.1$, we get

$$P(\text{machine works}) = P(0 \text{ failures}) + P(1 \text{ failure}) + P(2 \text{ failures})$$
$$= b(0; 14, 0.1) + b(1; 14, 0.1) + b(2; 14, 0.1)$$
$$= 0.229 + 0.356 + 0.257$$
$$= 0.842. \qquad \bullet$$

EXAMPLE 5 *Risks accompanying decision making* A drug manufacturing company is debating whether a vaccine is safe enough to be marketed. The company claims that the vaccine is 90 percent effective; that is, when tried on a person, the chance for that person to develop immunity is 0.9. The federal drug agency, however, believes that the claim is exaggerated and that the drug is 40 percent effective. To test the company claim, the following procedure is devised: The vaccine will be tried on ten people. If eight or more people develop immunity, the company claim will be granted. Find the probability that:

(a) the company claim will be granted incorrectly, that is, the company claim will be granted when the federal drug agency is correct in its assertion

(b) the company claim will be denied incorrectly, that is, the company claim will be denied when the vaccine is 90 percent effective.

SOLUTION   Let X denote the number of people among ten who develop immunity. If the company claim is valid, then the distribution of X is binomial with $n = 10$, $p = 0.9$. On the other hand, if the federal agency assertion is valid, then X has a binomial distribution with $n = 10$, $p = 0.4$.

(a)   The company claim will be granted when eight or more people develop immunity. The probability that the claim will be granted *incorrectly* is equal to the probability that eight or more people develop immunity when $p = 0.4$. Thus the probability is obtained from Table A-1 as

$$b(8; 10, 0.4) + b(9; 10, 0.4) + b(10; 10, 0.4)$$
$$= 0.011 + 0.002 + 0$$
$$= 0.013.$$

(b)   The company claim will be denied when less than eight people develop immunity. The probability that the claim is denied *incorrectly* is equal to the probability that less than eight people develop immunity when $p = 0.9$. Thus the probability in question is

$$\sum_{k=0}^{7} b(k; 10, 0.9) = 0.001 + 0.011 + 0.057$$
$$= 0.069.$$

Note: Some of the probabilities in the table are 0.   ●

In Example 5, there are two competing possible errors, the error of licensing a vaccine when it is not effective versus not approving it when it is effective. The federal drug agency has to have a balance between the risks involved with the two kinds of errors. By being overcautious the consumer may be denied the benefits of an effective vaccine. On the other hand, an overeager approach to approve could expose the population to serious consequences. In Example 5, the chance of approving the vaccine erroneously is about 1 in 100, whereas the chance of not approving it when it is effective is about 7 in 100. The federal drug agency will have to weigh the seriousness of the consequences when considering approval of the vaccine.

SIDEBAR   Toxic Spill and Miscarriages

Consider the brief item on the next page concerning miscarriages due to toxic spill, which appeared through the AP wire services in the *Times Standard* of Eureka, California. The article says that the average rate of miscarriage is 10 to 20 percent, meaning that p, the probability that a pregnant woman has a miscarriage, is between 0.1 and 0.2.

Let us compute the probability that 5 or more miscarriages will occur among 11 pregnant women, for two values of p: $p = 0.2$, $p = 0.1$. That is, defining X as the number of women having miscarriages in a sample of 11 pregnant women, we will compute $P(X \geq 5)$ for $p = 0.2$ and $p = 0.1$. Assuming independence, X has a binomial distribution with 11 trials.

Miscarriage Rate Increased after Toxic Spill

Pregnant women living along the Sacramento River near Dunsmuir had a higher than usual miscarriage rate after a railroad pesticide spill, state officials say.

But they say they won't ever know if the spilled chemical caused the miscarriages because of the statistically small number of women involved.

A door-to-door survey found 11 women at the beginning of pregnancies when a Southern Pacific train derailed in July 1991. Five of the women later reportedly miscarried. The average rate is 10 to 20 percent.

"At this point we can't say the rate is not due to the spill, or that it is due to the spill," said state research scientist Amy Casey. "We are very concerned about the rate, we are just limited by the science."

Suppose that indeed the rate is 20 percent. In that case, from Appendix III, Table A-1, with $n = 11$ and $p = 0.2$, we get

$$P(X \geq 5) = 0.039 + 0.010 + 0.002 = 0.051.$$

If the national average is 0.20, the chances of 5 or more miscarriages in 11 pregnancies are about 5 in 100, which cannot be characterized as practically impossible. The pesticide company could argue successfully that the phenomenon is not too uncommon and that the findings have nothing to do with the spill.

Next, let us compute $P(X \geq 5)$ when $p = 0.1$. From Table A-1, using $n = 11$ and $p = 0.1$, we have

$$P(X \geq 5) = 0.002.$$

The chances of 5 or more miscarriages in 11 pregnancies are very small—2 in 1000—if $p = 0.1$. It seems very unlikely to have 5 or more miscarriages if the national average is ten percent. The women could argue that the spill affected them. Further investigation would seem most advisable.

THE MEAN AND THE STANDARD DEVIATION OF A BINOMIAL DISTRIBUTION

Because the binomial distribution is a discrete distribution, the mean and the standard deviation of the distribution can be found by following the methods presented in Section 4-2. The algebraic manipulations become cumbersome, so we will illustrate the method for $n = 3$ and, from the results obtained, inductively state the general formula for arbitrary n. The probability distribution and

Source: "Miscarriage Rate Increased after Toxic Spill" from the Associated Press, March 23, 1993. Reprinted by permission of the Associated Press.

TABLE 4-15   Binomial Probability Distribution, $n = 3$

Number of Successes x	Probability $p(x)$	$x \cdot p(x)$	$x^2 \cdot p(x)$
0	q^3	$0 \cdot q^3$	$0^2 \cdot q^3$
1	$3pq^2$	$1 \cdot 3pq^2$	$1^2 \cdot 3pq^2$
2	$3p^2q$	$2 \cdot 3p^2q$	$2^2 \cdot 3p^2q$
3	p^3	$3 \cdot p^3$	$3^2 \cdot p^3$

related computations are presented in Table 4-15. Following the general rule described in Section 4-2, the mean μ of the distribution can be found by multiplying each of the values 0, 1, 2, 3 by the corresponding probability and then adding the terms.

$$\begin{aligned} \mu &= 0(q^3) + 1(3pq^2) + 2(3p^2q) + 3(p^3) \\ &= 3p(q^2 + 2pq + p^2) \\ &= 3p(p + q)^2 \\ &= 3p \quad \text{since } p + q = 1. \end{aligned}$$

The preceding result suggests a formula for μ in the general case when there are n trials. The formula for μ reduces to np. This should not come as a surprise. Wouldn't we expect a basketball player to make twelve out of twenty free throws, if we knew from past experience that the player makes 60 percent of the free throws? Notice that $20(0.6) = 12$.

To find the variance σ^2, we employ the formula

$$\sigma^2 = \sum x_i^2 p(x_i) - \mu^2$$

where the x_i values are 0, 1, 2, 3. The quantity $\sum x_i^2 \, p(x_i)$ is the sum of the entries in Column 4 of Table 4-15. Thus,

$$\begin{aligned} \sum x_i^2 \, p(x_i) &= 0^2(q^3) + 1^2(3pq^2) + 2^2(3p^2q) + 3^2(p^3) \\ &= 3pq^2 + 12p^2q + 9p^3. \end{aligned}$$

Therefore

$$\begin{aligned} \sigma^2 &= 3pq^2 + 12p^2q + 9p^3 - (3p)^2 \quad \text{because } \mu = 3p \\ &= 3pq^2 + 12p^2q - 9p^2(1 - p) \\ &= 3pq^2 + 3p^2q, \text{ since } 1 - p = q \\ &= 3pq(p + q) \\ &= 3pq. \end{aligned}$$

It turns out that in the case of arbitrary n, the formula for the variance is given as npq.

In summary we have the following results.

> **MEAN AND STANDARD DEVIATION OF A BINOMIAL DISTRIBUTION**
>
> The mean μ and the standard deviation σ of a binomial distribution consisting of n trials and probability of success p are given as
>
> $$\mu = np \quad \text{and} \quad \sigma = \sqrt{np(1 - p)} = \sqrt{npq}$$

EXAMPLE 6   The probability that a freshman entering a university will graduate after a four-year program is 0.6. If 1000 freshmen enroll at a university, find the expected number of graduates. Also find the standard deviation of the number of freshmen who graduate.

SOLUTION   Because $n = 1000$ and $p = 0.6$, we have

$$\mu = 1000(0.6) = 600$$

and

$$\sigma = \sqrt{1000(0.6)(0.4)} = 15.49.$$

Thus, the expected number of students who graduate is 600 with a standard deviation of 15.49.

_10/17/96_

Section 4-3 Exercises

1. Suppose the probability that a U.S. citizen picked at random feels that the president of the United States is doing a satisfactory job is 0.4. Tim, John, Mary, Brice, Julie, Nancy, and Ernesto are picked at random and asked the question "Is the president doing a satisfactory job?" If they respond independently, find the probability that
   (a) they will answer, respectively, agree, disagree, agree, disagree, disagree, agree, disagree
   (b) exactly three of the seven interviewed will agree.

2. The probability that a player makes a free throw is 0.6. In four free throws, find the probability that the player makes:
   (a) exactly three free throws   (b) at least one free throw
   (c) at most three free throws   (d) an even number of free throws.

3. A person who claims to have powers of extrasensory perception (ESP) is shown six cards, each of which could be black or red with equal probability. If the person really does not have ESP and guesses on each card, what is the probability that all the cards will be identified correctly?

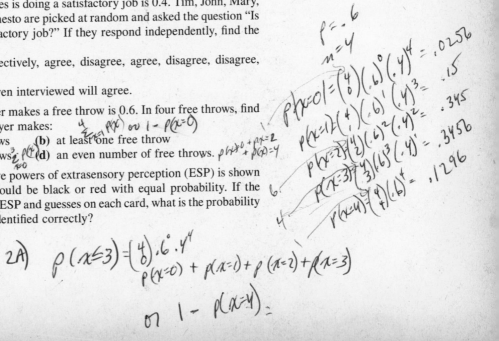

4. The results below are from a survey conducted by Yankelovich Partners Inc., in which 500 adult Americans were asked whether cloning of human embryos was a good thing. (Source: Study by Yankelovich Partners, Inc. from Time magazine, November 18, 1993. Copyright © 1993 Time Inc. Reprinted with permission.)

Do you think human cloning is a good thing?	
Yes	**14%**
No	**75%**

Suppose you conduct your own survey and interview 6 adult Americans selected at random. If the figures of the survey hold for the entire population, what is the probability that in your sample 1 will respond no?

5. Use the table of binomial probabilities (Table A-1) to find the following.
   (a) $b(3; 5, 0.4)$ (b) $b(8; 10, 0.7)$
   (c) $b(5; 12, 0.3)$ (d) $b(4; 8, 0.5)$

6. Use Table A-1 to find a value of p, the probability of success, if:
   (a) $b(5; 10, p) = 0.201$ (b) $(2; 15, p) = 0.267$
   (c) $b(8; 12, p) = 0.133$.

7. A machine manufactures a large number of bolts of a certain type. From past experience, it is known that if ten bolts are tested, the probability of finding four defective bolts is 0.111. Find the probability of finding six defective bolts among ten bolts tested, assuming that defects on bolts are independent.

   Hint: Use Table A-1 with the appropriate n to find p and then proceed.

8. Suppose the probability that a child is a son is 0.4. If there are eight children in a family, find the most probable number of sons and the corresponding probability.

   Hint: Use the appropriate binomial table and find the value x with the highest probability.

9. The Centers for Disease Control have determined that when a person is given a vaccine, the probability that the person will develop immunity to a virus is 0.8. If eight people are given this vaccine, find the probability that:
   (a) none will develop immunity
   (b) exactly four will develop immunity
   (c) all will develop immunity.

10. Suppose 10 percent of the tubes produced by a machine are defective. If six tubes from a huge lot are inspected at random, detemine the probability that:
    (a) three tubes are defective
    (b) at least two tubes are defective
    (c) at most five tubes are defective
    (d) at least three tubes are defective.

11. Suppose it can be assumed that 60 percent of adults favor capital punishment. Find the probability that on a jury of twelve members:
    (a) all will favor capital punishment
    (b) there will be a tie; that is, six members will favor capital punishment and six will oppose it.

12. *Quality control inspection*   In a factory that manufactures a large number of washers, the quality control department has adopted the following inspection plan for the production of the day: Ten washers are picked at random and inspected; if this sample contains less than three defective washers, the entire production is accepted as tolerable and shipped. What is the probability that the production of a day will be shipped when in fact it contains 50 percent defective washers?

13. A machine produces defective items at the rate of 1 in 10. If 100 items produced by this machine are inspected, what is the expected number of defective items? Also find the variance of the number of defective items.

$\mu = np = 100(\frac{1}{10}) = 10$
$\sigma^2 = npq = (100)(\frac{1}{10})(\frac{9}{10}) = 9$
$\sigma = \sqrt{9} = 3$

14. The probability that a door-to-door salesperson makes a sale when she visits a house is 0.2. If the salesperson visits sixty houses, find the following.
    (a) The expected number of houses where she makes sales
    (b) The variance and the standard deviation of the number of houses where she makes a sale

a) $60(.2) = 12$
b) $\sigma^2 = npq = (60)(.2)(.8) = 9.598$
$\sigma = \sqrt{9.598} = 3.098$

15. A large genetic pool contains 60 percent alleles of Type *A* and 40 percent of Type *a*. If fourteen alleles are picked at random, find the following.
    (a) The probability that there are exactly five alleles of Type *a*
    (b) The expected number of alleles of Type *A*
    (c) The variance of the number of alleles of Type *a*

a) $P(5) = \binom{14}{5}(0.4)^5(0.6)^9 = 0.207$
b) $(14)(.60) = 8.4$
c) $\sigma = \sqrt{(14)(0.4)(0.6)} = 1.833$ $\sigma^2 = 3.36$

16. A company received eight fluorescent tubes packed in a box. The probability that a tube is defective is 0.1. Assuming independence, find the following.
    (a) The probability that there are three defective tubes
    (b) The expected number of defective tubes
    (c) The standard deviation of the number of defective tubes

17. Suppose a company received a consignment of 200 boxes of fluorescent tubes, eight tube to a box. If the probability that a tube is defective is 0.1, how many of the boxes would you expect to have three defective tubes?

   Hint: See Part a in Exercise 16.

18. If the probability that a child is a son is 0.4, find the probability that in a family of four children there are:
    (a) two sons   (b) at least two sons   (c) all girls   (d) three girls.

19. Suppose the probability that a child is a son is 0.4. If 100 families, each with four children, are interviewed, find how many families you would expect to have; **(a)** two sons   **(b)** at least two sons   **(c)** all girls   **(d)** three girls.

    Hint: Refer to Exercise 18.

20. The following question and response to it appeared in *Parade,* the Sunday newspaper magazine, October 3, 1993, in the "Ask Marilyn" column by Marilyn Vos Savant. Justify the response of Ms. Vos Savant.

Ask Marilyn®

BY MARILYN VOS SAVANT

If you have four children, they may all be of one sex, there may be three of one sex and one of the other sex, or there may be two of each. Which is most likely?
—Adrian R. Beck, Joplin, Mo.

I know it sounds strange, but more families with four children have three of one sex and one of the other sex than any other combination. The chance of having all girls or all boys is 1 in 8; the chances of having two of each are 3 in 8; and the chances of having three girls and one boy (or three boys and one girl) are 4 in 8.

---

21. On the question of whether marijuana should be decriminalized, suppose the probability that a person favors decriminalization is 0.4, that a person opposes it is 0.5, and that a person has no opinion is 0.1. If six persons are interviewed and they respond independently, find the probability that:
    (a) five favor decriminalization
    (b) five oppose decriminalization
    (c) five have no opinion
    (d) more than four favor decriminalization
    (e) less than three oppose decriminalization.

22. The Republican majority in the House and Senate following the 1994 elections have declared an amendment to the Constitution to force a balanced federal budget as their top priority. In its January 23, 1995 issue, *Time* magazine stated that a public opinion poll recorded 80% in favor of the amendment. Suppose you interview nine individuals picked at random and they respond independently. Assuming the figure reported in *Time* to be a true reflection of the sentiment in the country, what is the probability that at least two-thirds of the nine individuals you interview will support the amendment?

23. A radar complex consists of eight units that operate independently. The probability that a unit detects an incoming missile is 0.90. Find the probability that an incoming missile will:
    (a) not be detected by any unit
    (b) be detected by at most four units.

---

24. A pediatrician reported that 30 percent of the children in the U.S. have cholesterol levels that should be considered above normal. If this is true, find the probability that in a sample of fourteen children tested for cholesterol more than six will have above-normal cholesterol levels.

25. The article "Nursing care of the patients with a pneumatic ventricular assist device" appeared in *Heart and Lung*, a journal of critical care nursing (July 1988, pp. 399–405). The article is followed by a multiple-choice exam consisting of ten questions. Each question has five options, of which only one is the correct response. Passing score is seven correct answers.
    (a) Suppose you don't read the article and merely guess randomly on each question. What are your chances of getting exactly seven correct answers? What is the probability that you pass the exam?
    (b) Suppose you read the article and are certain that five of your answers are correct. You are not sure about the other five questions, so you randomly guess on these questions. What are your chances of passing the exam?

26. A 1986 Gallup poll reported that 52 percent of American teenagers believe in the pseudoscience of astrology. Suppose this figure is true for the general teenage population. If a random sample of 6 teenagers is interviewed and they all answer independently, find the probability that a majority in the sample will believe in astrology.*

27. Consider the game of roulette described in Exercise 26 in Section 4-2. Suppose you make four successive bets on the red color.*
    (a) What is the probability that you win three times?
    (b) What is the probability that you win net
        i. two dollars
        ii. zero dollars
        iii. three dollars?

28. In the California weekly lottery, which involves picking 6 numbers without replacement from the numbers 1 to 51, the probability of picking 6 winning numbers is 1/18,009,460. Suppose 20 million individuals participate during a week. Assuming independence, find the following and comment.
    (a) Probability that there is no winner
    (b) Probability that there is at least one winner
    (c) Expected number of winners.

29. (*Investigative mathematics*) Consider the following values of n and p.
    (a) $n = 6, p = 0.3$ (b) $n = 15, p = 0.95$
    (c) $n = 11, p = 0.4$ (d) $n = 13, p = 0.05$

    In each case, compute the expected value and, from Table A-1, find the most probable number. Do you see any connection between the expected number of successes and the most probable number? Make a general statement.

---

*Hint: Use the binomial formula on page 250 to compute the probabilities.

Suppose a finite population consists of two kinds of items. When a random sample of items is picked *with replacement* from such a population, that is, an item is returned to the lot before picking the next one, all the conditions of the binomial experiment are satisfied. However, if the sampling is carried out *without replacement,* then the condition of independence of trials is not met.

Specifically, suppose there are N items in a lot of which d are defective and, consequently, $(N - d)$ are nondefective. Let X represent the number of defective items picked when a random sample of n items is drawn from the lot. Thus, "getting a defective on a draw" is classified as success and the number of trials is n.

If the sample is picked *with replacement,* then the n trials are independent and the probability of picking a defective item on any draw is d/N. That is, the probability of success is d/N. All the conditions necessary for applying the binomial distribution are satisfied and we have

$$P(X = x) = \binom{n}{x}\left(\frac{d}{N}\right)^x\left(1 - \frac{d}{N}\right)^{n-x}, \qquad x = 0, 1, \ldots, n$$

However, if the sampling is carried out *without replacement,* then the condition of independence is not met. For example, if the first item drawn is defective, then the probability of obtaining a defective item on the second draw is $(d - 1)/(N - 1)$. On the other hand, if the first item is nondefective, then the probability of a defective on the second draw is $d/(N - 1)$. In this situation, X, the number of defective items, has a distribution called the **hypergeometric distribution** and it is derived using the following argument developed in Section 3-4. If the sample is to contain x defective items, then we have the following setup.

$$
\begin{array}{ccc}
& N \text{ items} & \\
\swarrow & & \searrow \\
\text{defective} & & \text{nondefective} \\
d & & N - d \qquad \leftarrow \text{composition of lot} \\
\downarrow & & \downarrow \\
x & & n - x \qquad \leftarrow \text{composition of selection}
\end{array}
$$

where, naturally, $x \le d$ and $(n - x) \le (N - d)$. There are $\binom{d}{x}$ ways of picking x defective items out of d and $\binom{N - d}{n - x}$ ways of picking $(n - x)$ non-defectives out of $(N - d)$. Since there are $\binom{N}{n}$ possible samples of size n from a lot containing N items, we get

$$P(X = x) = \frac{\binom{d}{x}\binom{N - d}{n - x}}{\binom{N}{n}}, \qquad x = 0, 1, \ldots, n$$

If the lot size N is small and the items are picked without replacement, then the proper distribution to use is the hypergeometric distribution. But if the lot size is very large compared to the sample size n, then the conditions necessary for the binomial distribution are satisfied, approximately. For all practical purposes we have independence of trials and we can solve the problem as if it were a binomial experiment with probability of success on any draw equal to d/N. This is easy to see because, since the lot is large and we pick only a few items, its composition is not affected much as we remove items from it one by one.

It can be shown that the mean μ and the standard deviation σ for the hypergeometric distribution are given by

$$\mu = n\frac{d}{N} \quad \text{and} \quad \sigma = \sqrt{n\frac{d}{N}\left(1 - \frac{d}{N}\right)\left(\frac{N - n}{N - 1}\right)}$$

The term $(N - n)/(N - 1) \approx 1$ if N is large compared to n. In that case, we get the result

$$\sigma \approx \sqrt{n\left(\frac{d}{N}\right)\left(1 - \frac{d}{N}\right)} \qquad \text{if } N \text{ is large compared to } n$$

This result coincides with the standard deviation of the binomial distribution with $p = d/N$.

HYPERGEOMETRIC DISTRIBUTION

Suppose a sample of n items is picked *without replacememt* from a lot containing N items with d items of Type I and $N - d$ of Type II. Then, X, the number of items of Type I in the sample, has a distribution called the **hypergeometric distribution** given by

$$P(X = x) = \frac{\binom{d}{x}\binom{N - d}{n - x}}{\binom{N}{n}}, \qquad x = 0, 1, 2, \ldots, n$$

where $x \leq d$ and $(n - x) \leq (N - d)$. The mean μ and the standard deviation σ of X are

$$\mu = n\frac{d}{N} \quad \text{and} \quad \sigma = \sqrt{n\frac{d}{N}\left(1 - \frac{d}{N}\right)\left(\frac{N - n}{N - 1}\right)}$$

If N is large and n is relatively small, then

$$\sigma \approx \sqrt{n\frac{d}{N}\left(1 - \frac{d}{N}\right)}$$

EXAMPLE 1   A genetic pool contains 80 alleles of which 32 are of Type *A* and 48 are of Type *a*. Suppose 15 alleles are picked at random. Find the probability of getting four *A* alleles, the mean number of *A* alleles, and the standard deviation of the number of *A* alleles if the sample is picked

(a)   with replacement

(b)   without replacement.

SOLUTION   We will identify success with "picking an *A* allele" and let

$$X = \text{number of } A \text{ alleles in the sample.}$$

(a)   If the sampling is with replacement, we have a binomial distribution with 15 trials and probability of success $p = 32/80 = 0.4$. Hence,

$$P(X = 4) = b(4; 15, 0.4) = 0.127$$
$$\mu = np = 15(0.4) = 6$$
$$\sigma = \sqrt{np(1 - p)} = \sqrt{15(0.4)(0.6)} = 1.9$$

(b)   If the sampling is without replacement, we have a hypergeometric distribution and so

$$P(X = 4) = \frac{\binom{32}{4}\binom{48}{11}}{\binom{80}{15}} = 0.122$$

Observe that this answer is not much different from the one obtained in Part a above using the binomial formula. The reason is that $N (= 80)$, compared to $n(= 15)$, is fairly large.

$$\mu = 15(0.4) = 6$$
$$\sigma = \sqrt{15\left(\frac{32}{80}\right)\left(1 - \frac{32}{80}\right)\left(\frac{80 - 15}{80 - 1}\right)} = 1.72 \qquad \bullet$$

Section 4-4 Exercises

1. Suppose a box contains 100,000 beads, of which 30,000 are black and 70,000 are red. If ten beads are picked at random from this box, find the probability that there are:

(a) four black beads   (b) no black beads

(c) less than two red beads   (d) at least three red beads.

2. Solve Exercise 1 using the binomial approximation and compare these answers with those obtained using the hypergeometric distribution.

3. A pond contains 60 fish. Ten fish are caught, tagged, and relased in the same pond. After the tagged fish have mixed well among the rest of the fish, 6 fish are caught.
   (a) Find the probability distribution of X, the number of tagged fish among the 6 fish caught the second time.
   (b) Use the probability distribution in Part a to compute the mean number of tagged fish and the standard deviation.
   (c) Compute the mean number and the standard deviation using the formulas in this section. Verify that the answers in Part b agree with these answers.

4. Work out Exercise 3 using the binomial approximation.

5. An enterprising graduate student assembles PCs and sells them at a discount. He receives a shipment of 10 microprocessors every month for his computers and has no time to inspect them all to see if they are satisfactory. So, he has the following inspection plan. He inspects 3 microprocessors, and if all of them are satisfactory he keeps the shipment; otherwise, he returns it. What is the probability that he will accept a shipment if, in fact,
   (a) there are 6 defectives in the shipment
   (b) there is 1 defective in the shipment?

6. Suppose the final test for a statistics course is based on 18 different topics and the instructor announces in the class that the exam will have 6 questions from 6 different topics. A student decides to study only 12 topics.
   (a) What is the probability that, from the topics the student studies,
      i. no question is asked
      ii. two questions are asked
      iii. all six questions are asked?
   (b) What is the expected number of questions asked on the test from the topics the student studies? Interpret the answer.

7. A computer store received a shipment of 72 microchips. Suppose twenty microchips are picked at random and inspected. The shipment will be kept only if there are fewer than two defective microchips in the sample inspected. What is the probability that the shipment will be returned when in fact the shipment contains only 2 defective chips?

8. An investor has stocks of 18 different companies in her portfolio. Of these companies, 13 are industrial and 5 are utility. She decides to sell stocks of six of these companies. If she picks the companies at random, what is the probability that there will be equal numbers of industrial and utility companies in the selection?

9. A mail-order house received an order for 5 cameras and has 20 cameras in stock. Three of these cameras have a minor defect. If the salesperson picks 5 cameras at random and ships them, what is the probability that the shipment will include all the cameras with the minor defects?

CHAPTER 4 Summary

✓ *CHECKLIST: KEY TERMS AND EXPRESSIONS*

- ❑ random variable, page 227
- ❑ discrete random variable, page 229
- ❑ continuous random variable, page 229
- ❑ probability distribution, page 230
- ❑ probability function, page 230
- ❑ mean of a discrete random variable (or its distribution), page 234

- ❑ population mean, page 234
- ❑ standard deviation of a probability distribution, page 238
- ❑ variance of the distribution, page 238
- ❑ binomial distribution, page 246

- ❑ success, page 246
- ❑ failure, page 246
- ❑ binomial experiment, page 246
- ❑ binomial random variable, page 246
- ❑ hypergeometric distribution, page 262

KEY FORMULAS

mean of a discrete random variable

$$\mu = \sum x_i p(x_i)$$

standard deviation of a discrete random variable

$$\sigma = \sqrt{\sum (x_i - \mu)^2 p(x_i)} = \sqrt{\sum x_i^2 p(x_i) - \mu^2}$$

binomial probabilities

$$P(x \text{ successes}) = \frac{n!}{x!(n-x)!} p^x (1-p)^{n-x} \qquad x = 0, 1, \ldots, n.$$

binomial mean

$$\mu = np.$$

binomial standard deviation

$$\sigma = \sqrt{np(1-p)}.$$

hypergeometric probabilities

$$P(X = x) = \frac{\binom{d}{x}\binom{N-d}{n-x}}{\binom{N}{n}}, \qquad x = 0, 1, \ldots, n$$

hypergeometric mean

$$\mu = n\frac{d}{N}$$

hypergeometric standard deviation

$$\sigma = \sqrt{n\frac{d}{N}\left(1 - \frac{d}{N}\right)\left(\frac{N-n}{N-1}\right)}$$

CHAPTER **4** **Concept and Discussion Questions**

1. Explain the connection between the frequency distribution of the data on a variable and the probability distribution of the corresponding random variable.

2. What features characterize the binomial distribution?

3. How many parameters does a binomial distribution with a given number of trials have?

4. Why is independence so important in connection with the binomial distribution?

5. Suppose ten independent trials are performed with probability of success p on each trial. Explain intuitively why the probability of 3 successes is the same as the probability of 7 failures.

6. Describe clearly an experiment or a situation and give an example of a random variable that has a binomial distribution.

7. In a large school district 68 percent of the teachers are women. Suppose 12 teachers are selected at random. Which of the following are binomial random variables?
   (a) Genders of the 12 teachers
   (b) 8 women out of 12
   (c) Number of women teachers among the 12 teachers
   (d) Number of male and female teachers
   (e) Number of male teachers among the 12 teachers.

8. Compare and contrast the binomial and the hypergeometric distributions.

CHAPTER **4** **Review Exercises**

Note: Exercise with asterisk (*) is solved using counting techniques from Section 3-4.

1. Give examples of two random variables that are discrete and two that are continuous.

*2. Four defective transistors are mixed with five good transistors. Suppose three transistors are picked at random without replacement. If X represents the number of defective transistors, find the values X assumes together with the corresponding probabilities.

3. The number of cars that a salesperson sells on any day has the following probability function.

Number of cars	0	1	2	3	4	5
Probability	0.3	0.2	0.3	0.1	0.08	0.02

If the salesperson earns $80 for every car he sells, find the following:
(a) The probability that on any day he earns
     i. more than $260
     ii. $240
     iii. between $120 and $280
     iv. $200
     v. less than $280
     vi. no money
(b) The expected earnings of the salesperson on any day
(c) The variance of the earnings of the salesperson on any day

4. A lottery is set up with a first prize of $500 and a second prize of $100. If a total of 2000 tickets are sold, what should the price of a ticket be so that the lottery can be considered fair? (A game is considered fair if the expectation of net winnings is zero.)

5. A lottery is set up with one first prize of $5000, three second prizes of $2000, and ten third prizes of $1000 each. If 10,000 tickets are sold in all, what should the price of a ticket be so that the lottery can be considered fair?

$\mu = (5000-c)(1/1000) + (2000-c)(3/10,000)$
$+ (1000-c)(10/10,000) + (0-c)(\frac{9986}{10,000}) = 0$

$\mu = \frac{1}{2} * \frac{3}{5} + 1 - c = 0$
$c = \$2.10$

6. Mrs. Wilson sells vacuum cleaners. The following table includes information about the number of sales she makes during a day, with the corresponding probabilities.

Number of vacuum cleaners	0	1	2	3
Probability	0.05	0.15	0.30	0.40

Mrs. Wilson's commission is $100, $75, $55, $30, $0, respectively, if she sells 4 or more, 3, 2, 1, 0 vacuum cleaners. Find the expected amount of Mrs. Wilson's commission during a day.

7. When two fruit flies are crossed, the probabilities that there are 100, 110, 120, 140, or 150 flies in the progeny are, respectively, 0.15, 0.15, 0.20, 0.30, and 0.20. Find the expected number of flies in a cross and the variance of the number of flies.

8. Mr. Roberts bought 100 shares of a certain stock at $25 each. He has followed the stock's price fluctuations closely over a long period of time and has determined that the price of the stock on any day is $20, $22, $25, $32, $36 with respective probabilities 0.1, 0.35, 0.3, 0.15, 0.1. If Mr. Roberts sells the stock on a random day, what is his expected profit?

9. A billiards player is trying a trick shot. He has decided to make at most five attempts and will quit the first time he sinks the ball. He will also quit if he is not successful by the fifth attempt. The following table gives the probability distribution of the number of attempts when the player quits.

Number of Attempts	Probability
1	0.6000
2	0.2400
3	0.0960
4	0.0384
5	0.0256

(a) Find the probability that the player will quit in, at most, two attempts.

(b) Calculate the expected number of attempts when the player quits.

(c) Find the standard deviation of the number of attempts when the player quits.

10. Suppose you want to invest $20,000 in a venture capital fund. The possible returns on your investment are $50,000, $70,000, or $100,000 with respective probabilities 0.4, 0.15, and 0.05. The only other possibility is that you might lose all your investment.

(a) Prepare the probability function of the returns on your investment.

(b) What is the expected return on your investment?

(c) Prepare the probability function of the profit on your investment.

(d) What is the expected profit on your investment?

(e) Compare the answers in Parts b and d and comment.

11. As stated in Exercise 36 in Section 3-5, carriers of the sickle cell trait who do not show any symptoms of the disease have one A gene and one S gene. Such individuals are considered healthy. (The S gene is the culprit). Individuals who show symptoms of the disease have two S genes. Normal blood cells have two A genes. For parents who are carriers of the trait without any symptoms, the chance that their child will be a non-carrier is 1 in 4, the chance of their child being a carrier without the symptoms is 1 in 2, and the chance of their child showing the symptoms is 1 in 4. If 3 children are born to carrier parents, find the probability that

(a) all the children are normal, that is, are non-carriers

(b) all the children are healthy, that is, do not show any symptoms

(c) there is one child who shows the symptoms.

Also, interpret the results in parts a, b, and c above.

12. Ernesto is interested in the results of four baseball games (against the L. A. Dodgers, St. Louis Cardinals, Atlanta Braves, and New York Mets) that his home team, the San Francisco Giants, is scheduled to play. Suppose the following are the probabilities of his team winning: 0.6 against the L.A. Dodgers, 0.3 against the St. Louis Cardinals, 0.7 against the Atlanta Braves, and 0.5 against the New York Mets. What is the probability that Ernesto's team will win exactly three of the games? (You may assume that the outcome of one game does not influence that of the other.)

13. As in Exercise 12, Ernesto is interested in the results of the four games. However, this time suppose the probability of his team winning against any team is the same and equal to 0.6. What is the probability that Ernesto's team will win exactly three of the games? Once again, assume independence.

14. From past experience, an automobile dealer has determined that 20 percent of his customers buy compact cars. Find the probability that of ten customers who buy an automobile (independently):

(a) three will buy a compact

(b) at most three will buy a compact

(c) at least three will buy a compact.

15. Refer to Exercise 11 on page 269. Remember that a healthy child shows no symptoms of sickle cell anemia and could be a non-carrier, or could have one *S* gene. The expected number of healthy children in a family with 3 children born to carrier parents without the symptoms is 2.25. (Justify the answer.) For a carrier couple without the symptoms planning to have 3 children, which of the following interpretations are the correct ones?

 (a) They will have 2.25 healthy children
 (b) The chance that they will have a child with symptoms is 75 percent
 (c) They will have two healthy children and one child with the symptoms
 (d) They will have at least one child with the symptoms
 (e) It is impossible for them to have all children normal, that is, non-carriers
 (f) In a survey of a large number of families with 3 children born to carrier parents without the symptoms, the mean number of healthy children, will be approximately 2.25
 (g) In a survey of 1000 families with 3 children born to carrier parents without the symptoms, the total number of healthy children will be approximately 2250.

 ### GRAPHING CALCULATOR INVESTIGATIONS

Refer to the TI-82 Programs Appendix for instructions regarding the program BINOMIAL. Then use that program to answer the following questions.

 1. It has been reported that 25% of all births in Massachusetts are by Cesarean. In a random sample of 20 births, what is the probability that:
 (a) Exactly 6 are by Cesarean?
 (b) At most 4 are by Cesarean?
 (c) At least three are by Cesarean?
 (d) Draw the probability distribution for this experiment.

 2. It has been reported that 38% of all households own at least one dog. In a random sample of 30 households, find the probability that:
 (a) Between 10 and 15 households (inclusive) have at least one dog
 (b) Between 15 and 20 (inclusive) have at least one dog
 (c) More than 20 have at least one dog (Hint: Ignore really small probabilities)
 (d) What is the expected number of households that will have at least one dog?
 (e) What is the standard deviation for this binomial distribution?
 (f) Draw the probability distribution for this experiment.

 3. According to a recent study, 56% of the children who own bicycle helmets wear them regularly. If 25 children who own bicycle helmets are surveyed, what is the probability we will find:
 (a) 9 or 10 who wear them regularly?
 (b) Between 10 and 15 (inclusive) who wear them regularly?
 (c) Fewer than eight who wear them regularly (Hint: Ignore really small probabilities)?
 (d) What is the expected number of regular helmet wearers?
 (e) What is the standard deviation for this distribution?
 (f) Draw the probability distribution for this experiment.

4. The crime data for a particular large city indicated that assaults accounted for 18% of all criminal complaints. In a random sample of 35 criminal complaints, what is the probability that:

   (a) At most 4 were assault complaints?
   (b) Five or more were assault complaints?
   (c) What is the expected number of assault complaints?
   (d) What is the standard deviation?
   (e) Draw the probability distribution for this experiment.

5. In 1992 the Massachusetts Department of Education reported that 39% of the seniors attending a particular public school system performed at grade school level in mathematics. In a random sample of 24 students selected from this school system, what is the probability that:

   (a) At least 15 performed at grade level in math? *Hint:* Ignore really small probabilities.
   (b) Between 9 and 11 (inclusive) perform at grade level in math?
   (c) At least 6 perform at grade level in math?
   (d) What is the expected number of students in the sample performing at grade level in math?
   (e) What is the standard deviation?
   (f) Draw the probability distribution for this experiment.

6. Look back at the graphs for the probability distributions in Exercises 1 to 5. Which appear symmetrical? Which appear skewed? Do you have a conjecture regarding the conditions under which the graph of a binomial distribution will be symmetrical?

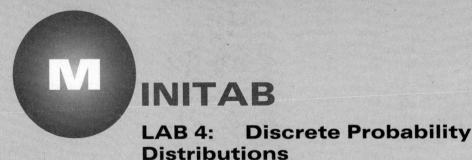

MINITAB

LAB 4: Discrete Probability Distributions

As you have seen in this chapter, a probability distribution provides a probability model describing a phenomenon. In what follows, given a probability model, we carry out simulation of the underlying experiment and collect data. We then compare the statistics computed from these data with the parameters of the model.

Log on and access MINITAB.

1 In the following we simulate from a population described by a discrete probability distribution. We consider the probability model of Example 3 in Section 4-2.

```
MTB >NOTE DISCRETE PROBABILITY DISTRIBUTION—THE
        MODEL
MTB >NOTE EXAMPLE 3 GARDENER'S EARNINGS SECTION  4-2
MTB >READ C1 C2
DATA>0   0.3
DATA>6   0.2
DATA>12  0.3
DATA>16  0.2
DATA>END
MTB >PRINT C1 C2
```

As a first step, compute the mean μ and the standard deviation σ of the distribution above. There are no MINITAB commands that compute these parameters directly. So we write a short program using the LET command.

```
MTB >NOTE MEAN OF THE DISTRIBUTION
MTB >LET C3 = C1 * C2
MTB >LET K1 = SUM (C3)
MTB >NOTE VARIANCE OF THE DISTRIBUTION
MTB >LET C4 = (C1—K1) ** 2
```

```
MTB >LET C5 = C2 * C4
```

```
MTB >LET K2 = SUM (C5)
```

```
MTB >NOTE STANDARD DEVIATION OF THE DISTRIBUTION
```

```
MTB >LET K3 = SQRT(K2)
```

```
MTB >NOTE KI IS MEAN K2 IS VARIANCE K3 IS STANDARD
        DEVIATION
```

```
MTB >PRINT K1-K3
```

You will see that the mean $\mu = 8$ and the standard deviation $\sigma = 6.13188$.

Next, we pick a random sample of 200 observations from the preceding distribution simulating gardener's earnings for 200 days. We plot the histogram for this sample, compute its mean $\bar{x}$, and standard deviation s.

```
MTB >NOTE SIMULATION OF A SAMPLE—EMPIRICAL ASPECT
```

```
MTB >RANDOM 200 VALUES PUT IN C10;
```

```
SUBC>DISCRETE VALUES IN C1 WITH PROBABILITIES IN C2.
```

```
MTB >LET K5 = MEAN(C10)
```

```
MTB >LET K6 = STAND(C10)
```

```
MTB >NOTE K5 IS X-BAR AND K6 IS S FOR SIMULATED
        SAMPLE
```

```
MTB >PRINT K5 K6
```

```
MTB >NOTE K1 IS MEAN AND K3 IS SIGMA FOR THE MODEL
```

```
MTB >PRINT K1 K3
```

```
MTB >HISTOGRAM C10
```

The command RANDOM and the subcommand DISCRETE that follow it were used to pick the 200 values from the given probability model. We elected to place these values in Column C10.

How do the values of the mean and the standard deviation computed from the sample compare with the corresponding parameters μ and σ? You do not expect them to be equal, but are they close?

2 We store the set of commands used in the preceding set of instructions in a file which we call 'PROBDIST'. This enables us to pick a sample of 150 observations from any probability distribution. For a different size sample, insert the appropriate number in place of 150 under the command RANDOM in the following set of instructions.

```
MTB >STORE 'PROBDIST'
```

```
STOR>NOECHO
```

```
STOR>LET C3 = C1 * C2

STOR>NOTE K1 IS MEW

STOR>LET K1 = SUM(C3)

STOR>PRINT K1

STOR>LET C4 = (C1 - K1) ** 2

STOR>LET C5 = C2 * C4

STOR>NOTE K2 IS VARIANCE, K3 IS SIGMA

STOR>LET K2 = SUM(C5)

STOR>LET K3 = SQRT(K2)

STOR>PRINT K2 K3

STOR>RANDOM 150 VALUES PUT IN C10;

STOR>DISCRETE VALUES IN C1 PROBABILITIES IN C2.

STOR>LET K5 = MEAN(C10)

STOR>LET K6 = STAND(C10)

STOR>PRINT K5 K6

STOR>HISTOGRAM C10

STOR>END
```

Once the commands are stored we have only to READ in the probability distribution in Columns C1 and C2, with the probabilities being entered in Column C2 and variable values in C1. Note that because the way the stored program is written it is important that the probability distribution be entered specifically this way in Columns C1 and C2. The command EXECUTE 'PROBDIST' will compute the parameters, carry out the simulation, and compute the statistics.

We now use the program PROBDIST in Example 6 in Section 4-2 in the text.

```
MTB >READ C1 C2

DATA>0 0.05

DATA>1 0.10

DATA>2 0.20

DATA>3 0.40

DATA>4 0.15

DATA>5 0.10
```

```
DATA>END

MTB >PRINT C1 C2

MTB >EXECUTE 'PROBDIST'
```

Assignment Consider the following probability distribution, which gives the net amount that a person can win in a raffle after paying $1 to purchase a ticket, and the corresponding probability.

x	-1	4	9	20
$p(x)$	0.70	0.15	0.1	0.05

(a) Compute the mean μ and the standard deviation σ.

(b) Simulate a sample of 500 from this distribution indicating that an individual participates 500 times in this type of raffle. Compute the sample mean and standard deviation and plot the histogram. Compare the parameters with the values of the corresponding statistics.

In the following we consider the binomial distribution.

3 To begin, we simulate a binomial experiment.

```
MTB >NOTE FIVE REPETITIONS; BINOMIAL EXPT 10 TRIALS
     P = 0.3

MTB >RANDOM 5 BINOMIAL EXPERIMENTS PUT IN C1;

SUBC>BINOMIAL N = 10, P = 0.3.

MTB >PRINT C1
```

The command RANDOM tells how many experiments are to be performed and where the results of the experiments are to be stored. The subcommand BINOMIAL states that the experiments are binomial and also stipulates *n* and the value of the parameter *p*. In the preceding simulation the experiment is binomial with ten independent trials and probability of success 0.3. The experiment is repeated five times, yielding a sample of five values placed in Column C1. Note that the sample values can be any of the numbers 0, 1, . . . , 10, as there are ten trials in each experiment.

4 Next, we investigate the empirical aspects associated with the binomial distribution. For this we repeat the experiment a certain (large) number of times and compare the mean and the standard deviation of the sample values with what the theory predicts. We store the program under the filename 'BNOMLEXP' so that we can use it with other binomial distributions obtained by changing *n* and *p*. The stored program will plot the histogram for the empirical data, and compute its mean and standard deviation.

```
MTB >STORE 'BNOMLEXP'

STOR>HISTOGRAM C1
```

```
STOR>MEAN OF C1

STOR>STANDARD DEVIATION OF C1

STOR>END
```

Now we generate data in Column C1 for $n = 10$ and $p = 0.3$ and execute the program. (The data must be stored in Column C1 since this is implied in the program.) We repeat the experiment 100 times below.

```
MTB >RANDOM 100 BINOMIAL EXPERIMENTS PUT IN C1;

SUBC>BINOMIAL N = 10, P = 0.3.

MTB >EXECUTE 'BNOMLEXP'
```

Because $p = 0.3$ and $n = 10$, the mean of the probability distribution is $10(0.3) = 3$ and its standard deviation is $\sqrt{10(0.3)(0.7)} = 1.4491$. See how the mean and the standard deviation computed from the sample of 100 values compare with the parameter values, $\mu = 3$ and $\sigma = 1.449$. They should be close.

Assignment   Carry out the simulation with $n = 12$ and $p = 0.65$, repeating the experiment 200 times.

5   MINITAB will print out the probabilities $b(x; n, p)$ under the command PDF, when it is followed by the subcommand that specifies the parameters of the distribution. We can ask for the entire column of probabilities or the probability for a specific value of x. For $n = 10$ and $p = 0.3$ we find $b(4; 10, 0.3)$ and the entire column of probabilities as follows.

```
MTB >NOTE PROBABILITY B(4;10,0.3)

MTB >PDF 4;

SUBC>BINOMIAL WITH N = 10, P = 0.3.

MTB >NOTE COLUMN OF PROBABILITIES B(X; N, P)

MTB >PDF;

SUBC>BINOMIAL WITH N = 10, P = 0.3.
```

Notice that except for rounding off, we get precisely the column of values in Appendix III Table A-1 corresponding to $n = 10$ and $p = 0.3$.

MINITAB will also produce cumulative probabilities $P(X \le x)$ when given the command CDF. This is illustrated for $n = 10$, $p = 0.3$ in the following.

```
MTB >CDF;

SUBC>BINOMIAL WITH N = 10, P = 0.3.
```

Assignment

    (a)   Find the binomial probabilities $b(x; n, p)$ when $n = 15$ and $p = 0.85$.

    (b)   Find $P(X \leq 9)$ when X is binomial with $n = 12$ and $p = 0.65$.

6   Having completed the MINITAB session, type STOP after the prompt.

```
MTB >STOP
```

Now log off.

Commands you have learned in this session:

```
RANDOM;
DISCRETE.

STORE
EXECUTE

RANDOM;
BINOMIAL.

PDF;
BINOMIAL.

CDF;
BINOMIAL.
```

CAREER *PROFILE*

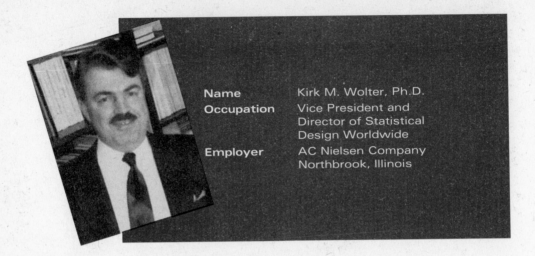

Name	Kirk M. Wolter, Ph.D.
Occupation	Vice President and Director of Statistical Design Worldwide
Employer	AC Nielsen Company Northbrook, Illinois

My introduction to statistics came when I was a senior math major at St. Olaf College in Minnesota wondering which courses to take my final year and what to do following graduation. By chance, I enrolled in a two-semester course in statistics, and as that year progressed, I became inspired by the opportunities in statistics for intellectual challenge and practical utility as well as personal enjoyment. This chance decision, made late in my college career, was to reshape the rest of my life.

I attended graduate school at Iowa State University and received an Ms and a PhD in statistics. I served as a teaching and research assistant helping students appreciate how statistics could be useful in almost all fields of human endeavor.

My career began in Washington, DC, at the Census Bureau. Contrary to some public opinion, the Census Bureau collects information about America, its people, and its industry every month of every year, not just once every ten years as part of the population and housing census. One of my most challenging and rewarding areas of work was the development of statistical methods to measure the number of people missed in a census.

After fourteen years, I left the Census Bureau to accept a new position as vice president with the AC Nielsen Company. Nielsen operates in two fundamental businesses, both in the United States and in 27 other countries worldwide. Television ratings is our most widely known business, measuring the number of Americans watching network programming, cable programming, and home video. The second component of our business is market research. We collect data on sales volumes, prices, and promotions from large samples of stores carrying consumer packaged goods. The data help businessmen decide what products to manufacture how many to make, how to market them, and where to market them. We produce data from a large national sample of households regarding the characteristics of the individual people buying consumer packaged goods.

My current work involves designing samples of stores and households, and estimation procedures for projecting data from a sample to an entire universe of stores or households. My colleagues and I study the relationships between the sales volume of a given product, the price of that product, and the promotional conditions (coupons, in-store displays, and advertising) under which the product is sold. Such relationships are helpful in formulating optimal strategies for marketing products. In addition, I work toward exporting statistical methods and innovations to other Nielsen countries around the world.

Quantitative literacy has become important for all Americans, especially the young who are beginning careers. It is a means to maintaining or improving our competitive position in an emerging global marketplace. Statistics is a rewarding career because it is the essence of both quantitative literacy and scientific method. It teaches how to approach thinking in confronting important national and global issues. It helps formulate the right questions to ask and how to solve or answer them. A good grounding in statistics and the scientific method gives you flexibility to work in many different areas of science, business, and government.

Continuous Probability Distributions

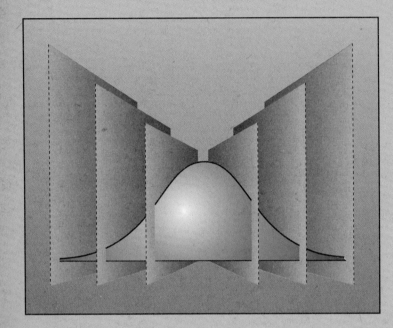

INTRODUCTION

I n the preceding chapter you were introduced to the general concept of random variables. We mentioned two types of random variables—discrete and continuous. We discussed discrete random variables at length. As you saw, in the discrete case the distribution can be described by giving the values that the random variable assumes along with the corresponding probabilities. In the continuous case, as we shall see now, the distribution is described by means of a smooth curve, and the probabilities are given in terms of appropriate areas under this curve. In our study of continuous distributions in this chapter we shall single out a very important continuous distribution called the *normal distribution,* since the probability laws of many physical phenomena can be modeled through it. The normal distribution is also called the *Gaussian distribution*, in honor of the great mathematician Karl Gauss (1777–1855). There are other important continuous distributions also, such as the *chi square* distribution, the *t* distribution, and so on, that we shall study in subsequent chapters.

In the last section of this chapter, we will see a connection between the normal and the binomial distribution and how, in some situations, the normal distribution can be used to compute, approximately, the probabilities of a binomial distribution.

5-1 CONTINUOUS RANDOM VARIABLES

A clue to understanding probability distributions of continuous random variables is in the concept of a histogram. The theoretical model for a phenomenon follows from the empirical evidence collected by observing it. Consider the following example where the random variable X represents the delay in the departure time of a plane scheduled to depart at a certain time, say 8 A.M. We assume that time is measured in units of hours. Suppose records are kept over 200 days and the information is summarized in a relative frequency distribution given in Columns 1 and 2 of Table 5-1.

We will now draw a histogram of the distribution as a modified version of the type presented in Chapter 1. Instead of using frequency densities, we use what are called the *relative frequency densities.* The **relative frequency density** of a class is obtained by dividing the relative frequency of the class by the class width.

$$\left(\begin{array}{c}\text{relative frequency density} \\ \text{of a class}\end{array}\right) = \frac{\text{relative frequency of the class}}{\text{width of the class}}.$$

In our example, the width of any class interval is 0.5. Therefore, the relative frequency density column in Table 5-1 is obtained by dividing the entries in Column 2 by 0.5. We now erect rectangles in the usual way with their bases centered at the class marks, except that the heights of the rectangles are taken

TABLE 5-1 Relative Frequency Distribution of the Delay Times

Delay-in-Departure Time*	Relative Frequency	Relative Frequency Density
$0.0 \leq t < 0.5$	0.10	0.20
$0.5 \leq t < 1.0$	0.35	0.70
$1.0 \leq t < 1.5$	0.20	0.40
$1.5 \leq t < 2.0$	0.18	0.36
$2.0 \leq t < 2.5$	0.10	0.20
$2.5 \leq t < 3.0$	0.05	0.10
$3.0 \leq t < 3.5$	0.02	0.04

*In Column 1, $1.5 \leq t < 2.0$, for example, means that the delay-in-departure time is at least 1.5 hours but less than 2 hours.

to correspond to the relative frequency densities. The resulting histogram is called a *relative frequency density histogram.* For the distribution in Table 5-1, the relative frequency density histogram is shown in Figure 5-1a. Notice that

$$\text{the area of a rectangle} = \text{base} \times \text{height}$$
$$= \left(\begin{array}{c}\text{width of the}\\ \text{class interval}\end{array}\right) \times \left(\begin{array}{c}\text{relative frequency}\\ \text{density}\end{array}\right)$$
$$= \left(\begin{array}{c}\text{width of the}\\ \text{class interval}\end{array}\right) \times \frac{\left(\begin{array}{c}\text{relative frequency}\\ \text{of the class}\end{array}\right)}{\left(\begin{array}{c}\text{width of the}\\ \text{class interval}\end{array}\right)}$$
$$= \text{relative frequency of the class.}$$

Thus, in geometric terms, the area of any rectangle erected over an interval represents the relative frequency of that class. Now, for example, the relative frequency of the flights of the plane delayed between 1.5 hours and 2.0 hours is precisely equal to the area of the darker shaded rectangle in Figure 5-1a. Also, because the sum of the relative frequencies is 1, it follows that *the area under a relative frequency density histogram is always equal to 1.*

With a larger sample we are in a position to take shorter and shorter class intervals, resulting in a refinement of the relative frequency density histogram as in Figure 5-1b, bringing into clearer focus the nature of the delay time. In Figure 5-1b, where the histogram is obtained with 1000 observations, the relative frequency of the flights delayed between 1.5 hours and 2.0 hours is equal to the sum of the areas of the three darker rectangles. Eventually, as we take larger and larger samples an idealization of the situation is the rendering

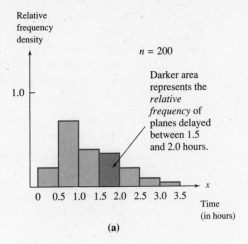

(a)

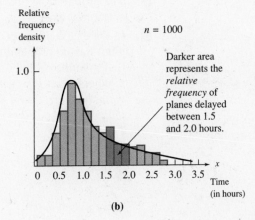

(b)

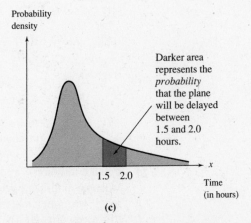

(c)

Figure 5-1 Delay in the Departure Time of a Plane

of the histogram into a smooth curve, as in Figure 5-1c. This curve is super-imposed on the histogram of Figure 5-1b. We call this curve the *probability density curve*. It represents the graph of a function called the **probability density function,** often abbreviated as *pdf*. The darker shaded area in Figure 5-1c represents the *probability* that the plane will be delayed between 1.5 hours and 2.0 hours.

Probability Density Curve; Probabilities as Areas under It

A few comments are in order at this point.

1   Because the area under each relative frequency density histogram is equal to 1, it can be seen, intuitively at least, that in the limiting situation, the area under the probability density curve is also equal to 1.

2   The graph of the probability density function is always above the horizontal axis.

3   In the graph of the probability density function, the height at any point has *no* probability interpretation. It is the *areas* under the graph that give the probabilities. Specifically, if *a* and *b* are any two numbers, then *the probability that the random variable assumes a value between a and b (written P(a < X < b)) is given as the area under the probability density curve between a and b, as shown in Figure 5-2.*

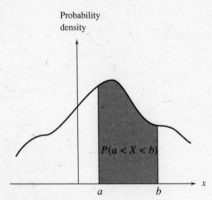

Figure 5-2
The shaded area under the curve is
$P(a < X < b)$.

4   For a continuous random variable, the probability that the random variable assumes any particular value is zero, that is, $P(X = a) = 0$ for any number *a*. This is because we can look at $P(X = a)$ as equal to $P(a \leq X \leq a)$, that is, as the area under the curve between *a* and *a*. But this area is zero, thus $P(X = a) = 0$. In our example, therefore, the probability that the plane will be delayed exactly 2.5 hours is zero.

5 Since $P(X = a) = 0$ for a continuous random variable, it does not matter whether we include or exclude equalities at the endpoints of the intervals, and we have

$$P(c < X < d) = P(c \le X < d) = P(c < X \le d) = P(c \le X \le d).$$

Note: In the discrete case this need not be true.

EXAMPLE 1 Suppose that an executive arrives in her office between 8 A.M. and 10 A.M. Records are kept of her arrival times (measured in hours) over 1000 days and the relative frequency density histogram given in Figure 5-3 is obtained.

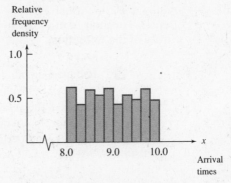

Figure 5-3 The Relative Frequency Density Histogram for the Arrival Times of an Executive

(a) Give an idealization that might best provide a probability model describing the distribution of the arrival times.

(b) What is the probability that the executive will arrive between 8:45 A.M. and 9:25 A.M.?

(c) What is the probability that she will arrive after 9:36 A.M.?

SOLUTION

(a) Since the vertical bars are approximately the same height, it seems reasonable that the graph given in Figure 5-4 on page 286 provides an appropriate probability model. Notice that the length of the interval 8:00 A.M to 10:00 A.M. is 2 units, and since we want the total area of the rectangle equal to 1, the probability density function at every point in the interval is 0.5. The area of a rectangle with these dimensions is 2(0.5), or 1.

(b) We want to find the probability that the executive will arrive between 8:45 A.M. and 9:25 A.M. Converting into hours, we want $P(8.75 < X < 9.42)$. This is equal to the area of the darker shaded rectangle in Figure 5-5. (See page 286.) Since the length of the rectangle is $9.42 - 8.75 = 0.67$, and its height is 0.5, the darker shaded area is equal to $(0.67)(0.5) = 0.335$. Therefore, the probability that the executive will arrive between 8:45 A.M. and 9:25 A.M. is 0.335.

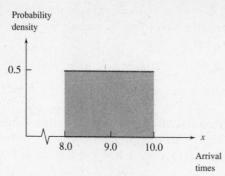

Figure 5-4 The Probability Density Function of the Arrival Times of the Executive

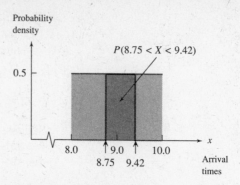

Figure 5-5 The Probability $P(8.75 < X < 9.42)$

(c)   Converting into hours, the probability of arriving after 9:36 A.M. is equal to $P(X > 9.6)$ and is obtained as the shaded area in Figure 5-6. As can be seen, this area is equal to $(10 - 9.6)(0.5)$, or 0.2. Thus, the probability that the executive will arrive after 9:36 A.M. is equal to 0.2. That is, over a long period of time, approximately 20 percent of the times the executive will arrive between 9:36 A.M. and 10:00 A.M.

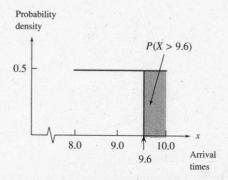

Figure 5-6 The Probability $P(X > 9.6)$

In Example 1 it was easy to compute the areas, as the graph of the probability density function was a familiar geometric figure, the rectangle. The distribution of the type displayed in Figure 5-4 is, therefore, called the **rectangular distribution.** It is also called the **uniform distribution.** In many situations, computations of the areas is an impossible task without using techniques of integral calculus or numerical analysis. Extensive tables of areas are available for important continuous distributions, and we will have occasion to use them in our study. Further, in the continuous case the definition of the mean and the standard deviation of a probability distribution require a background in calculus, so we will steer clear of this subject.

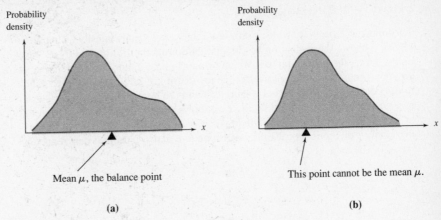

Figure 5-7
In Figure (a) the fulcrum is at the balance point; in Figure (b) it is not.

However, it is worth mentioning that if we regard the probability distribution (the shaded portion in Figure 5-7) as a thin plate of some material of uniform density, then the location of the mean along the horizontal axis will be where the plate balances.

Also, the magnitude of the standard deviation reflects the spread of the distribution about the mean. Distributions with small standard deviation σ have high, narrow curves while those with large standard deviation σ have low, flat curves. For example, the curve in Figure 5-8a has smaller standard deviation than the one in Figure 5-8b.

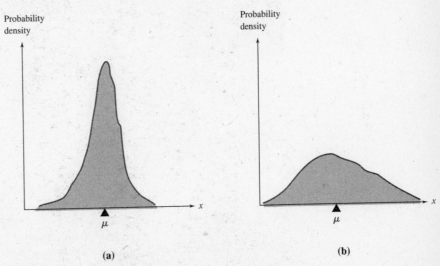

Figure 5-8
The curve in Figure (a) has smaller standard deviation than the one in Figure (b).

Section 5-1 Exercises

1. The following figures show three different relative frequency density histograms. Figure a shows the histogram of the distribution of the life of 10,000 electric bulbs. Along the horizontal axis we plot time (in hours), and the vertical bar at any point represents the relative frequency density of bulbs that stopped functioning during that time period. Figure b shows the histogram of the distribution of time needed to perform a task. Figure c shows the histogram of the distribution of the heights of 1000 adult individuals. For each case, approximate the histogram by a theoretical probability density curve that you think is best representative of the data.

Relative frequency
density of bulbs

(a)

Relative frequency
density of workers

(b)

Relative frequency
density of individuals

(c)

For the continuous random variables in Exercises 2–8, consider how a suitable histogram of each distribution would appear if you were to obtain a large number of measurements on the variable. From this infer the nature of the theoretical probability density curve.

2. The amount of rainfall during a year

3. The weight of an adult male

4. The burning time of an experimental rocket

5. The reaction time of a frog to a given electric stimulus

6. The time taken by a biological cell to divide in two

7. The length of time that a person has to wait in line to be served

8. After applying brakes, the distance that a car going at 40 miles per hour covers before it comes to a stop

9. The curves in the figures that follow represent theoretical probability density curves.
   (a) In Figure a, shade the appropriate region to represent the probability $P(175 < X < 250)$.

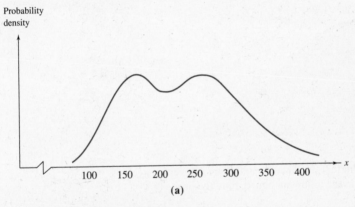

(a)

(b) In Figure b, shade the appropriate region to represent the probability $P(X \geq -2.5)$.

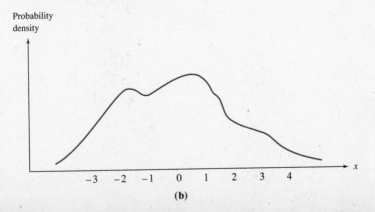

(b)

(c) In Figure c, shade the appropriate region to represent the probability that X does *not* lie between 10 and 20.

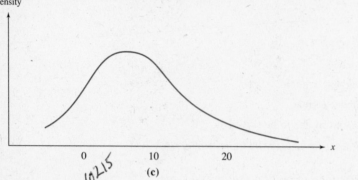

(c)

10. In the following figure, probabilities over some intervals are given as appropriate areas under the curve. Use this information to find the following probabilities.

(a) $P(X \le -6)$ **(b)** $P(-3.5 < X \le 3)$ **(c)** $P(X \ge 1)$

(d) Probability that X lies between -6 and -3.5, or between 1 and 3

(e) Probability that X does not lie between -6 and 1

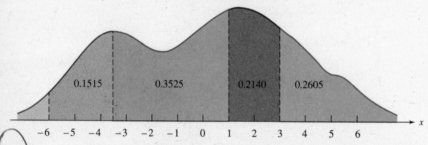

11. The probability distribution of the height (in feet) attained by a balloon is given by the probability density function shown in the following figure. Find the probability that a balloon will rise:

(a) beyond 1800 feet

(b) not more than 1400 feet

(c) between 1200 and 1600 feet.

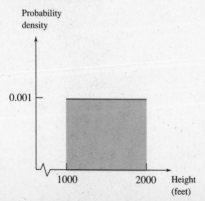

12. If the distribution of the height attained by a ballon is given by the following probability density curve, find the probability that a balloon will rise:
    (a) beyond 1800 feet
    (b) not more than 1400 feet
    (c) between 1200 and 1600 feet.

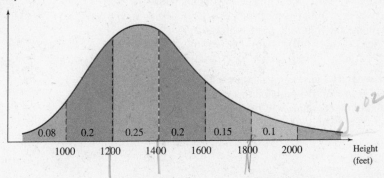

13. Of the two probability density curves in Exercises 11 and 12 describing the height attained by a balloon, which one depicts a more realistic situation? Explain.

14. Referring to the probability density curve in the figure for Exercise 12, find the value c for which:
    (a) $P(1200 < X < c) = 0.60$
    (b) $P(X > c) = 0.27$
    (c) $P(X < c) = 0.53$.

15. Consider a continuous random variable with a probability density function that is symmetric about the vertical axis at 0. If, as indicated in the shaded regions in the following figure, $P(-2 < X < 2) = 0.4$ and $P(X > 3) = 0.18$, compute the following probabilities.
    (a) $P(0 < X < 2)$ (b) $P(2 < X < 3)$
    (c) $P(-2 < X < 3)$ (d) $P(-3 < X < 0)$
    (e) $P(X < -2)$ (f) $P(X < 2)$

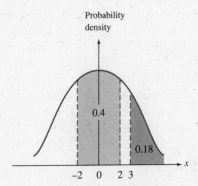

16. Suppose the probability density function of a continuous random variable X is known to be symmetric about the vertical axis at $x = 10$. If you are given that $P(X > 14) = 0.12$ and $P(12 < X < 14) = 0.18$, compute the following probabilities. *Hint:* Sketch a suitable probability density curve and shade the appropriate regions to indicate the given probabilities.

   (a) $P(X > 10)$ **(b)** $P(10 < X < 12)$ **(c)** $P(6 < X < 12)$

   (d) $P(8 < X < 12)$ **(e)** $P(X < 6)$ **(f)** $P(X < 12)$

17. Consider a continuous random variable X with a probability density function that is symmetric about the vertical axis at $x = 100$. Given that $P(X > 120) = 0.1$ and $P(80 < X < 90) = 0.22$, compute the following probabilities.

   (a) $P(110 < X < 120)$ **(b)** $P(90 < X < 110)$ **(c)** $P(X < 80)$

   (d) $P(80 < X < 110)$ **(e)** $P(90 < X < 100)$ **(f)** $P(X > 90)$

18. Suppose the life span of a whale, measured in years, has a probability density function given in the figure below. Based on the probabilities indicated by the shaded areas, compute the probability that a whale will live:

   (a) less than 20 years   **(b)** between 40 and 80 years   **(c)** past 60 years.

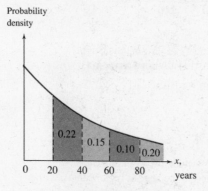

5-2   THE NORMAL DISTRIBUTION

For a large number of phenomena, a smooth, bell-shaped curve serves as a mathematical model to describe their probability distribution. Such a bell-shaped curve is called a **normal curve,** and the probability distribution is called a **normal distribution.*** It is by far the most widely used continuous distribution and plays a central role in statistical inference. The IQ of university students, the amount of annual rainfall, and the amount of error in filling a one-pound jar of coffee provide some examples. In Figure 5-9, we show how a cursory consideration of the histogram in each case gives an idealization by means of a smooth, symmetrical curve that tapers off at both ends and is peaked at the center. For example, consider the histogram and the corre-

---

* For the mathematically oriented, the shape of the curve is defined by

$$f(x) = \frac{1}{\sigma\sqrt{2\pi}}e^{-(x-\mu)^2/(2\sigma^2)}, \; -\infty < x < \infty$$

where μ is the population mean and σ is the population standard deviation.

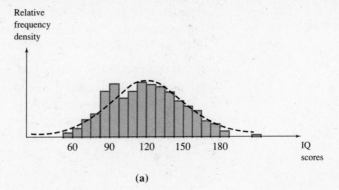

(a)

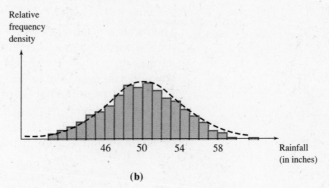

(b)

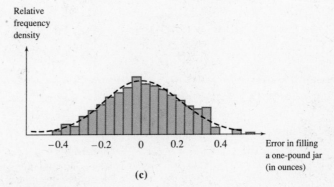

(c)

Figure 5-9

sponding smooth curve describing the distribution of error in filling one-pound jars of coffee. We assume that the error is positive if a jar is overfilled and negative if it is underfilled. If the process is under control, we expect that most of the jars will be neither overfilled nor underfilled by large amounts. Fewer and fewer jars will be overfilled by larger and larger amounts; thus the graph will taper off on the right. In the same way, fewer and fewer jars will be underfilled by larger and larger amounts; thus the graph will taper off on the left.

Properties of the Normal Distribution

The mathematical function of the normal curve was developed by DeMoivre in 1733. The following features characterize the normal distribution.

1   It is a continuous distribution described by a bell-shaped curve.

2   It is completely determined by its mean, which can be any number—positive or negative, and its standard deviation, which can be any positive number.

3   It is unimodal, and its curve is peaked at the center and symmetric about a vertical line at the mean.

4   The tails of the curve extend indefinitely in both directions from the center, getting closer and closer to the horizontal axis but never quite touching it.

5   The mean μ determines where the center of the curve is located and the standard deviation σ determines its flatness. As one traces a normal curve, there is the central portion of the curve which looks down like an outline of an inverted bowl (the dotted part in Figure 5-10) and the tail portions of the curve which look up (the solid parts in Figure 5-10). The x-coordinates of the points where the nature of the curvature changes are located precisely one standard deviation from the mean, that is, at $\mu - \sigma$ and $\mu + \sigma$.

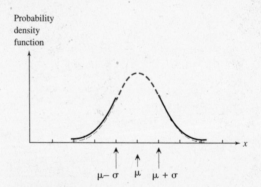

Figure 5-10

The points where the curvature changes are located one standard deviation from the mean.

    All of the curves in Figure 5-11 have the same mean of 50 but different standard deviations, with $\sigma = 0.5$, $\sigma = 1$, and $\sigma = 2$. Note that the greater the standard deviation, the flatter the curve. For example, the curve with $\sigma = 0.5$, is more peaked than the one with $\sigma = 2$.

    In Figure 5-12 we have three curves all having the same standard deviation, $\sigma = 2$, but different means, $\mu = -6$, $\mu = 0$, and $\mu = 8$. Because σ is the same, the curves have the same shape and size, but the one with $\mu = 8$ is to the right of the one with $\mu = -6$.

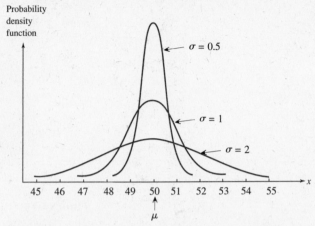

Figure 5-11 Curves of Normal Distribution with Mean 50 and σ = 0.5, 1, 2

The greater the standard deviation, the flatter the curve.

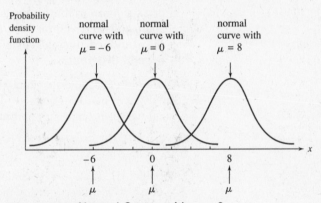

Figure 5-12 Normal Curves with σ = 2 and μ = −6, 0, 8

THE STANDARD NORMAL DISTRIBUTION

Because a normal distribution is a continuous distribution, the probabilities are given in terms of appropriate areas. Thus, the probability that a random variable X having a normal distribution will assume a value between two numbers a and b is equal to the area under the curve between $x = a$ and $x = b$, indicated by the shaded region in Figure 5-13. (See page 296.) Because of the shape of the curve, it is impossible to compute these areas geometrically. Tables of areas for normal distributions with all possible means and standard deviations are needed, which would of course be impractical, as there is an endless number of such curves. Fortunately, we need a table of areas for only one normal curve. The curve that is used is called the **standard normal curve.**

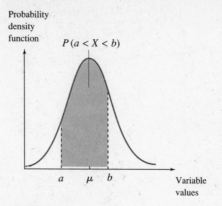

Figure 5-13

The shaded area represents the probability that
X assumes a value between *a* and *b*.

A standard normal curve describes the probability density function
of a random variable having a normal distribution with mean $\mu = 0$
and standard deviation $\sigma = 1$. The random variable itself is called
the standard normal variable and is said to have **the standard
normal distribution.**

We will always denote the standard normal variable by Z. Areas under any
normal curve can be obtained by comparing the curve with the standard
normal curve through an appropriate transformation, as we will see later.

Areas under Standard Normal Curve Using Table A-2 and Table A-3

*Appendix III, Table A-2, gives areas under the standard normal curve from
$z = 0$ to $z = a$, where *a* is any number 0.00, 0.01, 0.02, . . . , 3.09. To find
the area between $z = 0$ and $z = 1.73$, for example, we go to 1.7 in the column
and 0.03 in the row, both headed by *z*, and read the corresponding entry in
the body of the table as 0.4582, as indicated on the following page in the
excerpt reproduced from Table A-2.

Thus, the area between 0 and 1.73 is 0.4582 and

$$P(0 < Z < 1.73) = 0.4582.$$

This is shown by means of the shaded region in Figure 5-14.

---

* In later chapters dealing with statistical inference it will be necessary to look up areas in the
tails of the standard normal curve. For convenient reference we also provide Appendix III,
Table A-3. It is read exactly as Table A-2 is read, except that the entries in the body of the table
are areas in the right tail of the standard normal curve.

z	0.00	0.01	0.02	0.03	0.04	0.05	0.06	0.07	0.08	0.09
0.0	0.0000	0.0040	0.0080	0.0120	0.0160	0.0199	0.0239	0.0279	0.0319	0.0359
0.1	0.0398	0.0438	0.0478	0.0517	0.0557	0.0596	0.0636	0.0675	0.0714	0.0753
0.2	0.0793	0.0832	0.0871	0.0910	0.0948	0.0987	0.1026	0.1064	0.1103	0.1141
.										
.										
1.6	0.4452	0.4463	0.4474	0.4484	0.4495	0.4505	0.4515	0.4525	0.4535	0.4545
1.7	0.4454	0.4564	0.4573	0.4582	0.4591	0.4599	0.4608	0.4616	0.4625	0.4663
1.8	0.4641	0.4649	0.4656	0.4664	0.4671	0.4678	0.4686	0.4692	0.4699	0.4706

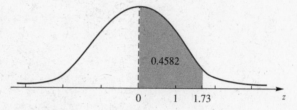

Figure 5-14

Area between $z = 0$ and $z = 1.73$ is 0.4582.

Notice that, because of the symmetry of the standard normal distribution about the vertical line at 0, we can say even without the table that

$$P(Z \leq 0) = P(Z \geq 0) = 0.5.$$

We will now use Tables A-2 and A-3 to find areas under the standard normal curve. Remember that the curve is symmetric with respect to the vertical axis through 0. It is helpful to draw the curves and identify the areas under the curve and the values along the horizontal axis. This procedure is highly recommended.

EXAMPLE 1   If $P(0 < Z < c) = 0.3944$, find c.

SOLUTION   Here the probability is given as 0.3944, that is, the entry in the body of the table is given as 0.3944 and we want to find the corresponding z value. Having determined that the row heading is 1.2 and the column heading is 0.05, we read $c = 1.25$. This reading is shown in Figure 5-15.   ●

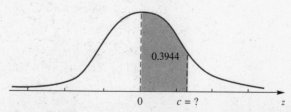

Figure 5-15

EXAMPLE 2   Find $P(-2.42 < Z < 0.8)$.

SOLUTION   This probability is shown in Figure 5-16 as the sum of the shaded regions between 0 and 0.8 and between -2.42 and 0. Notice that due to symmetry, the area between -2.42 and 0 is equal to the area between 0 and 2.42. Thus,

$$P(-2.42 < Z < 0.8) = P(0 < Z < 0.8) + P(0 < Z < 2.42)$$
$$= 0.2881 + 0.4922$$
$$= 0.7803.$$

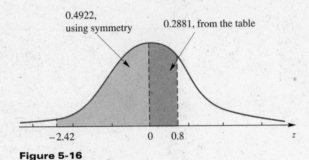

Figure 5-16

EXAMPLE 3   Find the following

(a) $P(1.8 < Z < 2.8)$ **(b)** $P(-2.8 < Z < -1.8)$

SOLUTION
(a) $P(1.8 < Z < 2.8)$ is the area between $z = 1.8$ and $z = 2.8$ and is equal to the area from 0 to 2.8 *minus* the area from 0 to 1.8, as can be seen from the shaded portion in the right tail of the curve in Figure 5-17. Thus,

$$P(1.8 < Z < 2.8) = 0.4974 - 0.4641$$
$$= 0.0333.$$

0.0333, using symmetry

0.0333, from the table

-2.8 -1.8 0   1.8   2.8 z

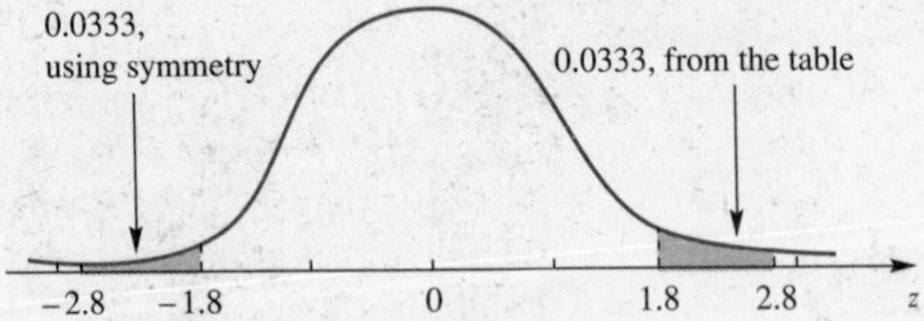

Figure 5-17

(b)   Due to the symmetry of the normal curve, as can be seen from the left tail of the curve in Figure 5-17,

$$P(-2.8 < Z < -1.8) = P(1.8 < Z < 2.8)$$
$$= 0.0333.$$

EXAMPLE 4 Find the following.

(a) $P(Z > -2.13)$ (b) $P(Z < -1.81)$

SOLUTION

(a) The area representing $P(Z > -2.13)$ is the shaded region in Figure 5-18 to the right of -2.13. This area is equal to the area between -2.13 and 0 *plus* the area to the right of 0. Now the area between -2.13 and 0 is the same as the area between 0 and 2.13 and is equal to 0.4834. Also, the area to the right of 0 is 0.5. Thus,

$$P(Z > -2.13) = 0.4834 + 0.5$$
$$= 0.9834.$$

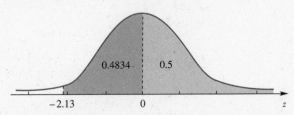

Figure 5-18

(b) $P(Z < -1.81)$ is equal to the area to the left of -1.81, which, due to symmetry, is equal to the area to the right of 1.81, as can be seen from Figure 5-19. This, in turn, is equal to the area to the right of 0 *minus* the area between 0 and 1.81. But the area between 0 and 1.81 is 0.4649. Thus,

$$P(Z < -1.81) = 0.5 - 0.4649$$
$$= 0.0351.$$

Incidentally, the area to the right of 1.81 can be obtained quite easily straight from Table A-3 in Appendix III. ●

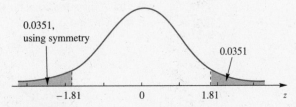

Figure 5-19

EXAMPLE 5 Suppose Z is a standard normal variable. In each of the following cases, find c for which:

(a) $P(Z \leq c) = 0.1151$

(b) $P(Z \leq c) = 0.8238$

(c) $P(1 \leq Z < c) = 0.1525$

(d) $P(-c < Z < c) = 0.8164.$

SOLUTION   In this example we are given the areas and want to read the values along the horizontal axis.

(a)   Because the area to the left of c is 0.1151, and it is less than 0.5, c must be located to the left of 0 as in Figure 5-20. Thus c is negative. Now, due to symmetry, the area to the right of $-c$ is 0.1151. But from Table A-3, the area to the right of 1.2, is 0.1151. Therefore, $-c = 1.2$, so that $c = -1.2$.

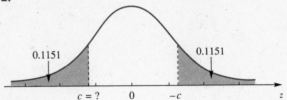

0.1151   0.1151

$c = ?$ 0 $-c$ z

Figure 5-20

(b)   Because $P(Z \le c) = 0.8238$, c is to the right of 0; that is, c is positive. Also, as can be seen from Figure 5-21, the area between 0 and c is equal to $0.8238 - 0.5 = 0.3238$. From Table A-2, the area between 0 and 0.93 is 0.3238. Therefore, $c = 0.93$.

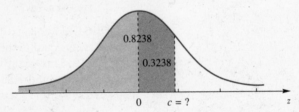

0.8238

0.3238

0        $c = ?$ z

Figure 5-21

(c)   Because $P(1 \le Z < c) = 0.1525$, the area between 1 and c is equal to 0.1525. Now the area between 0 and 1 is 0.3413. Thus,

$$P(0 < Z < c) = 0.3413 + 0.1525 = 0.4938.$$

We have the situation shown in Figure 5-22. Refering back to Table A-2, it follows that $c = 2.5$.

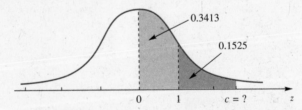

0.3413

0.1525

0 1 $c = ?$ z

Figure 5-22

(d) $P(-c < Z < c) = 0.8164$. Therefore, on account of symmetry,

$$P(0 < Z < c) = \frac{1}{2}(0.8164) = 0.4082.$$

From Table A-2, $c = 1.33$

EXAMPLE 6

Suppose the Greek letter α (alpha) represents a number between 0 and 0.5 and z_α represents the z value such that to its right the area under the standard normal curve is α as shown in Figure 5-23.

Find the following.

(a) α if

   i. $z_\alpha = 1.24$

   ii. $z_\alpha = 1.99$

   iii. $z_\alpha = 2.73$

(b) i. $z_{0.0043}$

   ii. $z_{0.0202}$

   iii. $-z_{0.0202}$

   iv. $z_{0.025}$

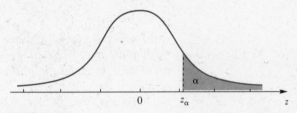

Figure 5-23

SOLUTION We could find all the answers in this example using Appendix III, Table A-2. However, since α is the area in the right tail it will be more convenient to use Appendix III, Table A-3.

(a) In this part of the example we are given z values and are expected to find α, the area in the right tail.

  i. To find α when $z_\alpha = 1.24$, we look for the entry in Table A-3 along the row with heading 1.2 and down the column with heading 0.04. The value of α in this case is 0.1075.

  ii. When $z_\alpha = 1.99$, the entry α with row heading 1.9 and column heading 0.09 is 0.0233.

  iii. When $z_\alpha = 2.73$, looking along row heading 2.7 and down the column heading 0.03, we find $\alpha = 0.0032$.

(b) In this part we do exactly the reverse of what we did in Part a. We locate the given number α in the body of the table and read the corresponding row and column headings to find the z value.

  i. Because $\alpha = 0.0043$, the row heading is 2.6 and the column heading is 0.03. Thus $z_{0.0043} = 2.63$.

  ii. When $\alpha = 0.0202$, we see that the row heading is 2.0 and the column heading is 0.05. Therefore $z_{0.0202} = 2.05$.

  iii. Obviously $-z_{0.0202} = -2.05$.

  iv. Here $\alpha = 0.025$. The corresponding row heading 1.9 and column heading is 0.06. Thus $z_{0.025} = 1.96$.   ●

Remark In dealing with inference problems, we will have occasion to use values such as $z_{0.05}$ and $z_{0.06}$. From Appendix III, Table A-3, $z_{0.05}$ is midway between 1.64 and 1.65. So we take $z_{0.05} = 1.645$. In the same way we take $z_{0.06} = 1.555$, as this value lies between 1.55 and 1.56.

NORMAL DISTRIBUTION WITH MEAN μ AND STANDARD DEVIATION σ

Having considered areas under the standard normal curve, we now turn our attention to the general case of a normal distribution with any mean μ and any standard deviation σ where, of course, $\sigma > 0$. It can be shown that if X is a normal variable with mean μ and standard deviation σ, then one can convert X into a standard normal variable Z by setting

$$Z = \frac{X - \mu}{\sigma}.$$

The result is accomplished in two steps, first by finding $X - \mu$ and then by dividing by σ. The effect of subtracting μ from X is that of shifting the curve with its center of symmetry at μ so that its new center is at 0. The new position is denoted by the dashed curve in Figure 5-24(a). This step gives us the deviation from the mean μ.

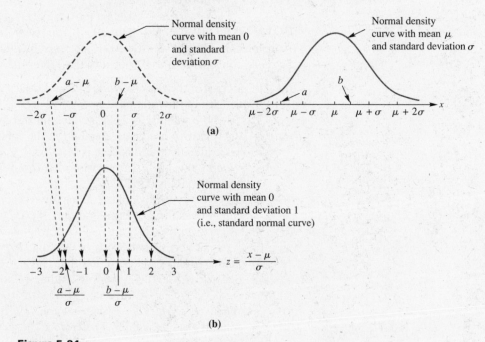

Figure 5-24

The solid curve in (a) is first shifted to the dashed curve. Then a change of scale is applied to produce the curve in (b).

The effect of dividing by σ is to change the scale of the dashed curve in the figure so that it coincides with the standard normal curve as shown in Figure 5-24(b). This step gives us the number of standard deviations from the mean.

How do we find $P(a < X < b)$? Notice that

$$\text{when } x = a, \text{ we have } z = \frac{a - \mu}{\sigma}$$

$$\text{and when } x = b, \text{ we have } z = \frac{b - \mu}{\sigma} .$$

This means that when X is between a and b, Z is between $\dfrac{a - \mu}{\sigma}$ and

$\dfrac{b - \mu}{\sigma}$. Thus we have the following result.

PROBABILITIES UNDER A GENERAL NORMAL CURVE

If X is a normally distributed random variable with mean μ and standard deviation σ, then

$$P(a < X < b) = P\left(\frac{a - \mu}{\sigma} < Z < \frac{b - \mu}{\sigma}\right)$$

In short, the area between a and b under a general normal curve with mean μ and standard deviation σ is equal to the area under the standard normal curve between $\dfrac{a - \mu}{\sigma}$ and $\dfrac{b - \mu}{\sigma}$.

The resulting situation can be shown conveniently as in Figure 5-25 on the following page. The important point is that *we subtract the mean of the random variable from the end points of the interval, divide by the standard deviation of the random variable, and refer, in Appendix III, to Table A-2 or Table A-3 of areas under the standard normal curve.*

The conversion of x values into z values as described here is called **standardization.** If x is a given value, then its **standard score** or **z-score** is determined by

$$z = \frac{x - \mu}{\sigma}.$$

This gives the number of standard deviations that the x value is from the mean μ.

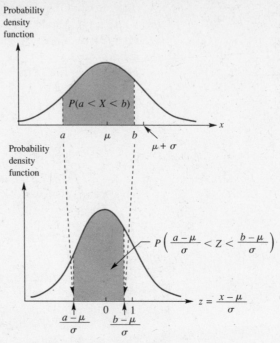

Figure 5-25

The shaded regions under the two curves have the same area. The center of the upper curve is at μ; its standard deviation σ. The center of the lower curve is at 0; its standard deviation 1.

EXAMPLE 7   Suppose X has a normal distribution with $\mu = 30$ and $\sigma = 4$. Find the following.

(a) $P(30 < X < 35)$

(b) $P(X > 40)$

(c) $P(X < 22)$

(d) $P(X > 20)$

(e) $P(X < 37)$

(f) $P(20 < X < 37)$

SOLUTION

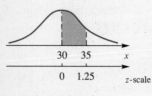

(a) $P(30 < X < 35) = P\left(\dfrac{30 - 30}{4} < Z < \dfrac{35 - 30}{4}\right)$

$= P(0 < Z < 1.25)$

$= 0.3944$, from Table A-2

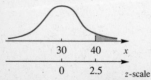

(b) $P(X > 40) = P\left(Z > \dfrac{40 - 30}{4}\right)$

$= P(Z > 2.5)$

$= 0.0062$, from Table A-3

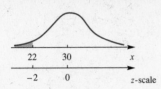

(c) $P(X < 22) = P\left(Z < \dfrac{22 - 30}{4}\right)$

$= P(Z < -2)$

$= P(Z > 2)$

$= 0.0228$, from Table A-3

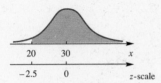

(d) $P(X > 20) = P\left(Z > \dfrac{20 - 30}{4}\right)$

$= P(Z > -2.5)$

$= P(-2.5 < Z < 0) + P(Z \geq 0)$

$= 0.4938 + 0.5$

$= 0.9938$

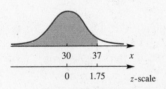

(e) $P(X < 37) = P\left(Z < \dfrac{37 - 30}{4}\right)$

$= P(Z < 1.75)$

$= P(Z \leq 0) + P(0 < Z < 1.75)$

$= 0.5 + 0.4599$

$= 0.9599$

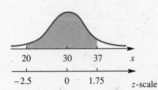

(f) $P(20 < X < 37) = P\left(\dfrac{20 - 30}{4} < Z < \dfrac{37 - 30}{4}\right)$

$= P(-2.5 < Z < 1.75)$

$= P(-2.5 < Z < 0) + P(0 < Z < 1.75)$

$= P(0 < Z < 2.5) + P(0 < Z < 1.75)$

$= 0.4938 + 0.4599$

$= 0.9537$ ●

EXAMPLE 8   Suppose X has a normal distribution with mean μ and standard deviation σ. Find the probability that X falls within 1, 2, or 3 standard deviations of the mean μ. That is, find the following.

(a) $P(\mu - \sigma < X < \mu + \sigma)$

(b) $P(\mu - 2\sigma < X < \mu + 2\sigma)$

(c) $P(\mu - 3\sigma < X < \mu + 3\sigma)$

SOLUTION   Notice that for any number k

$$P(\mu - k\sigma < X < \mu + k\sigma) = P\left(\dfrac{(\mu - k\sigma) - \mu}{\sigma} < Z < \dfrac{(\mu + k\sigma) - \mu}{\sigma}\right)$$

$$= P(-k < Z < k).$$

Note. What we did above is subtract the mean μ and divide by the standard deviation σ. Thus, for $k = 1, k = 2, k = 3$, we find, respectively, the following.

(a) $P(\mu - \sigma < X < \mu + \sigma) = P(-1 < Z < 1)$ **because $k = 1$**
$$= 2(0.3413)$$
$$= 0.6826$$

(b) $P(\mu - 2\sigma < X < \mu + 2\sigma) = P(-2 < Z < 2)$ **because $k = 2$**
$$= 2(0.4772)$$
$$= 0.9544$$

(c) $P(\mu - 3\sigma < X < \mu + 3\sigma) = P(-3 < Z < 3)$ **because $k = 3$**
$$= 2(0.4987)$$
$$= 0.9974$$

These probabilities are given in Table 5-2 and displayed in Figure 5-26. ●

TABLE 5-2   Probability That X Is within k Standard Deviations from the Mean, $k = 1, 2, 3$

k	Interval	Probability That X Is in the Interval	Area under the Normal Curve
1	$\mu - \sigma$ to $\mu + \sigma$	0.6826	68.26 percent
2	$\mu - 2\sigma$ to $\mu + 2\sigma$	0.9544	95.44 percent
3	$\mu - 3\sigma$ to $\mu + 3\sigma$	0.9974	99.74 percent

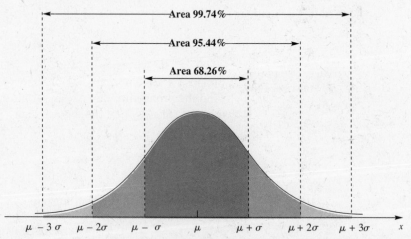

Figure 5-26 Area within k Standard Deviations from the Mean, $k = 1, 2, 3$

Remark   Theoretically a normal curve extends indefinitely in both directions, in that the range of the random variable is any number between minus infinity and plus infinity. However, in practice we approximate the distribution of such random variables as the heights of people by a normal distribution, although we are fully aware that the height of a person can never assume a negative value. In view of the fact that practically all the area under a normal curve is within three standard deviations from the mean (99.74 percent, as seen from Table 5-2), we need only be concerned with whether the normal curve provides a good model within three standard deviations from the mean. The behavior in the curve's tails beyond three standard deviations is inconsequential.

EXAMPLE 9

The amount of annual rainfall in a certain region is known from past experience to be a normally distributed random variable with a mean of 50 inches and a standard deviation of 4 inches. If the rainfall exceeds 57 inches during a year, it leads to floods. Find the probability that during a randomly picked year there will be floods.

SOLUTION Let X denote the amount of annual rainfall (in inches). The distribution of X is shown in Figure 5-27. The probability of a flood is equal to the probability that the rainfall exceeds 57 inches, that is, $P(X > 57)$. Since $\mu = 50$ and $\sigma = 4$,

$$P(X > 57) = P\left(Z > \frac{57 - 50}{4}\right)$$
$$= P(Z > 1.75)$$
$$= 0.0401, \text{ from Appendix III, Table A-3}$$

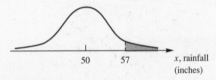

Figure 5-27

Over a long period of years, about once every 25 years there will be floods.

●

EXAMPLE 10

The weight of food packed in certain containers is a normally distributed random variable with a mean weight of 500 pounds and a standard deviation of 5 pounds. Suppose a container is picked at random. Find the probability that it contains:

(a) more than 510 pounds

(b) less than 498 pounds

(c) between 491 and 498 pounds.

SOLUTION Suppose X denotes the weight of food packed in the containers.

(a) We want to find $P(X > 510)$.

$$P(X > 510) = P\left(Z > \frac{510 - 500}{5}\right)$$
$$= P(Z > 2)$$
$$= 0.0228, \text{ from Appendix III, Table A-3}$$

(b) Here we want $P(X < 498)$.

$$P(X < 498) = P\left(Z < \frac{498 - 500}{5}\right)$$
$$= P(Z < -0.4)$$
$$= P(Z > 0.4)$$
$$= 0.3446, \text{ from Appendix III, Table A-3}$$

(c) We want $P(491 < X < 498)$.

$$P(491 < X < 498) = P\left(\frac{491 - 500}{5} < Z < \frac{498 - 500}{5}\right)$$
$$= P(-1.8 < Z < -0.4)$$
$$= P(0.4 < Z < 1.8) \quad \text{(Why?)}$$
$$= P(0 < Z < 1.8) - P(0 < Z < 0.4)$$
$$= 0.4641 - 0.1554 = 0.3087. \qquad \bullet$$

EXAMPLE 11 *Grading on the curve* Suppose the scores of students on a test are approximately normally distributed with a mean score of 70 points and a standard deviation of 10 points. It is decided to give A's to 10 percent of the students and B's to 23 percent of the students. Find what scores should be assigned A's and B's.

SOLUTION "The scores on a test" is in reality a discrete random variable, but experience shows that the normal distribution provides a good approximation. Denote the random variable by X. Suppose scores above u are assigned the grade A and those between v and u are assigned the grade B. Then the problem is reduced to finding two numbers u and v for which

$$P(X \geq u) = 0.1$$

and

$$P(v \leq X < u) = 0.23$$

as shown in Figure 5-28.

From the figure we see that the area to the right of u is 0.1 and the area to the right of v is $0.23 + 0.10 = 0.33$. Thus,

$$P(X \geq u) = 0.1 \quad \text{and} \quad P(X \geq v) = 0.33.$$

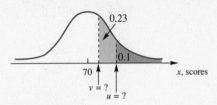

Figure 5-28

Because $\mu = 70$ and $\sigma = 10$, converting to the z scale we have

$$P\left(Z \geq \frac{u - 70}{10}\right) = 0.1$$

and

$$P\left(Z \geq \frac{v - 70}{10}\right) = 0.33.$$

From the standard normal table (Table A-3 in Appendix-III), it therefore follows that

$$\frac{u - 70}{10} = 1.28, \text{ that is, } u = 82.8$$

and

$$\frac{v - 70}{10} = 0.44, \text{ that is, } v = 74.4.$$

Thus, students with a score of 83 points or over should be given an A, and those with a score between 74 and 82 points should receive a B. ●

EXAMPLE 12   A special task force of a military unit requires that the recruits not be too tall or too short. Suppose 12 percent of the applicants are rejected because they are too tall and 18 percent because they are too short. If the height of an applicant is normally distributed with mean height 67 inches and standard deviation 3.5 inches, determine the heights below and above which the candidates are rejected.

SOLUTION   Let us denote the height of an applicant by X. Suppose an applicant above height A or below height B is rejected. Since 12 percent are rejected because they are too tall and 18 percent are rejected because they are too short, the numbers A and B should be such that

$$P(X > A) = 0.12 \quad \text{and} \quad P(X < B) = 0.18$$

as shown in Figure 5-29. Since $\mu = 67$ and $\sigma = 3.5$, converting to z-scale, A and B should satisfy

$$P\left(Z > \frac{A - 67}{3.5}\right) = 0.12 \quad \text{and} \quad P\left(Z < \frac{B - 67}{3.5}\right) = 0.18$$

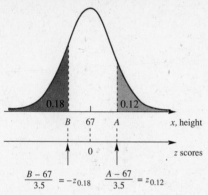

Figure 5-29

Now $z_{0.12}$ is midway between 1.17 and 1.18. So we take $z_{0.12} = 1.175$. Therefore, we must have

$$\frac{A - 67}{3.5} = 1.175$$

Solving this for A we get $A = 67 + (1.175)(3.5) = 71.1$ inches.

Also, $z_{0.18}$ is midway between 0.91 and 0.92; therefore, we take $z_{0.18} = 0.915$. Consequently, since the area 0.18 is in the left tail,

$$\frac{B - 67}{3.5} = -0.915$$

Solving for B, we have $B = 67 - (0.915)(3.5) = 63.8$ inches.

Thus, applicants shorter than 63.8 inches or taller than 71.1 inches are rejected. ●

Test of Normality Assumption

Some of the statistical test procedures that we develop in later chapters rely on the assumption that the population from which the data are collected is very nearly normally distributed. So, a question of interest is how we decide whether the sample data bear out this assumption about the population. Several methods are available, and we discuss below some that are easy to apply. The methods that we consider below, however, call for subjective judgment.

1. An analysis of the histogram of the data can provide information regarding the normality of the population. A cursory analysis of the histogram can shed light on the symmetry or the lack of it. Together with the way the tails of the distribution taper off, it can suggest whether a bell-shaped model would seem appropriate.
2. One can compute $\bar{x}$ and s from the data and construct the intervals $(\bar{x} - s, \bar{x} + s)$, $(\bar{x} - 2s, \bar{x} + 2s)$, and $(\bar{x} - 3s, \bar{x} + 3s)$. If the

percentages of observations in the data falling in each of these intervals are approximately in compliance with those given in Table 5-2, then a normal distribution of the population is very plausible.

3. A method that is widely used is a graphical method due to Shapiro and Wilks. It involves the sketch of a graphical plot called the **normal probability plot,** also called **normal scores plot.** We now describe this method in some detail.

First we order the data values from low to high and then compare the ordered sample values with their *expected locations* obtained assuming that the normality assumption is justified. These expected locations are called the **normal scores** or **nscores** and are given in Appendix III, Table A-8, for sample sizes $n = 5$ to $n = 16$. For example, for $n = 8$ the table shows the expected location of the 4th smallest observation is -0.152, that of the 7th smallest observation is 0.852, and so on. In general, the normal scores for any sample size n can be found by adopting the following steps.

i. Order the n data values from the lowest to the highest.

ii. For the ith smallest observation compute $\dfrac{i - 0.375}{n + 0.25}$. For example, if $n = 8$, for the 4th smallest observation we compute $\dfrac{4 - 0.375}{8 + 0.25} = 0.4394$. This value gives area in the left tail of the standard normal distribution.

iii. Compute the z value such that to its left there is area equal to $\dfrac{i - 0.375}{n + 0.25}$, computed in Step (ii). This z value is the nscore associated with the ith smallest value in the data. In our illustration, with $n = 8$ and $i = 4$, the computed value in step (ii) is 0.4394. Looking up in Appendix III, Table A-3, it can be verified that the z value is -0.152.

When the nscores are obtained we form pairs with the ith smallest observation in the data paired with the ith smallest normal score. These pairs are then plotted to get the normal probability plot. *If the points in the graph are approximately close to a straight line, then there is indication that the population is normally distributed.* The subjective aspect of our judgment is clearly evident because we have not specified what we would consider as being "close." Let us now consider the following example.

EXAMPLE 13

The data below record the number of hours a team of workers takes to assemble a custom-built motorcycle. The data are recorded for 10 different teams each assembling a motorcycle.

89 78 48 85 67 45 60 62 62 56

Use the normal probability plot to check the validity of normality assumption.

SOLUTION Since there are 10 sample values, the nscores are obtained from Appendix III Table A-8 for $n = 10$. The ordered sample values, the corresponding nscores, and areas in the left tail are given in the following table.

Ordered Values	Nscores	Probabilities in the Left Tail
45	−1.546	0.0610
48	−1.001	0.1585
56	−0.655	0.2561
60	−0.375	0.3537
62	−0.123	0.4512
62	0.123	0.5488
67	0.375	0.6463
78	0.655	0.7439
85	1.001	0.8415
89	1.546	0.9390

The normal probability plot is given in Figure 5-30 by plotting the 10 pairs (45, −1.546), (48, −1.001), . . . , (89, 1.546). As can be seen, we plot the nscores along the vertical axis drawn on the right and the data values along the horizontal axis. The corresponding probabilities in the left tail are reflected along the vertical axis on the left. Notice the scale along this vertical axis; equal lengths do not signify equal probability. Graph paper with such scale is available commercially and is called *normal probability paper*. Referring to Figure 5-30, we see that the points are rather close to the straight line in the figure. So, there is a reasonable basis to assume that the length of time to assemble the motorcycles is normally distributed.

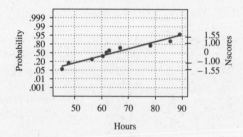

Figure 5-30 Normal Probability Plot for Number of Hours Required

There are easy-to-use computer software packages that give normal probability plots. The graph on the next page was obtained on the MINITAB software for the SYSBP data contained in Appendix IV, Table B-3.

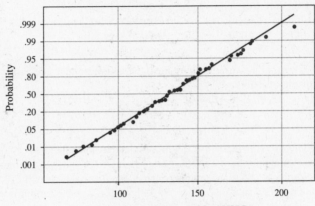

Normal probability plot for SYSBP,
Appendix IV, Table B-3

Average: 132.560
Std Dev: 21.2357
N of Data: 315

10/24/96

Section 5-2 Exercises

1. If Z is a standard normal variable, use Table A-2 or A-3 and find the following probabilities.
   (a) $P(0 \leq Z < 1.2)$ (b) $P(Z \geq 1.2)$
   (c) $P(Z > -1.2)$ (d) $P(-1.2 < Z \leq 0)$
   (e) $P(1 < Z < 2.5)$ (f) $P(-2.5 < Z < -1)$
   (g) $P(-1 < Z < 2.3)$ (h) $P(0.23 < Z \leq 0.68)$
   (i) $P(-0.18 < Z < 0.45)$ (j) $P(Z \geq -0.75)$
   (k) $P(Z \leq 0.68)$

2. Suppose Z is a standard normal variable. In each of the following cases, find the value of c if: $c = .48$
   (a) $P(0 < Z \leq c) = 0.1844$ (b) $P(Z \leq c) = 0.8729$ 1.14
   (c) $P(Z \leq c) = 0.1038$ -1.26   (d) $P(Z \geq c) = 0.7357$
   (e) $P(Z \geq c) = 0.0154$ (f) $P(-c \leq Z \leq c) = 0.8926$
   (g) $P(Z \geq 2c) = 0.0778$ (h) $P(0.5 \leq Z < c) = 0.2184$ *1.34
   (i) $P(c \leq Z \leq 0) = 0.4881$ (j) $P(-2 \leq Z \leq c) = 0.055$ -1.42
   (k) $P(c \leq Z \leq 2) = 0.64$

3. Find the following values of z where z_α is defined as in Example 6.
   (a) $z_{0.0918}$ (b) $z_{0.1446}$ (c) $-z_{0.0244}$
   (d) $-z_{0.0104}$ (e) $z_{0.3745}$ (f) $-z_{0.3897}$

4. Find α if:
   (a) $z_\alpha = 2.25$ (b) $z_\alpha = 1.39$ (c) $z_\alpha = 1.93$
   (d) $z_\alpha = 0.56$ (e) $z_{\alpha/2} = 1.74$ (f) $z_{\alpha/2} = 1.46$.

5. Suppose X is normally distributed with mean 50 and standard deviation 4. In each case indicate the given probability as an area under the normal curve, and determine its value.
   (a) $P(51 \leq X \leq 56)$ (b) $P(X \geq 47)$
   (c) $P(44 \leq X \leq 53)$ (d) $P(X \leq 48)$
   (e) $P(X \leq 52)$ (f) $P(X \geq 59)$

hardest problem of this type

.1628

.4772   .4772

C   on 2

.4772

from table for .1628

-.42

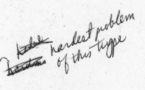

6. Suppose X has the normal distribution with mean 10 and standard deviation 2. Find the following probabilities.

   (a) $P(10 \leq X \leq 11.6)$ (b) $P(X \leq 12.8)$

   (c) $P(7 < X \leq 13)$ (d) $P(X \geq 8.6)$

   (e) $P(11 \leq X \leq 13)$ (f) $P(7.6 \leq X \leq 9.2)$

7. If X is normally distributed with mean -0.5 and standard deviation 0.02, find the following probabilities.

   (a) $P(-0.5 \leq X \leq -0.46)$ (b) $P(X \leq -0.47)$

   (c) $P(-0.53 \leq X \leq -0.45)$ (d) $P(X \geq -0.53)$

   (e) $P(-0.48 \leq X \leq -0.45)$ (f) $P(-0.55 \leq X \leq -0.52)$

8. The length of life of an instrument produced by a machine has a normal distribution with mean life of 12 months and standard deviation of 2 months. Find the probability that an instrument produced by the machine will last:

   (a) less than 7 months   (b) between 7 and 12 months.

9. Suppose the annual production of milk, per milk cow, has a normal distribution with mean amount $\mu = 11{,}200$ pounds and a standard deviation $\sigma = 280$ pounds. What fraction of the milk cows have an annual yield of more than 12,000 pounds?

10. The height of an adult male is known to be normally distributed with a mean of 69 inches and a standard deviation of 2.5 inches. How high should a doorway be so that 96 percent of the adult males can pass through it without having to bend?

11. The diameter of a lead shot produced by a machine has a normal distribution with a mean diameter equal to 2 inches and a standard deviation equal to 0.05 inches. Find approximately what the diameter of a circular hole should be so that only 3 percent of the lead shots will pass through it.

12. The nicotine content in a certain brand of king-size cigarettes has a normal distribution with a mean content of 1.8 mg and a standard deviation of 0.2 mg. Find the probability that the nicotine content of a randomly picked cigarette of this brand will be:

   (a) less than 1.45 mg

   (b) between 1.45 and 1.65 mg

   (c) between 1.95 and 2.15 mg

   (d) more than 2.15 mg.

13. In Exercise 12, what value is such that 80 percent of the cigarettes will exceed it in their nicotine content?

14. The diameter of a bolt produced by a machine is normally distributed with a mean diameter of 0.82 cm and a standard deviation of 0.004 cm. What percent of the bolts will meet the specification that they be between 0.816 cm and 0.826 cm in diameter?

15. The amount of sap collected from a tree has a normal distribution with a mean amount of 12 gallons and a standard deviation of 2.5 gallons.

   (a) Find the probability that a tree will yield more than 11 gallons.

   (b) If 1000 trees are tapped, approximately how many trees will yield more than 11 gallons?

16. The amount of coffee (in ounces) filled in a jar by a machine has a normal distribution with a mean amount of 16 ounces and a standard deviation of 0.2 ounces.
    (a) Find the probability that a randomly picked jar will contain.
        i. less than 15.7 ounces
        ii. more than 16.3 ounces.
    (b) In a random sample of 200 jars, find approximately how many jars will contain:
        i. less than 15.7 ounces
        ii. more than 16.3 ounces.

$$Z = \frac{X - \mu}{\sigma}$$

17. The demand for meat at a grocery store during any week is approximately normally distributed with a mean demand of 5000 pounds and a standard deviation of 300 pounds.
    (a) If the store has 5300 pounds of meat, what is the probability that it is overstocked?
    (b) How much meat should the store have in stock per week so as not to run short more than 10 percent of the time?

$$a) \frac{5300 - 5000}{300} = 1 \qquad (Chart\ A\text{-}2)$$
$$P(X < 5300) = P(Z < 1) = .5 + .3413$$
$$= .8413$$

18. The time that a skier takes on a downhill course has a normal distribution with a mean of 4.3 minutes and a standard deviation of 0.4 minutes. Find the probability that on a random run the skier will take between 4.1 and 4.5 minutes.

19. In Exercise 18, if the skier makes 6 runs, assuming independence, find the probability that:
    (a) two of the runs will be between 4.1 and 4.5 minutes
    (b) all of the runs will be between 4.1 and 4.5 minutes.
        Hint: Solve using the binomial formula.

20. Suppose X has a normal distribution with $\sigma = 4$. If $P(X \geq -8) = 0.0668$, find the mean μ.

21. Suppose the systolic resting blood pressure of an adult male is normally distributed with mean 138 mm of mercury and standard deviation 10 mm of mercury. If an adult male is picked at random, find the probability that his systolic blood pressure will be:
    (a) greater than 160 mm   (b) between 120 and 135 mm.

22. The mean annual income of a worker in a certain professional category is $23,200, with a standard deviation of $1,400. Assuming that the distribution of income is approximately normal, what is the probability that two workers picked at random will each earn less than $22,500?

23. According to the local transit authority, the mean length of time that a passenger spends waiting for a bus is 5 minutes with a standard deviation of 1.2 minutes. It may be assumed that the waiting time is normally distributed. If a passenger waits for a bus at two different times, what is the probability that this person will have to wait for more than 7 minutes on each occasion?

$$(Chart\ A\text{-}3)$$
$$\mu = 5 \quad \sigma = 1.2 \quad \frac{7-5}{1.2} = 1.67$$
$$P(X > 7) = P(Z > 1.67) = 0.0475$$
$$P(both\ times) = (0.0475)^2 = 0.002256$$

24. Water from the surface layer of the ocean was filtered through membrane filters with pore diameter of 0.32 μm. If the distribution of the width of the bacteria in the water is normal with mean width 0.28 μm and standard deviation 0.085 μm, what fraction of the bacteria will pass through the filter?

25. Suppose you board a plane in Los Angeles at 10 A.M. and have a connecting flight from New York to Paris 8 hours later. If the flight time from Los Angeles to New York is normally distributed with mean time of 6.2 hours and a standard deviation of 0.4 hours, what is the probability that you will land in New York with time to catch the connecting flight to Paris? You must take into account that it will take you half an hour to disembark the plane once it lands in New York and another half hour to get to the terminal where your flight to Paris is scheduled to take off.

26. A tire manufacturer claims that the steel-belted radial tire manufactured by the company has tire life that is normally distributed with a mean life of 32,000 miles and a standard deviation of 1800 miles. If the claim is valid, find the probability that a tire selected at random will last:
   (a) less than 28,000 miles   **(b)** more than 35,000 miles.

27. A manufacturer of TV sets wants to advertise that if a TV set manufactured by the company lasts for less than a certain number of years, the company will refund the full amount paid for the set. The company wants to pick the number of years for the advertisement in such a way that it will not have to give refunds on more than 4 percent of the sets. If, as a statistician for the company, you are asked to provide the number, what would be your answer? You may assume that the life of a TV set is normally distributed with a mean life of 8.5 years and a standard deviation of 1.8 years.

28. An applicant for a job with an insurance company has to take a standard psychological test. Suppose the time it takes to answer the test is normally distributed with mean time 60 minutes and standard deviation 8 minutes. Determine how much time an applicant should be given in general so that 97 percent of them will be able to complete the test.

29. A research university has stringent requirements that only students who score in the top 8 percent on the math SAT will be considered for admission as freshmen in the science program. Suppose scores on math SAT in the general population of students have a normal distribution with mean score 474 points and standard deviation 72 points. What is the minimum score that a student should obtain to be considered for admission?

30. A prison psychologist administers a profile test to the prison inmates. A prisoner is classified as "docile" if the score is less than 130 points, as "latently hostile" if the score is between 130 points and 210 points, and "dangerous, requiring close supervision" if the score is above 210 points. Through long experience the psychologist has determined that scores on such tests administered to prisoners have a normal distribution with mean 172 points and standard deviation 33 points. Find out what fraction of the prisoners will be classified as
   (a) docile   **(b)** latently hostile
   (c) dangerous, requiring close supervision.

31. A lawnmower manufacturing company has the policy that it will repair free of charge any lawnmower of a certain brand manufactured by them if it breaks down within one year of purchase. Suppose the length of time a lawnmower works without needing repairs is approximately normally distributed with mean time 4.8 years and standard deviation 1.8 years. What fraction of the lawnmowers will require the free repair?

32. The excerpt on the next page, describing investment risk, appeared in the *Investment Digest*, Winter 1994, published by the Variable Annuity Life Insurance

Company (VALIC). Provide a critical appraisal of the figure and the explanation presented for measuring investment risk.

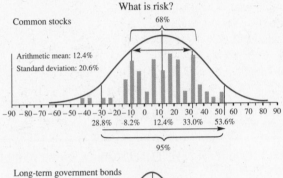

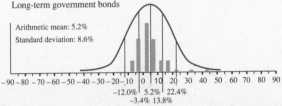

Measuring Investment Risk

Investment risk is often measured in terms of the volatility of an investment over time. A figure called the standard deviation is used to measure volatility. In general, as the standard deviation increases, so too does the degree of risk. As the standard deviation declines so too does risk and the higher the probability of achieving the expected performance in the future.

The diagram illustrates the historical standard deviation of common stocks and long-term government bonds. As you can see, the arithmetic mean or the average return for the common stocks asset class is 12.4%. This means that, on the average, stocks produced a return of 12.4% per year. A standard deviation of 20.6% means that 68% of the annual returns for com-

mon stocks have ranged from a low of −8.2% (12.4% − 20.6%) to a high of 33.0% (12.4% + 20.6%). Ninety-five percent of the returns for this category have been between −28.8% in the worst year and 53.6% in the best year.

The average (mean) return for long-term government bonds is lower, at 5.2%. The risk (standard deviation) is also much lower. Sixty-five percent of the annual returns fall between −3.4% and 13.8%. Ninety-five percent fall between −12.0% and 22.4%. In other words, by comparing the standard deviation of stocks and bonds, we see that stocks have historically provided a higher annual total return but it comes at the expense of higher volatility, as measured by standard deviation.

Source: Graphic "What is risk?" by Ibbotson, *Investment Digest,* Winter, 1994. Reprinted by permission of the Variable Annuity Life Insurance Company.

33. Use the normal probability plot to check the following data sets for the validity of normality assumption.

(a) 62   41   52   45   43   41   54   51

(b) 0.87   1.78   5.81   1.83   9.14   7.40   2.43   1.49
    4.31   6.51   1.60   3.62

(c) 106.2   104.8   97.8   101.4   98.9   106.0   94.3
    110.4   90.8   95.5

(d) 34.2   23.5   28.2   19.6   38.2   25.9   22.7   25.1

(e) 4.8   4.7   4.7   5.0   4.5   5.1   4.7   4.8   4.6   4.9

5-3   THE NORMAL APPROXIMATION TO THE BINOMIAL DISTRIBUTION

The arithmetic computations for finding the binomial probabilities can be both involved and time consuming, even for moderately small values of n.

If n is large, there is a convenient way to obtain an approximate value for binomial probabilities. This is accomplished by approximating the binomial distribution with n trials and probability of success p by means of a normal distribution that has a mean np and a standard deviation $\sqrt{npq}$, where $q = 1 - p$. We will consider the case when $n = 15$ and $p = 0.4$.

The *exact* probabilities $b(x; 15, 0.4)$ for various values of x, except for rounding, are reproduced in Table 5-3 from Table A-1 in Appendix III. Graphically, these probabilities can be represented as in Figure 5-31 by drawing line segments at each of the values 0, 1, 2, . . . , 15, or by means of a histogram, as in Figure 5-32, by erecting vertical bars. (See following page.)

TABLE 5-3   Binomial Probabilities When $n = 15$
and $p = 0.4$

x	$b(x; 15, 0.4)$
0	---
1	0.005
2	0.022
3	0.063
4	0.127
5	0.186
6	0.207
7	**0.177** $\longleftarrow \;\; = \dbinom{15}{7}(0.4)^7(0.6)^8$
8	0.118
9	0.061
10	0.024
11	0.007
12	0.002
13	---
14	---
15	---

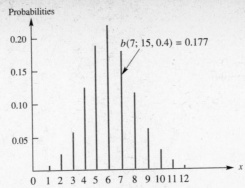

Figure 5-31 Graph of Binomial Probabilities for $n = 15$,
$p = 0.4$

From Table 5-3 we see that the exact probability of getting seven successes is equal to 0.177. In the discussion that follows, we will see how to obtain the approximate value for this probability by normal approximation.

Notice first of all that in Figure 5-32 the vertical bar centered at $x = 7$ has a base length of 1 unit extending from 6.5 to 7.5, and has a height equal to 0.177. Thus, the area of this vertical bar is equal to 0.177×1, or 0.177. In other words, $b(7; 15, 0.4)$ is *precisely* equal to the area of the darker rectangle in Figure 5-32. The important fact to realize is that the base of the rectangle extends from 6.5 to 7.5.

Under the normal approximation, the area of the rectangle will be approximated by the darker region in Figure 5-33, the area between 6.5 and 7.5 under the normal curve with $\mu = 15(0.4) = 6$ and $\sigma = \sqrt{15(0.4)(0.6)} = \sqrt{3.6} = 1.897$. Therefore, converting to the z scale, the probability of seven successes is approximately equal to the area between $z = \dfrac{6.5 - 6}{1.897}$ and $z = \dfrac{7.5 - 6}{1.897}$ under the standard normal curve. Letting the symbol $\approx$ stand

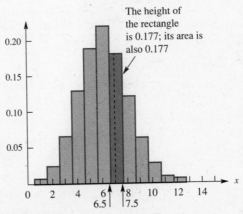

Figure 5-32

Histogram of binomial probabilities for $n = 15$, $p = 0.4$. Here, $b(7; 15, 0.4) = 0.177$ is *precisely* the area of the darker rectangle.

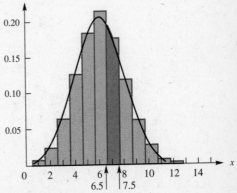

Figure 5-33

The darker area under the normal curve is *approximately* the probability $b(7; 15, 0.4)$.

319

for "approximately equal to," we have

$$b(7; 15, 0.4) \approx P\left(\frac{6.5 - 6}{1.897} < Z < \frac{7.5 - 6}{1.897}\right)$$

$$= P(0.26 < Z < 0.79)$$

$$= P(0 < Z < 0.79) - P(0 < Z < 0.26)$$

$$= 0.2852 - 0.1026$$

$$= 0.1826.$$

This approximate value, as can be seen, is remarkably close to the exact value of 0.177, off by only 0.0056.

We are now in a position to compute probabilities such as $P(5 \le X < 9)$, $P(5 < X \le 9)$, $P(X \ge 8)$, $P(X < 7)$, and so on. First, wherever there are strict inequalities $<$ or $>$, we rewrite the event equivalently using $\le$ or $\ge$. Thus, we rewrite $P(5 \le X < 9)$ as $P(5 \le X \le 8)$, $P(5 < X \le 9)$ as $P(6 \le X \le 9)$, and $P(X < 7)$ as $P(X \le 6)$.

It now follows that

i. $P(5 \le X < 9) = P(5 \le X \le 8)$

= sum of the areas of the rectangles *centered* at 5, 6, 7, and 8

$\approx$ area under the normal curve between 4.5 and 8.5 (See Figure 5-34a)

$$= P\left(\frac{4.5 - 6}{1.897} < Z < \frac{8.5 - 6}{1.897}\right)$$

$$= P(-0.79 < Z < 1.32)$$

$$= 0.4066 + 0.2852$$

$$= 0.6918$$

(The exact value from Table 5-3 is $0.186 + 0.207 + 0.177 + 0.118 = 0.688$.)

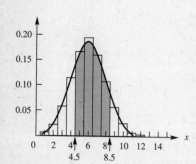

Figure 5-34a

ii. $P(5 < X \le 9) = P(6 \le X \le 9)$

= sum of the areas of the rectangles centered at 6, 7, 8, and 9

$\approx$ area under the normal curve between 5.5 and 9.5 (See Figure 5-34b)

$$= P\left(\frac{5.5 - 6}{1.897} < Z < \frac{9.5 - 6}{1.897}\right)$$

$$= P(-0.26 < Z < 1.85)$$

$$= 0.4678 + 0.1026 = 0.5704$$

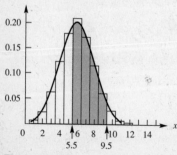

Figure 5-34b

iii. $P(X \ge 8) =$ sum of the areas of the rectangles centered at 8, 9, . . . , 15

$\approx$ area under the normal curve to the right of 7.5 (See Figure 5-34c)

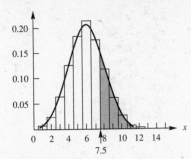

Figure 5-34c

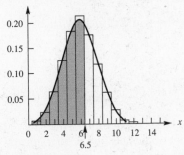

Figure 5-34d

$$= P\left(Z > \frac{7.5 - 6}{1.897}\right)$$

$$= P(Z > 0.79)$$

$$= 0.2148, \text{ from Table A-3 in Appendix III}$$

iv. $P(X < 7) = P(X \le 6)$

$\qquad = $ sum of the areas of the rectangles centered at $0, 1, \ldots, 6$

$\qquad \approx $ areas under the normal curve to the left of 6.5
(See Figure 5-34d)

$$= P\left(Z < \frac{6.5 - 6}{1.897}\right)$$

$$= P(Z < 0.26)$$

$$= 0.5 + 0.1026$$

$$= 0.6026$$

and so on.

In general, the approximation is carried out as follows: If n is large and p is not close to 0 or 1, the binomial probability $b(x; n, p)$ can be obtained approximately as the area from $x - \frac{1}{2}$ to $x + \frac{1}{2}$ under a normal curve where the mean of the distribution is np and the standard deviation is $\sqrt{npq}$. Converting to the z scale, we get the following result.

APPROXIMATION OF BINOMIAL PROBABILITIES

If n is large and p is not close to 0 or 1, then **the normal approximation to the binomial** is given by

$$b(x; n, p) \approx P\left(\frac{\left(x - \frac{1}{2}\right) - np}{\sqrt{npq}} < Z < \frac{\left(x + \frac{1}{2}\right) - np}{\sqrt{npq}}\right)$$

If i and j are integers,

$$P(i \le X \le j) \approx P\left(\frac{\left(i - \frac{1}{2}\right) - np}{\sqrt{npq}} < Z < \frac{\left(j + \frac{1}{2}\right) - np}{\sqrt{npq}}\right)$$

The approximation is satisfactory if both $np \ge 5$ and $n(1 - p) \ge 5$.

The approximation is rather close, even for n as low as 15, if p is in the neighborhood of 0.5. As a working rule, *the procedure will yield a satisfactory approximation to the binomial probabilities if both np and $n(1 - p)$ are greater than 5*. In the context of a normal approximation to the binomial distribution, this is what we mean when we say n is large. The addition and subtraction of $\frac{1}{2}$ is called the **continuity correction.**

EXAMPLE 1 Suppose the probability that a worker meets with an accident during a one-year period is 0.4. If there are 200 workers in a factory, find the probability that more than 75 workers will meet with an accident in one year. (Assume that whether a worker has an accident or not is independent of any other worker having one.)

SOLUTION Let X denote the number of workers who meet with an accident during a year. Then X has a binomial distribution with

$$\text{mean} = 200(0.4) = 80$$

and

$$\text{standard deviation} = \sqrt{200(0.4)(0.6)} = 6.928.$$

Now the probability that more than seventy-five workers meet with an accident is

$$P(X > 75) = P(X = 76) + P(X = 77) + \cdots + P(X = 200)$$

$$= \frac{200!}{76!124!}(0.4)^{76}(0.6)^{124} + \cdots + \frac{200!}{200!0!}(0.4)^{200}(0.6)^0$$

using the formula for binomial probabilities. The computation of the value of the expression above without a computer is out of the question. Under normal approximation, we approximate $P(X > 75)$ that is, $P(X \geq 76)$, as the area to the right of 75.5 under the normal curve with $\mu = 80$ and $\sigma = 6.928$, that is, as the area to the right of $\dfrac{75.5 - 80}{6.928}$ under the standard normal curve. Thus,

$$P(X > 75) = P(X \geq 76)$$

$$\approx P\left(Z > \frac{75.5 - 80}{6.928}\right)$$

$$= P(Z > -0.65)$$

$$= 0.5 + P(0 < Z < 0.65)$$

$$= 0.5 + 0.2422$$

$$= 0.7422. \qquad \bullet$$

EXAMPLE 2 A new vaccine was tested on 100 persons to determine its effectiveness. If the claim of the drug company is that a random person who is given the vaccine will develop immunity with probability 0.8, find the probability that:

(a) less than seventy-four people will develop immunity

(b) between seventy-four and eighty-five people, inclusive, will develop immunity.

SOLUTION Let X represent the number of people who develop immunity. Since $n = 100$ and $p = 0.8$,

$$\mu = 100(0.8) = 80 \quad \text{and} \quad \sigma = \sqrt{100(0.8)(0.2)} = 4.$$

(a)   We want $P(X < 74)$ that is, $P(X \leq 73)$. The approximate value of this probability is equal to the area to the left of 73.5 under the normal curve with $\mu = 80$ and $\sigma = 4$. Converting to the z scale, this is equal to $P\left(Z < \dfrac{73.5 - 80}{4}\right)$. Thus,

$$P(X < 74) \approx P\left(Z < \frac{73.5 - 80}{4}\right)$$
$$= P(Z < -1.62)$$
$$= P(Z > 1.62), \text{ due to symmetry}$$
$$= 0.0526, \text{ from Appendix III, Table A-3.}$$

(b) $P(74 \leq X \leq 85)$ can be approximated as the area between 73.5 and 85.5 under the normal curve with $\mu = 80$ and $\sigma = 4$, that is, as the area between $\dfrac{73.5 - 80}{4}$ and $\dfrac{85.5 - 80}{4}$ under the standard normal curve. Thus,

$$P(74 \leq X \leq 85) \approx P\left(\frac{73.5 - 80}{4} < Z < \frac{85.5 - 80}{4}\right)$$
$$= P(-1.62 < Z < 1.37)$$
$$= P(0 < Z < 1.37) + P(0 < Z < 1.62)$$
$$= 0.4147 + 0.4474$$
$$= 0.8621. \qquad \bullet$$

Remark   The discrepancy resulting from not using the continuity correction in a normal approximation to a binomial distribution will not be appreciable if n is very large.

10/29/96

Section 5-3 Exercises

1. Suppose the probability that a randomly selected college student favors strict environmental controls is 0.7. A sample of 200 students, selected at random, is interviewed, and suppose they respond independently. Let X be the number of students in the sample who favour environmental controls.
   (a) What is the exact distribution of X?
   (b) Compute the mean and the standard deviation of the number of students in the sample who favor strict controls.
   (c) Explain why the distribution of X can be approximated by the normal distribution. What are the parameters of this normal distribution?
   (d) Find the approximate probability that 140 students in the sample favor strict controls.
   (e) Find the approximate probability that between 130 and 146 students, inclusive, favor strict controls.

2. From past experience, a restaurant owner knows that about 40 percent of her customers order cold drinks. If 400 customers visit the restaurant, find the approximate probability that more than 180 customers will order cold drinks after justifying the approximation. (Assume that the customers order independently.)

3. The probability that a household owns a self-defrosting refrigerator is 0.9. If 190 households are interviewed, find the approximate probability that a self-defrosting refrigerator is owned by:
   (a) one hundred sixty households
   (b) more than one hundred sixty households
   (c) less than one hundred seventy households.
   State the assumptions and justify why approximation is permissible.

4. Seventy-three percent of the respondents in a Harris poll conducted November 11–15, 1993, agreed that the law should allow doctors to comply with the wishes of a dying patient in severe distress who asks to have his or her life ended. Assume that this figure reflects the general sentiment of the population. In a random sample of 82 people that you interview, find the probability that you will find
   (a) more than 70 such respondents
   (b) less than 50 such respondents
   (c) between 50 and 70 such respondents, both inclusive.

5. The city of Streeterville has a population of 50,000. Sixty percent of the voters in the upcoming mayoral election are believed to favor Mr. Nesbitt. What is the approximate probability that of the 140 people interviewed at random by a television network, less than 77 will favor Mr. Nesbitt? (Assume that the people respond independently.)

6. When a certain seed is planted, the probability that it will germinate is 0.1. If 1000 seeds are planted, find the approximate probability that:
   (a) more than 130 seeds will germinate
   (b) between ninety and ninety-five seeds, inclusive, will germinate.

7. **Risks accompanying decision making**   If a sample of eighty washers produced by a machine yields less than ten defective washers, the entire production of the day is accepted as good. Otherwise it is rejected. Find the probability that the production will be:
   (a) accepted when in fact the machine produces 20 percent defective washers
   (b) rejected when the machine produces 7 percent defective washers.

8. The probability that an electronic component functions beyond 1 hour is 0.8. If a machine uses 120 such components functioning independently, find the approximate probability that at the end of 1 hour:
   (a) at least ninety components will be functioning
   (b) between seventy and ninety components, inclusive, will be functioning.

9. **Risks accompanying decision making**   A drug company is debating whether a vaccine is effective enough to be marketed. The company claims that the vaccine is 90 percent effective. To test the company claim, the vaccine will be tried on 100 people. If eighty-two or more people develop immunity, the company claim will be granted. Find the approximate probability that the company claim will *not* be granted when in fact the drug is 90 percent effective, as the company claims.

10. The probability that a cathode tube manufactured by a machine is defective is 0.4. If a sample of sixty tubes is inspected, find the approximate probability that:
  (a) exactly half of the tubes will be defective
  (b) over half of the tubes will be defective.

11. The probability that a mail-order service receives between 400 and 500 orders on a day is 0.3. Find the approximate probability that during 90 days the mail-order service will receive between 400 and 500 orders on:
  (a) 32 days
  (b) more than 32 days.
  What assumption do you need to make?

$\mu = 90(.3) = 27$ $\sigma = \sqrt{90(.3)(.7)}$
$= \sqrt{18.9}$

$P(x=32) \approx P\left(\frac{31.5-27}{\sqrt{18.9}} < Z < \frac{32.5-27}{\sqrt{18.9}}\right)$

$= P(1.04 < Z < 1.27)$

$= 0.3980 - 0.3508$

$= 0.0472$

12. Suppose the probability that a person favors decriminalization of marijuana is 0.45. If 96 randomly selected people are interviewed, find the approximate probability that a majority will favor decriminalization. Assume independence.

13. A door-to-door salesperson visits 68 households. If the probability that the person makes a sale at a household is 0.2, find the approximate probability that he or she will make sales in more than 15 households. Assume independence.

14. The probability that a drunk driver will cause an accident is 0.2. If there are eighty drunk drivers on a freeway, assuming independence, find the approximate probability that:
  (a) fewer than fourteen will cause an accident
  (b) between fourteen and seventeen, inclusive, will cause an accident.

15. Seventy-two individuals picked at random responded independently when asked whether food stamps should be restricted to families who earn below the level of $12,000 per year for a family of four. A survey indicates that 70 percent of the population favors such a proposition. If the survey results are correct, what is the approximate probability that in the sample of 72 less than 44 will support the proposition?

16. A drug manufacturer claims that 75 percent of all doctors recommend aspirin over any other pain reliever. If this claim is indeed valid, what is the approximate probability that, in a sample of ninety-six doctors who were interviewed at random and who responded independently, less than sixty doctors will recommend aspirin? If in such a survey you actually found out that fifty-eight out of ninety-six preferred aspirin, what would you think about the nature of the claim?

17. A Gallup poll conducted in March 1993 reported that 70 percent of the U.S. population feels that laws covering the sale of firearms should be made more strict. Suppose you work as a statistician for an organization that wishes to challenge these findings and decide to conduct your own survey by interviewing 115 adults. Suppose 74 individuals respond that they would like more strict laws.
  (a) Is it highly unlikely to get a sample with this response if indeed the true percentage favoring strict laws is 70?
  (b) What is the probability of getting a sample with 74 or less wanting strict controls?
  (c) Do the results of your survey make a case for your organization to support its opposite point of view?

18. Answer questions in Exercise 17 if the survey results produced 66 respondents who favor strict controls.

✓ **19.** Recent medical studies have shown that 10 percent of women who are carriers of a certain errant gene are at risk for getting breast or ovarian cancer. A researcher has decided to follow the progress of 168 women who are carriers of the altered gene.

 (a) Assuming 10 percent to be the true percentage for the carriers to get the cancer,
   i. how many of the women in the study do you expect to get breast or ovarian cancer?
   ii. what is the approximate probability that between 8 and 26, both inclusive, will get it?

 (b) If it should turn out that 33 women eventually get breast or ovarian cancer, what would it indicate about the quoted rate of 10 percent?

 (c) Answer Question b if it turns out that 22 women get breast or ovarian cancer.

20. If a genotype Aa is self-fertilized, then the probability that an offspring is of AA genotype is $\frac{1}{4}$. When a genotype Aa was self-fertilized, there were 184 offspring. Assuming independence, what is the approximate probability that there will be more than sixty offspring of AA genotype?

21. In an article entitled "Unproven Remedies Tempt the Ailing" published in *The Wall Street Journal* (May 27, 1988), it was reported that 27 percent of U.S. adults have used one or more questionable health treatments at some point in their lives. If this is true, find the probability that in a sample of ninety individuals selected at random, between twenty-two and thirty, both inclusive, will have used questionable health treatments.

22. A study conducted by the Department of Health and Human Services reported that 15 percent of adults suffer from migraine headache. To test the validity of this statement, sixty-eight randomly selected adults are interviewed. Find the probability that the number of those having migraine headache is:

 (a) at least fifteen

 (b) less than seven

 (c) between seven and fifteen, both inclusive.

CHAPTER **5 Summary**

✓ *CHECKLIST: KEY TERMS AND EXPRESSIONS*

- ❑ relative frequency density, page 281
- ❑ probability density function, page 284
- ❑ rectangular distribution, page 286
- ❑ uniform distribution, page 286
- ❑ normal curve, page 292
- ❑ normal distribution, page 292
- ❑ standard normal curve, page 295
- ❑ standard normal variable, page 296
- ❑ standard normal distribution, page 296
- ❑ standardization, page 303
- ❑ standard score, page 303
- ❑ z-score, page 303
- ❑ normal probability plot, page 311
- ❑ normal scores plot, page 311
- ❑ normal scores, page 311
- ❑ nscores, page 311
- ❑ normal approximation to the binomial, page 321
- ❑ continuity correction, page 321

KEY FORMULAS

z-score for x

$$z = \frac{x - \mu}{\sigma}.$$

approximation of binomial probabilities

If n is large and p is not close to 0 or 1, then

$$b(x; n, p) \approx P\left(\frac{\left(x - \frac{1}{2}\right) - np}{\sqrt{npq}} < Z < \frac{\left(x + \frac{1}{2}\right) - np}{\sqrt{npq}} \right)$$

If i and j are integers,

$$P(i \le X \le j) \approx P\left(\frac{\left(i - \frac{1}{2}\right) - np}{\sqrt{npq}} < Z < \frac{\left(j + \frac{1}{2}\right) - np}{\sqrt{npq}} \right)$$

CHAPTER 5   Concept and Discussion Questions

1. Describe the features that characterize the normal distribution.

2. How many parameters does a normal distribution have? What does each parameter determine?

3. How large is the theoretical normal population?

4. Suppose a variable has a normal distribution with mean, say, 100. Describe the effect on the normal distribution curve if
   (a) the mean is increased   (b) the mean is decreased.

5. Suppose a variable has a normal distribution with standard deviation, say, 8. Describe the effect on the normal distribution curve if
   (a) the standard deviation is increased
   (b) the standard deviation is decreased.

6. What is the theoretical range of values of a normally distributed random variable?

7. A phenomenon cannot be modeled approximately by a normal distribution if the underlying variable assumes only positive values. True or false? Explain.

8. If you have been following the "Law and Probability" discussion introduced on page 204, we have already computed on page 252 that $P(X \ge 2) = 0.013$. This was done with the understanding that X has a binomial distribution with 2,000,000 independent trials and probability of success 1/12,000,000 on each trial. Here, the number of trials n is certainly very large. However, if we use the normal approximation to the binomial we get

$$P(X \ge 2) \approx P\left(Z \ge \frac{1.5 - 1/6}{\sqrt{1/6}} \right) = P(Z \ge 3.27).$$

From Table A-3, $P(Z \ge 3.27)$ is less than 0.001 and considerably different from the answer 0.013. Explain why.

CHAPTER **5** **Review Exercises**

1. The graph below gives the probability density curve of a continuous random variable X.
   Copy the graph and shade appropriate regions to describe the following probabilities:
   (a) $P(7 < X \le 9)$ **(b)** $P(X \le 6)$ **(c)** $P(X > 10)$

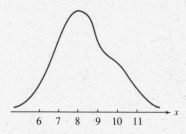

2. The following curve is the probability distribution of a region's annual rainfall. Copy the graph and shade appropriate regions to describe the following probabilities that:
   (a) the annual rainfall will exceed 40 inches
   (b) the annual rainfall will be between 20 and 40 inches
   (c) the annual rainfall will not be between 20 and 50 inches.

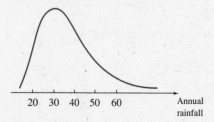

3. The following figure represents the probability density curve describing the distribution of the error (in ounces) in filling a one-pound bag of flour. (Error is

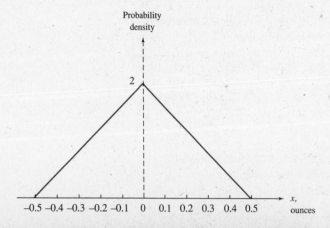

positive if the bag is overfilled, negative if underfilled.) Find the probability that a bag is:

(a) overfilled by more than 0.2 ounces

(b) underfilled by more than 0.3 ounces

(c) underfilled by more than 0.2 ounces or overfilled by more than 0.3 ounces.

4. Answer true or false: If X has a symmetric distribution with $P(X > 12) = 0.3$, then $P(X < -12) = 0.3$, also.

5. A continuous random variable X has a distribution that is symmetric about a vertical axis at $x = -8$. Suppose it is given that $P(-10 < X < -6) = 0.36$. Compute the following probabilities.

(a) $P(X > -6)$ (b) $P(X > -10)$ (c) $P(X < -10 \text{ or } X > -8)$

6. Describe the probability density curve of a normal distribution. Draw, on the same graph, probability density curves for two normal distributions that have the same variance but have means -3 and 5.

7. If Z is a standard normal random variable, use Table A-2 or A-3 to find the following probabilities.

(a) $P(1.2 < Z < 2.2)$ (b) $P(-1.2 < Z < 2.8)$

(c) $P(-2.8 < Z < -1.5)$ (d) $P(Z > -2.7)$

(e) $P(-1.2 < Z < 0.8)$ (f) $P(-1.5 < Z < -0.6)$

8. If Z is a standard normal random variable, find the value of c in each of the following cases.

(a) $P(Z < c) = 0.9495$ (b) $P(Z < c) = 0.1251$

(c) $P(Z > 3c) = 0.1038$ (d) $P(Z < 4c) = 0.0073$

(e) $P(c < Z < 2.5) = 0.9295$ (f) $P(c < Z < -0.8) = 0.1965$

9. Suppose X has a normal distribution with mean 150 and standard deviation 8. Compute the following probabilities.

(a) $P(X < 152.8)$ (b) $P(X > 140.8)$

(c) $P(160.4 < X < 168.4)$ (d) $P(144.4 < X < 148.0)$

10. If X has a normal distribution with mean 12.5 and standard deviation 5, determine c when:

(a) $P(X > c) = 0.0322$ (b) $P(X < c) = 0.1685$

(c) $P(X > c) = 0.9793$ (d) $P(X < c) = 0.6985$

11. The distance (in miles) that a brand of radial tire lasts is normally distributed with a mean distance of 40,000 miles and a standard deviation of 4000 miles. If the company guarantees free replacement of any tire that lasts less than 33,000 miles, what fraction of the tires will the company have to replace?

12. The breaking strength of cables manufactured by a company has the normal distribution with a mean strength of 1200 pounds and a standard deviation of 64 pounds.

(a) Find the probability that a randomly picked cable will have breaking strength:

   i. less than 1059 pounds

   ii. greater than 1136 pounds.

(b) If the breaking strengths of 500 randomly picked cables are measured, find approximately how many will have breaking strength:

   i. less than 1059 pounds

   ii. greater than 1136 pounds.

13. Dresswell Company is a large apparel manufacturer. A survey conducted by the company revealed that collar sizes of men's shirts in a certain region are normally distributed with a mean collar size of 17 and a standard deviation of 1.6. Suppose that an individual whose collar size is between 17.5 and 18.5 will purchase a shirt with collar size 18. If Dresswell manufactures shirts of all collar sizes to meet the needs of its customers, what proportion of shirts should be of size 18?

14. A company administers a battery of tests in its recruiting procedure. The first test is a verbal test. Suppose it is known from past experience that scores on the verbal test have a normal distribution with a mean score of 150 points and standard deviation of 30 points. A candidate will be considered for further screening only if his or her score on this test is 160 points or more. What fraction of the candidates will be eliminated at this stage?

15. A machine manufactures a very large number of resistors. Suppose that 10 percent of the resistors produced by the machine are defective. What is the approximate probability that a sample of 400 will contain 43 defective resistors?

16. The probability that an electric sander will last between 500 and 600 hours before it needs repairs is 0.7. In a school district, 200 sanders are in use in different woodworking classes. What is the approximate probability that more than 150 sanders will last between 500 and 600 hours?

17. According to an article published in *The Wall Street Journal* (May 27, 1988), about one out of seven adult Americans suffers from arthritis. Assuming this to be true, find the probability that in a sample of 340 individuals picked at random there will be:
    (a) at least fifty-five having arthritis
    (b) at most thirty-five having arthritis.

 ## WORKING WITH LARGE DATA SETS

Project: Refer to the SAT scores data in Appendix IV, Table B-2. Using MINITAB, compute the nscores for each of the variables: math scores, verbal scores, and high school GPA. Obtain the normal probability plots and determine for which variables normal distribution appears reasonable. Prepare a written report.

 ## GRAPHING CALCULATOR INVESTIGATIONS

Refer to the instructions given in the TI-82 Programs Appendix regarding the operation of the program ZPROG. Then use that program to answer the following questions.

1. Find the following probabilities:
   (a) $P(-1 < z < 1)$ (b) $P(-2 < z < 2)$ (c) $P(-3 < z < 3)$
   (d) $P(-4 < z < 4)$ (e) $P(z < -2.4)$ (f) $P(z < -1.12)$
   (g) $P(z < 2.89)$ (h) $P(-1.11 < z < -0.23)$

2. Read the section of the TI-82 Programs Appendix regarding the operation of the program NSNORMAL. Then redo the following textbook exercises using this program.
   Section 5-2: Problems 6, 7, 8, 9, 12, 15, 16, and 22.

3. Read the section of the Appendix entitled "Using the TI-82 as a Data Analysis Tool" to solve normal approximation to the binomial problems using the program NSNORMAL from the TI-82 Programs Appendix. Then redo the following textbook exercises:
   Section 5-3: Problems 3, 5, 7, 9, and 11.

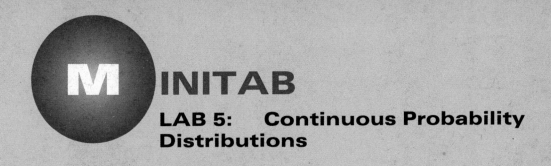

MINITAB

LAB 5: Continuous Probability Distributions

I n Lab 5, we learn how MINITAB can be used to find certain probabilities associated with the normal distribution. Also, we consider the important aspect of empirical verification by simulating experiments from this distribution.

1 The goal here is to sketch the normal density curves. First we sketch the standard normal curve.

```
MTB >NOTE PLOT OF THE STANDARD NORMAL CURVE

MTB >NOTE X-VALUES IN C1

MTB >SET C1

DATA>-3.2:3.2/0.5

DATA>END

MTB >NOTE HEIGHTS FOR X-VALUES IN CI PLACE IN C2

MTB >PDF AT C1 PUT IN C2;

SUBC>NORMAL MEAN 0 SIGMA 1.

MTB >PRINT C1 C2

MTB >PLOT C2 VERSUS C1
```

To plot the curve we first generate numbers to be plotted along the x-axis. These numbers are generated in Column C1 with the command SET C1 and the input DATA $> -3.2 : 3.2/0.5$, which generates consecutive points from -3.2 to 3.2, spaced 0.5 units apart. (The range and the spacing can be changed suitably, as desired.) The heights at these points are obtained with the command PDF* and are placed in Column C2. The subcommand NORMAL stipulates the underlying distribution, giving its mean and standard deviation. The PLOT command then sketches the curve.

---

* For a discrete distribution the command PDF gives the probabilities, whereas for a continuous distribution it gives the heights of the probability density curve for specified values of x.

The procedure for any other normal curve is similar, as can be seen from the following program.

```
MTB >NOTE PLOT OF THE NORMAL CURVE WITH MEAN 20
       SIGMA 6

MTB >NOTE X-VALUES PLACED IN C5

MTB >SET C5

DATA>2:38/3

DATA>END

MTB >NOTE HEIGHTS FOR X-VALUES IN C5 PLACE IN C6

MTB >PDF AT C5 PUT IN C6;

SUBC>NORMAL MEAN 20 SIGMA 6.

MTB >PLOT C6 VERSUS C5
```

In the above, the input DATA $>$ 2 : 38/3 covers the range 20–3(6) to 20 + 3(6) spaced 6/2, that is, 3 units apart.

Assignment   Sketch the normal probability curves having $\mu = 50$ and standard deviations 0.5, 1 and 2, respectively. Compare the plots with the curves in Figure 5-11. In each case use x values with the range $\mu - 3\sigma$ to $\mu + 3\sigma$, and spacing given by $\mu - 3\sigma : \mu + 3\sigma/(\sigma/2)$.

Assignment   Sketch the normal probability curves with $\sigma = 2$ and means $-6, 0$, and 8, respectively. Compare the plots with the curves in Figure 5-12.

2   MINITAB provides answers to questions dealing with probabilities associated with a normal distribution. For example, if X is normally distributed with mean 10 and sigma 4, MINITAB will give answers to probabilities such as $P(X \leq 7.3)$ or $P(X \leq 13.5)$. The command CDF provides the answers. Also, for example, given that $P(X \leq c) = 0.9463$, MINITAB will provide the value of c with the command INVCDF. This procedure is used to find answers to the following questions.

1. $P(Z \leq 1.73) = ?$ We want the probability, given the value along the x-axis.

```
MTB >NOTE AREA LEFT OF 1.73 UNDER STANDARD NORMAL
       CURVE

MTB >CDF AT 1.73;

SUBC>NORMAL MEAN 0 SIGMA 1.
```

(If you subtract 0.5 from the answer displayed, you will see that it agrees with the value you get from Appendix III, Table A-2.)

2. If X is normal with mean $= 20$ and sigma $= 6$, what is $P(X \leq 23.5)$ equal to?

```
MTB >NOTE AREA LEFT OF 23.5; NORMAL MEAN 20 SIGMA 6

MTB >CDF AT 23.5;

SUBC>NORMAL MEAN 20 SIGMA 6.
```

If you wish to use the value $P(Z \leq 23.5)$ for subsequent computations, you can store the answer as a constant as follows:

```
MTB >CDF AT 23.5 STORE AS K1;

SUBC>NORMAL MEAN 20 SIGMA 6.
```

For example, you can now compute $P(16 < X \leq 25)$ as follows:

```
MTB >CDF AT 25 AS K1;

SUBC>NORMAL MEAN 20, SIGMA 6.

MTB >CDF AT 16 AS K2;

SUBC>NORMAL MEAN 20, SIGMA 6.

MTB >LET K3 = K1-K2

MTB >PRINT K3
```

3. If $P(Z \leq c) = 0.9515$ what is c? We want the value along the x-axis, given the probability—a process inverse to finding CDF.

```
MTB  >NOTE VALUE FOR WHICH AREA TO ITS LEFT IS
        0.9515

MTB  >NOTE STANDARD NORMAL CURVE

MTB  >INVCDF AT 0.9515;

SUBC >NORMAL MEAN 0 SIGMA 1.
```

4. If X is normal with mean 20, sigma 6, and $P(X \leq c) = 0.9515$, what is c?

```
MTB >NOTE VALUE FOR WHICH AREA TO ITS LEFT IS
       0.9515

MTB >NOTE NORMAL CURVE MEAN 20 SIGMA 6

MTB >INVCDF AT 0.9515;

SUBC>NORMAL MEAN 20 SIGMA 6.
```

Assignment

(a) If *X* is normal with mean 60 and sigma 8 find the following probabilities.
   i. $P(X \leq 71.3)$
   ii. $P(X \leq 51.82)$
   iii. $P(X \geq 51.82)$
   iv. $P(X > 54.75)$
   v. $P(61.84 < X \leq 79.38)$

(b) If *X* is normal with mean -15 and sigma 7.5, find the value of *c* if:
   i. $P(X \leq c) = 0.8542$
   ii. $P(X \leq c) = 0.238$.

3 Next, MINITAB has the command NSCORES which can be used to obtain the *normal probability plot* for any given data. We illustrate this using the following set of commands to produce a normal probability plot of the data of Example 13 on page 311.

```
MTB >SET DATA IN C1

DATA>89 78 48 85 67 45 60 62 62 56

DATA>END OF DATA

MTB >PRINT C1

MTB >NSCORES OF C1 INTO C2

MTB >PLOT C2 VS C1
```

For the data in Column C1, we placed the normal scores in Column C2 using the NSCORES command. The plot command gives the plot.

Assignment   Work out Exercise 33 on page 318.

4 Finally, we consider simulation and empirical aspects when the underlying distribution is a normal distribution.

```
MTB >NOTE SIMULATION, 10 VALUES FROM NORMAL MEAN 20
      SIGMA 6

MTB >RANDOM 10 VALUES PUT IN C1;

SUBC>NORMAL MEAN 20 SIGMA 6.

MTB >PRINT C1
```

To discuss the empirical aspects we take a reasonably large sample of 150 values.

```
MTB >NOTE SAMPLE OF 150 FOR EMPIRICAL ANALYSIS

MTB >RANDOM 150 VALUES PUT IN C1;

SUBC>NORMAL MEAN 20 SIGMA 6.
```

```
MTB >MEAN C1

MTB >STANDARD DEVIATION C1

MTB >HISTOGRAM C1
```

Compare the sample mean of the simulated data with 20, the mean of the distribution, and the sample standard deviation with 6, the standard deviation of the distribution.

We now type STOP and conclude the MINITAB session using the logging off procedure.

Commands you have learned in this session.

```
PDF;
NORMAL.

CDF;
NORMAL.

INVCDF;
NORMAL.

RANDOM;
NORMAL.

NSCORES
```

Sampling Distributions and Designs

6

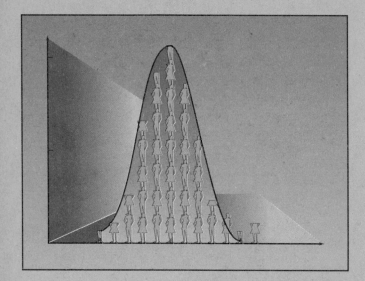

INTRODUCTION

This is the last chapter in our step-by-step development of probability theory. In the chapters that follow, we will study the inferential, or inductive, aspects of statistics. The field of inductive statistics is concerned with studying facts about populations. Specifically, our interest is in learning about population parameters, which involves picking a sample and computing the values of appropriate statistics. Now, the value of a statistic will depend on the observed sample values, which, more than likely, will differ from sample to sample. Thus, a statistic is a random variable. In keeping with the convention adopted in Chapter 4, we will denote a statistic by an uppercase letter. For example, $\bar{X}$ will stand for the statistic *sample mean*. Its value for a given set of sample values will be denoted by $\bar{x}$, the lowercase letter x with a bar. Similarly, the sample variance S^2 is a statistic whose value for a given sample data will be denoted as s^2. Although we will try not to overemphasize this notational encumbrance, it is important to be aware of the distinction between a statistic, which is a random variable, and its particular value, which depends on a given set of sample values.

Why Study Sampling Distribution?

By the **sampling distribution** of a statistic we mean the theoretical probability distribution of the statistic. Consider the following example: Suppose an advertising agency wants to find out the proportion of households in the United States that use Toothpaste *A*. The exact value of this proportion will not be known unless every household is interviewed. Suppose the agency decides to question 100 households and use the proportion in this sample as an indicator of the proportion in the population. It is possible that when 100 households are actually interviewed, 62 will say they use Toothpaste *A*, thus giving a value of the sample proportion as 0.62. If another sample of 100 is taken, we might get a response that 55 use Toothpaste *A*, that is, a sample proportion of 0.55. Every time a new group of 100 is interviewed, a new value for the sample proportion is obtained. Such variation from sample to sample is unavoidable because of the inherent variability in the population. At this point we might envision the totality of all the values of the sample proportion when groups of 100 households are interviewed. This totality of values will describe the distribution of the sample proportion.

We are interested in sampling distributions for a reason that is easy to understand. From the first sample that was drawn, the sample proportion obtained was 0.62. Is 0.62 the true proportion in the population? Maybe not. But before drawing the sample, we intuitively feel that the chance is high that we will find a value near the true population value of the proportion. Exactly how great is this chance? To be more specific, what is the probability that the sample proportion will differ from the true proportion by, perhaps, more than 0.05? We can answer this question and questions of this type only if we know the probability distribution of the sample proportions; for this reason we must learn about the sampling distribution of a sample proportion. In the same way, a need arises to discuss sampling distributions of other statistics, such as $\bar{X}$, S^2, and so on.

Empirical Meaning of a Sampling Distribution

The problem of determining the sampling distribution of a statistic is essentially a mathematical problem. Informally the following concept is involved: We conceive of *all the possible samples* of a given size n that can be drawn from the population under study. For each of these samples we compute the value of the statistic of interest. The sampling distribution of the statistic is then the distribution of the totality of these values. Consequently, a nonmathematical understanding of the theoretical distribution can be acquired by considering an experiment where a *large* number of samples of size n are drawn, every time computing the value of the statistic. From the values generated in this way we might obtain the relative frequencies and the histogram of the empirical distribution as discussed in Chapter 1. If a fairly large number of samples are drawn, the histogram so obtained will provide a fairly close approximation to the theoretical sampling distribution of the statistic.

After discussing distributions of the sample mean and the sample proportion in Section 1, we consider some basic sampling techniques in Section 2, and some aspects of designs of experiments in Section 3.

6-1   DISTRIBUTION OF THE SAMPLE MEAN AND SAMPLE PROPORTION

In this section we first concentrate entirely on the **sampling distribution of the mean,** or, to put it more descriptively, the distribution of the sample mean $\bar{X}$. For the purpose of illustration, consider the following rather artificial example where a sample of two (sample of size 2) is picked from a population consisting of four members whose weights (in pounds) are 148, 156, 176, and 184. The probability distribution of the weights in this population can be described as follows.

Weight x	148	156	176	184
Probability	$\frac{1}{4}$	$\frac{1}{4}$	$\frac{1}{4}$	$\frac{1}{4}$

Using the formulas in Chapter 4, we obtain the mean μ and the variance σ^2 of this population as follows.

$$\mu = (148)\frac{1}{4} + (156)\frac{1}{4} + (176)\frac{1}{4} + (184)\frac{1}{4}$$

$$= 166$$

$$\sigma^2 = (148 - 166)^2\frac{1}{4} + (156 - 166)^2\frac{1}{4} + (176 - 166)^2\frac{1}{4}$$

$$+ (184 - 166)^2\frac{1}{4}$$

$$= 212$$

The population described above is the parent population with which we start. Now suppose we pick a random sample of two. Sampling carried out in such a way that once an item is picked, it is not returned to the lot, is called **sampling without replacement.** On the other hand, if an item is returned to the lot before picking the next one, the sampling is called **sampling with replacement.** In an actual experiment the procedure will usually consist of picking items without replacement. However, to inject some important ideas into our discussion, we will not only consider sampling without replacement but also sampling with replacement. We begin with the latter case.

SAMPLING WITH REPLACEMENT

In sampling with replacement from a population of four items, we get 4^2, or 16, possible samples. The 16 samples are listed below in Table 6-1 along with the corresponding sample means and probabilities. When we, as experimenters, actually pick our sample, we will end up with *one* of these. By a *random sample* we mean that the probability of picking any one of these samples is the same, and equal to $\frac{1}{16}$. (We could have written each sample on a slip of paper, mixed the sixteen slips of paper thoroughly, and then picked one blindfolded. Later in Section 6-2 we will explain how random number tables can be used for this purpose.)

TABLE 6-1 Samples of Size 2 Drawn with Replacement, Their Means and Probabilities

Sample Number	Sample	Mean $\bar{x}$	Probability
1	148, 148	148	1/16
2	148, 156	152	1/16
3	148, 176	162	1/16
4	148, 184	166	1/16
5	156, 148	152	1/16
6	156, 156	156	1/16
7	156, 176	166	1/16
8	156, 184	170	1/16
9	176, 148	162	1/16
10	176, 156	166	1/16
11	176, 176	176	1/16
12	176, 184	180	1/16
13	184, 148	166	1/16
14	184, 156	170	1/16
15	184, 176	180	1/16
16	184, 184	184	1/16

TABLE 6-2 Sampling Distribution of $\overline{X}$ (Sampling with Replacement)

Sample Mean $\overline{x}$	Probability
148	1/16
152	2/16
156	1/16
162	2/16
166	4/16
170	2/16
176	1/16
180	2/16
184	1/16

Listing the distinct values of $\overline{x}$ and combining the corresponding probabilities, we get the probability distribution in Table 6-2. It is the so-called sampling distribution of $\overline{X}$.

Now that we have obtained the sampling distribution of $\overline{X}$, we will find the mean and the variance of this distribution. We will denote these quantities by using the subscript $\overline{x}$ as in $\mu_{\overline{X}}$ and $\sigma_{\overline{X}}^2$, respectively. This will enable us to distinguish them from the mean μ and variance σ^2 of the parent population, which are written without subscripts.

Applying the rule for finding the mean of a discrete random variable given in Chapter 4, we get

$$\mu_{\overline{x}} = (148)\frac{1}{16} + (152)\frac{2}{16} + (156)\frac{1}{16} + (162)\frac{2}{16} + (166)\frac{4}{16}$$
$$+ (170)\frac{2}{16} + (176)\frac{1}{16} + (180)\frac{2}{16} + (184)\frac{1}{16}$$
$$= 166.$$

We see that $\mu_{\overline{x}}$ is the same as the population mean μ, both equal to 166.

Similarly, using the formula for the variance appropriately, we get

$$\sigma_{\overline{X}}^2 = (148 - 166)^2\frac{1}{16} + (152 - 166)^2\frac{2}{16}$$
$$+ (156 - 166)^2\frac{1}{16} + (162 - 166)^2\frac{2}{16}$$
$$+ (166 - 166)^2\frac{4}{16} + (170 - 166)^2\frac{2}{16}$$
$$+ (176 - 166)^2\frac{1}{16} + (180 - 166)^2\frac{2}{16}$$
$$+ (184 - 166)^2\frac{1}{16}$$
$$= 106.$$

Notice from these computations that since $\sigma_{\overline{X}}^2 = 106$ and $\sigma^2 = 212$, we get

$$\sigma_{\overline{X}}^2 = \frac{\sigma^2}{2} = \frac{\text{variance of the parent population}}{\text{sample size}}.$$

Therefore,

$$\sigma_{\overline{X}} = \frac{\sigma}{\sqrt{2}} = \frac{\text{standard deviation of the population}}{\sqrt{\text{sample size}}}.$$

The standard deviation $\sigma_{\overline{X}}$ is smaller than the standard deviation σ of the parent population. This is not difficult to understand. For example, in the original population the value 184 at the upper extreme had a probability of $\frac{1}{4}$ and the value 148 at the lower extreme had a probability of $\frac{1}{4}$. But as can be seen from the distribution of $\overline{X}$ obtained in Table 6-2, the values at the extremes of 184 and 148 carry a probability of only $\frac{1}{16}$. This is reflected in the smaller value for the standard deviation $\sigma_{\overline{x}}$.

This connection between μ and $\mu_{\overline{x}}$ and between σ and $\sigma_{\overline{x}}$ that we have shown is always true, irrespective of the size of the population. We can make the following general statement.

MEAN AND STANDARD DEVIATION—SAMPLING WITH REPLACEMENT

If samples of size n are drawn with replacement from a population with mean μ and standard deviation σ, then

$$\mu_{\overline{x}} = \mu = \text{population mean}$$

and

$$\sigma_{\overline{x}} = \frac{\sigma}{\sqrt{n}} = \frac{\text{population standard deviation}}{\sqrt{\text{sample size}}}$$

SAMPLING WITHOUT REPLACEMENT

When sampling without replacement, we have 4×3, or 12, possible samples given in Table 6-3. By a *random sample,* we now mean that any sample has a probability of $\frac{1}{12}$ of being drawn. Combining the values of $\overline{x}$, we obtain the sampling distribution of $\overline{X}$ given in Table 6-4.

TABLE 6-3 Samples of Size 2 Drawn without Replacement, Their Means and Probabilities

Sample Number	Sample	Mean $\overline{x}$	Probability
1	148, 156	152	1/12
2	148, 176	162	1/12
3	148, 184	166	1/12
4	156, 148	152	1/12
5	156, 176	166	1/12
6	156, 184	170	1/12
7	176, 148	162	1/12
8	176, 156	166	1/12
9	176, 184	180	1/12
10	184, 148	166	1/12
11	184, 156	170	1/12
12	184, 176	180	1/12

TABLE 6-4 Sampling Distribution of $\overline{X}$ (Sampling without Replacement)

Sample Mean $\overline{x}$	Probability
152	1/6
162	1/6
166	1/3
170	1/6
180	1/6

The mean of the sampling distribution of $\overline{X}$ in this case is given as

$$\mu_{\overline{x}} = (152)\frac{1}{6} + (162)\frac{1}{6} + (166)\frac{1}{3} + (170)\frac{1}{6} + (180)\frac{1}{6}$$

$$= 166.$$

Thus, we see that $\mu_{\overline{x}}$ is equal to the mean of the parent population even when the sampling is carried out without replacement. This is always the case. *Irrespective of whether the sampling is with or without replacement, the mean of the sampling distribution of $\overline{X}$ is always equal to the mean μ of the parent population.*

Turning now to the variance $\sigma_{\overline{X}}^2$, since $\mu_{\overline{x}} = 166$, we have $\sigma_{\overline{X}}^2$ given as

$$\sigma_{\overline{X}}^2 = (152 - 166)^2\frac{1}{6} + (162 - 166)^2\frac{1}{6} + (166 - 166)^2\frac{1}{3}$$

$$+ (170 - 166)^2\frac{1}{6} + (180 - 166)^2\frac{1}{6}$$

$$= \frac{424}{6}.$$

This value of $\sigma_{\overline{X}}^2$ can be rewritten as

$$\sigma_{\overline{X}}^2 = \frac{212}{2} \cdot \frac{4 - 2}{4 - 1}.$$

This equality contains a clue as to what we might anticipate for the formula in the general case where we have a population of size N from which a sample of size n is drawn without replacement. It turns out that, in sampling without replacement,

$$\sigma_{\overline{X}}^2 = \frac{\sigma^2}{n} \cdot \frac{N - n}{N - 1}$$

so that

$$\sigma_{\overline{X}} = \sqrt{\frac{\sigma^2}{n} \cdot \frac{N - n}{N - 1}} = \frac{\sigma}{\sqrt{n}} \cdot \sqrt{\frac{N - n}{N - 1}}.$$

Notice that on the right-hand side of the formula there is not only the term $\sigma/\sqrt{n}$, which we had in the case of sampling with replacement, but also a factor $\sqrt{(N - n)/(N - 1)}$. This factor is called the **finite population correction factor.** If the population is very large compared to the sample size, then this factor will be very close to 1. For example, if $N = 10,000$ and $n = 40$, then

$$\sqrt{\frac{N - n}{N - 1}} = \sqrt{\frac{10,000 - 40}{10,000 - 1}} = \sqrt{0.9961} = 0.998$$

For all practical purposes this value is equal to 1. Thus, if the population size is large relative to the sample size, then $\sigma_{\bar{X}} \approx \sigma/\sqrt{n}$, and this is the same as the standard deviation of $\bar{X}$ when sampling with replacement. Intuitively, this is just what one might expect. After all, if the population is extremely large, its composition is not going to change appreciably if we do not return the objects to the lot as they are picked. Even if the sampling is without replacement, for all practical purposes we can regard it as a case of sampling with replacement.

In summary, we have the following results.

MEAN AND STANDARD DEVIATION—SAMPLING WITHOUT REPLACEMENT

When random samples of size n are drawn without replacement from a finite population of size N that has a mean μ and a standard deviation σ, the mean and the standard deviation of the sampling distribution of $\bar{X}$ are given by

$$\mu_{\bar{X}} = \mu$$

and

$$\sigma_{\bar{X}} = \frac{\sigma}{\sqrt{n}} \sqrt{\frac{N - n}{N - 1}}.$$

If the population size is large compared to the sample size,

$$\sigma_{\bar{X}} \approx \frac{\sigma}{\sqrt{n}},$$

where $\approx$ means "approximately equal to."

Remark Those of you who read the material in Chapter 4 on the hypergeometric and binomial distributions will recall that the standard deviation formula for the hypergeometric differed from that of the binomial by this same $\sqrt{(N - n)/(N - 1)}$ factor. Remember that the hypergeometric setup involves sampling without replacement. If sampling is carried out with replacement, the underlying distribution is binomial!

The standard deviation of the sampling distribution of $\bar{X}$ is commonly known as the **standard error of the mean.** So, it will be $\sigma/\sqrt{n}$ when sampling with replacement and

$$\frac{\sigma}{\sqrt{n}} \sqrt{\frac{N - n}{N - 1}}$$

when sampling without replacement from a finite population of size N. In the latter case the standard error of the mean is approximately $\sigma/\sqrt{n}$ if the population is very large compared to the sample size.

In our discussion, for the most part, we assume that the population is large enough that $\sigma/\sqrt{n}$ can be taken as the value of $\sigma_{\bar{x}}$ even when sampling without replacement. The standard error of the mean then depends on two quantities, σ and n. It will be large if σ is large, that is, if the scatter in the parent population is large, just as one would expect. On the other hand, since n appears in the denominator, the standard error will be small if the sample size is large. This would seem reasonable because with a larger sample we stand to acquire more information about the population mean μ and consequently less scatter of the sample mean about μ. The standard deviation of the parent population is usually not under the experimenter's control. Therefore, one sure way of reducing the standard error of the mean is by picking a large sample—the larger the better.

EXAMPLE 1   A sample of 36 is picked at random from a population of adult males. If the standard deviation of the distribution of their heights is known to be 3 inches, find the standard error of the mean if:

(a)   the population consists of 1000 males

(b)   the population is extremely large.

SOLUTION

(a)   We are given that $n = 36$, $N = 1000$, and $\sigma = 3$. Therefore, the standard error of the mean is

$$\frac{\sigma}{\sqrt{n}} \cdot \sqrt{\frac{N-n}{N-1}} = \frac{3}{\sqrt{36}} \cdot \sqrt{\frac{1000-36}{1000-1}}$$

$$= 0.5(0.982)$$

$$= 0.491$$

(b)   If the population size is extremely large (practically infinite), the standard error is given as $\sigma/\sqrt{n} = 3/\sqrt{36} = 0.5$.   ●

CENTRAL LIMIT THEOREM

So far we have concerned ourselves with the two parameters of the sampling distribution of $\bar{X}$, its mean and standard deviation. We now turn our attention to the distribution itself. The probability distribution of $\bar{X}$ will depend greatly on the distribution of the sampled population. In the particular example where we picked a sample of two from a population of four, it was fairly straightforward to obtain the distribution of $\overline{X}$. In general, the problem is not so simple as that. It is remarkable and most fortunate that if n, the sample size, is large, the distribution of $\overline{X}$ is close to a normal distribution, of course, with mean μ and standard deviation $\sigma/\sqrt{n}$. The statement of this result is contained in one of the most important theorems of probability theory, called the **central limit theorem.**

CENTRAL LIMIT THEOREM

The distribution of the sample mean $\overline{X}$ of a random sample drawn from practically any population with mean μ and standard deviation σ can be approximated by a normal distribution with mean μ and standard deviation $\sigma/\sqrt{n}$, provided the sample size is large.

The prime importance that the normal distribution derives in statistical theory is to be found in this extremely important theorem. Its basic essence is that if we observe almost any random phenomenon, then the means of repeated samples based on a large number of independent observations will tend to have, approximately, a normal distribution.

The theoretical aspect of the sampling distribution of the sample mean is summarized in the following schematic display in Figure 6-1.

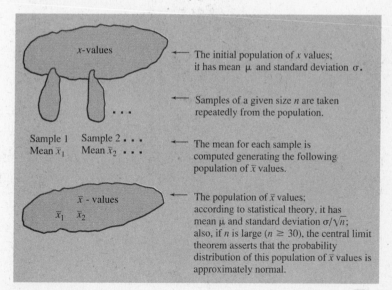

Figure 6-1 Schematic Description of Sampling Distribution of $\overline{X}$

EMPIRICAL INTERPRETATION OF THE CENTRAL LIMIT THEOREM

Simulation of Sampling Distribution of $\overline{X}$ from a Discrete Uniform Distribution

To understand what the central limit theorem signifies, suppose we have an extremely large population consisting of students of ages 5, 6, 7, 8, and 9 years where the proportion of students at each age level is 0.2. The probability distribution of this population is described in Figure 6-2. Using the formulas in Chapter 4, it can be verified that the mean age μ is 7 years and the standard deviation σ is $\sqrt{2} = 1.414$ years.

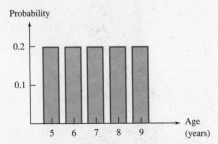

Figure 6-2 Probability Distribution of the Population

From this population we now pick 40 random samples, each sample composed of 8 students. The results of the simulated experiment are shown in Table 6-5. Individual ages are given in the middle column and the mean age for each sample is given in the last column. When we consider just the individual values we have a sample of 320 students. The mean of this sample is 6.82 years and the standard deviation is 1.377 years. As can be seen, both these values are fairly close to the corresponding values, $\mu = 7$ and $\sigma = 1.414$, in the population. Also, the histogram in Figure 6-3, based on these 320 values, looks more or less like the one in Figure 6-2.

So far nothing has been said about the central limit theorem. Let us now turn our attention to its significance. Since the central limit theorem is concerned with the distribution of the sample means, we have to examine the 40 sample means in the last column. The mean of these values is 6.82 and the standard deviation is 0.5272. You will notice that this mean is close to the population mean of 7. Also, the standard deviation is very close to $(\sigma/\sqrt{n}) = (\sqrt{2}/\sqrt{8}) = 0.5$. If you scrutinize the entries in the last column and compare them with those in the middle column, you will see the reason for the low variability of the sample means. The histogram of the sample

TABLE 6-5   Forty Simulated Samples Each of Size 8 from a Population with Values 5, 6, 7, 8, 9

Sample Number	Simulated Data								Sample Means
1	5	8	5	6	7	8	9	6	6.750
2	6	6	8	6	7	9	6	7	6.875
3	8	9	7	7	8	8	6	7	7.500
4	7	9	6	7	6	6	8	9	7.250
5	6	6	8	8	6	8	9	5	7.000
6	5	5	7	5	7	5	7	9	6.250
7	8	6	9	5	6	7	8	7	7.000
8	5	7	7	5	8	5	7	6	6.250
9	8	6	5	9	6	6	6	5	6.375
10	5	6	7	6	8	6	7	6	6.375
11	5	7	5	5	8	9	5	6	6.250
12	5	7	5	6	7	5	7	6	6.000
13	5	5	7	6	9	6	7	7	6.500
14	8	6	7	9	5	5	9	9	7.250
15	7	6	8	5	5	8	5	5	6.125
16	9	6	8	8	6	5	7	8	7.125
17	5	7	9	8	9	5	9	6	7.250
18	5	9	7	5	7	5	6	8	6.500
19	6	8	5	5	5	6	5	8	6.000
20	6	9	7	5	7	9	8	7	7.250
21	9	7	6	5	5	5	5	5	5.875
22	7	5	6	8	8	6	7	8	6.875
23	6	8	9	6	7	5	6	7	6.750
24	9	6	5	6	7	8	6	8	6.875
25	5	7	7	7	9	9	5	5	6.750
26	7	7	9	5	8	8	9	7	7.500
27	7	9	9	6	8	6	8	6	7.375
28	8	7	7	6	7	8	5	8	7.000
29	6	7	5	5	7	6	6	9	6.375
30	5	8	5	7	9	9	9	7	7.375
31	9	7	8	7	9	7	7	9	7.875
32	8	7	6	6	9	8	8	5	7.125
33	6	5	9	5	5	7	9	5	6.375
34	9	9	8	7	8	9	8	8	8.250
35	9	5	8	5	9	7	5	6	6.750
36	5	5	9	5	7	8	8	5	6.500
37	8	7	9	7	7	6	8	6	7.250
38	7	8	8	6	6	7	6	6	6.750
39	6	7	8	9	5	7	5	5	6.500
40	7	8	5	8	6	8	5	8	6.875

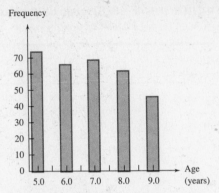

Figure 6-3 Histogram Based on the Collective Sample of 320 Students

means is shown in Figure 6-4(a). Even though the sample size of 8 is not all that large, the tendency toward symmetry and normality starts to show. The histogram in Figure 6-4(b) is based on a simulated experiment consisting of 40 samples each of size 60. You will notice that with larger size samples the closeness to the normal distribution is much better. This is the essence of the central limit theorem.

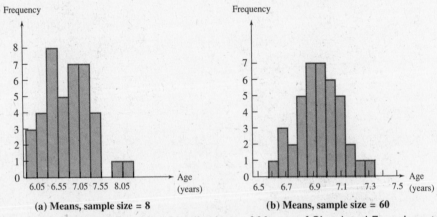

(a) Means, sample size = 8   **(b) Means, sample size = 60**

Figure 6-4 Histograms of the Distributions of Means of Simulated Experiments
(a) Means with sample size 8; (b) Means with sample size 60

Simulation of Sampling Distribution of $\overline{X}$ from Exponential Distribution

To appreciate the scope of the central limit theorem, consider yet another example where the population has a continuous distribution described by the density function given in Figure 6-5. The parent population has the so-called *exponential distribution* which, as you will agree, can hardly be regarded as normal. The distribution is far from being even symmetric. It has mean 1 and standard deviation 1.

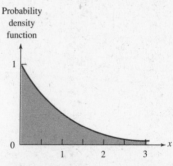

Distribution of the parent population

Figure 6-5 Exponential Distribution with Mean 1 and Standard Deviation 1

According to the central limit theorem, if the sample size n is large, the distribution of sample means will be approximately normal with mean 1 and, since $\sigma = 1$, standard deviation $1/\sqrt{n}$. The frequency distribution of $\overline{x}$ presented in Table 6-6 was actually obtained when 200 samples, each of size 30, were simulated on the computer. Thus, $n = 30$. The table also contains the relative frequency densities obtained using the fact that the width of each class is 0.1.

TABLE 6-6   Frequency Distribution of 200 Values of $\overline{x}$

Class Marks	Frequency	Relative Frequency	Relative Frequency Density (rel freq/0.1)
0.5	1	0.005	0.05
0.6	3	0.015	0.15
0.7	11	0.055	0.55
0.8	27	0.135	1.35
0.9	38	0.190	1.90
1.0	43	0.215	2.15
1.1	36	0.180	1.80
1.2	20	0.100	1.00
1.3	10	0.050	0.50
1.4	6	0.030	0.30
1.5	2	0.010	0.10
1.6	3	0.015	0.15

The relative frequency density histogram based on the 200 means is shown in Figure 6-6. Since $n = 30$, for the purpose of comparison we have also superimposed in Figure 6-6 a normal distribution with mean 1 and standard deviation $1/\sqrt{30}$. Notice how remarkably well the normal density curve fits the histogram.

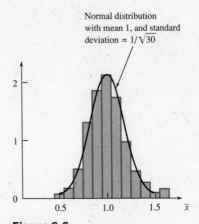

Figure 6-6

The relative frequency density histogram and the probability density curve of the normal distribution with mean 1 and standard deviation $1/\sqrt{30}$.

The central limit theorem tells us that the shape of the distribution of the sample mean is approximately the shape of a normal distribution. From our previous discussion, we already know that if the population has mean μ and standard deviation σ, then $\mu_{\bar{X}} = \mu$ and $\sigma_{\bar{x}} = \sigma/\sqrt{n}$. Converting to the z-scale, we can give an alternate version of the central limit theorem.

ALTERNATE VERSION OF THE CENTRAL LIMIT THEOREM

When the sample size is large, the distribution of

$$\frac{\bar{X} - \mu}{\sigma/\sqrt{n}}$$

is very much like that of the standard normal variable Z.

Recall that to convert to the z-scale the rule is: Subtract the mean and divide by the standard deviation of the random variable in question. Thus, for example,

$$P(a < \bar{X} < b) \approx P\left(\frac{a - \mu}{\sigma/\sqrt{n}} < Z < \frac{b - \mu}{\sigma/\sqrt{n}}\right)$$

if n is large.

Because the central limit theorem applies if the sample size is large, a natural question is: How large is large enough? This will depend on the nature of the sampled population. First of all, there is the following result of statistical theory.

If the parent population is normally distributed, then the distribution of $\overline{X}$ is exactly normal for any sample size.

If the parent population has a symmetric distribution, the approximation to the normal distribution will be reached for a moderately small sample size, in many cases as low as 10. In most instances the tendency toward normality is so strong that the approximation is fairly satisfactory with a sample of about 30. With larger samples, of course, the approximation is even more satisfactory.

EXAMPLE 2

The records of an agency show that the mean expenditure incurred by a student during 1992 was $8000 and the standard deviation of the expenditure was $800. Find the approximate probability that the mean expenditure of sixty-four students picked at random was:

(a)   more than $7820   **(b)**   between $7800 and $8120.

SOLUTION   Here $\mu = 8000$, $\sigma = 800$, and $n = 64$. Therefore,

$$\mu_{\overline{X}} = 8000$$

and

$$\sigma_{\overline{X}} = \frac{800}{\sqrt{64}} = 100.$$

(a)   We want to find $P(\overline{X} > 7820)$. Converting to the z-scale, we get

$$P(\overline{X} > 7820) \approx P\left(Z > \frac{7820 - 8000}{100}\right)$$
$$= P(Z > -1.8)$$
$$= 0.9641.$$

(b)   We want $P(7800 < \overline{X} < 8120)$, which, after we have converted to the z-scale, is given by

$$P(7800 < \overline{X} < 8120) \approx P\left(\frac{7800 - 8000}{100} < Z < \frac{8120 - 8000}{100}\right)$$
$$= P(-2 < Z < 1.2)$$
$$= 0.3849 + 0.4772$$
$$= 0.8621.$$

Interpretation: If we consider a large number of samples, each having sixty-four students, in approximately 86.2 percent of the samples the mean expenditure would be between $7800 and $8120.   ●

EXAMPLE 3   The length of life (in hours) of a certain type of electric bulb is a random variable with a mean of 500 hours and a standard deviation of 35 hours. What is the approximate probability that a random sample of forty-nine bulbs will have a mean life between 488 and 505 hours? Interpret the answer.

SOLUTION   We have $\mu = 500$, $\sigma = 35$, and $n = 49$. Thus,

$$\mu_{\bar{X}} = 500$$

and

$$\sigma_{\bar{X}} = 35/\sqrt{49} = 5.$$

We are interested in finding $P(488 < \bar{X} < 505)$, which, after we have converted to the z-scale, is given by

$$P(488 < \bar{X} < 505) \approx P\left(\frac{488 - 500}{5} < Z < \frac{505 - 500}{5}\right)$$
$$= P(-2.4 < Z < 1)$$
$$= 0.3413 + 0.4918$$
$$= 0.8331.$$

Suppose we sample repeatedly from the population of bulbs, every time picking a sample of 49 bulbs, and compute the mean life for each sample. Then, approximately 83 percent of the samples will have mean between 488 and 505 hours.   ●

EXAMPLE 4   The weight of food packed in certain containers is a random variable with a mean weight of 16 ounces and a standard deviation of 0.6 ounces. If the containers are shipped in boxes of 36, find, approximately, the probability that a randomly picked box will weigh over 585 ounces.

SOLUTION   Here $\mu = 16$, $\sigma = 0.6$, and $n = 36$. We want to find the probability that the total weight of 36 containers exceeds 585 ounces, that is, the mean weight $\bar{X}$ of a container exceeds $\frac{585}{36}$, or 16.25 ounces. Now

$$\mu_{\bar{X}} = 16$$

and

$$\sigma_{\bar{X}} = 0.6/\sqrt{36} = 0.1$$

Conversion to the z-scale gives

$$P(\bar{X} > 16.25) \approx P\left(Z > \frac{16.25 - 16}{0.1}\right)$$
$$= P(Z > 2.5)$$
$$= 0.0062.$$

The chance is extremely small, about 62 in 10,000.   ●

EXAMPLE 5

The amount of time (in minutes) devoted to commercials on a TV channel during any half-hour program is a random variable whose mean is 6.3 minutes and whose standard deviation is 0.866 minutes. What is the approximate probability that a person who watches 36 half-hour programs will be exposed to over 220 minutes of commercials?

SOLUTION   Here $\mu = 6.3$, $\sigma = 0.866$, and n represents the number of half-hour programs and is equal to 36. Assume that $n = 36$ is large enough that we can use the normal approximation for the distribution of $\overline{X}$, the mean amount of time devoted to commercials in 36 half-hour programs.

Now "the total exposure in 36 programs is over 220 minutes" implies that the mean exposure per program is over 220/36, or 6.111 minutes. Thus, we want to find $P(\overline{X} > 6.111)$ where $\mu_{\overline{x}} = 6.3$ and $\sigma_{\overline{x}} = 0.866/\sqrt{36} = 0.1443$. We get

$$P(\overline{X} > 6.111) \approx P\left(Z > \frac{6.111 - 6.3}{0.1443}\right)$$
$$= P(Z > -1.31)$$
$$= 0.9049 \qquad \bullet$$

A corollary to the central limit theorem   For the interested reader, there is an off-shoot to the central limit theorem that deals with the distribution of the *total* or *the sum* of sample values if the sample size is large. It states that, for samples of size n, the distribution of the total $\sum X_i$ is approximately normal with mean $n\mu$ and standard deviation $\sigma \sqrt{n}$ if n is large. As a consequence,

$$P\left(a < \sum X_i < b\right) \approx P\left(\frac{a - n\mu}{\sigma \sqrt{n}} < Z < \frac{b - n\mu}{\sigma \sqrt{n}}\right).$$

Applying this result, we can rework Example 5 as follows.

$$P\left(\sum X_i > 220\right) \approx P\left(Z > \frac{220 - 36(6.3)}{(0.866)\sqrt{36}}\right)$$
$$= P(Z > -1.31) = 0.9049$$

SAMPLING DISTRIBUTION OF THE PROPORTION

A need to know the *sampling distribution of the proportion* might arise if, for example, we were conducting a survey and wanted to find the probability that in a sample of 40 randomly picked people at least 70 percent will be in favor of a certain presidential candidate. Stated differently, we are interested in the probability that in a sample of 40 *the proportion of voters in favor of the candidate* will be greater than 0.7. To answer this, we need to know the distribution of the proportion of voters in favor of the candidate when samples of 40 are taken.

If p denotes the probability that a randomly picked person is in favor of the candidate, we already know that X, the number of voters among 40 who favor the candidate, has a binomial distribution. The proportion of voters in favor is simply the ratio of the number of voters in favor to the number of voters interviewed, which is just the random variable $X/40$. In general, when we pick a sample,

$$\binom{\text{sample proportion}}{\text{of an attribute}} = \frac{\binom{\text{number of items in the sample}}{\text{having the attribute}}}{\text{sample size}}.$$

So, if X is the number of items having a certain attribute in a sample of size n, then "the sample proportion having the attribute" is the random variable X/n. The probability distribution of this statistic is called the **sampling distribution of the proportion.** We can discuss the distribution of X/n, its mean, and its variance by using the known facts about the distribution of X that we learned in Chapter 4. However, we prefer to look at it here in the framework of the central limit theorem.

The method of counting votes essentially consists of recording a 1 if the voter is in favor of the candidate and a 0 if he or she is not in favor. In a typical survey, the sample proportion might be obtained as a value $(1 + 0 + 0 + 1 + \cdots + 0 + 1)/40$, or $29/40$, if there were twenty-nine 1s, that is, twenty-nine in favor. This shows that we can regard the sample proportion X/n as a mean of a sample of size n drawn from an infinite population consisting of 1s and 0s where the fraction of 1s is p and the fraction of 0s is $1 - p$. Thus, the sampled population can be described by the following probability distribution.

x	Probability
0	$1 - p$
1	p

The computations for finding the mean and the variance of this population are shown in Table 6-7.

TABLE 6-7   Mean and Variance of the Population

Values x	Probability $p(x)$	$x \cdot p(x)$	$x^2 \cdot p(x)$
0	$1 - p$	$0 \cdot (1 - p)$	$0^2 \cdot (1 - p)$
1	p	$1 \cdot p$	$1^2 \cdot p$
sum	1	p	p

From Column 3 in the table above, $\Sigma\, x \cdot p(x) = p$ and so the population μ is p. Thus,

population mean = p.

Because the sum in Column 4 is $\Sigma\, x^2 \cdot p(x) = p$, the variance σ^2 is given by

$$\sigma^2 = \sum x^2 \cdot p(x) - \mu^2$$
$$= p - p^2$$
$$= p(1 - p)$$

Thus,

population standard deviation = $\sqrt{p(1 - p)}$

Now recall that if the population has mean μ and standard deviation σ, then the distribution of the sample mean (based on a sample of size n) has mean μ and standard deviation $\sigma/\sqrt{n}$. In our present case X/n represents the mean of a sample of size n from a population with mean p and standard deviation $\sqrt{p(1 - p)}$. It follows, therefore, that the mean of the sampling distribution of X/n is p, the same as the parent population mean. The standard deviation of its sampling distribution is $\sqrt{p(1 - p)}/\sqrt{n}$, the parent population standard deviation divided by the square root of the sample size. The standard deviation of the distribution of X/n can be written as $\sqrt{p(1 - p)/n}$. Thus,

$$\mu_{X/n} = p \quad \text{and} \quad \sigma_{X/n} = \sqrt{\frac{p(1 - p)}{n}}$$

We now summarize our findings.

> **DISTRIBUTION OF THE SAMPLE PROPORTION**
>
> If n items are picked independently from a population where the probability of success is p (not very near 0 or 1), and if n is large, then the distribution of the sample proportion X/n is approximately normal with mean p and standard deviation $\sqrt{p(1 - p)/n}$.
>
> For most situations the approximation is considered satisfactory if $np \geq 5$ and $n(1 - p) \geq 5$.

For example, converting to the z-scale by subtracting the mean p from the endpoints and dividing by the standard deviation $\sqrt{p(1 - p)/n}$, we now get

$$P\left(a \leq \frac{X}{n} \leq b\right) \approx P\left(\frac{a - p}{\sqrt{\dfrac{p(1 - p)}{n}}} \leq Z \leq \frac{b - p}{\sqrt{\dfrac{p(1 - p)}{n}}}\right)$$

if n is large.

EXAMPLE 6   According to the Mendelian law of segregation in genetics, when certain types of peas are crossed, the probability that the plant yields a yellow pea is $\frac{3}{4}$ and that it yields a green pea $\frac{1}{4}$. For a plant yielding 400 peas, find the following.

(a)   The standard error of the proportion of yellow peas

(b)   The approximate probability that the proportion of yellow peas will be between 0.7 and 0.8.

SOLUTION   Here $p = \frac{3}{4}$, $n = 400$, and X = number of yellow peas.

(a)   The standard error of the proportion of yellow peas is

$$\sqrt{\frac{p(1-p)}{n}} = \sqrt{\left(\frac{3}{4} \cdot \frac{1}{4}\right) \Big/ 400}$$
$$= 0.0217.$$

(b)   We want to find $P(0.70 \le X/n \le 0.80)$. Converting to the z-scale, since $p = \frac{3}{4} = 0.75$, we get

$$p(0.70 \le X/n \le 0.80) \approx P\left(\frac{0.70 - 0.75}{0.0217} \le Z \le \frac{0.80 - 0.75}{0.0217}\right)$$
$$= P(-2.30 \le Z \le 2.30)$$
$$= 2(0.4893)$$
$$= 0.9786$$

You can be practically certain, with a chance of about 98 percent, that the sample proportion of yellow peas, based on a sample of 400, will be between 0.7 and 0.8. In an experiment you carry out, if you were to find that the proportion of yellow peas was 64 percent, you would question the truth of the Mendelian law or the way the experiment was conducted. Wouldn't you?   ●

Continuity correction   For computing $P(a \le X/n \le b)$ a continuity correction is recommended for better accuracy. This requires that we subtract $(0.5/n) = 1/(2n)$ from the left extremity a and add the same amount $1/(2n)$ to the right extremity b. (You might recall that we subtracted 0.5 and added 0.5 in Section 5-3 when using normal distribution to approximate the binomial probabilities for large n. Adding and subtracting $0.5/n$ here is closely tied to that concept. Remember, we are dealing with proportion here, and not the total as in Section 5-3.) So, we have the following result.

Using continuity correction,

$$P\left(a \le \frac{X}{n} \le b\right) \approx P\left(\frac{a - \dfrac{1}{2n} - p}{\sqrt{\dfrac{p(1-p)}{n}}} \le Z \le \frac{b + \dfrac{1}{2n} - p}{\sqrt{\dfrac{p(1-p)}{n}}}\right)$$

if n is large.

Since $n = 400$ in Example 6, the continuity correction factor will be equal to $\frac{1}{2(400)} = 0.00125$. Therefore,

$$P\left(0.7 \leq \frac{X}{n} \leq 0.8\right) \approx P\left(\frac{0.7 - 0.00125 - 0.75}{0.0217} \leq Z\right.$$
$$\left. \leq \frac{0.8 + 0.00125 - 0.75}{0.0217}\right)$$
$$= P(-2.36 \leq Z \leq 2.36)$$
$$= 2(0.4909)$$
$$= 0.9818$$

We get a slightly better accuracy.

The continuity correction is not significant if n is very large, because in that case $1/2n$ will be rather small. So, we will leave it out.

SIDEBAR   *Are the Results of Pollster Surveys Reliable?*

*I*t may seem rather foolhardy for the pollsters to venture statements about the preferences of such a large group as the U.S. adult population based on just a small sample of about 1000 individuals. This is especially so when we are aware that due to the intrinsic variability in the population, in repeated sampling the sample proportion of an attribute will differ from sample to sample. Can the results of the surveys be relied upon? The answer is in the affirmative. The important reason is that, even though the proportions differ from sample to sample, the variability is extremely small because the standard error of the sample proportions is less than

$$\sqrt{\frac{1}{4n}} = \sqrt{\frac{1}{4,000}} = 0.0158.$$ (We shall not be concerned with the "why?" here, as only the result is of interest to us.) As a result, it is extremely unlikely, with a chance of less than 2 in 10,000, that, if the population proportion is indeed 0.3, a survey will produce a sample proportion which is off by, say, more than 0.06, that is, a proportion which is less than 0.24 or greater than 0.36. The event is practically impossible. (See the computations in the box below.)

SIDEBAR[*]   *Computations for Reliability of Survey Results*

*C*onsider a pollster survey which involves interviews of 1000 individuals. Suppose p is the true proportion of an attribute in the population where $1000p \geq 5$ and $1000q \geq 5$. The measure of the variability of the sample

---

[*]The computations are somewhat involved.

proportions is the standard error which is $\sqrt{\dfrac{pq}{1000}}$. It can be shown, using algebra, that this quantity will have the maximum value when $p = q = \frac{1}{2}$. Thus, in our discussion the standard error will be less than $\sqrt{1/4000} = 0.0158$, a small value. Let us compute the probability that the sample proportion will differ from the true p value by more than 0.06. The probability that the sample proportion is within 0.06 of the true p is $P\left(-0.06 \leq \dfrac{X}{n} - p \leq 0.06\right)$. So, what we want is

$$1 - P\left(-0.06 \leq \frac{X}{n} - p \leq 0.06\right)$$

$$\approx 1 - P\left(-\frac{0.06}{\sqrt{\dfrac{pq}{1000}}} \leq Z \leq \frac{0.06}{\sqrt{\dfrac{pq}{1000}}}\right)$$

$$\leq 1 - P\left(-\frac{0.06}{0.0158} \leq Z \leq \frac{0.06}{0.0158}\right)$$

$$= 0.00015$$

which is less than 2 in 10,000. (For a good understanding of the computations above see Figure 6-7.)

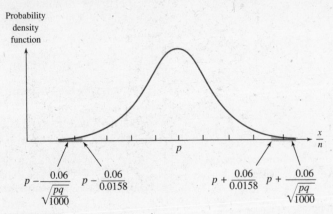

Figure 6-7

Section 6-1 Exercises

1. A sample of 8 items is picked at random from a population consisting of 100 items and a certain quantitative variable X is measured. If the population mean is 112.8, find the mean of the distribution of $\overline{X}$ if the sampling is carried out:
   (a) with replacement   (b) without replacement.

2. Suppose twelve items are picked at random from a population having eighty items. If the population variance of a variable X is 67.24, find the variance of the distribution of $\overline{X}$ if the sampling is:
   (a) with replacement   (b) without replacement.

3. A population consisting of 100 items has a mean of 60 and a standard deviation of 5. If a sample of 9 items is drawn without replacement from this population, what is the mean and the standard deviation of the distribution of the sample mean?

   p. 343
   $$\sigma_{\overline{x}}^2 = \frac{25}{9} \cdot \frac{100-9}{100-1} = 2.5533$$
   $$\sigma_{\overline{x}} = 1.6$$

4. Answer the question in Exercise 3 if the sample is drawn from the population with replacement.

   $\mu = 60 \quad \sigma = 5 \quad n = 9 \quad N = 100$
   $\mu_{\overline{x}} = 60 \qquad \sigma_{\overline{x}} = \frac{\sigma}{\sqrt{n}} = \frac{5}{\sqrt{9}} = 1.67$

5. Suppose it is known that the annual income of a family has a distribution with a mean of $22,200 and a standard deviation of $3,500. If a sample of sixty-four families is picked at random, find the standard error of the mean family income if the population:
   (a) consists of 400 families   (b) can be considered extremely large.

6. Assuming that the population is infinite, discuss the effect on the standard error of the mean if the sample size is changed:
   (a) from 25 to 225   (b) from 320 to 20.

7. Suppose a population consists of six values (all equally likely): 10, 8, 5, 6, 9, 13. Find the following:
   (a) The population mean   (b) The population standard deviation

8. A sample of four items is picked at random from the population described in Exercise 7. Use appropriate formulas to find the mean and the standard deviation of $\overline{X}$ if the four items are picked:
   (a) with replacement   (b) without replacement

9. Suppose a population has the distribution given by the following probability function.

x	−3	0	5	12
$p(x)$	0.3	0.2	0.4	0.1

   (a) Find the mean μ and the standard deviation σ of the population.
   (b) Find the mean and the standard deviation of $\overline{X}$, the sample mean based on a sample of 100 independent observations from the above population.
   (c) What is the approximate distribution of $\overline{X}$?
   (d) Find the approximate probability that:
      i. $\overline{X}$ will lie between 1.366 and 2.767.
      ii. $\overline{X}$ will exceed 3.234.
      iii. $\overline{X}$ will be less than 1.366.

10. The annual rainfall in a region has a distribution with a mean rainfall of 100 inches and a standard deviation of 12 inches. What is the approximate probability that the mean annual rainfall during 36 randomly picked years will exceed 103.8 inches?

11. A cigarette manufacturer claims that the mean nicotine content in its king-size cigarettes is 2 mg with the standard deviation equal to 0.3 mg. If this claim is valid, what is the approximate probability that a sample of 900 cigarettes will yield a mean nicotine content exceeding 2.02 mg?

12. *Quality control inspection*   A hardware store receives a consignment of bolts whose diameters have a normal distribution with a mean diameter of 1.2 inches and a standard deviation of 0.02 inch. The consignment will be considered substandard and returned if the mean diameter of a sample of 120 bolts is less than 1.197 inches or greater than 1.203 inches. Find the probability that the consignment will not be returned.

13. A farmer has divided his entire farm into a large number of small plots of the same size. The amount of yield on a plot has a mean of 100 bushels with a standard deviation of 8 bushels. If 64 of the plots are picked at random, find the approximate probability that the mean yield will:
    (a) exceed 102.5 bushels   (b) be less than 98.5 bushels
    (c) be between 98 and 102 bushels.

14. The maximum load that a ferry can carry is 16,375 pounds. Suppose on any trip the ferry carries 100 passengers and the distribution of the weights of the passengers has a mean of 160 pounds and a standard deviation of 15 pounds. Find the approximate probability that the ferry will exceed the limit on a trip.

15. The distance that Jane, a substitute teacher, travels to teach on a school day has the following probability function.

Distance traveled	0	6	10	15
Probability	0.1	0.2	0.4	0.3

During an academic year consisting of 225 days, find the approximate probability that Jane will travel to teach between 2082 and 2258 miles.

16. A fair die is rolled 200 times. On any roll, if an even number shows up, Bernice will win as many dollars as the number that shows up on the die, and if an odd number shows up, she will lose $3.60. Find the approximate probability that Bernice's total winnings will be:
    (a) more than $42   (b) between $39 and $42.

17. The driver of a truck loaded with 900 boxes of books will be fined if the total weight of the boxes exceeds 36,450 pounds. If the distribution of the weight of a box has a mean of 40 pounds and a standard deviation of 6 pounds, find the approximate probability that the driver will be fined.

18. The amount of ice cream in an ice-cream cone has a distribution with a mean amount of 3.2 ounces per cone and a standard deviation of 0.4 ounces. If there are 50 children at a birthday party, what is the approximate probability that more than 165 ounces of ice cream will be served?

19. A computer is programmed in such a way that the error involved in carrying out one computation has a distribution with mean 10^{-7} and standard deviation 10^{-8}. If 10,000 independent computations are made, find the approximate probability that the total error will be less than 0.001001.

20. The distribution of the test scores of students has an unknown mean μ and a standard deviation equal to 15 points. Suppose 160 students are picked at

random. What is the approximate probability that the mean score of these students will differ from their true mean score μ by at most 3 points?

21. Suppose 10 percent of the tubes produced by a machine are defective. If a sample of 100 tubes is inspected at random, find the following.
    (a) The expected proportion of defectives in the sample 0.1
    (b) The variance of the proportion of defectives in the sample $\sigma^2 = \frac{pq}{n} = \frac{(0.1)(0.9)}{100} = 0.0009$
    (c) The approximate distribution of the sample proportion $mean = .1 \ \sigma^2 = .0009$
    (d) The probability that the proportion of defectives will exceed 0.16

    [handwritten] $p = 0.1 \quad n = 100$

    $d) \ P\left(\frac{X}{n} > 0.16\right) \approx P\left(Z > \frac{0.16 - 0.1}{\sqrt{.0009}}\right) = P(2.2)$
    $= .0288$

22. It is believed that 40 percent of the people favor capital punishment. If 400 persons are interviewed at random, find the approximate probability that the proportion of individuals in the sample who favor capital punishment:
    (a) will exceed 0.43   (b) will be between 0.35 and 0.42.

23. If 60 percent of the population feels that the president of the United States is doing a satisfactory job, find the approximate probability that in a sample of 800 people interviewed at random, the proportion who share this view will:
    (a) exceed 0.65   (b) be less than 0.56.

24. It is hypothesized that the proportion of individuals in the population with blood type O is 0.3. If this hypothesis is correct, what is the approximate probability that in a random sample of 560, the proportion of blood type O individuals will be less than 0.25 or greater than 0.35?

25. The probability that a signal that originates as a dash will be received as a dot is 0.20. If there are 600 dashes in a message, find the approximate probability that the proportion of dashes received as dots will exceed 0.21.

26. It is felt that 80 percent of the population of the United States favors a reduction in the federal budget deficit. What is the approximate probability that in a random sample of sixty-four individuals interviewed, less than 70 percent will favor a reduction of the deficit?

27. It is believed that 40 percent of the adult population feels that Congress should provide housing facilities to the homeless. Find the approximate probability that in a sample of 120 adults interviewed, the proportion of individuals who feel this way is:
    (a) less than 0.3
    (b) greater than 0.35 and less than 0.43.

    [handwritten] $p = 0.4 \quad n = 120 \quad expected \ prop = 0.4$
    $\sigma^2 \ of \ prop. = \frac{(0.4)(0.6)}{120} = 0.002$
    $a) \ P\left(\frac{X}{n} < 0.3\right) \approx P\left(Z < \frac{0.3 - 0.4}{\sqrt{0.002}}\right) = P(Z < -2.24)$
    $= P(Z > 2.24)$
    $= 0.0125$

28. Suppose the annual production of milk, per milk cow, has a certain distribution with mean amount $\mu = 11,200$ pounds and standard deviation $\sigma = 280$ pounds. A farmer owns eighty-four milk-yielding cows. What is the probability that the mean annual yield of the cows will be between 11,145 and 11,240 pounds?

    [handwritten] $b) \ P\left(0.35 < \frac{X}{n} < 0.43\right) \approx P\left(\frac{.35 - .4}{\sqrt{.002}} < Z < \frac{.43 - .4}{\sqrt{.002}}\right)$
    $= P(-1.12 < Z < 0.67) = 0.3686 + 0.2486$
    $= 0.6172$

29. Consider the game of roulette described in Exercise 26 in Section 4-2. Suppose 100,000 one-dollar bets are made on the red color at a casino in one day. Find the probability that the house will lose any amount on a day. (*Hint:* If $\overline{X}$ denotes the gamblers' mean net winnings on 100,000 bets, compute $P(\overline{X} > 0)$.)

6-2   SAMPLING METHODS

A sampling method is a scientific and objective procedure that involves selecting units from a population to obtain a sample which, we expect, will produce results that are closely representative of the population. The process of picking the units constituting a sample has inherent in it the risk of a bias, whereby consciously or unconsciously certain types of units are favored over others. An effective way to avoid such a bias is to adopt a random process for selecting the units from the population. The concept of randomness implies that the investigation should be "honest," with no extraneous bias in the sampling procedure. When carried out properly, random selection gives the investigator no freedom to decide which units in the population are selected. However, it should be realized that, though random sampling gets rid of sampling bias, it does not eliminate variability. There are many approaches to picking a random sample, and we will discuss some of them below.

SIMPLE RANDOM SAMPLE

A simple random sample is the simplest type of sample involving a random process of selection, and it forms an important component of the overall approach for more complex sampling models. A **simple random sample** may be defined as a sample drawn in such a way that every member of the population has an equal chance of being included. This means that if the population is finite and we list all the possible samples of a given size, the probability of picking any one sample is the same. A mechanical procedure might entail writing slips, one for each sample; mixing them well; and then drawing one slip. It is, of course, important to realize that the term *random* refers to the mode of drawing the sample and in no way guarantees that the sample will be representative. However, it does minimize any bias on the part of the experimenter.

As can be seen, the preceding procedure could become extremely tedious and time-consuming even for moderately small populations and samples. An alternative is to use a computer software package that generates random samples from given populations, or to use **tables of random numbers.** The reader who has been using MINITAB has already seen how computers can be used to generate random samples of desired sizes.

The tables of random numbers consist of digits 0, 1, . . . , 9 generated by the computer using a random process and recorded randomly in rows and columns. Random number tables running literally into hundreds of pages are available. In Table A-7 in Appendix III a small portion of such a table is reproduced.

How is the table used? Suppose we want to pick a sample of 4 from a population consisting of 150 households in a residential block. As a first step we write down the 150 households in a serial listing as 001, 002, . . . , 150. Then we pick any arbitrary page of the table of random numbers—for example, the third page, an arbitrary row, such as the fourteenth, and three contiguous columns, such as Columns 18, 19, and 20. Starting with Row twenty-four, we then read down the page in the three columns. We obtain successively the following three-digit numbers.

691, **078,** 736, 272, 441, **093,** 227, 683, 317, **038,** 543, **002**

Discarding those that are over 150, we stop as soon as we have accomplished our objective of picking four numbers not exceeding 150. The first four such numbers that we obtain are 078, 093, 038, and 002. We would accordingly select the sample of households bearing these serial numbers.

If there were 5000 households in the population, the procedure would still be the same, except that we would make a serial listing as 0001, 0002, . . . , 5000, and we would, of course, use four columns reading down the page.

STRATIFIED SAMPLE

It often happens that a population can be subdivided into distinct subpopulations of various sizes because there is considerable internal homogeneity among units of each subpopulation but the subpopulations themselves differ greatly from one another. For example, suppose we are interested in estimating by sampling the annual income of statisticians. In this case we might stratify the population of statisticians into five groups based on their specialty, say, biostatistics, agricultural statistics, actuarial statistics, statistical quality control, and a fifth group that we might classify as "others." Of course, such a stratification will increase the accuracy to a marked extent if incomes differ substantially from specialty to specialty. In any case, stratification is highly desirable if estimates are required for each of the five groups.

Each subpopulation is called a *stratum*. A **stratified random sample** is obtained by picking a simple random sample in each of the strata. The drawing of the samples is carried out independently in different strata. Also, the number of units picked in a stratum may differ from stratum to stratum. A reasonable approach is to pick a sample size in each stratum proportional to the size of the stratum, in such a way that when the subsamples are pooled together we get a sample of the desired size n. This method is called the **proportional allocation method.** However, a larger sampling fraction is preferred in a stratum if the stratum is more variable internally or when sampling is cheaper in the stratum.

Stratification can lead to increased precision because the variability within each stratum is reduced due to internal homogeneity. Also, by dividing the population into strata and picking a sample from each stratum, we can make sure that each subpopulation is adequately represented.

> **S I D E B A R**   *Survey Polls*
>
> **A** typical sample in a survey poll involves about 500 to 1200 respondents. Pollsters sometimes conduct personal interviews, but today 90 percent of interviewing is done by telephone.
>
> National survey results presented in the *Gallup Poll Monthly* usually are based on interviews with about 1000 adult civilians in the United States. It is indeed remarkable that such a small sample is able to elicit information about such a large population of about 100 million adults with a fair degree of accuracy. The sample size is so insignificant that one might almost say that the information is obtained by interviewing just about nobody. The late George F. Gallup used to remark that the chances of a person being interviewed by one of his poll takers were about the same as being struck by lightning.

CLUSTER SAMPLE

Some populations exist or can be naturally divided into clusters or groups of basic units in the population. A **cluster sample** is obtained by picking at random some clusters from the entire list of clusters in the population and then picking some or all of the units from the selected clusters.

For example, an agency of the government interested in investigating energy consumption per household in the United States might divide the country into counties. Each county then constitutes a cluster. For the purpose of investigation a sample could be selected by first picking a random number of counties from the total list of counties and then selecting every household within each of the selected counties. A two-stage selection procedure is also possible, with a random number of households selected within each of the randomly selected counties.

As another example, suppose a retailer receives a shipment of 10,000 cathode tubes in 1000 boxes, 10 tubes per box. Suppose the retailer decides to test fifty tubes to see if they meet the quality standards. A simple random sample would not be feasible in this situation unless all the boxes were opened. Cluster sampling is most suitable here if each box can be regarded as a natural cluster. The retailer could pick five boxes at random and test every tube in each of the five boxes, or the retailer might pick a random sample of twenty-five boxes and from each box pick two tubes at random, or some other variation of this.

Cluster sampling is particularly appropriate when the listing of the units constituting clusters is not available, but the frame, that is, a list of the clusters themselves is available. This sampling technique is particularly useful when the population is scattered widely and some of the clusters are in remote locations. Once you get to a cluster it is relatively easy to sample units in the cluster, and the tendency is to take as many units as possible in that cluster. This is what makes it cost effective, though not necessarily statistically efficient.

SYSTEMATIC SAMPLE

Suppose the population contains N units arranged from 1 to N in some order and we want to pick a sample of size n. A **systematic sample** is a sample obtained by picking a unit at random from the first k units and every kth unit thereafter until a sample of n units is obtained. The number k itself is determined by dividing N by n. (If N/n is not an integer, then take k as the integer part of the number.)

As an example, suppose we wish to select a sample of 10 companies from the listing of the *Fortune* 500 companies using systematic sampling. Since $500/10 = 50$, we would first pick a company at random from the first 50 companies, say, by picking a random number between 01 and 50 and then include every 50th company from then on. Thus, if the number of the first company picked at random happens to be 42, then the next company included would bear the number 92, and so on. The sample would be comprised of companies with the following rankings.

42   92   142   192   242   292   342   392   442   492

A systematic sample is convenient because the procedure is such that it is much easier to organize selection of units in the field than is possible with a simple random sample. Also, a systematic sample generally provides a better spread of the sample and is therefore more representative of the population than a simple random sample.

There is, however, a disadvantage with systematic sampling when there are some periodic features in the list associated with the sampling interval. Thus, in estimating mean daily sales volume for a grocery store, we would probably underestimate the true mean if we sample every Wednesday, and overestimate it if we sample every Friday. To avoid this problem we might, for example, sample every ninth workday.

SIDEBAR   *Sloppy Sampling*

When we taste a dish being cooked, we see to it that it is stirred well before taking a sample from it. When a survey is conducted in a natural population, the concept of "stirring" it is meaningless and not feasible in most instances. We have to accept a population as it is and still conduct the survey so as to get a good representation of the population. Sloppy sampling procedures can lead to erroneous conclusions with sensational claims. Numerous examples of faulty sampling can be cited, and we mention two here.

An interview conducted on Main Street U.S.A. on a weekday regarding a national issue provides a good example of a nonrepresentative sample. Such a sample would leave out people in several walks of life. It would certainly leave out most factory workers, teachers, and college students.

Another example of a nonrepresentative sample is when the response is obtained through a call-in 900 number. By and large, most people who call in are those who feel strongly about the issue, that is, a respondent with an ax to grind. Also, these people do not particularly mind paying for the telephone toll.

A well-known example of a sloppy sampling procedure that led to an erroneous conclusion is the prediction by *Literary Digest* in the 1936 election that Alf Landon would win over Franklin Roosevelt. Roosevelt trounced Landon.

Section 6-2 Exercises

1. Suppose 250 students in a grade school are listed serially 1 through 250. Select a random sample of 8 students using Columns 7, 8, and 9 of the table of random numbers, and starting in Row 14 of the third page of the table.

2. Describe how, using the random numbers table, you would pick a random sample of six members from a club consisting of sixty members.

3. There are 1200 members attending a convention and you need to form a committee by picking 10 members at random. How would you go about picking such a committee using a table of random numbers?

6-3 DESIGN OF EXPERIMENTS* (OPTIONAL)

One of the objectives of a statistical investigation is to answer some theoretical aspects about a population. Accordingly, data should be collected in a manner that will shed light on the scope of the investigation. All too often the investigator will spend an inordinate amount of resources collecting the data only to realize that it was an exercise in wasted effort. Consider the following simple illustration.

EXAMPLE 1

Suppose there are two methods of instruction of a statistics course which we would like to compare for their effectiveness. As goals of the inquiry one could make statements such as "Method I is superior to Method II"; or "Methods I and II are similar"; and so on. Toward verifying such statements, what experiment are we contemplating? Suppose it is agreed to teach two sections of students, adopting Method I in the section meeting at 9 A.M. and Method II in the section meeting at 3 P.M. Suppose, also, that as part of the experiment it is agreed to administer some sort of a test at the end of the

---

*This section may be taken up in conjunction with Chapter 12 on analysis of variance.

session and to use scores on the test as the basis for comparison. The criterion such as "scores on the test" is called the *response variable* or the *dependent variable*. In a general setup one has to make sure that the response variable appropriately provides grounds for verifying the goals of the inquiry.

Suppose, at the end of the session, the mean score in the section where Method I was implemented turns out to be 85 points and that in the other section turns out to be 65 points. The difference in the mean scores certainly is large. But is this to be taken as evidence that Method I is superior to Method II? There are plenty of loopholes in the planning of the experiment. Among others, one is wide open to the rebuttal that maybe the students and the instructor alike are fresh and alert in the morning session in contrast to those in the late afternoon session.

We are not able to distinguish whether the mean score for the morning session is higher than that for the afternoon session because Method I is intrinsically better than Method II, or whether the outcome had to do with the scheduling of the classes, with the afternoon session being affected adversely. We have adopted a poor design. The effect of the teaching method is inextricably mixed up, that is, confounded, with the scheduling effect, thereby rendering the entire investigation worthless. This type of mixing up of the effects on the response variable is called *confounding of effects*.   ●

The above illustration describes a situation of confounding that is obviously discernible. Most people would be aware of it and would avoid the predicament. But there are more subtle ways in which confounding can occur when one is dealing with more complex investigations. Proper planning of the experiment is essential for avoiding such problems.

An experiment should be planned, or designed, in such a way that proper inferences can be drawn from the collected data and the questions that were posed can be answered satisfactorily. The questions about the population are formulated as statements called *hypotheses*. It may happen that data collected in one way might be suitable for verifying a hypothesis and for drawing valid inferences, but those collected in another way may not be suitable at all. The goal of a design of an experiment is basically to provide a pattern or a layout for collecting data so that the hypotheses can be verified by guarding against such pitfalls as confounding of effects and by taking into account the basic variability of the experimental units.

The variables that affect the response variable, that is, the variables on which the response depends, are called **independent variables.** In our illustration we are interested in the study of the independent variable "method of teaching" insofar as it influences the response variable "score on the test." So, the main independent variable of interest in the study is "the method of teaching." We will refer to such a variable which is manipulated to study its effect on the response variable, as a **factor.** It also is called an **explanatory variable** because it serves to explain the changes in the response variable. In our study there are two **levels of the factor** "method of teaching," namely, Method I and Method II.

Other independent variables also could have a bearing on the scores of the students. These could be the socioeconomic background of the students, their ages, their race, their nationality of origin, their field of major, and other variables that we have not even thought about. These variables which, per se, are not of direct or intrinsic interest to the investigator but might influence the response are called **extraneous variables.** As mentioned earlier, the mixing up of the effects of the independent variables obscuring the effect of the explanatory variable on the response is known as **confounding of the effects.**

In the planning stage of the experiment the investigator should be aware and take cognizance of the independent variables that influence the response. Depending on the goals of the investigation a decision might be made to intervene and manipulate at certain levels those factors whose effects on response we might wish to evaluate. The advantage of a planned experiment is that it will allow us to study the effect of such intervention.

Of the remaining variables—the extraneous ones—it might be possible to control some, and, if the experimenter chooses, these might be held rigidly constant while conducting the experiment. The influence of the extraneous variables that cannot be controlled is sought to be minimized by adopting a method of allocation called *randomization,* first introduced by the late Sir Ronald A. Fisher (1890–1962).

Fisher was a pioneer in the field of statistics and developed many of the far-reaching concepts of experimental designs. Much of the terminology, such as *treatment* and *plot,* that he coined was in connection with his work at the Rothamsted Agricultural Experiment Station in London, England. Designs of experiments now are used in many diverse fields such as psychology, business, economics, medicine, social sciences, and engineering. Also, terms such as *treatment* and *plot* are used in a much wider sense. Thus, a **plot** is simply any experimental unit. A **treatment** is any given experimental condition applied to the experimental units. The treatments involved in an experiment are derived from the combinations that can be formed from the different factors by combining their levels. The **design of an experiment** provides a pattern for applying treatments to the experimental units, or, equivalently, a pattern for assigning experimental units to the treatments. The objective of a well-designed experiment is to control the extraneous variables that can be controlled and to try to minimize the influence of those that cannot be controlled through the process of randomization.

Randomization is a simple idea, but a fundamental one for a well-planned statistical experimental design. It provides insurance against systematic influence on the response by the uncontrolled sources of variation. It refers to the method of allocation of the experimental units to the various treatment groups, with the basic intent that no bias should enter in the assignment.

Randomization is the process of allocating experimental units to the treatments at random.

Basically, randomization averages out the effects of variables over which we cannot exercise control.

We shall now briefly discuss two designs. They are the **completely randomized design (CRD)** and the **randomized complete block design (RCBD).**

COMPLETELY RANDOMIZED DESIGN

A completely randomized design has the simplest layout, where experimental units are assigned to the treatments entirely by chance. A random selection of experimental units of a given size is made to assign to one treatment; from the remaining, another random selection of a specified size is made to assign to the next treatment; and so on, until all the units have been assigned to the treatments in specified numbers.

EXAMPLE 2

Suppose an experimenter is interested in finding out the effects of three diuretic agents which induce production of urine in dogs. Each dog will be given a certain dose of one of the diuretics. The experiment will involve 15 male dogs. There are some extraneous variables such as the age of the dogs and the breed of the dogs that are easily controllable by choosing the 15 dogs to be of the same age and the same breed. Of these 15 dogs, four dogs will be administered Diuretic A, five dogs Diuretic B, and the remaining six dogs will be administered Diuretic C.

For a completely randomized design we pick simple random samples. We pick a simple random sample of 4 dogs from the 15 and assign them to Diuretic A. Next, from the remaining 11 dogs we pick a simple random sample of 5 and assign them to Diuretic B. The remaining 6 are assigned to Diuretic C.

One way to accomplish this would be to assign numbers 01, 02, . . . , 15 to the dogs. Then select two columns and a row in the random numbers table, and, going down along the columns, pick numbers 1 through 15 in the sequence in which they appear. Suppose we get the sequence

06 01 15 12 | 14 02 04 05 07 | 10 03 09 08 13 11

We would then allocate dogs bearing tag numbers 06, 01, 15, 12 to Diuretic A, those bearing tags 14, 02, 04, 05, 07 to Diuretic B, and the remaining six bearing tag numbers 10, 03, 09, 08, 13, 11 to Diuretic C. The resulting layout of the allocations is displayed in Table 6-8.

The analysis of data obtained from such a design will be considered in Chapter 12 on analysis of variance.

TABLE 6-8	Layout of the Completely Randomized Design		
Diuretic A	**Diuretic B**	**Diuretic C**
dog 06	dog 14	dog 10
dog 01	dog 02	dog 03
dog 15	dog 04	dog 09
dog 12	dog 05	dog 08
	dog 07	dog 13
		dog 11

RANDOMIZED COMPLETE BLOCK DESIGN

A completely randomized design is the basic design. All other randomized designs are its offshoots, where some restriction is placed on the method of randomization. A randomized complete block design is one such design. It is particularly effective when the experimental units are not homogeneous. It involves grouping together into blocks the units that are reasonably alike so that within each block the units are fairly homogeneous.

A **block** is a group of experimental units that are relatively homogeneous. There are as many experimental units in a block as there are treatments under consideration. Units in distinct blocks, however, may be dissimilar.

Since units within a block are more alike than units in different blocks, blocking serves the useful purpose of controlling the variability. The allocation is carried out in such a way that the treatments are assigned randomly within each block. Each treatment is assigned once and only once within each block, with a new randomization being carried out for each separate block. Because of homogeneity of units within blocks, the variation among experimental units can be attributed, by and large, to the variation between blocks instead of within blocks. You will recall that this idea was central in stratified sampling while forming the strata.

EXAMPLE 3

In Example 2 we did not make any distinction between the dogs. However, suppose the 15 dogs were picked from 5 litters, with 3 dogs being selected from each of the litters. The litters can be used to serve as blocks because the dogs within a litter are prone to be more alike. The method of allocation is as follows: From Litter 1, allocate the three dogs at random to the three diuretics; the same procedure is to be carried out for the other litters 2, 3, 4, and 5.

Specifically, suppose we identify the dogs in Litter 1 as $D_{L1,1}$, $D_{L1,2}$, $D_{L1,3}$, with the dogs in other litters being identified similarly. Then we would pick a dog at random, say, $D_{L1,2}$, and give it Diuretic A. From the remaining two dogs in the litter, pick the next dog at random, say, $D_{L1,3}$, and give it Diuretic B. Finally, give $D_{L1,1}$, the last dog in the litter, the Diuretic C. Thus, we would have the following allocation:

	Diuretic A	Diuretic B	Diuretic C
Litter 1	$D_{L1,2}$	$D_{L1,3}$	$D_{L1,1}$

If we carried out a similar procedure with Litter 2, 3, 4, and 5, we might end up with the randomized complete block allocation presented in Table 6-9.

The results, where the amount of urine discharged is recorded as x, could be presented as in Table 6-10.

TABLE 6-9   Layout of the Randomized Complete Block Design

	Diuretic A	Diuretic B	Diuretic C
Litter 1	$D_{L1,2}$	$D_{L1,3}$	$D_{L1,1}$
Litter 2	$D_{L2,3}$	$D_{L2,1}$	$D_{L2,2}$
Litter 3	$D_{L3,3}$	$D_{L3,2}$	$D_{L3,1}$
Litter 4	$D_{L4,2}$	$D_{L4,1}$	$D_{L4,3}$
Litter 5	$D_{L5,1}$	$D_{L5,2}$	$D_{L5,3}$

TABLE 6-10   The Yields for the Randomized Complete Block Design

	Diuretic A	Diuretic B	Diuretic C
Litter 1	x_{11}	x_{12}	x_{13}
Litter 2	x_{21}	x_{22}	x_{23}
Litter 3	x_{31}	x_{32}	x_{33}
Litter 4	x_{41}	x_{42}	x_{43}
Litter 5	x_{51}	x_{52}	x_{53}

Remark   In Example 3, litters were used as blocks. If dogs from the same litters are not available, one might choose to form blocks on the basis of the weights of the dogs, with Block one consisting of dogs of almost the same weight, Block 2 having dogs of similar weight, and so on. It is possible that a researcher might form blocks on the basis of the age of the dogs. In the final analysis, we use one of the important extraneous variables as the factor to form the blocks.

Blocking as described above is a common practice in agricultural experiments where the soil has fertility gradient. The land is divided into blocks consisting of plots that have similar soil conditions.

The simplest randomized complete block design is the *paired test design,* which will be studied in Chapter 9, Section 4.

Concluding Remarks: The study of a phenomenon, commonly referred to as research, requires close cooperation between the researcher and the statistician. The researcher has expertise in his or her field of study which the statistician may lack. The statistician can guide by providing a pattern for collecting the data so that it can be analyzed properly to shed light on the investigation. The following points deserve consideration when planning an experimental design:

1. The thrust of the study. Set the objectives and state them clearly.
2. Response variable to be observed. Is it appropriate for shedding light on the objectives?
3. Variables that influence the response
4. Factors that are to be manipulated so that their influence can be studied. This will be guided by the objectives of the study.
5. Extraneous variables that can be controlled, and those which cannot be controlled and are to be randomized
6. The method of randomization to be adopted
7. The appropriate mathematical model based on the nature of randomization and other considerations
8. The number of experimental units to be involved
9. The method of collecting data
10. The kind of analysis to be performed on the data
11. The reliability of the conclusions
12. Interpretation of the results so that they become meaningful to the layperson

Section 6-3 Exercises

1. FDA regulations state that a drug must retain at least 95% of its potency before its expiration date. A study on how much potency is retained in a certain drug after the date of its production is to be conducted using samples from 18 manufacturers. Name three extraneous variable that might influence the potency of the drug.

2. In Exercise 1 above suppose the following four classifications for the length of time after production are to be considered

   A_1: within 1 year after manufacture

   A_2: between 1 and 2 years after manufacture

   A_3: between 2 and 3 years after manufacture

   A_4: between 3 and 4 years after manufacture

   Prepare a completely randomized allotment with four samples tested for potency at the end of one year, five tested at the end of two years, six tested at the end of three years, and the rest tested at the end of four years.

3. The information on the following page regarding the salutary effects of taking vitamin E in protecting your heart was reported on AP wire service, December 14, 1993.

   (a) Based on the information provided, what kind of design is used by the investigators?

   (b) Name two extraneous variables that you would select to control.

   (c) Name two extraneous variables that would be handled by randomization.

Vitamin E May Protect Heart

DALLAS—Vitamin E proved more potent than either vitamin C or beta carotene in preventing a first step in the formation of the artery-clogging plaque that causes heart disease, a new study concludes.

In the study published in the December issue of the American Heart Association's journal Circulation, 36 men were divided into three groups of 12. The first group took only vitamin E supplements, the second took a "vitamin cocktail" of vitamins C, E and beta carotene, and the third received a placebo.

---

4. Suppose a chemical engineer wants to compare four different brands of industrial glues, G-I, G-II, G-III, G-IV, for their sealing strength (kg/25 cm^2) and hires you as a statistician. If twenty specimens are to be tested, prepare a completely randomized design so that the same number of specimens are tested with each brand.

5. The old growth redwood forest is a very important economic resource for the British Columbian government. The past decade of unusual rainy seasons has caused some concern among forestry officials because it has led to a very poor regeneration of seedlings, with a possibly serious economic impact on the region. To expedite the regeneration of the old growth, four different types of tree hormones were experimented on 8 trees of similar characteristics such as age, height, basal width, etc. Four healthy branches were selected on each of the eight trees. Using trees to form blocks and the branches as experimental units, make a randomized block allocation of the hormones to the branches.

6. A farmer has a farm with the following layout of plots (four rows and four columns) and wishes to test four fertilizers. On the east side of the farm there are tall trees that cast shade on the farm in the morning, but the shade gradually disappears as the sun rises. It is suspected that the shade of the trees might have an effect on the yield. Prepare a randomized block allocation of the fertilizers to the plots. Discuss your method.

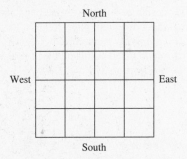

Source: From "Vitamin E May Protect Heart," from the Associated Press, December 14, 1993. Reprinted by permission of the Associated Press.

7. A cattle breeder wishes to compare three hormones—Hormone 1, Hormone 2, Hormone 3—which presumably induce production of milk in cows. There are 18 cows available, with 6 in each of the following age groups:

    G1: 4^+–6^- years   G2: 6^+–8^- years   G3: 8^+–10^- years

    Prepare a randomized block allocation of the cows to the hormones. You must realize that there will be two blocks for each age group.

8. In Exercise 7, what would happen if it was decided to test Hormone 1 with Group G1, Hormone 2 with Group G2, and Hormone 3 with Group G3?

9. A study is to be conducted to compare four diets: NomoreFat, ByeFat, LoseFat, Fatbegone. The experiment will subject individuals to diet regimens for a period of three months, and at the end of the period their loss in weight will be recorded. Sixteen volunteers are available, and they are classified into three groups

    E: Excessively overweight (8 volunteers)

    M: Moderately overweight (4 volunteers)

    S: Slightly overweight to normal (4 volunteers)

    Which factor would you use to form blocks? Make a randomized block assignment based on this stratification.

CHAPTER 6 Summary

✓ CHECKLIST: KEY TERMS AND EXPRESSIONS

❏ sampling distribution, page 337
❏ sampling distribution of the mean, page 338
❏ sampling without replacement, page 339
❏ sampling with replacement, page 339
❏ finite population correction factor, page 342
❏ standard error of the mean, page 343
❏ central limit theorem, page 345
❏ sampling distribution of the proportion, page 354
❏ simple random sample, page 362
❏ table of random numbers, page 362
❏ stratified random sample, page 363
❏ proportional allocation method, page 363
❏ cluster sample, page 364
❏ systematic sample, page 365

❏ independent variable, page 367
❏ factor, page 367
❏ explanatory variable, page 367
❏ levels of a factor, page 367
❏ extraneous variable, page 368
❏ confounding of effects, page 368
❏ plot, page 368
❏ treatment, page 368
❏ design of an experiment, page 368
❏ randomization, page 369
❏ completely randomized design (CRD), page 369
❏ randomized complete block design (RCBD), page 369
❏ block, page 370

KEY FORMULA

$$P(a < \overline{X} < b) \approx P\left(\frac{a - \mu}{\sigma/\sqrt{n}} < Z < \frac{b - \mu}{\sigma/\sqrt{n}} \right)$$

$$P\left(a < \sum X_i < b \right) \approx P\left(\frac{a - n\mu}{\sigma\sqrt{n}} < Z < \frac{b - n\mu}{\sigma\sqrt{n}} \right), \text{ where } \sum X_i \text{ is the sample total.}$$

$$P\left(a \le \frac{X}{n} \le b\right) \approx P\left(\frac{a - p}{\sqrt{\dfrac{p(1 - p)}{n}}} \le Z \le \frac{b - p}{\sqrt{\dfrac{p(1 - p)}{n}}}\right), \text{ where } \frac{X}{n} \text{ is the sample proportion.}$$

CHAPTER 6   Concept and Discussion Questions

1. Compare and contrast the distribution of a population and the distribution of the sample means from that population.

2. Which of the following statements is false? Explain.
   (a) A population parameter has a sampling distribution.
   (b) A statistic has a sampling distribution.

3. What aspect of the distribution of the sample means tells us that increasing the sample size leads to reduction of variability? Give a detailed discussion.

4. How is the standard deviation of the distribution of the sample means affected if the sample size is quadrupled?

5. The following figure is presented in a book on statistics to describe the distribution of $\overline{X}$ in relation to the distribution of X. Point out the fallacy.

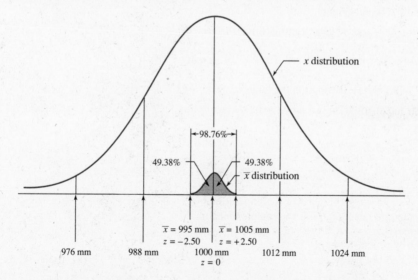

6. Is is true that the central limit theorem can be applied to the distribution of practically any random variable as long as the sample size is large?

7. Describe why the normal distribution enjoys the place of primacy in statistics.

8. Explain how the sample proportion can be regarded as the sample mean.

9. Outline the role of a designed experiment in a statistical study of a phenomenon.

CHAPTER 6   Review Exercises

1. The mean of the sampling distribution of $\overline{X}$ will depend on whether the sampling is with replacement or without replacement. Is this statement true or false?

2. It is said that the normal distribution forms the cornerstone of statistics. Explain this statement.

3. Suppose the population mean is μ. State whether the following statements are true or false.
   (a) The distribution of $\overline{X}$ is normal if the population is normally distributed.
   (b) The distribution of $\overline{X}$ is approximately normal if the population is large.
   (c) The distribution of $\overline{X}$ is approximately normal if the sample size n is large.
   (d) The mean of the sampling distribution of $\overline{X}$ is μ only if the sample size is large.
   (e) The mean of the sampling distribution of $\overline{X}$ is always μ.
   (f) The mean of the sampling distribution of $\overline{X}$ is approximately μ if the sample size n is large.

4. A sample of 10 is drawn from a population of 30. If the population variance is 16, find the variance of the distribution of $\overline{X}$ if the sampling is:
   (a) with replacement   (b) without replacement.

5. Which of the following statements is correct? The central limit theorem is important because it tells us that:
   (a) the mean and the variance of the distribution of $\overline{X}$ are μ and σ^2/n, respectively.
   (b) the distribution of $\overline{X}$ is close to a normal distribution for large n.

6. Suppose as infinite population has the distribution given by the following probability function.

x	-5	4	6
$p(x)$	0.3	0.5	0.2

   (a) Compute the mean μ and the standard deviation σ of the population.
   (b) Compute the mean and the standard deviation of $\overline{X}$, the sample mean based on a sample of 180 measurements from the above population.
   (c) What is the approximate distribution of $\overline{X}$?
   (d) Compute the approximate probability that:
      i. $\overline{X}$ will lie between 1.302 and 2.436
      ii. $\overline{X}$ will exceed 2.629
      iii. $\overline{X}$ will be less that 2.529
      iv. $\overline{X}$ will be less than 0.705 or exceed 2.695.

7. The amount of gasoline (in gallons) an automobile driver uses during a week has a distribution with a mean amount of 10 gallons per week and a standard deviation of 3.5 gallons. What is the approximate probability that during a 49-week period the driver will use more than 540 gallons?

8. A firm is manufacturing ball bearings with a diameter (in inches) that is normally distributed with a mean length of 0.5 inches and a standard deviation of 0.02 inches. If a sample of 25 is picked at random, what is the probability that the mean diameter will be less than 0.496 inches or greater than 0.512 inches?

9. According to available actuarial tables, 88 percent of the people of age 50 will reach age 60 and 12 percent will die before reaching that age. Suppose an insurance company has approved eighty applications for ten-year term insurance policies from people aged 50. Find the probability that the proportion of people reaching age 60 is:
   (a) less than 0.78
   (b) greater than 0.82.

10. Suppose the monthly consumption of electricity (kw hr) per residential consumer has a certain probability distribution with mean consumption of 900 kw hrs and standard deviation of 120 kw hrs. Find the probability that the mean monthly consumption of sixty residential customers will:
    (a) be between 880 and 912 kw hrs
    (b) exceed 885 kw hrs.

11. The number of orders that a mail-order service receives in a day has a probability distribution with mean number of orders equal to 500 and a standard deviation of 30 orders. Find the probability that the mean number of orders received, based on forty-five randomly picked days, will be:
    (a) more than 512 orders
    (b) less than 490 orders
    (c) between 487 and 512 orders.

12. Noskid Tire Company has introduced a new radial tire. The number of trouble-free miles it gives has a normal distribution with a mean of 35,000 miles and a standard deviation of 4,000 miles. Sixteen tires are tested in a simulated experiment.
    (a) What is the distribution of the mean trouble-free miles $\overline{X}$ based on sixteen tires? Is the distribution exact or approximate?
    (b) Find the probability that the mean trouble-free mileage based on sixteen tires will exceed 37,500 miles.

13. The time that Leonard takes to repair a machine has a probability distribution with a mean time of 2.5 hours and a standard deviation of 1.2 hours. If he has to repair thirty-six machines of this type, find the approximate probability that he will take:
    (a) more than 100 hours
    (b) between 82 and 100 hours.

WORKING WITH LARGE DATA SETS

Project: Refer to the real estate data in Appendix IV, Table B-1. There are 347 data entries in the table. Assume that this is the targeted population. Compute the mean μ and the standard deviation σ of the sale price of the houses. Use the random numbers table in Appendix III, Table A-7, or a computer program to pick a random sample of 20 houses. Compute the sample mean and sample standard deviation. Repeat the procedure for other variables. Write a report based on your findings.

 GRAPHING CALCULATOR INVESTIGATIONS

Read the section of the Appendix entitled "Using the TI-82 as a Data Analysis Tool" to solve central limit theorem problems using the program NSNORMAL from the TI-82 Programs Appendix. Then redo the following exercises from the text:

   Section 6-1: Problems 17, 18, 23, 25 and 27.

MINITAB

LAB 6: Sampling Distributions and Designs

The mathematical theory of the sampling distribution of a statistic is quite complicated. However, the empirical concept is easy to grasp. It involves taking random samples of the same size from the population, repeatedly. We then compute the value of the statistic under consideration for each of the samples, and from the values of the statistic thus obtained plot the histogram. If the number of samples is very large, then the histogram will closely resemble the theoretical probability density curve of the statistic. It is this aspect, an empirical understanding of the sampling distribution, that we focus on in this lab.

1 In the following we analyze the sampling distribution of the mean when the population is described by a discrete probability distribution.

```
MTB >NOTE SAMPLING DISTRIBUTION OF THE MEAN

MTB >NOTE PARENT DISTRIBUTION IS DISCRETE

MTB >READ DATA IN C1 C2

DATA>15   0.10

DATA>21   0.25

DATA>26   0.15

DATA>30   0.10

DATA>35   0.30

DATA>40   0.10

DATA>END

MTB >PRINT C1 C2
```

For the preceding distribution let us first find its mean μ and standard deviation σ.

```
MTB >NOTE PARAMETERS OF THE DISTRIBUTION

MTB >NOTE K1 GIVES MEAN OF THE DISTRIBUTION
```

```
MTB >LET C3 = C1*C2

MTB >LET K1 = SUM(C3)

MTB >NOTE K2 GIVES STANDARD DEVIATION OF THE
        DISTRIBUTION

MTB >LET C4 = C2*(C1-K1)**2

MTB >LET K2 = SQRT(SUM(C4))

MTB >PRINT K1 K2
```

You will see that the mean $\mu = 28.15$ and the standard deviation $\sigma = 7.69594$.

We can simulate random samples into rows just as we simulated them previously into columns. We do this by picking 200 values in each of the columns C5-C40. Now each sample is in a separate row and has 36 observations. There are thus 200 samples corresponding to the 200 rows.

```
MTB >NOTE PICK 200 SAMPLES EACH HAVING 36
        OBSERVATIONS

MTB >RANDOM 200 SAMPLES IN C5-C40;

SUBC>DISCRETE VALUES IN C1 WITH PROBABILITIES IN C2.

MTB >NOTE MEAN OF EACH SAMPLE PLACED IN C41

MTB >RMEAN C5-C40 PUT IN C41

MTB >NOTE MEAN OF SAMPLE MEANS

MTB >MEAN OF C41 PUT IN K1

MTB >NOTE STANDARD DEVATION OF SAMPLE MEANS

MTB >STANDARD DEVIATION OF C41 PUT IN K2

MTB >HISTOGRAM C41
```

The operations of summing, finding the mean, and so on that we have carried out previously with columns also can be performed with rows. We have to add an "R" before the command name. Thus, for example, RMEAN C5-C40 PUT IN C41 places the mean of each row in the corresponding row in Column C41. Column C41 now contains 200 sample means.

Since $\mu = 28.15$ and $\sigma = 7.69594$, based on theory, the mean of the sampling distribution of the mean is $\mu = 28.15$, and its standard deviation is $\sigma/\sqrt{n} = 7.69594/\sqrt{36} = 1.2827$. Compare $K1$, the mean of the 200 sample means, with 28.15, and $K2$, their standard deviation, with 1.2827. Are they close? Also, check the histogram to see if it displays the prominent features of a normal distribution. The central limit theorem states that the theoretical distribution is close to a normal distribution.

Assignment   Carry out the analysis and investigate the empirical aspect of the sampling distribution of the mean by picking 180 random samples, each of size 40, when the population is described by the following probability function.

x	−15	5	10	25	30	40
$p(x)$	0.25	0.15	0.10	0.25	0.15	0.10

2 Here we consider the sampling distribution of the mean when the population is described by an exponential distribution, which is a continuous distribution. Before we proceed with considering the distribution of the means, we simulate just one sample of 500 from this distribution and compare the histogram with its probability density curve, which is sketched in Figure 6-5.

```
MTB >RANDOM 500 OBSERVATIONS PUT IN C1;
SUBC>EXPONENTIAL PARAMETER 1.

MTB >HISTOGRAM C1
```

Does the histogram look, more or less, like the curve in Figure 6-5, with hardly any symmetry? In the following we see how the distribution of the sample means exhibits strong features of a normal distribution. We pick 200 samples, each one having thirty observations where, once again, each row represents a sample.

```
MTB >NOTE SAMPLING DISTRIBUTION OF THE MEAN

MTB >NOTE PARENT DISTRIBUTION CONTINUOUS—
     EXPONENTIAL

MTB>RANDOM 200 SAMPLES PUT IN C1-C30;
SUBC>EXPONENTIAL WITH PARAMETER 1.

MTB >NOTE C31 CONTAINS SAMPLE MEANS

MTB >RMEAN C1-C30 PUT IN C31

MTB >HISTOGRAM C31

MTB >NOTE MEAN OF SAMPLE MEANS

MTB >MEAN C31 PUT IN K1

MTB >NOTE STANDARD DEVIATION OF SAMPLE MEANS

MTB >STANDARD DEVIATION C31 PUT IN K2

MTB >PRINT K1, K2
```

For this exponential distribution the mean μ is 1 and the standard deviation σ is also 1. Therefore, the sampling distribution of the mean has mean 1 and standard deviation $\sigma/\sqrt{n} = 1/\sqrt{n} = 1/\sqrt{30} = 0.1826$. Check to see if K1, the mean of the 200 sample means, is close to 1, and if K2, their standard deviation, is close to 0.1826. Also, does the histogram resemble a bell-shaped curve?

* **3** The statistic considered below is $(n - 1)S^2/\sigma^2$. When the population has a normal distribution it has a very important distribution called the chi-square distribution with $(n - 1)$ degrees of freedom, where n is the sample size. It is discussed later on in Chapter 7, and its probability density function is described in Figure 7-12, p. 425, for different degrees of freedom.

We now simulate 200 samples, with each sample having six values placed in a row. Since each sample is represented by entries in a row, we add "R" before the command STDEV to find the standard deviations. They are placed in Column C10. Column C11 contains s^2 and Column C12 contains $(n - 1)s^2/\sigma^2$, where $n = 6$ and $\sigma = 8$.

```
MTB >RANDOM 200 SAMPLES IN C1-C6;

SUBC>NORMAL MEAN 30 SIGMA 8.

MTB >RSTDEV C1-C6 PUT IN C10

MTB >LET C11 = C10**2

MTB >LET C12 = 5*C11/8**2

MTB >HISTOGRAM C12

MTB >MEAN OF C12 PUT IN K1

MTB >STANDARD DEVIATION OF C12 PUT IN K2

MTB >PRINT K1, K2
```

Since $n = 6$, the chi-square distribution has 5 degrees of freedom. Compare the histogram with the graph of the probability density function for 5 degrees of freedom in Figure 7-12. Also, according to theory, the mean of the distribution should be $n - 1$, that is, 5 in this case, and the standard deviation should be $\sqrt{2(n - 1)}$, that is $\sqrt{10} = 3.1623$. Check for similarity between these values and the corresponding values of K1, K2 obtained from the 200 samples.

---

*The chi-square simulation may be skipped here and considered later in Chapter 7 when deemed appropriate.

Assignment   Pick 180 samples, each consisting of 16 observations, from a normal population with mean $\mu = 150$ and standard deviation $\sigma = 12$. Investigate the distribution of the statistic $(n-1)S^2/\sigma^2$ on the basis of the empirical information. Compare the histogram with the curve in Figure 7-12 for 15 degrees of freedom. Also, check whether the empirical mean and standard deviation are, respectively, close to the theoretical values 15 and $\sqrt{2 \times 15} = 5.4772$.

We now type STOP and conclude the MINITAB session using the logging off procedure.

Commands you have learned in this session:

RMEAN

RANDOM;

EXPONENTIAL.

RSTDEV

CAREER *PROFILE*

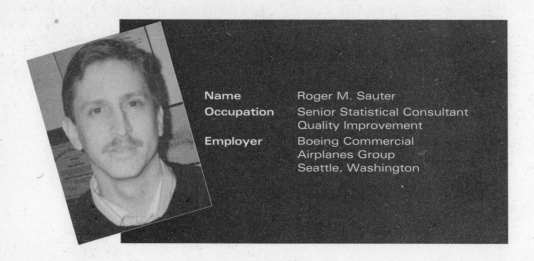

Name   Roger M. Sauter
Occupation   Senior Statistical Consultant
                  Quality Improvement
Employer   Boeing Commercial
                  Airplanes Group
                  Seattle, Washington

My introduction to the career of a statistician came in the ninth grade. In a guidance counseling class, we were assigned to pick a career that looked interesting. I chose to be a statistician because it was described as an applied form of mathematics and because of the huge salary (listed as a whopping $8000 in 1972). By the end of my senior year of college, I didn't know what I wanted to do with a major in mathematics. I didn't think I wanted to teach high school math or be a computer programmer, which seemed to be the only options open to me as a math major. I liked the two courses I had taken in statistics. The first was an introductory course in statistics taught by the psychology department. The professor challenged us with a final project that involved obtaining some real data and analyzing it. The other was an introduction to probability course from the math department. During the spring break of my senior year, I checked out the graduate program in statistics at a nearby university. I liked what the professor had to say and enrolled that fall.

Currently my job involves advising people how to collect and analyze data, and answering questions important to them. Many times analyzing the data is as simple as calculating summary statistics, like means and standard deviations, or plotting simple graphs like histograms, scatter plots, and control charts. Other times more in-depth analysis is needed. A current project involves forecasting the status of a performance/business measure based on its past behavior. A previous project involved designing an experiment to isolate a problem and finding the right combination of materials to help a system function properly. Because the human brain does not randomize, some combinations of variables are often overlooked. A designed experiment can aid tremendously in situations like this. The whole idea of statistical analysis is to use objective data to get to the facts rather than relying on subjective opinions and preconceived ideas.

I find that the most gratifying part of my job is helping clients arrive at solutions that solve their problems. The solution may appear obvious after considering the data (just using simple statistical techniques), but because clients are close to their problems they need the statistical evidence to lead them down the proper pathway, Learning about the area to which I am applying statistical techniques is one of the exciting parts about being a statistician. I chose Boeing Commercial Airplanes partly because I wanted to learn about the Boeing 747 airplane. I had never been in a 747 before working at Boeing.

Being a statistician is a rewarding and interesting occupation. Thinking "statistically" will aid you in whatever career you choose, even if you don't become a full-time statistician. It is one of the most important tools you can obtain from school, not just memorizing the formulas, but understanding the thought process that provides the reasoning behind the formulas. Even if math and/or statistics doesn't come easily for you, stick with it and you will benefit from it in the real work world.

Estimation—Single Population

7

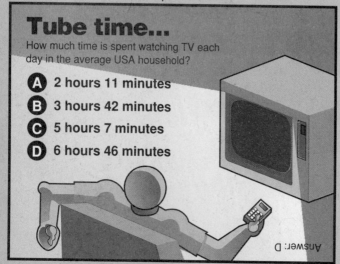

INTRODUCTION

We are entering the most important phase of a statistical investigation, called *statistical inference,* which is a procedure whereby, on the basis of observed experimental data in a particular sample, we generalize and draw conclusions about the population from which the sample is drawn. Among other things, this procedure is concerned with estimation of unknown parameter values and tests of hypotheses regarding these values. We will discuss the principles of estimation in this and the next chapter, deferring tests of hypotheses to Chapter 9.

The object of the study of an investigator is the population. As we are aware, the investigation of the entire population of interest may not be feasible for several reasons. The only recourse is to pick a sample of some size from the population. From the sample data that are collected, we compute the values of appropriate statistics. On the basis of the values of the statistics, we draw inferences about the corresponding population parameters that serve to explain the population. The overall procedure associated with the problem of statistical inference is shown in the following diagram.

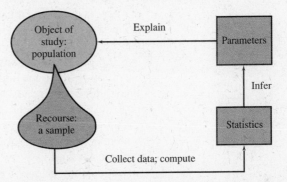

The *Thorndike-Barnhart Advanced Dictionary* defines the word *estimate* as "judgment or opinion of the approximate size, amount, etc." The practice of estimating is not foreign to us. For example, if we are taking a trip from San Diego to San Francisco, we might estimate the distance as 650 miles, the mileage per gallon as 22 miles, and the price per gallon as $1.38, and eventually we might put all the information together and estimate how much the entire trip might cost us. We might also estimate the distance as being between 600 and 700 miles, the mileage per gallon as being between 20 and 25 miles, and so on. In the first case we are estimating the distance, mileage, and price per gallon as specific values, or specific points. In the latter case we are estimating these quantities in terms of certain intervals. The first method of giving an estimate is referred to as *point estimation* and the second method is referred to as *interval estimation.*

The main objective of a statistical investigation is to acquire an understanding of the population, and this is often accomplished by studying the constants associated with the population. These constants are what we have called the population parameters. We have come across three parameters so far. One of these is p, the proportion of items of a certain type in the population, for example, the proportion of Democrats in a state, the proportion of defectives produced by a machine, the proportion of people of Type A blood, and so on. While discussing quantitative data, we came across the population mean μ and another parameter, the population standard deviation σ. Examples of these include the mean IQ of students and standard deviation of the IQ, and the mean nicotine content in a certain brand of cigarettes and the standard deviation of the nicotine content.

As in our example of the trip to San Francisco, there are two ways of giving an estimate of a parameter. One might give an estimate as a single value, that is, a point estimate, or one might give an estimate stating that the parameter is enclosed in an interval, described as an interval estimate.

S I D E B A R   *Capture-Recapture Method*

*T*here is a serious concern among environmentalists regarding the extinction of many animal species such as the rhino due to poaching, the blue whale due to whale fishing, the spotted owl due to logging, and so on. When the pros and cons about the survival of a threatened species are debated, one question is highly relevant for the discussion: "If the species is endangered and needs to be protected, how many are there in the population?" A full count is, of course, impossible, since the animals are not stationary in one location. We can only estimate the number. A fairly common practice for estimating the number is the so-called **capture-recapture method.**

For instance, suppose we wish to estimate the number of deer in a wilderness. As a first step we capture a certain number of deer—say, 100—tag them, and release them in the wilderness. In a few days, after these tagged deer have mixed well in the population, we make a second catch of, say, 80 deer from the population. If the number of tagged deer in this sample is 25, for example, then the number N in the population is estimated using the relation $25/80 = 100/N$. Solving for N, the estimated population size is 320.

7-1   GENERAL APPROACH TO ESTIMATION

As mentioned in the introduction, there are two methods of providing an estimate of a parameter. One method is *point estimation* and the other is *interval estimation.*

Survey results such as the following, which appeared in *Time,* April 5, 1993, are fairly common. As we all have experienced, opinion pollsters have a field day beaming a constant barrage of survey reports during presidential election years.

VOX POP

Do you favor or oppose
a tax of $1 on a pack of
cigarettes?

FAVOR	OPPOSE
63%	**32%**

From a telephone poll of 800 adult Americans
taken for TIME/CNN on March 18 by Yankelovich
Partners Inc. Sampling error is ± 3.5%.

In the *Time* survey report, the parameter of interest is the proportion p in the entire adult U.S. population who favor a tax of $1 on a pack of cigarettes. The value 63%, or 0.63, is a point estimate of p based on a sample of 800 adult individuals. It is a point estimate because it is a specific number.

The statement also says that the *sampling error* was ±3.5%, or ±0.035. Another name for sampling error is the *margin of error*. This way of presenting results by giving a point estimate and the margin of error is the method of interval estimation, because it implies that p is in the interval $(0.63 - 0.035, 0.63 + 0.035)$, that is, in the interval $(0.595, 0.665)$. Actually, this is one component of the method of interval estimation. An important component that is left out in the survey report is the level of reliability, or how confident we can be that the given interval contains p.

POINT ESTIMATION

The methodology of computing the value of a statistic from a sample in order to provide an approximate answer to the value of a parameter is called *estimation*. The statistic used for the purpose is called an **estimator.** It is a random variable which, in some sense, is relevant for estimating the parameter and is simply the form of a mathematical expression, that is, it is a formula. The computed value of the estimator based on a set of sample data is called a **point estimate,** or, simply, an *estimate* of the parameter. It provides a single numerical value as an assessed value of the parameter. In the following, an estimator will be denoted by an uppercase letter and the estimate based on it by the corresponding lowercase letter.

In the *Time* survey, the estimator for estimating p is the suggestive statistic X/n, the sample proportion. Presumably, there were 504 adults who favored a tax of $1 in the sample of 800 surveyed, giving an estimate of p as $504/800 = 0.63$. It must be realized that in another survey of 800 individuals we would, very likely, get a different value of X/n, giving another estimate of p. Thus, estimates of a parameter will differ from sample to sample. In general, if there are x individuals in the survey who are in favor, then the estimate of p will be x/n.

For estimating the population mean μ, intuitively, the natural estimator would seem to be $\overline{X}$, the sample mean. The intuition is backed by the following properties of $\overline{X}$.

1. Expected value of $\overline{X}$ is μ

2. Standard deviation of $\overline{X}$ is $\dfrac{\sigma}{\sqrt{n}}$.

Property 1 asserts that the distribution of $\overline{X}$ is centered about the parameter μ. Because of this property, we say that $\overline{X}$ is an **unbiased estimator*** of μ. From Property 2 it follows that as the sample size n gets larger, the standard deviation of $\overline{X}$ will become smaller since $\sqrt{n}$ is in the denominator. The distribution of $\overline{X}$ becomes narrower about the center μ, so that there is a higher probability that $\overline{X}$ will be close to μ. As a result, we stand to get a more precise estimate of μ if the sample size is large. Additionally, using the central limit theorem, the distribution of $\overline{X}$ is, approximately, normal. This consequence will be useful later in obtaining a confidence interval for μ.

If a sample yields values $x_1, x_2, \ldots, x_n$ and the mean $\overline{x}$ is computed, we will say that $\overline{x}$ is a point estimate of μ.

Our common sense would tell us that we should use the sample proportion X/n to estimate the population proportion p, as was done in the *Time* survey. Also, you have seen in Chapter 6 that the estimator X/n has expected value p and standard deviation $\sqrt{p(1 - p)/n}$. So, *X/n is an unbiased estimator of p* and its standard deviation becomes smaller as n increases. These properties of X/n support our intuitive choice.

For our purpose, we will also find it satisfactory to use the sample variance S^2 to estimate the population variance σ^2 and the sample standard deviation S to estimate σ. (It turns out that S^2 is an unbiased estimator of σ^2, but S is not an unbiased estimator of σ.)

EXAMPLE 1   The following readings give the weights of four bags of flour (in pounds) picked at random from a shipment: 102, 101, 97, and 96. Find estimates of the following.

(a)   The true (population) mean weight of all the bags

(b)   The true variance of weights of all the bags

(c)   The true standard deviation of weights

SOLUTION   The sample mean is

$$\overline{x} = \frac{396}{4} = 99$$

and

$$\sum (x_i - \overline{x})^2 = 3^2 + 2^2 + (-2)^2 + (-3)^2 = 26.$$

(a)   The estimate of the true mean is $\overline{x}$, that is, 99 pounds.

(b)   The estimate (in pounds²) of the true variance is

$$\sum (x_i - \overline{x})^2/(4 - 1) = 26/3 = 8.67.$$

(c)   The estimate (in pounds) of the true standard deviation is $\sqrt{8.67} = 2.9$. ●

_____

*For further discussion on properties of estimators, see *Elementary Statistics in a World of Applications,* Third Edition, by R. Khazanie, HarperCollins Publishers.

EXAMPLE 2

Antirrhinum majus (snapdragon) is a plant with saclike, two-lipped flowers. When a yellow-flowered plant which was known to be heterozygous was self-mated, it was found that there were two types of flowers, with 105 yellow flowers and 31 ivory flowers. Find a point estimate of the probability that a flower is yellow when the heterozygous variety is self-mated.

SOLUTION   We have qualitative information. Altogether there are 105 + 31, or 136, flowers, of which 105 are yellow. Thus, $x = 105$ and $n = 136$. Consequently, a point estimate of p is

$$\frac{x}{n} = \frac{105}{136} = 0.772.$$

●

INTERVAL ESTIMATION

The method of point estimation has some serious drawbacks because, for one thing, unless some additional information is given, there is no way to decide how good the estimate is. Certainly we would like to have some idea about the extent of departure of the estimator from the parameter it is supposed to estimate and the chances of such a departure, that is, about how close the computed estimate is to the parameter value and how confident we can be in our assertion.

As an example, suppose we wish to estimate the mean weight of miniature poodles in a certain population of poodles. Suppose when a sample is picked, the sample mean weight turns out to be 20.5 pounds. We will accept 20.5 as a point estimate of μ. How reliable is this estimate? Unless this information is supplemented with other facts, we have no way of knowing how much faith to place in this value. Suppose we are told that the population standard deviation is 8 pounds. In that case we would not have very much confidence in the estimate because of the large spread indicated by the fact that σ is 8. Our confidence in the estimate would improve considerably if the population standard deviation was, say, 1 pound. Again, suppose the mean of 20.5 was obtained by weighing just two poodles. We would be skeptical of an estimate that is computed from such a small sample. But had the estimate been obtained by weighing, say, 100 poodles, we would have more faith in the estimate. Also, the weakness in simply providing a point estimate of 20.5 is that there is no accompanying statement about the reliability of the answer.

An alternate approach for estimating a parameter which takes into account considerations of the preceding nature is the method of **interval estimation,** also called the method of **confidence intervals.**

An **interval estimate** provides an interval—a range of numerical values—that is believed to contain the unknown parameter, together with the degree of confidence in the assertion that the interval does contain the parameter.

CONFIDENCE INTERVAL OF μ WHEN σ IS KNOWN

To get started, let us show in detail how to construct a confidence interval for the parameter μ when the population standard deviation σ is known. As an example, we might wish to find a confidence interval for the population (or true) mean tar content μ in a certain brand of cigarettes when, from prior studies, it may be assumed that $\sigma = 0.6$ milligrams. The assumption that σ is known may seem rather artificial, since in reality it is more than likely that it will not be known. However, we will handle this situation in the next section. In the meantime, the ideas developed here will serve to guide our future work, which will present more realistic situations.

For obtaining a confidence interval for μ, in this section we make the following assumptions.

1. The sample consists of n independent observations, that is, any one observation does not influence any other.
2. The underlying population is normally distributed with mean μ, which is, of course, unknown.
3. The population standard deviation σ is known.

A 95 Percent Confidence Interval

In what follows we shall first construct a confidence interval where we can be '*95 percent confident*' that the interval brackets the mean μ. We shall then provide a procedure for other levels of confidence.

Recall from Chapter 6 that, under the assumptions we made above, the distribution of $\overline{X}$ is normal, with mean μ and standard deviation $\sigma/\sqrt{n}$. Standardizing, $\dfrac{\overline{X} - \mu}{\sigma/\sqrt{n}}$ has a standard normal distribution. To obtain a 95 percent confidence interval we must first determine the z value such that the area under the standard normal curve between $-z$ and z is 0.95. This results into leaving area 0.025 to the left of $-z$ and area 0.025 to the right of z. But we know, from Table A-3 in Appendix III that the area to the right of 1.96 is 0.025 and, due to symmetry, the area to the left of -1.96 is also 0.025. So the area between -1.96 and 1.96 is 0.95 as in Figure 7-1.

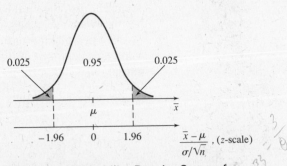

Figure 7-1 Probability Density Curve of Normal Distribution with Mean μ and Standard Deviation $\sigma/\sqrt{n}$

Areas in the left and right tail each equal 0.025.

Therefore, we can write

$$P\left(-1.96 < \frac{\bar{X} - \mu}{\sigma/\sqrt{n}} < 1.96\right) = 0.95$$

After some algebraic steps dealing with inequalities, we get

$$P\left(\bar{X} - 1.96\frac{\sigma}{\sqrt{n}} < \mu < \bar{X} + 1.96\frac{\sigma}{\sqrt{n}}\right) = 0.95.$$

According to the last statement the random interval $\left(\bar{X} - 1.96\frac{\sigma}{\sqrt{n}},\right.$ $\left.\bar{X} + 1.96\frac{\sigma}{\sqrt{n}}\right)$ encloses the mean μ with probability 0.95. The interval is called a **random interval** because its endpoints are random variables whose values depend on the sample values. If the mean from the collected data is $\bar{x}$, then we will say that

$$\left(\bar{x} - 1.96\frac{\sigma}{\sqrt{n}}, \bar{x} + 1.96\frac{\sigma}{\sqrt{n}}\right)$$

is a 95 percent confidence interval of μ. (For an empirical interpretation of this statement read the discussion given later in this section on page 397.)

Now suppose we examine a sample of twenty cigarettes for their tar content and find the mean tar content in this sample to be 10.8 milligrams. Taking into account that $\sigma = 0.6$, since $\bar{x} = 10.8$ and $n = 20$, the endpoints of the interval are computed as follows:

$$\bar{x} - 1.96\frac{\sigma}{\sqrt{n}} = 10.8 - 1.96\frac{0.6}{\sqrt{20}} = 10.54$$

$$\bar{x} + 1.96\frac{\sigma}{\sqrt{n}} = 10.8 + 1.96\frac{0.6}{\sqrt{20}} = 11.06$$

The resulting interval (10.54, 11.06) is a *95 percent confidence interval for the true mean* μ.

A $(1 - \alpha)100$ Percent Confidence Interval

Having constructed a 95 percent confidence interval for μ we can now consider confidence intervals with other levels of confidence. For a 95 percent confidence interval, that is $(1 - 0.05)100$ percent, we found the z value for which the area in the right tail is $0.05/2 = 0.025$. For a 98 percent confidence interval, that is $(1 - 0.02)100$ percent, we would need to find the z value for which area in the right tail is $0.02/2 = 0.01$. This value is 2.33. In general, for a $(1 - \alpha)100$ percent confidence interval we would need to find the z value for which the area in the right tail is $\alpha/2$. We denote this value by $z_{\alpha/2}$. Hence the area between $-z_{\alpha/2}$ and $z_{\alpha/2}$ is $(1 - \alpha)$, as shown in Figure 7-2.

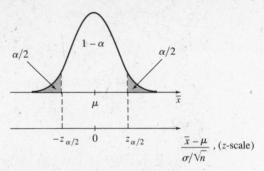

Figure 7-2 Probability Density Curve of Normal Distribution with Mean μ and Standard Deviation $\sigma/\sqrt{n}$
Shows area $\alpha/2$ in each of the two tails.

Thus, we can make the following probability statement.

$$P\left(-z_{\alpha/2} < \frac{\overline{X} - \mu}{\sigma/\sqrt{n}} < z_{\alpha/2}\right) = 1 - \alpha$$

Or, equivalently,

$$P\left(-z_{\alpha/2}\frac{\sigma}{\sqrt{n}} < \overline{X} - \mu < z_{\alpha/2}\frac{\sigma}{\sqrt{n}}\right) = 1 - \alpha.$$

That is, $\overline{X}$ will differ from μ by at most $z_{\alpha/2}\dfrac{\sigma}{\sqrt{n}}$ with probability $(1 - \alpha)$. The quantity $z_{\alpha/2}\dfrac{\sigma}{\sqrt{n}}$ is, therefore, called the *maximum error of estimate* of μ at the $(1 - \alpha)100$ percent level. The maximum error of estimate of μ is commonly called the **margin of error** or the **sampling error.**

> The **margin of error** of estimate of μ at the $(1 - \alpha)100$ percent level is the maximum error in estimating μ and is given by
>
> $$\text{margin of error} = z_{\alpha/2}\frac{\sigma}{\sqrt{n}}$$

With some additional algebraic steps dealing with inequalities, it can be shown that the statement

$$P\left(-z_{\alpha/2}\frac{\sigma}{\sqrt{n}} < \overline{X} - \mu < z_{\alpha/2}\frac{\sigma}{\sqrt{n}}\right) = 1 - \alpha$$

leads to

$$P\left(\overline{X} - z_{\alpha/2}\frac{\sigma}{\sqrt{n}} < \mu < \overline{X} + z_{\alpha/2}\frac{\sigma}{\sqrt{n}}\right) = 1 - \alpha.$$

The interval $\left(\bar{X} - z_{\alpha/2}\dfrac{\sigma}{\sqrt{n}}, \bar{X} + z_{\alpha/2}\dfrac{\sigma}{\sqrt{n}}\right)$ is a random interval and the probability that it covers the mean μ is $(1 - \alpha)$. The endpoints

$$L = \bar{X} - z_{\alpha/2}\frac{\sigma}{\sqrt{n}}$$

and

$$U = \bar{X} + z_{\alpha/2}\frac{\sigma}{\sqrt{n}}$$

of the interval are random variables. Their values depend on the value of $\bar{X}$ which, in turn, depends on the sample values.

Our premise is that σ is known. In any problem the level of confidence will be stipulated, and from this we should be able to find $z_{\alpha/2}$. When a sample of a given size is picked, we will obtain $\bar{x}$, the mean for this data. Thus, n and $\bar{x}$ will also be known to us. All this information will give the resulting $(1 - \alpha)100$ percent confidence interval as

$$\bar{x} - z_{\alpha/2}\frac{\sigma}{\sqrt{n}} < \mu < \bar{x} + z_{\alpha/2}\frac{\sigma}{\sqrt{n}}$$

We observe that the left extremity of the interval is obtained by subtracting $z_{\alpha/2}\dfrac{\sigma}{\sqrt{n}}$ from $\bar{x}$, and the right extremity is obtained by adding $z_{\alpha/2}\dfrac{\sigma}{\sqrt{n}}$ to $\bar{x}$. Therefore, a confidence interval for μ is often given simply by specifying the limits as

$$\bar{x} \pm z_{\alpha/2}\frac{\sigma}{\sqrt{n}}.$$

We have established the following result.

Commonly Used Confidence Levels and Corresponding Values of z

Confidence Level	Value of z
90%	1.645
95%	1.960
98%	2.330
99%	2.575

CONFIDENCE INTERVAL FOR μ; σ KNOWN
If the population has a normal distribution and σ is known, then a $(1 - \alpha)100$ percent confidence interval for μ is given by

$$\bar{x} - z_{\alpha/2}\frac{\sigma}{\sqrt{n}} < \mu < \bar{x} + z_{\alpha/2}\frac{\sigma}{\sqrt{n}}$$

where $\bar{x}$ is the sample mean based on n observations.

Remark   We have constructed the confidence interval above for μ with the assumption that the population is normally distributed. This assumption was necessary since it allowed us to proceed by stating that $\bar{X}$ (consequently, $\dfrac{\bar{X} - \mu}{\sigma/\sqrt{n}}$) has a normal distribution. If n is large (at least 30), the assumption of a normal distribution is not crucial, because the central limit theorem permits us to proceed by stating that $\dfrac{\bar{X} - \mu}{\sigma/\sqrt{n}}$ has *approximately* a normal distribution. We still get the same confidence limits given above, but now they are approximate limits.

EXAMPLE 3

When ten vehicles were observed at random for their speeds (in mph) on a freeway, the following data were obtained.

61	53	57	55	62	58	60	54	62	60

Suppose that from past experience it may be assumed that vehicle speed is normally distributed with $\sigma = 3.2$ mph. Construct a 90 percent confidence interval for μ, the true mean speed.

SOLUTION   In Figure 7-3 we have obtained the normal probability plot for vehicle speeds. You will notice that the points are close to the straight line, thereby providing a reasonable basis to assume that vehicle speed is normally distributed.

From the available data we can compute $\bar{x} = 58.2$. Further, we know that $n = 10$ and $\sigma = 3.2$. For a 90 percent confidence interval, $\alpha = 0.10$. Thus, $z_{\alpha/2} = z_{0.05} = 1.645$ and the confidence interval for μ is given by

$$58.2 - 1.645\left(\frac{3.2}{\sqrt{10}}\right) < \mu < 58.2 + 1.645\left(\frac{3.2}{\sqrt{10}}\right)$$

$$58.2 - 1.66 < \mu < 58.2 + 1.66$$

which can be simplified to $56.54 < \mu < 59.86$, giving the confidence interval (56.54, 59.86). Thus, with 90 percent confidence we estimate that the mean speed μ of the vehicles driving on the freeway is between 56.54 and 59.86 miles. Since the vehicle speed is normally distributed, one could infer with 90 percent confidence that a majority of the drivers on the freeway

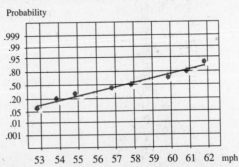

Figure 7-3 Normal Probability Plot for Vehicle Speeds

(that is, at least 50 percent) are in violation of the federally mandated speed limit of 55 mph. ●

EXAMPLE 4

Basing her findings on fifty successful pregnancies involving natural birth, Dr. Vivien McKnight found that the mean pregnancy term was 274 days. Suppose that from previous studies the standard deviation σ of a pregnancy term may safely be assumed to be 14 days. Construct a 98 percent confidence interval for the true mean pregnancy term μ.

SOLUTION The problem does not state that pregnancy term is normally distributed. However, $n = 50$, which is large enough that we can apply the central limit theorem. Of course, we realize that the confidence interval will be approximate.

From the given information, $\bar{x} = 274$, $\sigma = 14$, and $n = 50$. For a 98 percent confidence interval, $\alpha = 0.02$. Thus, $z_{\alpha/2} = z_{0.01} = 2.33$ so that, with 98 percent confidence, the approximate interval is as follows:

$$274 - 2.33\left(\frac{14}{\sqrt{50}}\right) < \mu < 274 + 2.33\left(\frac{14}{\sqrt{50}}\right)$$

$$274 - 4.6 < \mu < 274 + 4.6$$

$$269.4 < \mu < 278.6$$

So, a 98 percent confidence interval for the true mean pregnancy term involving natural birth is (269.4, 278.6). From this interval we infer, with 98 percent confidence, that the mean pregnancy term μ is between 269.4 and 278.6 days. ●

Empirical Interpretation

As an example of an empirical interpretation of a confidence interval statement, suppose we wish to construct a 95 percent confidence interval for the mean weight of a certain strain of monkeys. Based on past experience, we will assume that σ equals 4 pounds. Also, we agree to construct an interval by weighing sixty-four monkeys; that is, $n = 64$. Substituting for σ and n, the random interval becomes

$$\left(\bar{X} - 1.96\frac{4}{\sqrt{64}}, \bar{X} + 1.96\frac{4}{\sqrt{64}}\right),$$

that is, $(\bar{X} - 0.98, \bar{X} + 0.98)$. Thus,

$$P(\bar{X} - 0.98 < \mu < \bar{X} + 0.98) = 0.95.$$

Suppose we weigh a sample of sixty-four monkeys and find their mean weight to be 20.8 pounds. The random interval then reduces to the interval $(20.8 - 0.98, 20.8 + 0.98)$, that is, $(19.82, 21.78)$. Does this interval include μ? We do not know.

Next, suppose we take a second sample, once again consisting of sixty-four monkeys, and obtain a mean of 19.7 pounds. This time the random interval reduces to $(18.72, 20.68)$. Here, too, we do not know whether the interval encloses μ.

We can continue this process of taking samples of 64 and computing intervals. In Table 7-1 we give results for five samples, each of size 64. These intervals corresponding to each sample are plotted in Figure 7-4.

TABLE 7-1

Sample Number	Sample Mean $\bar{x}$	Intervals $(\bar{x} - 0.98, \bar{x} + 0.98)$
1	20.8	(19.82, 21.78)
2	19.7	(18.72, 20.68)
3	21.3	(20.32, 22.28)
4	20.2	(19.22, 21.18)
5	21.0	(20.02, 21.98)
⋮	⋮	⋮

Which of these intervals enclose μ and which do not we cannot actually say. Recall at this point the empirical notion of probability: *The probability of an event is interpreted as the relative frequency with which the event occurs in a large series of identical trials of an experiment.* Therefore, the probability statement

$$P(\bar{X} - 0.98 < \mu < \bar{X} + 0.98) = 0.95$$

states that approximately 95 percent of the intervals constructed in the manner above will encompass the population mean.

Of course, we pick samples of size n repeatedly only in theory. When we construct a confidence interval, no repetitive process is involved. We pick just one sample and compute the endpoints. Now the population mean is either included in this interval or it is not. We will never really know.

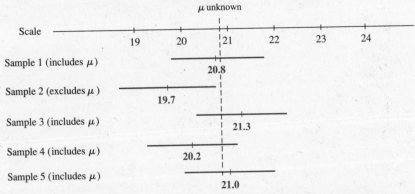

Figure 7-4 Confidence Intervals from Five Samples Each of Size Sixty-four
Though μ is unknown, it is shown in the figure for illustration. Only Sample 2 excludes this value of μ.

Since the first sample that we picked gave us an interval (19.82, 21.78), we claim that the population mean is contained in the interval 19.82 to 21.78 and write

$$19.82 < \mu < 21.78.$$

We do not contemplate taking any more samples. We attach a 95 percent confidence to the claim $19.82 < \mu < 21.78$ with the following rationale: The approach that we adopted for computing the confidence interval is such that of all the intervals obtained by computing $(\overline{X} - 0.98, \overline{X} + 0.98)$ by repeatedly drawing samples of sixty-four monkeys, 95 percent will include μ and 5 percent will not. We have a strong feeling and hope that the one interval that we have obtained, (19.82, 21.78), is from the former category. We quantify this hope by saying that we have a 95 percent confidence in the claim.

Note: It is definitely wrong to say

$$P(19.82 < \mu < 21.78) = 0.95$$

because it would imply that μ varies as a random variable. However, it would not be inappropriate to say

$$\text{conf}(19.82 < \mu < 21.78) = 0.95,$$

which means that the confidence is 95 percent.

EXAMPLE 5   A gas station sold a total of 8019 gallons of gas on nine randomly picked days. Suppose the amount sold on a day is normally distributed with a standard deviation σ of 90 gallons. Construct confidence intervals for the true mean amount μ sold on a day with the following confidence levels.

(a)   98 percent

(b)   80 percent

SOLUTION   We are given that $\Sigma\, x_i = 8019$, $n = 9$, and $\sigma = 90$. Therefore,

$$\overline{x} = \frac{8019}{9} = 891.$$

(a)   For a 98 percent confidence interval, $\alpha = 0.02$. So, $z_{\alpha/2} = z_{0.01} = 2.33$. Therefore, a 98 percent confidence interval for the true mean amount μ sold on a day is given by

$$\overline{x} - 2.33\,\frac{\sigma}{\sqrt{n}} < \mu < \overline{x} + 2.33\,\frac{\sigma}{\sqrt{n}}$$

Since $\overline{x} = 891$, $n = 9$, and $\sigma = 90$, we get

$$891 - 2.33\!\left(\frac{90}{\sqrt{9}}\right) < \mu < 891 + 2.33\!\left(\frac{90}{\sqrt{9}}\right)$$

which, after simplification, gives

$$821.1 < \mu < 960.9.$$

Therefore, a 98 percent confidence interval for the mean daily amount sold is (821.1, 960.9). The formula that we have used is such that 98 percent of the intervals computed using it will contain μ. With that in mind, we put faith in stating, with 98 percent confidence, that the mean amount sold on a day is between 821.1 and 960.9 gallons.

(b) In this part we wish to estimate μ with a much lower level of confidence than in Part a. If you think about it, now we are not that particular if the intervals computed by taking repeated samples miss the target parameter more often. In the long run, 80 percent of them will contain the mean μ. The interval that we obtain will be narrower, as you will see.

Since $\alpha = 0.20$, we get $z_{\alpha/2} = z_{0.1} = 1.28$. Consequently, with 80 percent confidence,

$$891 - 1.28\left(\frac{90}{\sqrt{9}}\right) < \mu < 891 + 1.28\left(\frac{90}{\sqrt{9}}\right).$$

After simplifying, an 80 percent confidence interval for μ is (852.6, 929.4). The interval is narrower now, which is good. But our confidence in saying that μ is in the interval is lower, which is not so good.   ●

THE LENGTH OF A CONFIDENCE INTERVAL

In providing confidence intervals, the most ideal situation would be to pinpoint the parameter with 100 percent certainty. The variability inherent in the population, however, limits us in accomplishing this. Short of this ideal target, *one should strive for the twin goals of a high level of confidence and a narrow interval.* But there is a seesaw relationship between these goals. Let us consider the following example as it pertains to the **length of a confidence interval,** also called its *width.*

EXAMPLE 6   From past records, the seasonal rainfall in a county when observed over sixteen randomly picked years yielded a mean rainfall of 20.8 inches. If it can be assumed from past experience that rainfall during a season is normally distributed with $\sigma = 2.8$ inches, construct confidence intervals for the true mean rainfall μ with the following confidence levels.

(a) 90 percent

(b) 95 percent

(c) 98 percent

SOLUTION   We are given that $\bar{x} = 20.8$, $\sigma = 2.8$, and $n = 16$. Also, we are granted the assumption that rainfall is normally distributed.

(a) For a 90 percent confidence interval, $\alpha = 0.10$. Thus, $z_{\alpha/2} = z_{0.05} = 1.645$, and the confidence interval is given by

$$20.8 - 1.645\left(\frac{2.8}{\sqrt{16}}\right) < \mu < 20.8 + 1.645\left(\frac{2.8}{\sqrt{16}}\right)$$

which can be simplified to $19.65 < \mu < 21.95$, or the interval (19.65, 21.95).

(b) Because $\alpha = 0.05$, $z_{\alpha/2} = 1.96$, and a 95 percent confidence interval is given by

$$20.8 - 1.96\left(\frac{2.8}{\sqrt{16}}\right) < \mu < 20.8 + 1.96\left(\frac{2.8}{\sqrt{16}}\right).$$

This, upon simplification, gives $19.43 < \mu < 22.17$, or the interval $(19.43, 22.17)$.

(c) Because $\alpha = 0.02$, $z_{\alpha/2} = 2.33$, and we get, with 98 percent confidence,

$$20.8 - 2.33\left(\frac{2.8}{\sqrt{16}}\right) < \mu < 20.8 + 2.33\left(\frac{2.8}{\sqrt{16}}\right).$$

That is, $19.17 < \mu < 22.43$, or μ is in the interval $(19.17, 22.43)$.
An important observation that we should register from this example is that as the level of confidence desired increases, the z value becomes larger, and this makes the interval wider. ●

If we exhibit the confidence intervals in Example 6 graphically, as in Figure 7-5, we observe that they become wider as the level of confidence we desire increases.

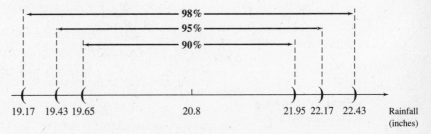

Figure 7-5 A Graphical Presentation Showing How the Length of the Confidence Interval Increases with the Level of Confidence

What are the other factors that affect the length of a confidence interval? Notice that the length of any interval from a to b is $b - a$. Consequently, the length of the confidence interval $\left(\bar{x} - z_{\alpha/2}\dfrac{\sigma}{\sqrt{n}}, \bar{x} + z_{\alpha/2}\dfrac{\sigma}{\sqrt{n}}\right)$ is

$$\left(\bar{x} + z_{\alpha/2}\frac{\sigma}{\sqrt{n}}\right) - \left(\bar{x} - z_{\alpha/2}\frac{\sigma}{\sqrt{n}}\right) = 2z_{\alpha/2}\frac{\sigma}{\sqrt{n}}.$$

Thus,

$$\begin{matrix}\textbf{the length of the confidence}\\ \textbf{interval for } \mu\end{matrix} = 2z_{\alpha/2}\frac{\sigma}{\sqrt{n}} = 2(\textbf{margin of error}).$$

We see that the length of the confidence interval for μ does not depend on $\bar{x}$, but it does depend on $z_{\alpha/2}$, σ, and n, in the following manner.

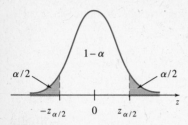

Figure 7-6

Standard normal curve; if $1 - \alpha$ increases, $z_{\alpha/2}$ moves to the right and becomes larger.

1 First, let us keep n fixed and find out how the level of confidence and the length of the interval influence each other.

Suppose we decide to have a high level of confidence. In that case $(1 - \alpha)$ will be large and this will make $z_{\alpha/2}$ large (see Figure 7-6). From the formula above, this will make the confidence interval wider.

On the other hand, if the interval is made narrower, then $z_{\alpha/2}$ will be smaller (from the formula). From Figure 7-6, this will make $1 - \alpha$ smaller and, consequently, will make the confidence level smaller.

This is the seesaw relationship mentioned earlier between confidence level and length of the confidence interval. *For a given sample size, if the confidence level is raised, the price is paid by making the interval wider, so, estimation is less precise. On the other hand, if the interval is made narrower so estimation is more precise, then the price is paid by lowering the confidence level.*

2 The quantity σ appears in the numerator of the formula above. Therefore, the length of the interval will be large if σ is large and small if σ is small. Notice, however, that σ is a population characteristic and the experimenter has no control over it.

3 The third factor affecting the length of the interval is the sample size n. Since n appears in the denominator of $2z_{\alpha/2}\dfrac{\sigma}{\sqrt{n}}$, the width will be small if n is large. Thus, *if a very high accuracy in estimating μ is desired, that is, a narrow confidence interval with a high degree of confidence is desired, then a way to accomplish this goal is by picking an appropriately large sample.* We discuss this aspect next.

How Large a Sample for Estimating μ?

The question of how large a sample should be investigated is an important consideration in designing an experiment. Suppose we want to decide how large a sample should be picked so that we can be $(1 - \alpha)100$ percent confident that the estimate $\bar{x}$ does not differ from μ by more than some given amount. For example, if an environmental engineer wants to estimate the mean dissolved oxygen concentration in a stream, the engineer might want to know how many measurements should be made so that one can be $(1-0.05)100$, or 95, percent confident that the estimate $\bar{x}$ will not differ from the true mean concentration μ by, say, more than 1 mg/liter. We will refer to the amount by which the estimate $\bar{x}$ differs from μ as the *error in estimation* and will denote the maximum error of estimate allowed by E.

Since the maximum error allowed in estimating μ is E and since we know that the maximum error of estimate of μ at the $(1 - \alpha)100$ percent level is $z_{\alpha/2}\dfrac{\sigma}{\sqrt{n}}$, we must have

$$E = z_{\alpha/2}\frac{\sigma}{\sqrt{n}}.$$

A slight simplification gives the following formula for the sample size n.

$$n = \left(\frac{z_{\alpha/2}\sigma}{E}\right)^2.$$

The value of n is rounded upward to the next higher integer and this guarantees that the desired goal of error tolerance and confidence level is met.

With a sample size as large as this, we can be $(1 - \alpha)100$ percent confident that the estimate will not be in error by more than an amount E. If the experimenter can take a sample larger than $\left(\frac{z_{\alpha/2}\sigma}{E}\right)^2$, so much the better.

We should not forget that with a larger sample, we always stand to gain more information about the population. Of course, one might have reached a point of diminishing returns, in which case the additional information acquired may not be commensurate with the added expenditure in time and money needed to investigate more units. In summary, we have established the following result.

SAMPLE SIZE FOR ESTIMATING μ

The **sample size** needed so as to be $(1 - \alpha)100$ percent confident that the estimate $\overline{x}$ does not differ from μ by more than a preassigned quantity E is

$$n = \left(\frac{z_{\alpha/2}\sigma}{E}\right)^2$$

where the population is normally distributed with known standard deviation σ. If n comes out fractional, then its value is rounded upward to the next higher integer.

Notice that n will be large if (1) σ is large, (2) $z_{\alpha/2}$ is large, which will be the case if we want a high level of confidence, or (3) E is small, that is, if the degree of accuracy desired is high. This is in keeping with what we might expect intuitively.

EXAMPLE 7

An electrical firm that manufactures a certain type of bulb wants to estimate its mean life. Assuming that the standard deviation σ is 40 hours, find how many bulbs should be tested so as to be:

(a)   95 percent confident that the estimate $\overline{x}$ will not differ from the true mean life μ by more than 10 hours

(b)   98 percent confident to accomplish accuracy in Part a.

(Assume a normal distribution of the life of a bulb.)

SOLUTION   We have $\sigma = 40$ and $E = 10$.

(a)   Because $\alpha = 0.05$, $z_{\alpha/2} = z_{0.025} = 1.96$. Therefore,

$$n = \left(\frac{1.96 \times 40}{10}\right)^2 = 61.5 \approx 62.$$

Rounding upward, approximately sixty-two bulbs should be tested if we wish to be 95 percent confident.

(b)   Because $\alpha = 0.02$, $z_{\alpha/2} = z_{0.01} = 2.33$. Therefore,

$$n = \left(\frac{2.33 \times 40}{10}\right)^2 = 86.9 \approx 87.$$

Rounding upward, approximately eighty-seven bulbs should be tested for a 98 percent confidence level to be attained.

Notice that a larger number of bulbs must be tested if we desire higher level of confidence.   ●

General Procedure

The general procedure for constructing a confidence interval for a parameter involves the following steps.

1 Devise two appropriate random variables, which we shall call L and U, thereby getting a **random interval** (L, U), or an **interval estimator.** The random interval (L, U) is arrived at in such a way that the probability that it covers the parameter is $1 - \alpha$, where α is a number between 0 and 1. That is,

$$P(L < \text{parameter} < U) = 1 - \alpha.$$

2 Pick a random sample of a certain size.

3 Compute the values of L and U from the sample data. If the value of L is l and that of U is u, give an interval estimate of the parameter as (l, u).

For example, in this section the parameter was μ, and L and U are given by

$$L = \bar{X} - z_{\alpha/2}\sigma/\sqrt{n} \quad \text{and} \quad U = \bar{X} + z_{\alpha/2}\,\sigma/\sqrt{n}.$$

Since L and U contain $\bar{X}$, a random variable, they are themselves random variables. When we compute the mean from given sample data we get $\bar{x}$, a specific number. When this value is substituted in the formulas we get l and u as specific numbers given by

$$l = \bar{x} - z_{\alpha/2}\sigma/\sqrt{n} \quad \text{and} \quad u = \bar{x} + z_{\alpha/2}\sigma/\sqrt{n}.$$

As a result we get a specific confidence interval (l, u).

The statistical solution to an estimation problem thus provides an interval (l, u) together with an accompanying statement regarding the measure of reliability.

The fraction $1 - \alpha$ associated with a confidence interval is called the **confidence coefficient.** It is the probability with which an interval estimator encloses the population parameter. The empirical interpretation is that when the interval is computed repeatedly a large number of times by picking samples of the same size, the fraction of such intervals that enclose the parameter will be, approximately, $1 - \alpha$.

When the confidence coefficient is expressed as a percentage as $(1 - \alpha)100$ percent, we get the **level of confidence.**

The left endpoint of the interval (l, u) is called the **lower confidence limit** and the right endpoint is called the **upper confidence limit.**

We will limit our discussion to the single population in this chapter and consider two populations in Chapter 8.

11/12/96

Section 7-1 Exercises

1. An aptitude test was administered to 100 of the 400 students in a freshman math class. The mean score for the 100 students was found to be 71.8 points. Which of the following statements are inferential?
   (a) The mean score of the 100 students is 71.8.
   (b) The mean score of the freshman math class is 71.8.
   (c) This year's freshmen are better than last year's.

2. The student records of a school extend from the day it opened to the present. State whether the mean height of all the students when they enrolled is a parameter or a statistic. (You are interested only in this school.)

3. In each of the following, identify the quantities as μ, p, $\bar{x}$, or x/n, if, with reference to Exercise 2, these facts were found.
   (a) 42 percent of the students enrolled in the 1950s graduated.
   (b) The mean weight of the students enrolled in 1960 was 115 pounds.
   (c) 25 percent of all the students entered a university after graduating.
   (d) 28 percent of all the students enrolled were blonde.
   (e) The mean IQ of all the students was 110.

4. In each of the following, obtain point estimates of the population mean, variance, and standard deviation.
   (a) $n = 12$, $\Sigma \, x_i = 1080$, $\Sigma \, (x_i - \bar{x})^2 = 6875$
   (b) $n = 10$, $\Sigma \, x_i = 185$, $s^2 = 1.764$

5. The following readings were obtained for the chlorophyll content (in mg/gm) of eight wheat leaves.

   2.3   2.1   2.6   2.2   1.9   2.7   2.4   2.2

   Obtain point estimates of the true mean, variance, and standard deviation of the chlorophyll content of wheat leaves.

6. Sylvia's grandmother deposits in a box whatever change she collects each day. In this way she has collected 50,000 coins of different denominations—pennies, nickels, dimes, and quarters. A random sample of 200 coins from this collection was found to contain 85 pennies, 52 nickels, 38 dimes, and 25 quarters. Estimate how much the collection of coins is worth.

7. A manuscript has 5000 pages. You have only 1 hour in which to come up with an approximate figure for the number of words in it. Devise a suitable approach.

8. Describe a suitable scheme to estimate the number of grains in a container that holds 100 pounds of rice.

9. Estimate the proportion of people who wear glasses if a random sample of 370 yielded 76 individuals who wore them.

10. When fifteen randomly picked people were asked about whether they smoked cigarettes, the following responses were obtained: yes, no, yes, no, no, yes, no, yes, no, yes, no, yes, no, no, no. Obtain a point estimate of the proportion of smokers in the population.

11. Of 100 fluorescent tubes tested in a laboratory, 68 lasted beyond 250 hours. Find a point estimate of the true proportion of fluorescent tubes that will last beyond 250 hours.

12. A fisherman cast his net five times. The following figures refer to the number of fish caught each time and the proportion of salmon in each catch.

Proportion of salmon	0.2	0.32	0.18	0.12	0.28
Number of fish caught	150	350	200	140	190

   (a) How many salmon did he catch altogether? $(.2)(150) + (.32)(350) + (.18)(200) + (.12)(140) + (.28)(190) = 248$.
   (b) What was the proportion of salmon in the entire catch? $248/1,030 = 0.241$
   (c) Obtain a *pooled estimate* of the true proportion of salmon, that is, an estimate by pooling together the results of the five catches.

13. Suppose a box contains 100,000 beads, of which some are red and others are blue. If you have limited time at your disposal and are short on patience, how would you go about estimating the number of red beads using a sample of beads from the box?

14. For the given level of confidence in each case determine α.
   (a) 96 percent   (b) 92 percent
   (c) 84 percent   (d) 97 percent

15. Determine the indicated value for the given level of confidence.
   (a) $z_{\alpha/2}$; 90 percent   (b) $z_{\alpha/2}$; 93.84 percent
   (c) $-z_{\alpha/2}$; 91.64 percent   (d) $-z_{\alpha/2}$; 96.16 percent
   (e) $z_{\alpha/2}$; 96.92 percent   (f) $-z_{\alpha/2}$; 97.42 percent

In Exercises 16–19, assuming that the population is normally distributed, find the margin of error at the indicated level of confidence.

16. $n = 20$, $\sigma = 4$; 90 percent

17. $n = 8$, $\sigma = 0.02$; 99 percent

18. $n = 16$, $\sigma = 12$; 95.4 percent

19. $n = 20$, $\sigma = 1.9$; 80 percent

In Exercises 20–22, assuming that the population is normally distributed, determine a confidence interval for the population mean at the specified level of confidence.

20. $n = 9, \bar{x} = 15.2, \sigma = 2.8$; 95 percent

21. $n = 16, \bar{x} = 0.8, \sigma = 0.12$; 90 percent

22. $n = 13, \bar{x} = 152.3, \sigma = 9.8$; 91.64 percent

23. A random sample of size $n = 16$ is drawn from a population having a normal distribution with standard deviation σ equal to 6. If the sample mean is $\bar{x} = 46$, find a 90 percent confidence interval for the population mean μ.

24. The time (in minutes) taken by a biological cell to divide into two cells has a normal distribution. Suppose the standard deviation σ can be assumed to be 3.5 minutes. When sixteen cells were observed, the mean time taken by them to divide was 31.2 minutes.
   (a) Find the margin of error in estimating the true mean time for cell division using a 98 percent confidence level.
   (b) Estimate the true mean time for cell division using a 98 percent confidence interval.

25. The mean yield of grain on ten randomly picked experimental plots of farm was found to be 150 bushels. If the yield per plot can be assumed to be normally distributed with a standard deviation of yield σ equal to 20 bushels, determine a 90 percent confidence interval to estimate the true mean yield μ per plot. Interpret the interval.

$$\left(\bar{x} - Z_{\alpha/2}\cdot\frac{\sigma}{\sqrt{n}} < \mu < \bar{x} + Z_{\alpha/2}\cdot\frac{\sigma}{\sqrt{n}}\right)$$

$n = 10 \quad \bar{x} = 150 \quad \sigma = 20 \quad \alpha = 0.10$

$Z_{\alpha/2} = 2.05 = 1.645 \left[\frac{20}{\sqrt{10}}\right] = 10.404$

$(139.596, 160.404)$

26. A random sample of nine full-grown crabs had a mean weight of 24 ounces. If the population standard deviation σ of the weight of a crab can be assumed to be 3 ounces, construct confidence intervals for the true mean weight of a crab with the following confidence levels.
   (a) 90 percent   **(b)** 95 percent   **(c)** 98 percent

Comment on the lengths of the confidence intervals obtained. You may assume that the weight of a full-grown crab has a normal distribution.

27. A paint manufacturer wants to introduce a new acrylic paint on the market. The following data were obtained for drying time (hours) when twelve samples of the paint were tested.

6.62	4.85	6.44	5.73	5.98	5.15
5.20	7.67	4.68	6.57	4.99	6.25

   (a) Assuming that the drying time is distributed with standard deviation $\sigma = 1.2$ hours, obtain a 98 percent confidence interval to estimate μ, the true mean drying time.
   (b) What assumptions are required to apply the formula of this section? Check whether the assumptions are reasonable.
   (c) Interpret the interval.
   (d) What would you recommend if it was desired to make the interval narrower without changing the level of confidence?

28. The data below contain the concentration of acid-soluble sulfides (mg/liter fresh ooze) in the bottom sediments of a lagoon. Twenty-five samples were analyzed.

998	938	971	949	1013	982	941
1000	1021	1002	995	956	976	1047
1010	1015	967	987	990	974	965
954	974	984	988			

   (a) Use the data to compute a point estimate of μ, the true mean concentration.
   (b) Construct a 95 percent confidence interval to estimate the true mean concentration. You may assume a normal distribution with $\sigma = 25$. (The normal probability plot will support the assumption of normality.)
   (c) Would the confidence interval be wider or narrower if the level of confidence was increased to a higher level?

29. In each case find the length of the given confidence interval.
   (a) $10.87 < \mu < 12.63$ (b) $-4.88 < \mu < 8.44$

30. The following confidence intervals are obtained for μ using the formula $\bar{x} \pm z_{\alpha/2}\dfrac{\sigma}{\sqrt{n}}$. In each case determine the sample mean $\bar{x}$.
   (a) $10.87 < \mu < 12.63$ (b) $-4.88 < \mu < 8.44$

   Hint: The sample mean is the midpoint of the interval.

31. In computing a confidence interval for a parameter, which of the following situations should we strive for?
   (a) To have as wide an interval as possible with a high level of confidence
   (b) To have as narrow an interval as possible with a high level of confidence

32. A 94 percent confidence interval for μ is (13.8, 16.2). Would a 97 percent confidence interval with the same data be wider or narrower?

33. A margin of error for a 96 percent confidence interval based on 120 observations is reported as 2.38. Would the margin of error increase or decrease if
   (a) the level of confidence was changed to 98 percent
   (b) the sample size was increased to 185?

34. A population has a normal distribution with variance 225. Find how large a sample must be drawn in order to be 95 percent confident that the sample mean will not differ from the population mean by more than 2 units.

35. It is desired to estimate the true mean of a certain characteristic that is normally distributed correct to within 1.2 units with 98 percent confidence level. From a preliminary survey it is safe to assume that $\sigma = 8.9$ units. Find the size of the sample that should be investigated to achieve the goal.

36. Suppose you wish to estimate the true mean weight of the liver of full-grown frogs of a certain variety correct to within 0.1 gm. Each frog costs $6 and there are $300 available for purchase. If frog liver weight is normally distributed with $\sigma = 0.35$ and you want to be 98 percent confident, is the amount allotted sufficient? If the amount is not sufficient, what compromises would you suggest?

37. Suppose the breaking strength of cables (in pounds) is known to have a normal distribution with a standard deviation σ equal to 6 pounds. Find how large a sample must be taken so as to be 90 percent confident that the sample mean breaking strength will not differ from the true mean breaking strength by more than 0.75 pounds.

38. To determine the mean weekly idle time of machines, the management has decided to study a sample of machines from the total number of machines in the factory. (It may be assumed that the total number of machines is large.) The mean idle time of a machine during a week is to be estimated within 1.12 hours of the true mean idle time with a 95 percent level of confidence. From past experience, it is safe to assume that the idle time of a machine during a week is normally distributed with a standard deviation of 4 hours. Find what size sample should be drawn.

39. Based on a random sample of 100 cows of a certain breed, a confidence interval for estimating the true mean yield of milk is given by $41.6 < \mu < 44.0$. If the yield of milk of a cow may be assumed to be normally distributed with σ equal to 6, what was the level of confidence used?

40. A manufacturer of ball bearings wants to have the true mean diameter of the bearings with an accuracy within 0.0026 cm. A random sample of 8 bearings gave a mean diameter of 0.502. Assuming the diameters are normally distributed and the standard deviation σ is 0.003 cm, how confident can the manufacturer be that the estimate of 0.502 is within 0.0026 cm of the true mean diameter?

41. A chemist has to estimate the true mean amount of an active ingredient in a pill to within 0.017 mg. When a random sample of 20 pills was analyzed for the active ingredient, the mean active ingredient was 0.52 mg. Suppose it is reasonable to assume that the amount of active ingredient in a pill is normally distributed with $\sigma = 0.04$ mg. How confident can the chemist be that the estimate 0.52 is within 0.017 mg of the true mean?

7-2   CONFIDENCE INTERVAL FOR μ WHEN σ IS UNKNOWN

The case where the population standard deviation is unknown is more realistic, since in practice it will hardly ever be known to the experimenter. *The method that we are about to develop is valid for any arbitrary sample size but is particularly important when the sample size is small.* (Recall that if the sample size is large we can use the normal approximation.) Because of this, some authors refer to the confidence interval discussed here as a **small sample confidence interval for μ.**

For setting a confidence interval for μ we make the following assumptions.

1.   The observations are picked from the population under study and are independent.
2.   The population has a normal distribution.

Now recall that for setting a confidence interval for μ when σ is known, we start with the fact that the sampling distribution of $\overline{X}$ is normal with mean μ and standard deviation $\sigma/\sqrt{n}$ or, equivalently, converting to the z-scale, that

$$Z = \frac{\overline{X} - \mu}{\sigma/\sqrt{n}}$$

has the standard normal distribution. In the present context, since σ is not known, we do the next best thing and use its point estimate. In other words, we start with the random variable

$$T = \frac{\overline{X} - \mu}{S/\sqrt{n}}.$$

Now it turns out that the random variable T does not have the standard normal distribution, but has a distribution called Student's t distribution, which will be studied next.

STUDENT'S t DISTRIBUTION

We notice immediately that for Z only the numerator changes from sample to sample (assuming n is fixed). But for T both the numerator and the denominator change from sample to sample, since $\overline{X}$ and S both depend on sample values. Thus, it is not surprising that the exact distribution of T is not the same as that of Z. If n is small, the departure of the distribution of T from the standard normal distribution is rather marked. William Gosset, an employee of an Irish brewery, realized this fact in his research and derived the exact distribution of T. Gosset published his work under the pseudonym of Student. Hence the distribution is commonly known as **Student's t distribution,** or, simply, as the **t distribution.** There are many other statistics that also have the t distribution. A typical value of a random variable that has the t distribution is denoted by t.

Properties of Student's t Distribution

The following features characterize the t distribution.

1.   It is a continuous distribution with mean 0.
2.   It is symmetric about the vertical axis at 0 and is bell-shaped like the normal distribution.
3.   Its shape depends on a parameter ν (which is related in some way to the sample size) called the **number of degrees of freedom,** or simply *degrees of freedom,* usually abbreviated df. There is one t distribution curve for each degree of freedom (t distributions with 2, 4, and 10 df are shown in Figure 7-7).
4.   The curves for the t distribution are lower at the center and higher at the tails than the standard normal curve (see Figure 7-7). That means results far from 0 are more probable than with the standard normal distribution.
5.   As the number of degrees of freedom becomes larger, the t distribution looks more and more like the standard normal distribution; the curves for the t distributions rise at the center, getting closer to the standard normal distribution (see Figure 7-7).

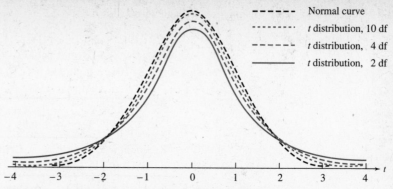

Figure 7-7 *t* Distributions with 2, 4, and 10 df

Also, the figure shows how the *t* distribution approaches the standard normal distribution as the degrees of freedom become large.

The particular statistic *T* that we have described above, namely,

$$T = \frac{\bar{X} - \mu}{S/\sqrt{n}}$$

has a *t* distribution with $n - 1$ degrees of freedom where n is the size of the sample.

Since the *t* distribution is continuous, probabilities are given as areas under the curve. Appendix III, Table A-4 gives such areas for different numbers of degrees of freedom. Because there is one curve for each number of degrees of freedom, for economy of space, the table is not as exhaustive as the one giving areas for the standard normal distribution. (We would need thirty tables for the first 30 df!) For each number of degrees of freedom ν given in the first column, the *t values are given along the row for a select set of probabilities contained in the right tail of the distribution.*

A general *t* value is denoted with a pair of subscripts where the first subscript indicates the degrees of freedom of the *t* distribution and the second subscript gives the area to the right of the *t* value. Thus, for example, $t_{\nu, \alpha}, t_{\nu, \alpha/2}$, $t_{\nu, 0.05}$ are the *t* values for a *t* distribution with ν degrees of freedom leaving, respectively, areas α, $\alpha/2$, 0.05 in the right tail. The value $t_{\nu, \alpha}$ is shown in Figure 7-8. Such *t* values are called the **critical values of the *t* distribution.**

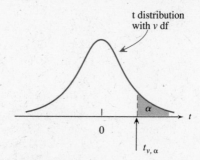

Figure 7-8

For ν df, a *t* value such that the area in the right tail is α.

We will show how to read Table A-4 specifically for $\nu = 10$. The relevant entries for $\nu = 10$ are included here in the partial reproduction of that table.

Degrees of Freedom ν	Area in the Right Tail				
	0.1	0.05	0.025	0.01	0.005
	Critical Values of t				
	$t_{\nu,0.1}$	$t_{\nu,0.05}$	$t_{\nu,0.025}$	$t_{\nu,0.01}$	$t_{\nu,0.005}$
1	3.078	6.314	12.706	31.821	63.657
2	1.886	2.920	4.303	6.965	9.925
3	1.638	2.353	3.182	4.541	5.841
⋮	⋮				
9	1.383	1.833	2.262	2.821	3.250
10	**1.372**	**1.812**	**2.228**	**2.764**	**3.169**
11	1.363	1.796	2.201	2.718	3.106
⋮	⋮				
29	1.311	1.699	2.045	2.462	2.756
∞	1.282	1.645	1.960	2.326	2.576

The values 1.372, 1.812, 2.228, 2.764, 3.169 represent the t values that we would ordinarily mark along the horizontal axis. The column headings are the areas (that is, the probabilities) under the curve to the right of the corresponding t value, as shown in Figure 7-9. Thus, to the right of 1.812 there is

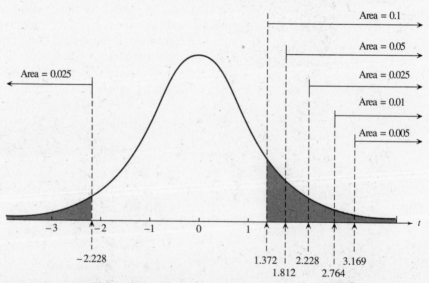

Figure 7-9 Probability Density Curve for the t Distribution with 10 df
Area to the right of 1.372 is 0.1, and so on.

an area of 0.05, to the right of 2.764 it is 0.01, and so on. Since the t distribution is symmetric about the vertical axis at $t = 0$, it is possible to obtain areas for certain other t values as well. For example, the area to the left of -1.812 is also 0.05, the area between -1.812 and 1.812 is 0.90, and so on.

As remarked earlier, the t distribution approaches the standard normal distribution as the degrees of freedom become large. The row with ∞ df in the preceding table on page 412 corresponds to the standard normal distribution and the entries in that row give the values of z. Notice how the t values obtained for 29 df are close to the corresponding values in the row shown below it. It is for this reason that the following rule of thumb is often pre-scribed: *One may use the standard normal distribution instead of the* t *distribution to obtain the* t *values, approximately, when the number of degrees of freedom is 30 or more.* The magic number 30 has no scientific basis and depends on the simple fact that about 30 rows, one for each degree of freedom, can be accommodated on a page of t tables while, at the same time, providing a reasonably satisfactory approximation.

It is not possible to give a complete answer to the question "How large is large enough?" since this will depend on what is going to be done with the answer. Sometimes a crude approximation is all that is desired. In such situations, following the rule of thumb, one may use values from the standard normal distribution in place of the t values with satisfactory approximation when the degrees of freedom are 30 or more. On the other hand, when a great deal of precision in the answer is needed one should consult tables where critical values of t are available for higher degrees of freedom. Some com-puter software packages such as MINITAB or SAS can also be used for the purpose.

EXAMPLE 1   Find the following.

(a) $t_{4,0.05}$ (b) $t_{21,0.025}$

(c) $-t_{17,0.1}$ (d) $t_{5,\alpha/2}$ if $\alpha = 0.01$

(e) $-t_{16,\alpha/2}$ if $\alpha = 0.05$

SOLUTION   Sketching the graph of the appropriate t distribution would help.

(a)   To find $t_{4,0.05}$, we consider t distribution with 4 degrees of freedom. Notice that the area is 0.05 to the right of the value. From Table A-4, in the row corresponding to 4 degrees of freedom under column heading 0.05, we find $t_{4,0.05} = 2.132$ (see Figure 7-10(a)).

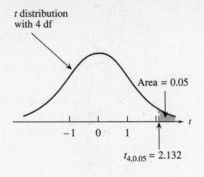

(a)

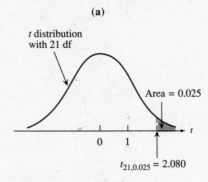

(b)

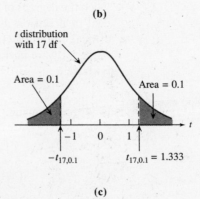

(c)

Figure 7-10

(b) As in Part a, we look in Table A-4 in the row corresponding to $\nu = 21$ and under column heading 0.025. The answer is $t_{21,0.025} = 2.080$ (see Figure 7-10(b)).

(c) To find $-t_{17,0.1}$, we first find $t_{17,0.1}$. From Table A-4 we see that $t_{17,0.1} = 1.333$. Thus, $-t_{17,0.1} = -1.333$ (see Figure 7-10(c)). Notice, incidentally, that $-t_{17,0.1}$ has area 0.9 to its right and area 0.1 to its left.

(d) Because $\alpha = 0.01$, $\alpha/2 = 0.005$. From Table A-4, we immediately get $t_{5,\alpha/2} = t_{5,0.005} = 4.032$.

(e) $\alpha = 0.05$; thus $\alpha/2 = 0.025$. As a result, $t_{16,\alpha/2} = t_{16,0.025} = 2.12$, and $-t_{16,\alpha/2} = -2.12$.

Confidence Interval

Having discussed the t distribution, we now return to our primary goal—that of setting a confidence interval for μ when σ is not known.

Because the statistic T has Student's t distribution with $n - 1$ degrees of freedom, we can write

$$P\left(-t_{n-1,\alpha/2} < \frac{\overline{X} - \mu}{S/\sqrt{n}} < t_{n-1,\alpha/2}\right) = 1 - \alpha$$

where $t_{n-1,\alpha/2}$ is the value from the table for a t distribution with $n - 1$ degrees of freedom such that the area in its right tail is $\alpha/2$. The value $-t_{n-1,\alpha/2}$ is the value located symmetrically on the other side of zero and is such that the area in the left tail is $\alpha/2$, as shown in Figure 7-11.

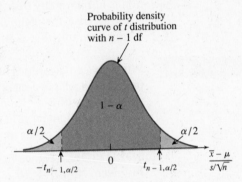

Figure 7-11

t values such that there is an area $\alpha/2$ in the right tail and an area $\alpha/2$ in the left tail of the distribution.

Through some algebraic steps, this leads to

$$P\left(\overline{X} - t_{n-1,\alpha/2}\frac{S}{\sqrt{n}} < \mu < \overline{X} + t_{n-1,\alpha/2}\frac{S}{\sqrt{n}}\right) = 1 - \alpha.$$

Thus, we get the following result.

CONFIDENCE INTERVAL FOR μ; σ UNKNOWN

A $(1 - \alpha)100$ percent confidence interval for μ when the population is normally distributed and σ is *not* known is given by

$$\overline{x} - t_{n-1,\alpha/2}\frac{s}{\sqrt{n}} < \mu < \overline{x} + t_{n-1,\alpha/2}\frac{s}{\sqrt{n}}$$

where $\overline{x}$ is the sample mean and s the sample standard deviation based on n observations.

You will notice that the formula for the confidence interval is pretty much the same as the one given in the previous section except for the following: We use the sample standard deviation s in place of the population standard deviation σ, which is not known. For that reason, we use t-values in place of the z-values.

Remark   We have mentioned earlier that $t_{n-1,\alpha/2}$ will be very near $z_{\alpha/2}$ if n is 30 or more. Thus, *for large samples,* meaning $n \geq 30$, the confidence interval for μ when σ is unknown becomes, approximately,

$$\bar{x} - z_{\alpha/2}\frac{s}{\sqrt{n}} < \mu < \bar{x} + z_{\alpha/2}\frac{s}{\sqrt{n}}.$$

Actually, if n is large, we can use this formula even if the parent population is not normally distributed. This comment and those made earlier make a strong case for planning an experiment properly and, if at all possible, taking a sample that is sufficiently large.

EXAMPLE 2   When sixteen cigarettes of a particular brand were tested in a laboratory for the amount of tar content, it was found that their mean content was 18.3 milligrams with $s = 1.8$ milligrams. Set a 90 percent confidence interval for the mean tar content μ in the population of cigarettes of this brand. (Assume that the amount of tar in a cigarette is normally distributed.)

SOLUTION   Notice that σ is not known but the sample standard deviation s is given. Now $n = 16$, $\bar{x} = 18.3$, $s = 1.8$, and $\alpha = 0.10$. Thus, the underlying t distribution has $16 - 1 = 15$ degrees of freedom and $t_{15,\alpha/2} = t_{15,0.05} = 1.753$. As a result, a 90 percent confidence interval is given by

$$18.3 - 1.753\left(\frac{1.8}{\sqrt{16}}\right) < \mu < 18.3 + 1.753\left(\frac{1.8}{\sqrt{16}}\right)$$

which can be simplified to $17.51 < \mu < 19.09$, or the interval (17.51, 19.09). From this interval we estimate, with 90 percent confidence, that the mean tar content μ of the cigarettes of the brand is between 17.51 and 19.09 mg. ●

EXAMPLE 3   In order to estimate the amount of time (in minutes) that a teller spends on a customer, a bank manager decided to observe sixty-four customers picked at random. The amount of time the teller spent on each customer was recorded. It was found that the sample mean was 3.2 minutes with $s = 1.2$. Find a 98 percent confidence interval for the mean amount of time μ.

SOLUTION   From the given information $\bar{x} = 3.2$, $s = 1.2$, $n = 64$, and $\alpha = 0.02$. The population standard deviation σ is not known, so we use its estimate s. Since n is large, the confidence interval is constructed, approximately, using the formula

$$\bar{x} - z_{\alpha/2}\frac{s}{\sqrt{n}} < \mu < \bar{x} + z_{\alpha/2}\frac{s}{\sqrt{n}}.$$

Now $z_{\alpha/2} = z_{0.01} = 2.33$. Therefore, a 98 percent confidence interval for μ is approximately

$$3.2 - 2.33\frac{1.2}{\sqrt{64}} < \mu < 3.2 + 2.33\frac{1.2}{\sqrt{64}}.$$

Simplifying, we get

$$2.85 < \mu < 3.55.$$

A 98 percent confidence interval for μ, the mean amount of time that the teller spends on a customer, is (2.85, 3.55). So, we can say with 98 percent confidence that the mean time spent on a customer is between 2.85 and 3.55 minutes. The width of the interval of estimation is 0.7 minutes. If this interval seems too wide, we could get a narrower interval by lowering the level of confidence. It is a trade-off, however. ●

EXAMPLE 4 Weight loss is a major manifestation of infection with the human immuno-deficiency virus (HIV). A study was carried out by Derek C. Macallan et al. to analyze weight change in 30 male subjects with Stage IV HIV infection related to different clinical events. One such clinical event is PCP, *Pneumocystis carinii* pneumonia. The following figures are obtained from the data reported by the investigators in the *American Journal of Clinical Nutrition* 1993; 58: 417–424 and give the weight loss per month of 9 Stage IV HIV patients infected with PCP. Use a 95 percent confidence interval to estimate μ, the mean weight loss per month of HIV male patients in Stage IV infected with PCP. Interpret the results.

Subject	Weight Loss (Kg/mo)
1	11.9
2	7.8
3	6.2
4	6.1
5	5.8
6	5.0
7	4.5
8	2.8
9	3.6
Mean $\bar{x}$	5.97
Std dev s	2.678

SOLUTION   The values for the sample mean and standard deviation given in the table can be easily verified. To apply the formula developed in this section we need to make the assumption that the population is normally distributed. The normal probability plot reasonably bears out this assumption.

There are 9 data values. So, the underlying t distribution has $9 - 1 = 8$ degrees of freedom. For a 95 percent confidence interval $\alpha = 0.05$. Therefore, $t_{8,\alpha/2} = t_{8,0.025} = 2.306$, and a 95 percent confidence interval for μ is given by

$$\bar{x} - t_{n-1,\alpha/2}\frac{s}{\sqrt{n}} < \mu < \bar{x} + t_{n-1,\alpha/2}\frac{s}{\sqrt{n}}$$

$$5.97 - 2.306\frac{2.678}{\sqrt{9}} < \mu < 5.97 + 2.306\frac{2.678}{\sqrt{9}}$$

$$3.91 < \mu < 8.03$$

Hence, a 95 percent confidence interval for μ is (3.91, 8.03). That is, with 95 percent confidence we estimate that the true mean weight loss of Stage IV HIV male patients infected with PCP is between 3.91 and 8.03 kg per month. ●

11/12/96

Section 7-2 Exercises

1. Find the following.
   (a) $t_{18,0.05}$ (b) $t_{13,0.025}$
   (c) $t_{6,0.01}$ (d) $t_{9,0.025}$

2. In each case find the indicated quantity for the given level of confidence.
   (a) $t_{6,\alpha/2}$; 95 percent   (b) $t_{10,\alpha/2}$; 98 percent
   (c) $t_{14,\alpha/2}$; 98 percent   (d) $t_{26,\alpha/2}$; 80 percent
   (e) $t_{16,\alpha/2}$; 99 percent   (f) $t_{12,\alpha/2}$; 99 percent

In Exercises 3–6, assuming that the population is normally distributed, determine a confidence interval for the corresponding population mean at the specified level of confidence.

3. $n = 9$, $\bar{x} = 15.2$, $s = 3.3$; 95 percent

4. $n = 16$, $\bar{x} = 0.8$, $s = 0.1$; 98 percent

5. $n = 13$, $\bar{x} = 152.3$, $s = 10.82$; 99 percent

6. $n = 21$, $\bar{x} = -1.23$, $s = 2.34$; 90 percent

7. A random sample of size $n = 16$ is drawn from a population having a normal distribution. If the sample mean and the sample standard deviation from the data are given, respectively, as $\bar{x} = 23.8$ and $s = 3.2$, find a 98 percent confidence interval for μ, the population mean.

8. When a skier took ten runs on the downhill course, the mean time was found to be 4.3 minutes with $s = 1.2$ minutes. Assuming a normal distribution, estimate the true mean time for a run of the skier using a 90 percent confidence interval.

9. In ten half-hour programs on a TV channel, Jill found that the number of minutes devoted to commercials were 6, 5, 5, 7, 5, 4, 6, 7, 5, and 5. Set a 95 percent confidence interval on the true mean time devoted to commercials during a half-hour program. Based on the normal probability plot, it seems reasonable to assume that the amount of time devoted to commercials is normally distributed.

10. A sample of eight workers in a clothing manufacturing company gave the following figures for the amount of time (in minutes) needed to join a collar to a shirt: 10, 12, 13, 9, 8, 14, 10, and 11. Assuming that the length of time is normally distributed, set a 90 percent confidence interval on the true mean time μ.

11. The following data present the sealing strength (kg/25 cm²) of an industrial glue at room temperature (10°C–25°C).

11.8	13.6	11.6	12.6	12.8	12.9
13.9	11.8	13.5	11.6	14.2	13.7

(a) Use the normal probability plot to check if the sealing strength of the glue may be assumed to be normally distributed.

(b) It is desired to estimate the true mean sealing strength of the glue. Use a 99 percent confidence interval to estimate it.

(c) How do you interpret the interval estimate?

12. A sensor detects an electric charge each time a turbine blade makes one rotation and then measures the amplitude of the electric current. The following eight readings were obtained for a certain setting of the speed of rotation of the blades and the gap between the blades.

26.9   30.1   28.9   25.5   23.6   29.9   24.7   30.0

(a) Use a 98 percent confidence interval to estimate the true mean amplitude of the electric current for the given setting.

(b) State the assumptions that are required, mentioning the variable under study.

(c) What would happen to the width of the confidence interval if the level of confidence is lowered?

13. A study on the amount of dye needed to get the best color for a certain type of fabric was conducted. The following data give the photometer readings for the color density of the fabric when the amount of dye used was 1% of the weight of the fabric.

13.2	11.5	12.9	13.0	11.7	10.4	12.1
12.1	11.5	10.3	11.7	12.3	11.2	

(a) Obtain a 90 percent confidence interval for the true mean color density of the fabric. State the underlying assumptions.

(b) Interpret the interval to comment on the true mean color density of the fabric.

14. Suppose you decide to buy a hotel. Before buying the business you would certainly like to get an idea of the past earnings of the hotel. So you go through the hotel account books and examine the earnings on 45 randomly picked days. Suppose the mean daily profit based on these 45 days is $1,825 with $s = \$318$.

(a) State whether "Profit on a day" is a discrete variable or a continuous variable.

(b) Is it appropriate to classify the variable as normally distributed?

(c) State why we may obtain approximate confidence intervals for the true mean daily profit using a normal approximation.

(d) Estimate the true mean daily profit using an interval so that you can be 98 percent confident.

(e) Interpret the interval.

In Exercises 15–26 below assume that the variable under study has a normal distribution. (When normal probability plots were obtained for data sets in Exercises 15–16 and 18–26, in each case this assumption was supported.) Also, provide an interpretation of the confidence interval for each exercise.

15. The following data were obtained when the lives (in hours) of 18 randomly picked dry cell batteries were tested in the laboratory.

109	101	112	107	115	79	129	125	119
134	106	108	128	95	107	91	106	92

Estimate the true mean life of a battery giving the margin of error at the 98 percent level.

16. The data given below represent the time (in minutes) taken by 14 randomly selected students to complete a national exam.

99	109	69	64	100	65	100
104	92	83	87	73	79	65

Based on these data, determine the true mean estimated time and the margin of error at the 95 percent level.

17. There is a general concern about the escalating costs of providing health care in the United States. One of the components contributing to the increasing costs is the length of the hospital stay of a patient. In a sample of 23 patients the mean time was $\bar{x} = 4.5$ days with standard deviation $s = 1.3$ days. Find a 95 percent confidence interval for the true mean length of hospital stay of a patient. You may assume that the hospital stay of a patient is normally distributed.

18. It is suspected that an industrial plant is polluting the water of a stream. To determine the extent of damage, the dissolved oxygen concentration was measured at a certain location along the stream on 13 randomly selected days. The data are presented below.

1.9	2.3	2.6	2.9	2.1	2.8	2.7
2.9	1.4	2.1	2.1	2.2	1.9	

Construct a 90 percent confidence interval for the true mean daily concentration.

19. The following figures give the percentage recovery of chemically reactive lysine in the presence of glucose by a modified trinitrobenzene sulphonic acid procedure.

57.2	79.7	82.7	66.3	64.7
78.4	70.2	66.1	55.6	67.3
72.1	80.5	62.8	74.7	64.8
74.0	71.2	71.5	70.7	70.0
73.9	69.9	73.3	70.9	61.4

Estimate the true mean percentage recovery using a 98 percent confidence interval.

$n = 45 \quad \bar{x} = \$1,825 \quad s = \$318$

$E_{98} = \bar{x} \pm t_{n-1, \alpha/2} \cdot \dfrac{s}{\sqrt{n}}$

$= 1,825 \pm t_{44, 0.01} \cdot \dfrac{318}{\sqrt{45}}$

$= 1,825 \pm 2.33 \cdot \dfrac{318}{\sqrt{45}}$

$= 1,825 \pm 110.45$

$(\$1,714.55, \ \$1,935.45)$

20. Because of their nutritional importance, an analysis was carried out to find the basic organic constituents of the tissues of the edible freshwater fish *Mystus vittatus*. Adult fish were freshly collected and biochemical estimations were made of the liver, muscle, and gill tissue of each. The total carbohydrate content of each was estimated using Anthrone reagent. The following data give the amount of total carbohydrates in the liver tissue of *Mystus vittatus*.

35.7	34.9	35.3	35.7	35.4
36.0	35.9	33.4	37.5	35.8
37.6	33.9	37.3	36.7	37.0

    Estimate the mean carbohydrate content using a 98 percent confidence interval.

21. A bank has 250,065 cardholders. A senior vice president calls upon the bank statistician to estimate the total outstanding amount charged by the cardholders. A random sample of 25 accounts is selected and it was found that the mean amount was $\bar{x} = 682$ dollars with $s = 102$ dollars. Assuming a normal distribution, use a 90 percent confidence interval to estimate
    (a) the true mean of the amount charged by a cardholder.
    (b) the total amount charged by the cardholders.

22. The recent failure of the Philadelphia police department to respond to 911 calls in an emergency where a youth was beaten to death has caused some concern among authorities in many cities in the United States. To determine whether the service in your area is satisfactory, suppose the police chief investigated 18 emergency calls, picked at random, that had been received during the previous year. The response time is measured as the time when the first call was received to the time the squad car leaves the station. The computer log showed the following times (in minutes).

11.6	7.9	9.1	7.7	6.8	8.5	2.5	10.1	11.0
3.6	10.5	6.3	5.3	7.2	5.8	12.6	4.3	2.9

    Determine a 99 percent confidence interval estimate of the true mean response time in your area. Interpret your answer.

23. In recent years the public has become increasingly aware of the importance of fiber in their diet. The following data present the amount of dietary fiber (g/100 g) in ten samples of white bread sold in the supermarket.

4.8	4.7	4.7	5.0	4.5
5.1	4.7	4.8	4.6	4.9

    Estimate the true mean amount of dietary fiber in white bread using a 90 percent confidence interval.

24. An important aspect that should be considered in developing a psychological profile of an individual is the person's attitude and general disposition. An attitude-measuring test was given to 21 randomly picked individuals. The following figures give their scores.

247	211	195	262	187	205	217
270	195	274	231	199	267	216
222	288	233	260	243	276	197

    Estimate the true mean of the attitude scores using a 95 percent confidence level.

$$s^2 = \frac{1}{n-1}\sum\left(x_i - \bar{x}\right)^2$$
or
$$s^2 = \frac{1}{n-1}\left\{\left(\sum x_i^2\right) - \frac{\left(\sum x_i\right)^2}{n}\right\}$$

25. Ten 17-week-old turkeys were slaughtered and the total haem pigments (mg/g tissue) were extracted with phosphate buffer (pH 6.8). The following data give the total haem pigments in breast and thigh meats of the ten birds.

Haem Pigment (mg/g)	
Breast Meat	**Thigh Meat**
0.57	2.11
0.51	2.21
0.60	2.20
0.57	2.19
0.56	2.00
0.58	2.22
0.65	2.13
0.68	2.10
0.57	2.17
0.59	1.92

(a) Construct a 95 percent confidence interval for the true mean haem pigment in turkey breast meat.

(b) Construct a 98 percent confidence interval for the true mean haem pigment in turkey thigh meat.

26. The following data contain information about thymol and carvacrol concentrations in essential oils from *Coridothymus capitatus,* a wild spice plant that grows in Israel and the West Bank of the Jordan River. Twelve thymol-type plants near Kfar Galim were selected for a random sample and the essential oil from individual *C. capitatus* plants was analyzed (Alexander Fleisher et al., "Chemovarieties of *Coridothymus capitatus I. Rchb.* growing in Israel"; *J. Sc. Food Agric. 35* (1984): 495–499).

Thymol (%)	Carvacrol (%)
58.8	6.2
58.0	6.9
61.0	5.6
54.8	3.9
66.2	6.3
67.0	6.4
52.2	4.8
63.0	7.2
55.0	5.9
50.5	5.1
50.8	4.7
58.5	6.6

(a) Estimate the true mean thymol concentration using a 95 percent confidence interval.

(b) Estimate the true mean carvacrol concentration with a 98 percent confidence interval.

In Exercises 27–34, construct a confidence interval for the corresponding population mean if the given information is available. Hint: Notice n is large in each case, so you may use z-values.

27. $n = 40, \bar{x} = 18.6, s = 9.486$; 95 percent

28. $n = 64, \bar{x} = 0.82, s = 0.12$; 90 percent

29. $n = 100, \bar{x} = 148.9, s^2 = 13.83$; 98 percent

30. In a laboratory experiment, sixty-four crosses of a certain type of beetle were made and the number of pupae produced were counted. It was found that the mean number of pupae were 85 with $s = 15$ pupae. Find a 95 percent confidence interval for the true mean number of pupae in a cross.

31. A laboratory tested thirty-six chicken eggs and found that the mean amount of cholesterol was 230 milligrams with $s = 20$ milligrams.
  (a) Determine a 90 percent confidence interval for the true mean cholesterol content of an egg using normal approximation.
  (b) Determine a 90 percent confidence interval for the true mean cholesterol content of an egg using $t_{35,0.05} = 1.6896$.
  (c) Comment on the widths of the two intervals.

32. A local telephone company has recently installed a new electronic switchboard and since then the company has been flooded with customer complaints about misdirected phone calls. As a result the company decided to estimate the true mean number of misdirected phone calls per 1000 calls placed. For this purpose, 36 random sets, each of one thousand phone calls, were randomly selected, and the misdirected phone calls were counted. The data below give the number of misdirected calls per 1000 calls.

4	6	7	12	7	6	5	5	8	7	10	6
6	7	5	6	8	9	8	7	6	5	8	9
6	4	4	11	8	4	6	5	8	8	6	5

Estimate the true mean number of misdirected phone calls per 1000 calls using a 95 percent confidence interval.

33. The daily paint production of a paint factory was recorded over 84 days and it was found that the mean daily production was $\bar{x} = 9910$ gallons with standard deviation $s = 1105$ gallons. Estimate the true mean daily production of paint with 90 percent confidence level.

34. A large state university is funded depending on the mean number of credit units a student takes. A preliminary survey involving 72 students yielded the mean number of $\bar{x} = 13.4$ units with a standard deviation $s = 2.1$ units.
  (a) Find a 99 percent confidence interval to estimate the true mean number of units.
  (b) Suppose there are 20,800 students enrolled in the university and the institution is funded at the rate of $300 per student unit. At the 99 percent confidence level, estimate the interval that covers the total funding of the university for the year.

7-3   CONFIDENCE INTERVAL FOR THE POPULATION VARIANCE

While discussing point estimation, we agreed to accept S^2 as providing a good estimate of σ^2. For setting a confidence interval for σ^2, we make the following assumptions.

1. The observations are independent.
2. The parent population has a normal distribution.

The statistic relevant for the purpose is

$$U = \frac{(n-1)S^2}{\sigma^2}.$$

Based on this definition of U, its value for any set of sample data $x_1, x_2, \ldots, x_n$ is given by

$$\frac{(n-1)s^2}{\sigma^2} = \frac{\sum (x_i - \bar{x})^2}{\sigma^2}.$$

Notice that only the numerator of the statistic U changes from sample to sample. The denominator is a fixed number σ^2.

THE CHI-SQUARE DISTRIBUTION

A quick analysis of the statistic reveals that we are comparing the product of $(n-1)$ and the sample variance with the population variance. It might seem more relevant that we compare the sample variance itself with the population variance, that is, study the statistic S^2/σ^2. This would be fine. It just happens that the statistic U not only allows us to accomplish our primary goal of inference, it also has a well-known distribution called the **chi-square distribution,** with $n-1$ degrees of freedom.

There are other statistics which also have the chi-square distribution. We are thus led to the study of yet another important continuous distribution.

Properties of the Chi-Square Distribution

A typical value of a random variable that has a chi-square distribution is denoted as χ^2, the square of the Greek letter chi (pronounced "kī").

The following features characterize the chi-square distribution.

1 It is a continuous distribution.

2 χ^2 values cannot be negative. As a result, curves describing the distributions are always to the right of the vertical axis at 0.

3 Its shape depends on a parameter ν (which is related to the sample size) called the number of degrees of freedom, or simply *degrees of freedom,* usually abbreviated df. There is one chi-square distribution curve for each number of degrees of freedom. (Chi-square distributions with 5, 10, 15, and 30 df are shown in Figure 7-12.)

4   When the number of degrees of freedom is small, the distribution is markedly skewed to the right (see Figure 7-12).

5   Skewness disappears rapidly as degrees of freedom increase. For $\nu \geq 30$, the distribution is approximately normal (see Figure 7-12).

Since a chi-square distribution is continuous, the probabilities are given as areas under the curve just as in the case of the t distribution. Appendix III, Table A-5 gives χ^2 values where specified areas are in the right tail of the distribution. *The degrees of freedom are given in the first column; for each degree of freedom, the χ^2 values are given along the corresponding row; and areas in the right tail of the distribution are indicated by the corresponding column headings.*

A general χ^2 value is denoted with a pair of subscripts where the first subscript indicates the degrees of freedom of the chi-square distribution and the second subscript gives the area to the right of the χ^2 value. Thus, for example, $\chi^2_{\nu,\alpha}$, $\chi^2_{\nu,\alpha/2}$, $\chi^2_{\nu,0.05}$ are the χ^2 values for a chi-square distribution with ν degrees of freedom leaving, respectively, areas α, $\alpha/2$, 0.05 in the right tail. The value $\chi^2_{\nu,\alpha}$ is shown in Figure 7-13. Such χ^2 values are called the **critical values of the chi-square distribution.**

As with the t distribution, since there is one curve for each degree of freedom, for economy of space *χ^2 values in our table are given only for a certain set of probabilities contained in the right tail.* This table is read exactly as the one for the t distribution. We reproduce a part of Table A-5 that includes $\nu = 10$ degrees of freedom.

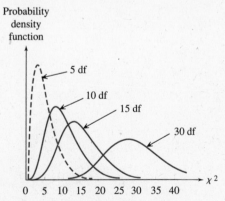

Figure 7-12 Probability Density Curves for Chi-Square Distributions with 5, 10, 15, 30 df

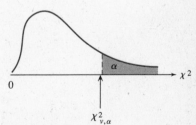

Figure 7-13

For ν df, χ^2 value such that area in the right tail is α.

Degrees of Freedom ν	Area in the Right Tail							
	0.995	0.99	0.975	0.95	0.05	0.025	0.01	0.005
	Critical Values of χ^2							
	$\chi^2_{\nu,0.995}$	$\chi^2_{\nu,0.99}$	$\chi^2_{\nu,0.975}$	$\chi^2_{\nu,0.95}$	$\chi^2_{\nu,0.05}$	$\chi^2_{\nu,0.025}$	$\chi^2_{\nu,0.01}$	$\chi^2_{\nu,0.005}$
1	0.0^4393	0.0^3157	0.0^3982	0.0^2393	3.841	5.024	6.635	7.879
2	0.0100	0.0201	0.0506	0.103	5.991	7.378	9.210	10.597
3	0.0717	0.115	0.216	0.352	7.815	9.348	11.345	12.838
⋮	⋮							
9	1.735	2.088	2.700	3.325	16.919	19.023	21.666	23.589
10	2.156	2.558	3.247	3.940	18.307	20.483	23.209	25.188
11	2.603	3.053	3.816	4.575	19.675	21.920	24.725	26.757
⋮	⋮							

For example, when $\nu = 10$, the area to the right of 2.156 is 0.995, to the right of 3.940 is 0.95, to the right of 20.483 is 0.025, and so on. The locations of these χ^2 values along the horizontal axis together with the corresponding probabilities are shown in Figure 7-14. It will be noticed that, unlike the table for the t distribution, the column headings in the table include not only the probabilities 0.05, 0.025, 0.01, and 0.005, but also the probabilities 0.95, 0.975, 0.99, and 0.995. Because the t distribution is symmetric, there was no need to provide the t values for this latter set of probabilities.

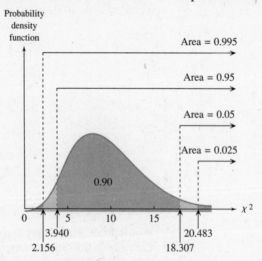

Figure 7-14 Probability Density Curve of the Chi-Square Distribution with 10 df
Area to the right of 2.156 is 0.995, and so on.

The chi-square distribution finds applications in many statistical tests, as you will soon discover. It will form the mainstay of the tests carried out in Chapter 10, where we discuss goodness of fit and contingency tables.

EXAMPLE 1

Find the following.

(a) $\chi^2_{4,0.05}$ **(b)** $\chi^2_{21,0.025}$

(c) $\chi^2_{17,0.95}$ **(d)** $\chi^2_{5,\alpha/2}$ if $\alpha = 0.1$

(e) $\chi^2_{16,1-\alpha/2}$ if $\alpha = 0.05$

SOLUTION   It will help us to draw a rough sketch of the graph of the appropriate chi-square distribution.

(a)   To find $\chi^2_{4,0.05}$, we consider the χ^2 distribution with 4 degrees of freedom. Notice that the area to the right of the desired value is 0.05. From Table A-5 in the row for 4 degrees of freedom under column heading 0.05, we find, in the body of the table, $\chi^2_{4,0.05} = 9.488$ (see Figure 7-15(a)).

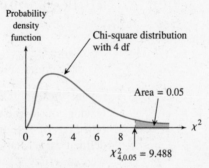

Figure 7-15(a)

(b)   As in Part a, we look in Table A-5 for the row corresponding to $\nu = 21$, under column heading 0.025. The value we find is $\chi^2_{21,0.025} = 35.479$ (see Figure 7-15(b)).

Figure 7-15(b)

(c) To find $\chi^2_{17,0.95}$ we are interested in a chi-square distribution with 17 degrees of freedom where the area to the right of the value is 0.95. From Table A-5 in the row corresponding to $\nu = 17$ under column heading 0.95, such a value is 8.672. Thus, $\chi^2_{17,0.95} = 8.672$. (Notice that the value is in the left tail of the distribution in Figure 7-15(c).)

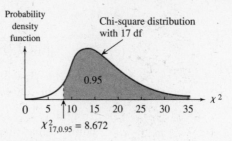

Figure 7-15(c)

(d) Because $\alpha = 0.1$, $\alpha/2 = 0.05$. From Table A-5, we immediately find $\chi^2_{5,\alpha/2} = \chi^2_{5,0.05} = 11.07$.

(e) Here $\alpha = 0.05$. Thus,

$$1 - \alpha/2 = 1 - 0.025 = 0.975$$

so that $\chi^2_{16,1-\alpha/2} = \chi^2_{16,0.975} = 6.908$ (in the row corresponding to $\nu = 16$, under column heading 0.975). ●

Confidence Interval

Having considered the chi-square distribution at length, we return to our main objective of setting a confidence interval for σ^2. Because the statistic U defined earlier has a chi-square distribution with $n - 1$ degrees of freedom, we can write

$$P\left(\chi^2_{n-1,1-\alpha/2} < \frac{(n-1)S^2}{\sigma^2} < \chi^2_{n-1,\alpha/2} \right) = 1 - \alpha$$

where $\chi^2_{n-1,1-\alpha/2}$ and $\chi^2_{n-1,\alpha/2}$ represent values from the chi-square distribution with $n - 1$ degrees of freedom such that they leave, respectively, areas of $1 - \alpha/2$ and $\alpha/2$ to their right, as shown in Figure 7-16.

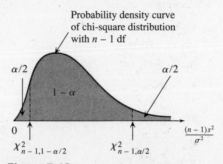

Figure 7-16
Chi-square values such that areas $1 - \alpha/2$ and $\alpha/2$ are to their right.

While setting a confidence interval on μ, we isolated it in the center between the inequalities. In the present case, we isolate σ^2. Remember that this is what the general procedure on page 404 stipulates.

It can be shown through a series of algebraic steps that the preceding probability statement leads to

$$P\left(\frac{(n-1)S^2}{\chi^2_{n-1,\alpha/2}} < \sigma^2 < \frac{(n-1)S^2}{\chi^2_{n-1,1-\alpha/2}}\right) = 1 - \alpha.$$

When a sample of values $x_1, x_2, \ldots, x_n$ is given, we compute s^2 from it and obtain the interval

$$\left(\frac{(n-1)s^2}{\chi^2_{n-1,\alpha/2}}, \frac{(n-1)s^2}{\chi^2_{n-1,1-\alpha/2}}\right),$$

which is called a $(1 - \alpha)100$ percent confidence interval for σ^2.

Thus, we arrive at the following result.

CONFIDENCE INTERVAL FOR σ^2

A $(1 - \alpha)100$ percent confidence interval for σ^2 is given by

$$\frac{(n-1)s^2}{\chi^2_{n-1,\alpha/2}} < \sigma^2 < \frac{(n-1)s^2}{\chi^2_{n-1,1-\alpha/2}}$$

where s^2 is the sample variance based on n observations from a normal population.

Note: Keep in mind that $\chi^2_{n-1,1-\alpha/2}$ is *not* equal to $-\chi^2_{n-1,\alpha/2}$. This is because the chi-square distribution is not symmetric. Also, chi-square values cannot be negative.

EXAMPLE 2   The following data represent the amount of enamel coating (in ounces) needed to paint six plates.

    8.1   8.7   7.6   7.8   8.5   7.9

Find the following.

(a)   A point estimate of σ^2, the variance of the amount of enamel coating needed to paint a plate

(b)   Confidence intervals for σ^2 and σ with the following confidence levels, assuming that the amount of enamel coating per plate is normally distributed

   i. 95 percent   ii. 98 percent

SOLUTION

(a) It can be easily checked that $\Sigma (x_i - \bar{x})^2 = 0.9$. A point estimate of σ^2 is s^2. Since

$$s^2 = \frac{\Sigma (x_i - \bar{x})^2}{6 - 1} = \frac{0.9}{5} = 0.18$$

a point estimate of σ^2 is 0.18 ounces squared.

(b) Because $n = 6$, the chi-square distribution has 5 degrees of freedom.

i. To find a 95 percent confidence interval we observe that $\alpha = 0.05$ so that $\alpha/2 = 0.025$ and $1 - \alpha/2 = 0.975$. Now $\chi^2_{5,0.025} = 12.832$ and $\chi^2_{5,0.975} = 0.831$. Because $(n - 1)s^2 = 0.9$, a 95 percent confidence interval is given by

$$\frac{0.9}{12.832} < \sigma^2 < \frac{0.9}{0.831}$$

That is,

$$0.07 < \sigma^2 < 1.08.$$

Thus, a 95 percent confidence interval for σ^2 is (0.07, 1.08).

Taking square roots of the upper and lower limits of the interval for σ^2 we get $(\sqrt{0.07}, \sqrt{1.08})$ or (0.26, 1.04) as a 95 percent confidence interval for σ. In other words, we estimate with 95 percent confidence that the standard deviation σ of the amount of enamel coating needed to paint a plate is greater than 0.26 and less than 1.04 ounces.

ii. In this case $\alpha = 0.02$. We note that $\chi^2_{5,0.01} = 15.086$ and $\chi^2_{5,0.99} = 0.554$, and we get a 98 percent confidence interval for σ^2 given by

$$\frac{0.9}{15.086} < \sigma^2 < \frac{0.9}{0.554}.$$

That is,

$$0.06 < \sigma^2 < 1.62.$$

A 98 percent confidence interval for σ is

$$\sqrt{0.06} < \sigma < \sqrt{1.62}$$
$$0.24 < \sigma < 1.27$$

A 98 percent confidence interval for σ^2 is (0.06, 1.62) and for σ is (0.24, 1.27). You will notice that the 98 percent confidence interval for σ (or σ^2) is wider than the 95 percent confidence interval. This trade-off is just as in the case of confidence interval for μ. The result that, given a fixed sample size, the interval becomes wider when the confidence level is increased is true when setting confidence interval for *any* parameter. This is, admittedly, in keeping with what one might expect.   ●

11/14/96

Section 7-3 Exercises

1. Find the following.
   (a) $\chi^2_{7, 0.05}$
   (b) $\chi^2_{12, 0.01}$
   (c) $\chi^2_{14, 0.95}$
   (d) $\chi^2_{7, 0.025}$
   (e) $\chi^2_{22, 0.99}$
   (f) $\chi^2_{19, 0.975}$

2. In each case find the indicated quantity for the given level of confidence.
   (a) $\chi^2_{5, \alpha/2}$; 90 percent
   (b) $\chi^2_{15, \alpha/2}$; 99 percent
   (c) $\chi^2_{7, \alpha/2}$; 98 percent
   (d) $\chi^2_{10, \alpha/2}$; 90 percent

3. Suppose you are constructing a 90 percent confidence interval. Determine $\alpha/2$, $1 - \alpha/2$ and find the indicated chi-square value.
   (a) $\chi^2_{16, 1-\alpha/2}$
   (b) $\chi^2_{16, \alpha/2}$
   (c) $\chi^2_{8, 1-\alpha/2}$
   (d) $\chi^2_{7, 1-\alpha/2}$

4. In each case, find the chi-square value for the indicated level of confidence.
   (a) $\chi^2_{14, \alpha/2}$; 95 percent
   (b) $\chi^2_{14, 1-\alpha/2}$; 95 percent
   (c) $\chi^2_{8, 1-\alpha/2}$; 98 percent
   (d) $\chi^2_{10, 1-\alpha/2}$; 99 percent

5. In each case, of the two values, determine which value is larger. (*Hint:* This can be done without using the tables by sketching graphs.)
   (a) $\chi^2_{5, 0.08}, \chi^2_{5, 0.94}$
   (b) $\chi^2_{16, 0.03}, \chi^2_{16, 0.02}$
   (c) $\chi^2_{20, 0.96}, \chi^2_{20, 0.92}$
   (d) $\chi^2_{10, 0.05}, \chi^2_{10, 0.88}$
   (e) $\chi^2_{12, 0.07}, \chi^2_{12, 0.03}$
   (f) $\chi^2_{4, 0.97}, \chi^2_{4, 0.90}$

In Exercises 6–12, assuming that the population is normally distributed, construct a confidence interval for the corresponding population variance σ^2 at the indicated confidence level if the given data summary is available.

6. $n = 7$, $s^2 = 105$; 98 percent

7. $n = 10$, $s = 1.22$; 98 percent

8. $n = 14$, $s = 0.4$; 95 percent

9. $n = 23$, $s^2 = 12.8$; 90 percent

10. $n = 12$, $\Sigma (x_i - \bar{x})^2 = 16.38$; 95 percent

11. $n = 18$, $\Sigma (x_i - \bar{x})^2 = 4.597$; 90 percent

12. $n = 7$, $\Sigma (x_i - \bar{x})^2 = 797.645$; 99 percent

13. The following data present the weight loss of 11 subjects treated in a weight control clinic over a certain period of time.

19	27	24	20	17	25
30	22	23	16	21	

    (a) Determine a 95 percent confidence interval for the true variance of the weight loss.
    (b) Determine a 95 percent confidence interval for the true standard deviation of the weight loss.
    (c) State the assumptions that are needed.

(Handwritten annotations:)

table p. 773

$s^2 = \dfrac{\Sigma(x_i - \bar{x})^2}{n-1} = 17.7636$

a) $\alpha = 0.05$ $n = 11$

$\dfrac{(n-1)s^2}{\chi^2_{n-1, \alpha/2}} = \dfrac{(11-1)\,17.7636}{20.483} = 8.6724$
$\quad\quad 10, .025$

$\dfrac{(n-1)s^2}{\chi^2_{n-1, 1-\alpha/2}} = \dfrac{(10)\,17.7636}{3.247} = 54.7077$
$\quad\quad 10, 0.975$

$\sigma^2_{\text{for } 95\%} = (8.6724, 54.7077)$

14. The braking distance of a car at a given speed is the distance it travels before it comes to a complete halt after the brakes are applied. A scientific experiment was run to measure the braking distance from 60 mph for ten cars of a new model. The data (in feet) are presented below

  161   153   175   179   173   169   187   171   172   181

(a) Estimate the true mean braking distance using a 90 percent confidence interval.

(b) Estimate the true variance of braking distance using a 90 percent confidence interval.

(c) State the assumptions that would be considered necessary.

(d) Give an interpretation of the confidence intervals obtained.

15. A grade-school teacher has obtained the following figures for the time (in minutes) that ten randomly picked students in her class took to complete an assigned task.

  15   10   12   8   11
  10   10   13   12   9

(a) Find a 95 percent confidence interval for:
  i. the true mean time taken to complete the task
  ii. the true standard deviation of the time to complete the task

(b) State the underlying assumptions.

(c) Interpret the confidence intervals in Part a.

In each of the Exercises 16–22, a normal probability plot will support the assumption of a normal distribution of the underlying population. You might like to check this to satisfy your curiosity. Also, provide an interpretation of the results obtained in each case.

16. A machine is set so that it will dispense close to 6 oz of coffee in a cup. If the machine overfills or underfills by large amounts, it will need to be adjusted. The measure of this variability is provided by the variance of the amount filled. In fifteen trials the following amounts of coffee were dispensed.

  5.3   6.4   6.4   5.6   5.7   6.2   6.5   5.1
  6.1   6.2   5.3   5.9   5.6   5.6   5.4

(a) Obtain a 98 percent confidence interval for the true mean amount of coffee dispensed in a cup.

(b) Obtain a 98 percent confidence interval for the true variance of the amount dispensed.

17. Food legumes contain a substance called saponin credited with lowering plasma cholesterol in animals. Twenty samples of a high-yielding variety of chick-peas (*Cicer arietinum*) were laboratory tested for total saponin (mg/100 g). The figures are presented as follows.

  3366   3337   3361   3410   3316   3387   3348
  3356   3376   3382   3377   3355   3408   3401
  3390   3428   3383   3374   3484   3390

Estimate the following.

(a) The true mean saponin using a 90 percent confidence interval

(b) The true standard deviation σ by constructing a 90 percent confidence interval

18. Fifteen individuals were sampled to determine their daily intake of wheat selenium (μg/day), which is considered essential as a constituent of erythrocyte glutathione peroxidase. The data are presented as follows.

83.2	91.8	84.4	77.1	80.0
85.7	78.9	84.6	87.8	78.9
87.3	83.8	92.5	86.0	87.8

(a) Estimate the mean daily selenium intake with 98 percent confidence interval.

(b) Find an interval estimate of the variance of daily selenium intake with a 98 percent confidence interval.

19. Fifteen 5- to 6-week-old male rats of approximately 100 grams body weight were fed a test diet containing casein treated with methanol (HCHO) solution containing 40 grams HCHO/liter. The rats were housed for 14 days in metabolism cages and fed 10 grams of fresh weight of diet per day. During the final 7-day period a total collection of feces and urine was carried out and the samples were analyzed for total nitrogen (N). The data that follow give figures for N digestibility of the rats.

0.91	0.91	0.94	0.95	0.93
0.93	0.92	0.92	0.91	0.92
0.94	0.92	0.92	0.93	0.92

Use a 95 percent confidence interval to estimate the standard deviation σ of nitrogen digestibility.

20. The total soluble sugar (g/100 ml) in a brand of canned soft drink was determined through high-performance liquid chromatography. The data from ten samples tested are presented as follows.

10.4	10.2	10.4	10.3	10.2
10.2	10.1	10.2	10.3	10.4

(a) Construct a 98 percent confidence interval for the true mean soluble sugar.

(b) Estimate the true standard deviation using a 98 percent confidence interval.

21. The suspended organic matter (SOM) (g/m^3) was determined in the northern and northwestern zones of Lake Onega, including Petrozavodsk and Kondopoga inlets, in the Soviet Union (*Hydrobiological Journal* 1982; 18: pp. 19–94). The concentration of SOM was determined at nineteen stations. The data are as follows.

SOM (g/m^3)
3.2
3.2
3.8
3.8
2.7
3.2
2.7
3.9
1.8
2.0
1.9

(continued)

SOM (g/m^3)
2.5
5.4
4.6
4.2
4.1
4.0
3.0
1.7

Using a 98 percent confidence interval, estimate the true variance of the SOM content.

22. Thirty pigs were fed a diet containing 1.8 percent linoleic acid (percent of dry feed) from 20 kilograms to 35 kilograms of their weight followed by 1.4 percent linoleic acid from 35 kilograms to slaughter at 85 kilograms live weight. The following figures relate to the fat (percent of wet weight) in the tissue composition of the back fat.

78.4	71.5	81.9	84.2	76.9	76.6	82.8	70.3	65.0	81.4
86.7	74.0	69.8	88.5	79.5	79.1	72.9	72.3	76.5	74.4
83.4	75.7	81.0	73.5	84.2	87.7	76.4	84.9	78.1	80.6

Use the data to estimate:
(a) the true mean fat in the tissue with a 95 percent confidence interval
(b) the true variance of the fat in the tissue with a 95 percent confidence interval.

7-4   CONFIDENCE INTERVAL FOR THE POPULATION PROPORTION

If p represents the proportion of an attribute in the population, we have already seen that the sample proportion X/n provides a good estimate of p. We will now give an interval estimate for p under the following assumptions.

1. The sample consists of n independent observations
2. The sample size is large
3. The population proportion is not too close to 0 or 1.

Under these assumptions, we know that the sampling distribution of X/n can be approximated by a normal distribution that has mean p and standard deviation $\sqrt{p(1-p)/n}$.

To construct a confidence interval for p, we can now adopt the same argument used to find a confidence interval for μ and write

$$P\left(\frac{X}{n} - z_{\alpha/2}\sqrt{\frac{p(1-p)}{n}} < p < \frac{X}{n} + z_{\alpha/2}\sqrt{\frac{p(1-p)}{n}}\right) = 1 - \alpha.$$

point estimator
of p

standard deviation
of the estimator

Thus, if x/n is the observed proportion from the sample, one might give a $(1 - \alpha)100$ percent confidence interval as

$$\frac{x}{n} - z_{\alpha/2} \sqrt{\frac{p(1 - p)}{n}} < p < \frac{x}{n} + z_{\alpha/2} \sqrt{\frac{p(1 - p)}{n}}.$$

But you may notice that there is something incongruous about this formula. We are giving a confidence interval for estimating p, but that very unknown quantity is used in computing the endpoints of the interval! Luckily, it turns out that we can substitute x/n for p in $\sqrt{p(1 - p)/n}$. If the sample size is large, this will not bring in much discrepancy. Thus, we get the following result.

LARGE SAMPLE CONFIDENCE INTERVAL FOR p

An approximate $(1 - \alpha)100$ percent confidence interval for the population proportion p is given by

$$\frac{x}{n} - z_{\alpha/2} \sqrt{\frac{\frac{x}{n}\left(1 - \frac{x}{n}\right)}{n}} < p < \frac{x}{n} + z_{\alpha/2} \sqrt{\frac{\frac{x}{n}\left(1 - \frac{x}{n}\right)}{n}}$$

if the sample size is large. Here x/n is the proportion of an attribute in a sample of n independent observations.

Remark   As a working rule, we will apply the formula only if the observed sample proportion x/n is not too close to 0 or 1 and the observed number of successes x and the observed number of failures $(n - x)$ both exceed 5. If these conditions are not met, the procedure is not recommended.

EXAMPLE 1   In a random sample of 400 people who were questioned regarding their participation in sports, 160 said that they did participate. Set an approximate 98 percent confidence interval for p, the proportion of people in the population who participate in sports.

SOLUTION   The number of people participating in sports is $x = 160$ in a sample of $n = 400$. Therefore, the sample proportion of those participating in sports is $x/n = 160/400 = 0.4$. So, a point estimate of p is 0.4. Consequently, we get

$$\sqrt{\frac{\frac{x}{n}\left(1 - \frac{x}{n}\right)}{n}} = \sqrt{\frac{(0.4)(0.6)}{400}} = 0.0245$$

For a 98 percent confidence interval, $\alpha = 0.02$ and, therefore, $z_{\alpha/2} = z_{0.01} = 2.33$. From the formula

$$\frac{x}{n} - z_{\alpha/2}\sqrt{\frac{\frac{x}{n}\left(1 - \frac{x}{n}\right)}{n}} < p < \frac{x}{n} + z_{\alpha/2}\sqrt{\frac{\frac{x}{n}\left(1 - \frac{x}{n}\right)}{n}}$$

an approximate 98 percent confidence interval for p is, therefore,

$$0.4 - 2.33(0.0245) < p < 0.4 + 2.33(0.0245)$$

that is,

$$0.34 < p < 0.46.$$

A 98 percent confidence interval for p is (0.34, 0.46). We can be 98 percent confident that the proportion of people in the population participating in sports is between 34 and 46 percent. The interval has width of 0.12. To get a narrower interval, while maintaining the confidence level at 98 percent, a larger sample would need to be interviewed. ●

Margin of Error

From the formula for the confidence interval for p, the margin of error is

$$\textbf{margin of error} = z_{\alpha/2}\sqrt{\frac{\frac{x}{n}\left(1 - \frac{x}{n}\right)}{n}}$$

Quite often, however, the margin of error is given anticipating the worst possible error that could result under all the possible values of p. It just happens that the maximum value of $z_{\alpha/2}\sqrt{\frac{p(1-p)}{n}}$ is $z_{\alpha/2}\sqrt{\frac{1}{4n}}$ and occurs when $p = 1/2$. We will call this the *safe* or *conservative* margin of error, whatever the actual value of p in the population. It is given by the following formula

$$(\textit{conservative})\ \textbf{margin of error} = z_{\alpha/2}\sqrt{\frac{1}{4n}}.$$

Values quoted by most pollsters when presenting the results of their surveys are computed using this formula.

In the survey report based on 800 adult Americans, discussed on page 389, the margin of error was quoted as ± 0.035. Let us verify this using $n = 800$.

The level of confidence is not specified. But it is easy to see that it must be 95 percent because, then $z_{\alpha/2} = z_{0.025} = 1.96$ and we get

$$\text{margin of error} = 1.96\sqrt{\frac{1}{4(800)}} = 0.035.$$

EXAMPLE 2   The following results of a survey conducted by Yankelovich Partners Inc., were published in *Time*, January 17, 1994. The sampling error, or the margin of error, is given as ±4.5%, that is, 0.045. Explain this.

> If it were possible, would you want to take a genetic test telling you which diseases you are likely to suffer from later in life?
>
> | YES | 50% |
> |-----|-----|
> | NO | 49% |
>
> If you or your spouse were pregnant, would you want the unborn child tested for genetic defects?
>
> | YES | 58% |
> |-----|-----|
> | NO | 39% |
>
> Do you think it should be legal for employers to use genetic tests in deciding whom to hire?
>
> | YES | 9% |
> |-----|-----|
> | NO | 87% |
>
> From a telephone poll of 500 adult Americans taken for TIME/CNN on Dec.2 by Yankelovich Partners Inc. Sampling error is ± 4.5%. "Not sures" omitted.

SOLUTION   There are three qualitative variables under consideration. The survey is attempting to estimate true proportions of those who

1. would want to take a genetic test for diseases.
2. would want the unborn child tested for genetic defects.
3. think it should be legal to use genetic tests.

The population proportions p for three attributes are considered here. As we know, the formula for the margin of error that will handle all the possible values of p is

$$\text{margin of error} = z_{\alpha/2} \sqrt{\frac{1}{4n}}$$

Source: Study by Yankelovich Partners, Inc. from *Time* magazine, January 17, 1994. Copyright © 1994 Time Inc. Reprinted with permission.

The level of confidence is not specified in the published report. So let us get the margin of error supposing a 95 percent confidence interval is intended. In that case $z_{\alpha/2} = z_{0.025} = 1.96$ and, since $n = 500$, we get

$$\text{margin of error} = 1.96\sqrt{\frac{1}{4(500)}} = 0.044.$$

This value is slightly less than 0.045. Presumably, the value 1.96 was rounded upward to 2, in which case the margin of error does turn out to be 0.045. This is a fairly common practice. In that case, the level of confidence is slightly higher at 95.44 percent.   ●

How Large a Sample for Estimating p?

The question of how large a sample to pick in order to attain a certain degree of precision was considered earlier in Section 7-1 in the context of constructing a confidence interval for μ when σ is known. A question of a similar nature also comes up when estimating a proportion p. For example, a pollster might wish to know how many people should be interviewed so that one can be 95 percent confident that the estimate will not be in error of the true proportion p in the population by more than 0.01.

Suppose the maximum possible error is denoted by E, as before. Arguing as in Section 7-1, we equate the maximum possible error with the margin of error. Thus, we set

$$E = z_{\alpha/2}\sqrt{\frac{1}{4n}}.$$

Rearranging the formula algebraically, we get

$$n = \frac{z_{\alpha/2}^2}{4E^2}$$

You should realize that this is a *conservative* sample size at the given confidence level since we are giving the maximum n which will attain the desired goal, no matter what the actual value of p turns out to be in a given population.

In conclusion, we obtain the following result.

> **SAMPLE SIZE FOR ESTIMATING p**
>
> A conservative sample size needed so as to be $(1 - \alpha)100$ percent confident that the estimate x/n is not in error of p by more than a quantity E is given by
>
> $$n = \frac{z_{\alpha/2}^2}{4E^2}.$$

Thus, in the case of the pollster who wants to be 95 percent confident that the estimate will not be in error of p by more than 0.01, a conservative value of n will be

$$n = \frac{(1.96)^2}{4(0.01)^2} = 9604.$$

A much smaller sample might give an estimate that is in error by less than 0.01 with 95 percent confidence. But this will depend on the actual value of p. In being conservative, we have given a value of n that will do the job no matter what the actual value of p is.

11/14/96

Section 7-4 Exercises

Exercises 1 and 2 are intended to exhibit that the discrepancy is not great if the population proportion p is replaced by the sample estimate x/n in $\sqrt{\dfrac{p(1 - p)}{n}}$.

1. Suppose $n = 400$. Compute

   (a) $\sqrt{\dfrac{p(1 - p)}{n}}$ if $p = 0.2$

   (b) $\sqrt{\dfrac{\frac{x}{n}\left(1 - \frac{x}{n}\right)}{n}}$ when $\dfrac{x}{n} = 0.26$

   (c) $\sqrt{\dfrac{\frac{x}{n}\left(1 - \frac{x}{n}\right)}{n}}$ when $\dfrac{x}{n} = 0.14$

   Observe that, when the answers in Parts b and c are compared with that in Part a, the differences are very small. Also, you may wish to verify that if, indeed, $p = 0.2$, then it is practically certain that X/n will be between 0.14 and 0.26 with a sample size as large as 400.

2. Repeat Exercise 1 when $n = 600$ and
   (a) $p = 0.75$ (b) $x/n = 0.80$ (c) $x/n = 0.70$

In Exercises 3–6, construct a confidence interval for the corresponding population proportion p at the indicated level of confidence if the following information is given.

3. $n = 60, x = 18$; 95 percent   4. $n = 78, x = 30$; 90 percent

5. $n = 100, x = 62$; 88 percent   6. $n = 80, x = 52$; 95 percent

7. In a study with a new vaccine involving fifty-three lung cancer patients, thirty-three survived five years after surgery. Obtain a 98 percent confidence interval for the true proportion of patients surviving five years after surgery with the new vaccine.

8. In a sample of 100 cathode tubes inspected, twelve were found to be defective. Set a 95 percent confidence interval on the true proportion of defectives in the entire production.

9. Of the 200 individuals interviewed, 80 said that they were concerned about fluorocarbon emissions in the atmosphere. Obtain a 99 percent confidence interval for the true proportion of individuals who are concerned.

10. A food marketing company wants to introduce a new spaghetti sauce on the market. The research and testing department of the company picked a random sample of 300 individuals to taste the sauce. Of these individuals, 208 said that they liked the sauce and would buy it. Find a 95 percent confidence interval for p, the true proportion of individuals who would buy the sauce if introduced to it.

11. In order to test the effectiveness of a flu vaccine, it was administered to 650 people who were selected at random. At the end of the flu season it was determined that 553 of the individuals did not get the flu. Using a 90 percent confidence interval, estimate the true proportion of people who would not get the flu if given the vaccine.

12. When a surgical procedure was tried on 212 patients, 186 recovered and were able to resume a normal life. Based on this information, obtain a 98 percent confidence interval for p, the true proportion of patients who would recover and be able to resume a normal life after the surgical procedure.

13. The following question was asked in a survey conducted by the Gallup organization in which 1003 individuals were interviewed: "In your opinion, has the federal government under President Clinton become a lot more liberal, a little more liberal, or has it not become more liberal at all?" The results of the survey were as follows:

	Percent
A lot more	24
A little more	43
Not more at all	24
No opinion	9

Obtain a 95 percent confidence interval for the true proportion of those who feel that the federal government has become
(a) a little more liberal
(b) not more liberal at all.

14. The transportation department in a metropolitan area interviewed 148 randomly selected commuters to determine if they used public transportation. If 18 responded that they did, construct a 95 percent confidence interval for the true proportion of commuters in the metropolis who use public transportation.

15. A new vaccine is to be tested on the market. Find how large a sample should be drawn if we want to be 95 percent confident that the estimate will not be in error of the true proportion of success by more than:
(a) 0.1   (b) 0.02   (c) 0.001.

Comment on your findings.

16. A survey is to be conducted to estimate the proportion of citizens who favor imposing trade restraints on Japan. Determine how large the sample should be so that, with 98 percent confidence, the sample proportion will not differ from the true proportion of those who favor restraints by more than 0.02.

17. An insurance company offering term insurance wants to estimate the true proportion of all 40-year-old men who die before reaching the age of 50. Find how many past records on survivability of men in the age group 40–50 should be analyzed to be 95 percent confident that the true proportion is estimated to within 0.012.

18. A coffee company wants to estimate the true proportion in the U.S. population that drink its brand. How many individuals should be surveyed to be 99 percent confident of having the true proportion of people drinking the brand estimated to within 0.018?

19. Should a certain nuclear power plant be reopened? A random sample of 500 adults gave the following results: 240 in favor, 180 opposed, 80 undecided. Using a 90 percent confidence interval, estimate the true proportion of individuals in the population:
(a) who favor reopening the plant
(b) who are undecided.

20. A pollster published the findings of a survey with the following comment:

> Results are based on telephone interviews with a randomly selected national sample of 1,000 adults, 18 years and older, ... For results based on samples of this size, one can say with 95 percent confidence that the error attributable to sampling and other random effects could be plus or minus 3 percentage points.

Justify the statement.

21. In the following results of the survey conducted by Yankelovich Clancy Shulman for TIME/CNN, explain the statement that the sampling error is plus or minus 2.8%.

VOX POP

Do you favor a federal law that would prohibit employers from hiring permanent replacements for striking workers?

YES	NO
29%	**60%**

From a telephone poll of 1,250 American adults taken for TIME/CNN on April 9 by Yankelovich Clancy Shulman. Sampling error is is ± 2.8%.

Source: "Vox Pop" survey from *Time* magazine, April 20, 1992. Copyright © 1992 Time Inc. Reprinted with permission.

22. A poll of 780 registered voters found that 42 percent of them believed that the biggest problems facing the country were crime and violence. What is the margin of error at the 98 percent level? Present the results in a way similar to that in the TIME/CNN report in Exercise 21.

23. Because of concern over fat and cholesterol, consumption of poultry has steadily increased in recent years. In a survey of 384 households, 114 reported eating more chicken and less beef than previously. Find an estimate of the true proportion of households eating more chicken and less beef. Also, give the margin of error if you wish to be 90 percent confident.

24. In an article entitled "The (Executive) Board Vs. the 'Babe'," *Time,* August 30, 1993, it was stated that " . . . sexual bias and harassment remain an entrenched fact of life at many U.S. companies. In a recent survey poll of 439 female executives, the recruiting firm Korn-Ferry found that 60% had been sexually harassed during their careers." Assuming that the female executives who were interviewed responded independently, estimate the true proportion of female executives harassed during their careers and give the sampling error if you wish to be 99 percent confident that the interval contains the true proportion.

25. The statistics presented below appeared in the weekly magazine *Time,* August 23, 1993, in the article "Danger in the Safety Zone," which described violence

Do you worry about being a victim of crime?	Do you favor the death penalty?
City Suburb Rural	
YES 59% 57% 43%	YES 77%
NO 39% 43% 54%	NO 17%

Has the amount of crime in your community increased in the past five years?	Should there be a five-day waiting period before anyone can buy a gun?
Increased 61%	YES 92%
Decreased 5%	NO 8%
Remained the same 30%	

How much will increasing the number of police reduce crime?	Are stricter gun-control laws a necessary part of any anticrime bill?
Great deal / fair amount 80%	YES 65%
Little or not at all 18%	NO 28%

From a telephone poll of 500 adult Americans taken for TIME/CNN on Aug. 12 by Yankelovich Partners Inc. Sampling error is ± 4.5%.

Would you favor building more prisons, even if your taxes would be raised?
YES 60%
NO 35%

Source: "Danger in the Safety Zone" from *Time* magazine, August 23, 1993. Copyright © 1993 Time Inc. Reprinted with permission.

in the United States. Consider the footnote: "From a telephone poll of 500 adult Americans taken for TIME/CNN on Aug. 12 by Yankelovich Partners Inc. Sampling error is ±4.5%" (that is, 0.045).

(a) Is it correct to use the sampling error of 0.045 for the first question "Do you worry about being a victim of crime"? Explain.

(b) For the remaining questions justify the footnote, stating the level of confidence.

26. Refer to Exercise 25 above. From the survey information and the sampling error provided, obtain a confidence interval for the true proportion of those who
    (a) believe that increasing the number of police will reduce crime a great deal/fair amount
    (b) believe that stricter gun control laws are a necessary part of any anticrime bill
    (c) favor the death penalty.

27. In Exercise 26, are the confidence intervals wider or narrower than the intervals you would get by using the formula on page 435? Explain your answer.

28. A student obtained the following 95 percent confidence interval for the population proportion p: $1.38 < p < 2.33$. Comment on its acceptability.

CHAPTER **7** **Summary**

✓ *CHECKLIST: KEY TERMS AND EXPRESSIONS*

❑ capture-recapture method, page 388
❑ estimator, page 389
❑ point estimate, page 389
❑ unbiased estimator, page 390
❑ interval estimation, page 391
❑ confidence interval, page 391
❑ interval estimate, page 391
❑ random interval, page 393
❑ margin of error, page 394
❑ sampling error, page 394

❑ length of confidence interval, page 400
❑ sample size, page 403
❑ random interval, page 404
❑ interval estimator, page 404
❑ confidence coefficient, page 405
❑ level of confidence, page 405
❑ lower (upper) confidence limits, page 405
❑ small sample confidence interval, page 409

❑ Student's t distribution, page 410
❑ number of degrees of freedom, page 410
❑ critical values of t distribution, page 411
❑ chi-square distribution, page 424
❑ critical values of chi-square distribution, page 425
❑ margin of error for p, page 436

KEY FORMULAS

Summary of Confidence Intervals (Single Population)

Nature of the Population	Parameter on Which Confidence Interval Is Set	Procedure	Limits of Confidence Interval with Confidence Coefficient $1 - \alpha$
Quantitative data; standard deviation σ is *known;* population normal	μ, the population mean	Draw a sample of size n and compute the value of $\bar{x}$, the estimate of μ.	$\bar{x} \pm z_{\alpha/2} \dfrac{\sigma}{\sqrt{n}}$
Quantitative data; standard deviation σ is *not* known; population normal; important when sample size is small ($n < 30$)	μ, the population mean	Draw a sample of size n and compute $\bar{x}$ and $$s = \sqrt{\dfrac{1}{n-1} \sum (x_i - \bar{x})^2}$$	$\bar{x} \pm t_{n-1,\,\alpha/2} \dfrac{s}{\sqrt{n}}$ $t_{n-1,\alpha/2}$ is the value obtained from the t distribution with $n - 1$ degrees of freedom.
Quantitative data; standard deviation σ is *not* known; population *not* necessarily normal; *sample size is large ($n \geq 30$)*	μ, the population mean	Draw a sample of size n and compute $\bar{x}$ and s.	$\bar{x} \pm z_{\alpha/2} \dfrac{s}{\sqrt{n}}$ Confidence interval is approximate.
Qualitative data; binomial case	p, the probability of success (the population proportion)	Draw a sample of size n and note x, the number of successes; obtain x/n, the estimate of p. Here n is assumed large.	$\dfrac{x}{n} \pm z_{\alpha/2} \sqrt{\dfrac{\dfrac{x}{n}\left(1 - \dfrac{x}{n}\right)}{n}}$ Confidence interval is based on the central limit theorem, thus, is approximate.
Quantitative data; population normal	σ^2, the population variance	Draw a sample of size n and compute s^2.	$\left(\dfrac{(n-1)s^2}{\chi^2_{n-1,\,\alpha/2}}, \dfrac{(n-1)s^2}{\chi^2_{n-1,\,1-\alpha/2}} \right)$

CHAPTER 7 Concept and Discussion Questions

1. Answer true or false: The interpretation of a confidence interval for a parameter refers to
   (a) all possible samples
   (b) all possible samples of the same size.

2. Answer true or false: When one makes a probability interpretation of a confidence interval at a given confidence level, it refers to
   (a) different intervals computed from different samples of the same size
   (b) different values of the parameter.

3. Suppose you are given (2.3, 3.4) as a confidence interval for p. Explain why you could not accept this interval.

4. A person interprets the 95 percent confidence interval (3.89, 6.72) for σ to mean $P(3.89 < \sigma < 6.72) = 0.95$. Why is the interpretation incorrect?

5. A university has a large student body, in excess of 30,000 students. A random sample of 60 students was taken to determine the mean amount spent by students for textbooks during the academic year. The sample mean was $628, with a 95 percent confidence interval for the true mean amount spent being (604, 652). Determine which of the following statements is (are) correct.
   (a) The probability that the true mean μ falls between 604 and 652 is 0.95.
   (b) Of all the sample means based on all possible samples of size 60, 95 percent fall in the interval.
   (c) 95 percent of the population means are inside the interval.
   (d) We cannot be sure that the population mean is in the interval.
   (e) 95 percent of the students spend between 604 and 652 dollars for textbooks.
   (f) Of all the confidence intervals constructed taking repeatedly samples of size 60, approximately 95 percent of the intervals will contain the true mean. To that extent we trust that the interval (604, 652) contains μ, the true mean amount spent.

6. Suppose you set a confidence interval for a parameter based on a sample. With the same sample data you decide to have a wider interval. How will this influence your confidence in the interval estimate? Have you gained or lost precision in estimating the parameter?

7. Discuss how the sample size needed to estimate μ with a given precision is affected by various factors.

8. Explain what is wrong with the following statement: A 95 percent confidence for σ is $(-12.3, -6.9)$.

9. Christmas is the time for borrowing and spending. Suppose an economist wishes to estimate the true mean of the amount that a household borrowed for Christmas-related shopping last season and seeks your advice. If the amount that a household borrowed cannot be assumed to be quite normally distributed, what would you recommend to circumvent this problem? Which formula would you use for estimating the true mean amount borrowed at the 95 percent level?

CHAPTER 7 Review Exercises

In Exercises 1–4, obtain point estimates of the population
   (a) mean   **(b)** variance   **(c)** standard deviation.

1. $n = 15, \Sigma x_i = 24.3, \Sigma (x_i - \bar{x})^2 = 0.824$

2. $n = 6, \Sigma x_i = -16, s^2 = 12.67$

3. $x_1 = 3.2, x_2 = 3.9, x_3 = 2.8, x_4 = 4.3, x_5 = 3.4$

4. $n = 7, \Sigma x_i = 26.M3, \Sigma x_i^2 = 128.03$

5. In a survey conducted to estimate the percentage of the population in the 20–40 year age bracket engaged in some kind of physical exercise, it was found that, of the 150 randomly picked individuals, 86 exercised. Estimate the percentage in the population.

6. Eight randomly picked individuals were asked the question, "Do you fear that religious fundamentalism is spreading around the world?" The following responses were obtained: yes, no, no, no, yes, no, yes, no. Based on these responses, estimate the proportion in the population who share the fear.

In Exercises 7–11, find a confidence interval for the indicated parameter at the specified level of confidence. Mention the assumptions that would be considered necessary in each case.

7. $\bar{x} = 13.8$, $n = 17$, $s = 6.2$; parameter μ; 98 percent

8. $n = 14$, $s = 1.3$; parameter σ^2; 95 percent

9. $\bar{x} = -3.4$, $n = 65$, $s = 7.6$; parameter μ; 90 percent

10. $n = 80$, $x = 27$ successes; parameter p; 88 percent

11. $\bar{x} = 7.6$, $n = 16$, $\sigma = 2.8$; parameter μ; 95 percent

12. In each case find the length of the given confidence interval.
    (a) $6.85 < \mu < 17.92$
    (b) $1.8 < \sigma^2 < 3.84$
    (c) $-3.13 < \mu < 6.84$
    (d) $0.36 < p < 0.84$

13. A random sample of sixteen servings of canned pineapple has a mean carbohydrate content of 49 grams. If it can be assumed that the population is normally distributed and that σ^2, the variance of the carbohydrate content of all the servings from which the sample was taken, is 4 (grams)^2, find a 98 percent confidence interval for the true mean carbohydrate content of a serving.

14. In a survey in which 100 randomly picked middle-class families were interviewed, it was found that their mean medical expenses during a year were $760 with a standard deviation s of $120. Find an 88 percent confidence interval for the true mean of the medical expenses of a middle-class family.

15. An archaeologist found that the mean cranial width of seventeen skulls was 5.3 inches with s equal to 0.5 inch. Using a 90 percent confidence level, set a confidence interval on:
    (a) the true mean cranial width
    (b) the true variance of cranial width.

    You may assume that cranial width has a normal distribution.

16. Do you think the president is doing a satisfactory job? The following response was obtained among forty randomly picked citizens (Y = yes, N = no, U = undecided).

N	U	U	N	Y	Y	N	Y	U	Y
N	Y	Y	N	U	Y	Y	Y	U	Y
N	Y	Y	U	N	Y	Y	Y	U	Y
Y	N	Y	Y	U	N	N	N	Y	U

    Estimate the true proportion of citizens who think the president is doing a satisfactory job.

17. It is important for timberland owners and managers to estimate the mean population site index with a high degree of precision since the site index represents the only basis for the prediction of the future yields from the trees. The following site index figures were reported for ten one-fifth acre permanent plots in a relatively homogeneous tract of young-growth redwood trees. (James Lundquist and Marshal Palley. Bulletin 796, California Agricultural Station.)

166	160	150	156	170
176	166	143	165	175

Estimate the mean population site index using a 98 percent confidence interval. Assume a normal distribution for site index.

18. It is suspected that a substance called actin is linked to various movement phenomena of nonmuscle cells. In a laboratory experiment when eight fertilized eggs were incubated for fourteen days the following amounts (mg) of total brain actin were obtained.

1.2	1.4	1.5	1.2	1.4	1.7	1.5	1.7

Assuming that the brain-actin amount after fourteen days of incubation is normally distributed, find a 95 percent confidence interval for:

(a) the true mean brain-actin amount
(b) the true variance of the brain-actin amount.

19. To investigate the effect of the insecticide lead arsenate on table wines, twenty-five bottles of a white wine from a region where the insecticide was used were sampled and the amount of lead (mg/l) was determined by carbon rod atomic absorption spectrophotometry. The findings were as follows.

0.44	0.50	0.45	0.42	0.46
0.34	0.47	0.44	0.34	0.39
0.41	0.43	0.37	0.41	0.47
0.37	0.36	0.45	0.47	0.34
0.44	0.39	0.37	0.30	0.40

Estimate the true mean lead concentration in the wine of the region using a 98 percent confidence interval and interpret the results. Assume a normal distribution for the amount of lead.

20. A dietician studied 240 school-age children selected at random. It was found that eighty-eight of the children had a high cholesterol count (in excess of 180). Construct a 98 percent confidence interval for p, the true proportion of school-age children with a high cholesterol count. Provide an interpretation.

21. A nutritionist interested in finding out what kind of food children eat interviewed 170 parents picked at random. In this survey 132 responded that they were influenced by their children in what they buy in the supermarket. On the basis of this information, estimate the true proportion of parents influenced by their children in what they buy in the supermarket, using a 99 percent confidence interval.

22. Sections of bacon, each weighing 100 grams and consisting of both back fat and lean muscle tissue, were homogenized in a food processor, chemically treated, and the copper content assayed by submitting the samples for copper analysis by flame atomic absorption spectroscopy. The data below give copper content of fat (μg/g).

0.55	0.46	0.72	0.32	0.45	0.47
0.51	0.39	0.36	0.62	0.71	

(a) Obtain a 98 percent confidence interval for the mean copper content of fat.
(b) Estimate the variance of the copper content of fat using a 98 percent confidence interval.

You may assume a normal distribution for the copper content.

23. Ten hens, approximately 64 weeks of age, were slaughtered and the pH values of breast and thigh meat were measured about 24 hours after slaughter. The data on pH values are as follows.

pH Values	
Breast Meat	**Thigh Meat**
5.9	6.6
6.1	6.4
6.0	6.6
5.9	6.5
6.0	6.3
6.1	6.1
5.9	6.7
6.1	6.5
6.0	6.3
6.3	6.0

(a) Construct a 95 percent confidence interval for the true variance of pH value in breast meat.
(b) Construct a 90 percent confidence interval for the true standard deviation of pH value in thigh meat.

Assume that the pH value is normally distributed.

24. Thirty milk-yielding Ayrshire cows were fed a diet of silage supplemented with a protein concentrate. The following figures refer to the milk fat concentration (g/kg) for each cow.

30.5	28.7	27.3	24.1	42.3
35.7	32.7	41.5	20.5	36.3
25.5	30.9	34.6	35.1	26.3
33.6	31.9	34.7	35.0	35.9
33.3	34.5	30.6	34.0	38.2
42.5	37.4	33.3	40.3	21.7

(a) Estimate the true mean milk fat concentration using a 90 percent confidence interval.
(b) Estimate the true standard deviation of the milk fat concentration with a 90 percent confidence interval.

25. Consider the following results of the survey about whether pennies should be removed from circulation.

VOX POP

Should the U.S. remove
pennies from circulation?

YES   NO

18%   74%

From a telephone poll of 1,250 American adults
taken for TIME/CNN on April 9 by Yankelovich
Clancy Shulman. Sampling error is ± 2.8%.

(a) Justify the value for the sampling error quoted in the report, stating the level of confidence.

(b) Use the sampling error provided, to determine the confidence interval for the true proportion of those who feel that pennies should be removed from circulation.

(c) Find a 95 percent confidence interval for the true proportion of those who feel that pennies should be removed from circulation, using the given point estimate of p to compute the margin of error. Is there much discrepancy between this interval and the one obtained in Part b?

 WORKING WITH LARGE DATA SETS

Project: Refer to the medical data from the diabetes study in Appendix IV, Table B-3. Obtain normal probability plots for each of the variables HTCM, WTKG, SYSBP, DIASBP and CHOLES and determine whether normality assumption is valid in each case. Assuming that the data are from a random sample of diabetes patients, estimate the population means of the variables HTCM, WTKG, SYSBP, DIASBP and CHOLES. You will notice that some of the variables do not support the normality assumption. Would that pose a serious problem in setting confidence intervals? Prepare a report of your investigation.

Project: Refer to a national newspaper which gives, in its business section, stock quotations for companies listed on the New York Stock Exchange. There are about 1200 companies on the list on any trading day. The last column of the quotations gives the net change in the stock price on a given trading day. (If the figure is not given for a stock, it means the change was 0.) See the illustration on the next page.

New York Stock Exchange Issues

52-week High	Low	Stock	Div	Yld %	P/E	Sales 100s	High	Low	Last	Chg
					A					
$17\frac{3}{8}$	$11\frac{7}{8}$	AAR	48	3.7	24	109	$13\frac{1}{2}$	13	13	$-\frac{1}{4}$
$36\frac{1}{8}$	$25\frac{3}{8}$	AbtLab	84h	2.4	19	7507	35	$34\frac{5}{8}$	$34\frac{7}{8}$	$+\frac{1}{8}$
$8\frac{1}{4}$	$3\frac{7}{8}$	Abex	...	...	...	255	$7\frac{3}{4}$	$7\frac{5}{8}$	$7\frac{3}{4}$	$+\frac{1}{4}$
$15\frac{1}{4}$	$11\frac{1}{2}$	Abtibi	...	...	...	910	$12\frac{1}{2}$	12	$12\frac{3}{8}$	$+\frac{1}{4}$
$24\frac{3}{8}$	$17\frac{5}{8}$	AMBlnd	60h	2.6	14	41	$22\frac{3}{4}$	$22\frac{5}{8}$	$22\frac{3}{4}$	0
18	$11\frac{1}{8}$	Acpls	...	...	11	43	$15\frac{1}{4}$	$15\frac{1}{8}$	$15\frac{1}{8}$	$-\frac{1}{4}$
$28\frac{7}{8}$	$20\frac{3}{4}$	AceLtD	44	1.9	...	1655	$23\frac{3}{4}$	$23\frac{1}{2}$	$23\frac{5}{8}$	$-\frac{1}{4}$

Select a trading day during the last week. Suppose you are interested in the change in stock prices on that day. What is your target population? Write the changes using decimals (for example, write $\frac{1}{8} = 0.125$) and compute the mean μ for this population. Computation of the mean μ involves a lot of tedious work. Describe a process to estimate it.

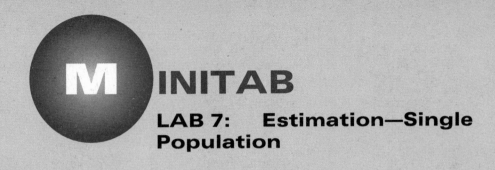

MINITAB

LAB 7: Estimation—Single Population

As explained in Section 7-1, the empirical meaning attached to a confidence interval with a given level of confidence is that it gives us some idea of the fraction of the intervals that will cover the parameter in question and the fraction of the intervals that will not cover it. Thus, for instance, a 95 percent confidence interval for the mean tells us that if we take samples repeatedly and use the formula for a 95 percent confidence interval for the mean, then approximately 95 percent of the intervals will cover the unknown mean and approximately 5 percent of them will not. The notion of taking repeated samples is only conceptual. One cannot hope to do this in practice.

MINITAB, with its ability to perform experiments artificially in relatively short time, provides a unique opportunity to put the theory to test.

1 In the following, we simulate forty samples, each of size 25, from a normal population with mean $\mu = 18$ and standard deviation $\sigma = 6.5$. Each of the forty columns, C1 to C40, represents a sample of 25. Now, even though we know the true mean μ, we compute 95 percent confidence intervals for it, based on each sample, as though it is not known. When we examine each interval we will know exactly which ones covered the true mean $\mu = 18$ and which ones failed to cover it.

```
MTB >RANDOM 25 OBSERVATIONS IN C1-C40;
SUBC>NORMAL MEAN 18 SIGMA 6.5.
MTB >ZINTERVAL 0.95, SIGMA 6.5, FOR COLUMNS C1-C40
```

Since the standard deviation is known and the distribution is normal, the command for finding a confidence interval for the mean is ZINTERVAL. The other information that needs to be provided with the command is the level of confidence, the value of σ, and the columns where the samples are located. If the level of confidence is not specified, then it is taken to be 0.95. In the output, count the number of intervals that cover the true mean, which happens to be 18. (It might be easier to count the intervals that fail to cover μ since there will be fewer such intervals.)

This method puts a limit on the number of samples we can have because of the limited number of columns available on a given computer system. An alternate way is to proceed as follows, with commands stored in a file which we name 'CONFZ'.

```
MTB >STORE 'CONFZ'

STOR>NOECHO

STOR>RANDOM 25 OBSERVATIONS IN C1;

STOR>NORMAL MEAN 18 SIGMA 6.5.

STOR>ZINTERVAL CONF LEVEL 0.95 SIGMA 6.5, DATA IN C1

STOR>END OF PROGRAM

MTB >EXECUTE 'CONFZ' 120 TIMES
```

As a result of the two preceding procedures, you will have obtained, altogether, 160 confidence intervals for μ. Find out what fraction of these intervals actually contain the true mean 18. Because the confidence level used is 0.95, we expect the fraction to be approximately 95 percent. The following two observations need to be made in this context. Whatever the actual fraction in your simulation, it is not likely to vary much from 95 percent. Also, when an interval does not cover μ it is not likely to miss μ by much. When you construct a confidence interval from your field data, you will not know whether it covers the true mean μ or not. But you should find these two observations very reassuring, in the way they allow you to rely on the confidence interval you might compute in your investigation.

2   Having considered the empirical aspect, we will now solve Exercise 27 in Section 7-1. Given that the population is normally distributed with standard deviation $\sigma = 1.2$, we use the command ZINTERVAL to construct the confidence interval, proceeding as follows.

```
MTB >NAME C1 'TIME'

MTB >SET DATA IN C1

DATA>6.62    4.85    6.44    5.73    5.98    5.15    5.20

DATA>7.67    4.68    6.57    4.99    6.25

DATA>END OF DATA

MTB >PRINT C1

MTB >ZINTERVAL CONF LEVEL 0.98, SIGMA 1.2, DATA IN C1
```

Assignment   Work out Exercise 28 in Section 7-1.

It should be mentioned that the ZINTERVAL command will work only on raw data. It will not work if the summary of the data is provided by giving the mean and the sample size. We can, however, write a program to handle this situation. We store the commands in a file we call ZINTSUM, meaning *z* interval for summary data.

```
MTB >STORE 'ZINTSUM'

STOR>NOECHO

STOR>NOTE K1 IS Z VALUE, K2 IS SIGMA, K3 IS SAMPLE SIZE

STOR>LET K6 = K1*K2/SQRT(K3)

STOR>NOTE THE CONSTANT K4 BELOW IS SAMPLE MEAN

STOR>NOTE LOWER CONFIDENCE LIMIT K11

STOR>LET K11 = K4 − K6

STOR>PRINT K11

STOR>NOTE UPPER CONFIDENCE LIMIT K12

STOR>LET K12 = K4 + K6

STOR>PRINT K12

STOR>END OF PROGRAM
```

All that we do now is execute the program after providing the *z* value, sample mean, σ, and *n*. We will work out Example 4 in Section 7-1 where $\alpha = 0.02$, the sample mean is 274, $\sigma = 14$, and $n = 50$.

```
MTB >NOTE EXAMPLE 4 SECTION 7-1

MTB >NOTE Z VALUE K1 WITH ALPHA 0.02

MTB >INVCDF 0.99 AS K1;

SUBC>NORMAL MEAN 0 SIGMA 1.

MTB >NOTE K2 IS SIGMA, K3 IS SAMPLE SIZE, K4 IS SAMPLE
      MEAN

MTB >LET K2 = 14

MTB >LET K3 = 50

MTB >LET K4 = 274

MTB >EXECUTE 'ZINTSUM'
```

Recall that INVCDF, together with the subcommand specifying the underlying distribution, gives the *x* value with specified area (probability) to its left. In this case, because $1 - \dfrac{\alpha}{2} = 1 - 0.01 = 0.99$, the area is 0.99. Check the upper and lower limits that you get here with those given in the solution for Example 4.

Assignment   Work out Exercise 24 in Section 7-1.

3   The following solves Exercise 22 in Section 7-2. We want a confidence interval for the mean μ of a normally distributed population when the standard deviation σ is not known. The command in this case is TINTERVAL, which requires that we specify the level of confidence, and the column where the data are located.

```
MTB >NAME C5 'RSP-TIME'

MTB >SET DATA IN C5

DATA>11.6    7.9  9.1  7.7  6.8  8.5  2.5  10.1  11.0

DATA>3.6   10.5  6.3  5.3  7.2  5.8  12.6  4.3   2.9

DATA>END OF DATA

MTB >PRINT C5

MTB >TINTERVAL CONFIDENCE LEVEL 0.99, DATA IN C5
```

Like the command ZINTERVAL, the command TINTERVAL will work only on raw data.

Assignment   Work out Exercise 26 in Section 7-2.

4   There is no direct command in MINITAB to construct confidence intervals for σ^2. But we can write a program to construct one, as the following shows. First, we store the commands in a file which we name 'CHI-INT'.

```
MTB >STORE 'CHI-INT'

STOR>NOECHO

STOR>NOTE K1 IS SAMPLE SIZE, K2 STD DEV FOR DATA IN C1

STOR>LET K1 = COUNT (C1)

STOR>LET K2 = STDEV(C1)

STOR>LET K3 = (K1 - 1)*K2**2

STOR>NOTE K4, K5 BELOW ARE TABLE CHI-SQUARE VALUES

STOR>NOTE LOWER CONFIDENCE LIMIT

STOR>LET K10 = K3/K4
```

```
STOR>PRINT K10

STOR>NOTE UPPER CONFIDENCE LIMIT

STOR>LET K11 = K3/K5

STOR>PRINT K11

STOR>END OF PROGRAM
```

5 The preceding program requires that the data be entered in Column C1. The constant K3 gives $(n - 1)s^2$. Also, the constants K4 and K5 refer to the chi-square values which will have to be supplied. The chi-square values can be obtained either from Appendix III Table A-5 or with MINITAB commands as shown in the following, which works out Example 2b.i. in Section 7-3.

```
MTB >NOTE ENTER DATA FOR EXAMPLE 2 SECTION 7-3

MTB >NAME C1 'ENAMEL'

MTB >SET DATA IN C1

DATA>8.1    8.7    7.6    7.8    8.5    7.9

DATA>END OF DATA

MTB >PRINT C1

MTB >NOTE CHI-SQUARE VALUES FOR 95 PERCENT INTERVAL

MTB >INVCDF FOR PROBABILITY 0.025 AS K5;

SUBC>CHISQUARE WITH 5 DF.

MTB >INVCDF FOR PROBABILITY 0.975 AS K4;

SUBC>CHISQUARE WITH 5 DF.

MTB >EXECUTE 'CHI-INT'
```

Assignment   Work out Exercise 22 in Section 7-3.

We now type STOP and conclude the MINITAB session using the logging off procedure.

Commands you have learned in this session:

```
ZINTERVAL

TINTERVAL
```

Estimation—Two Populations

USA SNAPSHOTS®

A look at statistics that shape our lives

Student aid covers less

College costs have gone up

$7,470

$5,854

'83–'84 '92–'93

While student aid has declined

$3,583 $3,261

'83–'84 '92–'93

(average amounts in 1992 dollars)

▲ Source: Graphic "Student Aid Covers Less" by Patti Stang and Web Bryant, *USA Today*. Data from The College Board, Education Department. Reprinted by permission of *USA Today*, December 2, 1993, Section D, p. 1.

INTRODUCTION

In a statistical investigation, it is often necessary to study two populations and compare a parameter of one with the corresponding parameter of the other. This might be done by comparing means of two populations, or by comparing proportions of objects having a given attribute in two populations, or by comparing variances of two populations. For example, we might wish to compare the mean diameter of bolts produced by one machine with the mean of those produced by another machine, or we might want to compare the proportion of smokers among men with that among women, or we might want to compare the variance of altimeter readings on one instrument with the variance of readings on another instrument.

There are two common ways to determine how near two quantities are to each other. One way is to take the difference: If the difference is nearly zero, then the quantities are very near each other; if the difference is very far removed from zero, then the two quantities are far apart. The other way is to take the ratio of the two quantities: If the ratio is nearly equal to 1, then two quantities are near each other; departure of the ratio from 1 will provide a measure of how far apart the two quantities are. In this chapter we compare means of two populations and also proportions in two populations. For considerations in theoretical statistics, we use the difference between quantities as a measure. It is for this reason that we will construct confidence intervals for the **difference of means** and **difference of proportions** in two populations.

Whether we are comparing population means or population proportions, our basic approach will be to take two random samples, one from each population, and then compare corresponding sample analogues. We will be making a very important assumption throughout in this context. We assume that the two samples are independent; that is, the outcomes in one sample have no influence on the outcomes in the other.

8-1   CONFIDENCE INTERVAL FOR THE DIFFERENCE OF POPULATION MEANS WHEN POPULATION VARIANCES ARE KNOWN

Suppose we wish to compare two brands of 9 volt batteries, Brand 1 and Brand 2. Specifically we would like to compare the mean life for the population of batteries of Brand 1 and the mean life for the population of batteries of Brand 2. To obtain a meaningful comparison we shall estimate the difference of the two population means by picking samples from the two populations. The data on page 458 are based on two samples and give the length of useful life (in hours) before the batteries went dead, when twelve batteries of Brand 1 and nine batteries of Brand 2 were tested. The data were simulated from normal distributions with standard deviation 1.4 hours for Brand 1 and 1.2 hours for Brand 2.

	Brand 1	Brand 2
	7.7	7.2
	7.5	7.4
	7.3	8.2
	6.5	7.1
	4.3	8.5
	7.1	7.8
	9.1	9.3
	8.3	5.5
	6.1	9.0
	9.4	
	6.4	
	5.8	
Mean	7.13	7.78

The 12 batteries of Brand 1 represent a sample from the population of 9 volt batteries of Brand 1. Likewise, the 9 batteries of Brand 2 are a sample from the population of 9 volt batteries of Brand 2. Also, notice that the two samples are obviously independent.

From the summary of the data, a *point estimate* of the population mean for Brand 1 is 7.13 hours and a *point estimate* of the population mean for Brand 2 is 7.78 hours. It would, therefore, be natural for us to take a *point estimate* of the difference of the two population means as $7.13 - 7.78 = -0.65$ hours.

Having found a point estimate of the difference of the population means, our next goal is to determine a confidence interval for it. Before doing that we shall develop a general setting that deals with problems of this type.

In a general setting we consider two populations, say, Population 1 and Population 2. Suppose the mean of Population 1 is μ_1 and its standard deviation is σ_1 with the corresponding quantities for Populations 2 being μ_2 and σ_2. In order to construct a confidence interval for $\mu_1 - \mu_2$ we pick two samples, one from each population, and compute their means.

Suppose the size of the sample from Population 1 is m and the sample mean is $\overline{x}$. Let n be the size and $\overline{y}$ the mean of the sample from Population 2. (Since in the alphabet the letter m precedes n and x precedes y, we associate m and x with Population 1 and n and y with Population 2.) Schematically, we can show the situation as in Figure 8-1.

The method developed in this section for finding a confidence interval is valid if the following assumptions are justified.

1. The two populations are normally distributed. This assumption is not crucial if both sample sizes are at least 30.
2. The standard deviations of the two populations are known. (This assumption is somewhat artificial, because if these quantities are

Population 1, mean μ_1, standard deviation σ_1

Population 2, mean μ_2, standard deviation σ_2

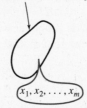

$(x_1, x_2, \ldots, x_m)$

$(y_1, y_2, \ldots, y_n)$

Sample 1, mean $\bar{x}$, size m Sample 2, mean $\bar{y}$, size n

Figure 8-1 Schematic Display Showing Two Samples Picked From Two Populations, and the Notation Used

known, then we would surely know the means. But there are instances when reliable estimates of the standard deviations are available from previous studies. Also, the discussion assuming knowledge of the standard deviations prepares a basis for the more realistic situation when they are not known.)

3. Two random samples are picked, one from each population. They are independent, that is, any outcome in one sample does not influence any outcome in the other sample.

4. Within each sample the outcomes are independent.

We know that a point estimate of μ_1 is $\bar{x}$ and a point estimate of μ_2 is $\bar{y}$. Intuitively, then, we would expect a point estimate of $\mu_1 - \mu_2$ to be $\bar{x} - \bar{y}$. This turns out to be true. (If we were interested in a point estimate of $\mu_2 - \mu_1$, then we would take $\bar{y} - \bar{x}$.)

To construct a confidence interval for $\mu_1 - \mu_2$, we have to know the distribution of $\bar{X} - \bar{Y}$, the statistic that is used to estimate $\mu_1 - \mu_2$.

DISTRIBUTION OF $\bar{X} - \bar{Y}$

Suppose two populations are normally distributed with respective means μ_1 and μ_2 and standard deviations σ_1 and σ_2. Then, the sampling distribution of $\bar{X} - \bar{Y}$ is normal, with expected value (that is, the mean) and the standard deviation given by

$$E(\bar{X} - \bar{Y}) = \mu_1 - \mu_2$$

$$sd(\bar{X} - \bar{Y}) = \sqrt{\frac{\sigma_1^2}{m} + \frac{\sigma_2^2}{n}}$$

when two random and independent samples are drawn, one from each population.

We will accept this fact without any elaboration. The nature of the distribution is indicated in Figure 8-2.

Now recall how, in Section 7-1, we set a confidence interval for μ using the statistic $\overline{X}$ once it was established that the statistic had a normal distribution. Using that as the guide, we obtain the following result.

$$P\left(\overline{X} - \overline{Y} - z_{\alpha/2}\sqrt{\frac{\sigma_1^2}{m} + \frac{\sigma_2^2}{n}} < \mu_1 - \mu_2 < \overline{X} - \overline{Y} + z_{\alpha/2}\sqrt{\frac{\sigma_1^2}{m} + \frac{\sigma_2^2}{n}}\right) = 1 - \alpha$$

point estimator of $\mu_1 - \mu_2$ standard deviation of the point estimator

This result leads to the following upper and lower limits for the confidence interval for $\mu_1 - \mu_2$.

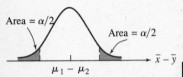

Figure 8-2 Normal Probability Density Curve of the Distribution of $\overline{X} - \overline{Y}$ with Mean $\mu_1 - \mu_2$ and Standard Deviation $\sqrt{\dfrac{\sigma_1^2}{m} + \dfrac{\sigma_2^2}{n}}$

CONFIDENCE INTERVAL FOR $\mu_1 - \mu_2$: STANDARD DEVIATIONS KNOWN

If the populations are normally distributed and the population standard deviations are known, then the upper and lower limits of a $(1 - \alpha)100$ percent confidence interval for $\mu_1 - \mu_2$ are given by

$$(\overline{x} - \overline{y}) \pm z_{\alpha/2}\sqrt{\frac{\sigma_1^2}{m} + \frac{\sigma_2^2}{n}}$$

where $\overline{x}$ and $\overline{y}$ are the sample means based, respectively, on samples of sizes m and n, and σ_1 and σ_2 are the population standard deviations.

The margin of error E is

$$E = z_{\alpha/2}\sqrt{\frac{\sigma_1^2}{m} + \frac{\sigma_2^2}{n}}$$

We now apply these results to determine a 95 percent confidence interval for the difference of the means of the two brands of batteries. Identifying Brand 1 with Population 1 and Brand 2 with Population 2, we have $m = 12$ and $n = 9$. Also, recall that $\sigma_1 = 1.4$ and $\sigma_2 = 1.2$. So, the margin of error is given by

$$\text{margin of error} = z_{\alpha/2}\sqrt{\frac{\sigma_1^2}{m} + \frac{\sigma_2^2}{n}}$$

$$= 1.96\sqrt{\frac{1.4^2}{12} + \frac{1.2^2}{9}}$$

$$= 1.115$$

Since we have seen earlier that $\bar{x} - \bar{y} = -0.65$, the endpoints of a 95 percent confidence interval for $\mu_1 - \mu_2$ are -0.65 ± 1.115. That is, a 95 percent confidence interval for $\mu_1 - \mu_2$ is $(-1.765, 0.465)$. Since the interval includes 0, there is no clear-cut evidence that the mean life of one brand is higher than that of the other brand. One can be 95 percent confident that the mean life for Brand 1 is less than the mean for Brand 2 by up to 1.765 hours or greater by up to 0.465 hours. There is slight indication to suggest that Brand 2 has higher mean life. But to ascertain this, further investigation may be necessary with larger sample sizes.

EXAMPLE 1

Tests were carried out to compare strengths of two types of yarn. When fifteen random tests were carried out with Yarn A, it was found that the mean strength was 140 pounds/square inch. Also, ten random tests carried out with Yarn B yielded a mean of 145 pounds/square inch. Assume that the strength of yarn has a normal distribution for both yarns and that the variance σ_1^2 for Yarn A is 25 (pounds/square inch)^2 while the variance σ_2^2 for Yarn B is 30 (pounds/square inch)^2. Construct an 80 percent confidence interval for the difference $\mu_1 - \mu_2$, where μ_1 and μ_2 represent, respectively, the true means for Yarn A and Yarn B.

SOLUTION   Let x and y represent the strengths of Yarn A and Yarn B, respectively. Notice that instead of standard deviations we are given the variances as $\sigma_1^2 = 25$ and $\sigma_2^2 = 30$. Also, we have $\bar{x} = 140$, $m = 15$, $\bar{y} = 145$, $n = 10$, and $\alpha = 0.2$. Therefore, $z_{\alpha/2} = z_{0.1} = 1.28$. Let us first compute the margin of error E.

$$
\begin{aligned}
E &= z_{\alpha/2} \sqrt{\frac{\sigma_1^2}{m} + \frac{\sigma_2^2}{n}} \\
&= 1.28 \sqrt{\frac{25}{15} + \frac{30}{10}} \\
&= 2.765
\end{aligned}
$$

As a result, an 80 percent confidence interval for $\mu_1 - \mu_2$ is given by

$$
(140 - 145) - 2.765 < \mu_1 - \mu_2 < (140 - 145) + 2.765
$$
$$
-7.77 < \mu_1 - \mu_2 < -2.24
$$

We have obtained $(-7.77, -2.24)$ as an 80 percent confidence interval for $\mu_1 - \mu_2$. Equivalently, an 80 percent confidence interval for $\mu_2 - \mu_1$ is $(2.24, 7.77)$, obtained by changing the signs of the endpoints and switching them around. We can use either of these intervals for interpretation. Let us interpret from the confidence interval for $\mu_2 - \mu_1$. We can say with 80 percent confidence that the mean strength for Yarn B is more than the mean strength for Yarn A by an amount at least as much as 2.24 pounds/square inch and up to as much as 7.77 pounds/square inch. So, with 80 percent confidence, we might infer that Yarn B is stronger than Yarn A. ●

We have assumed that the populations are normally distributed. This assumption is not vital if the sample sizes are large (as a rule of thumb, if m and n are both at least 30). Actually, in this case even the population variances need not be known. We can use s_1^2 and s_2^2, the sample variances, to provide reasonable estimates of the respective population variances σ_1^2 and σ_2^2. We have the following result.

LARGE SAMPLES CONFIDENCE INTERVAL FOR $\mu_1 - \mu_2$

If the **sample sizes m and n both are large** and s_1 and s_2 represent the sample standard deviations, then the upper and lower limit of a $(1 - \alpha)100$ percent confidence interval for $\mu_1 - \mu_2$ are, approximately,

$$\bar{x} - \bar{y} \pm z_{\alpha/2} \sqrt{\frac{s_1^2}{m} + \frac{s_2^2}{n}}$$

The approximate margin of error is

$$E = z_{\alpha/2} \sqrt{\frac{s_1^2}{m} + \frac{s_2^2}{n}}$$

EXAMPLE 2

At the end of a crash dieting program administered to fifty men and forty women, the following information was obtained about the loss of weight (in pounds).

	Men	Women
Mean loss of weight	$\bar{x} = 16.5$	$\bar{y} = 13.2$
s, sample standard deviation	$s_1 = 5$	$s_2 = 4$

Suppose μ_1 and μ_2 represent, respectively, the true mean losses for men and women. Obtain an approximate 85 percent confidence interval for the difference in the mean losses, $\mu_1 - \mu_2$.

SOLUTION Because the sample sizes are large, we can treat s_1 and s_2 as though they are σ_1 and σ_2.

Since $\alpha = 0.15$, we get $z_{\alpha/2} = z_{0.075} = 1.44$. Therefore,

$$E = z_{\alpha/2} \sqrt{\frac{s_1^2}{m} + \frac{s_2^2}{n}}$$

$$= 1.44 \sqrt{\frac{25}{50} + \frac{16}{40}}$$

$$= 1.366$$

Consequently, since $\bar{x} = 16.5$ and $\bar{y} = 13.2$, an approximate 85 percent confidence interval for $\mu_1 - \mu_2$ is given by

$$(16.5 - 13.2) - 1.366 < \mu_1 - \mu_2 < (16.5 - 13.2) + 1.366$$
$$1.93 < \mu_1 - \mu_2 < 4.67$$

Our 85 percent confidence interval for $\mu_1 - \mu_2$ is (1.93, 4.67). So, with 85 percent confidence we can assert that the mean weight loss for men is greater than the mean weight loss for women by an amount which is somewhere between 1.93 and 4.67 pounds.   ●

Section 8-1 Exercises

1. Find a point estimate of $\mu_1 - \mu_2$ if:
   (a) $\bar{x} = 13.75$ and $\bar{y} = 10.7$
   (b) $\Sigma\, x_i = 105$, $\Sigma\, y_i = 150$. *Note: m = 10 and n = 15.*

In Exercises 2–4, find a point estimate of $\mu_2 - \mu_1$.

2. $x_1 = 2.2$, $x_2 = 3.2$, $x_3 = 4.1$, $x_4 = 1.8$, $x_5 = 2.4$ and $y_1 = 3.5$, $y_2 = 4.1$, $y_3 = 3.0$, $y_4 = 3.7$

3. $\Sigma\, x_i = 36.18$ and $\Sigma\, y_i = 42.96$ if $m = 6$ and $n = 8$

4. $\bar{x} = 1.85$ and $\bar{y} = -0.63$

Suppose there are two normally distributed populations. Assuming that two independent random samples are picked from the two populations, determine a confidence interval for estimating $\mu_1 - \mu_2$ in Exercises 5–7 at the indicated level of confidence.

5. $\bar{x} = 8.0$, $m = 20$, $\sigma_1 = 3.2$ and $\bar{y} = 7.2$, $n = 24$, $\sigma_2 = 3.8$; 95 percent

6. $\bar{x} = 1.85$, $m = 14$, $\sigma_1^2 = 20$ and $\bar{y} = -0.63$, $n = 9$, $\sigma_2^2 = 14$; 90 percent

7. $x_1 = 2.2$, $x_2 = 3.2$, $x_3 = 4.1$, $x_4 = 1.8$, $x_5 = 2.5$ with $\sigma_1 = 0.883$ and $y_1 = 3.5$, $y_2 = 4.1$, $y_3 = 3.0$, $y_4 = 3.7$ with $\sigma_2 = 0.529$; 98 percent

8. Suppose there are two normally distributed populations. One population has variance $\sigma_1^2 = 10$, and when a sample of size 20 was picked from it, the mean was $\bar{x} = 8$. The second population has variance $\sigma_2^2 = 12$, and a sample of 24 yielded a mean $\bar{y} = 7.2$. Assume that the two samples are independent.
   (a) Compute a point estimate of the difference $\mu_1 - \mu_2$.
   (b) Determine the maximum error of estimate of $\mu_1 - \mu_2$ at the 95 percent level of confidence.
   (c) Construct a 95 percent confidence interval for estimating $\mu_1 - \mu_2$.

9. The mean height of ten adult Japanese women was found to be $\bar{x} = 63.8$ inches, and that of fifteen adult Chinese women was found to be 62.2 inches. The heights of adult women for the two nationalities may be assumed to be normally distributed with a standard deviation σ_1 of 2.8 inches for Japanese women and σ_2

of 3.1 inches for Chinese women. Let μ_J represent the (population) mean height for Japanese women and μ_C for Chinese women.

(a) Find a 92 percent confidence interval for $\mu_J - \mu_C$.

(b) Find a 92 percent confidence interval for $\mu_C - \mu_J$.

(c) Compare the intervals in Parts a and b and comment.

10. An engineering student measured the daily dissolved oxygen concentration (in milligrams/liter) of two streams. The mean daily concentration of Stream 1 when observed during 20 days was found to be $\bar{x} = 4.58$ mg/1 and that of Stream 2 when observed during 14 days was found to be $\bar{y} = 3.68$ mg/1. Suppose it is reasonable to assume that daily concentration is normally distributed for each stream and that the concentration of one stream does not influence that of the other. Also, it may be assumed that the standard deviation of the concentration for Stream 1 is $\sigma_1 = 0.8$ mg/1 and that for Stream 2 is $\sigma_2 = 1.1$ mg/1.

(a) Determine a 98 percent confidence interval for $\mu_1 - \mu_2$ where μ_1 is the (population) mean concentration for Stream 1 and μ_2 that for Stream 2.

(b) Interpret the interval.

(c) How will the width of the confidence interval be affected if the confidence level is lowered?

Suppose two independent random samples are picked from two populations. In Exercises 11–13, determine a confidence interval for estimating $\mu_1 - \mu_2$ at the indicated level of confidence.

Hint: Notice *m* and *n* are large in each case.

11. $\bar{x} = 6.2$, $m = 120$, $s_1 = 1.122$ and $\bar{y} = 3.8$, $n = 80$, $s_2 = 1.523$; 95 percent

12. $\bar{x} = 0.19$, $m = 60$, $s_1 = 1.3$ and $\bar{y} = -0.32$, $n = 68$, $s_2 = 1.6$; 90 percent

13. $\bar{x} = 123.6$, $m = 48$, $s_1^2 = 8.82$ and $\bar{y} = 98.7$, $n = 64$, $s_2^2 = 5.62$; 98 percent

14. When 100 king-size cigarettes of Brand A and 152 king-size cigarettes of Brand B were tested for their nicotine content (in milligrams), the following information was obtained.

	Sample Mean	Sample Standard Deviation
Brand A	1.2	0.2
Brand B	1.4	0.25

(a) Compute a point estimate of the difference in true mean nicotine contents for the two brands.

(b) Determine the margin of error in estimating the difference in true mean nicotine contents at the 90 percent level.

(c) Estimate the difference in true mean nicotine contents for the two brands at the 90 percent level of confidence.

(d) Interpret the interval to compare the two brands of cigarettes.

15. Sixty-nine children aged 2–5 years convalescing from shigellosis were divided into two groups in a random selection. Thirty-six of the children were fed a high protein diet and thirty-three were fed a standard protein diet. At the end of 21 days the change in height (cm) was measured for the children in each group.

The following summary gives the results for change in height for the children fed the high protein diet and those fed the standard protein diet. (Iqbal Kabir et al in the *American Journal of Clinical Nutrition* 1993: *57,* 3:441–5)

	High Protein Diet	Standard Protein Diet
Sample size	36	33
Mean	1.02	0.69
Standard deviation	0.44	0.34

(a) Describe the two populations that are being studied in the investigation.
(b) Determine a 95 percent confidence interval for the difference in mean changes of height for those children fed the high protein diet and those fed the standard protein diet. Interpret the interval to evaluate effectiveness of the high protein diet.
(c) Compute the width of the interval.

16. The data below present the summary of SAT scores in mathematics for 228 white students and 105 Asian-American students in 1992–93.

	White	Asian-American
Sample size	228	105
Mean	498	529
Standard deviation	68	73

(a) Estimate the difference in true mean scores for the two groups using a 99 percent confidence interval.
(b) Interpret the interval and make a statement about the relative performance of the two groups.

17. In a test to measure the performance of two comparable models of cars, Cheetah and Pioneer, 128 cars of each model were driven on the same terrain with ten gallons of gasoline in each car. The mean number of miles for Cheetah was 340 miles with s_1 of 10 miles; the number of miles for Pioneer was 335 miles with s_2 of 16 miles.
(a) Obtain a 98 percent confidence interval for the difference between the true mean numbers of miles for the two models of cars.
(b) Interpret the interval.

18. A rancher tried Feed A on 256 cattle and Feed B on 144 cattle. The mean weight of cattle given Feed A was found to be 1350 pounds with s_1 of 180 pounds. On the other hand, the mean weight of the cattle given Feed B was found to be 1430 pounds with s_2 of 210 pounds.
(a) Set a 95 percent confidence interval on the difference in the true mean weights of cattle given the two feeds.
(b) If a narrower interval is desired for estimating the difference in true means, would the confidence level be increased or lowered?

19. A company has two branches, one on the East Coast and the other on the West Coast. The mean daily sales of the company on the East Coast, when observed on 100 days, were found to be $32,000 with s_1 of $2400. For the company on the

West Coast, when observed on 200 days, the mean daily sales were $36,000 with s_2 of $2700.

(a) Construct a 95 percent confidence interval for the difference in true mean daily sales of the two branches.

(b) Interpret the interval.

20. A laboratory study was carried out to investigate the physical and chemical characteristics of Sri Lankan nutmeg oil to determine the norms for nutmeg oil from this source. The chemical characteristics, myristicin content and elemicin content, were determined using gas-liquid chromatography.

   The following information summarizes the data for the laboratory-distilled oil and industrial samples for each chemical characteristic mentioned.

	MYRISTICIN CONTENT	
	Laboratory	Industrial
Mean	1.94	2.29
Standard deviation	1.63	1.21
Number	36	40

	ELEMICIN CONTENT	
	Laboratory	Industrial
Mean	0.913	1.186
Standard deviation	0.58	0.70
Number	33	39

   For each chemical characteristic, obtain a confidence interval at the indicated level of confidence for the difference between the true means for laboratory distilled oil and industrial samples.

(a) Myristicin content at the 90 percent level

(b) Elemicin content at the 98 percent level

8-2 CONFIDENCE INTERVAL FOR THE DIFFERENCE OF POPULATION MEANS WHEN THE VARIANCES ARE UNKNOWN BUT ARE ASSUMED EQUAL

In most fields of investigation, it is more realistic to suppose that the variances of the two populations are not known. If the sample sizes are large, we know from the previous section, how to obtain a confidence interval for $\mu_1 - \mu_2$ even when the population variances are not known. So, *the method that we are going to develop in this section is particularly relevant when the sample sizes are small.* It is appropriate if the following assumptions are satisfied.

1. The two populations are normally distributed.
2. The population variances σ_1^2 and σ_2^2, though unknown, are the same, say, each equal to σ^2.
3. Two random and independent samples are picked, one from each population.

Our immediate task is to find an estimate of the common variance σ^2. Since the variances of the two populations are assumed to be equal, a combined estimate of the variance is obtained by pooling the sums of squared deviations, $\Sigma (x_i - \bar{x})^2$ and $\Sigma (y_i - \bar{y})^2$, for the two samples. The total is then divided by $(m - 1) + (n - 1) = m + n - 2$. Denoting the pooled estimate of σ^2 by s_p^2,

$$s_p^2 = \frac{\Sigma (x_i - \bar{x})^2 + \Sigma (y_i - \bar{y})^2}{m + n - 2}.$$

Because

$$\Sigma (x_i - \bar{x})^2 = (m - 1)s_1^2 \quad \text{and}$$

$$\Sigma (y_i - \bar{y})^2 = (n - 1)s_2^2$$

the pooled estimate of σ^2 may also be written as

$$s_P^2 = \frac{(m - 1)s_1^2 + (n - 1)s_2^2}{m + n - 2}.$$

You will find this formula for s_p^2 more useful because electronic calculators have a built-in key which gives sample standard deviations. Also, results published in research journals will usually provide a summary that includes sample standard deviations and sample sizes.

For example, suppose the sample standard deviations and the sample sizes in an investigation are reported as

$$s_1 = 12.8 \quad \text{with } m = 13, \quad \text{and } s_2 = 16.9 \quad \text{with } n = 15.$$

Let us assume that the population variances are equal. (There is a statistical test which would support this for the given values of the sample standard deviations and the sample sizes.) Then, a pooled estimate s_p^2 of the common variance is

$$s_p^2 = \frac{(13 - 1)12.8^2 + (15 - 1)16.9^2}{13 + 15 - 2} = 229.41.$$

So, a pooled estimate s_p of the common standard deviation σ is $s_p = \sqrt{229.41} = 15.15$.

Now, if $\sigma_1^2 = \sigma_2^2 = \sigma^2$, the upper and lower limits of the confidence interval for $\mu_1 - \mu_2$ given on page 460 will reduce to

$$(\bar{x} - \bar{y}) \pm z_{\alpha/2} \sigma \sqrt{\frac{1}{m} + \frac{1}{n}}$$

with the margin of error E equal to

$$E = z_{\alpha/2}\sigma\sqrt{\frac{1}{m} + \frac{1}{n}}.$$

Since σ is not known, we take the obvious and expedient step of replacing it with its pooled estimate s_p. But the consequence is that now we cannot use z values from the normal table. Rather, we are required to use values from the t distribution with $m + n - 2$ degrees of freedom. This is particularly important if the degrees of freedom are less than 30 because then the t values will be markedly different from the z values. Thus, as a result of the substitution, the margin of error changes to

$$E = t_{m+n-2,\alpha/2}s_p\sqrt{\frac{1}{m} + \frac{1}{n}}$$

and we get a confidence interval with upper and lower limits given as follows.

CONFIDENCE INTERVAL FOR $\mu_1 - \mu_2$; EQUAL BUT UNKNOWN VARIANCES

The upper and lower limits of a $(1 - \alpha)100$ percent confidence interval for $\mu_1 - \mu_2$, when two populations are normally distributed and have the same (but unknown) variance, are given by

$$(\overline{x} - \overline{y}) \pm t_{m+n-2,\alpha/2}s_p\sqrt{\frac{1}{m} + \frac{1}{n}}.$$

Here $\overline{x}$ and $\overline{y}$ are sample means based, respectively, on m and n observations, and s_p is the **pooled estimate of σ,** given by

$$s_p = \sqrt{\frac{(m-1)s_1^2 + (n-1)s_2^2}{m+n-2}} = \sqrt{\frac{\sum(x_i - \overline{x})^2 + \sum(y_i - \overline{y})^2}{m+n-2}}$$

Note: We omit the case involving *small* samples picked from two normal populations having *unknown, unequal variances,* which involves a complicated formula for computing the degrees of freedom.

EXAMPLE 1   Two diets, Diet X and Diet Y, were fed to two groups of randomly picked dogs of a certain breed. Diet X was fed to ten dogs. Their mean weight $\overline{x}$ was 50 pounds, and the sample standard deviation s_1 was 6 pounds. Diet Y was fed to fourteen dogs. Their mean weight $\overline{y}$ was 56 pounds, and the sample standard deviation s_2 was 8 pounds. Find a confidence interval for the difference

---

*The interested reader is referred to page 126 of *Introduction to Statistical Analysis,* 4th edition, by W. J. Dixon and F. J. Massey, McGraw Hill Publishers.

of the means with a 90 percent confidence level. (Assume that the weights of the dogs given Diet X and the weights of dogs given Diet Y are normally distributed with the same variance.)

SOLUTION Because $s_1 = 6$, $s_2 = 8$, $m = 10$, and $n = 14$,

$$s_p = \sqrt{\frac{(m-1)s_1^2 + (n-1)s_2^2}{m+n-2}}$$

$$= \sqrt{\frac{9(6)^2 + 13(8)^2}{10 + 14 - 2}}$$

$$= 7.249.$$

Also there are $m + n - 2 = 22$ degrees of freedom, and since $\alpha = 0.1$, we get $t_{m+n-2, \alpha/2} = t_{22, 0.05} = 1.717$. Thus, when computing a 90 percent confidence interval for $\mu_1 - \mu_2$, the margin of error is

$$E = 1.717 s_p \sqrt{\frac{1}{m} + \frac{1}{n}}$$

$$= 1.717(7.249) \sqrt{\frac{1}{10} + \frac{1}{14}}$$

$$= 5.153.$$

Consequently, because $\bar{x} = 50$ and $\bar{y} = 56$, a 90 percent confidence interval for $\mu_1 - \mu_2$ is given by

$$(50 - 56) - 5.153 < \mu_1 - \mu_2 < (50 - 56) + 5.153$$

which, upon simplification, gives the confidence interval

$$-11.15 < \mu_1 - \mu_2 < -0.85.$$

From the confidence interval $(-11.15, -0.85)$ for $\mu_1 - \mu_2$ we are able to deduce with 90 percent confidence that the mean μ_1 for Diet X is less than the mean μ_2 for Diet Y by as much as 0.85 pounds and possibly by as much as 11.15 pounds. It would seem reasonable to draw the inference that dogs fed Diet X have less mean weight than the dogs fed Diet Y. ●

EXAMPLE 2

A phosphate fertilizer was applied to five plots and a nitrogen fertilizer to six plots. The yield of grain on each plot was recorded (in bushels) as presented in Table 8-1.

Construct a 98 percent confidence interval for $\mu_1 - \mu_2$. (μ_1 represents the mean for all the conceivable plots to which the phosphate fertilizer might be applied, with a similar interpretation for μ_2.) Suppose it may be assumed that the populations are normally distributed and the variances are equal.

SOLUTION In Table 8-2, we find $\Sigma (x_i - \bar{x})^2$ and $\Sigma (y_i - \bar{y})^2$, which will be used to obtain a pooled estimate of the common variance.

TABLE 8-1 Yield of Grain on Each Plot

Phosphate	Nitrogen
40	50
49	41
38	53
48	39
40	40
	47

TABLE 8-2 Computations of $\Sigma (x_i - \bar{x})^2$ and $\Sigma (y_i - \bar{y})^2$

x	$(x - \bar{x})^2$	y	$(y - \bar{y})^2$
40	9	50	25
49	36	41	16
38	25	53	64
48	25	39	36
40	9	40	25
		47	4

$$\bar{x} = \frac{215}{5} \quad \Sigma (x_i - \bar{x})^2 \quad \bar{y} = \frac{270}{6} \quad \Sigma(y_i - \bar{y})^2$$
$$= 43 \qquad = 104 \qquad = 45 \qquad = 170$$

Because $m = 5$ and $n = 6$, the pooled estimate of σ^2 is

$$s_p^2 = \frac{104 + 170}{5 + 6 - 2} = \frac{274}{9} = 30.44.$$

Therefore, $s_p = 5.517$.

Now $\alpha = 0.02$ and the t distribution has $5 + 6 - 2 = 9$ degrees of freedom.

Therefore, $t_{m+n-2,\alpha/2} = t_{9,0.01} = 2.821$ and the margin of error E is

$$E = (2.821)s_p \sqrt{\frac{1}{m} + \frac{1}{n}}$$

$$= (2.821)(5.517) \sqrt{\frac{1}{5} + \frac{1}{6}}$$

$$= 9.424$$

Since $\bar{x} = 43$ and $\bar{y} = 45$, a 98 percent confidence interval for $\mu_1 - \mu_2$ is

$$(43 - 45) - 9.424 < \mu_1 - \mu_2 < (43 - 45) + 9.424.$$

After simplifying, we get

$$-11.42 < \mu_1 - \mu_2 < 7.42.$$

The confidence interval $(-11.42, 7.42)$ points out with 98 percent confidence that the mean yield μ_1 for the phosphate fertilizer could be less than the mean yield μ_2 for the nitrogen fertilizer (as judged from the interval $(-11.42, 0)$, or it could be greater than the mean yield for the nitrogen fertilizer (as judged from the interval $(0, 7.42)$). It could be less by as much as 11.42 pounds or greater by as much as 7.42 pounds. So, we are unable to recommend that one fertilizer is better than the other.   ●

Section 8-2 Exercises

If two independent random samples are picked from two populations, both having the same variance σ^2, find a pooled estimate of σ^2 in Exercises 1–6.

1. $\Sigma (x_i - \bar{x})^2 = 8.47$ with $m = 8$ and $\Sigma (y_i - \bar{y})^2 = 6.84$ with $n = 6$

2. $s_1^2 = 10$ with $m = 16$ and $s_2^2 = 14$ with $n = 11$

3. $\Sigma (x_i - \bar{x})^2 = 0.189$ with $m = 10$ and $\Sigma (y_i - \bar{y})^2 = 01.68$ with $n = 7$

4. $\Sigma (x_i - \bar{x})^2 = 26.24$ with $m = 5$ and $s_2^2 = 8.62$ with $n = 6$

5. $x_1 = 2.2$, $x_2 = 3.2$, $x_3 = 4.1$, $x_4 = 1.8$, $x_5 = 2.4$ and $y_1 = 3.5$, $y_2 = 4.1$, $y_3 = 3.0$, $y_4 = 3.7$

6. $x_1 = 16.1$, $x_2 = 18.3$, $x_3 = 14.6$, $x_4 = 13.6$ and $y_1 = 11.8$, $y_2 = 9.6$, $y_3 = 8.2$, $y_4 = 9.2$, $y_5 = 10.8$

7. Assume that we have two populations, both normally distributed with a common variance σ^2 that is unknown. If $\bar{x} = 10.2$, $\bar{y} = 8.5$, $m = 10$, $n = 12$, $s_1 = 1.67$, and $s_2 = 1.79$, assuming the samples are independent, find the following.
   (a) A pooled estimate, s_p
   (b) The degrees of freedom
   (c) A point estimate of $\mu_1 - \mu_2$
   (d) The margin of error in estimating $\mu_1 - \mu_2$ at the 90 percent level
   (e) A 90 percent confidence interval for $\mu_1 - \mu_2$

8. In order to evaluate the degree of suspension of a polyethylene, its gel contents (gel proportion) are determined after extraction using a solvent. The method is called the gel proportion estimation method. A study was run to compare two solvents, ethanol and toluene, using extraction time of 8 hours. The data are presented below.

Ethanol	Toluene
94.7	96.6
96.4	95.5
93.0	95.9
95.4	95.0
96.4	96.1
94.6	95.3
94.8	95.4
96.9	96.8
94.8	95.2
94.8	96.1

   (a) Obtain a point estimate of the difference in true mean gel contents extracted using the two solvents.
   (b) What is the margin of error in estimating the difference when 95 percent level of confidence is used?
   (c) Obtain a 95 percent confidence interval for the difference in true mean gel contents extracted using the two solvents.
   (d) Give the assumptions, with reference to the variables in the problem, that you would need to make to apply the formula in this section.
   (e) Interpret the interval for comparing the two solvents.

9. Two models of cars that were similar in specifications were compared for their acceleration from 0 to 60 mph. Acceleration tests were carried out from standstill with engine idling at start. Also, all the runs were done with gears shifted to best advantage. The results giving the number of seconds are presented below.

Model 1	15	12	10	12	10	11	12	12	14	16
Model 2	14	16	14	12	16	12	13	15	13	

   (a) Compute a point estimate of the difference in true mean acceleration times for the two models.
   (b) Find the margin of error in estimating the difference in mean acceleration times at the 98 percent level.
   (c) Construct a 98 percent confidence interval to estimate the difference in true mean acceleration times for the two models.
   (d) What assumptions did you make? State them in terms of the variables involved.
   (e) Interpret the interval.
   (f) What would you recommend if the investigator wanted to reduce the width of the confidence interval?

10. The quality of an air purifier is measured by the cubic feet of fresh air it provides per minute. Two models of air purifiers were compared for smoke removal and the following data were obtained giving the amount of fresh air provided per minute.

SmokeBgone	230	251	277	269	206	234	237	247	251	241	266
SmokeGuard	222	217	223	200	195	204	272	231	215	242	194

   (a) Using a 95 percent confidence interval, estimate the difference in the true mean rates for the two purifiers.
   (b) State the assumptions you had to make.
   (c) Give an interpretation of the interval you have obtained.

11. Exercise 15 in Section 8-1 involves computation of a 95 percent confidence interval for the difference in mean changes in heights of children using normal approximation because the sample sizes were greater than 30.
   (a) Construct a 95 percent confidence interval for the difference using the method of this section, given that $t_{67,0.025} = 1.996$.
   (b) Compute the width of the interval and compare it with the width obtained in Exercise 15. What is the percent error when the approximation was used?

In Exercises 12–21, assume that the two populations are normally distributed with the same, but unknown, variance. In each case interpret the interval.

12. The mean length of fourteen trout caught in Clear Lake was 10.5 inches with s_1 of 2.1 inches, and the mean length of eleven trout caught in Blue Lake was 9.6 inches, with s_2 of 1.8 inches. Construct a 90 percent confidence interval for the difference in the true mean lengths of trout in the two lakes.

13. A random sample of ten light bulbs manufactured by Company A had a mean life of 1850 hours and a standard deviation s_1 of 130 hours. Also, a random sample of twelve light bulbs manufactured by Company B had a mean life of 1940 hours with a standard deviation s_2 of 140 hours. Construct a 95 percent confidence interval for the difference in the true means.

14. In ten half-hour morning programs, the mean time devoted to commercials was 6.8 minutes with s_1 of 1 minute. In twelve half-hour evening programs, the mean time was 5.6 minutes with s_2 of 1.3 minutes. Estimate the difference in the true mean times devoted to commercials during the morning and the evening half-hour programs, using a 90 percent confidence interval.

15. A psychologist measured the reaction times (in seconds) of eight individuals who were not given any stimulant and six individuals who were given an alcoholic stimulant. The results are as follows.

Reaction time without stimulant	3.0	2.0	1.0	2.5	1.5	4.0	1.0	2.0
Reaction time with stimulant	5.0	4.0	3.0	4.5	2.0	2.5		

(a) Describe the two populations that are being compared.

(b) Determine a 95 percent confidence interval for the difference in true mean reaction times for the two populations.

(c) What would you recommend if the psychologist wanted a narrower interval without altering the confidence level?

16. Information on peptides identified in the brains of different animals is of interest to biologists. An experiment was carried out to compare the content of peptides (μ mole/g fresh tissue) in the brains of two species of fish from the Black Sea: dogfish (*Squalus acanthias*) and the skate (*Raja clavata L.*). Brains were quickly removed from fish freshly caught, and the peptides in the brain determined by electropherosis. The values are given in the following table.

PEPTIDE CONTENT (μ Mole/g Tissue)

Skate	Dogfish
1.00	0.51
0.53	0.74
0.41	0.84
0.70	0.80
0.70	0.60
0.59	0.74
0.82	0.82
0.80	0.73
0.53	0.98
0.70	0.59
0.68	0.74

(a) Estimate the difference in the true means of peptide content in the brain tissues for the two species, using a 90 percent confidence interval.

(b) If the confidence level is altered to 95 percent, how would it affect the width of the confidence interval? Answer without actually computing the confidence interval.

17. Casein, a phosphoprotein, is one of the chief constituents of milk and the basis of cheese. A change in the cow's diet is believed to affect milk composition and, in particular, its casein content.

A random sample of ten Jersey cows was fed a typical summer diet and another random sample of twelve cows was fed a typical winter diet. Milk samples were collected for each cow and the casein content of milk was estimated. The data are presented below.

CASEIN CONTENT	
Summer Diet	Winter Diet
3.69	3.20
3.59	3.20
3.75	3.07
3.54	3.49
3.69	3.13
3.60	3.14
3.69	3.15
3.71	3.24
3.57	3.43
3.68	3.22
	3.66
	3.61

From the given data, construct a 95 percent confidence interval for the difference in the true means of casein content under the two diets.

18. The values presented in the following table refer to the apparent disappearance of aspartic acid (an amino acid) from the small intestine of sheep. Eight sheep received an infusate of untreated casein at pH 2.5, and ten sheep received it with untreated casein at pH 6.8.

APPARENT DISAPPEARANCE OF ASPARTIC ACID	
pH 2.5	pH 6.8
0.82	0.78
0.82	0.85
0.79	0.77
0.82	0.88
0.88	0.82
0.81	0.81
0.79	0.82
0.83	0.85
	0.77
	0.75

Use the data to obtain a 98 confidence interval for the difference in the true means of apparent disappearance of aspartic acid at the two pH levels. Comment on the interval obtained to compare the true mean aspartic acid disappearance at the two pH levels.

19. The following data summarize the total concentration (pg/ml) of serum vitamin B_{12} in two groups of sheep. One group of three sheep was fed a high forage (HF) diet and the other group of three sheep was fed an HF diet with 33 mg/kg of added monensin (M). The results of the experiment are summarized in the following table.

	SERUM VITAMIN B_{12} (pg/ml)	
	HF Diet	**HF Diet + M**
Mean	1358	1868
Standard deviation	349	402
Number	3	3

(a) Find an estimate of the difference in the true means for serum vitamin under the two diets, using a 95 percent confidence interval.

(b) What options are there if the experimenter wanted to reduce the width of the confidence interval?

20. Two types of soft drinks, a cola and a non-cola, were tested for the amount of glucose (g/100 ml). The findings are summarized in the following table.

	GLUCOSE (g/100 ml)	
	Cola	**Non-cola**
Mean	4.10	4.63
Standard deviation	0.06	0.09
Number	15	18

Find an estimate of the difference between the true means of glucose amount in the two kinds of drinks, using a 90 percent confidence interval.

21. Fifteen hens, approximately 62 weeks of age, and ten turkeys, approximately 17 weeks of age, were slaughtered and the total haem pigments (mg/g of tissue) in the breast meat measured (*Journal of the Science of Food and Agriculture* 1986; 37: pp. 1236–1240). The data are summarized in the following table.

	HAEM CONTENT IN BREAST MEAT	
	Hens	**Turkeys**
Mean	0.64	0.58
Standard deviation	0.046	0.052
Number	15	10

Use a 98 percent confidence interval to estimate the difference between the true mean of haem content in the breast meat of hens and that in the breast meat of turkeys. Use the interval obtained to compare the true mean haem content in the breast meat of the two birds.

8-3   CONFIDENCE INTERVAL FOR THE DIFFERENCE OF POPULATION PROPORTIONS

Consider the following example, in which we compare the proportion of women favoring the death penalty with the proportion of men. The collection of all women represents one population. We denote the proportion favoring the death penalty in this population by p_1. The collection of all men represents the other population, where the proportion is p_2. The values of p_1 and p_2 are not known to us, and we intend to construct a confidence interval for $p_1 - p_2$. The first step is to pick two random samples, one from each population, taking care that the two samples are independent.

Suppose we interview m women. Let X denote the number among them who favor the death penalty. Then, from what we learned earlier, X/m is a good point estimator of p_1. Similarly, if we interview n males and Y denotes the number among them favoring the death penalty, then Y/n will be a point estimator of p_2. It would then seem natural that we use $X/m - Y/n$ as a point estimator of $p_1 - p_2$. For example, if among 100 women interviewed, 30 are in favor, and among 200 men, 72 are in favor, then a point estimate of $p_1 - p_2$ is $30/100 - 72/200 = -0.06$.

Next, we construct a confidence interval for $p_1 - p_2$ under the following assumptions.

1. The population proportions p_1 and p_2 are not too close to 0 or 1.
2. Two random samples are taken, one from each population, and the two samples are independent.
3. Sample sizes m and n are large.

Our first order of business is to find the distribution of the estimator $\dfrac{X}{m} - \dfrac{Y}{n}$.

THE DISTRIBUTION OF $\dfrac{X}{m} - \dfrac{Y}{n}$

The expected value and the standard deviation of the distribution of $(X/m) - (Y/n)$ are given by

$$E\left(\frac{X}{m} - \frac{Y}{n}\right) = p_1 - p_2$$

$$sd\left(\frac{X}{m} - \frac{Y}{n}\right) = \sqrt{\frac{p_1(1 - p_1)}{m} + \frac{p_2(1 - p_2)}{n}}$$

Moreover, the distribution of $(X/m) - (Y/n)$ is approximately normal if m and n are large and neither p_1 nor p_2 is near 0 or 1 in such a way that all the four quantities mp_1, $m(1 - p_1)$, np_2, $n(1 - p_2)$ are greater than 5.

Therefore, using the argument adopted in Section 7-4 while setting a confidence interval for the proportion p of a single population, an approximate $(1 - \alpha)100$ percent confidence interval for $p_1 - p_2$ would be given by the endpoints

$$\left(\frac{x}{m} - \frac{y}{n}\right) \pm z_{\alpha/2} \sqrt{\frac{p_1(1 - p_1)}{m} + \frac{p_2(1 - p_2)}{n}}$$

where x/m is an estimate of p_1 based on a sample of size m from Population 1 and y/n is an estimate of p_2 based on a sample of size n from Population 2. From the expression above, the margin of error E is

$$E = z_{\alpha/2} \sqrt{\frac{p_1(1 - p_1)}{m} + \frac{p_2(1 - p_2)}{n}} \, .$$

But this formula for E involves p_1 and p_2, which we do not know. However, it turns out that not much error results if we use x/m and y/n, respectively, in their place. Hence, the margin of error is taken as

$$E = z_{\alpha/2} \sqrt{\frac{\frac{x}{m}\left(1 - \frac{x}{m}\right)}{m} + \frac{\frac{y}{n}\left(1 - \frac{y}{n}\right)}{n}} \, .$$

In summary, we have the following result.

LARGE SAMPLES CONFIDENCE INTERVAL FOR $p_1 - p_2$

The upper and lower limits of a $(1 - \alpha)100$ percent confidence interval for $p_1 - p_2$ are approximately given by

$$\left(\frac{x}{m} - \frac{y}{n}\right) \pm z_{\alpha/2} \sqrt{\frac{\frac{x}{m}\left(1 - \frac{x}{m}\right)}{m} + \frac{\frac{y}{n}\left(1 - \frac{y}{n}\right)}{n}}$$

where we assume that both m and n are large (at least 30). Here x/m and y/n are the sample proportions of the attribute in the two populations.

The margin of error is

$$E = z_{\alpha/2} \sqrt{\frac{\frac{x}{m}\left(1 - \frac{x}{m}\right)}{m} + \frac{\frac{y}{n}\left(1 - \frac{y}{n}\right)}{n}} \, .$$

EXAMPLE 1   A survey was conducted February 8–9, 1993, by the Gallup organization, involving 503 adult whites and 315 adult blacks 18 years or older. The following question was asked: "Do you think that police officers treat criminal suspects differently in low-income neighborhoods than in middle- or high-income neighborhoods?" The survey resulted in 312 whites responding yes and 252 blacks responding yes.

(a) Obtain a point estimate of $p_1 - p_2$, the difference in the proportions, where p_1 is the proportion among all white adults who would respond yes and p_2 is the proportion among all black adults who would respond yes.

(b) Obtain an approximate 95 percent confidence interval for $p_1 - p_2$.

SOLUTION

(a) The data provide the following information.

$$x = 312, \ m = 503; \ y = 252, \text{ and } n = 315.$$

Therefore,

$$\frac{x}{m} = \frac{312}{503} = 0.62 \quad \text{and} \quad \frac{y}{n} = \frac{252}{315} = 0.80.$$

A point estimate of $p_1 - p_2$ is $\dfrac{x}{m} - \dfrac{y}{n} = 0.62 - 0.80 = -0.18.$

(b) Since $\alpha = 0.05$, we have $z_{\alpha/2} = 1.96$. Let us next compute the margin of error.

$$E = z_{\alpha/2} \sqrt{\frac{\dfrac{x}{m}\left(1 - \dfrac{x}{m}\right)}{m} + \frac{\dfrac{y}{n}\left(1 - \dfrac{y}{n}\right)}{n}}$$

$$= (1.96) \sqrt{\frac{(0.62)(0.38)}{503} + \frac{(0.80)(0.20)}{315}}$$

$$= 0.061$$

We have seen that $\dfrac{x}{m} - \dfrac{y}{n} = -0.18$. Therefore, a 95 percent confidence interval for $p_1 - p_2$ is

$$-0.18 - 0.061 < p_1 - p_2 < -0.18 + 0.061.$$

Simplifying, we find

$$-0.241 < p_1 - p_2 < -0.119.$$

We can be 95 percent confident that $p_1 - p_2$ is between -0.241 and -0.119, that is, it is negative and considerably less than 0. This tells us a lot about the relative consensus among white and black adults. First of all we can be 95 percent confident that the proportion among whites who would respond yes is less than the corresponding proportion among blacks. It is less by at least as much as 11.9 percent, up to 24.1 percent.   ●

Section 8-3 Exercises

*Suppose two independent random samples are picked from two populations and
we wish to compare proportions of an attribute in them. In Exercises 1–3:*
*a. determine the margin of error in estimating $p_1 - p_2$ at the indicated level of
confidence*
b. obtain a confidence interval for $p_1 - p_2$ using the margin of error in Part a.

1. $x = 42$, $m = 110$ and $y = 25$, $n = 60$; 95 percent

2. $x = 17$, $m = 48$ and $y = 36$, $n = 112$; 98 percent

3. $x = 35$, $m = 86$ and $y = 62$, $n = 156$; 90 percent

4. A certain proposition was being debated in the Congress of the United States. In
   a countrywide survey, in a sample of 154 randomly selected Democrats, 92 were
   in favor of the proposition. In a sample of 118 randomly selected Republicans, 79 were in favor.
   (a) Compute a point estimate of $p_D - p_R$ where p_D and p_R are, respectively, the
       proportions among Democrats and Republicans in the population who favor
       the proposition.
   (b) Determine the margin of error in estimating $p_D - p_R$ at the 90 percent level.
   (c) Estimate $p_D - p_R$, the difference in the proportions of Democrats and Re-
       publicans, at the 90 percent level of confidence.
   (d) Interpret the interval.

5. A salesperson contacted 98 women and 148 men about a utility product. He sold
   the product to 43 women and 72 men.
   (a) Using a 95 percent level of confidence, estimate the difference in the true
       proportions among males and females who will buy the product if contacted.
   (b) Provide an interpretation of the interval.

6. In a survey to compare the true proportions of blacks and whites who favor
   government assistance for child care, it was found that among 128 blacks, 90
   favored assistance, and among 82 whites, 49 favored assistance. Suppose p_B and
   p_W are, respectively, the true proportions among blacks and whites who favor
   assistance.
   (a) Determine a 98 percent confidence interval for $p_B - p_W$.
   (b) Determine a 98 percent confidence interval for $p_W - p_B$.
   (c) Compare the intervals in Parts a and b and comment.

7. The records of the Internal Revenue Service showed that of 212 business executives picked at random from its files, 51 had defaulted on tax payments, and
   of 108 white-collar workers, also picked at random, 19 had defaulted.
   (a) Determine a 90 percent confidence interval for the following difference in
       true proportions: the true proportion of defaulting business executives *minus*
       the true proportion of defaulting white-collar workers.
   (b) Interpret the interval to compare the true proportions among the two groups.
   (c) Will the interval be narrower or wider if the confidence level is increased?

8. When a chemist tested Dynamite spray insecticide on 283 insects, 191 were
   killed. When she tested Quick spray on 198 insects, 113 were killed. Compare
   the two sprays by estimating the difference in true proportions killed by the two
   insecticides using a 95 percent confidence interval.

9. In a laboratory test, of 92 electric bulbs of Brand A, 37 lasted beyond 800 hours, and of 110 bulbs of Brand B, 33 lasted beyond 800 hours.

 (a) Determine a 90 percent confidence interval to estimate the difference in population proportions of the bulbs that last beyond 800 hours for the two brands.

 (b) What would you recommend if it was desired to increase the level of confidence and have a narrower interval than in Part a?

10. In a Gallup poll consisting of 504 males and 503 females, it was reported that 61 percent of the males wanted more strict controls and 77 percent of the females wanted more strict controls when the following question was asked: "In general, do you feel the laws covering the sale of firearms should be made more strict, less strict, or kept as they are now?" Construct a 95 percent confidence interval for the difference in proportions among males and females who want more strict controls. Interpret the interval and comment on the relative proportions among males and females.

CHAPTER 8   Summary

CHECKLIST: KEY TERMS AND EXPRESSIONS

❑ difference of means, page 457

❑ difference of proportions, page 457

❑ pooled estimate of σ, page 468

KEY FORMULAS

TABLE   Summary of Confidence Intervals (two populations)

Nature of the Population	Parameter on Which Confidence Interval Is Set	Procedure	Limits of Confidence Interval with Confidence Coefficient $1 - \alpha$
Quantitative data; variances σ_1^2 and σ_2^2 are *known*; both populations are normally distributed.	$\mu_1 - \mu_2$, the difference of the population means	Draw samples of sizes m and n from the two populations, get the respective means $\bar{x}$ and $\bar{y}$, and find the value of $\bar{x} - \bar{y}$, an estimate of $\mu_1 - \mu_2$.	$(\bar{x} - \bar{y}) \pm z_{\alpha/2} \sqrt{\dfrac{\sigma_1^2}{m} + \dfrac{\sigma_2^2}{n}}$
Quantitative data; σ_1^2 and σ_2^2 are *not* known but assumed *equal;* both populations are normally distributed.	$\mu_1 - \mu_2$, the difference of the population means	Draw samples of sizes m and n from the two populations; compute $\bar{x} - \bar{y}$ and $$s_p = \sqrt{\dfrac{(m-1)s_1^2 + (n-1)s_2^2}{m+n-2}}.$$	$(\bar{x} - \bar{y}) \pm t_{m+n-2,\,\alpha/2}\, s_p \sqrt{\dfrac{1}{m} + \dfrac{1}{n}}$

Nature of the Population	Parameter on Which Confidence Interval Is Set	Procedure	Limits of Confidence Interval with Confidence Coefficient $1 - \alpha$
Quantitative data; σ_1^2 and σ_2^2 are *not* known and *not* assumed equal; populations may *not* be normal; *sample sizes m and n are large.*	$\mu_1 - \mu_2$, the difference of the population means	Draw samples of sizes m and n; compute $\bar{x} - \bar{y}$; compute s_1^2 and s_2^2.	$\bar{x} - \bar{y} \pm z_{\alpha/2} \sqrt{\dfrac{s_1^2}{m} + \dfrac{s_2^2}{n}}$ Confidence interval is approximate.
Qualitative data; p_1 is the proportion in Population 1 and p_2 in Population 2; sample sizes m and n are assumed large.	$p_1 - p_2$, the difference of the population proportions	Draw two samples, one from each population, of sizes m and n; find x/m and y/n, the estimates of p_1 and p_2 respectively; then get $x/m - y/n$.	$\left(\dfrac{x}{m} - \dfrac{y}{n}\right) \pm$ $z_{\alpha/2} \sqrt{\dfrac{\dfrac{x}{m}\left(1 - \dfrac{x}{m}\right)}{m} + \dfrac{\dfrac{y}{n}\left(1 - \dfrac{y}{n}\right)}{n}}$ Confidence interval is based on the central limit theorem, thus is approximate.

CHAPTER 8 Concept and Discussion Questions

1. Suppose two surveyors measure the height of a monument. The first surveyor makes 16 measurements in June and the second surveyor make 12 measurements in August.
   (a) Is it reasonable to assume that the measurements of each surveyor are independent?
   (b) If the surveyors have no access to each other's measurements, are the measurements of the two surveyors independent?

2. Suppose it is desired to compare two instruments used for measuring heights. Sixteen surveyors recorded the height of a monument using one instrument and twelve surveyors recorded it using the second instrument.
   (a) Is it reasonable to assume that the readings on each instrument are independent?
   (b) Are the measurements from the two instruments independent?

3. Give an example where data are picked from two populations and the resulting samples are independent.

4. Give an example where data are collected from two populations but, based on your experience, the two samples are dependent.

5. A researcher wishes to compare the means of two populations by estimating the difference in the population means. If the researcher comes to you for advice, what questions would you ask the researcher so that, depending on the validity of assumptions in any context, you can provide appropriate guidance? (Think of yourself as a medical doctor who elicits information from the patient about the symptoms in order to prescribe proper medication.)

In Exercises 1–5, find a confidence interval for the indicated difference of parameters at the specified level of confidence. Wherever appropriate, mention the assumptions that would be considered necessary.

1. 98 percent for $\mu_1 - \mu_2$ if $\bar{x} = 6.8$, $m = 6$, $s_1^2 = 2.7$ and $\bar{y} = 4.5$, $n = 10$, $s_2^2 = 3.9$

2. 95 percent for $\mu_2 - \mu_1$ if $\bar{x} = 1.8$, $m = 10$, $\sigma_1^2 = 1.69$ and $\bar{y} = -3.0$, $n = 16$, $\sigma_2^2 = 2.32$

3. 98 percent for $p_1 - p_2$ if $x = 28$, $m = 72$ and $y = 40$, $n = 90$

4. 92 percent for $p_2 - p_1$ if $x = 42$, $m = 100$ and $y = 53$, $n = 125$

5. 90 percent for $\mu_1 - \mu_2$ if $\bar{x} = -3.2$, $m = 48$, $s_1^2 = 7.6$ and $\bar{y} = -6.4$, $n = 60$, $s_2^2 = 5.9$

6. Eight fertilized eggs were incubated for 14 days and six fertilized eggs were incubated for 18 days. When a determination was made of the amount (in milligrams) of total brain actin, the following figures were obtained.

Incubated 14 days	1.2	1.4	1.5	1.2	1.4	1.7	1.5	1.7
Incubated 18 days	1.3	1.8	1.9	2.0	1.5	2.2		

   (a) Assuming a normal distribution of total brain actin for the two incubation periods, find a 95 percent confidence interval to estimate $\mu_{14} - \mu_{18}$, where the subscripts stand for number of days of incubation.

   (b) Besides the assumption that the populations are normally distributed, what other assumptions did you need to make?

7. The following information was collected on the amount of insulin released when pancreatic tissues of some animals were treated in a laboratory with two drugs, Drug A and Drug B. Seven tissues were treated with Drug A and eight with Drug B.

	Sample Mean	Sample Variance
Drug A	5.24	0.813
Drug B	7.06	1.572

   (a) Assuming that the amount of insulin released is normally distributed with the same variance for both drugs, set a 90 percent confidence interval for $\mu_A - \mu_B$.

   (b) Give an interpretation of the interval you have obtained.

   (c) If the confidence level is desired to be increased, would the width of the interval be increased or decreased?

8. The following table gives calorific values of dry organic matter ashed with and without benzoic acid.

Ashing with Benzoic Acid (cal/gm)	Ashing without Benzoic Acid (cal/gm)
4729	4327
5416	5233
5318	4383
4626	4675
4781	4428
5508	
4962	

(a) Assume that calorific values are normally distributed in both situations. Suppose μ_B and μ_N are, respectively, true mean calorific values when ashed with and ashed without benzoic acid. Find a 98 percent confidence interval for estimating $\mu_B - \mu_N$.

(b) Are there any other assumptions that you need to make?

(c) Interpret the confidence interval to compare the calorific values of dry organic matter with and without benzoic acid.

(d) If it is desired to have a narrower confidence interval without lowering the level of confidence, what would you recommend?

9. Out of 400 Europeans interviewed, 123 were of blood type O, and out of 200 Japanese, 57 were of that blood type. Set a 90 percent confidence interval for the difference in true proportions of O blood types for the two groups.

10. Five blue-collar workers and six white-collar workers, interviewed regarding the amount of time they spent watching television during a weekend, gave the following figures (in hours).

Blue-collar workers	10	9	14	4	8	
White-collar workers	13	7	8	6	10	8

(a) Determine a 95 percent confidence interval to estimate the difference in the true mean times for the two types of workers.

(b) State the underlying assumptions in your approach.

(c) Recommend a confidence level so that the width of the resulting confidence interval will be narrower than in Part a.

11. A social worker who wanted to compare racial tolerance among rural dwellers and city dwellers interviewed 52 rural dwellers and 58 city dwellers. Each person answered 20 questions in a questionnaire. On each question the respondent was assigned a score of 0, 1, 2, . . . , or 5 on an increasing scale, 0 showing low tolerance and 5 showing high tolerance. The summary of the response was as follows.

	Rural Dwellers	City Dwellers
Mean	71.5	74.2
Standard deviation	7.2	8.9

Determine a 95 percent confidence interval to estimate the difference $\mu_R - \mu_C$ where μ_R and μ_C are, respectively, the true mean tolerance scores for rural and city dwellers. Interpret the interval to compare the tolerance among the two groups.

In Exercises 12–16, state clearly the underlying assumptions when applying a particular formula to determine the confidence interval. Also, provide an interpretation of the confidence interval that you find.

12. The saponin content of a variety of black gram, a legume, was determined by two methods: cooking of soaked seeds and cooking of unsoaked seeds. The following summary figures were reported.

	SAPONIN (mg/100 g)	
	Soaked Seeds	**Unsoaked Seeds**
Mean	2275	2318
Standard deviation	56	67
Number	16	14

Use the information to construct a 98 percent confidence interval for the difference in the true means of saponin under the two methods.

13. Urea that occurs naturally as a minor component of milk has a significant effect on its heat stability. The following experiment was conducted to compare two diets for their effect on the urea concentration of milk. Twenty-two cows that had calved approximately a month earlier were selected for the experiment. Diet 1 was given to ten cows for four weeks and Diet 2 was given to the remaining twelve cows for the same length of time. The urea concentration in the milk (g/kg) was measured for each cow at the end of the period.

UREA CONCENTRATION (g/kg)	
Diet 1	**Diet 2**
0.45	0.36
0.42	0.37
0.40	0.38
0.49	0.43
0.45	0.40
0.43	0.42
0.50	0.43
0.40	0.39
0.44	0.35
0.48	0.41
	0.37
	0.45

Set a 98 percent confidence interval for the difference between the true means of urea concentration under the two diets.

14. The following figures refer to the weight (g) of natural food zooplankton in the intestines of fifteen carp grown in a lake and twelve grown in a bay.

ZOOPLANKTON WEIGHT (g)	
Lake	**Bay**
212	181
196	138
186	163
185	156
171	158
173	146
182	175
193	168
160	167
192	164
148	155
191	168
188	
178	
197	

Obtain a 95 percent confidence interval to estimate the difference in the true means of intake of zooplankton by carp grown in the lake and in the bay.

15. To investigate whether cultivation of cocoa leads to a degradation of topsoil, soil samples from forest areas and areas under cocoa plantation were analyzed for the soil properties *(Journal of Environmental Management* 1987; 25: pp. 61–70). Magnesium (mg/100 g soil) was one of the ingredients considered. The findings of the analysis are summarized in the following table.

	MAGNESIUM (mg/100 g soil)	
	Forest	**Cocoa**
Mean	3.8	2.2
Standard deviation	0.929	0.989
Number	15	13

Form a 90 percent confidence interval for the difference between the true mean magnesium content of the forest soil and that of the soil under cocoa cultivation.

16. The following data contain the summary of total lipid content (g) of smoke dried beef depending on whether the beef was cooked in water or pressure cooked prior to smoke drying.

	LIPID CONTENT (g)	
	Pressure Cooked	**Water Cooked**
Mean	8.50	9.58
Standard deviation	0.6	0.8
Number	15	12

Estimate the difference in the true means of lipid content of smoke dried beef under the two methods, using a 95 percent confidence interval.

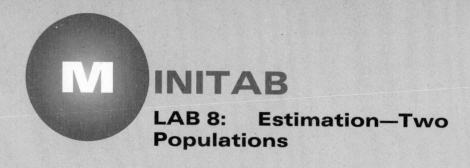

MINITAB

LAB 8: Estimation—Two Populations

In this lab we will consider confidence intervals for the difference of population parameters.

Log on and access MINITAB.

1 First we consider confidence intervals for the difference of population means using the z statistic. MINITAB does not have a command that directly computes the confidence interval in this case. So we write a program to compute one when the summary of the data provides the sample sizes, the sample means, and the population variances.

```
MTB >STORE 'ZMUDIFF'

STOR>NOECHO

STOR>NOTE K1 IS X-BAR, K2 IS Y-BAR

STOR>NOTE K3 IS SIGMA 1, K4 IS SIGMA 2

STOR>NOTE K5 IS M, K6 IS N, K7 IS Z VALUE

STOR>NOTE VALUES K1-K7 SHOULD BE PROVIDED

STOR>LET K11 = K1 - K2

STOR>LET K12 = SQRT (K3 ** 2/K5 + K4 ** 2/K6)

STOR>NOTE K15 IS LOWER CONFIDENCE LIMIT

STOR>LET K15 = K11 - K7 * K12

STOR>PRINT K15

STOR>NOTE K16 IS UPPER CONFIDENCE LIMIT

STOR>LET K16 = K11 + K7 * K12

STOR>PRINT K16

STOR>END OF PROGRAM
```

We now use the preceding program to solve Example 1 in Section 8-1. We enter the constants K1–K7 as follows.

```
MTB >LET K1 = 140

MTB >LET K2 = 145

MTB >LET K3 = SQRT(25)

MTB >LET K4 = SQRT(30)

MTB >LET K5 = 15

MTB >LET K6 = 10

MTB >INVCDF 0.9 AS K7;

SUBC>NORMAL MEAN 0 SIGMA 1.

MTB >EXECUTE 'ZMUDIFF'
```

Note that K3 and K4 in the stored program stand for the standard deviations, and because we are given the variances in Example 1, we had to enter them as square roots.

The output will display the upper and lower limits of the confidence interval. Check that these values are the same as those presented in the solution of Example 1.

Assignment   Work out Exercises 9 and 20 in Section 8-1.

2   Next, we deal with the more realistic case, in which the population variances are not known.

The basic command to compute a confidence interval for the difference in population means when the variances are not known is TWOSAMPLE-T. Under this command we have to specify the level of confidence and the columns informing where the samples from the two populations are located. When the command is executed, MINITAB prints out the confidence interval without assuming that the variances are equal. The formula for the degrees of freedom is rather complicated, but it is handled internally and the result for the confidence interval is presented.

If, however, the variances may be assumed to be equal, so that we can perform pooled analysis, MINITAB has the subcommand POOLED. This is the type of problem we have treated in Section 8-2. The MINITAB procedure for solving Example 2 in Section 8-2 is as follows. We use the subcommand POOLED with the main command TWOSAMPLE-T. Notice that under the main command we specify the level of confidence and the location of the two samples.

```
MTB >NOTE EXAMPLE 2 IN SECTION 8-2

MTB >NOTE ASSUMPTION VARIANCES ARE EQUAL

MTB >SET DATA IN C1

DATA>40    49    38    48    40

DATA>END OF DATA

MTB >SET DATA IN C2

DATA>50    41    53    39    40    47

DATA>END OF DATA

MTB >PRINT C1    C2

MTB >TWOSAMPLE-T 0.98 FOR SAMPLES IN C1 C2;

SUBC>POOLED.
```

You will see that the confidence interval agrees with the answer provided in the solution to Example 2, Section 8-2.

Assignment   Work out Exercises 10 and 17 in Section 8-2.

We now type STOP and conclude the MINITAB session using the logging off procedure.

Commands you have learned in this session:

```
TWOSAMPLE-T;

 POOLED.
```

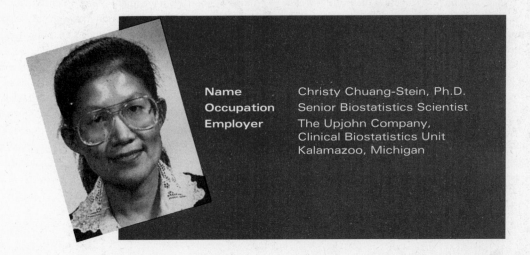

Name Christy Chuang-Stein, Ph.D.
Occupation Senior Biostatistics Scientist
Employer The Upjohn Company,
Clinical Biostatistics Unit
Kalamazoo, Michigan

Both of my parents are educators and enjoy great respect from their children. At a very young age, I decided to grow up to be just like them and became a teacher, too. Thanks to my mother who spent numerous hours with me during my summer vacations in grade school, I became not only comfortable with numbers but also very fond of them. My interest in mathematics led to a major in the subject. Toward the latter part of my college years, I grew more interested in the applications of math, especially the inferential power of mathematical thinking. This, plus my fondness for numbers, influenced my graduate study in statistics. As the years went by, my aspiration to become a teacher narrowed to a career as a university professor. I patted myself on the back the day I began teaching statistics at the University of Rochester (Rochester, New York) for the fulfillment of my childhood career dream.

My career path took a turn when I joined The Upjohn Company in 1985. The Upjohn Company is a drug company whose primary mission is to develop safe and effective drugs. Before a drug can be marketed for general public consumption, it must undergo various tests on animals and human subjects. Testing on humans in a clinic setting is called a clinical trial. My job at The Upjohn Company consists of helping physicians design clinical trials, ensuring that data necessary for the objectives of the trials are collected, and finally analyzing the data when the trials are completed. Clinical trials have different objectives. Some are to study the safety of the drug and others to find the optimal dose and schedule. Large clinical trials usually include a standard therapy for the underlying disease. We randomize patients to receive either the test drug or the standard therapy. The goal is to evaluate the safety and the efficacy of the test drug in a controlled environment. Depending on the study designs, statistical techniques ranging from the paired t test and analysis of variance to more sophisticated response surface modeling are used.

My current job has a research component. I find it gratifying to be able to develop new statistical methods that can facilitate the drug development process. The methods can be efficient study designs that require fewer study participants or new analytical proceduces that result in more powerful tests. I also find it exciting to work on a promising drug because of its potential impact on human beings.

I have taught elementary statistics to students like you. Some of you might react to the course the same way that many people do when they find out what I do for a living. Unfortunately, *number scare* has deprived many students of the joy of learning what statistics is about. In addition, *computer scare* has diminished the chance for many to realize that numbers can be quite manageable and fun. For those of you who have spent hours practicing your swing on a golf course or your turn on a downhill ski slope, you know very well that only practice can bring familiarity, which in turn will bring confidence, fun, and pride. There is no myth about numbers and there is no myth about computers. They are tools to help us do our jobs better. Enroll in a statistics course with an open mind, attend the lectures regularly, be diligent in your practice, and find ways to apply your newly learned knowledge in the real world. You will find that statistics is not as formidable as many people think. In the end, you are the one who will benefit from your own effort.

Tests of Hypotheses

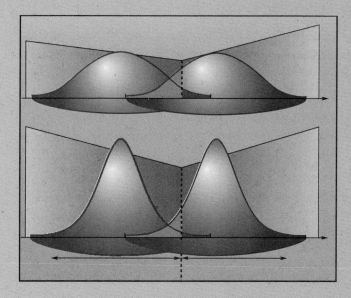

INTRODUCTION

I n Chapters 7 and 8 we considered one aspect of statistical inference, the estimation of population parameters. We will now turn our attention to another aspect called *hypothesis testing*. The main objective here is to formulate rules that lead to decisions culminating in acceptance or rejection of statements about the population parameters.

On a cloudy day, a man looks out the window and says, "It is going to rain today." What that man has done is make a statement forecasting the weather for that day. We will say that he has made a hypothesis regarding the weather. He is now faced with the following decision problem. Should he or should he not take his raincoat? Whatever his action, it is going to result in precisely one of the following four situations: He takes his raincoat and it rains (a wise decision); he does not take his raincoat and it does not rain (again, a wise decision); he takes his raincoat and it does not rain (a wrong decision); and, finally, he does not take his raincoat and it rains (a wrong decision). The last two situations are not desirable in that the actions result in erroneous decisions. Ideally, what this man would like is to have his raincoat with him if it does rain and not to have it with him if it does not rain. However, he will not know if it will rain when the day begins, and he must make his decision before the day begins. Before starting from home, he will have to take into account such factors as how overcast the sky is, his past experience, and so on, and balance the consequences of erroneous decisions. If he is highly susceptible to catching colds, he might prefer to err on the side of taking the raincoat with him even if it does not rain that day. On the other hand, if he is physically fit and prone to misplacing his raincoat, he might prefer to leave the raincoat home.

Ideas of this nature crop up in connection with statistical hypothesis testing. Fortunately, our intuition will guide us well in our discussion.

9-1   THE JARGON OF STATISTICAL TESTING

WHAT IS A HYPOTHESIS?

An important goal of a research investigation is to verify a theory. In connection with a statistical investigation, it amounts to verifying some aspects of a population under study.

A **statistical hypothesis** is a statement, assertion, or claim about the nature of a population. It is basic to any experimental inquiry.

Numerous examples of a person hypothesizing about a situation can be cited. Anyone who has watched commercial TV cannot fail to be aware of the

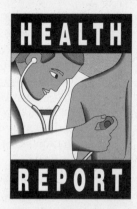

The Good News

√Cranberry juice really does protect against urinary-tract infections, as women have long believed and a scientific study now confirms. As it happens, the study was funded by Ocean Spray, which provided the juice.

√Children who drink caffeinated soda show significant improvements in tests of attention and manual dexterity, a new study reports. Not surprisingly, the kids also said the caffeine made them feel less sluggish—and more anxious.

√Fat consumption in the U.S. continues to fall—down to 34% of daily calories in 1990 from 42% in the mid–1960s. But Americans are eating more rather than fewer calories, and they grow heavier every year.

The Bad News

√The number of new AIDS cases surged unexpectedly last year, more than doubling, owing largely to a jump in infections among heterosexuals. The numbers had been expected to go up because of a change in reporting procedures, but not by this much. The increase was greater among women (151%) than among men (105%). The biggest increases of all were among teens and young adults.

√People who wear contact lenses overnight are more than eight times as likely to get eye infections as those who don't, even if the lenses are the so-called extended-wear type, according to a new study. Nearly 75% of those cases could be prevented by taking the lenses out at night.

constant barrage of claims. Brand X detergent will wash white clothes sparkling white; with a supergasoline, your car will get more miles to the gallon than before; a majority of dentists prefer Brand A toothpaste, and so on. A regular feature in the magazine *Time* is health news replete with claims such as those contained in the clipping which appeared in the issue of March 21, 1994. (See page 493.)

In the framework of statistical investigation, very often a hypothesis takes the form of stipulating the values of the unknown parameters of the population being studied. For example, consider the following.

a.   A coin is unbiased. If P (head) $= p$, the hypothesis is that $p = \frac{1}{2}$.

b.   A tire manufacturer claims that the new radial tire produced by the company will give more than 40,000 miles of trouble-free driving. If μ represents the mean number of miles, the hypothesis that the manufacturer might have in mind is $\mu > 40,000$.

c.   A social scientist asserts that students with an urban background perform better than those with a rural background. Suppose the measure used is the percent of students who graduate from high school. If p_U and p_R represent, respectively, the population proportions of students with urban and rural backgrounds who graduate, the hypothesis is that $p_U > p_R$.

d.   An engineer claims that a machine manufactures bolts within specifications. If she uses the standard deviation of the diameter of the bolts as a measure, the hypothesis may be that $\sigma < 0.1$ inch.

An investigator involved in the study of a research problem will often hypothesize a position. This hypothesized position is the *research hypothesis* that the investigator believes to be true. In statistical jargon the research hypothesis is called the *alternative hypothesis* and is commonly denoted as H_A. For example, an agronomist might believe and claim that a new technique of cross fertilization will give increased grain production. Specifically, suppose the claim is that μ, the mean grain production per acre, will be more than 180 bushels per acre. We take this claim as the alternative hypothesis and write this formally as

$$H_A: \mu > 150.$$

The hypothesis that counters the claim of the researcher and negates it is called the *null hypothesis*. It is a common convention to denote it as H_0. So, with regard to the agronomist's claim, the null hypothesis will be that the new technique of cross fertilization does not increase grain production. We write this as

$$H_0: \mu \leq 150.$$

The null hypothesis and the alternative maintain two mutually exclusive positions and so cannot both be true at the same time. The null hypothesis declares, in essence, that there is no merit to the researcher's claim. To put it simply, the alternative hypothesis conjectures that something out of the ordinary is happening, whereas the null hypothesis asserts that nothing unusual is happening.

The only way to establish the absolute truth or falsity of a particular hypothesis is by examining the entire population. Because usually this is not possible, we collect sample data for drawing conclusions. It is the fond hope of any researcher that the collected data will provide strong evidence supporting the research hypothesis, thereby leading to the rejection of the null hypothesis. But as will be expalined later, the philosophy of testing hypotheses is such that the null hypothesis will be rejected only under compelling evidence against it.

The hypothesis that a researcher believes to be true and expects to establish on the basis of collected data is called the **alternative hypothesis** or the **research hypothesis** and is denoted as H_A.

The hypothesis that negates the claim laid under the alternative hypothesis is called the **null hypothesis** and is denoted as H_0. Rejection of this hypothesis leads to the acceptance of the alternative hypothesis.

Important   The null and the alternative hypotheses are statements about population parameters and not sample statistics. For example, it would not make any sense to write H_0: $\overline{X} = 55$.

EXAMPLE 1

The promoters of a diet plan advertise that a conscientiously followed plan of their diet will result in mean loss of weight of at least 35 pounds in 3 months. State the null and the alternative hypotheses for testing whether the diet is effective.

SOLUTION   Let μ be the mean weight loss per person in 3 months. The claim of the promoter is that $\mu > 35$ pounds. So this is the alternative hypothesis. The null hypothesis negates the claim under the alternative hypothesis and, as a result, we write the two hypotheses as

$$H_0: \mu \leq 35.$$
$$H_{A:} \mu > 35.$$

●

EXAMPLE 2

Ten years ago twenty percent of high school teenagers smoked cigarettes. Recently, a government official in the health department in Washington, D.C., made the following statement: "Today, after ten years, smoking is on the rise among high school teenagers." State the null and alternative hypotheses for testing the official's statement.

SOLUTION   The parameter of interest is p, the proportion of high school teenagers who smoke cigarettes today. The research hypothesis is $p > 0.2$, since this is the official's contention which she believes will be supported by the data. We give the two hypotheses as

$$H_0: p \leq 0.2.$$
$$H_A: p > 0.2. \qquad \bullet$$

Since the null and the alternative hypotheses hold to two mutually exclusive positions, there is no overlap between the set of parameter values stipulated under the null hypothesis and those stipulated under the alternative hypothesis. Also, notice that in all the above situations the null hypothesis includes the equality sign. Actually, for our purpose of testing hypotheses, it is relevant that we state the null hypothesis with equality, giving a specific value which is nearest to those given under H_A.

For example, when the alternative hypothesis is $H_A: \mu < 40$, the values of μ that are not included in H_A are $\mu \geq 40$. One would usually state the null hypothesis as $H_0: \mu \geq 40$. However, it is sufficient for us to write $H_0: \mu = 40$. The rationale, in this case, is that if we go along with $H_A: \mu < 40$ by rejecting H_0 when it states that $\mu = 40$, then we could certainly go along with $H_A: \mu < 40$ if H_0 states that μ has a value greater than 40.

The research hypothesis is often couched in terms of a question. For instance, in Example 1 above, the question could be "Will the diet plan result in mean loss of weight of more than 35 pounds in 3 months?"; in Example 2, the question could be "Is smoking on the rise among high school teenagers today?" In each case, the null hypothesis denies the claim posed in the question. Thus, it makes good sense to pose the question appropriately, depending upon what the researcher is interested in establishing.

EXAMPLE 3

A cable manufacturer has to supply cables to a company that manufacturers electric motors. It is required that the cables should have mean cross-sectional diameter of 0.4 inches. If the mean diameter is different from 0.4 inches, then the settings will need to be adjusted. After initial production, the production engineer wonders: "Do the settings need to be adjusted?" Give the null and the alternative hypotheses.

SOLUTION   Let μ be the mean cross-sectional diameter of the cables. In view of the question, the alternative hypothesis is "the settings need to be adjusted" and the null hypothesis is "the settings do not need to be adjusted."

Now, the settings have to be adjusted if μ is less than or greater than 0.4 inches. So we set the null and the alternative hypotheses as

$$H_0: \mu = 0.4$$
$$H_A: \mu \neq 0.4$$

In summary, the formulation of H_0 and H_A is guided as follows:

FORMULATION OF H_0 AND H_A

When we wish to establish a statement about the population with evidence obtained from the sample, the negation of the statement is what we take as the null hypothesis H_0. The statement itself constitutes the alternative hypothesis H_A.

In approaching the problem of testing a statistical hypothesis, our attitude will be to assume initially that the null hypothesis H_0 is correct. As might be expected, any decision must be based on the evidence provided by the sample data. But a common feature of any collected data is that they are subject to variability, and even when H_0 is true, the data will not, in general, be completely in conformity with the hypothesis. Based on the assumption that H_0 is true, we calculate how probable the results of the experimental data are. If the results are highly improbable, we have strong indication that the data are not compatible with H_0. We take the evidence as grounds to reject H_0 in favor of H_A. If the results of the experimental data are reasonably probable under H_0, the status quo prevails in that we have no reason to believe differently. Our decision will be to not reject H_0. It is the extreme nature of the experimental or field data that will lead us to act and take a decision to reject H_0.

A **test of the null hypothesis** H_0 is a rule or procedure that leads to a decision to reject or not to reject H_0 when the sample values are analyzed. In any given situation the rule is based on the value of an appropriate statistic called the **test statistic**. The rule often is referred to as a *decision rule* or a *test of significance*.

TYPE I AND TYPE II ERRORS

In any hypothesis-testing problem, because we take action based on incomplete information, there is a built-in danger of an erroneous decision. A statistical test procedure based on sample data will lead to precisely one of the following four situations. Two of these situations will entail correct decisions and the other two, incorrect decisions.

1. H_0 is true and H_0 is not rejected—a correct decision.
2. H_0 is true and H_0 is rejected—an incorrect decision.
3. H_0 is false and H_0 is not rejected—an incorrect decision.
4. H_0 is false and H_0 is rejected—a correct decision.

Rejection of the null hypothesis when in fact it is true is called a **Type I error** or a *rejection error*. The probability of committing this error is denoted by the Greek letter α (alpha) and is referred to as the **level of significance** of the test.

Non-rejection of H_0 when it is false is called a **Type II error** or an *acceptance error*. The probability of making this error is denoted by the Greek letter β (beta).

The four possibilities mentioned are summarized in Table 9-1.

TABLE 9-1   Four Possibilities Based on the Decision Taken and the Truth or Falsity of H_0

	TRUE STATE OF NATURE (UNKNOWN)	
Test Procedure Conclusion	H_0 **Is True**	H_0 **Is False** **(H_A Is True)**
Do not reject H_0	Correct decision; probability is $1 - \alpha$	Incorrect decision; Type II error; probability is β
Reject H_0 (accept H_A)	Incorrect decision; Type I error; probability is α	Correct decision; probability is $1 - \beta$

Ideally, we would like to have both α and β very small so that a true hypothesis H_0 is rejected as rarely as possible, while, at the same time, a false hypothesis H_0 is rejected as often as possible. In fact, if it were possible, we would eliminate both these errors and set their probabilities equal to zero. However, once the sample size is agreed upon, there is no way to exercise simultaneous control over both errors. The only way to accomplish a simultaneous reduction of both the errors is to increase the sample size, and if we want both α and β equal to zero, to explore the entire population.

To understand the basic approach to hypothesis testing, we might consider the familiar presumption under our judicial system, "The accused is innocent until proven guilty beyond a reasonable doubt." Is the accused guilty? That is the question. We state the null hypothesis as

H_0: The accused is not guilty.

The alternative hypothesis is

H_A: The accused is guilty.

It is up to the prosecution to provide evidence to destroy the null hypothesis. If the prosecution is unable to provide such evidence, the accused goes free. If the null hypothesis is refuted, we accept the alternative hypothesis and declare that the accused is guilty. Bear in mind that if the accused goes free, it does not mean that the accused is indeed innocent. It simply means that there was not enough evidence to find the accused guilty. Nor, if the accused is convicted, does it mean that the accused did indeed commit the crime. It simply means that the evidence was so overwhelming that it is highly improbable that the accused is innocent. Only the accused knows the truth.

In this context, suppose the accused is innocent, in fact, but is found guilty. Then a Type I error has been made because the null hypothesis has been rejected erroneously. Thus, the probability of convicting the innocent would be α, and we would like to keep this value rather low. On the other hand, if a guilty person is declared not guilty, a Type II error has been made with probability β.

When we reject the null hypothesis, we have not proved that it is false, for no statistical test can give 100 percent assurance of anything. However, if relative to the body of evidence provided by the data, we reject H_0 with a small probability α of Type I error, then we are able to assert that H_0 is false and H_A is true *beyond a reasonable doubt*. Thus, in any test procedure, it makes good sense to let α be small. The magnitude of α should depend on how serious the consequences of making an error of Type I are.

A further understanding of the roles of the null and alternative hypotheses will be developed in the following Section 9-1 Exercises.

Section 9-1 Exercises

In Exercises 1–6 identify the null hypothesis and the corresponding alternative hypothesis.

1. A computer company claims that less than one in one thousand of its chips are defective and can establish it using sample data. Use the parameter p, which stands for true proportion of defective chips.

2. An environmentalist organization believes that it can establish statistically that a company that produces PCB for electric insulation is discharging, on the average, more than 6 parts per million of PCB in a water stream. Use the parameter μ which represents the true mean amount of PCB discharged.

3. A medical researcher claims that she can establish on the basis of data that a new treatment can extend beyond 3 years the true mean survival time μ of patients having a particular type of cancer.

4. A factory produces timing devices that are supposed to have a true mean timing advance of 0 seconds in a certain period of time. A quality control engineer working for the factory disagrees with the production engineer and believes that he can establish that the production process needs more fine tuning. Use the parameter μ, the true mean timing advance.

5. The president of a large corporation claims that he can establish that at least 60 percent of the shareholders are in favor of its acquisition of a company. Use the parameter p, the true proportion of shareholders in favor of acquisition.

6. A research biologist working for a bioengineering firm claims that, statistically, the majority of the U.S. population is indifferent to the use of hormones to induce milk production in cows. Use the parameter p, the true proportion of the U.S. population indifferent to the use.

7. Suppose a drug company has developed a new serum that could help prevent a disease. The question is whether the serum is effective. The course of action open is whether to market the serum or not to market it.
   (a) What are the two errors that could be committed?
   (b) As far as the drug company is concerned, which error is more serious?
   (c) As far as the Federal Drug Administration (which protects consumer interests) is concerned, which error is more serious?

8. Referring to Exercise 7, in deciding whether to market the serum, suppose the company does not want to miss the opportunity of marketing an effective drug. Explain why the company would state the hypotheses as follows.

   H_0: The serum is effective.

   H_A: The serum is ineffective.

9. Referring to Exercise 7, if the Federal Drug Administration is interested in testing the claim of the drug company, state how the null and the alternative hypotheses should be formulated.

10. Mr. and Mrs. Diaz have inherited some money and plan to invest it in a business venture. It is possible that the business will flourish. By the same token, it is quite possible that it might fail. Mr. and Mrs. Diaz formulate the null and alternative hypotheses as follows.

    H_0: The business venture will not flourish.

    H_A: The business venture will flourish.

    State the Type I and Type II errors in this context. From the viewpoint of Mr. and Mrs. Diaz, which error do you suppose would be more serious?

11. Suppose you are a cigarette smoker who wants to minimize your risk of cancer. You feel that if the mean nicotine content is less than 1 milligram, it is safe for you to smoke that brand. If a new cigarette is introduced on the market and you wish to decide by statistical testing whether it is safe for you to switch to this brand, how would you formulate the null and the alternative hypotheses? (Remember, you are not too eager to switch to a new brand.)

12. Suppose the probability of Type I error is 0.08. Decide whether the following statements are true or false.
    (a) The probability of rejecting the null hypothesis is 0.08.
    (b) The probability of rejecting the null hypothesis when it is true is 0.08.
    (c) The probability of Type II error is 0.92.
    (d) The probability of Type II error is a number between 0 and 1, both inclusive.

(e) The probability of Type II error will depend on the alternative hypothesis.

(f) The probability of accepting the null hypothesis is 0.92.

(g) The probability of accepting the null hypothesis when it is true is 0.92.

13. For each of the following, state what action would constitute a Type I error and what action would constitute a Type II error. (In each case, take H_A to be the negation of the statement under H_0.)

(a) H_0: The sulphur dioxide content in the atmosphere is at a dangerous level.

(b) H_0: The new process produces a better product than the old process.

(c) H_0: Most Americans favor gun control.

(d) H_0: The mean velocity of light in a vacuum is 299,795 km/sec.

(e) H_0: Eighty percent of doctors recommend aspirin for headaches.

(f) H_0: A geographic basin with a certain type of rock formation is oil-bearing.

(g) H_0: The life expectancy of people living in a certain geographic region is 52 years.

(h) H_0: Supersonic transport will affect the ozone level in the stratosphere.

(i) H_0: The old factory should be scrapped and a new factory should be installed.

(j) H_0: Nuclear power does not pose hazards to the public.

(k) H_0: The diet will help reduce weight.

14. An industrial engineer tested two types of steel wires to test H_0: (the mean strengths of the two types of steel wires are the same) against H_A: (the mean strengths are different). It is revealed subsequently that in coming to his conclusion, the engineer committed a Type II error. What is the actual state of affairs and what was the decision taken?

15. In Exercise 14, suppose in the long run the engineer does not mind if 10 percent of the times he decides erroneously that the two types of wires are different when in fact they are not. What type of error are we alluding to? What is its approximate probability?

16. An educational testing service wanted to compare the mean IQ of university students in a certain geographic region with that of the rest of the nation. The null hypothesis is given as follows:

H_0: The mean IQ of the region is less than or equal to the mean for the rest of the nation.

After carrying out the statistical analysis, the following conclusion is reached: At the 5 percent level of significance, the mean IQ of the region is higher. Interpret the meaning of the conclusion and the stated level of significance.

17. Widespread skepticism resulted from the report, in June 1985, that the body which had been exhumed from a graveyard in Embu, Brazil, was that of Joseph Mengele, the infamous Nazi doctor. After a team of international forensic experts examined the skeletal remains, Dr. Lowell Levine, a forensic specialist from the U.S. Justice Department, announced the experts' unanimous conclusion: "The skeleton is that of Joseph Mengele within a *reasonable scientific certainty.*" Comment on this announcement in light of the likely null hypothesis, alternative hypothesis, and the level of significance.

9-2   THE MECHANICS OF CARRYING OUT TESTS OF HYPOTHESES

In this section we shall formulate decision rules based on the nature of the alternative hypothesis and the levels of the two types of errors.

Let us consider the following situation: Suppose a factory manufactures electric bulbs. An engineer employed by the factory believes that she has developed a new manufacturing technique that will increase the life of the bulbs which currently are believed to have a mean life of 450 hours. We shall assume that there is evidence that the length of time that a bulb lasts before burning out is normally distributed and that the standard deviation is 60 hours and has not changed under the engineer's invention.

Putting the problem formally in a statistical framework, we have two hypotheses. The engineer is making a claim, so let her provide the evidence; it is the research hypothesis. We will take the stand that there is no improvement under the new technique. Our attitude is that the engineer should provide reasonably convincing evidence that will result in rejecting this stand. Thus, we set up the hypotheses as follows.

Null hypothesis H_0: $\mu = 450$, that is, the new technique does not result in improvement

Alternative hypothesis H_A: $\mu > 450$, that is, the new technique *does* result in improvement

To test the engineer's claim, let us agree to test nine bulbs produced using the new technique and record the number of hours these bulbs last before burning out. Suppose the following results are obtained.

583   518   464   384   445   534   549   563   376

If we examine the data, 5 of the bulbs lasted well beyond the hypothesized 450 hours under H_0. On the other hand, there were 3 bulbs that lasted less than 450 hours, with two bulbs lasting considerably less than 450 hours. How do we interpret these data and take a decision? The mean life of the bulbs is $\bar{x} = 490.67$ and is certainly greater than 450. Could this be due to random fluctuations? In other words, are the bulbs really from a population of bulbs where the mean life *is* 450 hours, but our sample yields, by chance, a value 490.6 for the sample mean? Or, is the observed mean large because the new technique really is superior and produces longer-lasting bulbs?

P-value

Let us find how probable it is to get a sample mean of 490.67 or greater if, in fact, H_0 is true, that is, the population mean is 450 hours. The above data were artificially simulated from a normal distribution with standard deviation 60.

So, if H_0 is true, the distribution of $\bar{X}$ is normal with mean 450 and standard deviation $\sigma/\sqrt{n} = 60/\sqrt{9} = 20$ hours. Therefore,

$$P(\bar{X} > 490.67) = P\left(Z > \frac{490.67 - 450}{20}\right)$$
$$= P(Z > 2.03)$$
$$= 0.0212$$

We find that the probability of getting a sample mean 490.67 or larger, if H_0 is true, is about 2 percent. This probability is called the *P*-value and will be explained in detail later in this section.

The small value 0.0212 of the probability casts doubts on the truth of the null hypothesis H_0. But is this probability small enough that the management should opt for the engineer's claim and switch over to the new technique? The decision will, of course, have to depend also on practical considerations such as costs involved in switching over, the share of the market that the company enjoys and how that share might be affected by the introduction of a better bulb, the strength of the competition, and so on.

DECISION RULE

There is another approach to the problem of testing hypotheses. Quite often the statistician will take into account the risk factors and provide a rule for arriving at a decision. The rule is called a **decision rule.** Let us discuss the approach in light of the above illustration.

Since the alternative hypothesis states that $\mu > 450$, it seems intuitively reasonable that a large value of $\bar{x}$ would cause us to lean in favor of rejecting H_0 and going along with H_A. The larger the value of $\bar{x}$, the stronger the leaning. Suppose we decide, rather arbitrarily at this stage, to choose a cutoff point of the sample mean at 475 and adopt the following decision rule.

Decision rule: Reject H_0 if the sample mean $\bar{x}$ is greater than 475 hours.

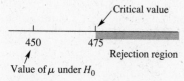

Figure 9-1

Critical value and rejection region

We will call 475 a *critical value.* The values to the right of 475 will constitute what is called a *rejection region,* also called a *critical region* (see Figure 9-1). Since the sample mean 490.67 of the 9 observations is greater than 475, the decision would have to be to reject H_0 and go along with the engineer's claim. Had the sample mean turned out to be, say, 471 hours, the decision would have been not to reject H_0. Let us now determine the risks involved when the above decision rule is adopted.

TYPE I AND TYPE II ERRORS

We have agreed that if the sample mean yields a value greater than 475, we will reject H_0 and grant the engineer's claim. However, due to sampling fluctuations, it is very possible that the bulbs are indeed from a population where the mean life is 450, but that our sample yields a mean greater than 475. If we grant the engineer's claim (that is, reject H_0) when she is wrong (that is, when H_0 is true), we will be committing the Type I error. Since the test is based on the value of $\bar{x}$, the probability of this error will depend on the distribution of $\bar{X}$.

The sampling distribution of $\bar{X}$ is normal because the population is normally distributed. Its mean is 450 if H_0: $\mu = 450$ is true, and its standard deviation is $\sigma/\sqrt{n} = 60/\sqrt{9} = 20$. The distribution is shown by the curve on the left in Figure 9-2.

The alternative hypothesis contemplates a wide range of possible values (greater than 450) for the parameter μ. Suppose, for definiteness, the alternative hypothesis is specifically H_A: $\mu = 500$. The distribution of $\bar{X}$ in this case is normal with mean 500 and standard deviation $\sigma/\sqrt{n} = 20$. This distribution is shown by the curve on the right in Figure 9-2.

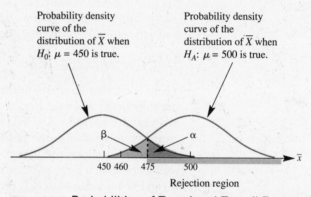

Figure 9-2 Probabilities of Type I and Type II Errors with Values Greater Than 475 Constituting the Critical Region

Type I Error

The probability of a Type I error, α, is the probability of rejecting H_0 when H_0 is true, that is, it is the probability of granting the engineer's claim when she is wrong. Now, we reject H_0 when $\bar{x} > 475$ and the statement "H_0 is true" means that $\mu = 450$. Thus, we want

Probability $\bar{X} > 475$ when $\mu = 450$

or, equivalently, we want

Probability $\bar{X} > 475$ when the distribution of $\bar{X}$ is given by the curve on the left in Figure 9-2.

This is simply the shaded area to the *right* of 475. Since the standard deviation of $\bar{X}$ is 20, conversion to the z-scale shows that the shaded area to the right of 475 is equal to the area to the right of $(475 - 450)/20 = 1.25$. From

Appendix III, Table A-3, it is equal to 0.11 (approximately). Thus, α, the level of significance, is 0.11. This means that in a long series of identical experiments in which nine bulbs are tested from the production, in approximately 11 out of 100 experiments the engineer's claim will be granted when the engineer is wrong in her claim. That is the risk we are running with the decision procedure we have adopted.

Type II Error

Next we will find β, the probability of not granting the engineer's claim when she is correct and μ is in fact 500. (We are assuming that 500 is the value of μ stipulated under H_A.) According to our decision rule, we do not grant the engineer's claim if $\bar{x}$ is less than 475. Thus, we want to find

Probability $\bar{X} < 475$ when $\mu = 500$.

Equivalently, we want

Probability $\bar{X} < 475$ when the distribution of $\bar{X}$ is given by the curve on the right in Figure 9-2.

This is the shaded area to the left of 475. After converting to the z-scale, this is equal to the area to the left of $(475 - 500)/20 = -1.25$, and is approximately equal to 0.11.

Interplay between α and β

Suppose the company is well established and conservative in its decisions and feels that granting the engineer's claim when she is wrong is fraught with much more serious consequences for the company than not granting her claim when she is right. In other words, instead of having a decision rule with $\alpha = \beta = 0.11$, the company wants a decision rule for which α is smaller. An easy way to accomplish this is by moving the critical point to the right of 475, as we have done in Figure 9-3. Suppose we pick arbitrarily 490 as the new critical point. The area shaded to the right of this critical point is now reduced, but the area shaded to the left has increased.

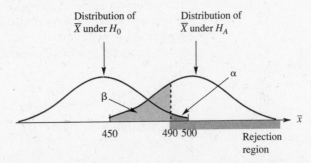

Figure 9-3 Probabilities of Type I and Type II Errors with Values Greater than 490 as the Critical Region

Thus, α has gone down but β has gone up. In a like manner, if we were to move the critical point to the left, β would go down but α would go up. This shows that we cannot control both errors simultaneously once the sample size is determined.

In any decision-making process, the chances of errors that an experimenter (whether businessperson, student, or scientist) is prepared to tolerate is basically up to the individual. However, as we mentioned in the previous section, it is most desirable to exercise control over the risk α of erroneously rejecting correct H_0 and to keep it small. The value of α chosen is usually between 0.01 and 0.1, the most common value being 0.05. *If, for instance, the level of α used is 0.05, then we say that the test is carried out at the 5 percent level of significance.* In the rest of this section we make the following assumptions just as we did when initiating the topic of confidence intervals in Section 7-1.

1. The sample consists of n independent observations.
2. The underlying population is normally distributed with mean μ, which is unknown. (If the sample size is large, that is, at least 30, we can dispense with the requirement that the population must be normally distributed.)
3. The population standard deviation σ is known.

FORMULATION OF THE DECISION RULE

Continuing with our example, suppose we wish to carry out a test of

$$H_0: \mu = 450 \quad \text{against} \quad H_A: \mu > 450$$

at the level of significance α. How do we formulate a decision rule with this significance level in mind?

From our earlier discussion, in view of the nature of the alternative hypothesis, the decision rule will be

$$\text{Reject } H_0 \text{ if } \bar{x} > c$$

(that is, reject H_0 for large values of $\bar{x}$) where the critical value c is to be determined in such a way that the probability of rejecting H_0 when H_0 is true must be α.

In terms of z-scores, the decision rule takes the following form (see Figure 9-4):

$$\text{Reject } H_0 \text{ if } \frac{\bar{x} - 450}{20} > z_a.$$

In general, if we have

$$H_0: \mu = \mu_0 \quad \text{against} \quad H_A: \mu > \mu_0$$

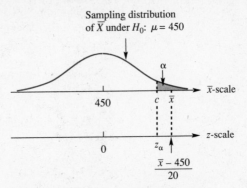

Figure 9-4 $\bar{x}$ is to the right of c, the critical value. On the z scale, $\dfrac{\bar{x} - 450}{20}$ is to the right of z_α.

and $\bar{x}$ is the mean based on n observations from a normal population with standard deviation σ, then the decision rule at the level of significance α would be

$$\text{Reject } H_0 \text{ if } \frac{\bar{x} - \mu_0}{\sigma/\sqrt{n}} > z_a.$$

Note: μ_0 is the value of μ stipulated under H_0. A statistic such as

$$\frac{\bar{X} - \mu_0}{\sigma/\sqrt{n}}$$

used in formulating the decision rule is called a **test statistic.**

The set of values of the test statistic that lead to the rejection of the null hypothesis is called the **rejection region** or the **critical region.**

A **test of a statistical hypothesis** is completely determined by providing the relevant test statistic and the rejection region.

ONE-TAIL AND TWO-TAIL TESTS

The nature of the critical region for a statistical test procedure depends on the alternative hypothesis. In the following discussion we will consider three cases of the alternative hypothesis: (1) H_A: $\mu > \mu_0$, (2) H_A: $\mu < \mu_0$, and (3) H_A: $\mu \neq \mu_0$, where μ_0 is a given specific value (the value 450 in our illustration).

1. A *right-tail test*. In discussing the engineer's claim, we have considered the principle of testing the null hypothesis

$$H_0: \mu = \mu_0$$

against the alternative hypothesis

$$H_A: \mu > \mu_0.$$

The decision rule we formulated there at the level of significance
α was

$$\text{Reject } H_0 \text{ if } \frac{\overline{x} - \mu_0}{\sigma/\sqrt{n}} > z_\alpha.$$

It is the one-sided nature of the alternative hypothesis
(greater than, $>$) that prompts the rejection of H_0 if the value of
the statistic falls in the right tail of its distribution. The test is
therefore called a **one-tail test**, specifically, a **right-tail test**.

2. A *left-tail test*. Suppose the null and alternative hypotheses are
given as

$$H_0: \mu = \mu_0$$
$$H_A: \mu < \mu_0.$$

Once again, the alternative hypothesis is one-sided (less
than, $<$). In this situation, common sense would tell us that we
should prefer to reject H_0 for smaller values of $\overline{x}$, leading to the
rejection of H_0 if the value falls in the left tail of the distribution
of $\overline{X}$ as indicated in Figure 9-5. This gives a one-tail test that is
specifically a **left-tail test**. In terms of the standard scale, the decision rule takes the following form.

$$\text{Reject } H_0 \text{ if } \frac{\overline{x} - \mu_0}{\sigma/\sqrt{n}} < -z_\alpha.$$

3. A *two-tail test*. Finally, a test leads to a two-tail test if the alternative hypothesis is two-sided. Consider the following example.
Suppose a machine is adjusted to manufacture bolts to the specification of a 0.5-inch diameter, and we state the null hypothesis
and the alternative hypothesis as

$$H_0: \mu = 0.5$$
$$H_A: \mu \neq 0.5$$

where μ is the mean diameter. If the sample mean of the diameters was too far off on either side of 0.5, we would favor rejecting
H_0. In other words, if the value of $\overline{x}$ falls in either tail of the distribution of $\overline{X}$, we will reject H_0.

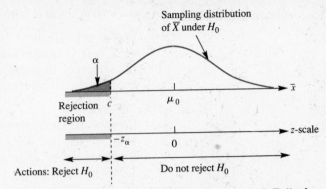

Figure 9-5 A Critical Region That Is in the Left Tail of the Distribution

If the test is carried out at the level of significance α, we distribute α equally between the two tails as in Figure 9-6 and so have two critical values c_1 and c_2. In terms of standardized scores the decision rule is then given by

Reject H_0 if $\dfrac{\bar{x} - \mu_0}{\sigma/\sqrt{n}}$ is less than $-z_{\alpha/2}$ or greater than $z_{\alpha/2}$.

In general, a **two-tail test** is one where the rejection region is located in the two tails of the distribution and results when the alternative hypothesis is two-sided.

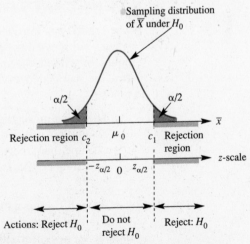

Figure 9-6 A Two-tail Critical Region for Testing H_0: $\mu = \mu_0$ Against H_A: $\mu \neq \mu_0$ with Level of Significance α

EXAMPLE 1   Suppose a population is normally distributed with $\sigma = 6$.

(a) Based on a sample of 16, describe a suitable test procedure at the 10 percent level of significance to test H_0: $\mu = 50$ against H_A: $\mu < 50$.

(b) If a sample of 16 yields a mean of 48.5, what decision would be appropriate at the 10 percent level of significance?

SOLUTION

(a) Because the alternative hypothesis is one-sided (less than), we have a one-tail (left-tail) test. We are given that $\mu_0 = 50$. (It is the value stipulated under H_0.) Also, $z_\alpha = z_{0.1} = 1.28$, $\sigma = 6$, and $n = 16$. Therefore,

$$\frac{\bar{x} - \mu_0}{\sigma/\sqrt{n}} = \frac{\bar{x} - 50}{6/\sqrt{16}} = \frac{\bar{x} - 50}{1.5}.$$

Because the test is a left-tail test, the decision rule is

$$\text{Reject } H_0 \text{ if } \frac{\bar{x} - 50}{1.5} < -1.28.$$

(b) The sample mean is $\bar{x} = 48.5$. From Part a the computed value of the test statistic is

$$\frac{\bar{x} - 50}{1.5} = \frac{48.5 - 50}{1.5} = -1.$$

Because this value is not less than -1.28, we do not reject H_0.   ●

We will see more of this approach in Sections 9-3 and 9-4.

SIGNIFICANCE PROBABILITY (P-VALUE)

The reader was introduced to the concept of P-value earlier in this section when we discussed the engineer's claim based on the sample mean of 490.6 hours. In general, the approach provides the value of the relevant test statistic and the associated P-value and is widely used in journals and technical reports.

> The **P-value** or the **significance probability** is the probability of obtaining a value of the test statistic as extreme as, or even more extreme than, the one computed from the data, *assuming* H_0 *is true*. The smaller the P-value, the stronger the justification for rejecting H_0.

A small P-value will indicate that the value of the test statistic is far in one of the tails of the distribution under H_0, thereby implying less likelihood of the data being in conformity with H_0. We would be more disposed to rejecting H_0. Hence, the smaller the P-value, the stronger the justification for rejection of H_0. A researcher is always delighted to report a small P-value because this leads to the rejection of H_0 and the consequent acceptance of the research hypothesis H_A.

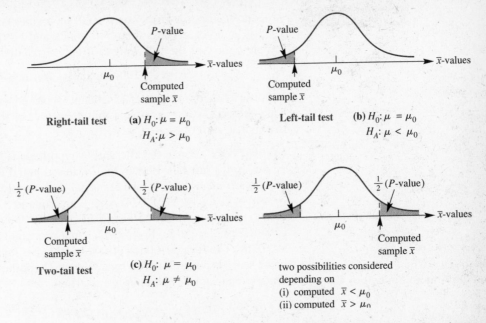

Figure 9-7 *P*-values for Tests of H_0: $\mu = \mu_0$ Against Different Alternative Hypotheses
Shaded areas give *P*-values. The curves describe the distribution of $\overline{X}$ when H_0 is true.

In Figure 9-7, we illustrate the *P*-values for right-, left-, and two-tail tests of the mean. When calculating it, we should keep the alternative hypothesis H_A in perspective, as the nature of H_A (less than, greater than, not equal to) determines the tail(s) in which the values of the test statistic are significant.

There is an advantage to this method of reporting results by providing the *P*-values. The readers are free to pick their own level of significance in arriving at a decision. Suppose, for instance, that the *P*-value in a certain situation is 0.083. Then, at the 5 percent level of significance, we would not reject H_0; however, we would reject it at the 10 percent level or at the 8.5 percent level. In fact, 0.083 is the smallest value of α for which we would reject H_0. For any value of α lower than 0.083 we would not reject H_0.

We now summarize these observations.

THE *P*-VALUE OR SIGNIFICANCE PROBABILITY

The *P*-value (or significance probability) associated with the test of a hypothesis is the smallest level of significance α for which the observed data would call for rejection of the null hypothesis in favor of the alternative. For a given level of significance,

1. reject H_0 if the *P*-value $\leq$ level of significance α
2. do not reject H_0 if the *P*-value $>$ level of significance α.

EXAMPLE 2

An educator feels that the mean SAT scores in mathematics of students in 1994 is significantly less than 500 points and he can establish it based on sample data. So, the following null and alternative hypotheses are set up.

$$H_0: \mu = 500 \quad \text{against} \quad H_A: \mu < 500.$$

Suppose a random sample of 15 students is picked and the mean score is 455. Calculate the P-value assuming that the scores of the students are normally distributed with $\sigma = 70$.

SOLUTION   Since the alternative hypothesis is *less than*, the critical region is in the left tail. Hence, the P-value is equal to $P(\overline{X} < 455)$ when the population mean is 500. The standard deviation of the distribution of $\overline{X}$ is $\sigma/\sqrt{n} = 70/\sqrt{15} = 18.074$. After converting to z-scale, $P(\overline{X} < 455)$ is the area to the left of $(455 - 500)/18.074 = -2.49$ under the standard normal curve (see Figure 9-8). From the normal probability tables, this probability is 0.0064. The P-value 0.0064 is so small that it is highly unlikely that the true mean score is 500 points. We reject the null hypothesis H_0. It is very unlikely that the educator is wrong in his assertion.   ●

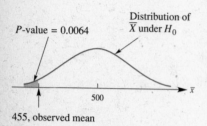

Figure 9-8 The P-value Shown as the Area to the Left of 455

EFFECT OF THE SAMPLE SIZE

It can be shown that for testing

$$H_0: \mu = 450 \quad \text{against} \quad H_A: \mu = 500$$

at the 5 percent level of significance with $\sigma = 60$ and a sample of size 9, the critical value on the $\overline{x}$ scale is 482.5. The resultant value of β turns out to be 0.192 (the area in the left tail of the standard normal distribution, to the left of $(482.5 - 500)/20$, or -0.875). We have discussed the interplay between α and β and are aware that once the sample size is fixed, there is no way of decreasing β below 0.192 without interfering with α and increasing its value beyond 0.05, and vice versa. The only way to reduce both α and β simultaneously is by increasing the sample size. For example, suppose instead of carrying out the test with a sample of 9, we pick a sample of 25. The distribution of $\overline{X}$ has less spread now, since the standard deviation of its distribution is $60/\sqrt{25}$, or 12. The distributions of $\overline{X}$ under H_0 and H_A are given in Figure 9-9 for $n = 9$ and $n = 25$.

You can see by comparing the respective shaded regions in Figure 9-9 that *both* α and β are reduced when the sample size is increased from 9 to 25. Actual computations will show that α is reduced from 0.05 to 0.0032 and β from 0.192 to 0.0764. Thus, it is clear that the accuracy of a test can be improved by taking a larger sample.

A CONNECTION BETWEEN HYPOTHESIS TESTING AND CONFIDENCE INTERVALS

Suppose we test

$$H_0: \mu = \mu_0$$

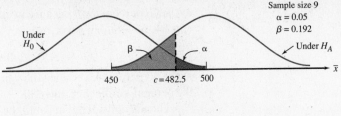

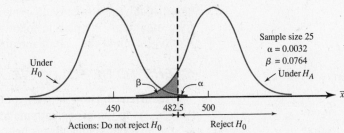

Figure 9-9 Effect on Type I and Type II Errors When Sample Size Is Increased from 9 to 25

against

$$H_A: \mu \neq \mu_0.$$

(Notice that the alternative hypothesis is two-sided.) As we have seen, the region where we would not reject H_0 at the α level of significance is given by

$$-z_{\alpha/2} < \frac{\bar{x} - \mu_0}{\sigma/\sqrt{n}} < z_{\alpha/2}.$$

We would reject H_0 outside this region. Now, resorting to some algebraic manipulations, we can say equivalently that we would not reject H_0 if

$$\bar{x} - z_{\alpha/2} \frac{\sigma}{\sqrt{n}} < \mu_0 < \bar{x} + z_{\alpha/2} \frac{\sigma}{\sqrt{n}}.$$

We recognize at once that the endpoints of the preceding interval are those we found previously for a $(1 - \alpha) 100$ percent confidence interval for μ (see page 395). *Thus, we do not reject* H_0 *at the α level of significance if μ_0 stipulated under it is inside the $(1 - \alpha)100$ percent confidence interval for μ and we do reject it if μ_0 is outside the confidence interval.*

Thus, suppose we have $(56.54, 59.86)$ as a 90 percent confidence interval for μ, as in Example 3 in Section 7-1. Based on the data of that example, we would not reject any of the null hypotheses

$$H_0: \mu = 57.3 \quad \text{against} \quad H_A: \mu \neq 57.3$$

$$H_0: \mu = 58.5 \quad \text{against} \quad H_A: \mu \neq 58.5$$

$$H_0: \mu = 59 \quad \text{against} \quad H_A: \mu \neq 59$$

at the 10 percent level because in each case μ_0 is inside the confidence interval (56.54, 59.86). However, we would reject

$$H_0: \mu = 55 \quad \text{against} \quad H_A: \mu \neq 55$$

at the 10 percent level since the value 55, stipulated under H_0, falls outside the 90 percent confidence interval.

Having considered in detail the factors that enter our considerations when a statistical test in devised, we now outline the general procedure in the following steps. We plan to follow these steps scrupulously in the next two sections, even at the risk of seeming repetitious.

STEPS TO BE FOLLOWED FOR TESTING A HYPOTHESIS

1. State the alternative hypothesis H_A. (This is the research hypothesis in the investigation and specifies a range of possible values for the parameter that is being tested. It is important in deciding whether the test is one-tail or two-tail. Rejection of H_0 leads to the acceptance of H_A.)

2.* State the null hypothesis. Under H_0 give a specific value of the parameter.

3. Pick an appropriate test statistic. The choice of the test statistic is determined by the conventional point estimator of the parameter in question.

4. Stipulate the level of significance α, the probability of rejecting H_0 wrongly. The critical point(s) are determined by the value of α. Together with Step 1, formulate the decision rule, that is, determine the values of the test statistic that will lead to the rejection of H_0 (the critical region). If the value of α is not stipulated, provide the P-value in Step 6 below.

5. Take a random sample and compute the value of the test statistic.

6. Make the decision in light of the decision rule formulated in Step 4. Keep in mind that you either reject H_0 or do not reject it. It is important to translate the conclusions into nonstatistical language for the benefit of the uninitiated.

Section 9-2 Exercises

1. Determine in which of the following we have acceptable sets of null and alternative hypotheses.
   (a) $H_0: \bar{x} = 50$ $H_A: \bar{x} = 62$
   (b) $H_0: \mu = 50$ $H_A: \mu \geq 62$
   (c) $H_0: \mu = 50$ $H_A: \mu \leq 62$
   (d) $H_0: \dfrac{x}{n} = 0.3$ $H_A: \dfrac{x}{n} \neq 0.3$

---

*The order of Steps 1 and 2 can be reversed.

(e) H_0: There is no difference between Pepsi and Coke.
   H_A: Pepsi is better than Coke.

(f) H_0: $p > 0.7$ H_A: $p \leq 0.6$

(g) H_0: $p \geq 0.7$ H_A: $p < 0.6$

(h) H_0: $p = 0.7$ H_A: $p \neq 0.7$

(i) H_0: $\mu = 30$ H_A: $\mu \neq 25$

2. For testing a null hypothesis H_0 against an alternative hypothesis H_A, suppose the critical region is as described in Figure 9-10. Copy the curves and shade appropriate regions under them to represent the probability of Type I error and the probability of Type II error.

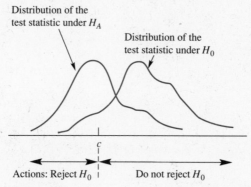

Distribution of the test statistic under H_A

Distribution of the test statistic under H_0

c

Actions: Reject H_0 Do not reject H_0

Figure 9-10

3. Figure 9-11 on page 516 gives critical regions for testing H_0 against H_A. In each case shade appropriate regions under the curves to represent the probabilities of Type I and Type II errors. In which of the cases would you say that the choice of the critical region was not prudently made.

4. Peggy is given a jar containing five chips and told that either the chips are marked 1, 3, 5, 7, and 9 or they are marked 2, 4, 6, 8, and 10. She is allowed to pick one chip at random. If Peggy sets the null hypothesis as H_0: The chips are marked 1, 3, 5, 7, and 9, what is the most appropriate decision rule that she can devise? Find the corresponding probabilities of Type I and Type II errors.

5. As a slight variation of Exercise 4, suppose Peggy is given a jar containing five chips and told that either the chips are marked 1, 2, 3, 4, and 5 or they are marked 5, 6, 7, 8, and 9. Suppose H_0 states that the chips are marked 1, 2, 3, 4, and 5. Peggy is allowed to pick one chip at random and has to decide according to the following decision rule: If the observed chip is one of the chips 5, 6, 7, 8, or 9, then reject H_0. Find the probabilities of Type I and Type II errors.

6. In Exercise 5 devise a decision rule such that the probability of Type I error is zero and the probability of Type II error is a minimum.

7. A population is known to be normally distributed with standard deviation 3. However, there is a dispute as to whether the mean μ is 50 or 60. Suppose you have to decide by picking one observation whether H_0: $\mu = 50$ is true. The following decision rule is adopted: *Reject H_0 if the observed value is greater than 56.* Find the probabilities of Type I and Type II errors.

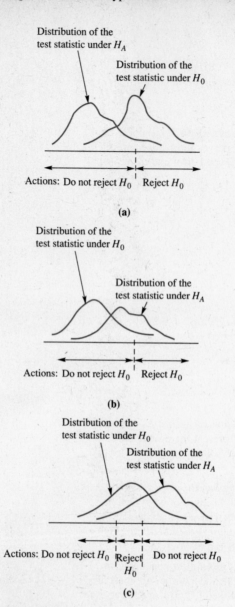

Figure 9-11

8. Suppose you cannot reject a null hypothesis at the 5 percent level of significance. Would you reject it at the 3 percent level? Would you reject it at any level less than 5 percent? Explain.

9. Suppose we know that a population is normally distributed with the variance σ^2 equal to 16, but we have doubts as to whether the mean is 40 or greater than 40. Suppose we wish to test

$$H_0: \mu = 40 \quad \text{against} \quad H_A: \mu > 40.$$

What decision would you take at the 10 percent level of significance if a sample of 25 yielded a mean equal to 42?

10. Suppose we want to test

$$H_0: \mu = 100 \quad \text{against} \quad H_A: \mu \neq 100$$

for a population that is normally distributed with a standard deviation of 3. What decision would you take, at the 10 percent level of significance, if a sample of 16 yields a mean of 97?

11. A population is known to be normally distributed with variance 17.64 but unknown mean μ. Suppose we wish to test

$$H_0: \mu = 100 \quad \text{against} \quad H_A: \mu < 100.$$

   (a) What decision would you take, at the 4 percent level of significance, if a sample of 9 yields a mean of 97.6?
   (b) In Part a determine the P-value.

12. Suppose we have a normal population with $\sigma = 4$ and wish to test

$$H_0: \mu = 20 \quad \text{against} \quad H_A: \mu < 20.$$

A sample of 9 items yields a mean $\bar{x} = 18.6$.
   (a) Compute the value of an appropriate test statistic.
   (b) Determine the P-value.

13. Determine P-values in the following cases.
   (a) $H_0: \mu = 12.5$ against $H_A: \mu > 12.5$
      If $\bar{x} = 14.8$, $\sigma = 6$, and $n = 25$
   (b) $H_0: \mu = 0.5$ against $H_A: \mu > 0.5$
      if $\bar{x} = 0.62$, $\sigma = 0.2$, and $n = 20$
   (c) $H_0: \mu = 100$ against $H_A: \mu \neq 100$
      if $\bar{x} = 82.8$, $\sigma = 24$, and $n = 15$
   (d) $H_0: \mu = -7$ against $H_A: \mu \neq -7$
      if $\bar{x} = -5.8$, $\sigma = 2$, and $n = 18$

14. Mr. Montoya is running for election in a certain district. Suppose p represents the proportion of voters who favor him. The following hypotheses are set up.

$$H_0: p = 0.6 \quad \text{against} \quad H_A: p = 0.4$$

It is agreed that a sample of 8 individuals picked at random will be interviewed, and if 6 or more people declare in favor of Mr. Montoya, then H_0 will be accepted. Otherwise, H_0 will be rejected. Find the probability of the following.
   (a) Type I error
   (b) Type II error
      Hint: Use the binomial tables, Table A-1, in Appendix III.

15. A stockbroker claims to be able to tell correctly whether a stock will go up or not during a trading session. She asserts that she is correct 80 percent of the time. Of course, if she really is guessing, her probability of being correct is 50 percent. The question is, "Does the stockbroker have the ability that she claims to have?"
   (a) Set up the null hypothesis and the alternative hypothesis.
   (b) What decisions would constitute Type I and Type II errors?

16. In Exercise 15, suppose in order to test the claim of the broker, she is shown twelve unrelated stocks and it is agreed that if the broker is correct nine times or

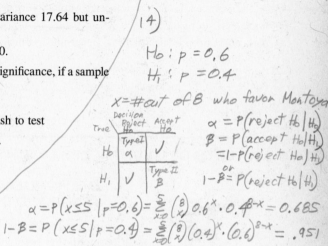

more, her claim will be granted. Otherwise the claim will not be granted. Determine α and β.

   Hint: Use the binomial tables, Table A-1, in Appendix III.

17. Suppose you are interested in testing a null hypothesis H_0 against the alternative hypothesis H_A. In carrying out the test, suppose the P-value turns out to be 0.043. At the 2 percent level of significance would you reject H_0?

18. Suppose, based on a sample you have collected, you obtain (23.9, 26.7) as a 98 percent confidence interval for μ. Determine in which of the following cases you would reject H_0.
   (a) H_0: $\mu = 24$ against H_A: $\mu \neq 24$
   (b) H_0: $\mu = 27$ against H_A: $\mu \neq 27$

9-3   TESTS OF HYPOTHESES (A SINGLE POPULATION)

THE POPULATION MEAN WHEN THE POPULATION STANDARD DEVIATION IS KNOWN

In the preceding section, we considered tests of a population mean when the population variance was known. A basic assumption about the population in such a case is that it is normally distributed. In the absence of a normally distributed population, we will require that the sample size be large, that is, at least 30. The level of significance is then approximately 100α percent.

   As we have seen, the relevant statistic in this case is

$$\frac{\overline{X} - \mu_0}{\sigma/\sqrt{n}}.$$

A summary of the test criteria developed in the last section to test H_0: $\mu = \mu_0$ against the three forms of alternative hypotheses is given in Table 9-2.

TABLE 9-2   Procedure to Test H_0: $\mu = \mu_0$ When σ Is Known

Alternative Hypothesis	The Decision Rule Is to Reject H_0 If the Computed Value Is
1. $\mu > \mu_0$	greater than z_a
2. $\mu < \mu_0$	less than $-z_a$
3. $\mu \neq \mu_0$	less than $-z_{a/2}$ or greater than $z_{a/2}$

   We will now solve some examples following the steps outlined at the end of the previous section.

EXAMPLE 1

After taking a refresher course, a salesperson found that sales (in dollars) on nine random days were as follows.

| 1280 | 1250 | 990 | 1100 | 880 | 1300 | 1100 | 950 | 1050 |

Has the refresher course had the desired effect, in that the mean sale is now more than 1000 dollars? Assume $\sigma = 100$, and the probability of erroneously saying that the refresher course is beneficial should not exceed 0.01. Also, assume that sales are normally distributed.

SOLUTION   The approach is outlined as follows.

1. H_0: $\mu = 1000$ (actually $\mu \leq 1000$)
2. H_A: $\mu > 1000$
3. The appropriate test statistic is $\dfrac{\overline{X} - 1000}{\sigma/\sqrt{n}}$, because $\mu_0 = 1000$.
4. The alternative hypothesis is one-sided (right-sided). Also, $\alpha = 0.01$ so that $z_a = 2.33$. Therefore, the decision rule is: *Reject* H_0 *if the computed value of the test statistic is greater than 2.33.*
5. Because $\sigma = 100$, $n = 9$, and $\overline{x} = 1100$, the computed value of the test statistic is

$$\frac{\overline{x} - 1000}{\sigma/\sqrt{n}} = \frac{1100 - 1000}{100/3} = 3. \text{ (See Figure 9-12.)}$$

6. Because the computed value is greater than 2.33, we reject H_0. Incidentally, because the computed value is 3, from Table A-3 the *P*-value is 0.0013.

Conclusion: At the 1 percent level of significance, there is evidence indicating that the refresher course was beneficial to the salesperson. It is indeed very unlikely (probability less than 0.01) that we would get a sample with such an extreme value of $\overline{X}$ if the course were not beneficial.   ●

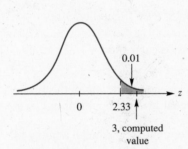

Figure 9-12

0.01

0   2.33

3, computed value

EXAMPLE 2

In advertising a brand of king-size cigarettes, a tobacco company says that the customer could switch down to lower tar by buying the brand. A random sample of 12 cigarettes was selected from the brand and the cigarettes were tested for their tar content. The results (in mg) were as follows.

| 8.4 | 8.1 | 6.6 | 7.9 | 9.3 | 6.1 |
| 7.2 | 9.1 | 7.1 | 7.0 | 8.7 | 7.9 |

Suppose it is reasonable to believe that the tar content for this brand is normally distributed with a standard deviation σ of 1 mg. At the 10 percent level, do the results of the study support the contention that the true mean tar content for this brand is less than 8 mg?

SOLUTION   In view of the question posed, the null hypothesis will stipulate the mean tar content is not less than 8 mg. Thus, we have the following.

1. H_0: $\mu = 8$ (actually $\mu \geq 8$)
2. H_A: $\mu < 8$
3. The test statistic is $\dfrac{\bar{X} - 8}{\sigma/\sqrt{n}}$.
4. $\alpha = 0.10$ so that $z_\alpha = 1.28$. Because the alternative hypothesis is one-sided (on the left), the decision rule is: *Reject* H_0 *if the computed value of the test statistic is less than* -1.28.
5. The sample mean is $\bar{x} = 7.783$, $\sigma = 1$, and $n = 12$. Therefore, the computed value is

$$\frac{\bar{x} - 8}{\sigma/\sqrt{n}} = \frac{7.783 - 8}{1/\sqrt{12}} = -0.75. \text{ (See Figure 9-13.)}$$

6. Because the computed value -0.75 is greater than -1.28, we do not reject H_0 (*P*-value 0.2266).

Conclusion: There is no evidence that the true mean tar content of this brand of cigarettes is less than 8 mg at the ten percent level of significance. ●

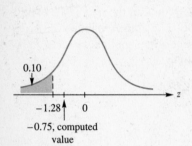

0.10

-1.28 0

-0.75, computed value

Figure 9-13

EXAMPLE 3

A machine can be adjusted so that when under control, the mean amount of sugar filled in a bag is 5 pounds. From past experience, the standard deviation of the amount filled is known to be 0.15 pound. To check if the machine is under control, a random sample of sixteen bags was weighed and the mean weight was found to be 5.1 pounds. At the 5 percent level of significance, is the adjustment out of control? (Assume a normal distribution of the amount of sugar filled in a bag.)

SOLUTION   The adjustment is out of control if it overfills or underfills. Thus, the alternative hypothesis is two-sided.

1. H_0: $\mu = 5$
2. H_A: $\mu \neq 5$
3. The test statistic is $\dfrac{\bar{X} - 5}{\sigma/\sqrt{n}}$, since $\mu_0 = 5$.
4. The alternative hypothesis is two-sided. Therefore, we have a two-tail test. Because $\alpha = 0.05$, $z_{\alpha/2} = z_{0.025} = 1.96$. The decision rule is: *Reject* H_0 *if the computed value is less than* -1.96 *or greater than* 1.96.
5. We have $\bar{x} = 5.1$, $\sigma = 0.15$, and $n = 16$. Therefore, the computed value of the test statistic is

$$\frac{\bar{x} - 5}{\sigma/\sqrt{n}} = \frac{5.1 - 5}{0.15/\sqrt{16}} = 2.67. \text{ (See Figure 9-14.)}$$

6. Because the computed value is greater than 1.96, we reject H_0 (*P*-value 2(0.0038), or 0.0076).

Conclusion: There is strong evidence indicating that the adjustment is out of control. ●

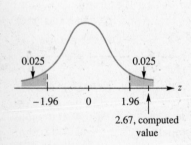

0.025   0.025

-1.96 0   1.96

2.67, computed value

Figure 9-14

THE POPULATION MEAN WHEN THE POPULATION STANDARD DEVIATION IS UNKNOWN

In tests of hypotheses regarding the population mean, the situation where the population standard deviation is not known is more realistic. *The procedure we are about to develop is particularly important when the sample size is small.* We wish to test the following null hypothesis.

$$H_0: \mu = \mu_0$$

In the case where σ was known, we used the test statistic

$$\frac{\overline{X} - \mu_0}{\sigma/\sqrt{n}}.$$

Since σ is not known, we will use its estimator S. Thus, the appropriate test statistic in this case is

$$T = \frac{\overline{X} - \mu_0}{S/\sqrt{n}}.$$

At this point, we make the following assumptions similar to those we made when setting confidence interval of μ when σ is unknown.

1. The observations are independent.
2. The parent population has a normal distribution.

Under these assumptions, the statistic T has Student's t distribution with $n - 1$ degrees of freedom, *if H_0 is true.* So we get the decision rules given in Table 9-3, depending upon the particular alternative hypothesis.

TABLE 9-3   Procedure to Test $H_0: \mu = \mu_0$ When σ Is Not Known

Alternative Hypothesis	The Decision Rule Is to Reject H_0 If the Computed Value of T Is
1. $\mu > \mu_0$	greater than $t_{n-1,\alpha}$
2. $\mu < \mu_0$	less than $-t_{n-1,\alpha}$
3. $\mu \neq \mu_0$	less than $-t_{n-1,\alpha/2}$ or greater than $t_{n-1,\alpha/2}$

If σ is unknown but n is large (that is, $n \geq 30$), then, even if the population is not normally distributed, the test statistic

$$\frac{\overline{X} - \mu_0}{S/\sqrt{n}}$$

will have, approximately, a standard normal distribution. So we can use the test procedure described in Table 9-2, with the only difference being that we have S in place of σ in the test statistic. Of course, we have to bear in mind that the level of significance will be, approximately, 100α percent.

EXAMPLE 4

A car manufacturer wants to determine whether a luxury model gives satisfactory mileage so that the overall EPA requirement on mileage per car manufactured can be met. The particular model will be considered satisfactory if the true mean mileage μ is greater than 20 miles per gallon. A field experiment was conducted in which ten cars were driven under almost identical conditions and the miles per gallon were computed for each. The results (in miles) were as follows.

| 23 | 18 | 22 | 19 | 19 | 22 | 18 | 18 | 24 | 22 |

Based on the data, is there sufficient evidence for the manufacturer to decide that the model is satisfactory? The manufacturer is prepared to run a risk of Type I error of 0.05. It may be assumed that mileage per gallon is normally distributed.

SOLUTION

1. H_0: $\mu = 20$
2. H_A: $\mu > 20$
3. The standard deviation of the distribution is not known. We will have to estimate it from the sample. Hence, the test statistic is
$$\frac{\bar{X} - 20}{S/\sqrt{n}}.$$
4. Here $\alpha = 0.05$ and, because $n = 10$, we have $10 - 1$, or 9 degrees of freedom. Thus, from the table for Student's t distribution (Table A-4, Appendix III), we find $t_{9,0.05} = 1.833$. The decision rule is: *Reject* H_0 *if the computed value of the test statistic is greater than 1.833.*
5. It can be easily verified from the data that
$$\bar{x} = 20.5 \quad \text{and} \quad s = 2.32.$$
Thus, the computed value of the test statistic is
$$\frac{\bar{x} - 20}{s/\sqrt{n}} = \frac{20.5 - 20}{2.32/\sqrt{10}} = 0.682. \text{ (See Figure 9-15.)}$$
6. We do not reject H_0, since the computed value is less than 1.833 (*P*-value greater than 0.1).

Conclusion: At the 5 percent level, the evidence at hand does not bear out that the true mean mileage per gallon is greater than 20 miles.   ●

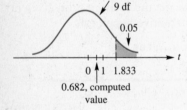

t-distribution
9 df
0.05
0 1 1.833
0.682, computed value

Figure 9-15

EXAMPLE 5

To investigate the effect of the insecticide lead arsenate on table wines, twenty-five bottles of a white wine from a certain region where the insecticide was used were sampled and the amount of lead (mg/l) was determined by carbon rod atomic absorption spectrophotometry. The sample mean was 0.409 and the sample standard deviation was 0.051. Based on the given data, is the true mean lead content in the white wines from this region significantly less than 0.42, as the local wineries claim? Use $\alpha = 0.05$. (Assume a normal distribution.)

SOLUTION

1. H_0: $\mu = 0.42$ (actually $\mu \geq 0.42$)
2. H_A: $\mu < 0.42$
3. The test statistic is $\dfrac{\overline{X} - 0.42}{S/\sqrt{n}}$.
4. $\alpha = 0.05$ and the number of degrees of freedom is $25 - 1 = 24$. Therefore, the decision rule is: *Reject* H_0 *if the computed value of the test statistic is less than* $-t_{24,0.05} = -1.711$.
5. The computed value of the test statistic is

$$\frac{\overline{x} - 0.42}{s/\sqrt{n}} = \frac{0.409 - 0.42}{0.051/\sqrt{25}} = -1.078. \text{ (See Figure 9-16.)}$$

6. Because the computed value -1.078 is greater than -1.711, we do not reject H_0 (P-value greater than 0.1).

Conclusion: There is no conclusive evidence at the 5 percent level of significance to justify the claim of the local wineries. ●

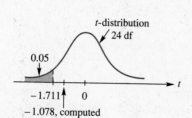

t-distribution 24 df

0.05

-1.711 0

-1.078, computed value

Figure 9-16

EXAMPLE 6

A machine can be adjusted so that when under control, the mean amount of sugar filled in a bag is 5 pounds. To check if the machine is under control, six bags were picked at random and their weights (in pounds) were found to be as follows.

5.3   5.2   4.8   5.2   4.8   5.3

At the 5 percent level of significance, is it true that the machine is not under control? (Assume a normal distribution for the weight of a bag.)

SOLUTION σ is not known and so will have to be estimated from the sample.

1. H_0: $\mu = 5$
2. H_A: $\mu \neq 5$
3. The test statistic is $\dfrac{\overline{X} - 5}{S/\sqrt{n}}$.
4. We have a two-tail test. Because $\alpha = 0.05$ and $n = 6$, $t_{n-1,\alpha/2} = t_{5,0.025} = 2.571$. Therefore, the decision rule is: *Reject* H_0 *if the computed value is less than* -2.571 *or greater than* 2.571.
5. $\overline{x} = \dfrac{30.6}{6} = 5.1$ and $s^2 = \dfrac{0.28}{5} = 0.056$.

   Thus,

$$\frac{\overline{x} - 5}{s/\sqrt{n}} = \frac{5.1 - 5}{\sqrt{0.056}/\sqrt{6}} = 1.04. \text{ (See Figure 9-17.)}$$

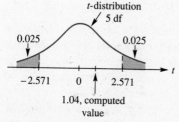

t-distribution 5 df

0.025   0.025

-2.571 0   2.571

1.04, computed value

Figure 9-17

6. Because the computed value is between -2.571 and 2.571, we do not reject H_0 (P-value greater than 2(0.1), or 0.2).

Conclusion: There is no significant evidence at the 5 percent level to indicate that the machine is not under control, so there is no need to adjust the machine. ●

THE POPULATION VARIANCE

While testing hypotheses concerning the population mean, we used the population variance when it was available and estimated it when it was not available. So we are already aware of cases in which there is a need for some information about the population variance σ^2. Knowledge about σ^2 is also important in many practical problems, such as in quality control, where the magnitude of the variance will be very crucial in determining whether the process is under control. For example, suppose a machine packs a mean amount of 5 pounds per bag exactly as the specifications require. In spite of this, we would regard the machine not under control if the standard deviation should turn out to be 1 pound. It is therefore easy to see that very often the variance of the population will be the main object of the researcher's interest and study.

We are interested in testing a null hypothesis that postulates that the population variance has a specified value σ_0^2 against various alternative hypotheses. In other words, we want to test

$$H_0: \sigma^2 = \sigma_0^2 \quad \text{(or, equivalently, } H_0: \sigma = \sigma_0).$$

For testing hypotheses concerning the population variance σ^2, it would be natural to start with its sample counterpart, the sample variance,

$$s^2 = \frac{\sum (x_i - \bar{x})^2}{n - 1}.$$

It turns out that this leads to the test statistic

$$\frac{(n - 1)S^2}{\sigma_0^2}$$

on the basis of which decision rules are formulated.

At this time we will have to know the distribution of $(n - 1)S^2/\sigma_0^2$. In this connection we make the following assumptions.

1. The observations are independent.
2. The parent population has a normal distribution.

We know from Chapter 7 that, if the population is normally distributed with variance σ^2, then the distribution of $(n - 1)S^2/\sigma^2$ is chi-square with $n - 1$ degrees of freedom, where n represents the sample size. Therefore, *if* H_0 *is true,* that is, if the population variance is σ_0^2, then the distribution of $(n - 1)S^2/\sigma_0^2$ will be chi-square with $n - 1$ degrees of freedom.

Knowing the distribution of the test statistic, in Table 9-4 we now provide test procedures for testing $H_0: \sigma^2 = \sigma_0^2$, depending upon whether the alternative hypothesis is right-sided, left-sided, or two-sided. (Recall that $\chi_{n-1,\alpha}^2$ is the table value of a chi-square distribution with $n - 1$ degrees of freedom such that the area to its right is α.)

TABLE 9-4 Procedure to Test H_0: $\sigma = \sigma_0$

Alternative Hypothesis	The Decision Rule Is to Reject H_0 If the Computed Value of $\dfrac{(n-1)S^2}{\sigma_0^2}$ Is
1. $\sigma > \sigma_0$	greater than $\chi^2_{n-1,\alpha}$
2. $\sigma < \sigma_0$	less than $\chi^2_{n-1,1-\alpha}$
3. $\sigma \neq \sigma_0$	less than $\chi^2_{n-1,1-\alpha/2}$ or greater than $\chi^2_{n-1,\alpha/2}$

EXAMPLE 7

The following measurements (in ounces) were obtained on the weights of six laboratory mice picked at random; 12, 8, 7, 12, 14, and 13. At the 5 percent level of significance, is the population variance of the weights of such mice greater than 2.25?

SOLUTION

1. H_0: $\sigma = \sqrt{2.25} = 1.5$
2. H_A: $\sigma > 1.5$
3. The test statistic is $\dfrac{(n-1)S^2}{2.25}$.
4. The alternative hypothesis is right-sided. Because $n = 6$, the chi-square has $6 - 1$ or 5 degrees of freedom. From Table A-5, $\chi^2_{5,0.05} = 11.07$. Thus, the decision rule is: *Reject* H_0 *if the computed value of the test statistic exceeds 11.07.*
5. It can be verified that $\bar{x} = 11$ and $(n-1)s^2 = \Sigma (x_i - \bar{x})^2 = 40$. Therefore, the computed value of the test statistic is

$$\frac{(n-1)s^2}{2.25} = \frac{40}{2.25} = 17.8. \text{ (See Figure 9-18.)}$$

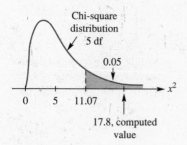

Chi-square distribution 5 df

0.05

0   5   11.07

17.8, computed value

Figure 9-18

6. Because the computed value 17.8 is greater than 11.07, we reject H_0 (*P*-value less than 0.005).

Conclusion: At the 5 percent level, there is evidence indicating that the population variance exceeds 2.25.   ●

EXAMPLE 8

A construction company received a very large consignment of bolts. The shipment will be accepted if the standard deviation σ of the diameters of the bolts is less than 0.1 inch. When a random sample of 11 bolts from the shipment was inspected, it was found that the sample standard deviation was 0.07.

Do the data provide sufficient evidence to the construction company to decide that the bolts meet their requirements? The company is willing to take a chance of saying that the shipment is satisfactory when in fact it is not with probability 0.05. Also, it may be assumed that the diameters are normally distributed.

SOLUTION

1. H_0: $\sigma = 0.1$
2. H_A: $\sigma < 0.1$
3. The test statistic is $\dfrac{(n-1)S^2}{(0.1)^2}$.
4. In saying that the shipment is satisfactory when in fact it is not, the construction company would be committing a Type I error. Since the probability of this error is given to be 0.05, we have $\alpha = 0.05$.

   Because $n = 11$, there are 10 degrees of freedom. The test is a left-tail test and $\chi^2_{10,0.95} = 3.94$. Therefore, the decision rule is: *Reject H_0 if the computed value of the test statistic is less than 3.94.*
5. $s = 0.07$. Thus, the computed value of the test statistic is

$$\frac{(n-1)s^2}{0.01} = \frac{10(0.07)^2}{0.01} = 4.9. \text{ (See Figure 9-19.)}$$

6. Because the computed value is greater than 3.94, we do not reject H_0 (*P*-value greater than 0.05).

Conclusion: There is no evidence at the 5 percent level that the bolts meet the stipulation of the construction company.   ●

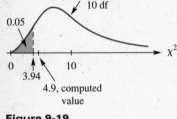

Chi-square distribution 10 df

0.05

0   3.94   10

4.9, computed value

Figure 9-19

EXAMPLE 9   A random sample of fifteen bulbs produced by a machine was picked, and each bulb was tested for the number of hours it lasted before it burned out. If $s = 13$ hours, at the 5 percent level of significance, is the population standard deviation different from 10 hours?

SOLUTION

1. H_0: $\sigma = 10$
2. H_A: $\sigma \neq 10$
3. The test statistic is $\dfrac{(n-1)S^2}{10^2}$.
4. The alternative hypothesis is two-sided so that the test is two-tailed. Because $n = 15$ and $\alpha = 0.05$,

$$\chi^2_{n-1,\alpha/2} = \chi^2_{14,0.025} = 26.119 \quad \text{and}$$
$$\chi^2_{n-1,1-\alpha/2} = \chi^2_{14,0.975} = 5.629.$$

   The decision rule is: *Reject H_0 if the computed value of the test statistic is less than 5.629 or greater than 26.119.*
5. Because $s = 13$, the computed value of the test statistic is

$$\frac{(n-1)s^2}{100} = \frac{(14)(169)}{100} = 23.66. \text{ (See Figure 9-20.)}$$

6. The computed value is between 5.629 and 26.119. We do not reject H_0.

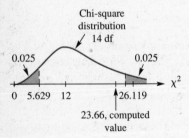

Chi-square distribution 14 df

0.025   0.025

0   5.629   12   26.119

23.66, computed value

Figure 9-20

Conclusion: There is insufficient evidence to warrant that $\sigma \neq 10$.   ●

THE POPULATION PROPORTION FOR QUALITATIVE DATA

So far in this chapter we have concerned ourselves with treating data where the observed variable can be measured on a numerical scale. We will now treat the case of a qualitative variable, where the data are recorded as tall-short, black-green, defective-nondefective, and so forth. Our objective will be to test hypotheses regarding the proportion p of a certain attribute in the population. We will specifically consider the problem of testing the null hypothesis

$$H_0: p = p_0$$

where p_0 is a number between 0 and 1, against various alternative hypotheses. For example, we might be interested in the proportion of defective items produced by a machine and wish to test

$$H_0: p = 0.2 \quad \text{against} \quad H_A: p > 0.2$$

or

$$H_0: p = 0.2 \quad \text{against} \quad H_A: p < 0.2$$

or

$$H_0: p = 0.2 \quad \text{against} \quad H_A: p \neq 0.2.$$

To carry out a test of hypothesis regarding the population proportion, we pick a sample of observations and take the proportion of the attribute in the sample as the statistic on which the test is based. At this stage we make the following assumptions.

1. The sample consists of n independent observations.
2. The sample size is large.
3. The hypothesized population proportion p_0 is not too close to 0 or 1 and is such that np_0 and $n(1 - p_0)$ are both greater than 5.

Now, if p is the proportion in the population, then, as we know, the sample proportion X/n has a sampling distribution with mean p and standard deviation $\sqrt{p(1 - p)/n}$. Furthermore, if the sample is large in such a way that both np and $n(1 - p)$ are greater than 5, the distribution of X/n is approximately normal. Consequently, if the null hypothesis is true, that is, if the population proportion is p_0, then X/n will have a distribution that is approximately normal with mean p_0 and standard deviation $\sqrt{p_0(1 - p_0)/n}$, since we have assumed that both np_0 and $n(1 - p_0)$ are greater than 5.

We now have a situation analogous to the one where we tested hypotheses regarding the population mean when σ was known. The role of $\overline{X}$ is played by X/n, that of μ_0 by p_0, and that of $\sigma/\sqrt{n}$ by $\sqrt{p_0(1 - p_0)/n}$. The test statistic is

$$\frac{(X/n) - p_0}{\sqrt{\dfrac{p_0(1 - p_0)}{n}}}.$$

TABLE 9-5 Procedure to Test H_0: $p = p_0$ When Sample Size Is Large So That $np_0 \geq 5$ and $n(1 - p_0) \geq 5$

Alternative Hypothesis	The Decision Rule Is to Reject H_0 If the Computed Value of $\dfrac{(X/n) - p_0}{\sqrt{\dfrac{p_0(1 - p_0)}{n}}}$ Is
1. $p > p_0$	greater than z_α
2. $p < p_0$	less than $-z_\alpha$
3. $p \neq p_0$	less than $-z_{\alpha/2}$ or greater than $z_{\alpha/2}$

We list in Table 9-5 the three cases based on the nature of the alternative hypothesis.

It is worth keeping in mind that the present test procedure is based on the application of the central limit theorem. Thus, the significance level is *approximately* 100α percent. This should be understood even when it is not mentioned in a specific problem.

EXAMPLE 10 When a coin was tossed 200 times, it showed heads 120 times. Is the coin biased in favor of heads? Carry out the test at the 5 percent level of significance.

SOLUTION Let p = probability of getting a head on a toss. The question posed is "Is $p > 0.5$?" Under null hypothesis, we state that this is not the case.

1. H_0: $p = 0.5$ (actually $p \leq 0.5$)
2. H_A: $p > 0.5$
3. Because $p_0 = 0.5$, the test statistic is $\dfrac{(X/n) - 0.5}{\sqrt{(0.5)(0.5)/n}}$.
4. $\alpha = 0.05$, therefore the decision rule is: *Reject* H_0 *if the computed value is greater than* $z_{0.05} = 1.645$.
5. Because $x = 120$ and $n = 200$, the computed value of the statistic is

$$\frac{(x/n) - 0.5}{\sqrt{(0.5)(0.5)/n}} = \frac{\dfrac{120}{200} - 0.5}{\sqrt{0.25/200}} = 2.83. \text{ (See Figure 9-21.)}$$

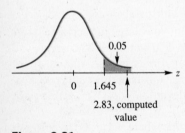

0.05

0   1.645

2.83, computed value

Figure 9-21

6. The computed value 2.83 is greater than 1.645. So, we reject H_0 (*P*-value 0.0023).

Conclusion: At the 5 percent level of significance, there is strong evidence that the coin is biased in favor of heads. ●

EXAMPLE 11

A machine is known to produce 30 percent defective tubes. After repairing the machine, it was found that it produced 22 defective tubes in the first run of 100. Is the true proportion of defective tubes reduced after the repairs? Use $\alpha = 0.01$.

SOLUTION Let p denote the proportion of defective tubes.

1. $H_0: p = 0.3$
2. $H_A: p < 0.3$
3. The test statistic is $\dfrac{(X/n) - 0.3}{\sqrt{(0.3)(0.7)/n}}$.
4. Because $\alpha = 0.01$, the decision rule is: *Reject* H_0 *if the computed value is less than* $-z_{0.01} = -2.33$.
5. Since $x = 22$ and $n = 100$, the computed value is

$$\frac{(x/n) - 0.3}{\sqrt{(0.3)(0.7)/n}} = \frac{0.22 - 0.3}{\sqrt{0.21/100}} = -1.746. \text{ (See Figure 9-22.)}$$

6. The computed value -1.746 is greater than -2.33. So, we do not reject H_0 (P-value 0.0401).

Conclusion: There is no evidence (at the 1 percent level) that the proportion of defective tubes produced by the machine is reduced. ●

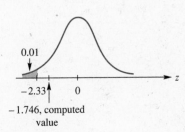

0.01

-2.33 0   z

-1.746, computed value

Figure 9-22

EXAMPLE 12

Suppose the proportion of families in the U.S. that took a vacation of at least 1 week last year was 20 percent. To find the attitude of families on traveling this year, 120 families picked at random were interviewed and, of these, 18 said they would take such a vacation. Has the attitude changed from last year? Use $\alpha = 0.10$.

SOLUTION We will interpret "change of attitude" to mean change in the true proportion of families. Let p represent the proportion of those who would take a vacation of at least 1 week.

1. $H_0: p = 0.2$
2. $H_A: p \neq 0.2$
3. Because $p_0 = 0.2$, the test statistic is $\dfrac{(X/n) - 0.2}{\sqrt{(0.2)(0.8)/n}}$.
4. Since $\alpha = 0.10$, $\alpha/2 = 0.05$, and $z_{\alpha/2} = 1.645$, the decision rule is: *Reject* H_0 *if the computed value is less than* -1.645 *or greater than* 1.645.
5. Because $x = 18$ and $n = 120$,

$$\frac{(x/n) - 0.2}{\sqrt{(0.2)(0.8)/n}} = \frac{0.15 - 0.2}{\sqrt{0.16/120}} = -1.37. \text{ (See Figure 9-23.)}$$

6. The computed value -1.37 is between -1.645 and 1.645, and so we do not reject H_0 (*P*-value 2(0.0853), or 0.1706).

Conclusion: At the 10 percent level of significance, there is no evidence to indicate that the attitude this year is different from that of last year. ●

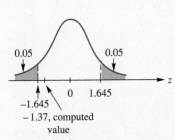

0.05   0.05

-1.645 0   1.645   z

-1.37, computed value

Figure 9-23

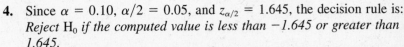

Section 9-3 Exercises

In Exercises 1–6, determine appropriate test statistics and critical regions that you would use in carrying out tests of the hypotheses. In each case provide the decision rule. You may assume that the populations are normally distributed.

1. $H_0: \mu = 10.2$ against $H_A: \mu > 10.2$ when $\sigma = 1.3, n = 16,$ $\alpha = 0.05$

2. $H_0: \mu = 110$ against $H_A: \mu < 110$ when $\sigma = 4.32, n = 25,$ $\alpha = 0.025$

3. $H_0: \mu = 32.8$ against $H_A: \mu < 32.8$ when $\sigma = 6.8, n = 14,$ $\alpha = 0.01$

4. $H_0: \mu = 17.8$ against $H_A: \mu > 17.8$ when $\sigma = 2.8, n = 20,$ $\alpha = 0.02$

5. $H_0: \mu = -3.3$ against $H_A: \mu \neq -3.3$ when $\sigma = 4.8, n = 20,$ $\alpha = 0.02$

6. $H_0: \mu = 11.6$ against $H_A: \mu \neq 11.6$ when $\sigma = 3.2, n = 12,$ $\alpha = 0.1$

[handwritten margin note: $\dfrac{\bar{X} - \mu}{\sigma/\sqrt{n}}$ Reject H_0 if value is greater than $Z_{.05} = 1.645$]

7. The time (in minutes) taken by a biological cell to divide into two cells has a normal distribution. From past experience, the population standard deviation σ can be assumed to be 3.5 minutes. When sixteen cells were observed, the mean time taken by them to divide into two was 31.6 minutes. At the 5 percent level of significance, test the following.
   (a) $H_0: \mu = 30$ against $H_A: \mu \neq 30$
   (b) $H_0: \mu = 30$ against $H_A: \mu > 30$

8. A breeder of rabbits claims that he can breed rabbits yielding a mean weight of greater than 58 ounces. Suppose the standard deviation σ is known to be 3 ounces. A random sample of sixteen rabbits had a mean weight of 59.2 ounces. At the 5 percent level of significance, is the breeder's claim justified? Assume a normal distribution.

9. The label on a can of pineapple slices states that the mean carbohydrate content per serving of canned pineapple is over 50 grams. It may be assumed that the standard deviation of the carbohydrate content σ is 4 grams. A random sample of twenty-five servings has a mean carbohydrate content of 52.3 grams. Is the company correct in its claim? Use $\alpha = 0.05$. Assume a normal distribution.

10. From his long-standing experience, a farmer believes that the mean yield of grain per plot on his farm is 150 bushels. When a new seed introduced on the market was tried on sixteen randomly picked experimental plots, the mean yield was 158 bushels. Suppose the yield per plot can be assumed to be normally distributed with a standard deviation of yield σ of 20 bushels. Is the new seed significantly better? Use $\alpha = 0.02$.

11. The contention of factory management is that the mean weekly idle time per machine is over 5 hours. The supervisor does not agree with this. When a random sample of twenty-five machines was observed, the mean weekly idle time was found to be 5.8 hours. From past experience, it is safe to assume that the idle time of a machine during a week is normally distributed with a standard deviation of 2 hours. At the 3 percent level of significance, is the contention of the management justified?

In Exercises 12–17, determine appropriate test statistics and critical regions for carrying out the tests of hypotheses. In each case, provide the decision rule. You may assume that the populations are normally distributed with unknown variance σ^2 having estimator S^2.

12. H_0: $\mu = 10.2$ against H_A: $\mu > 10.2$ when $n = 16$, $\alpha = 0.05$

13. H_0: $\mu = 110$ against H_A: $\mu < 110$ when $n = 25$, $\alpha = 0.025$

14. H_0: $\mu = 32.8$ against H_A: $\mu < 32.8$ when $n = 14$, $\alpha = 0.01$

15. H_0: $\mu = 17.8$ against H_A: $\mu > 17.8$ when $n = 20$, $\alpha = 0.025$

16. H_0: $\mu = -3.3$ against H_A: $\mu \neq -3.3$ when $n = 20$, $\alpha = 0.02$

17. H_0: $\mu = 11.6$ against H_A: $\mu \neq 11.6$ when $n = 12$, $\alpha = 0.1$

18. A random sample of size $n = 16$ is drawn from a pupulation having a normal distribution. The sample mean and the sample variance are given, respectively, as $\bar{x} = 23.8$ and $s^2 = 10.24$. At the 5 percent level of significance, test the following.
    (a) H_0: $\mu = 25$ against H_A: $\mu \neq 25$
    (b) H_0: $\mu = 25$ against H_A: $\mu < 25$

19. A ski coach claims that she can train beginning skiers for 3 weeks so that at the end of the program they will finish a certain downhill course in less than 13 minutes. It was found that when a random sample of ten skiers was given the training their mean time was 12.3 minutes with $s = 1.2$ minutes. On the basis of the evidence, is the true mean time significantly less than 13 minutes? Use $\alpha = 0.025$. Assume a normal distribution.

20. A sample of eight workers in a clothing-manufacturing company gave the following figures for the amount of time (in minutes) needed to join a collar to a shirt.

    10   12   13   9   8   14   10   11

    At the 5 percent level of significance, is the true mean time to join a collar more than 10 minutes? Assume a normal distribution.

21. A manufacturer of automobile tires claims that the mean number of trouble-free miles given by a new tire introduced by the company is more than 36,000. When sixteen randomly picked tires were tested, the mean number of miles was 38,900, with the sample standard deviation s equal to 8200 miles. At the 5 percent level of significance, is the manufacturer's claim justified? Assume a normal distribution.

22. The following figures refer to the amount of coffee (in ounces) filled by a machine in six randomly picked jars.

    15.7   15.9   16.3   16.2   15.7   15.9

    Is the true mean amount of coffee filled in a jar less than 16 ounces? Use $\alpha = 0.025$. Assume a normal distribution.

23. A company manufactures dry cell batteries. When 18 randomly picked batteries were tested in the laboratory, the mean life was 55.9 hours with a standard deviation of 6.5 hours. At the 1 percent level of significance, based on the information, is the true mean life of the batteries significantly less than 60 hours? Assume a normal distribution.

24. A school system has assumed that the true mean time to complete a particular exam is 90 minutes and, as a result, has allotted time for completing the exam equal to the assumed true mean time plus 30 minutes, that is, 120 minutes. One of the administrators, however, feels that the true mean time is less than 90 minutes and the time allotted should be reduced. In a sample of 14 randomly selected students it was found that the mean time to complete the exam was $\bar{x} = 84.93$ with a standard deviation $s = 16.0$. At the 5 percent level, is there evidence, based on the data, that the time allotted should be reduced? Assume a normal distribution.

25. Suppose the company you work for has decided to buy an industrial glue if there is evidence to support that the true mean sealing strength (kg/25 cm²) is greater than 12 kg/25 cm². The following data present the sealing strength at room temperature (10°C–25°C) when 12 specimens were tested

11.8	13.6	11.6	12.6	12.8	12.9
13.9	11.8	13.5	11.6	14.2	13.7

At the 1 percent level of significance, would you recommend that the company buy the brand of glue? Assume a normal distribution.

26. A sensor detects an electric charge each time a turbine blade makes one rotation and then measures the amplitude of the electric current. The following eight readings were obtained for a certain setting of the speed of rotation of the blades and gap between the blades.

26.9	30.1	28.9	25.5	23.6	29.9	24.7	30.0

Suppose you are willing to risk a Type I error of 2.5 percent. At this level, do the data justify that the true amplitude at the setting is less than 30? Assume a normal distribution.

27. A study on the amount of dye needed to get the best color for a certain type of fabric was conducted. The following data give the photometer readings for the color density of the fabric when the amount of dye used was 1% of the weight of the fabric.

13.2	11.5	12.9	13.0	11.7	10.4	12.1
12.1	11.5	10.3	11.7	12.3	11.2	

Do the data demonstrate that the true mean photometer reading for the color density is significantly different from 12? Use 10 percent as the level of significance. Assume a normal distribution.

28. The quality of water streams and lakes is determined by the concentration (mg/l) of dissolved oxygen in the water. It is suspected that an industrial plant might be polluting the water of a stream. When the dissolved oxygen concentration was measured along the stream on 13 randomly selected days, the following readings were obtained.

2.9	3.3	3.6	3.9	3.1	3.8	3.7
3.9	2.4	3.1	3.1	3.2	2.9	

Suppose it would be a matter of concern if the true mean daily concentration was less than 3.5 mg/l. At the 5 percent level, is concern warranted? Assume a normal distribution.

29. The following data are the room-temperature injectate cardiac outputs of twenty-nine patients in a cardiac intensive care unit (*Heart and Lung,* May 1987, p. 296).

2.60	5.16	6.18	3.22	4.99	3.62	5.40	5.93
3.31	4.11	5.24	4.27	3.42	4.70	5.90	4.11
5.42	5.36	2.63	3.70	5.39	5.44	4.44	2.64
3.86	6.68	5.35	3.26	4.06			

(a) Construct a 95 percent confidence interval for the true mean cardiac output for patients of this type.

(b) At the 5 percent level, is the true mean cardiac output significantly greater than 4.0? Assume a normal distribution for cardiac output.

30. Food legumes contain a substance called saponin which has been credited with lowering plasma cholesterol in animals. Twenty samples of a high-yielding variety of chick-peas (*Cicer arietinum*) were laboratory tested for their total saponin (mg/100g). The figures are presented as follows.

3366	3337	3361	3410	3316	3387	3348
3356	3376	3382	3377	3355	3408	3401
3390	3428	3383	3374	3484	3390	

At the 2.5 percent level, is the true mean saponin content less than 3400? Assume a normal distribution.

31. Fifteen individuals were sampled to find out their daily intake of wheat selenium (μg/day) which is considered essential as a constituent of erythrocyte glutathione peroxidase. The data are presented as follows.

83.2	91.8	84.4	77.1	80.0	87.3	83.8	92.5
85.7	78.9	84.6	87.8	78.9	86.0	87.8	

At the 10 percent level of significance, is the true mean daily selenium intake different from 85? Assume a normal distribution.

32. The data below contain the concentration of acid-soluble sulfides (mg/l fresh ooze) in the bottom sediments of a lagoon. Twenty-five samples were analyzed.

998	938	971	949	1013	982	941
1000	1021	1002	995	956	976	1047
1010	1015	967	987	990	974	965
954	974	984	988			

Assuming a normal distribution, at the 2 percent level of significance, is the true mean concentration of acid-soluble sulfides different from 1000?

33. The total soluble sugar (g/100 ml) in a brand of canned soft drink was determined through high performance liquid chromatography. The data from ten samples tested are presented as follows.

10.4	10.2	10.4	10.3	10.2
10.2	10.1	10.2	10.3	10.4

Is the true mean soluble sugar significantly different from 10.2? Test using $\alpha = 0.05$. Assume a normal distribution.

In Exercises 34–37, determine appropriate test statistics and critical regions for carrying out the test of hypotheses. In each case provide the decision rule.
   Hint: Notice n is a large in each case.

34. $H_0: \mu = 9.8$ against $H_A: \mu < 9.8$ when $n = 46$, $\alpha = 0.05$

35. $H_0: \mu = 82.0$ against $H_A: \mu \neq 82.0$ when $n = 64$, $\alpha = 0.02$

36. $H_0: \mu = -3.0$ against $H_A: \mu > -3.0$ when $n = 80$, $\alpha = 0.1$

37. $H_0: \mu = 0$ against $H_A: \mu \neq 0$ when $n = 50$, $\alpha = 0.1$

38. In a survey in which 100 randomly picked middle-class families were interviewed, it was found that their mean medical expenses during a year were \$770 with $s = \$120$. Are the true mean medical expenses during the year significantly greater than \$750? Use $\alpha = 0.025$.

39. A biologist claims that the mean cholesterol content of chicken eggs can be reduced by giving a special diet to the chickens. When thirty-six eggs obtained in this way were tested, it was found that the mean cholesterol content was 245 mg with a sample standard deviation equal to 20 mg. Is the true mean cholesterol content significantly less than 250 mg? Use $\alpha = 0.05$.

40. Suppose you consider buying a hotel. Before buying the business you would certainly like to get an idea of the past earnings of the hotel. So you go through the hotel account books and examine the earnings of 45 randomly picked days. The mean daily profit based on these 45 days is \$1,825 with $s = \$318$. Suppose you would buy the hotel if the evidence showed that the true mean daily profit was more than \$1600. Would the data induce you to buy the business? Use 5 percent level of significance.

41. The president is concerned about the escalating costs of providing health care in U.S. One of the components contributing to the increasing costs is the length of the hospital stay of a patient. In a sample involving 63 patients the mean time was $\bar{x} = 4.5$ days with standard deviation $s = 1.3$ days. Do these results demonstrate that the true mean stay of a patient is significantly longer than 4 days? Carry out tests using (i) 5 percent level of significance, (ii) 1 percent level of significance. Also, determine the P-value.

In Exercises 42–47, determine appropriate test statistics and critical regions for carrying out the tests of hypotheses. In each case provide the decision rule. You may assume that the populations are normally distributed.

42. $H_0: \sigma^2 = 8.2$ against $H_A: \sigma^2 \neq 8.2$ when $n = 16$, $\alpha = 0.05$

43. $H_0: \sigma^2 = 0.36$ against $H_A: \sigma^2 < 0.36$ when $n = 7$, $\alpha = 0.025$

44. $H_0: \sigma = 6.8$ against $H_A: \sigma > 6.8$ when $n = 14$, $\alpha = 0.05$

45. $H_0: \sigma^2 = 10$ against $H_A: \sigma^2 \neq 10$ when $n = 18$, $\alpha = 0.1$

46. $H_0: \sigma^2 = 0.78$ against $H_A: \sigma^2 < 0.78$ when $n = 8$, $\alpha = 0.01$

47. $H_0: \sigma^2 = 9$ against $H_A: \sigma^2 > 9$ when $n = 11$, $\alpha = 0.05$

48. A random sample of size $n = 11$ drawn from a normal population gave $s^2 = 2.45$. At the 5 percent level of significance, test the following.
   (a) $H_0: \sigma^2 = 7$ against $H_A: \sigma^2 \neq 7$
   (b) $H_0: \sigma^2 = 7$ against $H_A: \sigma^2 < 7$

49. An archaeologist found that the mean cranial width of twenty-one skulls was 5.3 inches with the sample standard deviation equal to 0.55 inch. Is the true standard deviation of cranial width significantly greater than 0.5? Use $\alpha = 0.025$. Assume a normal distribution.

50. A grade-school teacher has obtained the following figures for the time (in minutes) that ten randomly picked students in her class took to complete an assigned task.

    15   10   12   8   11   10   10   13   12   9

    At the 5 percent level of significance, is the true variance of the time different from 2.4 minutes squared? Assume a normal distribution.

51. The following data represent the weight loss (in pounds) of 11 subjects treated in a weight control clinic over a certain period of time.

    19   27   24   20   17   25
    30   22   23   16   21

    At the 5 percent level of significance, is the population standard deviation different from 5 pounds? You may assume that the loss of weight is normally distributed.

52. The braking distance of a car at a given speed is the distance through which it travels before it comes to a complete halt after the brakes are applied. A scientific experiment was run to measure the braking distance from 60 mph for ten cars of a new model. The data (in feet) are presented below.

    161   153   175   174   173   169   187   171
    172   181

    Do these results indicate that the true standard deviation of the braking distance is significantly greater than 8? Assume that the braking distance is normally distributed.

53. A machine is set to that it will dispense close to 6 oz of coffee in a cup. If the machine overfills or underfills by large amounts, it will need to be adjusted. The measure of this variability is provided by the standard deviation of the amount filled. In fifteen trial runs the following amounts of coffee were dispensed.

    5.3   6.4   6.4   5.6   5.7   6.2   6.5   5.1
    6.1   6.2   5.3   5.9   5.6   5.6   5.4

    Assume that the amount of coffee dispensed in a cup is normally distributed.
    (a) At the 2 percent level is the true mean significantly different from 6 oz?
    (b) At the 5 percent level is the true standard deviation less than 0.7?

54. Eleven sections of bacon each weighing 100 grams and consisting of both back fat and lean muscle tissue were homogenized in a food processor and then chemically treated. Their copper content was then assayed by submitting the samples for copper analysis by flame atomic absorption spectroscopy. The following data give the copper content of the fat (μg/g).

    0.55   0.46   0.72   0.32   0.45   0.47
    0.36   0.62   0.71   0.51   0.39

    (a) At the 2 percent level, is the true standard deviation significantly different from 0.12?
    (b) Test the claim that the true mean copper content is greater than 0.5. Use $\alpha = 0.01$. Assume a normal distribution.

55. Thirty milk-yielding Ayrshire cows were fed a diet of silage which was supplemented with a protein concentrate. The following figures refer to the milk fat concentration (g/kg) for each cow.

30.5	28.7	27.3	24.1	42.3	35.7	32.7
41.5	20.5	36.3	25.5	30.9	34.6	35.1
26.3	33.6	31.9	34.7	35.0	35.9	33.3
34.5	30.6	34.0	38.2	42.5	37.4	33.3
40.3	21.7					

(a) Based on the data, is the true mean milk fat concentration significantly less than 35? Carry out the test at the 5 percent level.

(b) At the 5 percent level of significance, is the true standard deviation different from 5.0? Assume a normal distribution.

56. Thirty pigs were each fed a diet containing 1.8 percent linoleic acid (percent of dry feed) from 20 kilograms to 35 kilograms of their weight followed by 1.4 percent linoleic acid from 35 kilograms to slaughter at 85 kilogram live weight. The figures that follow relate to the fat (percent of wet weight) in the tissue composition of the back fat.

78.4	71.5	81.9	84.2	76.9	76.6	82.8
70.3	65.0	81.4	86.7	74.0	69.8	88.5
79.5	79.1	72.9	72.3	76.5	74.4	83.4
75.7	81.0	73.5	84.2	87.7	76.4	84.9
78.1	80.6					

Use the data to determine the following.

(a) Is the true mean fat in the tissue significantly less than 80? Use $\alpha = 0.05$.

(b) Is the true standard deviation of the fat in the tissue significantly less than 5? Use $\alpha = 0.05$. Assume a normal distribution.

In Exercises 57–62, determine appropriate test statistics and critical regions for carrying out the tests of hypotheses. In each case provide the critical region.

57. $H_0: p = 0.4$ against $H_A: p \neq 0.4$ when $n = 60$, $\alpha = 0.05$

58. $H_0: p = 0.2$ against $H_A: p < 0.2$ when $n = 80$, $\alpha = 0.025$

59. $H_0: p = 0.7$ against $H_A: p > 0.7$ when $n = 36$, $\alpha = 0.05$

60. $H_0: p = 0.5$ against $H_A: p \neq 0.5$ when $n = 64$, $\alpha = 0.1$

61. $H_0: p = 0.8$ against $H_A: p < 0.8$ when $n = 120$, $\alpha = 0.01$

62. $H_0: p = 0.3$ against $H_A: p > 0.3$ when $n = 96$, $\alpha = 0.05$

63. A drug company claims that more than 80 percent of the people given a vaccine will develop immunity to a disease. Of 100 randomly picked people who were given the vaccine, 88 developed immunity. On the basis of this evidence, is the claim of the drug company valid? Use $\alpha = 0.03$.

64. In a sample of 160 cathode tubes inspected, 22 were found to be defective. If the true proportion of defectives is significantly higher than the 8 percent that the company considers tolerable, repairs on the machine are in order. At the 5 percent level of significance, does the machine need repairs?

65. Based on the data from the Second United States Health and Nutritional Examination Survey, it was reported (*Am. J. Public Health 76* (1986): pp. 287–289) that 35 percent of the U.S. population 18 to 74 years of age take vitamin/mineral supplements regularly. An investigator interviewed 300 randomly picked individuals in the age group and found that 125 took the supplements. Is there significant evidence to reject $H_0: p = 0.35$ in favor of $H_A: p > 0.35$? Use $\alpha = 0.025$.

66. A poll of 780 registered voters found that 42 percent of them believed that the biggest problems facing the country were crime and violence. At the 5 percent level, is the proportion in the population who share the belief significantly greater than 40 percent?

67. A wire service flashed the following news item: "The credibility of Congress is down to less than 20 percent." In a survey in which 280 random individuals were interviewed, 49 thought the performance of Congress was satisfactory. Is the wire service correct in its news flash? Use $\alpha = 0.10$.

68. Should a national lottery be instituted? A random sample of 360 adults gave the following results: 130 in favor, 230 opposed. Is it true that less than 40 percent of the population favor a national lottery? Use $\alpha = 0.05$.

69. In a magazine article, "The (Executive) Board Vs. the 'Babe'," *Time*, August 30, 1993, wrote " . . . sexual bias and harassment remain an entrenched fact of life at many U.S. companies. In a recent survey poll of 439 female executives, the recruiting firm Korn-Ferry found that 60% had been sexually harassed during their careers." Assume that the female executives who were interviewed responded independently. At the 2.5 percent level, are the majority of female executives in the U.S. harassed during their careers?

70. In Exercise 69, what would your decision be if, instead of having 439 female executives in the survey, there were 80 females? Comment on your findings.

71. The following question was asked in a survey conducted by the Gallup organization in which 1003 individuals were interviewed: "In your opinion, has the federal government under President Clinton become a lot more liberal, a little more liberal, or has it not become more liberal at all?" The results of the survey were as follows:

	Percent
A lot more	0.24
A little more	0.43
Not more at all	0.24
No opinion	0.09

At the 5 percent level, is it evident that at least one in five feel that the federal government has become a lot more liberal under President Clinton?

72. Mars, Inc., a candy company, claims that the color ratios in a bag of plain M&Ms are as shown on the next page. In a sixth grade class the teacher gave the students a project to verify if the claim was reasonable. In a sample of 683 M&M candies, the following distribution was obtained.

	Brown	Red	Yellow	Orange	Green	Tan
Number	226	152	119	52	57	77
Percent	33.1	22.3	17.4	7.6	8.3	11.3

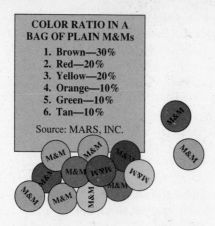

**COLOR RATIO IN A
BAG OF PLAIN M&Ms**
  1. **Brown—30%**
  2. **Red—20%**
  3. **Yellow—20%**
  4. **Orange—10%**
  5. **Green—10%**
  6. **Tan—10%**

Source: MARS, INC.

The students were ecstatic, declaring that all the percentages were off from the claimed percentages and that the company was not living up to its claim. At the 5 percent level of significance, is the discrepancy for each color due to chance or is it due to real difference, with the company not living up to its claim? You will have to perform six tests. (See also, Chapter 3, Example 5, page 158.) (*Comment.* In Chapter 10 we will discuss a better procedure where just one test will be needed for carrying out the analysis.)

9-4   TESTS OF HYPOTHESES (TWO POPULATIONS)

In this section we will test hypotheses concerning the difference of means of two populations and the difference of proportions of an attribute in two populations. The setup in each case regarding the nature of the sampled populations (notations included) is as defined in the corresponding counterpart in Chapter 8.

DIFFERENCE IN POPULATION MEANS WHEN THE VARIANCES ARE KNOWN

The null hypothesis under test is

$$H_0: \mu_1 = \mu_2, \text{ that is } \mu_1 - \mu_2 = 0,$$

and the test statistic appropriate for the purpose is

$$\frac{\bar{X} - \bar{Y}}{\sqrt{\dfrac{\sigma_1^2}{m} + \dfrac{\sigma_2^2}{n}}}.$$

The decision rules for various forms of alternative hypotheses are given in Table 9-6.

TABLE 9-6   Procedure to Test $H_0: \mu_1 = \mu_2$ When the Populations Are Normally Distributed and Variances Are Known

Alternative Hypothesis	The Decision Rule Is to Reject H_0 If the Computed Value Is
1. $\mu_1 > \mu_2$	greater than z_α
2. $\mu_1 < \mu_2$	less than $-z_\alpha$
3. $\mu_1 \neq \mu_2$	less than $-z_{\alpha/2}$ *or* greater than $z_{\alpha/2}$

The assumption of normal distribution of populations is not crucial if sample sizes are large, that is, as a general rule, *m* and *n* are both at least 30. Actually, in this case, even the population variances need not be known. We can use the sample variances s_1^2 and s_2^2 in place of the respective population variances σ_1^2 and σ_2^2. The test procedure is the same as in Table 9-6 except that we use the test statistic

$$\frac{\bar{X} - \bar{Y}}{\sqrt{\dfrac{S_1^2}{m} + \dfrac{S_2^2}{n}}}.$$

EXAMPLE 1

For a sample of fifteen adult European males picked at random, the mean weight $\bar{x}$ was 154 pounds, whereas for a sample of eighteen adult males in the United States, the mean weight $\bar{y}$ was 162 pounds. Assume that the variance of weight in Europe σ_1^2 is 100 and in the United States σ_2^2 is 169. Is it true that there is a significant difference between mean weights in the two places? Use $\alpha = 0.05$. Assume that the weights are normally distributed.

SOLUTION

1. $H_0: \mu_1 = \mu_2$, that is, $\mu_1 - \mu_2 = 0$
2. $H_A: \mu_1 \neq \mu_2$
3. The test statistic is $\dfrac{\bar{X} - \bar{Y}}{\sqrt{\dfrac{\sigma_1^2}{m} + \dfrac{\sigma_2^2}{n}}}.$
4. Because $\alpha = 0.05$, $z_{\alpha/2} = 1.96$. The decision rule is: *Reject* H$_0$ *if the computed value falls outside the interval* -1.96 *to* 1.96.
5. $\bar{x} = 154$, $\bar{y} = 162$, $\sigma_1^2 = 100$, $\sigma_2^2 = 169$, $m = 15$, and $n = 18$. Therefore, the computed value is

$$\frac{\bar{x} - \bar{y}}{\sqrt{\dfrac{\sigma_1^2}{m} + \dfrac{\sigma_2^2}{n}}} = \frac{154 - 162}{\sqrt{\dfrac{100}{15} + \dfrac{169}{18}}}$$

$$= \frac{-8}{\sqrt{16.056}} = -2. \text{ (See Figure 9-24.)}$$

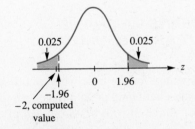

0.025   0.025

-1.96 0   1.96 z

-2, computed value

Figure 9-24

6. Because the computed value is less than -1.96, we reject H_0 (P-value 0.0456).

Conclusion: At the 5 percent level, there is a difference between the true mean weight of male adults in the United States and those in Europe.   ●

EXAMPLE 2   To compare two brands of cigarettes, Brand A and Brand B, for their tar content, a sample of 60 was inspected from Brand A and a sample of 40 from Brand B. The results of the tests are summarized as follows.

Brand A	$\bar{x} = 15.4 \quad s_1^2 = 3$
Brand B	$\bar{y} = 16.8 \quad s_2^2 = 4$

At the 5 percent level of significance, do the two brands differ in their mean tar content?

SOLUTION   Because the two samples are large, we can use s_1^2 in place of σ_1^2 and s_2^2 in place of σ_2^2. Let $\mu_{\text{Brand A}}$ represent the mean for Brand A with a similar interpretation for $\mu_{\text{Brand B}}$.

1. H_0: $\mu_{\text{Brand A}} = \mu_{\text{Brand B}}$
2. H_A: $\mu_{\text{Brand A}} \neq \mu_{\text{Brand B}}$
3. The test statistic is $\dfrac{\bar{X} - \bar{Y}}{\sqrt{\dfrac{S_1^2}{m} + \dfrac{S_2^2}{n}}}$.

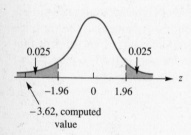

0.025   0.025

−1.96   0   1.96

−3.62, computed value

Figure 9-25

4. Because $\alpha = 0.05$, $z_{\alpha/2} = 1.96$. The decision rule is: *Reject H_0 if the computed value is less than -1.96 or greater than 1.96.*
5. The computed value of the statistic is

$$\frac{\bar{x} - \bar{y}}{\sqrt{\dfrac{s_1^2}{m} + \dfrac{s_2^2}{n}}} = \frac{15.4 - 16.8}{\sqrt{\dfrac{3}{60} + \dfrac{4}{40}}} = -3.62. \text{ (See Figure 9-25.)}$$

6. The computed value -3.62 is less than -1.96. We therefore reject H_0 (P-value less than 0.002).

Conclusion: There is a strong reason to conclude that the two brands of cigarettes differ in their tar content.   ●

DIFFERENCE IN POPULATION MEANS WHEN THE VARIANCES ARE UNKNOWN BUT ARE ASSUMED EQUAL

The following test procedure is particularly suited for the case when *small* independent samples are drawn from normally distributed populations both having the *same variance*.

We are interested in testing the null hypothesis

$$H_0: \mu_1 = \mu_2.$$

When the variances were known, we used the statistic

$$\frac{\bar{X} - \bar{Y}}{\sqrt{\dfrac{\sigma_1^2}{m} + \dfrac{\sigma_2^2}{n}}}.$$

But we are assuming that the variances are equal. So suppose $\sigma_1^2 = \sigma_2^2$ and let σ^2 represent the common value. The preceding test statistic then reduces to

$$\frac{\bar{X} - \bar{Y}}{\sigma \sqrt{\dfrac{1}{m} + \dfrac{1}{n}}}.$$

Because σ is not known, we will use its pooled estimate s_p, where

$$s_p = \sqrt{\frac{(m-1)s_1^2 + (n-1)s_2^2}{m+n-2}}.$$

Therefore, the test statistic appropriate for carrying out the test is

$$\frac{\bar{X} - \bar{Y}}{S_p \sqrt{\dfrac{1}{m} + \dfrac{1}{n}}}.$$

When H_0 is true, this statistic has Student's t distribution with $m + n - 2$ degrees of freedom. The test procedures for the various forms of the alternative hypothesis are given in Table 9-7.

TABLE 9-7 Procedure to Test $H_0: \mu_1 = \mu_2$ When the Variances Are Unknown but Are Assumed Equal

Alternative Hypothesis	The Decision Rule Is to Reject H_0 If the Computed Value Is
1. $\mu_1 > \mu_2$	greater than $t_{m+n-2,\alpha}$
2. $\mu_1 < \mu_2$	less than $-t_{m+n-2,\alpha}$
3. $\mu_1 \neq \mu_2$	less than $-t_{m+n-2,\alpha/2}$ *or* greater than $t_{m+n-2,\alpha/2}$

EXAMPLE 3

A nitrogen fertilizer was used on ten plots and the mean yield per plot $\bar{x}$ was found to be 82.5 bushels with s_1 equal to 10 bushels. On the other hand, fifteen plots treated with phosphate fertilizer gave a mean yield $\bar{y}$ of 90.5 bushels per plot with s_2 equal to 20 bushels. At the 5 percent level of significance, are the two fertilizers significantly different? It may be assumed that the yields for the two fertilizers are normally distributed with the same variance. (There exists a test which shows that with $s_1 = 10$, $m = 10$ and $s_2 = 20$, $n = 15$ we cannot reject that $\sigma_1 = \sigma_2$.)

SOLUTION

1. $H_0: \mu_1 = \mu_2$
2. $H_A: \mu_1 \neq \mu_2$
3. The test statistic is $\dfrac{\bar{X} - \bar{Y}}{S_p \sqrt{\dfrac{1}{m} + \dfrac{1}{n}}}$.
4. $m = 10$ and $n = 15$. Therefore, the degrees of freedom are $m + n - 2 = 23$. Because $\alpha = 0.05$, $t_{23, \alpha/2} = t_{23, 0.025} = 2.069$. The decision rule is: *Reject H_0 if the computed value is less than -2.069 or greater than 2.069.*
5. $\bar{x} = 82.5$, $\bar{y} = 90.5$; $s_1 = 10$, $s_2 = 20$; $m = 10$, $n = 15$. Therefore,

$$s_p^2 = \frac{(m - 1)s_1^2 + (n - 1)s_2^2}{m + n - 2} = \frac{9(10)^2 + 14(20)^2}{10 + 15 - 2}$$

$$= 282.6$$

so that $s_p = \sqrt{282.6} = 16.81$. The computed value of the test statistic is

$$\frac{\bar{x} - \bar{y}}{s_p \sqrt{\dfrac{1}{m} + \dfrac{1}{n}}} = \frac{82.5 - 90.5}{16.81 \sqrt{\dfrac{1}{10} + \dfrac{1}{15}}}$$

$$= \frac{-8}{6.86} = -1.166. \text{ (See Figure 9-26.)}$$

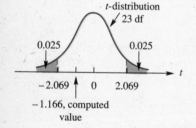

t-distribution 23 df

0.025   0.025

-2.069 0   2.069 t

-1.166, computed value

Figure 9-26

6. Because the computed value -1.166 is between -2.069 and 2.069, we do not reject H_0 (*P*-value greater than 0.05).

Conclusion: The data do not support the contention that the fertilizers are significantly different. ●

DIFFERENCE IN POPULATION MEANS USING THE PAIRED *t* TEST

The method of a **paired *t* test** reduces the problem of comparing the means of two populations to that of a one-sample *t* test. Suppose we wish to compare two treatments, which we will call Treatment 1 and Treatment 2. The two treatments could be two stimuli, two diets, two fertilizers, two methods of teaching, and so on. The idea involves taking experimental units in pairs so that within each pair, the units are as homogeneous as possible.* This method of collecting data eliminates the influence of extraneous factors. Treatment 1 is then assigned to one of the members of the pair at random and Treatment 2 to the other member.

For example, if we wish to compare two teaching methods we might pick ten pairs of students, such that within each pair both students are of the same socioeconomic background, same age, same IQ. (Pairs of twins might turn out to be ideal.) We then assign one student in a pair to be taught by one method and the other student by the second method.

A common method of obtaining matched pairs is to match an experimental unit with itself, leading to what is widely referred to as a "before-after" study.

The data can be recorded as ordered pairs (x, y), where x is the response to Treatment 1 and y is the response to Treatment 2. If there are n pairs that are independent of each other, we get a set of n independent pairs of values $(x_1, y_1), (x_2, y_2), \ldots, (x_n, y_n)$. *When data are collected in this way, obviously the two samples of x values and y values are not independent.* (If one member of a pair performs well (poorly) on a test, then we might reasonably expect the other one to do well (poorly) also.) Our previous test for comparing means was based on the strict assumption of independence of samples and so cannot be applied here. Therefore, we devise a different test which is based on the difference of responses within pairs.

Let us denote the difference of the pair (x_i, y_i) by $d_i = x_i - y_i$. The n differences $d_1, d_2, \ldots, d_n$ now constitute a sample of n independent observations from a population of "within-pair differences D." For carrying out analysis of the data, we require that the differences D be normally distributed. This assumption is particularly important if n is small. If μ_1 is the mean for Treatment 1 and μ_2 that for Treatment 2, then it can be shown that μ_D—the population mean of paired differences—is equal to $\mu_1 - \mu_2$. We are interested in testing

$$H_0: \mu_1 = \mu_2.$$

This is the same as testing the null hypothesis

$$H_0: \mu_D = 0.$$

At this stage we have all the ingredients appropriate for applying the one-sample t test. We have a sample of n independent observations d_1, $d_2, \ldots, d_n$ from a normal population whose variance is unknown, and we wish to test the hypothesis about its mean μ_D. We use the one-sample t test

---

* If you studied designs of experiments in Section 6-3, you saw that the design using paired data is a simple example of a randomized complete block design where each block has two experimental units.

with $n - 1$ degrees of freedom to test H_0: $\mu_D = 0$ against the various alternative hypotheses. The value of the test statistic is

$$\frac{\bar{d} - 0}{s_d/\sqrt{n}}, \qquad \text{that is,} \qquad \frac{\bar{d}\sqrt{n}}{s_d},$$

where

$$\bar{d} = \frac{\sum d_i}{n} \quad \text{and} \quad s_d^2 = \frac{1}{n-1} \sum (d_i - \bar{d})^2.$$

We have the decision rule given in Table 9-8.

TABLE 9-8 Procedure for Paired Difference Test of a Hypothesis H_0: $\mu_1 = \mu_2$

Alternative Hypothesis	The Decision Rule Is to Reject H_0 If the Computed Value of $\dfrac{\bar{d}\sqrt{n}}{s_d}$ Is
1. $\mu_D > 0$, that is, $\mu_1 > \mu_2$	greater than $t_{n-1,\alpha}$
2. $\mu_D < 0$, that is, $\mu_1 < \mu_2$	less than $-t_{n-1,\alpha}$
3. $\mu_D \neq 0$, that is, $\mu_1 \neq \mu_2$	less than $-t_{n-1,\alpha/2}$ or greater than $t_{n-1,\alpha/2}$

Since $d_1, d_2, \ldots, d_n$ constitute a sample of size n from a normally distributed population with unkown variance, we can apply the formula from Section 7-2 to give the following confidence interval for μ_D.

CONFIDENCE INTERVAL FOR μ_D

A $(1 - \alpha)100$ percent confidence interval for μ_D, that is, for $\mu_1 - \mu_2$ using paired observations is given by

$$\bar{d} - t_{n-1,\alpha/2}\frac{s_d}{\sqrt{n}} < \mu_D < \bar{d} + t_{n-1,\alpha/2}\frac{s_d}{\sqrt{n}}.$$

EXAMPLE 4

For studying human energy requirements and obesity, different methods are available for estimating total body fat. The following data were reported by Annica Sohlstrom et al. in *Am J Clin Nutr* 1993; *58:* 830–8, where total body fat of 20 subjects was estimated by underwater weighing (UWW) and magnetic resonance imaging (MRI).

(a) Are the two methods of estimating total body fat significantly different? Use a 5 percent level of significance.

(b) Compute a 95 percent confidence interval for the difference in true mean estimates of total body fat using the two methods.

Subject	UWW x	MRI y	Difference $d = x - y$
1	22.7	25.5	−2.8
2	25.1	24.3	0.8
3	31.1	30.1	1.0
4	30.9	28.3	2.6
5	28.2	29.6	−1.4
6	26.5	27.7	−1.2
7	22.2	24.6	−2.4
8	17.1	16.2	0.9
9	33.7	30.6	3.1
10	22.8	22.8	0
11	24.6	22.9	1.7
12	36.3	30.7	5.6
13	24.0	25.5	−1.5
14	35.6	32.7	2.9
15	30.7	27.1	3.6
16	29.7	30.8	−1.1
17	37.3	31.5	5.8
18	26.9	28.0	−1.1
19	29.5	23.2	6.3
20	26.0	20.7	5.3

SOLUTION The data are not from two independent samples. We have paired data. Column 4 of the table above gives the difference d for each of the pairs. Let μ_1 be the mean for UWW and μ_2 the mean for MRI. We know that $\mu_D = \mu_1 - \mu_2$.

(a) The procedure to carry out the test is as follows.

   i. $H_0: \mu_1 = \mu_2$ (or, equivalently, $\mu_D = 0$).

   ii. $H_A: \mu_1 \neq \mu_2$ (or, equivalently, $\mu_D \neq 0$).

   iii. The test statistic is $\dfrac{\bar{d}\sqrt{n}}{s_d}$.

   iv. The level of significance is $\alpha = 0.05$ and there are 19 degrees of freedom since the sample size is $n = 20$. Because $t_{19,0.025} = 2.093$, the decision rule is: *Reject* H_0 *if the computed value of the test statistic is greater than 2.093 or less than* -2.093.

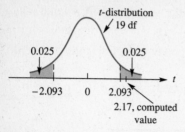

Figure 9-27

v. From Column 4 of the table, it can be verified that $\bar{d} = 1.4$ and $s_d = 2.884$. Therefore, the computed value of the test statistic is

$$\frac{\bar{d}\sqrt{n}}{s_d} = \frac{1.4\sqrt{20}}{2.884} = 2.17. \text{ (See Figure 9-27.)}$$

vi. The computed value is greater than 2.093, and so we reject H_0. (P-value between 0.01 and 0.025).

Conclusion: At the 5 percent level we conclude that the true means of estimated total body fat using UWW method and MRI method are significantly different. The two methods are different.

(b) Since we want a 95 percent confidence interval, the area in each tail is 0.025 and so the t value is the same as the one in Part a. Therefore,

$$t_{n-1,\alpha/2}\,\frac{s_d}{\sqrt{n}} = 2.093\frac{2.884}{\sqrt{20}} = 1.35$$

giving a 95 percent confidence interval for μ_D as

$$1.4 - 1.35 < \mu_D < 1.4 + 1.35, \quad \text{that is, } 0.05 < \mu_D < 2.75.$$

From the confidence interval (0.05, 2.75), we can be 95 percent confident that the mean estimate μ_1 by UWW is in excess of the mean estimate μ_2 by MRI by an amount which is between 0.05 and 2.75. ●

DIFFERENCE IN POPULATION PROPORTIONS

Very often we are interested in deciding whether the observed difference between two sample proportions is due to sampling error or due to the fact that the proportions in the parent populations from which the samples are taken are inherently different. As an example, we will consider the following experiment, where Vaccine A is given to 200 and Vaccine B to 300 randomly picked people.

	Vaccine A	Vaccine B
Infected	80	105
Not infected	120	195
Total	200	300

Of the 200 people who were given Vaccine A, 80 were infected, that is, 40 percent were infected. On the other hand, with Vaccine B, only 105 out of 300 were infected, that is, 35 percent. The difference in the sample proportions is $0.40 - 0.35$, or 0.05. We might ask whether this difference is large enough to suggest that the two vaccines are different in their effectivenes; or is it just possible that the difference is due to sampling error, in which case we should not read much into it? In short, if p_1 and p_2 represent proportions of those infected in the two populations, we want to test the null hypothesis

$$H_0: p_1 = p_2, \text{ that is } p_1 - p_2 = 0.$$

If H_0 is true and $p_1 = p_2$, we denote this common value as p. Suppose among the m people given Vaccine A, X are infected, and among the n people given Vaccine B, Y are infected. We can then obtain a pooled estimate of p using the estimator

$$\hat{p} = \frac{X + Y}{m + n}.$$

Writing p_e to represent the estimate, in our example we get

$$p_e = \frac{80 + 105}{200 + 300} = 0.37.$$

It can be shown that the appropriate test statistic in this case is

$$\frac{\dfrac{X}{m} - \dfrac{Y}{n}}{\sqrt{\hat{p}(1 - \hat{p})\left(\dfrac{1}{m} + \dfrac{1}{n}\right)}}.$$

Because $p_e = 0.37$, the value of the test statistic in our present case is

$$\frac{0.40 - 0.35}{\sqrt{0.37(1 - 0.37)\left(\dfrac{1}{200} + \dfrac{1}{300}\right)}} = \frac{0.05}{\sqrt{0.001942}} = 1.134.$$

If m and n are large, then the test procedures for various forms of the alternative hypothesis are given in Table 9-9.

TABLE 9-9 Procedure to Test $H_0: p_1 = p_2$ When Samples Are Large

Alternative Hypothesis	The Decision Rule Is to Reject H_0 If the Computed Value of the Test Statistic Is
1. $p_1 > p_2$	greater than z_α
2. $p_1 < p_2$	less than $-z_\alpha$
3. $p_1 \neq p_2$	less than $-z_{\alpha/2}$ or greater than $z_{\alpha/2}$

Because the tests are based on the application of the central limit theorem, the level of significance is *approximately* 100α percent.

EXAMPLE 5   A jar containing 120 flies was sprayed with an insecticide of Brand A and it was found that 95 of the flies were killed. When another jar containing 145 flies of the same type was sprayed with Brand B, 124 flies were killed. At the 2 percent level of significance, do the two brands differ in their effectiveness?

SOLUTION   Notice that the data are qualitative in nature.

1. $H_0: p_1 = p_2$, that is $p_1 - p_2 = 0$
2. $H_A: p_1 \neq p_2$

3. The test statistic is $\dfrac{\dfrac{X}{m} - \dfrac{Y}{n}}{\sqrt{\hat{p}(1 - \hat{p}) \left(\dfrac{1}{m} + \dfrac{1}{n} \right)}}$.

4. Because $\alpha = 0.02$, $z_{\alpha/2} = z_{0.01} = 2.33$. Thus, the decision rule is: *Reject H_0 if the computed value is less than -2.33 or greater than 2.33.*

5. Because $x = 95$, $y = 124$, $m = 120$, and $n = 145$,

$$p_e = \frac{x + y}{m + n} = \frac{95 + 124}{120 + 145} = 0.826.$$

Therefore, the computed value of the test statistic is

$$\frac{\dfrac{95}{120} - \dfrac{124}{145}}{\sqrt{0.826(1 - 0.826)\left(\dfrac{1}{120} + \dfrac{1}{145} \right)}} = \frac{0.792 - 0.855}{\sqrt{(0.144)(0.0152)}}$$

$$= -1.35. \text{ (See Figure 9-28.)}$$

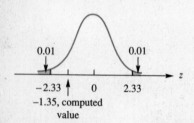

0.01 .... 0.01

−2.33   0   2.33
−1.35, computed
value

Figure 9-28

6. The computed value -1.35 is between -2.33 and 2.33. There is no reason to discard H_0 (P-value 0.177).

Conclusion: There is no evidence to indicate that one spray is significantly different from the other (at the 2 percent level).   ●

Section 9-4 Exercises

1. Suppose there are two normally distributed populations. One population has variance σ_1^2 of 10, and when a sample of size 20 was picked from it, the mean $\bar{x}$ was 9.2. The second population has variance σ_2^2 of 12, and a sample of 16 yielded a mean $\bar{y}$ of 7.2. Assuming that the samples are independent, at the 5 percent level of significance, test the null hypothesis $H_0: \mu_1 = \mu_2$ against the following alternative hypotheses.
   (a) $\mu_1 \neq \mu_2$ (b) $\mu_1 > \mu_2$

2. A placement exam in mathematics was given to fifteen students who had a modern math background and to ten students who had a traditional math background. The mean score of the modern math students was 88 points and that of the traditional math students was 82 points. Suppose it may be assumed that the variances of the score for modern math and traditional math are known and are,

respectively, $\sigma_1^2 = 20$ and $\sigma_2^2 = 12$. At the 5 percent level of significance, do the true mean scores for the two groups differ significantly? Assume that scores in the two populations are normally distributed.

3. Suppose there are two populations. When a sample of 100 items was picked from one population, it was found that $\bar{x}$ is 14 and s_1^2 is 1.2. A sample of 200 items drawn independently from the other population yielded $\bar{y} = 14.4$ with $s_2^2 = 1.4$. At the 10 percent level of significance, test the null hypothesis H_0: $\mu_1 = \mu_2$ against the following alternative hypotheses.
   (a) $\mu_1 \neq \mu_2$ (b) $\mu_1 < \mu_2$

4. A company has two branches, one on the East Coast and the other on the West Coast. The mean daily sales of the company on the East Coast, when observed on 100 days, were found to be $32,000 with s_1 of $2400. For the company on the West Coast, when observed on 200 days, the mean daily sales were $36,000 with s_2 of $2700. At the 5 percent level of significance, do the two branches differ in their true mean daily sales?

5. In a test to compare the performance of two models of cars, Thunderball and Fireball, 128 cars of each model were driven on the same terrain with ten gallons of gasoline in each car. The mean number of miles for Thunderball was 240 miles with $s_1 = 10$ miles; the mean number of miles for Fireball was 248 miles with $s_2 = 16$ miles. At the 2 percent level of significance, are the two models of cars significantly different?

6. When 100 raw commerical apples of Variety A and 150 of Variety B were assayed for their iron content (in milligrams), the following information was obtained.

	Sample Mean (in Milligrams)	Sample Standard Deviation s
Variety A	0.32	0.08
Variety B	0.29	0.05

At the 5 percent level of significance, do the two varieties of apples differ in their iron content?

7. A rancher tried Feed A on 256 cattle and Feed B on 144 cattle. The mean weight of cattle given Feed A was found to be 1350 pounds with s_1 of 180 pounds. On the other hand, the mean weight of the cattle given Feed B was found to be 1430 pounds with s_2 of 210 pounds. At the 5 percent level of significance, is Feed B significantly better than Feed A?

8. The data below present the summary of SAT scores in mathematics for 228 white students and 105 Asian-American students in 1992–93.

	White	Asian-American
Sample size	228	105
Mean	498	529
Standard deviation	68	73

At the 1 percent level, is the performance of white students lagging behind that of the Asian-American students?

9. A bank manager has the unpleasant task of eliminating a position of teller. Of two tellers who were the last hired, she has decided to let go one. Both tellers are equally satisfactory with regard to their personality and cordiality with the customers. So, the manager has decided to use statistical evidence to compare their efficiency. Toward this, unknown to the tellers, the time (in minutes) that each teller takes to serve a customer was recorded. For each teller 60 customers were observed. The table below gives the summary data.

	Teller 1	Teller 2
Mean time	2.3	2.6
Standard deviation	0.7	0.5

(a) Suppose the manager argues that the mean time for Teller 1 is less than that for Teller 2 and so Teller 2 should be given the pink slip. How would you counter this line of reasoning?

(b) Is there evidence that Teller 2 is slower than Teller 1? Carry out the test at the 5 percent level.

10. The myristicin content and elemicin content in Sri Lankan nutmeg oil samples were determined with gas-liquid chromatography. The following information summarizes the data for laboratory distilled oil and industrial samples for each chemical characteristic.

	MYRISTICIN CONTENT	
	Laboratory	Industrial
Mean	1.94	2.29
Standard deviation	1.63	1.21
Number	36	40

	ELEMICIN CONTENT	
	Laboratory	Industrial
Mean	0.913	1.186
Standard deviation	0.58	0.70
Number	33	39

For each chemical characteristic, test for the equality of means against specified alternative hypothesis H_A, at the indicated level of significance.

(a) Myristicin content, at the 10 percent level, where $H_A: \mu_{lab} \neq \mu_{ind}$

(b) Elemicin content, at the 2 percent level, where $H_A: \mu_{lab} < \mu_{ind}$

11. Assume that we have two populations, both normally distributed with a common variance σ^2 that is unknown. If $\bar{x} = 10.2$, $\bar{y} = 8.5$, $m = 10$, $n = 12$, $s_1^2 = 2.8$, and $s_2^2 = 3.2$, assuming the samples are independent, test the hypothesis $H_0: \mu_1 = \mu_2$ against the following alternative hypotheses at the 5 percent level.

(a) $\mu_1 \neq \mu_2$

(b) $\mu_1 > \mu_2$

In Exercises 12–29, assume that the two populations are normally distributed with the same, but unknown, variance.

12. The mean length of fourteen trout caught in Clear Lake was 10.5 inches with s_1 of 2.1 inches, and the mean length of eleven trout caught in Blue Lake was 9.6 inches with s_2 of 1.8 inches. Is the true mean length of trout in Clear Lake significantly greater than the true mean length of trout in Blue Lake? Use $\alpha = 0.1$.

13. A random sample of ten light bulbs manufactured by Company A had a mean life of 1850 hours and a sample standard deviation of 130 hours. A random sample of twelve light bulbs manufactured by Company B had a mean life of 1940 hours with a sample standard deviation of 140 hours. Is the true mean life of bulbs manufactured by Company B significantly different from the true mean life of bulbs manufactured by Company A? Use $\alpha = 0.05$.

14. In twelve half-hour morning programs, the mean time devoted to commercials was 6.4 minutes with s_1 of 1 minute. In fourteen half-hour evening programs, the mean time was 5.6 minutes with s_2 of 1.3 minutes. At the 10 percent level of significance, is the true mean time devoted to commercials in the evening significantly less than the true mean time devoted to commercials in the morning?

15. A sociologist wants to compare the dexterity of children living in a well-to-do neighborhood with that of children living in a poor neighborhood. A random sample is selected of five children from the well-to-do neighborhood and six children from the poor neighborhood, and the time that each child takes to assemble a certain toy is recorded. The following figures give the times.

Well-to-do	6	8	7	6	10	
Poor	4	10	9	7	11	8

At the 5 percent level of significance, are the children from the well-to-do neighborhood more adept at assembling the toy than those in the poor neighborhood?

16. In order to evaluate the degree of suspension of a polyethylene, its gel contents (gel proportion) are determined after extraction using a solvent. The method is called the gel proportion estimation method. A study was run to compare two solvents, ethanol and toluene, using extraction time of 8 hours. The data are presented below:

Ethanol	Toluene
94.7	96.6
96.4	95.5
93.0	95.9
95.4	95.0
96.4	96.1
94.6	95.3
94.8	95.4
96.9	96.8
94.8	95.2
94.8	96.1

At the 5 percent level of significance, is the true mean gel content extracted different for the two solvents?

17. Two models of cars that were similar in specifications were compared for their acceleration from 0 to 60 mph. Acceleration tests were carried out from standstill with engine idling at start. Also, all the runs were done with gears shifted to best advantage. The results giving the number of seconds are presented below.

Model 1	15	12	10	12	10	11	12	12	14	16
Model 2	14	16	14	12	16	12	13	15	13	

The promoter of Model 1 claims that the model is superior to Model 2 for acceleration. At the 5 percent level of significance, do the data bear out the claim?

18. A psychologist measured reaction times (in seconds) of eight individuals who were not given any stimulant and six individuals who were given an alcoholic stimulant. The figures are given as follows.

Reaction time without stimulant	3.0	2.0	1.0	2.5	1.5	4.0	1.0	2.0
Reaction time with stimulant	5.0	4.0	3.0	4.5	2.0	2.5		

At the 5 percent level of significance, is the true mean reaction time increased significantly by alcoholic stimulant?

19. The following figures refer to the amounts of carbohydrates (in grams) per serving of two varieties of canned peaches.

Variety A	32	32	27	38	29		
Variety B	27	28	33	30	26	29	28

Based on this sample data, at the 5 percent level of significance, are the two varieties significantly different in their true mean carbohydrate content?

20. Ten voters were picked at random from those who voted in favor of a certain proposition, and twelve voters were picked at random from those who voted against it. The following figures give their ages.

In favor	28	33	27	31	29	25	50	30	25	41		
Against	31	43	49	32	40	41	48	30	29	39	42	36

At the 5 percent level of significance, is the true mean age of those voting against the proposition significantly different from the true mean age of those voting for it?

21. In an experiment to investigate the effect of precooking of beef in water and pressure cooking of beef prior to smoke drying, a cut of hindquarters of beef was cut into rectangular pieces each 3 inches thick. Fifteen pieces were pressure cooked and twelve pieces were precooked in water. The thickness of each piece was measured. The following table gives the summary of the thickness data.

	THICKNESS (INCHES)	
	Pressure-Cooked Meat	**Water-Cooked Meat**
Mean	2.77	2.90
Standard deviation	0.32	0.23

At the 2 percent level, are the mean thicknesses for the two methods of cooking significantly different?

22. Two types of soft drinks, a cola and a noncola, were tested for their amount of glucose (g/100 ml). The findings are summarized as follows.

	GLUCOSE (g/100 ml)	
	Cola	Noncola
Mean	4.10	4.63
Standard deviation	0.06	0.09
Number	15	18

At the 5 percent level, is the true mean amount of glucose for noncola significantly greater than the true mean amount for cola?

23. Urea, which occurs naturally as a minor component of milk, has a significant effect on milk's heat stability. The following experiment was conducted to compare two diets for their effect on the urea concentration of milk. Twenty-two cows which had calved approximately a month earlier were selected for the experiment. Diet 1 was given to ten cows for 4 weeks, and Diet 2 was given to the remaining twelve cows for the same length of time. The urea concentration in the milk (g/kg) was measured for each cow at the end of the period.

UREA CONCENTRATION
(g/kg)

Diet 1	Diet 2
0.45	0.36
0.42	0.37
0.40	0.38
0.49	0.43
0.45	0.40
0.43	0.42
0.50	0.43
0.40	0.39
0.44	0.35
0.48	0.41
	0.37
	0.45

Is the true mean urea concentration significantly different for the two diets? Use $\alpha = 0.10$.

24. Three fish-rearing ponds were fertilized with fodder yeast and mineral fertilizer containing superphosphate and ammonium nitrate. (*Hydrobiological Journal* 1982; *18:* pp. 58–61). Five determinations of dissolved oxygen concentration

(mg/l) were made in two of the ponds and four determinations were made in the third pond.

DISSOLVED OXYGEN (mg/l)		
Pond 1	Pond 2	Pond 3
5.5	5.3	5.8
5.8	5.2	5.9
5.6	5.4	5.7
5.5	5.1	5.6
5.4	5.3	

Carry out the following tests of hypotheses, at the 5 percent level, to test whether the true mean dissolved oxygen content of one pond is significantly different from that of the other.

(a) Pond 1 versus Pond 2

(b) Pond 2 versus Pond 3

(c) Pond 1 versus Pond 3

25. It is believed that eels do not utilize soybean meal very efficiently. European eels (*Anguilla anguilla*) obtained from the coastal waters of France were studied to explore this hypothesis. A batch of eighteen eels was fed a diet consisting of 0% level of soybean meal and another batch of eighteen eels was fed 20% level of soybean meal. The following data give the body protein on a dry weight basis.

	Diet 1 0% Soybean	Diet 2 20% Soybean
Mean	59.67	55.79
Standard deviation	3.58	2.30
Number	18	18

(a) Use the data to obtain a 98 percent confidence interval for the difference in the true mean body protein with the two diets.

(b) On the basis of the information provided, is the true mean body protein at 20% level of soybean meal significantly less than the true mean body protein at 0% level of soybean meal? Test using $\alpha = 0.05$.

26. The figures below relate to the weights (g) of natural food zooplankton in the intestines of fifteen carp grown in a lake and twelve grown in a bay. At the 2.5 percent level of significance, is the true mean intake of the bay carp less than that of the lake carp?

ZOOPLANKTON WEIGHT(g)	
Lake	Bay
212	181
196	138
186	163
185	156

(continued)

ZOOPLANKTON WEIGHT(g)	
Lake	Bay
171	158
173	146
182	175
193	168
160	167
192	164
148	155
191	168
188	
178	
197	

27. The conventional method of acid hydrolysis of different types of solanum steroid glycoalkaloids is not suitable for the simultaneous hydrolysis of the different types of glycoalkaloids, especially solanine and tomatine. A new hydrolysis technique was developed using a two-phase system where the hydrolysis medium consists of aqueous acid and an apolar organic solvent, resulting in two immiscible phases. (*Journal of the Science of Food and Agriculture* 1984; *35:* pp. 487–494).

Glycoalkaloids are natural constituents of potatoes. To determine whether the new hydrolysis technique is superior, ten samples of fresh potato tubers of the cultivar *Bintje* were analyzed for solanine content (mg/20 g) by conventional acid hydrolysis (CAH) and a equal number of samples were analyzed using the two-phase hydrolysis (TPH). The folowing data give the solanine determination in the samples under the two methods.

Solanine Determination	
CAH	TPH
0.73	0.84
0.59	0.99
0.67	0.86
0.79	0.79
0.80	0.79
0.49	0.92
0.54	1.08
0.52	0.61
0.73	0.94
0.68	0.89

At the 5 percent level of significance, is the TPH method of hydrolysis superior to the CAH method?

28. Fifteen hens, approximately 62 weeks of age, and ten turkeys, approximately 17 weeks of age, were slaughtered and the total haem pigments (mg/g of tissue) in the breast meats measured. The data are summarized as follows.

	HAEM CONTENT IN BREAST MEAT	
	Hens	Turkeys
Mean	0.64	0.58
Standard deviation	0.046	0.052
Number	15	10

Test the claim that the true mean haem content in the breast meat of hens is significantly different from that in the breast meat of turkeys. Use $\alpha = 0.05$.

29. The following data contain the summary of total lipid content (g) of smoke dried beef depending on whether the beef was cooked in water or pressure cooked prior to smoke drying.

	LIPID CONTENT (G)	
	Pressure Cooked	Water Cooked
Mean	8.50	9.58
Standard deviation	0.6	0.8
Number	15	12

On the basis of the information provided, is the true mean lipid content when the beef is pressure cooked significantly different from the true mean lipid content when it is water cooked? Test using $\alpha = 0.05$.

30. Suppose you are interested in comparing means of two populations by using paired differences. The following data are available.

Pair	1	2	3	4	5
x	17	25	36	16	14
y	19	18	30	21	10

(a) Find the differences d_i, that is, $x_i - y_i$, and compute $\bar{d}$ and s_d.
(b) Suppose you wish to test

$$H_0: \mu_D = 0 \quad \text{against} \quad H_A: \mu_D > 0.$$

Determine the critical region at the 5 percent level.
(c) Compute the value of the test statistic.
(d) What conclusion do you arrive at using the 5 percent level of significance?

31. Suppose you want to compare means of two populations using paired observations. If $\bar{d} = -2.85$, $s_d = 3.72$, and $n = 15$, test the hypothesis $H_0: \mu_1 = \mu_2$ against the following alternative hypotheses using $\alpha = 0.05$.
(a) $\mu_1 \neq \mu_2$ (that is, $\mu_D \neq 0$)
(b) $\mu_1 < \mu_2$ (that is, $\mu_D < 0$)

32. A dealer of a certain Japanese car claims that it is easier to change spark plugs on a four-cylinder model that he sells than on a comparable U.S. model. To test the claim, eight mechanics of proven ability who were not employees of the dealership nor of the U.S. model being tested each were assigned to change spark plugs on the two models. Their times (in minutes) are recorded as follows.

Mechanic	Japanese Model	U.S. Model
Toni	15	19
Adam	21	24
Susan	22	19
Henry	13	16
Kenji	18	16
Carlos	21	23
Bob	20	19
Winnie	25	26

(a) Are the two samples independent? Why?

(b) At the 1 percent level of significance, is the dealer's claim valid?

(c) State the underlying assumptions.

33. Scores of nine students before and after a special coaching program are given as follows.

Student	Before Coaching	After Coaching
1	91	82
2	78	80
3	47	62
4	37	49
5	64	55
6	54	73
7	43	59
8	33	58
9	53	43

(a) Explain why matched pair analysis is appropriate in this case.

(b) At the 5 percent level of significance, does the special coaching benefit students in that student scores are higher after coaching?

(c) What assumptions did you make?

34. Is it safe for an individual to drink two beers and drive? To test whether this amount of beer impairs driving ability, the department of motor vehicles conducted a simulated experiment with seven volunteers, each of whom drove through an obstacle course prior to consuming any alcohol and again after drinking two beers. Driving ability was measured by recording the total reaction time (in seconds) of a driver in avoiding the various obstacles on the course. The data are presented as follows.

Driver	1	2	3	4	5	6	7
Without alcohol	8.82	6.62	9.84	10.81	9.20	11.67	4.42
With alcohol	10.61	8.18	12.24	13.45	8.88	10.88	5.38

At the 5 percent level of significance, is it true that consumption of two beers impairs driving ability? State the underlying assumptions.

35. Glutamic acid is one of the amino acids commonly found in ripe fruits. The amount of glutamic acid (mg/100 g fresh tissue) was determined spectrophotometrically in the pericarp tissue and the core tissue of ten ripe mangoes of a certain variety. The data are presented as follows.

GLUTAMIC ACID

Pericarp Tissue	Core Tissue
9.52	9.50
10.16	10.02
11.33	12.17
10.86	10.51
8.88	11.43
10.30	11.35
11.16	12.49
9.89	12.52
11.82	10.96
10.03	10.61

At the 5 percent level of significance, test to see whether the true mean glutamic acid in the core tissue is greater than that in the pericarp tissue.

36. Brandstetter et al. (*Heart and Lung,* March 1988, pp. 170–172) wanted to study the concern that meal-induced hypoxemia may occur in asymptomatic patients with chronic obstructive pulmonary disease (COPD). The research group did a study on eleven patients in which arterial blood gas (ABG) samples were drawn on each patient just before and 30 minutes after the initiation of bolus feeding through an NG tube. In each ABG sample the oxygen partial pressure (PaO_2) was measured (mm Hg). The data were as follows.

Patient	First PaO_2	Second PaO_2
1	60	62
2	83	85
3	87	85
4	52	50
5	86	84
6	95	93
7	59	58
8	101	99
9	80	77
10	91	88
11	89	82

On the basis of these data, at the 5 percent level of significance, is there a decrease in PaO_2 as a result of feeding? (That is, is meal-induced hypoxemia in patients with COPD a clinical problem?)

37. Determine a 90 percent confidence interval for μ_D in Exercise 32.

38. Determine a 95 percent confidence interval for μ_D in Exercise 33.

39. Determine a 95 percent confidence interval for μ_D in Exercise 34.

40. A psychology professor assigned a project to two students, Carol and Russell. They were to work on the project independently. The objective of the study was to evaluate whether good grooming increased self-esteem of students in middle school. Self-esteem was measured on a numerical scale based on the sum of scores on 8 criteria.

Carol studied 10 students before good grooming and again, at the end of a period, after good grooming.

Russell studied two sets of ten students. In one set the students were well groomed and in the other set they were not.

Discuss how the two approaches differ and the type of analysis that would be appropriate in each case.

41. A study was conducted by Daniela C. Wallace et al. to study "the effects of iced and room temperature injectate on cardiac output measurements in critically ill patients with low and high cardiac outputs". The analysis presented in the journal is reproduced below. ("Study of the Effects of Iced and Room Temperature Injectate on Cardiac Output Measurements in Critically Ill Patients with Low and High Cardiac Outputs" by Daniela C. Wallace et al. from *Heart & Lung*, January/February 1993, vol. 22, no. 1. Reprinted by permission of Mosby-Year Book, Inc. and the author.)

Summary of Differences between Room Temperature and Iced Injectate Cardiac Output (CO) Values in Subjects with Low and High Cardiac Outputs

Subject No.	Room Temp. CO (L/min)	Iced Temp. CO (L/min)	Difference*	% Difference[†]
Low CO Group				
1	1.7	1.3	0.4	24
2	3.4	3.3	0.1	3
3	3.2	2.9	0.3	9
4	3.7	2.7	1.0	27
5	3.3	3.2	0.1	3
6	3.2	2.8	0.4	13
7	3.3	2.9	0.4	12
8	3.5	3.4	0.1	3
9	3.5	3.1	0.4	11
10	2.4	1.8	0.6	25
11	2.3	2.0	0.3	13
Mean	3.0	2.7	.37	13

Subject No.	Room Temp. CO (L/min)	Iced Temp. CO (L/min)	Difference*	% Difference[†]
High CO Group				
12	10.0	9.4	0.6	6
13	9.3	8.7	0.6	6
14	13.7	10.7	3.0	22
15	8.2	7.8	0.4	5
16	9.3	8.9	0.4	4
17	9.4	6.7	2.7	29
18	8.9	8.6	0.3	3
19	12.2	10.3	1.9	16
20	9.5	8.7	0.8	8
21	9.3	8.3	1.0	11
Mean	10.0	8.8	1.17	11

*Note: The difference is calculated by subtracting the iced temperature CO from the room temperature CO.

[†] % Difference from room temperature values.

Group	N	t Value	p Value
Low CO	11	−4.75	0.001*
High CO	10	−3.90	0.004*

*$p < 0.05$.

Carry out a thorough critical analysis of the presented results.

42. Suppose we wish to compare proportions of an attribute in two populations. When a sample of 80 was picked from one population it was found that $x = 36$ had the particular attribute, whereas a sample of 60 picked from the other population showed that $y = 18$ individuals had the attribute. Assuming that the two samples are independent, at the 2 percent level of significance, test the hypothesis $H_0: p_1 = p_2$ against the following alternative hypotheses.
(a) $H_A: p_1 \neq p_2$ (b) $H_A: p_1 > p_2$

43. Of 150 Democrats interviewed, 90 favor a proposition, and of 120 Republicans interviewed, 80 favor it. Is the true proportion of individuals in favor of the proposition significantly larger among Republicans than among Democrats? Use $\alpha = 0.01$.

44. A survey was conducted to compare the proportions of males and females who favor government assistance for child care. It was found that among 64 males interviewed, 40 favored assistance, and among 100 females, 70 favored assistance. At the 5 percent level of significance, are the true proportions among males and females in the population who favor government assistance for child care significantly different?

45. Two hundred students in a class are divided into two groups of 100. One of the groups was given a vaccine against the common cold. The other group was not given any vaccine. It was found that in the group that was given the vaccine, forty-five got colds. In the other group, fifty-four got colds. At the 2 percent level of significance, is the vaccine effective against the common cold?

46. In the course of his campaign, a presidential candidate made a "foot-in-the-mouth" statement. When a random sample of 400 were interviewed to see if this had affected the candidate's standing, 212 said that they favored the candidate. However, during the week before this unfortunate episode, when a random sample of 600 had been interviewed, 348 had voted in favor. At the 5 percent level of significance, did the statement hurt the candidate?

47. A survey included 191 past drinkers (PD) of alcohol and 215 life-long abstainers (LLA) (Reported by Ann Edward et al. in *American Journal of Public Health* 1986; *76;* pp. 68–70). Among the PD group, 95 responded that they attended church one or more times a week, whereas in the LLA group 131 did. At the 5 percent level, does the true proportion of church attendants among PD differ significantly from that among LLA?

48. A clinical study was made to determine the side effects of Terazosin, a recently developed drug used for control of hypertension. The study consisted of 859 patients on Terazosin and 506 on a placebo. During the controlled time period each person was asked to keep track of general body ailments (headache, backache, asthenia). Of the 859 on Terazosin therapy, 139 complained of general ailments. Of the 506 taking a placebo, 80 complaints were registered. Should a warning of general body ailments as a side effect be issued to patients on Terazosin? Justify your answer.

49. *Do you worry about being a victim of crime?* The following results were obtained from a telephone poll of 500 adult Americans taken for *Time* on August 23, 1993, by Yankelovich Partners, Inc.

	City	Suburb	Rural
Yes	59%	57%	43%

Since the number of adults in each category in the poll are not given, assume that there were 140 adults from city, 160 from suburban, and 200 from rural areas.

(a) At the 2 percent level of significance, are the true proportions of those who worry about being a victim different for city and suburb?

(b) At the 5 percent level of significance, is the true proportion of those in city who worry significantly greater than that in the rural area?

50. A sample of 503 whites and 315 blacks in a Gallup poll survey yielded the results shown when they were asked the following question: *Do you think that police officers treat criminal suspects differently in low-income neighborhoods than in middle- or high-income neighborhoods?*

	PERCENT	
	Whites	**Blacks**
Yes	62	80

At the 5 percent level of significance, are the true proportions among whites and blacks who would say yes to the question significantly different?

✓ *CHECKLIST: KEY TERMS AND EXPRESSIONS*

- ❏ statistical hypothesis, page 492
- ❏ alternative hypothesis, page 495
- ❏ null hypothesis, page 495
- ❏ test of the null hypothesis, page 497
- ❏ test statistic, page 497
- ❏ Type I error, page 498
- ❏ level of significance, page 498

- ❏ Type II error, page 498
- ❏ decision rule, page 503
- ❏ test statistic, page 507
- ❏ rejection region, page 507
- ❏ critical region, page 507
- ❏ test of a statistical hypothesis, page 507
- ❏ one-tail test, page 508

- ❏ right-tail test, page 508
- ❏ left-tail test, page 508
- ❏ two-tail test, page 509
- ❏ *P*-value, page 510
- ❏ significance probability, page 510
- ❏ paired *t* test, page 542

KEY FORMULAS

TABLE Summary of Tests of Hypotheses (Single Population)

Nature of the Population	Parameter	Null Hypothesis	Alternative Hypothesis	Test Statistic	Decision Rule: Reject H_0 If the Computed Value Is
Quantitative data; variance σ^2 is known; population normally distributed	μ	$\mu = \mu_0$	1. $\mu > \mu_0$ 2. $\mu < \mu_0$ 3. $\mu \neq \mu_0$	$\dfrac{\bar{X} - \mu_0}{\sigma/\sqrt{n}}$	1. greater than z_α 2. less than $-z_\alpha$ 3. less than $-z_{\alpha/2}$ or greater than $z_{\alpha/2}$
Quantitative data; variance σ^2 is *not* known; population normally distributed; sample size small ($n < 30$)	μ	$\mu = \mu_0$	1. $\mu > \mu_0$ 2. $\mu < \mu_0$ 3. $\mu \neq \mu_0$	$\dfrac{\bar{X} - \mu_0}{S/\sqrt{n}}$	1. greater than $t_{n-1,\alpha}$ 2. less than $-t_{n-1,\alpha}$ 3. less than $-t_{n-1,\alpha/2}$ or greater than $t_{n-1,\alpha/2}$
Quantitative data; variance σ^2 is *not* known; population not necessarily normal; sample large ($n \geq 30$)	μ	$\mu = \mu_0$	1. $\mu > \mu_0$ 2. $\mu < \mu_0$ 3. $\mu \neq \mu_0$	$\dfrac{\bar{X} - \mu_0}{S/\sqrt{n}}$	1. greater than z_α 2. less than $-z_\alpha$ 3. less than $-z_{\alpha/2}$ or greater than $z_{\alpha/2}$ The level of significance is approximately 100α percent.
Quantitative data; population normally distributed	σ	$\sigma = \sigma_0$	1. $\sigma > \sigma_0$ 2. $\sigma < \sigma_0$ 3. $\sigma \neq \sigma_0$	$\dfrac{(n-1)S^2}{\sigma_0^2}$	1. greater than $\chi^2_{n-1,\alpha}$ 2. less than $\chi^2_{n-1,1-\alpha}$ 3. less than $\chi^2_{n-1,1-\alpha/2}$ or greater than $x^2_{n-1,\alpha/2}$
Qualitative data; binomial case	p	$p = p_0$	1. $p > p_0$ 2. $p < p_0$ 3. $p \neq p_0$	$\dfrac{(X/n) - p_0}{\sqrt{\dfrac{p_0(1 - p_0)}{n}}}$ Here n is assumed large.	1. greater than z_α 2. less than $-z_\alpha$ 3. less than $-z_{\alpha/2}$ or greater than $z_{\alpha/2}$ The level of significance is approximately 100α percent.

TABLE Summary of Tests of Hypotheses (Two Populations)

Nature of the Population	Parameter	Null Hypothesis	Alternative Hypothesis	Test Statistic	Decision Rule: Reject H_0 If the Computed Value Is
Quantitative data; variances σ_1^2 and σ_2^2 are known; both populations normally distributed.	$\mu_1 - \mu_2$	$\mu_1 = \mu_2$	1. $\mu_1 > \mu_2$ 2. $\mu_1 < \mu_2$ 3. $\mu_1 \neq \mu_2$	$\dfrac{\bar{X} - \bar{Y}}{\sqrt{\dfrac{\sigma_1^2}{m} + \dfrac{\sigma_2^2}{n}}}$	1. greater than z_α 2. less than $-z_\alpha$ 3. less than $-z_{\alpha/2}$ or greater than $z_{\alpha/2}$
Quantitative data; σ_1^2, σ_2^2 unknown; populations not necessarily normal; sample sizes are large	$\mu_1 - \mu_2$	$\mu_1 = \mu_2$	1. $\mu_1 > \mu_2$ 2. $\mu_1 < \mu_2$ 3. $\mu_1 \neq \mu_2$	$\dfrac{\bar{X} - \bar{Y}}{\sqrt{\dfrac{S_1^2}{m} + \dfrac{S_2^2}{n}}}$	1. greater than z_α 2. less than $-z_\alpha$ 3. less than $-z_{\alpha/2}$ or greater than $z_{\alpha/2}$
Quantitative data; σ_1^2 and σ_2^2 are not known but are assumed to be equal; both populations are normally distributed.	$\mu_1 - \mu_2$	$\mu_1 = \mu_2$	1. $\mu_1 > \mu_2$ 2. $\mu_1 < \mu_2$ 3. $\mu_1 \neq \mu_2$	$\dfrac{\bar{X} - \bar{Y}}{S_p \sqrt{\dfrac{1}{m} + \dfrac{1}{n}}}$ where $S_p = \sqrt{\dfrac{(m-1)S_1^2 + (n-1)S_2^2}{m + n - 2}}$	1. greater than $t_{m+n-2,\alpha}$ 2. less than $-t_{m+n-2,\alpha}$ 3. less than $-t_{m+n-2,\alpha/2}$ or greater than $t_{m+n-2,\alpha/2}$
Paired observations	$\mu_1 - \mu_2$ $(= \mu_D)$	$\mu_D = 0$ i.e., $\mu_1 = \mu_2$	1. $\mu_D > 0$ 2. $\mu_D < 0$ 3. $\mu_D \neq 0$	$\dfrac{\bar{d}\sqrt{n}}{s_d}$ where $\bar{d} = \sum d_i/n$ and $s_d = \sqrt{\dfrac{1}{n-1} \sum (d_i - \bar{d})^2}$	1. greater than $t_{n-1,\alpha}$ 2. less than $-t_{n-1,\alpha}$ 3. less than $-t_{n-1,\alpha/2}$ or greater than $t_{n-1,\alpha/2}$
Qualitative data; binomial case	$p_1 - p_2$	$p_1 = p_2$	1. $p_1 > p_2$ 2. $p_1 < p_2$ 3. $p_1 \neq p_2$	$\dfrac{\dfrac{X}{m} - \dfrac{Y}{n}}{\sqrt{\hat{p}(1 - \hat{p})\left(\dfrac{1}{m} + \dfrac{1}{n}\right)}}$ where $\hat{p} = \dfrac{X + Y}{m + n}$ Both m and n are assumed large.	1. greater than z_α 2. less than $-z_\alpha$ 3. less than $-z_{\alpha/2}$ or greater than $z_{\alpha/2}$ The level of significance is approximately 100α percent.

CHAPTER 9 Concept and Discussion Questions

1. A company is contemplating whether the ingredients of the soup it produces should be changed. If 50 percent or less prefer the new formula, the company will retain the old formula. If more than 50 percent prefer the new formula, the company will change to the new formula. Suppose p is the proportion in the population who would prefer the new formula, and the null and the alternative hypotheses are set as

$$H_0: p \leq 0.5$$
$$H_A: p > 0.5.$$

Does this formulation of the two hypotheses indicate that the company is eager to change the ingredients or to keep the old formula?

2. Suppose a survey of 200 male students and 100 female students gave the following summary regarding their verbal score on the SAT test.

Gender	Number	Verbal Scores
Male	200	384
Female	100	408

Is it possible to test statistically whether the true mean score for females is higher than that for males? Explain your answer.

3. Suppose a survey of all the male and female students in a county who took the SAT test in 1994 gave the following summary regarding their scores on the verbal test.

Gender	Number	Verbal Scores
Male	812	392
Female	741	408

Do you need a statistical test to decide whether female students in the county had a higher mean verbal test score than male students in 1994? If yes, what hypothesis would you be testing? If not, give your reason.

4. For testing a null hypothesis against an alternative hypothesis, data were collected and the value of the test statistic computed. If the value was not significant at the given level of significance, which of the following two interpretations is more appropriate?
   (a) Accept the null hypothesis.
   (b) There is insufficient evidence to reject the null hypothesis.

5. Suppose the level of significance for carrying out the test of a hypothesis is lowered from 5 percent to 1 percent. As a result, has the likelihood of rejecting the null hypothesis when it is the true increased or decreased?

6. Suppose you wish to test H_0: there is no difference in true mean IQ of males and females at the 2 percent level of significance. Interpret the "two percent level of significance" with reference to the hypothesis regarding IQ.

   After computing the relevant statistic, suppose the P-value is 0.001. Interpret the P-value. What decision would you take?

7. To which type of error does the level of significance of a test refer?

8. Can one make a Type I error and a Type II error on the same decision?

9. Since the Type I and Type II errors have an inverse relation, what venue does the experimenter have to exercise simultaneous control on both types of errors?

CHAPTER 9   Review Exercises

1. Define the following terms; *null hypothesis, alternative hypothesis, Type I error, Type II error, level of significance, critical region.*

2. In carrying out the test of a hypothesis, why is it so important to state the alternative hypothesis?

3. Suppose the null hypothesis states: Vitamin C is effective in controlling colds. State what action will constitute a Type I error and what action will constitute a Type II error.

4. Suppose we have H_0: $\sigma^2 = 25$ against H_A: $\sigma^2 = 9$. State which of the following statements are true.

   (a) $\dfrac{(n-1)S^2}{25}$ has a chi-square distribution.

   (b) $\dfrac{(n-1)S^2}{9}$ has a chi-square distribution.

   (c) $\dfrac{(n-1)S^2}{9}$ has a chi-square distribution if H_0 is false.

   (d) $\dfrac{(n-1)S^2}{25}$ has a chi-square distribution if H_0 is true.

5. Suppose we have H_0: $p = 0.4$ against H_A: $p = 0.3$. A sample of 100 is picked. Find the approximate distribution of $X/100$, the sample proportion:
   (a) when H_0 is true   (b) when H_A is true.

6. Rosemary Farm of Santa Maria, California, a poultry farm, says it has a secret formula for chicken feed that makes its eggs low in cholesterol (UPI report, October 1988). As a cautious consumer who wants to test the claim, how would you state the null hypothesis and the alternative hypothesis?

7. The following figures refer to weights (in ounces) of ten one-pound cans of peaches distributed by a company.

16.3	15.7	16.8	16.5	15.5	16.0	16.2	15.8
15.4	16.8						

At the 5 percent level of significance, decide the following.

(a) Is the true mean of all the cans distributed by the company different from 16 ounces?

(b) Is the true variance of all the cans distributed by the company different from 0.16 ounces squared?

8. Thirty experimental launches are made with a rocket-launching system. The outcomes are given as follows. (S is recorded if successful and U if unsuccessful.)

S	S	U	U	S	S	S	U	S	S
U	S	S	S	S	U	U	S	S	U
S	S	S	S	U	S	U	S	S	S

At the 5 percent level of significance, is the true proportion of successful launches among all possible launches different from 0.8?

9. The suspended organic matter (SOM) (g/m^3) was determined in the northern and northwestern zones of Lake Onega, including Petrozavodsk and Kondopoga inlets in the Soviet Union (*Hydrobiological Journal* 1982; *18:* pp. 91–94). The concentration of SOM was determined at nineteen stations. The data are as follows.

SOM (g/m^3)
3.2
3.2
3.8
3.8
2.7
3.2
2.7
3.9
1.8
2.0
1.9
2.5
5.4
4.6
4.2
4.1
4.0
3.0
1.7

On the basis of the data presented, assuming a normal distribution, is the true standard deviation of the SOM content significantly different from 1? Test using $\alpha = 0.05$.

10. In recent years the public has become increasingly aware of the importance of fiber in the diet. The following data present the amount of dietary fiber (g/100 g) in ten samples of white bread sold in the supermarket.

    4.8   4.7   4.7   5.0   4.5   5.1   4.7   4.8   4.6   4.9

    At the 5 percent level of significance, is the true mean dietary fiber in white bread significantly less than 5, as a nutritionist claims? Assume a normal distribution.

11. The following figures give the β-lactoglobuline in twelve samples of cow's milk.

    9.6   9.7   10.0   8.8   9.5   9.4   9.2   9.8   9.5
    8.5   10.1   9.3

    It is claimed that the true mean β-lactoglobuline is greater than 9.2. Test the claim at the 2.5 percent level of significance. Assume a normal distribution.

12. Because of their nutritional importance, an analysis was carried out to find the basic organic constituents of the tissue of the edible freshwater fish *Mystus vittatus*. The following give the amount of total carbohydrates in the liver tissue of the fish.

    35.7   34.9   35.3   35.7   35.4   36.0   35.9   33.4
    37.5   35.8   37.6   33.9   37.3   36.7   37.0

    Test the claim that the true mean carbohydrate contents is significantly greater than 35. Find the *P*-value and use $\alpha = 0.02$ to arrive at the decision. Assume a normal distribution for the amount of carbohydrate in the liver tissue.

13. The saponin contents of variety UH-80-7 of black gram, a legume, were determined by two methods: (1) cooking of soaked seeds; and (2) cooking of unsoaked seeds. The following summary figures were reported.

	SAPONIN (mg/100 g)	
	Soaked Seeds	Unsoaked Seeds
Mean	2275	2318
Standad deviation	56	67
Number	16	14

    Do the two ways of cooking yield significantly different mean saponin? Carry out the test at the 2 percent level. Assume a normal distribution of saponin under both methods.

14. Information on peptides identified in the brain of different animals is of interest to biologists. An experiment was carried out to compare the content of peptides (μmole/g fresh tissue) in the brains of two species of fish from the Black Sea: dogfish (*Squalus acanthias*) and the skate (*Raja clavata L.*). Brains were quickly removed from fish freshly caught and the peptides in the brain determined by electropherosis (*Hydrobiological Journal;* Vol. 5: pp. 81–83). The values are given as follows.

PEPTIDE CONTENT (μmole/g TISSUE)	
Skate	**Dogfish**
1.00	0.51
0.53	0.74
0.41	0.84
0.70	0.80
0.70	0.60
0.59	0.74
0.82	0.82
0.80	0.73
0.53	0.98
0.70	0.59
0.68	0.74

At the 10 percent level of significance, is there a difference in the true mean peptide content in the brain tissues of the two species? Assume that peptide content is normally distributed for the two species with the same variance.

15. To investigate whether cultivation of cocoa leads to a degradation of topsoil, soil samples from forest areas and areas under cocoa cultivation were analyzed for the soil properties. Magnesium (mg/100 g soil) was one of the ingredients considered. The findings of the analysis are summarized as follows.

	MAGNESIUM (mg/100 g SOIL)	
	Forest	**Cocoa**
Mean	3.8	2.2
Standard deviation	0.929	0.989
Number	15	13

Is there a significant difference in the true mean magnesium content of the two topsoils? Carry out the test using 0.02 as the level of significance. Assume that amount of magnesium is normally distributed in both areas with the same variance σ^2.

16. The following data summarize the total concentration (pg/ml) of serum vitamin B_{12} of sheep where one group of three sheep was fed a high forage (HF) diet and the other group of three sheep was fed an HF diet with 33 mg/kg added monensin (M). The results of the experiment are summarized as follows.

	SERUM VITAMIN B_{12} (pg/ml)	
	HF Diet	**HF Diet + M**
Mean	1358	1868
Standard deviation	349	402
Number	3	3

Is the true mean serum vitamin B_{12} significantly different under the two diets? Use a 5 percent level of significance. Assume normal distributions in both cases with the same variance.

17. The data summarized as follows present the total carbohydrates in the liver tissue of two types of edible freshwater fish, *Channa punctatus* and *Mystus vittatus*.

	TOTAL CARBOHYDRATE	
	Channa Punctatus	**Mystus Vittatus**
Mean	43.34	35.67
Standard deviation	1.23	1.46
Number	20	25

Is the true mean carbohydrate content in the liver tissue of the two types of fish significantly different? Carry out the test at the 2 percent level. Assume that the total carbohydrate content is normally distributed with the same variance.

18. A sociologist wanted to investigate whether sports activities and TV viewing have different effects on scholastic performance of grade school students. Twelve pairs of identical twins who did not show pronounced preference to any of the two activities were selected in the study. One of the twins in a pair, selected at random, was allowed three hours of sports activities while the other was allowed three hours of viewing TV. Each activity was allowed to the exclusion of the other. At the end of six months the twenty-four students were given a scholastic test. Their scores are recorded in the following table.

	SCORES	
Pair#	Sports	TV
1	82	92
2	62	65
3	69	64
4	68	74
5	79	87
6	75	80
7	73	67
8	76	73
9	70	77
10	75	86
11	66	77
12	90	83

At the 5 percent level do the two activities influence the scholastic performance of the students differently? How about at the 10 percent level?

19. A random sample of ten Jersey cows was fed a typical summer diet and another random sample of twelve cows was fed a typical winter diet. Milk samples were collected for each cow and the casein content of the milk was estimated. The data are presented as follows.

CASEIN CONTENT	
Summer Diet	**Winter Diet**
3.69	3.20
3.59	3.20
3.75	3.07
3.54	3.49
3.69	3.13
3.60	3.14
3.69	3.15
3.71	3.24
3.57	3.43
3.68	3.22
	3.66
	3.61

At the 5 percent level, is there a significant difference in the true mean casein content under the two diets?

20. A precision tool company needs plates of uniform thickness. A manufacturer claims that it can supply plates of thickness with $\sigma < 0.1$ inch. When a random sample of thirty plates was examined, it was found that $s = 0.095$. At the 5 percent level of significance, is the claim of the manufacturing company valid?

21. Sprouts grow from dormant buds around root collars and along the sides of many stumps of cut redwood tress. An investigation was carried out to determine if sprouts on smaller redwood stumps (diameter less than 60 inches) grow differently from those on larger stumps (diameter greater than 60 inches). Suppose the data that follow give the summary regarding height growth at the end of five years of unthinned stumps (Kenneth N. Boe, Research Note PSW-290, USDA Forest Service, 1974).

Stump Diameter	Sample Mean (in feet)	Sample Standard Deviation (in feet)	Sample Size
Less than 60 inches	12.4	1.38	12
Greater than 60 inches	11.5	1.62	15

At the 5 percent level of significance, do the true mean height growths for the two classifications of stump diameters differ significantly?

22. Of 200 college students interviewed, 150 favored strict environmental controls, while, of 300 college graduates interviewed, 195 favored them. At the 5 percent level of significance, do college students tend to be more in favor of environmental controls than college graduates?

23. A study was conducted to compare two brands of batteries. When ten batteries of Brand A were tested, it was found that the mean life was 1200 hours with s_1 of 120 hours. On the other hand, eight batteries of Brand B yielded a mean life of 1300 hours with s_2 of 150 hours. At the 5 percent level of significance, are the

two brands different? Assume that battery life is normally distributed for both brands.

24. When a salesperson contacted 100 women about a utility product, he sold the product to 43 of them. On the other hand, when he contacted 150 men, he sold the product to 78 of them. At the 10 percent level of significance, is the true proportion among men who buy the product when contacted different from the true proportion among women?

25. Eight breasts of 7- to 8-week-old broiler chickens weighing 0.9 to 1.2 kilograms were taken at random from a poultry-processing plant. The concentration of sarcoplasmic protein (mg N/g tissue) in the breast muscles of these specimens was measured before and after they were irradiated with 10 kGy. The data are presented below.

| | SARCOPLASMIC PROTEIN (mg/g) | |
Specimen #	Before	After
1	14.6	12.8
2	14.0	13.5
3	13.9	13.6
4	14.3	13.3
5	13.6	13.9
6	13.7	13.9
7	14.1	13.3
8	14.3	13.5

At the 2.5 percent level, is there a significant decrease in the true mean sarcoplasmic protein after irradiation?

26. The quality of an air purifier is measured by the cubic feet of fresh air it provides per minute. Two models of air purifiers were compared for smoke removal and the following data were obtained giving the amount of fresh air provided per minute.

SmokeBgone	230	251	277	269	206	234	237	247	251	241	266
SmokeGuard	222	217	223	200	195	204	272	231	215	242	194

Are the two models significantly different? Use 0.05 as the level of significance.

27. The following piece of news was published in *Time,* March 3, 1994.

The Good News

✓One in four children born to women who harbor the AIDS virus becomes infected while still in the womb. However, a federally funded study has shown that mothers who take AZT after their 14th week of pregnancy reduce to 8% the risk of transmitting the virus to their babies.

Source: "The Good News" from *Time* magazine, March 3, 1994. Copyright © 1994 Time Inc. Reprinted with permission.

Since the number of pregnancies involved in the study are not known, it is hard to say how reliable the published statistics are. Suppose 75 pregnancies were involved. At the 1 percent level, is AZT effective in reducing the risk of transmitting the AIDS virus to the babies?

 WORKING WITH LARGE DATA SETS

Project: Go to your favorite grocery store. Weigh about 20 loaves of bread of one certain kind from a specific bakery. Test whether the data collected are reasonably in keeping with the weight claimed by the company on the label. Submit a written report.

Project: Interview a random sample of 220 students on your campus asking whether a student is left-handed, right-handed, or ambidextrous. Are the data qualitative or quantitative? Obtain the proportion of students who are left-handed in your sample. Would this value serve to estimate the proportion of left-handed people in the U.S. at large, or would it provide an estimate only for your campus? Estimate the proportion in the population using a confidence interval.

Next suppose you are interested in finding out whether the proportion of left-handed people is the same among males and females. How would you proceed? Carry out this test by collecting new sets of data appropriately.

MINITAB

LAB 9: Tests of Hypotheses

MINITAB provides commands for tests of hypotheses in the following cases.

a. Test for μ when σ is known
b. Test for μ when σ is not known
c. Pooled t-test for comparing means of two populations assuming equal variances
d. Paired t-test using differences of paired data

In each instance, MINITAB prints out the P-value, which serves as the basis for arriving at a decision depending on the level of significance adopted.

1. Before solving examples, we will provide an empirical meaning to what we mean by the Type I error associated with a test and the level of significance α of a test. We do this by simulating samples from a known population. Specifically, we will draw samples of size 15, repeatedly, from a normal population with mean 150 and standard deviation 7.5. Then, as though the population mean is not known, we test

$$H_0: \mu = 150 \quad \text{against} \quad H_A: \mu \neq 150$$

each time a sample is drawn. Since σ^2 is known and the population from which the sample is drawn is normal, the MINITAB command to perform the test is ZTEST. We must specify in the command the value of μ_0, the value of σ, and the location of the sample data.

When the alternative hypothesis is not specified, it is assumed to be two-sided, so that the test is a two-tail test. If the alternative hypothesis is one-sided, we specify it under the subcommand ALTERNATIVE in the following way.

◄ Alternative Hypothesis	◄ Subcommand
$H_A: \mu < \mu_0$	ALTERNATIVE IS -1.
$H_A: \mu > \mu_0$	ALTERNATIVE IS 1.

We will now proceed with our simulation. We store the commands in a file which we name 'HYPTSTZ'. Notice that we do not specify the alternative hypothesis, as it is two-sided.

```
MTB >STORE 'HYPTSTZ'

STOR>NOECHO

STOR>RANDOM 15 OBSERVATIONS IN C1;

STOR>NORMAL MU = 150 SIGMA = 7.5.

STOR>ZTEST WITH MU = 150,SIGMA = 7.5, DATA IN C1

STOR>END OF PROGRAM

MTB >EXECUTE 'HYPTSTZ' 40 TIMES
```

Each time the stored program is executed, the output will include the P-value, which is the attained level of significance based on that sample. The P-values will differ from sample to sample. Suppose we use a 5 percent level of significance to carry out the test, that is, $\alpha = 0.05$. If the displayed P-value is less than 0.05, then naturally we will reject the null hypothesis H_0: $\mu = 150$. This would, of course, result in an incorrect decision because we certainly know that the population mean is 150, having simulated from such a population. The incorrect decision entails a Type I error.

What fraction of your 40 samples led to incorrectly rejecting H_0? If you were to execute the program many times, the fraction of the times you would reject the null hypothesis H_0 incorrectly, that is, commit a Type I error, would be very close to the α value we have chosen, 0.05.

2 We will now solve Examples 1 and 2 from Section 9-3. In both cases the population variances are known and the populations are normally distributed. So we use the command ZTEST.

```
MTB >NOTE EXAMPLE 1 IN SECTION 9-3

MTB >NAME C1 'SALES'

MTB >SET DATA IN C1

DATA>1280 1250 990 1100 880 1300 1100 950 1050

DATA>END OF DATA

MTB >PRINT C1

MTB >NOTE ALTERNATIVE HYPOTHESIS IS 'GREATER THAN'

MTB >ZTEST WITH MU 1000, SIGMA 100, DATA IN C1;

SUBC>ALTERNATIVE IS 1.
```

You will notice that the displayed P-value 0.0014 is slightly different from the answer in the text. This discrepancy is a result of rounding.

Next, we work out Example 2, in which the alternative hypothesis is left-sided.

```
MTB >NOTE EXAMPLE 2 IN SECTION 9-3

MTB >NAME C1 'TAR'

MTB >SET DATA IN C1

DATA>8.4 8.1 6.6 7.9 9.3 6.1 7.2 9.1 7.1 7.0 8.7 7.9

DATA>END OF DATA

MTB >PRINT C1

MTB >NOTE ALTERNATIVE HYPOTHESIS IS 'LESS THAN'

MTB >ZTEST WITH MU 8, SIGMA 1, DATA IN C1;

SUBC>ALTERNATIVE IS -1.
```

The displayed z value will be -0.75 with P-value of 0.23, as in the solution.

3 If the population variance is not known, then the command to use is TTEST. All the details are exactly as in the case of the command ZTEST except that there is no σ to specify.

We will now consider Examples 4 and 6 from Section 9-3. First we consider Example 4.

```
MTB >NOTE EXAMPLE 4 IN SECTION 9-3

MTB >NAME C1 'MILEAGE'

MTB >SET DATA IN C1

DATA>23  18  22  19  19  22  18  18  24  22

DATA>END OF DATA

MTB >PRINT C1

MTB >NOTE ALTERNATIVE HYPOTHESIS IS 'GREATER THAN'

MTB >TTEST MU 20, DATA IN C1;

SUBC>ALTERNATIVE IS 1.
```

The printout will include the computed t-value and the P-value. Observe that $t = 0.68$ and the P-value $= 0.26$. Since the P-value is greater than 0.1 we do not reject H_0 at the 10 percent level of significance.

In Example 6, solved as follows, the alternative hypothesis is two-sided. So there is no need to give the subcommand ALTERNATIVE.

```
MTB >NOTE EXAMPLE 6 IN SECTION 9-3

MTB >NAME C1 'WEIGHT'

MTB >SET DATA IN C1

DATA>5.3  5.2  4.8  5.2  4.8  5.3

DATA>END OF DATA

MTB >PRINT C1

MTB >NOTE ALTERNATIVE HYPOTHESIS IS TWO-SIDED

MTB >TTEST MU 5, DATA IN C1
```

Assignment Work out Exercises 29, 30, and 31 in Section 9-3.

4 Next, we compare the means of two populations using the pooled t-test. Recall that the basic assumptions are that the populations are normally distributed and have equal variances. The command in this case is TWOSAMPLE-T TEST, together with the subcommand ALTERNATIVE when the alternative is one-sided, followed by the subcommand POOLED.

We will now work out Exercise 18 in Section 9-4. Observe that the alternative hypothesis $H_A: \mu_1 < \mu_2$ is left-sided.

```
MTB >NOTE EXERCISE 18 SECTION 9-4

MTB >NAME C1 'NOSTIM'

MTB >SET DATA IN C1

DATA>3.0  2.0  1.0  2.5  1.5  4.0  1.0  2.0

DATA>END OF DATA

MTB >PRINT C1

MTB >NAME C2 'WITHSTIM'

MTB >SET DATA IN C2

DATA>5.0  4.0  3.0  4.5  2.0  2.5

DATA>END OF DATA

MTB >PRINT C2

MTB >TWOSAMPLE-T TEST FOR SAMPLES IN C1 C2;

SUBC>ALTERNATIVE IS -1;

SUBC>POOLED.
```

Observe that the printed *P*-value is 0.019, which would lead us to reject H_0 at the 5 percent level.

Assignment   Work out Exercises 23, 26, and 27 in Section 9-4.

5 Finally, we consider a test for the difference in population means based on paired data, that is, a test using paired differences. To carry out the test, we first obtain the difference for each pair and then use the command TTEST (see Part 3 of this lab) to test the null hypothesis that $\mu_D = 0$. While testing the hypothesis regarding μ_D, we can also compute a confidence interval for it using the command TINT introduced in Lab 7. We will work out Exercise 33 in Section 9-4 to see how this can be done. Observe that the alternative hypothesis is H_A: $\mu_D < 0$.

```
MTB >NOTE EXERCISE 33 IN SECTION 9-4

MTB >NAME C1 'BEFORE' C2 'AFTER'

MTB >READ DATA IN C1-C2

DATA>91    82

DATA>78    80

DATA>47    62

DATA>37    49

DATA>64    55

DATA>54    73

DATA>43    59

DATA>33    58

DATA>53    43

DATA>END OF DATA

MTB >PRINT C1  C2

MTB >NOTE C3 GIVES PAIRED DIFFERENCES, 'BEFORE MINUS
      AFTER'

MTB >LET C3 = C1 — C2

MTB >NAME C3 'DIFF'

MTB >PRINT C1-C3

MTB >NOTE CONFIDENCE INTERVAL

MTB >TINTERVAL WITH 0.95 CONFIDENCE LEVEL DATA IN C3
```

```
MTB >NOTE TEST OF HYPOTHESIS
MTB >TTEST MU = 0, DATA IN C3;
MTB >ALTERNATIVE IS −1.
```

Assignment   Work out Exercises 35 and 36 in Section 9-4.

There are no direct MINITAB commands to test for the population variance or proportion. But one can write programs to perform such tests.

We now type STOP and conclude the MINITAB session using the logging off procedure.

Commands you have learned in this session.

```
ZTEST ;
ALTERNATIVE IS .
TTEST ;
ALTERNATIVE IS .
TWOSAMPLE-T ;
ALTERNATIVE IS ;
POOLED .
```

Goodness of Fit and Contingency Tables

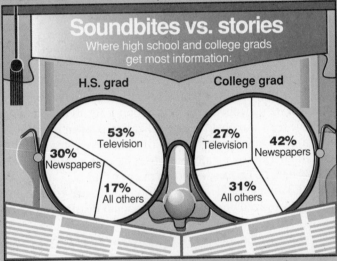

USA SNAPSHOTS®

A look at statistics that shape the nation

Soundbites vs. stories
Where high school and college grads
get most information:

H.S. grad

30% Newspapers
53% Television
17% All others

College grad

27% Television
42% Newspapers
31% All others

▲ Source: Graphic "Soundbites Vs. Stories" by Patti Stang and Marcia Staimer, *USA Today*. Data from The Harris Poll, Louis Harris & Associates. Reprinted by permission of *USA Today*, October 7, 1993, Section A, p. I.

INTRODUCTION

In the preceding chapter, we considered tests of hypotheses regarding the mean of a population and also comparison of the means of two populations. In this connection, the variables measured were quantitative in nature, and in most cases we had to assume that the populations were normally distributed. We also discussed tests of hypotheses about a population proportion and comparison of proportions of an attribute in two populations. The variables in these cases were essentially qualitative. We now extend these ideas, with the qualitative case to be covered in this chapter and the quantitative case in Chapter 12, Analysis of Variance.

10-1   A TEST OF GOODNESS OF FIT

In the discussion of tests of hypotheses about a population proportion, the items that were inspected were classified into one of two categories. For example, a coin could land heads or tails, a person could be a smoker or a nonsmoker, an item could be defective or nondefective, and so on. The independent choice of n items from such a population leads to the binomial distribution, as we have seen. An obvious generalization is when the population can be separated into more than two mutually exclusive categories. For example, a die could land showing up any one of six faces; a person might be a Democrat, a Republican, or neither; a person might be an A, B, O, or AB blood type; and so on. If n independent observations are made from such a population, we get a generalized concept of the binomial distribution called the *multinomial distribution*.

The preceding chapter equipped us to test the following null hypothesis.

H_0: The proportion of Democrats in the United States is 0.60 (of course, this implies that the proportion of non-Democrats is 0.40).

Now, as a generalization, this section will, for example, enable us to test a null hypothesis of the following type.

H_0: In the United States, the proportion of Democrats is 0.55, the proportion of Republicans is 0.35, and the proportion of voters belonging to neither party is 0.10.

To test the preceding hypothesis, suppose we interview 1000 people picked at random. On the basis of the stipulated null hypothesis, we would *expect* 550 Democrats, 350 Republicans, and 100 others. If we actually *observe* 568 Democrats, 342 Republicans, and 90 others in this sample, we might be willing to go along with the null hypothesis. On the other hand, if the sample yields 460 Democrats, 400 Republicans, and 140 others, we would be reluctant to accept H_0. Thus, in the final analysis the statistical test will have to be based on how good a fit there is between the observed numbers and the numbers that one would expect from the hypothesized distribution. A test of this type, which determines whether the sample data are in conformity with the hypothesized distribution, is called a **test of goodness of fit,** as it literally tests how good the fit is.

For the purpose of indicating the procedure, consider the following example. Suppose we want to test whether a die is a fair die. Let p_1 represent the probability of getting a 1, p_2 that of getting a 2, and so on. Then the null hypothesis H_0 can be stated as follows.

$$H_0: p_1 = \tfrac{1}{6}, p_2 = \tfrac{1}{6}, p_3 = \tfrac{1}{6}, p_4 = \tfrac{1}{6}, p_5 = \tfrac{1}{6}, p_6 = \tfrac{1}{6}$$

This hypothesis will be tested against the alternative hypothesis H_A, stated as

H_A: The die is not fair.

To test the hypothesis, suppose we roll the die 120 times and obtain the data given in Table 10-1. If the die is a fair die, we would *expect* each face to show up $120(\tfrac{1}{6})$, or 20, times. Any departure from this could be simply due to sampling error or due to the fact that the die is indeed a loaded die. A radical departure from the expected frequency of 20 would definitely make the die suspect in our judgment and the discrepancy hardly attributable to chance. We are thus faced with two questions. First, how do we measure this departure? Second, for how large a departure would we consider rejecting H_0? In short, what test criterion do we use? The test criterion is provided by a statistic **X** whose value for any sample is given as a number χ^2 defined as follows.

$$\chi^2 = \sum \frac{(O_i - E_i)^2}{E_i}$$

Here O_i represents the **observed frequency** of the face marked i on the die and E_i the corresponding **expected frequency** obtained by assuming that the null hypothesis is true. The idea is that for each face on the die, we find the quantity [(observed $-$ expected)^2/expected] and then add these quantities for all the faces.

TABLE 10-1   Observed Frequencies in 120 Tosses of a Die

Number on the die	1	2	3	4	5	6
Frequency	17	25	17	23	15	23

Notice that the observed frequencies of the outcomes will change from sample to sample. That is, if we were to roll our die another 120 times, we would probably get a different set of observed frequencies from those given in Table 10-1. Thus, **X** is a random variable that provides a measure of departure of the observed frequencies, because $(O - E)^2$ will be small if the observed and expected frequencies are close, and it will be large if they are far apart.

The procedure for computing the χ^2 value for a given sample is shown in Table 10-2. The total in Column 6 gives the χ^2 value as equal to 4.30. The significance of the number 4.30 in determining whether the die is fair or not is related to the distribution of **X**. It turns out that *if none of the expected frequencies is less than 5, then the distribution of* **X** *can be approximately represented by a chi-square distribution.* The next question is, how many degrees of freedom does the chi-square distribution have? The number of degrees of freedom associated with the chi-square is $6 - 1$, or 5. The rationale is that although we are filling 6 cells, or entries in the table, in reality we have freedom to fill only 5 cells. The number in the sixth cell is determined automatically due to the constraint that the total should be 120.

TABLE 10-2   Computation of the χ^2 Value for 120 Tosses of a Die

Outcome	Observed Frequency O	Expected Frequency E	$(O - E)$	$(O - E)^2$	$(O - E)^2/E$
1	17	20	-3	9	0.45
2	25	20	5	25	1.25
3	17	20	-3	9	0.45
4	23	20	3	9	0.45
5	15	20	-5	25	1.25
6	23	20	3	9	0.45
Total	120		0		4.30

Intuition tells us that we should reject the null hypothesis that the die is fair if the computed χ^2 value is large. In other words, we have a one-tail test with the critical region consisting of the right tail of the chi-square distribution with 5 degrees of freedom, as indicated in Figure 10-1.

At the 5 percent level of significance, for 5 degrees of freedom the table value of chi-square from Table A-5 is 11.07. Because the computed value of 4.30 does not exceed 11.07, we do not reject the null hypothesis. Thus, at the 5 percent level of significance, there is no reason to believe that the die is loaded.

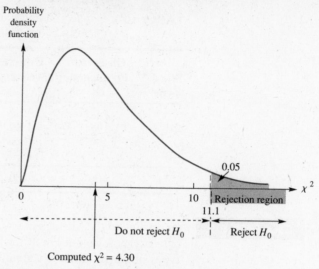

Figure 10-1 Probability Density Curve for the Chi-Square Distribution with 5 df and the Decision Rule

EXAMPLE 1   It is believed that the proportions of people with A, B, O, and AB blood types in the population are, respectively, 0.4, 0.2, 0.3, and 0.1. When 400 randomly picked people were examined, the observed numbers of each type were 148, 96, 106, and 50. At the 5 percent level of significance, test the null hypothesis that these data bear out the stated belief.

SOLUTION   Let p_A be the probability that a person has Type A blood, and so on. Then

$$H_0: p_A = 0.4, P_B = 0.2, p_O = 0.3, p_{AB} = 0.1.$$

We will state the alternative hypothesis H_A as

$$H_A: \text{The population proportions of blood types are not as stated in } H_0.$$

Our first goal should be to find the expected frequency in each cell. Since $n = 400$ and $p_A = 0.4$, the expected number of people with Type A blood is $400(0.4)$, or 160. Other expected frequencies are obtained similarly. The computations for finding the χ^2 value are presented in Table 10-3. Since there are four cells, the chi-square distribution has $4 - 1$, or 3, degrees of freedom. From Table A-5, $\chi^2_{3, 0.05} = 7.815$. Because the computed value of 8.23 is larger than 7.815, we reject the null hypothesis. Thus, there is strong evidence that the belief stated regarding the distribution of blood types is not correct.

TABLE 10-3 Observed and Expected Frequencies of Blood Types and Computation of the χ^2 Value

Blood Type	Observed Frequency O	Expected Frequency E	$(O - E)$	$(O - E)^2$	$(O - E)^2/E$
A	148	160	−12	144	0.90
B	96	80	16	256	3.20
O	106	120	−14	196	1.63
AB	50	40	10	100	2.50
Total	400	400	0		8.23

EXAMPLE 2

According to the Mendelian law of segregation in genetics, when two doubly heterozygous walnut comb fowls are crossed, the offspring consist of four types of fowls, namely, walnut, rose, pea, and single, with frequencies expected in the ratio $9 : 3 : 3 : 1$. Among 320 offspring, it was found that there were 160 walnut, 70 rose, 65 pea, and 25 single. Based on the data, are the probabilities for each type not in keeping with the Mendelian law?

SOLUTION Let p_{walnut} represent the probability that the offspring is walnut, and so on. Because the frequencies are expected in the ratio $9 : 3 : 3 : 1$, the null hypothesis is given as

$$H_0: p_{walnut} = \tfrac{9}{16}, \, p_{rose} = \tfrac{3}{16}, \, p_{pea} = \tfrac{3}{16}, \, p_{single} = \tfrac{1}{16}.$$

The alternative hypothesis H_A is

H_A: The probabilities are not given by the Mendelian law.

If the null hypothesis is true, then we expect $320(\tfrac{9}{16})$, or 180, walnut types. Other expected frequencies can be obtained similarly. The computation of the χ^2 value is presented in Table 10-4.

Because there are four comb types, the chi-square distribution has 3 degrees of freedom. The table value of $\chi^2_{3, 0.05}$ is 7.815. The computed value 5.56 is less than 7.815, and so we do not reject H_0. Thus, there is no reason to suspect that there is a disagreement with the Mendelian ratio $9 : 3 : 3 : 1$.

TABLE 10-4 Observed and Expected Frequencies of Comb Types and Computation of the χ^2 Value

Comb Type	Observed Frequency O	Expected Frequency E	$(O - E)^2$	$(O - E)^2/E$
Walnut	160	180	400	2.22
Rose	70	60	100	1.67
Pea	65	60	25	0.42
Single	25	20	25	1.25
Total	320	320		5.56

The essence of the previous examples can be summarized as follows.

1 The population is divided into k categories (or classes), $c_1, c_2, \ldots, c_k$.

2 The null hypothesis stipulates that the probability that an individual belongs to category c_1 is p_1, that it belongs to category c_2 is p_2, and so on.

3 To test this hypothesis, a random sample of n individuals is picked. The observed frequencies of the categories are recorded as O_1, $O_2, \ldots, O_k$.

4 If the null hypothesis is true, then the expected frequencies E_1, $E_2, \ldots, E_k$ are obtained as follows.

$$E_1 = np_1, \; E_2 = np_2, \ldots, E_k = np_k$$

5 The departure of the observed frequencies from those expected is measured by means of a statistic $\mathbf{X}$ whose value χ^2 is given by

$$\chi^2 = \frac{(O_1 - E_1)^2}{E_1} + \frac{(O_2 - E_2)^2}{E_2} + \cdots + \frac{(O_k - E_k)^2}{E_k}.$$

The computations leading to the evaluation of this statistic are displayed in Table 10-5.

TABLE 10-5 Observed and Expected Frequencies and Computation of the χ^2 Value

Category	Observed Frequency O	Expected Frequency E	$(O - E)^2/E$
c_1	O_1	$E_1 = np_1$	$(O_1 - E_1)^2/E_1$
c_2	O_2	$E_2 = np_2$	$(O_2 - E_2)^2/E_2$
$\vdots$	$\vdots$	$\vdots$	$\vdots$
c_k	O_k	$E_k = np_k$	$(O_k - E_k)^2/E_k$
Total	n	n	$\sum (O_i - E_i)^2/E_i$

6 *If n, the number of independent repetitions is large, H_0 is true, and none of the expected frequencies is less than 5, the distribution of $\mathbf{X}$ can be approximated very closely by a chi-square distribution. Since there are k categories, the number of degrees of freedom associated with the chi-square is $k - 1$.*

7 If the null hypothesis is true, then the departure of the observed frequencies from the expected frequencies will be small. In that case, the χ^2 value will be small. However, if there is a large discrepancy between the observed and expected frequencies, then we tend to question the null hypothesis. Such a large discrepancy will be reflected in a large χ^2 value. The critical region for a given level of significance will therefore consist of the right tail of the chi-square distribution with $k - 1$ degrees of freedom. The decision rule is as follows.

Reject H_0 if the computed χ^2 value is greater than the table value $\chi^2_{k-1,\,\alpha}$. (See Figure 10-2.)

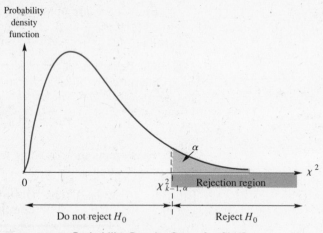

Figure 10-2 Probability Density Curve for Chi-Square Distribution with $k - 1$ df and the Decision Rule

It should be emphasized that the distribution of the statistic $\mathbf{X}$ employed here is only approximately chi-square. It should not be used if one or more of the expected frequencies is less than 5.

11/21/96

Section 10-1 Exercises

1. In a cross-fertilization experiment yielding 172 seedlings, a botanist found that there were 38 yellow and 134 green seedlings. At the 5 percent level of significance, test the null hypothesis that the segregation ratio of green seedlings to yellow is $3:1$. Use the following tests.
   (a) The chi-sqaure test developed in this section
   (b) A test concerning population proportion given in Section 9-3.

2. According to genetic theory, a cross between a homozygous fruit fly and a heterozygous fruit fly is expected to yield homozygous and heterozygous offspring in the ratio $1:1$. In an experimental cross, there were 27 homozygous offspring in a progeny of 68. At the 5 percent level, test the null hypothesis that the true ratio is $1:1$.

3. Suppose, according to a survey conducted in 1960, the *probability* distribution of the age of an adult over twenty years of age is as given in the first two columns of the following table. In 1993, when a sample of 1000 adults over twenty years of age was interviewed, the frequency distribution given in Column 3 was obtained. Test the null hypothesis that the probability distribution in 1993 is the same as that in 1960. Use $\alpha = 0.05$.

Age Group	Probability Distribution	Observed Sample Frequencies in 1988
20–29	0.25	270
30–39	0.23	202
40–49	0.20	180
50–59	0.15	160
60–69	0.08	88
70 and over	0.09	100

4. The number of accidents in a factory during the months of a year are as follows.

Jan	Feb	Mar	Apr	May	June	Jul	Aug	Sep
25	28	24	18	17	27	9	18	22

Oct	Nov	Dec
14	12	26

Using a 5 percent level of significance, test the null hypothesis that the number of accidents does not depend on the month of the year. *Hint:* If a person has an accident, then under the null hypothesis the probability of that accident taking place in any month is 1/12.

5. The Mendelian law of segregation in genetics states that when certain types of peas are crossed, the following four varieties—round and yellow (*RY*), round and green (*RG*), wrinkled and yellow (*WY*), and wrinkled and green (*WG*)— are obtained in the ratio $9:3:3:1$. In a greenhouse experiment, an agriculturist found that there were 652 *RY*, 200 *RG*, 185 *WY*, and 83 *WG*. Using a 5 percent level of significance, test the null hypothesis that these observed frequencies are consistent with those expected in theory.

6. A physicist theorizes that the probabilities that a radioactive substance emits 0, 1, 2, 3, or 4 particles of a certain kind during a one-hour period are, respectively, 0.05, 0.3, 0.3, 0.25, and 0.1. A record of 200 one-hour periods gave the following distribution of the number of particles emitted.

Number of particles	0	1	2	3	4
Number of times	16	68	51	46	19

At the 5 percent level of significance, test the null hypothesis that there is no departure from what the physicist theorizes.

7. Suppose the probability that a newborn baby is a boy is 1/2.
   (a) In a family with 4 children, find the probabilities of 0, 1, 2, 3, or 4 sons.
   (b) If 160 families, each with 4 children, are interviewed, find the expected number of families with 0, 1, 2, 3, and 4 sons.

(c) When 160 families, each with 4 children, were interviewed, the following information was obtained.

Number of sons	0	1	2	3	4
Number of families	12	35	53	44	16

At the 5 percent level of significance, test the null hypothesis that the probability that a child is a son is 1/2.

8. A machine is supposed to mix three types of candy—caramel, butterscotch, and coconut-chocolate—in the proportion $4:3:2$. A sample of 270 pieces of candy was found to contain 135 caramel, 70 butterscotch, and 65 coconut-chocolate. At the 5 percent level of significance, test the null hypothesis that the machine is mixing the candy in the proportion $4:3:2$.

9. Five hundred cars observed on a certain freeway were categorized as follows: 160 GM, 100 Ford, 80 Chrysler, 25 European, and 135 Japanese. Using a 5 percent level of significance, test the hypothesis that the expected frequencies of GM, Ford, Chrysler, European, and Japanese on that freeway are in the ratio $30:15:10:5:20$.

10. The distribution of the number of clients that an accountant had on each day during 400 randomly picked days is as follows.

Number of clients	0	1	2	3	4	5	6
Number of days	50	112	125	60	28	16	9

At the 5 percent level of significance, is there departure from the theoretical distribution given below?

Number of clients	0	1	2	3	4	5	6
Probability	0.10	0.27	0.27	0.18	0.09	0.05	0.04

11. The following table gives the distribution of child occupants, ages 1–5, killed in motor vehicle crashes, 1976–1980 (*Am. J. Public Health* 1986; 76: pp. 31–34).

Age (years)	Number of Children
1	698
2	747
3	594
4	538
5	513

Test the null hypothesis that a child occupant killed in a crash is equally likely to be of age 1, 2, 3, 4, or 5. Use $\alpha = 0.05$.

12. Refer to Exercise 72 in Section 9-3 of Chapter 9. Based on the data collected by the students, are the true proportions for different colors significantly different from the proportions advertised by Mars, Inc.? Answer the question considering different levels of significance.

10-2   A TEST OF INDEPENDENCE

The problem that we consider next has no previous counterpart. It differs from the earlier situations in which we conspicuously observed only one characteristic of any individual. For example, in classifying an individual as A, B, O, or AB blood type, we observed the characteristic "blood type"; in classifying a person as smoker or nonsmoker, we observed the "smoking habit." However, there are times when we might be interested in observing more than one variable on each individual to find if a relationship exists between these variables. As an example, for each person we might observe both blood type and eye color and investigate if these characteristics are related in any way. In short, our goal is a **test of independence,** specifically, to find whether two observed attributes of members of a population are independent.

As a first step, we pick a sample of size n and classify the data in a two-way table on the basis of the two variables. Such a table for determining whether the distribution according to one variable is contingent on the distribution of the other is appropriately called a **contingency table.** A table with r rows and c columns is an $r \times c$ contingency table (read this as *r by c contingency table*).

Going back to our example of blood type and eye color, we now investigate to determine if there is evidence indicating a relationship between these two variables. For this purpose, suppose we collect information on 400 randomly picked people and classify them according to blood type and eye color. The data recording the observed frequencies are contained in Table 10-6 which is a 2×4 contingency table, as there are 2 rows and 4 columns.

TABLE 10-6   A 2×4 Contingency Table Giving the Observed Frequencies

	BLOOD TYPE				
Eye Color	**A**	**B**	**O**	**AB**	**Total**
Blue	95	40	80	25	240
Brown	65	50	40	5	160
Total	160	90	120	30	400

There are 8 cells corresponding to the 2 rows and 4 columns. The row and column totals are referred to as the *marginal totals,* as they appear in the margins. The null hypothesis and the alternative hypothesis can be stated as follows.

H_0: The blood type of a person and the eye color of that person are independent variables.

H_A: The two attributes are not independent.

To test the *null* hypothesis, our objective will be the same as before, to determine the size of the discrepancy between the observed frequencies and the frequencies that we would expect assuming the null hypothesis is true, that is, assuming the two attributes are independent of each other. Therefore, we obtain the expected frequencies in the 8 cells under the assumption of independence.

As an illustration of the method used to obtain the expected frequencies, we will find the number of people that we expect to have Type A blood and blue eyes. Consider two events E and F defined as follows.

E: A person in the population has blue eyes.

F: A person in the population has Type A blood.

There are 240 people with blue eyes in a sample of 400. Therefore, we can use $240/400$ as a *point estimate* of the probability, $P(E)$. Similarly, because there are 160 people with Type A blood among 400 people in the sample, a point estimate of $P(F)$ is $160/400$. If, as per H_0, blood type and eye color are independent variables, then we must have $P(E \text{ and } F) = P(E) \cdot P(F)$. (See Chapter 3, page 199.) Consequently, we can obtain a point estimate of the probability that a person has blue eyes *and* is of Type A blood as $\left(\dfrac{240}{400} \right)\left(\dfrac{160}{400} \right)$. Because there are 400 people in all in the sample, the *expected* frequency of people with blue eyes and Type A blood is

$$400\left(\frac{240}{400} \right)\left(\frac{160}{400} \right) = \frac{(240)(160)}{400}.$$

A close examination will show that this is simply equal to

$$\frac{(\text{Total of Row 1})(\text{Total of Column 1})}{n}$$

where n is the total number in the sample.

In general, it is true that the expected frequency in any cell is obtained by multiplying the corresponding marginal row and column totals and then dividing this product by n. That is, the expected frequency in the cell that represents the intersection of Row i and Column j, which we denote by E_{ij}, is given by

$$E_{ij} = \frac{(\text{Total of Row } i)(\text{Total of Column } j)}{n}.$$

Thus, the expected number of people with brown eyes and Type O blood is $\dfrac{(160)(120)}{400}$, or 48, that of people with brown eyes and Type AB blood is $\dfrac{(160)(30)}{400}$, or 12, and so on. All the expected frequencies are given in Table 10-7.

TABLE 10-7   Expected Frequencies Assuming Eye Color and Blood Type are Independent Variables

	BLOOD TYPE				
Eye Color	**A**	**B**	**O**	**AB**	**Total**
Blue	96	54	72	18	240
Brown	64	36	48	12	160
Total	160	90	120	30	400

Remember that the expected frequencies are obtained under the assumption that eye color and blood type are independent variables. If there is a large discrepancy between these and the corresponding observed frequencies, it will cast doubt on the validity of the null hypothesis. Thus, once again we use a statistic $\mathbf{X}$ whose value χ^2 for any sample is given by

$$\chi^2 = \sum \frac{(O - E)^2}{E}$$

where the symbol Σ means that we are summing over all the cells. In each cell we divide the square of the difference between the observed and expected frequencies by the expected frequency of that cell and then add these quantities over all the cells. These computations are shown in Table 10-8.

If *n is large so that each of the expected frequencies is at least 5*, then the statistic $\mathbf{X}$ has a distribution that is close to a chi-square distribution, and for

TABLE 10-8   Observed Frequencies, Expected Frequencies Assuming Eye Color and Blood Type are Independent, and the χ^2 Value

Eye Color	Blood Type	Observed Frequency O	Expected Frequency E	$(O - E)$	$(O - E)^2$	$(O - E)^2/E$
Blue	A	95	96	−1	1	0.01
Blue	B	40	54	−14	196	3.63
Blue	O	80	72	8	64	0.89
Blue	AB	25	18	7	49	2.72
Brown	A	65	64	1	1	0.02
Brown	B	50	36	14	196	5.44
Brown	O	40	48	−8	64	1.33
Brown	AB	5	12	−7	49	4.08
Total		400	400	0		18.12

the particular problem we are discussing, it has 3 degrees of freedom. We can explain why there are 3 degrees of freedom as follows: There are 8 cells to be filled. Toward this goal we had to estimate the probabilities of three blood types. (The estimate of the probability of the fourth blood type is determined automatically, as the sum of all the four estimated probabilities must be 1.) Also, we had to estimate the probability of one of the two eye colors, say, the probability that the eye color is blue. (The estimate of the probability that the eye color is brown is determined automatically since the sum of the two probabilities should be 1.) Thus, we had to estimate four parameters. Also, a further constraint is that the total of all the expected frequencies has to be equal to n, the sample size. The degrees of freedom are given by the following.

$$\begin{pmatrix} \text{degrees of} \\ \text{freedom} \end{pmatrix} = \begin{pmatrix} \text{number} \\ \text{of cells} \end{pmatrix} - \begin{pmatrix} \text{number of} \\ \text{estimated} \\ \text{parameters} \end{pmatrix} - \begin{pmatrix} \text{number of} \\ \text{constraints} \end{pmatrix}$$
$$= 8 - 4 - 1 = 3$$

Notice that we can write the 3 degrees of freedom as $(2 - 1)(4 - 1)$. In general, *if there are r rows and c columns in the contingency table, then the corresponding chi-square distribution has $(r - 1)(c - 1)$ degrees of freedom.*

The critical region consists of the values in the right tail of the chi-square distribution. The decision rule for a level of significance α is as follows.

Reject H_0 if the computed χ^2 value is greater than the table value, $\chi^2_{(r-1)(c-1),\,\alpha}$.

To test the hypothesis of independence for the data in our example, take $\alpha = 0.05$. There are 3 degrees of freedom, and from the chi-square table, $\chi^2_{3,\,0.05} = 7.815$. Because the computed value of 18.12 is larger than the table value of 7.815, we reject the null hypothesis. The data do not support the hypothesis of independence on which the calculations are based. There is reason to believe that eye color and blood type are related.

To see what our conclusion means, examine Column 5 in Table 10-8, which gives the difference $O - E$. There is an indication that blue eyes and Type O blood, blue eyes and Type AB blood, and brown eyes and Type B blood associate together more often, and that blue eyes and Type B blood, brown eyes and Type O blood, and brown eyes and Type AB blood associate together less often than what one might expect if eye color and blood type were independent.

EXAMPLE

In a certain community, 360 randomly picked people were classified according to their age group and political leaning. The data are presented in Table 10-9.

TABLE 10-9   Observed Frequencies Giving the Distribution of Political Leaning and Age Group of People in a Community

Political Leaning	AGE GROUP 20–35	36–50	Over 50	Total
Conservative	10	40	10	60
Moderate	80	85	45	210
Liberal	30	25	35	90
Total	120	150	90	360

Test the null hypothesis that a person's age and political leaning are not related (use $\alpha = 0.05$).

SOLUTION

1. The null and alternative hypotheses are given as follows.

    H_0: A person's age and political leaning are independent.

    H_A: The two variables are not independent.

2. The expected frequency in each cell is obtained by multiplying the corresponding row and column totals and then dividing the product by 360, the total number observed. These frequencies are given in Table 10-10.

TABLE 10-10   Expected Frequencies Assuming Political Leaning and Age Level Are Independent

Political Leaning	AGE GROUP 20–35	36–50	Over 50	Total
Conservative	20	25.0	15.0	60
Moderate	70	87.5	52.5	210
Liberal	30	37.5	22.5	90
Total	120	150	90	360

3. The value χ^2 of the test statistic is $\chi^2 = \sum \dfrac{(O - E)^2}{E}$. The computations are given in Table 10-11, where we obtain the computed χ^2 value as 29.35.

4. The contingency table has 3 rows and 3 columns. Thus, the chi-square distribution has $(3 - 1)(3 - 1)$, or 4, degrees of freedom. For $\alpha = 0.05$, the table value of chi-square is $\chi^2_{4, 0.05} = 9.488$.

TABLE 10-11   Observed Frequencies, Expected Frequencies Assuming
Independence, and Computation of the χ^2 Value

Political Leaning	Age Group	Observed Frequency O	Expected Frequency E	$(O - E)^2$	$(O - E)^2/E$
Conservative	20–35	10	20.0	100	5.00
Conservative	36–50	40	25.0	225	9.00
Conservative	over 50	10	15.0	25	1.67
Moderate	20–35	80	70.0	100	1.43
Moderate	36–50	85	87.5	6.25	0.07
Moderate	over 50	45	52.5	56.25	1.07
Liberal	20–35	30	30.0	—	—
Liberal	36–50	25	37.5	156.25	4.17
Liberal	over 50	35	22.5	156.25	6.94
	Total	360	360		29.35

5. The computed value, 29.35, is greater than the table value, 9.488,
   and so we reject H_0.
6. The magnitude of the computed value indicates rather emphati-
   cally that age group and political leaning are strongly related.   ●

Note that if we do conclude that two attributes are dependent, it does not
imply any cause and effect relationship. For example, if we conclude that eye
color and blood type are related, it should not be construed to mean that eye
color causes blood type, or vice versa.

As a final comment, since the distribution of **X** is approximately chi-
square, we emphasize once again that n should be large, and the criterion for
the test of independence of two attributes is not recommended if any expected
frequency is less than 5.

Section 10-2 Exercises

1. A survey was conducted to investigate whether alcohol drinking and smoking are
related. The following information was compiled for 600 individuals.

	Smoker	Nonsmoker
Drinker	193	165
Nondrinker	89	153

Using a 2.5 percent level of significance, test the null hypothesis that alcohol
drinking and smoking are not related.

2. A store asked 500 of its customers whether they were satisfied with the service or not. The response was classified according to the sex of the customer and is contained in the following table.

	Male	Female
Satisfied	140	231
Not satisfied	39	90

At the 5 percent level of significance, test the null hypothesis that whether a customer is satisfied or not does not depend on the sex of the customer.

3. In a survey, 600 people were classified with respect to hypertension and heart ailment. The following data were obtained.

	HEART CONDITION	
Hypertension	**Ailment**	**No Ailment**
Constant	51	89
Occasional	72	280
Never	19	89

Handwritten annotations: Constant: 51 [33.13], 89 [106.87], 140; Occasional: 72 [83.31], 280 [268.69], 352; Never: 19 [25.8], 89 [82.44], 108; column totals 142, 458, 600.

$$\left(\frac{140}{600}\right)(142) \qquad \left(\frac{352}{600}\right)(142)$$

At the 1 percent level of significance, test the null hypothesis that hypertension and heart ailment are not related.

4. The following table gives the distribution of students according to the type of music they prefer and their IQ.

	IQ		
Music Preferred	**High**	**Medium**	**Low**
Classical	45	32	13
Semiclassical	58	62	30
Rock	87	126	87

Using a 5 percent level of significance, test the null hypothesis that music taste and IQ level are independent attributes.

5. The office of the dean has compiled the following information on 500 students regarding their grade point averages and how they ranked their instructor.

Handwritten: $\alpha = 0.05$ $\chi^2_{.05} = 16.919$ 9 deg. of freedom $\frac{(8-75)^2}{41.25} = 1.85$

	GRADE POINT AVERAGE			
Rating	**Above 3.5**	**3.0–3.5**	**2.0–2.9**	**Below 2.0**
Outstanding	40	50	55	20
Good	25	35	45	20
Average	10	25	50	36
Poor	10	15	40	24

Handwritten annotations: Outstanding row 50 [46.25], 20 [33], total 165; Good 35 [31.25], 20 [25], total 125; Average 25 [31.25], total 121; Poor 15 [25], total 89; column totals 85 125, 500.

Test the null hypothesis that a student's grade point average and evaluation of the instructor are not related. Use a 5 percent level of significance.

6. The data in the following table refer to achievement in high school and family economic level of 400 students.

	FAMILY ECONOMIC LEVEL		
Achievement	High	Medium	Low
High	15	80	20
Average	40	100	50
Low	25	30	40

At the 5 percent level of significance, is there a relationship between achievement in high school and family economic level?

7. The following table gives the distribution of weight and blood pressure of 300 subjects.

	BLOOD PRESSURE		
Weight	High	Normal	Low
Overweight	40	34	18
Normal	36	77	27
Underweight	16	33	19

At the 2.5 percent level of significance, are weight and blood pressure related?

8. The following figures give information regarding eye color and a certain eye trait of 300 individuals.

	EYE COLOR			
Trait	Blue	Brown	Green	Other
Present	50	69	45	34
Not present	25	24	30	23

At the 5 percent level of significance, is there any association between eye color and the eye trait?

9. To investigate women's attitudes toward cotton and other fibers used in clothing, the following question was posed: Which one of these materials—cotton, cotton and synthetics, others (including polyester, nylon, acrylic, etc.)—do you like best for your ready-made blouses? The data are summarized in the following table according to age groups.

	MATERIAL		
Age	Cotton	Cotton/Synthetic	Others
18–29	180	86	79
30–39	128	91	85
40–49	116	81	70
50 +	97	56	90

Does preference for fabric depend on the age group of women. Use $\alpha = 0.025$.

10-3   A TEST OF HOMOGENEITY

In Chapter 9 we saw how to test two populations for equality of proportions of some characteristic. With that background we are in a position, for example, to test whether the proportion of Democrats in California is the same as the proportion of Democrats in New York, assuming we are given the data in Table 10-12, which were collected by interviewing 200 people in New York and 300 in California.

TABLE 10-12

Party Affiliation	STATE	
	New York	**California**
Democrat	122	168
Non-Democrat	78	132
Total	200	300

We might generalize this procedure in several directions. For example, we might want to compare the proportions of Democrats in four states, such as New York, California, Indiana, and Florida. In other words, we might want to compare the proportions of a characteristic in more than two populations. Another generalization might be if we considered three states, for example, New York, California, and Indiana, to test whether, in these three states, the proportions of Republicans are the same, whether the proportions of Democrats are the same, and whether the proportions of voters belonging to neither party are the same. In short, what we are interested in is whether the three states are *homogeneous* with respect to the party affiliations of their residents. A test that deals with problems of this type is called a **test of homogeneity.**

Consider the last case. As has always been our practice, we pick a sample. Because each of the states has to be represented, we pick a certain number of individuals from each state. Suppose we interview 200 people in California, 200 in New York, and 100 in Indiana, and the survey produces the results given in Table 10-13 on page 598.

Notice that when we considered independence of attributes, we picked a random sample of a given size and then *both* the column and row totals were random numbers determined by chance. In our present discussion, the number of individuals that are to be interviewed in each state is predetermined; that is, the column totals are fixed.

TABLE 10-13   Observed Frequencies of Democrats, Republicans, and Others in California, New York, and Indiana in the Samples

| | STATE | | | |
Party	California	New York	Indiana	Total
Democrats	95	105	40	240
Republicans	80	60	50	190
Others	25	35	10	70
Total	200	200	100	500

In Table 10-14, suppose p_{11}, p_{12}, and so on, represent *theoretical probabilities* of the cells. In testing for homogeneity, we are actually setting the following null hypothesis.

$$H_0: \begin{cases} p_{11} = p_{12} = p_{13} & \text{The proportions of Democrats are equal.} \\ p_{21} = p_{22} = p_{23} & \text{The proportions of Republicans are equal.} \\ p_{31} = p_{32} = p_{33} & \text{The proportions of others are equal.} \end{cases}$$

The alternative hypothesis will, of course, be as follows.

H_A: There is no homogeneity.

We will begin by determining the expected number of Democrats in each of the states, assuming $p_{11} = p_{12} = p_{13}$.

TABLE 10-14   Theoretical Probabilities of the Cells

| | STATE | | |
Party	California	New York	Indiana
Democrats	p_{11}	p_{12}	p_{13}
Republicans	p_{21}	p_{22}	p_{23}
Others	p_{31}	p_{32}	p_{33}

Because there are 240 Democrats in the sample of 500, an *estimate* of the probability that a person is a Democrat is 240/500. (If there is homogeneity, as H_0 stipulates, we are entitled to obtain an estimate by pooling for all the states.) There are 200 people representing the state of California. Therefore, the expected number of Democrats from this state is $200\left(\dfrac{240}{500}\right)$. The same reasoning gives $200\left(\dfrac{240}{500}\right)$ and $100\left(\dfrac{240}{500}\right)$ as the expected number of

Democrats from New York and Indiana, respectively. Actually, once we have found the expected number of Democrats for any two states, the expected number of Democrats in the third state is determined automatically, in view of the fact that the total number of Democrats in the sample is 240. A careful observation shows that *the expected frequency in any cell is obtained by multiplying together the corresponding row and column totals and then dividing the product by n*, the same as in the test for independence of attributes. The expected numbers of Republicans and other voters who are neither Republican nor Democrat in the three states are computed in a similar way. These results are presented in Table 10-15. Notice, for example, that once we have determined that there are 96 Democrats and 76 Republicans in California, the number of other voters in that state has to be 28, as there are a total of 200 Californians in the sample.

TABLE 10-15   Expected Frequencies Assuming Homogeneity

Party	STATE			Total
	California	New York	Indiana	
Democrats	96	96	48	240
Republicans	76	76	38	190
Others	28	28	14	70
Total	200	200	100	500

Once again, the measure of departure from homogeneity is provided by a statistic **X** whose value for any sample is given by

$$\chi^2 = \sum \frac{(O - E)^2}{E}.$$

The distribution of the statistic is approximately chi-square with degrees of freedom equal to $(r - 1)(c - 1)$, where r represents the number of rows and c the number of columns. The approximation is satisfactory if none of the expected frequencies is less than 5.

In our particular example, there are $(3 - 1)(3 - 1)$, or 4, degrees of freedom, and the rejection region consists of the right tail of the distribution. Using a 5 percent level of significance, we would reject H_0 if the computed value is greater than the table value, $\chi^2_{4,0.05} = 9.488$. The computations are shown in Table 10-16.

The computed value of the statistic, 12.77, is greater than the table value of 9.488, and so we reject H_0. If we examine Column 5, which gives the difference $O - E$, and Column 7, which gives $(O - E)^2/E$, there are too few Republicans in New York and too many Republicans in Indiana for compatibility with the hypothesis of homogeneity.

TABLE 10-16 Computation of the χ^2 Value for Test of Homogeneity

State	Party Affiliation	Observed Frequency O	Expected Frequency E	$O - E$	$(O - E)^2$	$(O - E)^2/E$
California	Democrat	95	96	-1	1	0.01
California	Republican	80	76	4	16	0.21
California	Others	25	28	-3	9	0.32
New York	Democrat	105	96	9	81	0.84
New York	Republican	60	76	-16	256	3.37
New York	Others	35	28	7	49	1.75
Indiana	Democrat	40	48	-8	64	1.33
Indiana	Republican	50	38	12	144	3.80
Indiana	Others	10	14	-4	16	1.14
	Total	500	500	0		12.77

In summary, the approach for the test of homogeneity is the same as for the test of independence of variables, although the conclusions drawn are couched slightly differently.

EXAMPLE

In order to investigate whether the distribution of blood types in Europe is the same as in the United States, information was collected on 200 randomly picked people in Europe and 300 randomly picked people in the United States. From the data, which are summarized in Table 10-17, is it true that the distributions of blood types in Europe and the United States are significantly different?

TABLE 10-17 Observed Frequencies of the Distribution of Blood Types in Europe and the United States

| Blood Type | LOCATION | | Total |
	Europe	United States	
A	95	125	220
B	50	70	120
O	45	90	135
AB	10	15	25
Total	200	300	500

SOLUTION

1. The null and alternative hypotheses are given as follows.

H_0: The distribution of blood types in Europe is the same as in the United States.

H_A: There is no homogeneity.

2. The expected frequencies are obtained by multiplying appropriate row and column totals together and then dividing the product by $n = 500$. The computations to find the χ^2 value are given in Table 10-18.

TABLE 10-18 Computation of the χ^2 Value

Location	Blood Type	Observed Frequency O	Expected Frequency E	$O - E$	$(O - E)^2$	$(O - E)^2/E$
Europe	A	95	88	7	49	0.55
Europe	B	50	48	2	4	0.08
Europe	O	45	54	-9	81	1.50
Europe	AB	10	10	—	—	—
U.S.	A	125	132	-7	49	0.37
U.S.	B	70	72	-2	4	0.06
U.S.	O	90	81	9	81	1.00
U.S.	AB	15	15	—	—	—
	Total	500	500	0		3.56

3. The chi-square distribution has $(4 - 1)(2 - 1)$, or 3, degrees of freedom.
4. Using a 5 percent level of significance, the decision rule is to reject H_0 if the computed value is greater than $\chi^2_{3,0.05} = 7.815$.
5. Since the computed value 3.56 is not greater than the table value of 7.815, we do not reject H_0.

Conclusion: There is no reason to believe that the distribution of blood types in Europe is any different from the distribution in the United States. ●

Section 10-3 Exercises

1. In a laboratory experiment involving two crosses of fruit flies, the following data were obtained regarding the number of progeny with curly wings for the two crosses.

	Cross 1	Cross 2
Number of progeny with curly wings	63	40
Number of progeny	180	144

At the 5 percent level, is there a significant difference in the true proportions of curly-winged progeny for the two crosses? Use the following.
(a) The chi-square test developed in this section
(b) A test for comparing two population proportions given in Section 9-4

2. The following data give information regarding eighty patients, fifty of whom were administered a certain drug and thirty of whom were administered a placebo.

	Drug	Placebo
Cured	35	17
Not cured	15	13

Do the data indicate that the true proportion cured by the drug is different from that cured using a placebo? Use $\alpha = 0.05$.

3. Three launching pads are each fitted with a different device. The following data give the number of successful launches along with the total number of launches from each pad.

	Pad 1	Pad 2	Pad 3
Number of attempts	50	80	60
Number of successful attempts	30	52	33

At the 5 percent level of significance, is there a significant difference in the true proportions of successful launches from the three pads?

4. In a certain area, 200 Democrats, 150 Republicans, and 100 voters who belong to neither party were asked if they supported the welfare programs. The following figures give their responses.

	Support	Do Not Support
Democrats	133	67
Republicans	82	68
Other voters	59	41

At the 2.5 percent level of significance, test the null hypothesis that the support is the same among the three groups.

5. To find the attitude of people on the issue of gun control, 100 people in each of the four regions, the East, the West, the Midwest, and the South, were interviewed. The following sample data were obtained.

	East	West	Midwest	South
In favor	42	44	30	36

Using a 2.5 percent level of significance, test the null hypothesis that the sentiment in favor of gun control is the same in the four regions.

6. Four jars, each containing 100 insects, were sprayed with four different insecticides. The number of insects killed were 46, 70, 60, and 52. At the 5 percent level of significance, test the null hypothesis that there is no significant difference between the insecticides.

7. Professor Kieval is convinced that the traditional approach is best for teaching his beginning course in calculus. He feels that the modern approach does not provide any insight and, certainly, does not equip a student with the basic skills. To test the validity of this assertion, a section of sixty students was taught using the traditional approach and another section of eighty students was taught using the modern approach. At the end of the quarter, the same test (insight-oriented) was given to the two sections and the performance of the students graded as excellent, satisfactory, or poor. The following table presents the results.

	Traditional	Modern
Excellent	31	28
Satisfactory	21	27
Poor	8	25

At the 2.5 percent level of significance, would you say that Professor Kieval is justified in his assertion?

8. In a course in elementary statistics, there were forty chemistry majors, fifty forestry majors, and sixty fisheries majors. The following figures refer to the number of students who passed the course.

	Chemistry	Forestry	Fisheries
Number passed	34	35	44

At the 5 percent level, test the null hypothesis that the true proportions of students passing are the same for the three majors.

9. Three pain relievers, Extrarelief, Superrelief, and Mightyrelief, were each tried on 100 patients. The following figures refer to the extent of relief (recorded as excellent, moderate, or poor).

	Extrarelief	Superrelief	Mightyrelief
Excellent	42	51	55
Moderate	38	27	17
Poor	20	22	28

At the 5 percent level of significance, test the null hypothesis that the three pain relievers are basically the same in their effectiveness.

10. The following table gives the marital status of 200 blue-collar workers, 300 white-collar workers, and 150 university professors.

	Blue-Collar Workers	White-Collar Workers	University Professors
Married	105	147	90
Single (never married)	45	66	18
Divorced	50	87	42

At the 5 percent level of significance, test the null hypothesis that the true distribution of the marital status among the three groups is the same.

11. A single-dose study to compare three drugs was conducted over a four-hour period. The experiment involved 180 patients who were all suffering from one of the following conditions: sprain, strain, dislocation, fracture, or postsurgical pain. Each drug was tested on sixty patients allotted to it at random. The figures below give the number of patients reporting effective pain relief and those requesting additional medication.

	Drug A	Drug B	Drug C
Relief	46	33	40
No relief	14	27	20

At the 1 percent level of significance, are the true proportions of patients getting relief different for the three drugs?

12. The following data were obtained when high school students in grades 10–12 were interviewed regarding the use of smokeless tobacco. There were 104, 110, and 80 students from grades 10, 11, and 12, respectively.

	GRADE LEVEL		
Status	10	11	12
User	15	24	19
Nonuser	89	86	61

Is there a significant difference in the prevalence of smokeless tobacco use among the three grades? Use $\alpha = 0.01$.

13. The following table presents the blood alcohol level (mg percent) in suicide victims, classified by the sex of the victim. A sample of 345 male victims and 192 female victims was considered in the investigation.

	SEX	
Blood Alcohol Level	Male	Female
0	167	134
1–99	60	21
100–199	65	16
200–299	41	12
300 +	12	9

At the 2.5 percent level, is there a significant difference in blood alcohol levels of suicide victims among the two sexes?

CHAPTER **10** **Summary**

✔ *CHECKLIST: KEY TERMS AND EXPRESSIONS*

- ❏ test of goodness of fit, page 581
- ❏ observed frequency, page 581
- ❏ expected frequency, page 581
- ❏ test of independence, page 589
- ❏ contingency table, page 589
- ❏ test of homogeneity, page 597

KEY FORMULAS

For goodness of fit, expected frequency E_i is

$$E_i = np_i$$

where n is the sample size and p_i is the probability of category c_i hypothesized under H_0.

For tests of independence and homogeneity, in the cell that represents the intersection of Row i and Column j, expected frequency E_{ij} is

$$E_{ij} = \frac{(\textbf{Total of Row } i)(\textbf{Total of Column } j)}{n}$$

where n is the total number in the sample.

The test criterion for goodness of fit, independence, and homogeneity is

$$\chi^2 = \sum \frac{(O - E)^2}{E}.$$

The sum is found over all the cells. Here O represents the observed frequency and E, the expected frequency.

CHAPTER **10** **Concept and Discussion Questions**

1. Indicate which of the following are correct statements. The chi-square tests in this chapter deal with
   (a) measurements made on variables that are quantitative
   (b) counts of the number of cases falling into different categories
   (c) frequency of cases falling into different categories
   (d) apply equally well for data with quantitative measurements and frequency data.

2. Give two examples in your experience, one where chi-square analysis would not be appropriate and another where chi-square analysis would be appropriate.

3. Toss a coin 5 times. Record the number of heads in the 5 tosses. Perform this experiment a total of 60 times. Prepare a table as follows listing the number of experiments where 0 heads, 1 head, . . . , 5 heads occurred.

Number of heads in 5 tosses	0	1	2	3	4	5
Observed frequency in 60 repetitions						

Test the null hypothesis that the observed frequencies are consistent with the binomial distribution with $n = 5$ and $p = 0.5$.

4. When is a chi-square test for a two-way contingency table a test for homogeneity?

CHAPTER 10 Review Exercises

1. What null hypothesis do we stipulate in chi-square tests of goodness of fit? What is the alternative hypothesis?

2. What type of null hypothesis do we make when we carry out chi-square tests of independence of classification? What is the alternative hypothesis?

3. Eighty children were asked to pick one of the four numbers 1, 2, 3, 4—whichever came to their minds. The following figures give the distribution of the numbers picked.

Number picked	1	2	3	4
Number of children	25	14	15	26

Use a 5 percent level of significance to test the hypothesis that the numbers were picked at random.

4. The table below gives the distribution of eye color and hair color of 200 individuals picked at random.

	HAIR COLOR	
Eye color	Blonde	Brunette
Blue	32	40
Green	40	24
Brown	10	54

At the 5 percent level, are eye color and hair color related?

5. According to the Hardy-Weinberg law in genetics, if the proportion of A alleles in the population is p and that of a alleles is q (that is, $1 - p$), and the population is *mating at random*, then the proportions of AA, Aa, and aa genotypes in the progeny will be in the ratios $p^2 : 2pq : q^2$. Out of 200 individuals from a random

mating population, it was found that there were 26 *AA* genotypes, 90 *Aa* genotypes, and 84 *aa* genotypes. Test the null hypothesis that the proportion of *A*
alleles in the parent population is 0.4.

6. A survey was conducted in Chicago, Denver, and St. Louis. When 200 people
were interviewed in each of these cities as to whether they favored mandatory
car-safety regulations, the following responses were obtained.

	Chicago	Denver	St. Louis
Favor	60	86	70
Oppose	105	76	83
No opinion	35	38	47

At the 5 percent level, are the three cities significantly different regarding their
attitude toward mandatory car-safety regulations?

 ## GRAPHING CALCULATOR INVESTIGATIONS

*Before beginning these exercises, read the portion of the Appendix entitled "Using the
TI-82 as a Data Analysis Tool," on testing for independence in a contingency table.
Then do these exercises using your TI-82.*

1. A recent study investigated the relationship between gender and the time spent
on Home/Child Care on working days for households where both parents work
full time.

| | Time Spent on Home/Child Care on Working Days | | | |
	< 1 Hour	1–2 Hours	2–3 Hours	> 3 Hours
Male	36	26	20	10
Female	2	10	14	44

*Use the chi-square test for association to determine if a relationship exists between
genders on time spent on home/child care on weekdays. What is the calculated
chi-square? Is it significant, and at what level? Look at the expected values and
discuss how they compare to the observed values.*

2. On October 27, 1989, the *Journal of the American Medical Association* (JAMA) reported the results of a communitywide bicycle helmet campaign that took place in Seattle, Washington. Results on helmet use are summarized below. Observations were made two weeks before the campaign's start, and at 4, 12, and 16 months after the campaign's start.

	Number of Children Observed Wearing Helmets	**Number of Children Observed Not Wearing Helmets**
Before campaign	905	15,550
4 months	1,213	20,842
12 months	1,259	10,731
16 months	1,563	8,392

Test the null hypothesis that there is no association between helmet use and time period. (That is, test the null hypothesis that helmet use remained constant over the studied time period.) Report the calculated chi-square and examine the expected value matrix.

3. This JAMA study discussed in the previous problem also looked for an association between helmet use and gender. The data are as follows:

	Number of Children Observed Wearing Helmets	**Number of Children Observed Not Wearing Helmets**
Male	399	6985
Female	194	2226

Test for an association between gender and helmet use. What is the value of chi-square? Is it significant and, if so, at what level? Which gender, if either, has a higher percentage of helmet wearers?

4. Another factor the JAMA study considered was the companions of the bicycle riders.

	Riding Alone	**With Other Children, No Helmets**	**With Adults, No Helmets**	**With Other Children, Helmets**	**With Adults, Helmets**
Wearing helmets	196	41	48	115	194
Not wearing helmets	4580	4037	478	67	71

Test for an association between helmet use and riding companion. What is the value of chi-square? Is it significant and, if so, at what level? Examine the expected frequencies and comment on exceptional differences with observed frequencies.

Source for Exercise 1: Table from "Career Mobility: Does Gender Matter?" by Rose Bell from *Thought and Action,* Volume 8, 1, Spring 1992. Reprinted by permission of National Education Association.

Source for Exercises 2-4: Adapted from the *Journal of the American Medical Association,* October 27, 1989, Volume 262, pp. 2256-2261. Copyright © 1989, American Medical Association. Reprinted by permission.

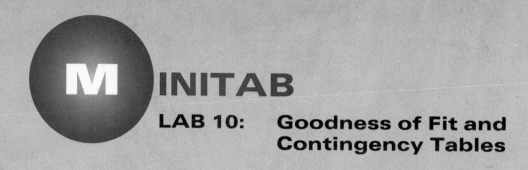

MINITAB

LAB 10: Goodness of Fit and Contingency Tables

In this lab we use the chi-square statistic to test for goodness of fit, for independence of attributes, and for homogeneity of probability distributions.

1 We start with goodness of fit. In several previous labs we have generated random observations from specified probability distributions. The goodness of fit test allows us to test whether the data obtained in this way are indeed in conformity with the prescribed distribution.

As an example, we will simulate rolls of a die as we did in Lab 3, and use the chi-square test for "goodness of fit" to test whether the data could have indeed resulted from rolling a normal die. To carry out the test we must develop the program, as it is not supported directly by MINITAB. We go about it as follows.

```
MTB >RANDOM 200 INTEGERS PUT IN C1;
SUBC>INTEGERS 1 TO 6.
```

This command simulates 200 rolls of a die. Count the number of 1s, 2s, . . . , 6s using the following procedure.

```
MTB >COPY C1 INTO C11;
SUBC>USE C1 = 1.
MTB >COPY C1 INTO C12;
SUBC>USE C1 = 2.
MTB >COPY C1 INTO C13;
SUBC>USE C1 = 3.
MTB >COPY C1 INTO C14;
SUBC>USE C1 = 4.
```

```
MTB >COPY C1 INTO C15;
SUBC>USE C1 = 5.
MTB >COPY C1 INTO C16;
SUBC>USE C1 = 6.
```

The command COPY together with the subcommand USE places the 1s, 2s, . . . , 6s in Columns C11, C12, . . . , C16, respectively. For example, COPY C1 INTO C11; USE C1 = 1. will place all the 1s from Column C1 into Column C11. Next, we count the frequency of each number.

```
MTB >LET K1 = COUNT (C11)
MTB >LET K2 = COUNT (C12)
MTB >LET K3 = COUNT (C13)
MTB >LET K4 = COUNT (C14)
MTB >LET K5 = COUNT (C15)
MTB >LET K6 = COUNT (C16)
MTB >PRINT K1-K6
```

K1, K2, . . . , K6 represent the observed frequencies. We insert these counts in Column C2 with the following COPY command.

```
MTB >COPY K1-K6 INTO C2
MTB >PRINT C2
```

The printout of C2 should match that of K1–K6. The chi-square analysis for goodness of fit is then carried out as follows.

```
MTB >NOTE NULL HYPOTHESIS P = 1/6 FOR EACH NUMBER
MTB >NOTE EXPECTED FREQUENCIES
MTB >LET C3 = SUM(C2) * (1/6)
MTB >NOTE DIFFERENCES, OBSERVED MINUS EXPECTED
MTB >LET C4 = C2 — C3
MTB >NOTE DIFFERENCES SQUARED OVER EXPECTED
MTB >LET C5 = C4 ** 2/C3
MTB >NAME C2 'OBS FR' C3 'EXP FR' C4'O MIN E'
      C5 'C4 SQ/C3'
MTB >PRINT C2-C5
MTB >NOTE CHI-SQUARE VALUE
MTB >SUM C5
```

The sum of Column C5 gives the chi-square value. Compare this computed value with the table value of chi-square with 5 degrees of freedom and a given level of α. Is the result significant?

2   Next, we solve Exercise 6 in Section 10-1 using MINITAB.

```
MTB >READ DATA IN C1-C3
DATA>0  0.05  16
DATA>1  0.30  68
DATA>2  0.30  51
DATA>3  0.25  46
DATA>4  0.10  19
DATA>END
MTB >PRINT C1-C3
MTB >NOTE EXPECTED FREQUENCIES
MTB >LET C4 = C2 * SUM(C3)
MTB >NOTE DIFFERENCES, OBSERVED MINUS EXPECTED
MTB >LET C5 = C3 − C4
MTB >NOTE DIFFERENCES SQUARED OVER EXPECTED
MTB >LET C6 = C5 ** 2/C4
MTB >NAME C3 'OBS FR' C4 'EXP FR' C5 'O MIN E'
     C6 'C5 SQ/C4'
MTB >PRINT C3-C6
MTB >NOTE CHI-SQUARE VALUE
MTB >SUM C6
```

Assignment   Consider Example 3, in Section 4-2 of Chapter 4 giving the landscape gardener's hourly earnings with the following probability distribution.

x	0	6	12	16
$p(x)$	0.3	0.2	0.3	0.2

Simulate the gardener's earnings on 150 days and test whether the data obtained are in fact in keeping with the given probability function.

Assignment   Work out Exercise 3 in Section 10-1 in the text.

3 We now consider contingency table analysis. It is supported by MINITAB software, and so we do not have to write a program for this purpose. Also, the procedure is the same for test of independence and test for homogeneity.

We will perform contingency table analysis on the data in the Example in Section 10-2. First, we read the three columns of data in Columns C1–C3. (You could pick any other set of three columns to enter the data.)

```
MTB >READ TABLE INTO C1-C3

DATA>10  40  10

DATA>80  85  45

DATA>30  25  35

DATA>END OF DATA

MTB >PRINT C1-C3
```

Once the data are read in, we enter the command CHISQUARE, properly identifying the columns where the data are placed, as follows.

```
MTB >CHISQUARE FOR DATA IN C1-C3
```

With this command MINITAB presents a detailed analysis giving, among other things, the expected frequencies, the value of $(O - E)^2/E$ for each cell, the chi-square value, and the appropriate number of degrees of freedom.

Assignment   Carry out contingency analysis for data in Exercise 9 in Section 10-3.

We now type STOP and conclude the MINITAB session using the logging off procedure.

Commands you have learned in this session:

```
COPY;

USE.

COUNT

CHISQUARE
```

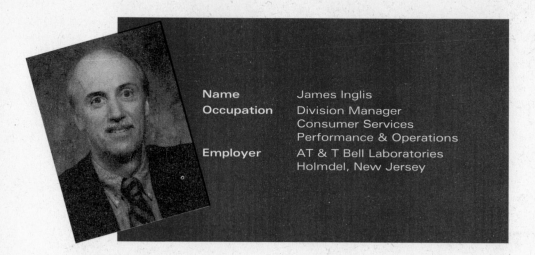

Name   James Inglis
Occupation   Division Manager
              Consumer Services
              Performance & Operations
Employer   AT & T Bell Laboratories
              Holmdel, New Jersey

I became interested in statistics as a mathematics major in college in the mid-60s. The problems in the probability and mathematical statistics courses I took had a complexity and an *association with the real world* that were fascinating. In graduate school, I became more interested in applied statistics, seeing how the statistical way of thinking and statistical techniques could make sense out of seemingly confused and messy, real-world data.

In my career, I've helped the chaplain of my Army battalion in Vietnam, who was dealing with a complaint about racial prejudice, to understand the racial mix of our unit. I've helped a magazine interpret data on the "power" of Broadway theater newspaper critics to infuence theater attendance with their praising and panning of shows. I've developed an equation to estimate the weight of babies, while still in the uterus, to aid in the management of high-risk pregnancies. I've helped a telephone company and a state public utilities commission understand the effects of rate changes on the demand for telephone service. The opportunities for meaningful data analysis are limitless.

In the course of my career, I've been an Army officer and a professor of statistics and biostatistics. Since 1978, I've been at AT & T Bell Laboratories, starting as an applied statistical researcher and consultant, and then working through two promotions as a technical manager.

Currently, I'm Manager of the Consumer Services Performance & Operations Division (a division of about 80 people) at Bell Laboratories' Holmdel location. I lead and support systems engineering work on new telecommunications services for the consumer market. Being involved with projects that use many kinds of data (customer needs, performance and reliability, and operations) is very satisfying.

In all my jobs, statistical thinking and an understanding of data analysis have been important. In my current job, I see many kinds of data—both managerial (budget, personnel) and technical project data. The data are in many forms: tables, histograms, graphs. Understanding how to look at data and how to draw conclusions from data is part of my job.

Statistics is a fascinating subject with many uses in many fields. A few useful suggestions: learn about computers and their abilities in data analysis—they are already heavily used for computations, and their role will only increase. Learn the principles of statistical inference and the difference between statistical correlation and physical causality. Most importantly, understand the fundamental role variability plays in any analysis—from surveys reported in the newspaper to complex medical studies, the inherent variability of people and nature is what applied statistics strives to help us understand.

Linear Regression and Linear Correlation

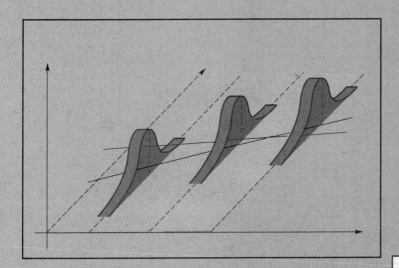

INTRODUCTION

For the most part, thus far we have restricted our study to statistical methods analyzing a single variable. Such a discussion about a single variable is commonly known as the **univariate case.** However, in dealing with problems in social sciences, business, biology, or economics, often we are interested in determining whether a discernible relationship exists between two or more variables. Recall that in Chapter 10 we considered only briefly a situation in which two response variables are observed on each member. The variables considered were qualitative, and using contingency tables, we went on to discuss whether they were related. In this chapter we turn our attention to the case in which two *numerical* variables are measured, resulting in a pair of measurements for each member. This is called the **bivariate case.** For example, an educator might be interested in finding out a relationship, if any, between the number of hours a student studies and his or her score on a scholastic test; a marketing specialist might wish to find a relationship between the amount spent on advertising and the volume of sales; a social scientist might be interested in knowing whether the income of a family and the amount spent on entertainment are related in some way; and so on.

In cases of this kind, many questions occur. In the first place, is there a reason to suspect that a mathematical relationship exists between the variables? If there is a relation, how can we describe it effectively? An important reason for establishing a relation between two variables is to make predictions. How reliable are such predictions?

We will treat the simplest case, in which the relation is linear. That is, the association can be expressed graphically by means of a straight line relation on the coordinate system. Specifically, we will discuss two concepts that are extensively used in scientific research, **linear correlation** and **linear regression.** It should be realized that in investigating an association between variables, these two concepts, though interrelated, serve different purposes. The role of linear correlation is to find out if there exists a *linear* relation between the variables. If a linear relation exists, the correlation coefficient will indicate how close to a straight line such a relation is. This closeness is often referred to as the strength of their association. The regression theory, on the other hand, goes on to describe the specific linear relation in quantitative terms, assuming that there does exist a linear relation. Once the equation describing the straight-line relation is established, we may use it for making predictions about one variable based on our knowledge of the other. The variable that is predicted is called the *dependent* variable (also called the *output* variable or *response* variable). The variable used for making the prediction is called the *independent* variable (also called the *input* variable). If, for example, the IQ is used as a predictor of performance on a scholastic test, then IQ is the independent variable and performance on the test, the dependent variable.

11-1   A LINEAR RELATION

WHAT IS A LINEAR RELATION?

As mentioned in the introduction, a linear relation between two variables is one that can be expressed graphically by means of a straight line. How do we express a straight-line relation by means of an equation? Consider the following examples.

Suppose a taxi driver charges 50¢ per mile. If x represents the number of miles traveled and y the total amount charged (in dollars), then we would write

$$y = 0.5x.$$

Now suppose the driver says that the initial charge is $2 just for turning on the meter and that each mile costs 50¢. Then we would write

$$y = 2 + 0.5x.$$

For different numbers of miles traveled x, we can compute the amount charged y as shown in Table 11-1. When the points that represent the pairs of values (x, y) computed in this way are plotted on the coordinate system as in Figure 11-1, we see that they all lie on a straight line that rises from left to right.

TABLE 11-1   Number of Miles Traveled and Amount Charged

Number of Miles x	Amount Charged (in Dollars) y
0	2.0
1	2.5
2	3.0
3	3.5
4	4.0

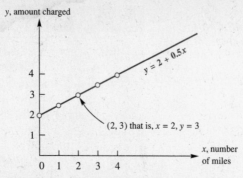

Figure 11-1 A Straight Line Rising From Left to Right

TABLE 11-2 Number of Miles Traveled and Amount Left

Number of Miles	Amount Left
x	y
0	40
2	39
4	38
10	35

As another example, suppose a person with $40 in his pocket hires a taxi and is charged at a rate of 50¢ a mile. If x represents the number of miles traveled and y the amount that the passenger has left after traveling x miles, then clearly

$$y = 40 - 0.5x.$$

The values of y for different values of x are given in Table 11-2. As can be seen from Figure 11-2, all the points (x, y) lie on a straight line that falls from left to right.

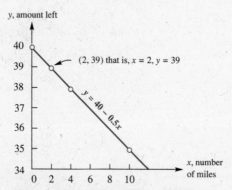

Figure 11-2 A Straight Line Falling From Left to Right

In general, a straight-line relation between two variables x and y can be expressed mathematically by means of an equation as

$$y = a + bx$$

where a and b are the constants that determine the nature of the particular straight line.

If b, the coefficient of x, is positive, the straight line rises from left to right as in Figure 11-1, and if b is negative, it falls from left to right as in Figure 11-2.

When $x = 0$, y is equal to a. We call a the **y-intercept.**

To attach a meaning to the constant b, consider a unit increase in x. That is, x changes to $x + 1$. Correspondingly, suppose y changes to y_1. We then have

$$y = a + bx$$

and

$$y_1 = a + b(x + 1).$$

Taking the difference, we get

$$y_1 - y = b.$$

Thus, b represents the change in y for a *unit* increase in x and is called the **slope** of the line. The y-intercept and the slope of a typical straight line are shown in Figure 11-3.

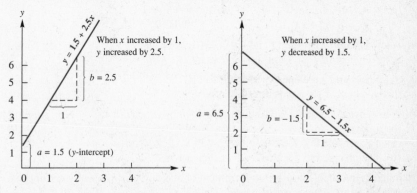

Figure 11-3 The y-intercepts a and Slopes b for Two Straight Lines $y = 1.5 + 2.5x$ and $y = 6.5 - 1.5x$

Thus, we summarize as follows.

EQUATION OF A STRAIGHT LINE

The **equation of a straight line** may be written in the form

$$y = a + bx$$

where the constant a represents the *y-intercept* and the constant b represents the *slope,* which is the change in y for a unit increase in x. If b is positive, then y increases as x increases; if b is negative, then y decreases as x increases; and if $b = 0$, then y is constant and $y = a$.

Section 11-1 Exercises

1. Suppose two variables x and y are related by the linear relation $2y = 3 - 4x$. Complete the following table.

x	3	0	-2
y	—	—	—

2. Determine which of the following points lie on the straight line given by $y = 7 - 3x$.
   (a) $(1, 4)$ **(b)** $(3, 7)$ **(c)** $(-3, 16)$
   (d) $(10, -23)$ **(e)** $(4, 5)$ **(f)** $(-2, 1)$

3. Draw the lines given by the following equations.
   (a) $y = 5 - 3x$ **(b)** $y = 7.5 + 2x$ **(c)** $y = 0.5 - 2.5x$

4. State whether y increases or decreases as x increases, in each case.
   (a) $y = -5 + 4x$ **(b)** $y = -2.5 - 8.7x$ **(c)** $y = 7.5 - 12.4x$

5. Without plotting the graphs, find the y-intercepts and the slopes of these lines.
   (a) $y = -10 - 15x$ **(b)** $y = -5 + 4x$
   (c) $y = 5.8 + 4.9x$ **(d)** $y = 15.2 - 24.9x$

6. A linear relation between x and y is given by the equation $y = 2.5 + 3.8x$. What is the change in y for a unit increase in x?

7. Suppose x and y are linearly related through the equation $y = 1.3 - 7.5x$. Find the change in y for a unit increase in x.

8. A linear relation between x and y is described by the equation $y = 6 - 3x$. Find the following.
   (a) The change in y for a unit increase in x
   (b) The change in y for two units of increase in x

9. What is the change in y for an increment of one unit in x, if x and y are linearly related by the equation $y = -0.875 + 0.75x$?

10. Consider the linear equation $y = -12.89 - 1.793x$. If x is changed from 18 to 19, what will the change in y be?

11-2 COMPUTATION OF A LINEAR CORRELATION COEFFICIENT AND FITTING OF A REGRESSION LINE

In this section we consider the twin problems of computing the correlation coefficient and fitting the regression line when the sample data are provided.

THE LINEAR CORRELATION COEFFICIENT

Consider the data given in Table 11-3, which represent the IQ and grade point averages of ten students. Designating IQ by x and grade point average by y, we have ten ordered pairs (x, y): $(100, 3.0)$, $(120, 3.8)$, . . . , $(90, 2.4)$. We call these pairs *ordered pairs* with the understanding that the first compo-

TABLE 11-3 IQ and Grade Point Averages of Ten Students

IQ x	Grade Point Average y
100	3.0
120	3.8
110	3.1
105	2.9
85	2.6
95	2.9
130	3.6
100	2.8
105	3.1
90	2.4

nent will refer to the value of x and the second component to the value of y. In treating problems involving bivariate data, as a preliminary step, it is a good idea to plot the data as points in the xy-coordinate system. The pictorial representation thus obtained is called a **scatter diagram.** When the points are plotted, a visual pattern may evolve suggesting a particular type of relation, such as linear, curvilinear, and so on. The scatter diagram for the data in Table 11-3 is shown in Figure 11-4.

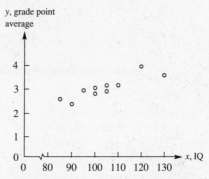

Figure 11-4 Scatter Diagram for IQ and Grade Point Average Data Given in Table 11-3

We see from the scatter diagram that students with higher IQs appear to obtain higher grade point averages and students with lower IQs tend to have lower grade point averages. Because high values of x tend to be associated with high values of y, and, conversely, low values of x tend to be associated with low values of y, we say that our data show a *positive relationship* between IQ and grade point average. On the other hand, if it should turn out that high values of x are associated with low values of y, and low values of x are associated with high values of y, we say that a *negative,* or an *inverse, relationship* exists. An example of a negative relationship is that of the price of a used car and the number of years it has been used.

From the scatter diagram in Figure 11-4 we see that the points are scattered in such a way that, more or less, a straight-line pattern is indicated. The measure characterizing the strength of this relationship, closeness to a straight line, is provided by the **coefficient of linear correlation,** also called the **correlation coefficient.** It is commonly designated by the letter r, and if (x_1, y_1), (x_2, y_2), . . . , (x_n, y_n) are n pairs of values, it is defined by the following formula.

$$r = \frac{\sum (x_i - \bar{x})(y_i - \bar{y})}{\sqrt{\sum (x_i - \bar{x})^2 \cdot \sum (y_i - \bar{y})^2}}$$

This coefficient is also called the *Pearson product-moment correlation coefficient*. For computational purposes, an alternate version is the following.

COEFFICIENT OF LINEAR CORRELATION

If (x_1, y_1), (x_2, y_2), . . . , (x_n, y_n) are n pairs of values, then the coefficient of linear correlation is defined by

$$r = \frac{n \sum x_i y_i - \sum x_i \cdot \sum y_i}{\sqrt{\left[n \sum x_i^2 - \left(\sum x_i\right)^2\right]\left[n \sum y_i^2 - \left(\sum y_i\right)^2\right]}}.$$

Thus, in order to compute r, we need $\sum x_i$, $\sum y_i$, $\sum x_i^2$, $\sum y_i^2$, $\sum x_i y_i$, and n, the number of paired observations. (Many handheld calculators are equipped to provide all these quantities by pressing a few keys. Some calculators also give r directly.

EXAMPLE 1   Obtain the coefficient of linear correlation between IQ and grade point average based on the data of Table 11-3.

SOLUTION   The data are reproduced in the first two columns of Table 11-4, and the relevant computations are carried out in Columns 3, 4, and 5.

TABLE 11-4   Computations for Obtaining Correlation Coefficient between IQ and Grade Point Average Based on the Data of Table 11-3

	x	y	x^2	y^2	xy
	100	3.0	10,000	9.00	300.0
	120	3.8	14,400	14.44	456.0
	110	3.1	12,100	9.61	341.0
	105	2.9	11,025	8.41	304.5
	85	2.6	7,225	6.76	221.0
	95	2.9	9,025	8.41	275.5
	130	3.6	16,900	12.96	468.0
	100	2.8	10,000	7.84	280.0
	105	3.1	11,025	9.61	325.5
	90	2.4	8,100	5.76	216.0
Total	1040	30.2	109,800	92.80	3187.5

Because $\Sigma\, x_i = 1040$, $\Sigma\, y_i = 30.2$, $\Sigma\, x_i^2 = 109{,}800$, $\Sigma\, y_i^2 = 92.8$, and $\Sigma\, x_i y_i = 3187.5$, substituting in the formula we get

$$r = \frac{10(3187.5) - (1040)(30.2)}{\sqrt{[10(109{,}800) - (1040)^2][10(92.80) - (30.2)^2]}}$$

$$= \frac{467}{\sqrt{(16{,}400)(15.96)}}$$

$$= 0.913.$$

How does r measure the strength of a linear relationship? What does a value of $r = 0.913$ in Example 1 mean? We will consider these questions later when we consider correlation analysis in Section 11-4. At this time, we simply list some of the properties of r.

Properties of the Correlation Coefficient

1. The coefficient of correlation always yields a value between -1 and 1, inclusive. That is, $-1 \le r \le 1$.

2. A value of $r = 1$ means that there is a perfect positive correlation and that in the scatter diagram all the points lie on a *straight line* rising upward to the right, as shown in Figure 11-5(a) on the next page.

3. A value of $r = -1$ means that there is a perfect negative correlation and that all the points lie on a *straight line* falling downward to the right, as in Figure 11-5(b). Any value of r in the vicinity of $+1$ or -1, such as 0.95 or -0.9, indicates that the points are scattered closely around a straight line.

4. A value of r near zero means either of the following.
    a. No clear relationship exists between the two variables, as, for example, is indicated by the scatter diagram in Figure 11-5(c). There is no definite pattern of dependence between x and y and the relation could be shown by a *horizontal line*.
    b. The variables are related, but *not* by a linear relation. For example, for Figure 11-5(d), the value of r would be very near zero. However, there is obviously a definite relation that is *curvilinear*.

5. A positive value of r indicates that the *linear trend* is upward from left to right, as in Figures 11-5(a) and (e). A negative value of r implies that the *linear trend* is downward from left to right, as in Figures 11-5(b) and (f).

FITTING A REGRESSION LINE TO DATA

From the magnitude of r we are able to decide whether two variables are linearly related and also to determine the closeness of this relationship. However, a correlation coefficient, even when it indicates a rather strong linear relationship, does not tell us anything about the way the two variables are

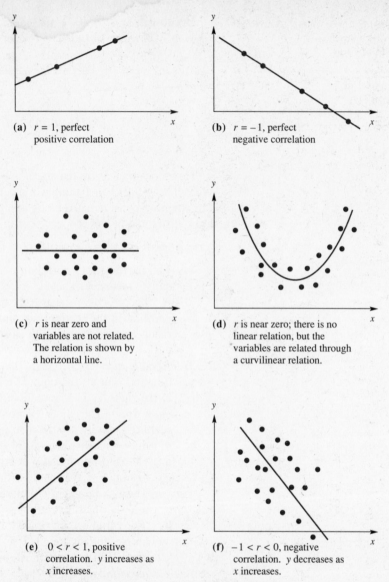

(a) $r = 1$, perfect positive correlation

(b) $r = -1$, perfect negative correlation

(c) r is near zero and variables are not related. The relation is shown by a horizontal line.

(d) r is near zero; there is no linear relation, but the variables are related through a curvilinear relation.

(e) $0 < r < 1$, positive correlation. y increases as x increases.

(f) $-1 < r < 0$, negative correlation. y decreases as x increases.

Figure 11-5 Some Possible Scatter Diagrams

linearly related, since r by itself gives us neither the slope nor the y-intercept. Therefore, the mere knowledge of r is inadequate for prediction purposes. What we need now is the theory of regression. We tacitly assume that there is a straight-line relation and proceed to determine the constants that describe it. In regression problems it is conventional to denote the independent variable as x and to plot its values along the horizontal axis. The dependent variable is denoted by y, and its values are plotted along the vertical axis.

TABLE 11-5 Yield of Grain and the Amount of Fertilizer Applied

Amount of Fertilizer x	Yield y
30	43
40	45
50	54
60	53
70	56
80	63

Consider the following example, in which we wish to predict the yield of grain (bushel per acre) as a linear function of the amount of fertilizer applied (pounds per acre).* Because yield is the predicted variable, it is the dependent variable. The amount of fertilizer applied is the variable used for the purpose of making a prediction; thus it is the independent variable.

Table 11-5 gives the observed yields of grain for corresponding applications of fertilizer. These pairs of values are plotted in Figure 11-6 to obtain the scatter diagram. Clearly a linear relation is indicated. We may therefore want to fit a straight line to bring out this linear relation. If the points were exactly on a straight line, there would be no problem about drawing the line. Because the points are not on a straight line, theoretically the number of lines that can be drawn is unlimited. Thus, given a set of data, our problem concerns drawing the particular straight line that best reflects the linear trend indicated by the points. In other words, the problem is simply that of determining appropriate constants a and b associated with the straight line.

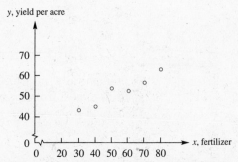

Figure 11-6 Scatter Diagram of the Yield of Grain and Amount of Fertilizer Applied

The principle involved in obtaining "the line of best fit" is called the *method of least squares* and was developed by Adrien-Marie Legendre (1752–1833). We will now explain the central idea behind this principle.

Suppose $\hat{y} = a + bx$ ($\hat{y}$ is read *y hat*) is the line that best fits the data consisting of the n pairs of values $(x_1, y_1), (x_2, x_2), \ldots, (x_n, y_n)$. We call this the **prediction equation**. For example, $\hat{y}_i = a + bx_i$ represents the predicted value of the dependent variable when the value of the independent variable is x_i. Of course, y_i is the corresponding observed y-value in the data. The quantities $y_1 - \hat{y}_1, \ y_2 - \hat{y}_2, \ldots, \ y_n - \hat{y}_n$, that is, $y_1 - (a + bx_1)$,

---

* This example will be referred to later in this section and in Section 11-3.

$y_2 - (a + bx_2), \ldots, y_n - (a + bx_n)$, give deviations of the observed y-values from the corresponding predicted values and each is called the **residual,** or **error,** denoted e. This information is summarized in Table 11-6, where $e_i = y_i - \hat{y}_i = y_i - (a + bx_i)$.

TABLE 11-6   The Observed x- and y-Values Together with the Predicted y-Values and the Corresponding Residuals

	y-VALUES		Residual
x	Observed	Predicted	(or Error)
x_1	y_1	$\hat{y}_1 = a + bx_1$	$y_1 - \hat{y}_1 = e_1$
x_2	y_2	$\hat{y}_2 = a + bx_2$	$y_2 - \hat{y}_2 = e_2$
$\vdots$	$\vdots$	$\vdots$	$\vdots$
x_i	y_i	$\hat{y}_i = a + bx_i$	$y_i - \hat{y}_i = e_i$
$\vdots$	$\vdots$	$\vdots$	$\vdots$
x_n	y_n	$\hat{y}_n = a + bx_n$	$y_n - \hat{y}_n = e_n$

Graphically, e_i, which is $y_i - (a + bx_i)$, denotes the vertical distance (plus or minus) from the observed value y_i to the line of best fit, as shown in Figures 11-7(a) and (b).

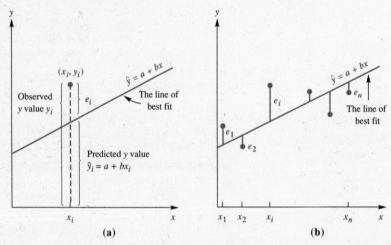

Figure 11-7 Deviations of the Observed y-values From the Predicted y-values $\hat{y} = a + bx$

The principle of the **method of least squares** is stipulated as follows.

METHOD OF LEAST SQUARES

Of all the possible straight lines that can be drawn on a scatter diagram, choose as the **line of best fit** the one for which the sum of the squares of the deviations of the observed y-values from the predicted y-values is a minimum. That is, determine the constants a and b in such a way that

$$\sum e_i^2 = \sum [y_i - (a + bx_i)]^2$$

is a minimum.

The straight line obtained using this criterion is called the **least squares regression line.** The values of a and b that determine the line of best fit can be obtained from the following formulas. (These formulas are derived using methods of calculus.)

THE LEAST SQUARES FORMULAS

The *slope* of the line of best fit is given by

$$b = \frac{n \sum x_i y_i - \sum x_i \cdot \sum y_i}{n \sum x_i^2 - \left(\sum x_i\right)^2}$$

and its *y-intercept* by

$$a = \bar{y} - b\bar{x}$$

where n is the number of pairs of observations in the data, $\bar{y}$ is the mean of the observed y-values, and $\bar{x}$ is the mean of the x-values.

The slope b given by the method of least squares is referred to as the **sample regression coefficient.** The specific line

$$\hat{y} = a + bx$$

obtained in this way is called the line of **regression of y on x.**

EXAMPLE 2   For the data on fertilizer application and yield of grain considered earlier, find the regression of yield of grain on the amount of fertilizer applied. The data are given in the first two columns of Table 11-7.

TABLE 11-7   Computations for Finding the Regression of Yield of Grain on the Amount of Fertilizer Applied

Fertilizer x	Yield y	x^2	xy	y^2
30	43	900	1,290	1,849
40	45	1,600	1,800	2,025
50	54	2,500	2,700	2,916
60	53	3,600	3,180	2,809
70	56	4,900	3,920	3,136
80	63	6,400	5,040	3,969
Total 330	314	19,900	17,930	16,704

SOLUTION   First we should compute a and b.* The quantities that are needed for this purpose are $\Sigma x_i, \Sigma y_i, \Sigma x_i^2, \Sigma x_i y_i$, and n, the number of paired observations. These computations are displayed in Columns 3, 4, and 5 of Table 11-7.

Because $n = 6$, $\Sigma x_i = 330$, $\Sigma y_i = 314$, $\Sigma x_i^2 = 19,900$, $\Sigma x_i y_i = 17,930$, we obtain the following value for b.

$$b = \frac{n \sum x_i y_i - \sum x_i \sum y_i}{n \sum x_i^2 - \left(\sum x_i\right)^2}$$

$$= \frac{6(17,930) - (330)(314)}{6(19,900) - (330)^2}$$

$$= 0.377$$

Because $\bar{x} = \dfrac{330}{6} = 55$ and $\bar{y} = \dfrac{314}{6} = 52.33$, we get

$$a = \bar{y} - b\bar{x}$$

$$= 52.33 - (0.377)(55)$$

$$= 31.593.$$

----

*Some calculators will give a and b directly.

Finally, the line of regression of y on x (that is, the prediction equation) is given by

$$\hat{y} = 31.593 + (0.377)x$$

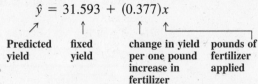

Predicted yield   fixed yield   change in yield per one pound increase in fertilizer   pounds of fertilizer applied

The value of $b = 0.377$ signifies that for an increase of 1 pound per acre of fertilizer, there is a corresponding increase of 0.377 bushels of grain. ⬤

Once the equation of the regression line is obtained, it is a simple matter to draw it. We need only two points in the plane to draw a straight line. Thus, we pick any two values of x and for these values obtain corresponding $\hat{y}$-values using the equation. In Example 2,

$$\text{for } x = 50, \hat{y} = 31.593 + 0.377(50) = 50.443$$

and

$$\text{for } x = 60, \hat{y} = 31.593 + 0.377(60) = 54.213.$$

Joining the two points (50, 50.443) and (60, 54.213), we get the line in Figure 11-8.

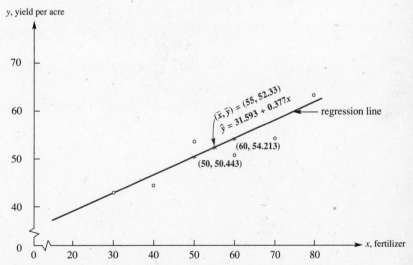

Figure 11-8 The Line of Regression of Grain Yield on the Amount of Fertilizer Applied

It is interesting to note from the formula for the y-intercept that $\bar{y} = a + b\bar{x}$. It therefore follows that *the regression line always passes through the point $(\bar{x}, \bar{y})$.*

EXAMPLE 3   The scores x of ten students on the midterm exam and their scores y on the final exam are given in the first two columns of Table 11-8.

TABLE 11-8   Scores of Ten Students on the Midterm and Final Exams and Other Computations

Scores on Midterm x	Scores on Final y	x^2	y^2	xy
58	55	3,364	3,025	3,190
64	75	4,096	5,625	4,800
80	82	6,400	6,724	6,560
74	63	5,476	3,969	4,662
40	82	1,600	6,724	3,280
70	83	4,900	6,889	5,810
72	78	5,184	6,084	5,616
26	50	676	2,500	1,300
80	65	6,400	4,225	5,200
74	57	5,476	3,249	4,218
Total 638	690	43,572	49,014	44,636

(a)   Plot the scatter diagram.

(b)   Compute the coefficient of correlation.

(c)   Obtain the regression of the scores in the final exam on the scores in the midterm.

(d)   Draw the least squares regression line.

SOLUTION
(a)   The scatter diagram is given in Figure 11-9.

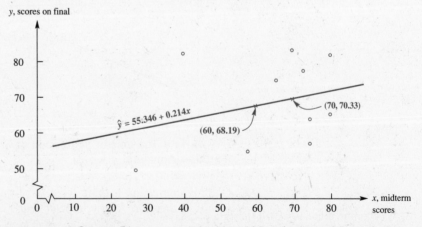

Figure 11-9 Scatter Diagram and the Line of Regression of Final Scores on the Midterm Scores

(b) The necessary computations are shown in Table 11-8. Since $n = 10$, $\Sigma\, x_i = 638$, $\Sigma\, y_i = 690$, $\Sigma\, x_i^2 = 43{,}572$, $\Sigma\, y_i^2 = 49{,}014$, and $\Sigma\, x_iy_i = 44{,}636$, from the formula for r we get

$$r = \frac{10(44{,}636) - (638)(690)}{\sqrt{[10(43{,}572) - (638)^2][10(49{,}014) - (690)^2]}}$$

$$= 0.306.$$

(c) The regression coefficient b is given from the formula as

$$b = \frac{10(44{,}636) - (638)(690)}{[10(43{,}572) - (638)^2]}$$

$$= 0.214$$

Because $\bar{x} = 638/10 = 63.8$, and $\bar{y} = 690/10 = 69$, we get

$$a = \bar{y} - b\bar{x}$$
$$= 69 - (0.214)(63.8)$$
$$= 55.346.$$

Therefore, the line of regression of final scores on the midterm scores is

$$\hat{y} = 55.346 + 0.214x.$$

(d) To draw the least squares regression line we pick any two convenient values of x and find the corresponding $\hat{y}$-values.

For $x = 60$, $\hat{y} = 55.346 + (0.214)(60) = 68.19$.

For $x = 70$, $\hat{y} = 55.346 + (0.214)(70) = 70.33$.

Joining the two points $(60, 68.19)$ and $(70, 70.33)$, we get the line in Figure 11-9. (We could have used $(\bar{x}, \bar{y})$ or $(63.8, 69)$ as one of the points.) ●

12/3/96

Section 11-2 Exercises

1. The following are the heights x (in inches) and weights y (in pounds) of seven athletes.

Heights x	69	66	68	73	71	74	71
Weights y	163	153	185	186	157	220	190

   (a) Find $\Sigma\, x_i$, $\Sigma\, y_i$, $\Sigma\, x_i^2$, $\Sigma\, y_i^2$, and $\Sigma\, x_iy_i$.
   (b) Computer r, the coefficient of correlation between heights and weights.

2. What does the value of $r = 0$ mean?

3. What is the value of r if all the points in a scatter diagram lie:
   (a) on a straight line sloping upward
   (b) on a straight line sloping downward?

4. Of the two correlation coefficients, $r = 0.3$ and $r = -0.8$, based on the same number of pairs of observations, which would you say suggests a stronger linear relation?

5. Suppose the following nine data points are given.

x	7	9	14	16	9	16	11	10	9
y	248	226	218	250	241	235	213	273	197

   (a) Plot the scatter diagram.
   (b) Compute the coefficient of correlation. Does the value of r indicate a linear relation between x and y?

6. Consider the data points with the following coordinates.

x	5.9	−1.1	12.5	5.5	12.9	23.5
y	41	59	53	46	35	27

   Compute the coefficient of correlation and determine whether the linear trend is upward or downward.

7. Consider the following pairs of data.

x	22	25	13	19	23	24	16	21
y	24	17	28	−25	74	80	32	28

   (a) Plot the scatter diagram.
   (b) Compute the coefficient of correlation.
   (c) Does the magnitude of r indicate a linear relation? If so, is the relation positive or negative?

8. The scores on a scholastic test of ten randomly picked students and the number of hours they devoted to preparing for it are as follows.

Number of hours x	18	27	20	10	30	24	32	27	12	16
Scores y	68	82	77	90	78	72	94	88	60	70

   Compute the coefficient of correlation r.

9. The following data give the average daily temperature on fifteen days and the quantity of soft drinks (in thousands of gallons) sold by a company on each of those days.

Temperature	70	75	80	90	93	98	72	75	75	80	90	95	98	91	98
Quantity	30	28	40	52	57	54	27	38	32	46	49	51	62	48	58

   (a) Plot the scatter diagram.
   (b) Compute the coefficient of correlation.

10. Compute the coefficient of correlation between the annual income (in thousands of dollars) and the amount of life insurance (in thousands of dollars) of eight families of the same size.

Annual income	30	33	35	38	41	44	47	50
Amount of insurance	150	120	250	200	250	300	350	320

11. This is the same situation as in Exercise 10, but now the annual income is doubled and the insurance amount is tripled.

Annual income	60	66	70	76	82	88	94	100
Amount of insurance	450	360	750	600	750	900	1050	960

Compare the value of the coefficient of correlation obtained here with that obtained in Exercise 10. Can you venture a guess, in general, as to what happens to the coefficient of correlation when the x-values are multiplied by a constant and the y-values are multiplied by a constant?

12. This is the same situation as in Exercise 10, but now the annual income is increased by $5000 and the amount of insurance is increased by $12,000 for each family. Find the coefficient of correlation and compare it with the answers in Exercises 10 and 11. Make a general comment.

13. In a research problem where two variables x and y were measured on each of forty items, the following data were obtained.

$$\sum x_i = 293 \qquad \sum y_i = 357$$

$$\sum x_i^2 = 2685 \qquad \sum x_i y_i = 3667$$

(a) Find the slope and the y-intercept of the line of best fit.
(b) Give the least squares regression line, $\hat{y} = a + bx$.

14. In a regression problem, the following information is available.

$$n = 50 \qquad \sum x_i = 170 \qquad \sum y_i = -282$$

$$\sum x_i^2 = 4350 \qquad \sum y_i^2 = 37{,}610$$

$$\sum x_i y_i = -12{,}590$$

(a) Find a and b and obtain the least squares regression line $\hat{y} = a + bx$.
(b) Draw the regression line.

15. Consider the following pairs of data points.

x	0.5	1.0	1.5	2.0	2.5	3.0
y	95	88	95	90	87	74

Find the least squares regression line for the data. Visually, does the line provide a satisfactory representation for the data?

16. Consider the following six pairs (x, y).

x	0.59	0.88	0.22	1.14	−0.35	−0.30
y	−0.4	9.4	6.0	−2.2	16.3	9.1

(a) Plot the scatter diagram.
(b) Compute the correlation coefficient. Does it indicate a linear relation between x and y?
(c) Obtain the line of regression of y on x and sketch it.

17. The following pairs of values (x, y) are given.

x	5	10	15	20	25
y	22	32	38	59	67

(a) Plot the scatter diagram.
(b) Find the least squares regression line for the data and sketch it.
(c) Using your visual judgment, does the line provide an adequate representation for the data?

18. The following pairs of values (x, y) are given.

x	−8	−4	0	4	8	12
y	−29	−18	−6	27	38	47

(a) Plot the scatter diagram.
(b) Find the least squares regression line.
(c) Graph the regression line. Does the line provide a good fit for the data?

19. Consider the following pairs of data.

x	2.5	3.5	4.5	6.0	8.0	10.0	12.5
y	−7	−9	−31	−52	−59	−64	−96

(a) Find the equation of the regression line $\hat{y} = a + bx$
(b) Using the regression line, what value $\hat{y}$ would you predict if $x = 5.5$?

20. The following pairs of values (x, y) are given.

x	7	6	5	3	9	5	2	10
y	5	1	−1	−5	3	4	0	7

(a) Plot the scatter diagram.
(b) Find the regression equation, $\hat{y} = a + bx$.
(c) Draw the regression line.

21. Consider the following bivariate data set.

x	4	6	8	9	12
y	12	16	25	30	39

(a) Find the equation of the regression line $\hat{y} = a + bx$.
(b) Using this regression line, what value $\hat{y}$ would you predict if $x = 7.2$?

22. Suppose the fitted linear regression of y, the score on a test, and x, the number of hours a student studies, is given by the relation

$$\hat{y} = 55.28 + 4.1x.$$

Predict a score for a student who studies:
(a) 1 hour   (b) 4 hours.

23. The scores on the final tests in mathematics, physics, and English for eight randomly picked students are as follows.

Mathematics	50	58	67	70	75	82	86	92
Physics	62	54	63	78	81	78	88	90
English	79	82	70	74	67	62	64	56

(a) Find the regression of scores in physics on scores in mathematics.

(b) Find the regression of scores in English on scores in mathematics.

(c) Plot the scatter diagram for the physics-mathematics scores and draw the fitted regression line.

(d) Do the same as in Part c for the English-mathematics scores.

24. Using the data in Exercise 23:

(a) Find the correlation coefficient between scores in mathematics and scores in physics.

(b) Find the correlation coefficient between scores in mathematics and scores in English.

(c) Comment on the signs of the correlation coefficients in Parts a and b in light of the signs of the corresponding regression coefficients in Exercise 23.

25. Nineteen sorghum samples were ground and analyzed for their protein content and the absorbance values. The data are presented in the following table.

Absorbance	Protein (%)
0.50	5.8
0.50	4.5
0.50	6.5
0.55	6.1
0.55	5.5
0.60	6.7
0.60	6.3
0.60	6.2
0.65	8.6
0.65	8.3
0.70	9.7
0.70	8.4
0.70	8.3
0.75	9.1
0.75	9.6
0.80	10.6
0.80	10.6
0.90	11.0
0.95	13.4

(a) Plot the scatter diagram.

(b) Find the regression of protein on absorbance.

(c) Calculate the correlation coefficient.

26. The data on page 636 contain information on fat thickness (mm) and percentage linoleic acid in the lipid of inner backfat of fifteen pigs fed a special diet containing 1.1% linoleic acid from 20 kilograms to 35 kilograms of their initial weight followed by 1.2% linoleic acid from 35 kilograms to slaughter at 85 kilograms live weight.

Fat Thickness (mm)	Linoleic Acid (%)
23	9.0
19	10.5
18	9.4
18	10.5
17	9.4
17	9.8
17	11.2
16	11.1
15	10.6
15	14.2
14	12.0
14	12.9
13	11.4
12	11.7
12	12.8

(a) Plot the scatter diagram.
(b) Obtain the regression of linoleic acid on fat thickness.
(c) Compute the correlation coefficient.

27. The table below represents the optical rotation (°) and the sabinese content (%) of nutmeg oil obtained by analyzing 20 samples. (*Journal of the Science of Food and Agriculture* 1985; *36*: pp. 93–100.)

Sabinese Content (%)	Optical Rotation (°)
28	22
28	25
38	29
38	32
43	27
43	32
45	28
45	35
49	35
49	38
55	31
55	32
55	35
55	40
57	50
60	40
65	50
65	52
65	40
70	60

(a) Plot the scatter diagram.
(b) Obtain the regression of the optical rotation on the sabinese content.
(c) Calculate the correlation coefficient.

28. The following data refer to the amount spent (in thousands of dollars) on advertising and the volume of sales (in millions of dollars) from 1986 to 1993.

Year	1986	1987	1988	1989	1990	1991	1992	1993
Advertising	30	35	50	40	60	80	60	75
Sales	0.8	1.1	1.3	1.8	1.2	2.2	3.1	1.8

  (a) Find the regression of the volume of sales during a year on the amount spent on advertising during that year.

  (b) Find the regression of the volume of sales on the amount spent on advertising during the previous year. That is, find the regression for the following data.

x	30	35	50	40	60	80	60
y	1.1	1.3	1.8	1.2	2.2	3.1	1.8

  (c) Comment on your findings in Parts a and b.

29. In Exercise 9, obtain the least squares regression line, $\hat{y} = a + bx$, where x represents the average temperature during a day and y the quantity of soft drinks sold. Estimate the quantity of soft drinks sold if the average temperature on a day is 85 degrees.

30. Based on your experience, guess in each of the following cases if there is a positive correlation, a negative correlation, or no correlation.
  (a) The price of a commodity and the demand for it
  (b) The supply of a certain food item and its price
  (c) The number of unemployed and the volume of exports
  (d) The height and weight of a person
  (e) The collar size of an individual and the amount of milk that person consumes
  (f) The number of years a machine is in use and the number of yearly repairs
  (g) The price of gold per ounce (in dollars) and the exchange rate of the dollar
  (h) The resale value of a machine and the number of years it is in use
  (i) The waist size of a person and the person's bank balance
  (j) The number of automobile accidents per year and the age of the driver (up to the age of 60 years)

31. A transportation agency interested in finding a relation between the highway speed of a car and mileage per gallon conducted the following experiment. Three test runs were made with a car of a certain make at each of the five speeds (measured to the nearest 5 miles) 45, 50, 55, 60, 65 over a distance of 100 miles, and the mileage per gallon was noted for each run. The following data were obtained.

Speed (mph)	45	50	55	60	65
Mileage per gallon	24.0	27.0	28.0	26.0	23.5
	23.0	26.5	29.0	26.5	23.5
	24.5	26.0	27.5	26.0	24.0

  (a) Draw the scatter diagram.
  (b) Draw a free-hand curve that might best bring out the relation between speed and mileage per gallon.
  (c) Estimate which speed will give the best mileage per gallon for a car of the above make.

32. For the data in Exercise 31, compute the correlation coefficient between speed and mileage per gallon. What explanation can you provide for a value of r so close to zero?

11–3   REGRESSION ANALYSIS

THE LINEAR MODEL

In the preceding section, using the method of least squares, we computed the straight line that best represents a set of n pairs of observations. The n pairs, of course, constitute a sample drawn from a much larger population. We therefore call that specific regression line the *sample regression line.*

Our primary interest in the sample regression line is to investigate the corresponding counterpart in the population. What we have in mind is a **linear model** in the population; that is, we feel that a straight line best depicts the relation between x and y in the population. This is not to say that we envision a precise relation of the type

$$Y = A + Bx.$$

(While we used lowercase letters a and b to represent the y-intercept and the slope, respectively, of the sample regression line, we will use uppercase letters A and B for the corresponding analogues in the population.) If this was the case, then any given value of x would always *determine* exactly the same y-value and we would have what is called a **deterministic model.** Of course, when we consider the fertilizer application-grain yield type of problems, we realize that this does not happen. The same amount of fertilizer applied to different lots of the same size can give different amounts of grain yield. To explain this phenomenon, we need a model that illustrates the overall linear trend in the population and also takes into account the random error. Thus, we think of a response with a basic linear component added to a random error component so that

response = (linear component)

+ (random error component).

A model of this type is called a **linear statistical model.** If Y represents the response variable, then our model stipulates that

$$Y = (A + Bx) + \text{(random error component)}$$

where A and B are population constants, that is, parameters. The assumptions that we make regarding the random error component are that, for any x, its mean is 0, its variance is σ^2, and its distribution is normal. This, in turn, implies that, for any x, the distribution of Y is normal with mean $A + Bx$ and variance σ^2.

To be more specific, we now consider the example in the preceding section where we obtained the regression of the amount of yield of grain on the amount of fertilizer applied. The levels of fertilizer are under the experimenter's control. Therefore, we assume that the levels of fertilizer are fixed, in that they are not decided by chance. It is the yield Y of grain that is a random variable, since one acre of land will not give the same yield as another acre, even with the same amount of fertilizer.

Now suppose we think about the totality of all the acres to which 50 pounds of fertilizer might be applied. The yields of these plots will constitute a subpopulation having a certain distribution. In keeping with our linear statistical model, we assume that the distribution is normal and has a mean equal to $A + 50B$ and a variance of σ^2. In a similar way, we can think of subpopulations of yields for different amounts of fertilizer applications. For any level x of the fertilizer, we assume that the subpopulation of yields is normally distributed with the *same* variance σ^2 and a mean yield equal to $A + Bx$. In short, we assume that the means of the subpopulations fall on a straight line, with the y-intercept equal to A and the slope equal to B. If we denote by $\mu_{Y|x}$ the mean yield of the subpopulation when x pounds of fertilizer are applied, then

$$\mu_{Y|x} = A + Bx.$$

Last, we assume that the yield of one plot is not influenced by the yield of another plot. That is, yields of plots are independent.

Summarizing, we now state formally the basic assumptions underlying a simple linear regression model.

Assumptions

1 The values of the independent variable x are fixed. They are non-random.

2 For each x, Y is normally distributed with mean $A + Bx$, where A and B are constants. That is, $\mu_{Y|x} = A + Bx$. We call the line given by $\mu_{Y|x} = A + Bx$ the **population regression line** or the **true regression line.** (In Figure 11-10 we sketch two straight lines, one representing the sample regression line based on the given data and the other, the population regression line generally unknown but drawn here for the sake of illustration.)

3 For each x, the variance of the distribution of Y is the same, say σ^2. The implication of this is that the sizes of the normal curves are the same for different values of x, as in Figure 11-11.

4 The random variables representing the response variable are independently distributed.

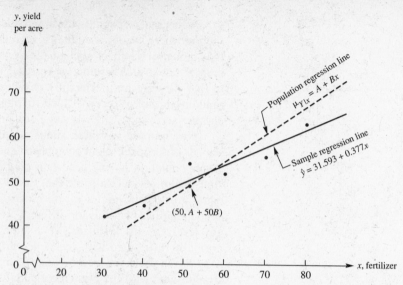

Figure 11-10 Sample Regression Line for the Fertilizer-grain Yield Data and the Population Regression Line

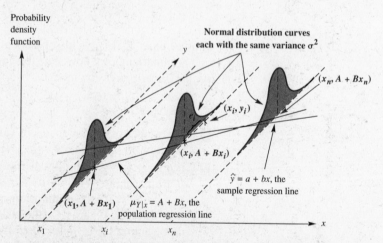

Figure 11-11 A Three-Dimensional Display Showing the Distribution of Y for Given Values of x, the True Regression Line, and the Sample Regression Line

By and large, the population regression line will not be known to the experimenter. In fact, it will be the experimenter's goal to investigate it. Also, we should be aware that this line is a fixed straight line and will not be the same as the sample regression line, which changes from sample to sample, just as we recognize, for example, that the population mean μ is a fixed constant and will not be the same as the sample mean $\bar{x}$, which changes from sample to sample. Now recall at this point that when we considered the univariate case, we first discussed the sample mean $\bar{x}$ and the sample variance s^2 because we were interested in the corresponding parameters μ and σ^2. Subsequently, we developed methods to set confidence intervals and to test hypotheses regarding these parameters. In a similar way, having considered the sample regression line in the preceding section, we now propose to investigate the population insofar as the linear trend is depicted in the population.

In view of the setup of the linear regression model, the statistical problem of regression analysis involves the following aspects:

1   Obtain estimates of the two important parameters A and B of the population regression line. From a practical standpoint, even more useful are inferences regarding $\mu_{Y|x_0} = A + Bx_0$, the mean value of Y for a given x value. For example, we might be interested in the true mean amount of yield Y of grain when x_0 pounds of fertilizer are applied per acre. Also, of paramount importance is the estimation of σ^2 which, as you will soon find out, is involved in the variances of the various estimators.

2   Construct confidence intervals for the parameters above.

3   Finally, test hypotheses regarding the parameters.

It can be shown that the least squares regression line $\hat{y} = a + bx$, which was fitted to the sample data in Section 11-2, provides an unbiased estimate for the population regression equation. Also, a and b provide unbiased estimates of A and B, respectively. In fact, writing $\hat{A}, \hat{B}, \hat{Y}_0 (= \hat{A} + \hat{B}x_0)$ for the point estimators of A, B, $\mu_{Y|x_0}$, it can be shown that all these estimators are normally distributed and that their expected values and variances are as follows.

EXPECTATIONS AND VARIANCES OF THE ESTIMATORS

$\hat{A}, \hat{B}$, and $\hat{Y}_0$ are normally distributed and

$$E(\hat{A}) = A \quad \text{and} \quad \text{Var}(\hat{A}) = \left[\frac{1}{n} + \frac{(\bar{x})^2}{\sum (x_i - \bar{x})^2}\right]\sigma^2$$

$$E(\hat{B}) = B \quad \text{and} \quad \text{Var}(\hat{B}) = \frac{\sigma^2}{\sum (x_i - \bar{x})^2}$$

$$E(\hat{Y}_0) = A + Bx_0 \quad \text{and} \quad \text{Var}(\hat{Y}_0) = \left[\frac{1}{n} + \frac{(x_0 - \bar{x})^2}{\sum (x_i - \bar{x})^2}\right]\sigma^2.$$

Since $\hat{A}, \hat{B}$, and $\hat{Y}_0$ are estimators, they represent random variables. Their values, a, b, and $\hat{y}_0$, computed from a given sample, will provide estimates of A, B, and $\mu_{Y|x_0}$, respectively.

VARIANCE ESTIMATES

As can be seen from the preceding formulas for the variances of the estimators, $\hat{A}$, $\hat{B}$, and $\hat{Y}_0$, they all involve the parameter σ^2 which, for the most part, will not be known to the experimenter. So before we can discuss the inferential aspects, we should obtain an estimate of σ^2. We will use this estimate, in turn, to estimate the variances of $\hat{A}$, $\hat{B}$, and $\hat{Y}_0$.

One of the assumptions of the linear regression model is that for any x, the distribution of Y has mean $A + Bx$ and the *same* variance σ^2, independent of x. Because of this, it can be shown that an unbiased estimate of σ^2, denoted by s_e^2, is given by the following formula.

> **ESTIMATE OF σ^2**
>
> In a linear regression model involving n pairs of observations, an unbiased estimate of σ^2 is provided by
>
> $$s_e^2 = \frac{\sum (y_i - \hat{y}_i)^2}{n - 2}.$$

The reason for dividing by $n - 2$ instead of by n is that to obtain $\hat{y}_i$, which is used in $\Sigma(y_i - \hat{y}_i)^2$, we had to estimate two parameters A and B. Thus, we have lost two degrees of freedom.

EXAMPLE 1   Find the estimate of σ^2 for the data on fertilizer application and yield of grain in the preceding section (see the first two columns of Table 11-9).

TABLE 11-9   Computations for Finding Estimate of σ^2 for the Grain Yield-Amount of Fertilizer Data in Table 11-5

Amount of Fertilizer x	Yield y	$\hat{y} = 31.593 + 0.377x$	$(y - \hat{y})^2$
30	43	42.903	0.010
40	45	46.673	2.800
50	54	50.443	12.652
60	53	54.213	1.472
70	56	57.983	3.932
80	63	61.753	1.555
Total			22.421

SOLUTION   It will be recalled that in this case in Example 2 in the preceding section, we obtained the sample regression line as $\hat{y} = 31.593 + 0.377x$. Thus, the computations to obtain s_e^2 could be carried out as shown in Table 11-9.

From Column 4 of the table, $\Sigma(y_i - \hat{y}_i)^2 = 22.421$. Therefore, applying the formula, the estimate of σ^2 is given as

$$s_e^2 = \frac{\sum (y_i - \hat{y}_i)^2}{n - 2}$$

$$= \frac{22.421}{6 - 2}$$

$$= 5.61.$$

In order to be able to use the above formula for computing s_e^2, we have to obtain the predicted value $\hat{y}_i$ for each x_i. It is clearly unsuitable if there are a large number of observations. The following alternate version of the formula is more manageable from the computational point of view.

AN ALTERNATE FORMULA FOR ESTIMATING σ^2

$$s_e^2 = \frac{1}{n - 2}\left[\sum y_i^2 - a\sum y_i - b\sum x_i y_i\right]$$

There are $n - 2$ degrees of freedom associated with s_e^2 because we used the n observations to compute two quantities, a and b, which are involved in the formula for s_e^2 (see the similar rationale given on page 105 in the context of sample variance).

Note: In using the alternate version, compute a and b to enough decimal places to compensate for the magnitudes of $\Sigma\ y_i$ and $\Sigma\ x_i y_i$. Remember that the products $a\Sigma\ y_i$, $b\Sigma\ x_i y_i$ are involved in the formula.

We now compute s_e^2 for the data in Example 1 using the alternate formula. In Section 11-2 (page 628) we have already computed all of the quantities involved in the formula but here they are given with a and b expressed to more decimal places.

$$n = 6 \qquad a = 31.5905 \qquad b = 0.377143$$

$$\sum y_i = 314 \qquad \sum y_i^2 = 16{,}704 \qquad \sum x_i y_i = 17{,}930$$

Therefore,

$$s_e^2 = \frac{1}{4}[16{,}704 - (31.5905)(314) - (0.377143)(17{,}930)]$$

$$= 5.61.$$

Thus, we get the same answer for s_e^2 using the alternate formula.

We denote the estimates of the variances of $\hat{A}$, $\hat{B}$, and $\hat{Y}_0$ by s_a^2, s_b^2, and $s_{\hat{y}_0}^2$, respectively. They are obtained directly from the corresponding expressions for the variances of the estimators (given on page 641) simply by replacing σ^2 by s_e^2. Therefore, we have the following formulas for the various variance estimates.

> **ESTIMATES OF THE VARIANCES OF $\hat{A}$, $\hat{B}$, AND $\hat{Y}_0$**
>
> $$s_a^2 = \left[\frac{1}{n} + \frac{(\bar{x})^2}{\sum (x_i - \bar{x})^2} \right] s_e^2 \; ; \qquad s_b^2 = \frac{s_e^2}{\sum (x_i - \bar{x})^2}$$
>
> and, if $\hat{y}_0 = a + bx_0$, then
>
> $$s_{\hat{y}_0}^2 = \left[\frac{1}{n} + \frac{(x_0 - \bar{x})^2}{\sum (x_i - \bar{x})^2} \right] s_e^2 .$$

EXAMPLE 2

In Example 1 we have already found a point estimate of σ^2 as $s_e^2 = 5.61$ for the fertilizer application-yield of grain data. Now compute the variance estimates s_a^2, s_b^2, and $s_{\hat{y}}^2$ when $x = 45$ lb.

SOLUTION Because $n = 6$, $\bar{x} = \dfrac{330}{6} = 55$, $\sum x_i = 330$, and $\sum x_i^2 = 19{,}900$, we get

$$\sum (x_i - \bar{x})^2 = \sum x_i^2 - \frac{\left(\sum x_i \right)^2}{n}$$

$$= 19{,}900 - \frac{(330)^2}{6} = 1750.$$

Therefore,

$$s_a^2 = \left[\frac{1}{n} + \frac{(\bar{x})^2}{\sum (x_i - \bar{x})^2} \right] s_e^2$$

$$= \left[\frac{1}{6} + \frac{55^2}{1750} \right] (5.61)$$

$$= 10.6323$$

and

$$s_b^2 = \frac{s_e^2}{\sum (x_i - \bar{x})^2}$$

$$= \frac{5.61}{1750}$$

$$= 0.00321.$$

Because $x_0 = 45$, the estimate of the variance of the estimated yield is

$$s_{\hat{y}}^2 = \left[\frac{1}{n} + \frac{(x_0 - \bar{x})^2}{\sum (x_i - \bar{x})^2} \right] s_e^2$$

$$= \left[\frac{1}{6} + \frac{(45 - 55)^2}{1750} \right] (5.61)$$

$$= 1.2556.$$

Taking square roots, we get $s_a = 3.261$ and $s_b = 0.0566$. Also, $s_{\hat{y}} = 1.121$, where the yield is estimated for 45 pounds of fertilizer application. ●

CONFIDENCE INTERVALS AND PREDICTION INTERVAL

Our first objective is to construct confidence intervals for the parameters A, B, and $\mu_{Y|x_0}$. We will then discuss the prediction interval.

Confidence Intervals

The upper and lower limits of the confidence intervals can be obtained using the following rule, which was established in Chapter 7 for the case in which the estimator is normally distributed but its variance must be estimated. The limits are given by

$$\begin{pmatrix} \text{Estimate of} \\ \text{the parameter} \end{pmatrix} \pm \begin{pmatrix} \text{Appropriate value} \\ \text{from Student's} \\ t \text{ distribution} \end{pmatrix} \cdot \begin{pmatrix} \text{Estimated} \\ \text{standard} \\ \text{deviation} \end{pmatrix}.$$

In the present case, the t distribution has $n - 2$ degrees of freedom because there are $n - 2$ degrees of freedom associated with s_e^2. Thus, we have confidence intervals for A, B, and $\mu_{Y|x_0}$ as shown.

CONFIDENCE INTERVALS FOR A, B, $\mu_{Y|x_0}$

1. A $(1 - \alpha)$ 100 percent confidence interval for A is given by

$$a - t_{n-2, \alpha/2} s_a < A < a + t_{n-2, \alpha/2} s_a.$$

   Or, equivalently, the limits of this interval are

$$a \pm t_{n-2, \alpha/2} s_e \sqrt{\frac{1}{n} + \frac{(\bar{x})^2}{\sum (x_i - \bar{x})^2}}.$$

2. A $(1 - \alpha)$ 100 percent confidence interval for B is given by

$$b - t_{n-2, \alpha/2} s_b < B < b + t_{n-2, \alpha/2} s_b.$$

   Or, equivalently, the limits of the interval are

$$b \pm t_{n-2, \alpha/2} \frac{s_e}{\sqrt{\sum (x_i - \bar{x})^2}}.$$

> 3. A $(1 - \alpha)$ 100 percent confidence interval for $\mu_{Y|x_0}$
> $(= A + B_{x_0})$, the mean of Y corresponding to a value x_0 of the
> independent variable, is given by
>
> $$(a + bx_0) - t_{n-2,\alpha/2} s_{\hat{y}_0} < A + Bx_0 < (a + bx_0) + t_{n-2,\alpha/2} s_{\hat{y}_0}.$$
>
> Or, equivalently, the limits of the interval are
>
> $$(a + bx_0) \pm t_{n-2,\alpha/2} s_e \sqrt{\frac{1}{n} + \frac{(x_0 - \bar{x})^2}{\sum(x_i - \bar{x})^2}}.$$

EXAMPLE 3   Construct 95 percent confidence intervals for the parameters A, B, and $\mu_{Y|45}$ involved in our fertilizer-yield of grain problem discussed earlier.

SOLUTION   In Example 2 in the preceding section, we found $a = 31.593$ and $b = 0.377$. Also, in Example 2 of this section we found $s_a = 3.261$, $s_b = 0.0566$, and $s_{\hat{y}_0} = 1.121$ when $x_0 = 45$. Now, because $n = 6$, there are 4 degrees of freedom and, because $\alpha = 0.05$, we have $t_{n-2,\alpha/2} = t_{4,0.025} = 2.776$. Therefore, 95 percent confidence intervals are obtained as follows.

1. Because $a = 31.593$ and $s_a = 3.261$, a 95 percent confidence interval for A is

$$a - t_{4,0.025} s_a < A < a + t_{4,0.025} s_a$$
$$31.593 - (2.776)(3.261) < A < 31.593$$
$$+ (2.776)(3.261)$$
$$22.540 < A < 40.646.$$

2. Because $b = 0.377$ and $s_b = 0.0566$, a 95 percent confidence interval for B is

$$b - t_{4,0.025} s_b < B < b + t_{4,0.025} s_b$$
$$0.377 - (2.776)(0.0566) < B < 0.377$$
$$+ (2.776)(0.0566)$$
$$0.220 < B < 0.534.$$

3. As a first step toward constructing a confidence interval for the mean yield when 45 pounds of fertilizer are applied (that is, for $\mu_{Y|45}$), we compute the estimated yield when $x_0 = 45$. This is given by

$$\hat{y}_0 = a + bx_0 = 31.593 + 0.377(45) = 48.558.$$

Now, a 95 percent confidence interval for $\mu_{Y|45}$ is

$$(a + bx_0) - t_{4,0.025} s_{\hat{y}_0} < \mu_{Y|45} < (a + bx_0) + t_{4,0.025} s_{\hat{y}_0}.$$

Because $s_{\hat{y}_0} = 1.121$ and $a + bx_0 = 48.558$, we get

$$48.558 - (2.776)(1.121)$$
$$< \mu_{Y|45} < 48.558 + (2.776)(1.121)$$
$$45.446 < \mu_{Y|45} < 51.670. \qquad \bullet$$

Prediction Intervals

As we know well, the expression *confidence interval* is used in the context of estimating intervals of population parameters. Very often it is quite important for the experimenter to predict a y-value for a given value of x, for example, x_0. We realize that we can get different y-values for the same x_0. Thus, we are interested in estimating an interval for a value of a random variable Y when $x = x_0$. For example, we might be interested in estimating an interval for the amount of grain yield when x_0 pounds of fertilizer are applied per acre (as opposed to the *mean* amount of grain yield $\mu_{Y|x_0}$, which is a specific constant). The expression **prediction interval** is commonly used in the context of setting an interval for the value of a random variable.

As before, suppose that corresponding to the value x_0 of the independent variable the value from the sample regression line is $\hat{y}_0$, so that $\hat{y}_0 = a + bx_0$. Then it can be shown that the prediction interval is as follows.

> **PREDICTION INTERVAL FOR y-VALUE WHEN $x = x_0$**
>
> The limits for a $(1 - \alpha)100$ percent prediction interval for a single y-value corresponding to a value x_0 of the independent variable are given by
>
> $$(a + bx_0) \pm t_{n-2,\alpha/2} s_e \sqrt{1 + \frac{1}{n} + \frac{(x_0 - \bar{x})^2}{\sum (x_i - \bar{x})^2}}.$$

EXAMPLE 4

Obtain a 95 percent prediction interval for yield y_0 of grain when 45 pounds of fertilizer are applied with reference to the fertilizer-yield of grain problem we have been discussing.

SOLUTION   We know that $x_0 = 45$, and we have seen earlier that $n = 6$, $\bar{x} = 55$, $\sum (x_i - \bar{x})^2 = 1750$, $s_e = \sqrt{5.61} = 2.368$, and $\hat{y}_0 = 48.558$. Also, $t_{n-2,\alpha/2} = t_{4,0.025} = 2.776$. Therefore, after substituting in the formula, the limits of a 95 percent prediction interval for the amount of yield of grain y_0 are given by

$$48.558 \pm (2.776)(2.368) \sqrt{1 + \frac{1}{n} + \frac{(45 - 55)^2}{1750}}.$$

That is, the interval is

$$48.558 - 7.272 < y_0 < 48.558 + 7.272$$
$$41.286 < y_0 < 55.830. \qquad \bullet$$

TEST OF HYPOTHESIS REGARDING B

Because the main function of the regression theory is to estimate the mean value of the dependent variable from the knowledge of the independent variable, a question of considerable importance is whether the independent variable is at all relevant to the prediction process. This would lead to our testing H_0: $B = 0$. (Notice that if $B = 0$, we would get $\mu_{Y|x} = A + 0x$, that is, $\mu_{Y|x} = A$, implying that the mean value of Y for any x is not dependent on x.)

The problem of a general nature is to test

$$H_0: B = B_0$$

where B_0 is some given number. For example, in the fertilizer-grain yield example we might wish to test the null hypothesis that increase in grain yield for an increase of 1 pound of fertilizer application is 1.5 bushels. In this case $B_0 = 1.5$ and we would be testing H_0: $B = 1.5$.

The alternative hypothesis could be one-sided or two-sided. Recall that this is what will decide whether the test is one-tail or two-tail. The decision rule is based on Student's t distribution with $n - 2$ degrees of freedom and depends on the value of $(b - B_0)/s_b$. It is carried out as shown in Table 11-10.

TABLE 11-10   Procedure to Test H_0: $B = B_0$

Alternative Hypothesis	The Decision Rule Is to Reject H_0 If the Computed Value of $\dfrac{b - B_0}{s_b}$ Is
$B > B_0$	greater than $t_{n-2, \alpha}$
$B < B_0$	less than $-t_{n-2, \alpha}$
$B \neq B_0$	less than $-t_{n-2, \alpha/2}$ or greater than $t_{n-2, \alpha/2}$

EXAMPLE 5   For the fertilizer-yield of grain problem, at the 2.5 percent level of significance test the null hypothesis

$$H_0: B = 0$$

against the alternative hypothesis

$$H_A: B > 0.$$

SOLUTION   At the 2.5 percent level of significance, because $t_{n-2, \alpha} = t_{4, 0.025} = 2.776$, the decision rule is: *Reject* H_0 *if the computed value* $(b - 0)/s_b$, *or* b/s_b, *is greater than* 2.776 (see Figure 11-12). Now $b = 0.377$ and $s_b = 0.0566$. Therefore,

$$\frac{b}{s_b} = \frac{0.377}{0.0566} = 6.66.$$

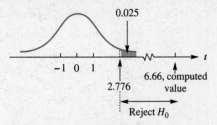

Figure 11-12 *t* Distribution with 4 Degrees of Freedom and the Critical Region

Because the computed value is greater than 2.776, we reject the hypothesis that $B = 0$ and conclude that the yield of grain depends on the level of fertilizer application. ●

For a given set of bivariate data, problems of statistical inference involve lengthy computations. At the very least, one needs the estimates a, b, and s_e^2. The following example illustrates, step by step, the general procedure involved.

EXAMPLE 6

Table 11-11 gives the resale values of a certain model of car (in hundreds of dollars) and the number of years each car has been in use (along with other necessary computations).

TABLE 11-11 Resale Value of Used Cars and Related Computations

Number of Years x	Resale Value (in Hundreds of Dollars) y	x^2	xy	y^2
1	30	1	30	900
1	32	1	32	1024
2	25	4	50	625
2	27	4	54	729
2	26	4	52	676
4	16	16	64	256
4	18	16	72	324
4	16	16	64	256
4	20	16	80	400
6	10	36	60	100
6	8	36	48	64
6	11	36	66	121
Total 42	239	186	672	5475

(a) Plot the scatter diagram.

(b) Obtain the regression of resale value on the number of years a car is used.

(c) Find the estimated resale value at the end of 5 years.

(d) Obtain an estimate of the variance σ^2.

(e) Find s_a, s_b, and $s_{\hat{y}_0}$ for $x = 5$ years.

(f) Construct 95 percent confidence intervals for A, B, and $\mu_{Y|5}$.

(g) Construct a 95 percent prediction interval for resale value y_0 when $x = 5$ years.

(h) Test the hypothesis that the annual rate of depreciation is $400. Use $\alpha = 0.05$.

SOLUTION

(a) The scatter diagram is given in Figure 11-13 and, as can be seen, a linear trend is evident.

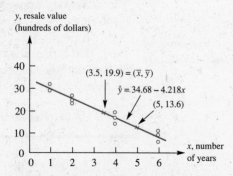

Figure 11-13 The Scatter Diagram, and the Line of Regression of Resale Value of a Car on the Number of Years in Use

(b) The necessary computations are given in Table 11-11. Because $n = 12$, $\Sigma x_i = 42$, $\Sigma y_i = 239$, $\Sigma x_i^2 = 186$, $\Sigma x_i y_i = 672$, and $\Sigma y_i^2 = 5475$ we get

$$b = \frac{n \sum x_i y_i - \sum x_i \sum y_i}{n \sum x_i^2 - \left(\sum x_i \right)^2}$$

$$= \frac{12(672) - (42)(239)}{12(186) - (42)^2}$$

$$= -4.218$$

and

$$a = \bar{y} - b\bar{x}$$

$$= \frac{239}{12} - (-4.218)\left(\frac{42}{12}\right)$$

$$= 34.68.$$

Thus, the regression of resale value on the number of years is given by $\hat{y} = 34.68 - 4.218x$.

The regression line is drawn in Figure 11-13. From the regression equation, because $b = -4.218$, the depreciation of a car per year is $421.80.

(c) The estimated resale value at the end of 5 years is

$$\hat{y}_0 = 34.68 - (4.218)(5)$$

$$= 13.59.$$

That is, the estimated resale value after 5 years is $1359.

(d) We have

$$s_e^2 = \frac{1}{n-2}\left[\sum y_i^2 - a\sum y_i - b\sum x_i y_i\right]$$

$$= \frac{1}{12-2}[5475 - (34.68)(239) - (-4.218)(672)]$$

$$= 2.098.$$

(e) Because $\sum (x_i - \bar{x})^2 = \sum x_i^2 - \frac{(\sum x_i)^2}{n} = 186 - \frac{42^2}{12} = 39$, and $\bar{x} = \frac{42}{12} = 3.5$, we get

$$s_a^2 = \left[\frac{1}{n} + \frac{(\bar{x})^2}{\sum (x_i - \bar{x})^2}\right]s_e^2$$

$$= \left[\frac{1}{12} + \frac{(3.5)^2}{39}\right](2.098) = 0.8338$$

$$s_b^2 = \frac{s_e^2}{\sum (x_i - \bar{x})^2}$$

$$= \frac{2.098}{39} = 0.0538$$

and, for $x_0 = 5$ years,

$$s_{\hat{y}_0}^2 = \left[\frac{1}{n} + \frac{(x_0 - \bar{x})^2}{\sum (x_i - \bar{x})^2}\right]s_e^2$$

$$= \left[\frac{1}{12} + \frac{(5 - 3.5)^2}{39}\right](2.098) = 0.2959.$$

Thus, $s_a = 0.913$, $s_b = 0.232$, and $s_{\hat{y}_0} = 0.544$.

(f) At this stage we need to make some assumptions. We assume that for any given number of year x a car has been in use, its resale value is normally distributed. Also, we assume that the variance of the distribution of resale value is the same for any x. Because $n = 12$, the t distribution has 10 degrees of freedom and $t_{n-2,\alpha/2} = t_{10,0.025} = 2.228$.

i. $a = 34.68$ and $s_a = 0.913$. Therefore, a 95 percent confidence interval for A is

$$a - t_{n-2,\alpha/2}\, s_a < A < a + t_{n-2,\alpha/2}\, s_a$$
$$34.68 - 2.228(0.913) < A < 34.68 + 2.228(0.913)$$
$$32.646 < A < 36.714.$$

ii. $b = -4.218$ and $s_b = 0.232$. Therefore, a 95 percent confidence interval for B is

$$b - t_{n-2,\alpha/2}\, s_b < B < b + t_{n-2,\alpha/2}\, s_b$$
$$-4.218 - 2.228(0.232) < B < -4.218 + 2.228(0.232)$$
$$-4.735 < B < -3.701.$$

iii. From Part c, $\hat{y}_0 = 13.59$. Also, $s_{\hat{y}_0} = 0.544$. Therefore, a 95 percent confidence interval for $\mu_{Y|5}$ is

$$\hat{y}_0 - t_{n-2,\alpha/2}\, s_{\hat{y}_0} < \mu_{Y|5} < \hat{y}_0 + t_{n-2,\alpha/2}\, s_{\hat{y}_0}$$
$$13.59 - 2.228(0.544) < \mu_{Y|5} < 13.59 + 2.228(0.544)$$
$$12.378 < \mu_{Y|5} < 14.802.$$

(g) We have $n = 12$, $x_0 = 5$, $\bar{x} = 3.5$, $\Sigma\,(x_i - \bar{x})^2 = 39$,

$$s_e = \sqrt{2.098} = 1.448,\ \hat{y}_0 = 13.59,\ \text{and}\ t_{n-2,\alpha/2} = 2.228.$$

Consequently, substituting in the formula

$$\hat{y}_0 \pm t_{n-2,\alpha/2}\, s_e \sqrt{1 + \frac{1}{n} + \frac{(x_0 - \bar{x})^2}{\Sigma\,(x_i - \bar{x})^2}}$$

we find the limits of a 95 percent prediction interval for y_0 when $x = 5$ years are

$$13.59 \pm (2.228)(1.448)\sqrt{1 + \frac{1}{12} + \frac{(5 - 3.5)^2}{39}}.$$

This can be simplified to yield the interval

$$10.144 < y_0 < 17.036.$$

(h) We want to test the null hypothesis that the annual rate of depreciation is \$400. In other words, we want to test $H_0\colon B = -4$.

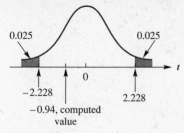

Figure 11-14

Suppose the alternative hypothesis is given as H_A: $B \neq -4$. Because $b = -4.218$ and $s_b = 0.232$, the computed value of the test statistic is

$$\frac{b - B_0}{s_b} = \frac{-4.218 - (-4)}{0.232}$$

$$= -0.94. \text{ (See Figure 11-14.)}$$

Now $t_{10, 0.025} = 2.228$, and because the computed value is between -2.228 and 2.228, we do not reject the null hypothesis.   ●

Section 11-3 Exercises

In the following exercises, in the context of setting confidence intervals and testing hypotheses, for each x assume that the response variable Y is normally distributed with the same variance.

1. The following table gives x, the age (in months) of a certain breed of calf in the tropics, and y, the mean length of its horns (in inches).

Age x	12	16	16	24	24	30	36	36
Mean length y	2.0	3.0	2.5	4.0	4.5	5.0	6.0	5.5

(a) Find n, $\Sigma\, x_i$, $\Sigma\, y_i$, $\Sigma\, x_i^2$, $\Sigma\, y_i^2$, and $\Sigma\, x_i y_i$.
(b) Find a and b, and obtain the least squares regression line $\hat{y} = a + bx$.
(c) Construct a table giving the values $\hat{y}$ for the different values of x, and obtain

$$\Sigma\, (y_i - \hat{y}_i)^2.$$

(d) Find s_e^2, an unbiased estimate of σ^2, using the result in Part c.
(e) Alternatively, obtain s_e^2 by using the formula

$$s_e^2 = \frac{1}{n-2}\left[\Sigma\, y_i^2 - a\Sigma\, y_i - b\Sigma\, x_i y_i\right].$$

2. The following pairs of values (x, y) are obtained in a laboratory experiment.

x	2	4	6	8	12	16
y	3.8	18.4	27.2	45.8	62.3	100.3

(a) Find n, $\Sigma\, x_i$, $\Sigma\, y_i$, $\Sigma\, x_i^2$, $\Sigma\, y_i^2$, and $\Sigma\, x_i y_i$.
(b) Find a and b.
(c) Find s_e^2.

3. Consider the following bivariate data.

x	6	8	10	11	13
y	10	14	23	28	37

Find an unbiased estimate of σ^2.

4. Consider the following bivariate data.

x	-4	-2	0	2	4	6
y	47	10	12	-13	-30	-28

Compute the following.
   (a) a and b **(b)** s_e **(c)** s_a **(d)** s_b

5. The following data points (x, y) are given.

x	10.0	12.5	15.0	17.5	20.0	22.5	25.0
y	80	86	118	115	130	134	144

Compute the following.
   (a) a and b **(b)** s_e **(c)** s_a **(d)** s_b

6. In Exercise 1, obtain the following.
   (a) s_a^2
   (b) A 90 percent confidence interval for A

7. In Exercise 1, obtain the following.
   (a) s_a^2
   (b) A 90 percent confidence interval for B

8. In Exercise 1
   (a) estimate the mean length of the horns of a calf of 20 months
   (b) set a 90 percent confidence interval for $\mu_{Y|20}$.

9. From a sample of 12 paired observations (x, y), the following information was obtained.

$$\sum x_i = 18 \qquad \sum x_i^2 = 38 \qquad s_e^2 = 1.21$$

Find the following.
   (a) $\sum (x_i - \bar{x})^2$ **(b)** s_a^2 and s_b^2

10. The following data are given on two variables x (independent) and y (dependent) measured on 20 experimental units.

$$\sum x_i = 92 \qquad \sum y_i = 183 \qquad \sum x_i^2 = 460$$

$$\sum y_i^2 = 2184 \qquad \sum x_i y_i = 717$$

   (a) Find the least squares regression equation.
   (b) Obtain the variance estimates s_e^2, s_a^2, and s_b^2.
   (c) Set 95 percent confidence intervals on A and B.

11. The following pairs of observations are given.

x	5	10	15	20	25	30
y	30	16	13	2	-4	-25

Obtain the following.
   (a) The least squares regression equation
   (b) s_e, s_a, s_b
   (c) A 95 percent confidence interval for A
   (d) A 95 percent confidence interval for B
   (e) A 95 percent confidence interval for $\mu_{Y|18}$

12. In Exercise 28 in Section 11-2, at the 5 percent level of significance, test the null hypothesis $B = 0$ against H_A: $B \neq 0$, where
    (a) B is the regression coefficient of the volume of sales during a year on the amount spent on advertising during that year
    (b) B is the regression coefficient of the volume of sales on the amount spent on advertising during the previous year.

13. The following data refer to the stopping distance y (in feet) of an automobile traveling at speed x (in miles per hour) in a simulated experiment.

Speed x	25	25	30	35	45	45	55	55	65	65
Stopping distance y	63	56	84	107	153	164	204	220	285	303

   (a) Plot the scatter diagram. Do you think that a linear model would be appropriate?
   (b) Fit a least squares regression line to the data.
   (c) Estimate the stopping distance for a car traveling at 50 mph.
   (d) Obtain an estimate of σ^2.
   (e) Find s_a^2 and s_b^2.
   (f) Test H_0: $B = 5$ against H_A: $B \neq 5$, at the 5 percent level.
   (g) Set a 95 percent confidence interval for $\mu_{Y|50}$, the mean stopping distance for a car traveling at 50 mph.

14. The following data present the hardness (in Rockwell units) of an alloy when 8 samples were treated at 500°C for different periods of time (in hours).

Time	6	9	9	12	12	15	18	21
Hardness	50	53	54	60	57	62	67	71

   (a) Plot the scatter diagram.
   (b) Obtain the regression of hardness on the time.
   (c) Find the estimated hardness at the end of 10 hours.
   (d) Obtain an estimate of the variance σ^2.
   (e) Find s_a, s_b, and $s_{\hat{y}}$ for $x = 10$ hours.
   (f) Construct 98 percent confidence intervals for A, B, and $\mu_{Y|10}$.
   (g) Test the null hypothesis that the hourly rate of increase of hardness is 1.5 Rockwell units against a suitable alternative hypothesis. (Use $\alpha = 0.10$.)

15. The following figures give the temperature x (in degrees Celsius) and the amount y (in grams) of a chemical substance extracted from 1 pound of a mineral soil.

Temperature x	210	250	270	290	310	340
Amount y	2.1	5.8	8.1	10.8	12.8	14.3

   (a) Find the regression of the amount of the chemical substance extracted on the temperature.
   (b) Find the estimated yield at 300°C.
   (c) Obtain an estimate of σ^2.
   (d) Find s_a, s_b, and $s_{\hat{y}}$ when $\hat{y}$ is the estimated yield at 300°C.
   (e) Construct 95 percent confidence intervals for A, B, and $\mu_{Y|300}$.
   (f) Construct a 95 percent prediction interval for the amount of chemical substance extracted when the temperature is 300°C.
   (g) Test the null hypothesis, that per degree Celsius increase in temperature, the yield of the extract is increased by 0.1 gram, against a suitable alternative hypothesis. (Use $\alpha = 0.05$.)

16. The following figures represent concentrations of standard manganese solutions and filter photometer readings.

Concentration of Standard Solution, mg Mn/250 ml x	Photometer Reading y
0.5	57
1.0	110
2.0	205
3.0	298
5.0	492

  (a) Set up a simple linear regression of photometer readings on the concentration of solution and determine the constants in the regression equation.

  (b) At the 5 percent level of significance, is the true increase in photometer reading per milligram increase in concentration different from 100?

17. A botanist has obtained the following information regarding y, the heights of seedlings (in inches), and x, the number of weeks after planting.

Number of weeks x	2	3	4	5	6	8
Height y (in inches)	2.5	4.8	8.0	10.5	13.0	18.0

  (a) Find the regression of the seedling height on the number of weeks after planting.

  (b) Set a 95 percent confidence interval on B.

  (c) Test the null hypothesis that $B = 2.4$ against $B \neq 2.4$.

  (d) Set a 95 percent confidence interval for $\mu_{Y|5}$, the mean seedling height at the end of 5 weeks.

  (e) Construct a 95 percent prediction interval for the height of the seedlings at the end of 5 weeks.

18. The following figures give the resting potential (in millivolts) of frog voluntary muscle fibers and body temperature (in degrees Celsius). Two readings of the resting potential were obtained at each of the temperatures.

Temperature x	5	10	15	20	25	30
Resting potential y	61	63	64	68	69	72
	62	65	66	68	70	72

Fit the regression equation of resting potential on temperature.
At the 5 percent level of significance, is the true regression coefficient B different from 0.4?

19. In order to study the effect of maturity on the starch content in pepper (*Piper nigrum L.*), spikes were collected from the plants by hand at 17, 19, 22.5, 26, 29.5, and 33 weeks after pollination. At each stage of maturity, sample peppercorns were separated from the stalks and the material dried in a forced-draft oven at 50°C for 30 hours, then analyzed for starch. The following table gives the time of harvest (weeks) and the starch content (g/100 g). (*Journal of the Science of Food and Agriculture* 1984; *35*: pp. 41–46.)

Time of Harvest (weeks)	Starch (g/100 g)
17.0	13.8
19.0	18.0
22.5	29.3
26.0	30.6
29.5	42.1
33.0	53.8

(a) Obtain the regression of starch on the time of harvest.

(b) Estimate the population regression coefficient B using a 90 percent confidence interval.

(c) Construct a 90 percent confidence interval for $\mu_{Y|20}$, the mean starch at the end of 20 weeks.

(d) Construct a 90 percent prediction interval for the amount of starch at the end of 20 weeks.

11-4   CORRELATION ANALYSIS

COEFFICIENT OF CORRELATION AS A MEASURE OF LINEAR ASSOCIATION

We will now explain why the coefficient of correlation is taken as a measure of the strength of linear association. If the values of y were considered without any reference to their dependence on some variable x, we would find the total variation in the y values in the usual way as $\sum (y_i - \bar{y})^2$. This is often called the **total sum of squares** and is commonly denoted by **SST**.

$$SST = \sum (y_i - \bar{y})^2$$

As can be seen from Figure 11-15, the deviation of y_i from $\bar{y}$ consists of two parts: one that represents the deviation of the observed value from the predicted value, $y_i - \hat{y}_i$, and the other that represents the deviation of the predicted value from the mean $\bar{y}$, $\hat{y}_i - \bar{y}$. Algebraically, this is expressed by the relation

$$y_i - \bar{y} = (y_i - \hat{y}_i) + (\hat{y}_i - \bar{y}).$$

Using these deviations, one could compute $\sum (y_i - \bar{y})^2$, $\sum (y_i - \hat{y}_i)^2$, and $\sum (\hat{y}_i - \bar{y})^2$. A remarkable algebraic fact is that

$$\sum (y_i - \bar{y})^2 = \sum (y_i - \hat{y}_i)^2 + \sum (\hat{y}_i - \bar{y})^2.$$

This relation shows that the total sum of squares is made up of two components, $\sum (y_i - \hat{y}_i)^2$, and $\sum (\hat{y}_i - \bar{y})^2$.

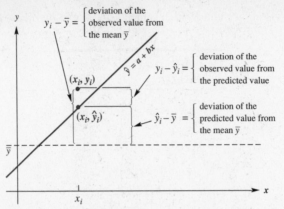

Figure 11-15 The Deviation $y_i - \bar{y}$ Expressed as the Sum of the Deviations $y_i - \hat{y}_i$ and $\hat{y}_i - \bar{y}$

The component $\Sigma \, (\hat{y}_i - \bar{y})^2$ is called the **explained sum of squares,** meaning that it explains or accounts for the portion of the total variation through the linear relationship of y and x. If the linear relationship is perfect, all the observed values will be equal to the predicted values; that is, $y_i = \hat{y}_i$. Consequently, $\Sigma \, (y_i - \hat{y}_i)^2 = 0$ and we get $\Sigma \, (y_i - \bar{y})^2 = \Sigma \, (\hat{y}_i - \bar{y})^2$, showing that *all* of the total variation is explained by means of the linear relationship. The explained sum of squares is also called the **regression sum of squares** and is denoted by **SSR.**

$$SSR = \sum (\hat{y}_i - \bar{y})^2$$

The other component, $\Sigma \, (y_i - \hat{y}_i)^2$, is the sum of squares of the deviations from the linear regression relation because, for each i, $y_i - \hat{y}_i$ denotes a deviation of an observed value from the corresponding predicted value. It measures the residual variation left unexplained by the regression line and is therefore called the **unexplained sum of squares.** This component is also called the **residual sum of squares** or the **error sum of squares** and is denoted by **SSE.**

$$SSE = \sum (y_i - \hat{y}_i)^2$$

If the relationship is perfect, all the observed values fall on the regression line, in which case $\Sigma \, (y_i - \hat{y}_i)^2 = 0$ and there is no unexplained variation.

Summarizing, we have accomplished the partitioning of the total variation as follows.

$$\sum (y_i - \bar{y})^2 \;\; = \;\; \sum (\hat{y}_i - \bar{y})^2 \;\; + \;\; \sum (y_i - \hat{y}_i)^2$$

$$\uparrow \qquad\qquad\qquad \uparrow \qquad\qquad\qquad \uparrow$$

$$\begin{pmatrix} SST, \text{ total sum} \\ \text{of squares} \end{pmatrix} = \begin{pmatrix} SSR, \text{ regression} \\ \text{sum of squares} \\ \text{(explained)} \end{pmatrix} + \begin{pmatrix} SSE, \text{ error sum} \\ \text{of squares} \\ \text{(unexplained)} \end{pmatrix}$$

From the foregoing discussion, it is obvious that the quantity $\Sigma \, (\hat{y}_i - \bar{y})^2$ expressed as a fraction of the total sum of squares $\Sigma \, (y_i - \bar{y})^2$ would play a prominent role in providing a measure of the degree of linear association between x and y. This fraction,

$$\frac{\sum (\hat{y}_i - \bar{y})^2}{\sum (y_i - \bar{y})^2} = \frac{\text{explained sum of squares}}{\text{total sum of squares}},$$

is called the **coefficient of determination** and represents the fraction of the total variation that can be ascribed to the linear relation. Through some algebraic manipulations, it can be shown that this quantity is precisely the square of the Pearson product-moment correlation coefficient r that we studied in Section 11-2. Thus,

$$r^2 = \frac{\sum (\hat{y}_i - \bar{y})^2}{\sum (y_i - \bar{y})^2} = \textbf{coefficient of determination.}$$

Notice that the explained variation cannot exceed the total variation. Thus $r^2 \leq 1$. Also, r^2 is a ratio of two positive quantities. Therefore, r^2 is greater than or equal to zero and $0 \leq r^2 \leq 1$, or, equivalently, $-1 \leq r \leq 1$.

As we remarked earlier, if all the observed values fall on the regression line, then $\Sigma \, (y_i - \bar{y})^2 = \Sigma \, (\hat{y}_i - \bar{y})^2$, in which case we will have $r^2 = 1$, that is, $r = 1$ or $r = -1$. On the other hand, $r^2 = 1$ will mean that $\Sigma \, (y_i - \hat{y}_i)^2 = 0$, showing that all the observed values fall on the straight line. (Why?) Thus, $r = +1$ or -1 means that there is a perfect linear relation among the observed pairs in the sample. If the observed values are close to the fitted line, the ratio r^2 will be close to 1. It will deteriorate and get closer to zero as the scatter gets larger.

Remark We now have two methods of computing the coefficient of correlation, one that was given in Section 11-2, and the other, just introduced, as the square root of the coefficient of determination. It should be mentioned that, for computational purposes, the formula introduced in Section 11-2 is preferred. Should one choose to compute r by using the relation

$$r = \pm \sqrt{\frac{\sum (\hat{y}_i - \bar{y})^2}{\sum (y_i - \bar{y})^2}}$$

how does one decide whether r is positive or negative? It turns out that the sign of r is always the same as that of the regression coefficient b.

EXAMPLE 1   For the data in Table 11-4 of Section 11-2, we have already obtained the coefficient of correlation between IQ and grade point average as equal to 0.913. Now we will find it as the square root of the coefficient of determination.

SOLUTION   In order to obtain the coefficient of determination, we must compute the various components of variation. In Table 11-4 we computed the following quantities.

$$n = 10 \qquad \sum x_i = 1040 \qquad \sum y_i = 30.2$$

$$\sum x_i^2 = 109{,}800 \qquad \sum y_i^2 = 92.80 \qquad \sum x_i y_i = 3187.5$$

Therefore, we can proceed as follows.

(a)   The regression coefficient is

$$b = \frac{10(3187.5) - (1040)(30.2)}{10(109{,}800) - (1040)^2}$$

$$= 0.0285.$$

(b)   Because $\bar{y} = 3.02$ and $\bar{x} = 104$, the y-intercept is

$$a = \bar{y} - b\bar{x}$$

$$= 3.02 - (0.0285)(104)$$

$$= 0.056.$$

(c)   The total sum of squares is

$$\sum (y_i - \bar{y})^2 = \sum y_i^2 - \frac{\left(\sum y_i\right)^2}{n}$$

$$= 92.80 - \frac{(30.2)^2}{10}$$

$$= 1.596.$$

(d)   Using the formula $\sum (y_i - \hat{y}_i)^2 = \sum y_i^2 - a \sum y_i - b \sum x_i y_i$, we find the unexplained sum of squares is

$$\sum (y_i - \hat{y}_i)^2 = 92.80 - 0.056(30.2) - (0.0285)(3187.5)$$

$$= 0.265.$$

From Parts c and d, the explained sum of squares $\sum (\hat{y}_i - \bar{y})^2$ is given by

$$\sum (\hat{y}_i - \bar{y})^2 = \sum (y_i - \bar{y})^2 - \sum (y_i - \hat{y}_i)^2$$

$$= 1.596 - 0.265$$

$$= 1.331.$$

TABLE 11-12   Components of Variation

Component	Sum of Squares
Explained (regression)	1.331
Unexplained (error)	0.265
Total	1.596

A summary of the components of variation is given in Table 11-12. The coefficient of determination is therefore given as

$$r^2 = \frac{\text{explained variation}}{\text{total variation}}$$

$$= \frac{1.331}{1.596}$$

$$= 0.834.$$

Because b is positive, we take the positive square root of 0.834 and obtain $r = \sqrt{0.834} = 0.913$, the same value we had obtained rather easily earlier. The coefficient of determination of 0.834 signifies that 83.4 percent of the variation in the grade point averages can be explained through the linear regression relation.   ●

EXAMPLE 2   For the data in Example 3 in Section 11-2, find the following.

(a)   The different components of variation

(b)   The coefficient of determination

(c)   The coefficient of correlation

SOLUTION

(a)   We have the following information from Example 3, Section 11-2.

$$n = 10 \qquad \sum y_i = 690 \qquad \sum y_i^2 = 49{,}014$$

$$\sum x_i y_i = 44{,}636 \qquad a = 55.346 \qquad b = 0.214$$

Thus, the total sum of squares is

$$\sum (y_i - \bar{y})^2 = \sum y_i^2 - \frac{\left(\sum y_i\right)^2}{10}$$

$$= 49{,}014 - \frac{(690)^2}{10}$$

$$= 1404$$

and the unexplained sum of squares is

$$\sum (y_i - \hat{y}_i)^2 = \sum y_i^2 - a \sum y_i - b \sum x_i y_i$$
$$= 49{,}014 - (55.346)(690)$$
$$- (0.214)(44{,}636)$$
$$= 1273.16.$$

Thus, we get Table 11-13, giving the components of variation. The explained component in the table is obtained by subtracting the unexplained component from the total variation.

TABLE 11-13   Components of Variation

Component	Sum of Squares	Ratio, Explained/Total
Explained (regression)	130.84	0.09319
Unexplained (error)	1273.16	
Total	1404.00	

(b) The coefficient of determination is given in Column 3 of Table 11-13. The small value of 0.09319 tells us that only 9.3 percent of the variation in the final scores can be explained through linear regression on the midterm scores.

(c) Since b is positive, the coefficient of correlation is given as the positive square root of the coefficient of determination and therefore is equal to $\sqrt{0.09319} = 0.306$, the value we obtained in Example 3 in Section 11-2.   ●

TEST OF HYPOTHESIS REGARDING POPULATION CORRELATION COEFFICIENT

We have discussed the meaning and significance of the correlation coefficient computed from a sample of n pairs of observations. To be more specific, we should refer to r as the *sample* **correlation coefficient.** Different samples of size n drawn from the population will give rise to different values of r. Thus, conceptually we can think of the distribution of r. (Notice our transgression in using r to denote both a random variable and its value. We take the liberty to avoid introducing new symbols.) The parameter that provides a measure of association between two variables in the population analogous to the way r does in the sample is called the **population correlation coefficient** and is denoted by the Greek letter ρ (rho).

Suppose we obtain a certain value of r from a given set of data. What does it suggest regarding ρ? We will consider only the simple case in which the null hypothesis we are interested in is

$$H_0: \rho = 0$$

meaning that there is no linear relationship between the two variables in the population. We test this against suitable alternative hypotheses.

$$H_A: \rho \neq 0 \qquad H_A: \rho < 0 \qquad H_A: \rho > 0$$

It is at this point that we need to make some assumptions regarding the distributions of the two variables. *Whereas in regression analysis one of the variables is nonrandom and the other is random, in correlation analysis we require that both the variables be random variables. Their joint distribution is given by what is called a bivariate normal distribution.* In such a case, $\rho = 0$ *does* imply that the two variables are independent.

THE TEST STATISTIC

The test statistic to carry out the test of the null hypothesis H_0: $\rho = 0$ is

$$r\sqrt{\frac{n-2}{1-r^2}}.$$

If H_0 is true, then this statistic has the Student's t distribution with $n - 2$ degrees of freedom.

If, for example, the alternative hypothesis is $H_A: \rho \neq 0$, then the decision rule is: *Reject* H_0 *if the computed value of the test statistic is less than* $-t_{n-2,\alpha/2}$ *or greater than* $t_{n-2,\alpha/2}$.

EXAMPLE 3   Consider the following data from a bivariate normal distribution.

x	6	13	18	24	30
y	12	16	14	23	20

Test $H_0: \rho = 0$ against $H_A: \rho \neq 0$ at the 5 percent level.

SOLUTION   First we compute r using the formula in Section 11-2. Because $n = 5$, $\Sigma \, x_i = 91$, $\Sigma \, x_i^2 = 2005$, $\Sigma \, y_i = 85$, $\Sigma \, y_i^2 = 1525$, and $\Sigma \, x_i y_i = 1684$ we get

$$r = \frac{5(1684) - (91)(85)}{\sqrt{[5(2005) - (91)^2][5(1525) - (85)^2]}} = 0.82.$$

Then, the computed value of the test statistic is

$$r\sqrt{\frac{n-2}{1-r^2}} = (0.82)\sqrt{\frac{5-2}{1-(0.82)^2}} = 2.481.$$

At the 5 percent level of significance $t_{n-2,\alpha/2} = t_{3,0.025} = 3.182$. The hypothesis that $\rho = 0$ cannot be rejected and so there is no evidence that the variables are dependent (in view of the bivariate normal distribution assumption). ●

Remarks   1. It can be shown that testing $H_0: \rho = 0$ is tantamount to testing $H_0: B = 0$. So, in essence, we have dealt with this problem in Section 11-3. A statistical test is available to test $H_0: \rho = \rho_0$ where ρ_0 is not equal to zero. But to carry out this test one has to perform what is called a *logarithmic transformation,* which is not presented here.

2. On first glance it may seem from Example 3 that a value of $r = 0.82$ is rather high and should be statistically significant. A closer scrutiny reveals that about 67.2 percent of the total variation can be ascribed to the linearity relation. This fraction in itself may not be negligible, but coupled with the fact that the sample size is very small, it explains why the test in Example 3 showed the result not significant. If we were to obtain the same r-value (that is, 0.82) from a sample of, for example, 12 pairs of observations, then the result would be highly significant since the computed value would be $0.82\sqrt{10}/\sqrt{1-0.82^2}$, or 4.53, and $t_{n-2,0.025} = t_{10,0.025} = 2.228$. Thus, once again we see how the sample size plays a prominent role in a decision-making process.

EXAMPLE 4   Concerning the IQ and grade point average data given in Example 1, we obtained $r = 0.913$ based on a sample with $n = 10$. Test the null hypothesis $H_0: \rho = 0$ against $H_A: \rho \neq 0$ at the 5 percent level of significance.

SOLUTION   The computed value of the test statistic is

$$r\sqrt{\frac{n-2}{1-r^2}} = (0.913)\sqrt{\frac{10-2}{1-(0.913)^2}}$$

$$= 6.33.$$

At the 5 percent level of significance $t_{n-2,\alpha/2} = t_{8,0.025} = 2.306$. We reject H_0, since the computed value does not fall in the interval -2.306 to 2.306, and we conclude that IQ and grade point average are linearly related. ●

EXAMPLE 5   In Example 3 in Section 11-2 we considered the data giving scores of ten students on their midterm and final exams. The value of r was computed to be 0.306. Test $H_0: \rho = 0$ against $H_A: \rho \neq 0$ at the 5 percent level of significance.

SOLUTION To test hypothesis H_0: $\rho = 0$, we assume that scores on the midterm exam and the final exam have a bivariate normal distribution and we compute the value of the statistic

$$r\sqrt{\frac{n-2}{1-r^2}}.$$

The computed value is

$$(0.306)\sqrt{\frac{10-2}{1-(0.306)^2}} = 0.909.$$

The null hypothesis is tested against the alternative hypothesis H_A: $\rho \neq 0$. At the 5 percent level of significance, $t_{n-2,\alpha/2} = t_{8,0.025} = 2.306$. We cannot reject H_0, since the computed value does not fall in the critical region. There is no evidence to indicate that the relation between the final scores and the midterm scores is linear.   ●

Section 11-4 Exercises

In the following exercises, for the purpose of testing hypotheses regarding ρ, assume that the two variables have a bivariate normal distribution.

1. The scores on a scholastic test and the number of hours that ten randomly picked students devoted preparing for it are as follows.

Number of hours x	18	27	20	10	30	24	32	27	12	16
Scores y	68	82	77	90	78	72	94	88	60	70

Determine the following.
   (a) The slope b
   (b) The y-intercept, $a = \bar{y} - b\bar{x}$
   (c) The total sum of squares $\Sigma(y_i - \bar{y})^2$
   (d) The unexplained sum of squares using the formula
   $$\sum (y_i - \hat{y}_i)^2 = \sum y_i^2 - a \sum y_i - b \sum x_i y_i$$
   (e) A table showing the partitioning of the total sum of squares
   (f) The coefficient of determination

2. If the value of r is given as 0.8 and the unexplained component of variation is given as 50.3, find the other components of variation.

3. Forty-nine percent of the total variation in y is explained by the linear regression of y on x. What is the coefficient of correlation?

4. If the coefficient of correlation between x and y is -0.6, what percent of the total variation is explained by the straight-line relation? If the total variation is 33.2, find the unexplained component of variation.

5. The following information is available regarding the annual yield of wheat per acre y and the annual rainfall x.

$$n = 9$$

$$\sum (x_i - \bar{x})^2 = 91.5$$

$$\sum (y_i - \bar{y})^2 = 2066.9$$

$$\sum (x_i - \bar{x})(y_i - \bar{y}) = 310$$

(a) Compute r.

(b) Compute the coefficient of determination.

(c) What does the coefficient of determination signify insofar as the variation in wheat yield is concerned?

6. The following figures pertain to the annual disposable income x and consumption expenditure y of five families.

Disposable income x (in thousands of dollars)	18.5	21.6	24.6	30.6	32.8
Consumption expenditure y (in thousands of dollars)	17.6	23.2	22.8	26.9	28.6

(a) Find the total sum of squares of y's.

(b) Find the explained and unexplained components of variation.

(c) Compute the coefficient of determination r^2.

(d) What does r^2 measure?

In Exercises 7–11, at the 5 percent level of significance, test the null hypothesis $\rho = 0$ against the alternative hypothesis $\rho \neq 0$.

7. $r = -0.5, n = 66$

8. $r = -0.28, n = 51$

9. $r = -0.5, n = 11$

10. $r = 0.3, n = 198$

11. The coefficient of determination is 0.49 and $n = 38$.

12. An experiment is conducted to determine the effect of pressure in a certain brand of tires on the number of miles the tire lasts. The following data are obtained.

Pressure x	27	27.5	28	28.5
Miles y	12,928	14,036	14,700	14,919

Pressure x	29	29.5	30
Miles y	14,694	14,024	12,910

(a) Plot a scatter diagram.

(b) Obtain the coefficient of correlation.

(c) Comment on the magnitude of ρ in view of the scatter diagram.

13. The following are heights and weights of seven athletes.

Heights x (in inches)	69	66	68	73	71	74	71
Weights y (in pounds)	163	153	185	186	157	220	190

(a) Follow Parts a–f in Exercise 1. (Construct a table showing the partitioning of the total sum of squares of weights.)

(b) Find the coefficient of correlation.

(c) Test the null hypothesis that $\rho = 0$ at the 5 percent level against the alternative hypothesis $H_A: \rho \neq 0$.

14. The following data refer to y, the level of cholesterol (in milligrams per hundred milliliters) of ten males, and x, their average daily intake of saturated fat (in grams).

Fat intake x	28	32	35	40	42	48	52	56	58	60
Cholesterol y	190	185	195	200	230	200	270	220	250	240

(a) Find the regression of the level of cholesterol on the intake of saturated fat.

(b) Find the components of variation explained and unexplained by the linear regression.

(c) Find the coefficient of determination. Interpret the value obtained in light of the components of variation.

(d) Find the sample coefficient of correlation r.

15. A survey of sixty-six families was conducted to find the relationship between x, the annual family income, and y, the amount spent on entertainment. It was found that the total variation was 409,600 and the unexplained component of variation was 77,824.

(a) What percentage of the variation in the amount spent on entertainment is accounted for by the linear regression relation?

(b) Compute the coefficient of correlation r.

(c) Test the null hypothesis that ρ, the coefficient of correlation for all potential families, is zero.

16. Seventeen phytoplankton samples were otabined in an upwelling environment off a coastal region. The amount of cellular carbon ($\mu g/l$) and the chlorophyl a content ($\mu g/l$) were determined for each sample, yielding the following data.

Chlorophyl a ($\mu g/l$)	Cellular Carbon ($\mu g/l$)
0.30	10
0.35	12
0.40	28
0.48	18
0.46	25
0.70	41
0.85	55
0.90	48
0.96	50
1.10	67
1.20	64
1.30	94
1.40	88
1.45	106
1.48	122
1.50	119
1.53	126

(a) Plot the scatter diagram.
(b) Compute the coefficient of correlation.
(c) Based on the data, is the true population coefficient of correlation greater than zero? Use $\alpha = 0.05$.

17. Liquid water in Colorado winter mountain clouds is an important component of the precipitation process. Due to the high elevations in the Rocky Mountains, nearly all wintertime clouds are entirely at temperatures below 0°C. Consequently, virtually all cloud liquid water is supercooled. The data present the liquid water content (g/m^3) in the clouds and the mountain top temperatures (°C).

Liquid Water Content (g/m^3)	Mountaintop Temperature (°C)
0.10	−18
0.08	−17
0.12	−16
0.15	−13
0.25	−9
0.21	−8
0.28	−7
0.24	−7

(a) Obtain the regression of the liquid water content on the temperature.
(b) Compute the coefficient of correlation.
(c) At the 2.5 percent level of significance, is the true population correlation coefficient greater than zero?

CONCLUDING REMARKS

Given any n pairs of observations, we can always compute the regression line and the linear correlation coefficient. It is when the inferential nature of the investigation is involved that we have to make some assumptions regarding the distributions of X and Y. For regression analysis the values of the independent variable are fixed, and for each value of x the subpopulation of y values has a normal distribution. For correlation analysis, X and Y are both random variables.

Regression technique is valuable for making predictions. But the predictions should be limited to the range of the values of the independent variable that led to obtaining the prediction equation in the first place. As an example, if a regression equation for predicting the yield of grain is fitted by considering levels of fertilizer ranging from 30 to 80 pounds per acre, it would be foolhardy to use the equation to predict the yield for 100 pounds of fertilizer, as crops are known to die with excessive use of fertilizer. And yet, in some situations, such as economic forecasts, some amount of extrapolation beyond the range is almost a necessity. For example, if a prediction equation is set up on the basis of the economic activity from 1980 to 1994, an economist may want to forecast what is in store in later years, say in 1997 and 1998. Whenever this is done, one should exercise utmost caution and be fully aware of other related factors that could invalidate the conclusions.

Finally, it should be clearly understood that a relationship between variables does not mean that there is a cause-and-effect relation. Even though we use terms such as dependent and independent variables, there is no implication that one variable is the cause of the other. There may exist a causal relation and there may not. For example, in considering the relationship between the amount of fertilizer applied and the amount of grain produced, we clearly have a causal relation in mind. But consider the following situation: During the 1950s production of cars increased every year in the United States and so did the population of Latin America. A mathematical relation could perhaps be established between the number of cars produced in the United States and the population of Latin America during that period. However, it would be ridiculous to conclude that more cars in the United States caused more people to be in Latin America or vice versa. We must be careful in deciding whether a relation is causal or simply mathematical.

CHAPTER **11 Summary**

✔ *CHECKLIST: KEY TERMS AND EXPRESSIONS*

- univariate case, page 616
- bivariate case, page 616
- linear correlation, page 616
- linear regression, page 616
- y-intercept, page 619
- slope, page 619
- equation of a straight line, page 619
- scatter diagram, page 621
- coefficient of linear correlation, page 621
- correlation coefficient r, page 621
- prediction equation, page 625

- residual (or error), page 626
- method of least squares, page 627
- line of best fit, page 627
- least squares regression line, page 627
- sample regression coefficient, page 627
- regression of y on x, page 627
- linear model, page 638
- deterministic model, page 638
- linear statistical model, page 638
- population, or true, regression line, page 639

- prediction interval, page 647
- total sum of squares, page 657
- explained, or regression, sum of squares, page 658
- unexplained, residual, or error sum of squares, page 658
- coefficient of determination, page 659
- sample correlation coefficient, page 662
- population correlation coefficient ρ, page 662

KEY FORMULAS

correlation coefficient

$$r = \frac{n \sum x_i y_i - \sum x_i \cdot \sum y_i}{\sqrt{\left[n \sum x_i^2 - \left(\sum x_i \right)^2 \right]\left[n \sum y_i^2 - \left(\sum y_i \right)^2 \right]}}$$

line of best fit

$$\hat{y} = a + bx$$

slope

$$b = \frac{n \sum x_i y_i - \sum x_i \cdot \sum y_i}{n \sum x_i^2 - \left(\sum x_i\right)^2}$$

y-intercept

$$a = \bar{y} - b\bar{x}$$

variance estimates

$$s_e^2 = \frac{1}{n-2}\left[\sum y_i^2 - a\sum y_i - b\sum x_i y_i\right] \qquad s_a^2 = \left[\frac{1}{n} + \frac{(\bar{x})^2}{\sum(x_i - \bar{x})^2}\right]s_e^2$$

$$s_b^2 = \frac{s_e^2}{\sum(x_i - \bar{x})^2} \qquad\qquad s_{\hat{y}_0}^2 = \left[\frac{1}{n} + \frac{(x_0 - \bar{x})^2}{\sum(x_i - \bar{x})^2}\right]s_e^2$$

confidence intervals

$$a - t_{n-2,\alpha/2}s_a < A < a + t_{n-2,\alpha/2}s_a$$
$$b - t_{n-2,\alpha/2}s_b < B < b + t_{n-2,\alpha/2}s_b$$
$$(a + bx_0) - t_{n-2,\alpha/2}s_{\hat{y}_0} < \mu_{Y|x_0} < (a + bx_0) + t_{n-2,\alpha/2}s_{\hat{y}_0}$$

prediction interval limits

$$(a + bx_0) \pm t_{n-2,\alpha/2}s_e \sqrt{1 + \frac{1}{n} + \frac{(x_0 - \bar{x})^2}{\sum(x_i - \bar{x})^2}}$$

Procedure used to test $H_0: B = B_0$

Alternative Hypothesis	The Decision Rule Is to Reject H_0 If the Computed Value of $\dfrac{b - B_0}{s_b}$ Is
$B > B_0$	greater than $t_{n-2,\alpha}$
$B < B_0$	less than $-t_{n-2,\alpha}$
$B \neq B_0$	less than $-t_{n-2,\alpha/2}$ or greater than $t_{n-2,\alpha/2}$

total sum of squares

$$SST = \sum(y_i - \bar{y})^2$$

regression sum of squares

$$SSR = \sum(\hat{y}_i - \bar{y})^2$$

error sum of squares $$SSE = \sum (y_i - \hat{y}_i)^2$$

coefficient of determination $$r^2 = \frac{SSR}{SST}$$

test statistic to test $H_0: \rho = 0$ $$\frac{r\sqrt{n-2}}{\sqrt{1-r^2}}$$

CHAPTER 11   Concept and Discussion Questions

1. Why is it useful to plot a scatter diagram?

2. What does the term *linear* mean in *linear correlation*?

3. Distinguish between regression and correlation with regard to the variables involved.

4. What do we assume about the population for setting confidence intervals for A and B?

5. What assumptions do we make about the variables x and y when testing a hypothesis about the population correlation coefficient ρ?

6. If you are given the coefficient of correlation between two variables, how do you know whether the relationship between the variables is positive or inverse?

7. Suppose a scatter diagram is given where the points fall exactly on a straight line. What could be the possible values of r?

8. Answer *true* or *false*.
   (a) An objective of regression analysis is to give estimates of the dependent variable from the values of the independent variable.
   (b) An objective of correlation analysis is to measure the strength of association between two variables.

9. Suppose you collect paired data (x, y) from situations described below. Based on your experience, state whether the linear correlation between the variables will be positive or negative, or close to zero.
   (a) Annual net income and annual expenditure on consumer durables
   (b) The level of cholesterol and blood pressure
   (c) Prime rate that the banks charge and the Dow Jones industrial averages index
   (d) Living area in a house and its sale price
   (e) Price of a car and the age of the car
   (f) Annual income of an employee and the number of years employed
   (g) Number of bank counters open and the waiting time in a queue

10. Suppose the sample coefficient of correlation is found to be 0.8. What is the significance of this value?

Chapter 11   Review Exercises

1. Describe the roles of linear correlation and linear regression in statistical analysis of bivariate data.

2. An electronics company has found the following relation between sales y (in millions of dollars) and advertising x (in thousands of dollars).

$$y = 20.18 + 0.42x$$

when x is between 60 and 80. What does the coefficient 0.42 of x mean?

3. List the basic properties of the coefficient of linear correlation.

4. In a research problem, two variables x and y were measured on each of seven items. The data are as follows.

x	0.15	0.20	0.20	0.35	0.45	0.55	0.55
y	18	20	19	28	42	50	58

(a) Plot the scatter diagram.
(b) Find the regression of y on x.
(c) Find the coefficient of correlation.

5. A research article reported findings for which the linear regression equation of y on x and the correlation coefficient r were given as follows.

$$y = 4.009 - 0.026x$$
$$r = 0.329$$

This information is inconsistent. Explain why.

6. Suppose the following nine data points are given.

x	2	4	9	11	4	11	6	5	4
y	60	44	24	27	51	22	32	57	33

(a) Plot the scatter diagram.
(b) Compute the coefficient of correlation. Does the value of r indicate a linear relation? If so, is the linear trend upward or downward?
(c) Sketch the line of best fit.

7. The following data give the total length (in millimeters) of a certain species of fish (*Epiplatys bifasciatus*) inhabiting a swampy region and its head length (in millimeters) based on a sample of eight fish.

Head length	3.8	4.1	4.0	4.3	4.8	4.6	5.1	6.0
Total length	20	25	27	30	34	38	40	45

(a) Find the regression of head length on the total length.
(b) Find the coefficient of correlation.

8. A study has asserted that based on 78 observations, the regression of age y (in years) on diameter at breast height (dbh) x (in inches), and correlation coefficient were as follows.

$$y = 103.95 + 5.27x$$
$$r = 0.81$$

(Kenneth N. Boe, "Growth and Mortality After Regeneration Cuttings in Old-growth Redwood," USDA Forest Service Research Paper PSW-104.)

(a) What fraction of the total variation in age can be explained through linear regression on dbh?

(b) At the 5 percent level of significance test the null hypothesis $H_0: \rho = 0$ against $H_A: \rho \neq 0$.

(c) Estimate the age of a tree when dbh is 30 inches.

9. The following measurements were obtained (in feet) for the length and the girth of ten full-grown sea cows. Find the coefficient of correlation between girth and length and, at the 5 percent level, test the null hypothesis that the population coefficient of correlation is zero against $H_A: \rho \neq 0$.

Length x	10.5	12.2	11.5	9.6	13.3	12.0	10.7	11.0	12.6	12.9
Girth y	7.6	8.3	8.1	7.2	8.7	8.4	7.3	7.9	8.2	8.5

10. The following data give the increase in weight (in grams) of male rats during a 1-month period when they were fed different doses of vitamin A (in milligrams).

Dose	0.3	0.6	0.9	1.2	1.5
Increase in weight	33	36	40	41	45
	29	34	37		

There were two rats at each of the doses 0.3, 0.6, and 0.9, and one rat at each of the doses 1.2 and 1.5. Estimate B, the population regression coefficient of weight increase on vitamin A dosage, and test the null hypothesis that it is equal to 12 against $H_A: B \neq 12$. (Use $\alpha = 0.05$.)

11. The volume of a tree is commonly regarded as a function of many variables, such as height, basal area, age, and so on. The following data give net 10-year volume stand growth y (in cubic feet) for trees larger than 4.5 inches at breast height and their basal area per acre x (in square feet) where stand basal area is a measure expressing the degree of land occupation by trees (James Lindquist and Marshal N. Palley, "Prediction of Stand Growth of Young Redwood," California Agriculture Experimental Station Bulletin 831).

Volume stand growth	610	1,034	1,753	2,387	2,972	3,522	4,046	4,549
Basal area per acre	50	100	200	300	400	500	600	700

(a) Obtain the regression of volume stand growth on the basal height.

(b) Find a 90 percent confidence interval for B.

(c) Test the null hypothesis $H_0: B = 5.5$ against the alternative hypothesis $H_A: B \neq 5.5$. Use $\alpha = 0.10$.

(d) Find a 90 percent confidence interval for $\mu_{Y|400}$.

(e) Find a 90 percent prediction interval for volume stand growth when basal area is 400 square feet.

12. Unlike most organisms, crustaceans such as crabs are enclosed in a rigid exterior skeleton. Through a process called molting, the entire exoskeleton is shed. Only then, before the formation of the new rigid shell, can the animal grow. The following data give the specific molt increment y (in millimeters) and the premolt

crab size x (in millimeters) for Dungeness crab found off the coast of Northern California (Nancy Diamond, Master's Thesis, Humboldt State University, Arcata, California).

Premolt size (mm)	126.2	131.0	131.7	132.8	136.2	136.8	138.3	138.3	140.0	143.9
Molt increment (mm)	17.1	16.5	15.1	15.9	14.8	15.1	14.4	12.8	11.9	11.7

(a) Plot the scatter diagram.
(b) Find the regression of molt increment on premolt size.
(c) Find the coefficient of correlation.
(d) Test $H_0: \rho = 0$ against $H_A: \rho \neq 0$. Use $\alpha = 0.05$.

13. *Paramysis lacustris* is a mysid (small, shrimplike crustacean) which, due to its high growth rate, fecundity, and relatively large size, is a valuable fodder organism. The study of the mechanism of the utilization of food by this mysid is of definite practical interest. The mysid feeds on silts containing a large number of substances that it can easily assimilate (*Hydrobiological Journal* 1982; *18:* pp. 72–75).

The following data represent the dry weight of the animals (mg) and the assimilated food (calories/day) determined in a controlled experiment.

Dry Weight (mg)	Assimilated Food (cal/day)
0.2	0.127
0.5	0.159
1.0	0.197
1.5	0.243
2.0	0.312
6.0	0.989

(a) Plot the scatter diagram.
(b) Obtain the regression of assimilated food on the dry body weight.
(c) Find the coefficient of correlation.
(d) Based on the data, is the population coefficient of correlation significantly greater than zero? Carry out the test at the 5 percent level of significance.

14. Suppose bivariate data are obtained on 15 experimental units under study. How high would r have to be so that you would reject $H_0: \rho = 0$ in favor of $H_A: \rho > 0$, at the 5 percent level of significance?

15. Suppose bivariate data are obtained on 18 units under study. How small would r have to be so that you would reject $H_0: \rho = 0$ in favor of $H_A: \rho < 0$, at the 2.5 percent level of significance?

16. Suppose bivariate data are obtained on 12 units under study. For what values of r would you consider rejecting $H_0: \rho = 0$ in favor of $H_A: \rho \neq 0$, at the 5 percent level?

17. The following data were collected on twelve randomly selected students at a university. They give their scores on the math placement exam and a subsequent calculus course they took.

SCORES	
Placement Exam	**Calculus**
106	77
142	84
111	69
94	64
130	87
105	72
140	85
139	72
110	78
106	54
186	97
152	78

At the 5 percent level of significance, is there any relation between student performance on the placement exam and the calculus course? *Hint:* You may use the result of Exercise 16.

MINITAB

LAB 11: Linear Regression and Linear Correlation

This chapter involved the most complex computations used so far in this book. We had to plot the scatter diagram, compute the coefficient of correlation, find the regression line, and obtain various variance estimates. MINITAB handles all these difficult calculations with just three simple commands: PLOT, REGRESS, and CORRELATION.

1   First, we simulate 80 pairs of observations, with x-values in Column C1, and y-values in Column C2. Here any two entries in the same row in Columns C1 and C2 form a pair. Once the data are obtained, we plot the scatter diagram and also compute the coefficient of correlation for this set of pairs.

```
MTB >RANDOM 80 OBSERVATIONS IN C1;

SUBC>NORMAL MEAN 150 SIGMA 20.

MTB >RANDOM 80 OBSERVATIONS IN C2;

SUBC>NORMAL MEAN 120 SIGMA 15.

MTB >PLOT C2 AGAINST C1

MTB >CORRELATION COEFFICIENT BETWEEN C1 C2
```

The command PLOT displays the scatter diagram and the command CORRELATION computes and prints out the value of r.

Because the pairs are formed by random association there is no reason to expect any kind of association between the values in C1 and corresponding values in C2. We anticipate the sample correlation coefficient to be close to zero. Is this the case with your sample r?

Next, we form new pairs of data using the values in Columns C1 and C2 and study the nature of the scatter plot and the resulting coefficient of correlation.

```
MTB >LET C3 = C1 + C2

MTB >PLOT C3 AGAINST C1

MTB >CORRELATION BETWEEN C1 C3
```

Column C3 contains the sum of the corresponding entries in Column C1 and C2. Because the values in Column C3 are obtained by using the corresponding values in Column C1 we have reason to expect an association between paired values in Columns C1 and C3.

The scatter diagram shows that, though the points in the plot are not exactly on a straight line, there is a pattern indicating an upward trend from left to right. This signals a positive correlation in the population. In fact, theory predicts the correlation in the population to be 0.8. Is your sample value of r near this value?

Finally, we consider another set of pairs from Columns C1 and C2 as follows.

```
MTB >LET C4 = C2 — C1

MTB >PLOT C4 AGAINST C1

MTB >CORRELATION BETWEEN C1 C4
```

The corresponding entries in the same row in Columns C1 and C4 form the pairs. In the scatter plot there is a downward trend from left to right reflecting a negative association between entries in Column C1 and corresponding entries in Column C4. This is easy to see, as Column C4 is obtained by subtracting Column C1 from C2, and as a result, generally, large values in C1 are likely to be associated with small values in C4 and vice versa. Is your answer for r near -0.8?

2 Next, we simulate a linear regression model in which the linear component is $150 + 0.6x$, so that $A = 150$ and $B = 0.6$, and the random component has a normal distribution with mean $\mu = 0$ and standard deviation $\sigma = 6.5$. First we must pick a set of values for the fixed variable x. In the following program, we select 25 values of x and place them in Column C1. There are three values of the variable x, each equal to 4.5, which is indicated by 3(4.5). There are four values equal to 8.5, indicated by 4(8.5), and so on.

```
MTB >SET X-VALUES IN COLUMN C1

DATA>3(4.5) 4(8.5) 3(12.5) 5(16.5) 2(20.5)

DATA>4(24.5) 4(28.5)

DATA>END OF DATA

MTB >PRINT C1

MTB >NOTE TWENTY FIVE VALUES OF THE RANDOM COMPONENT

MTB >RANDOM 25 VALUES IN C2;
```

```
SUBC>NORMAL MEAN 0 SIGMA 6.5.

MTB >NOTE ENTER A AND B

MTB >LET K1 = 150

MTB >LET K2 = 0.6

MTB >NOTE RANDOM RESPONSE Y PLACED IN COLUMN C3

MTB >LET C3 = K1 + K2*C1 + C2

MTB >NAME C1 'X INPUT' C3 'Y OUTPUT'

MTB >PLOT C3 AGAINST C1

MTB >REGRESS C3 ON 1 PREDICTOR IN C1
```

Notice that when giving the REGRESS command we must specify the number of predictors. Because we are dealing with a simple linear regression model, there is one predictor x which is in Column C1. Also, the column containing the y values must be mentioned first in the PLOT command.

Upon execution of the REGRESS command MINITAB presents an abridged version of the regression analysis. The output includes the equation of the regression line, estimates s_e, s_a, s_b of the standard deviations, the value of r^2, and the table of the components of variation similar to Table 11-12 on page 661. Compare the values of a and b from your simulation with $A = 150$ and $B = 0.6$. A more detailed regression analysis is available if the command BRIEF 3 is entered before giving the REGRESS command. The command BRIEF 1 will revert the output mode to the abridged version.

Assignment   Simulate a regression model where the linear component is $142.5 - 3.5x$, and the random component has a normal distribution with $\mu = 0$ and $\sigma = 10$. Take the values of x as 5, 10, 15, 20, 25, 30, with each value repeated three times. Thus, you should have 18 pairs of values altogether. Plot the simulated data, and carry out the regression analysis for the simulated data. Compare the value of a with $A = 142.5$ and the value of b with $B = -3.5$.

3   Having considered simulation aspects, we now tackle the fertilizer-yield of grain problem, which was worked out very painstakingly on pages 642–647 in the three examples. We want a detailed analysis here. So we enter the command BRIEF 3 and proceed.

```
MTB >BRIEF 3

MTB >NOTE FERTILIZER-YIELD OF GRAIN PROBLEM
```

```
MTB >NAME C1 'FERT' C2 'YIELD'

MTB >READ DATA IN C1 C2

DATA>30 43

DATA>40 45

DATA>50 54

DATA>60 53

DATA>70 56

DATA>80 63

DATA>END

MTB >PRINT C1 C2

MTB >REGRESS 'YIELD' ON 1 PREDICTOR 'FERT'
```

The printout resulting from the REGRESS command is reproduced on pages 680–681 with annotations. Except for rounding off, the values in the table should be the same as the corresponding quantities in the three examples on pages 642–647.

The preceding output gives the predicted values of y (yield) for different x (fert) values in Column C1. But if we want a predicted value of y for any other value of x, a quick way is to use the subcommand PREDICT with the command REGRESS. To get the prediction interval only, without the regression analysis that accompanies BRIEF 3, we should first type BRIEF. Use PREDICT for fertilizer level $x = 45$.

```
MTB >BRIEF

MTB >REGRESS 'YIELD' ON 1 PREDICTOR 'FERT';

SUBC >PREDICT FOR X = 45.
```

The PREDICT subcommand not only gives the predicted value, but also the estimated standard deviation of the predicted y, a 95 percent confidence interval for $\mu_{Y|x}$, and a 95 percent prediction interval.

The MINITAB output gives the values of a, b, s_a, and s_b, but it does not give confidence intervals for A or B. We can obtain them easily with a few LET commands. For example, in the printout on page 680, $a = 31.59$, $b = 0.37714$, $s_a = 3.259$, and $s_b = 0.05659$. So, the 95 percent confidence intervals for A and B can be computed as follows, taking note that here $t_{n-2,\alpha/2} = t_{4,0.025} = 2.776$.

```
MTB >NOTE CONFIDENCE LIMITS FOR A

MTB >LET K1 = 31.59 − 2.776*3.259
```

```
MTB >LET K2 = 31.59 + 2.776*3.259

MTB >PRINT K1 K2

MTB >NOTE CONFIDENCE LIMITS FOR B

MTB >LET K3 = 0.37714 − 2.776*0.05659

MTB >LET K4 = 0.37714 + 2.776*0.05659

MTB >PRINT K3 K4
```

Assignment   For the simulated regression model, with $A = 150$ and $B = 0.6$, find 95 percent confidence intervals for A and B based on your simulated sample. Do the respective intervals include A and B? Remember that the probability that a random interval includes the parameter in question is 0.95. There is a probability of 0.05 that the interval will not contain the parameter in question.

Assignment   Use the BRIEF 3 command and carry out the regression analysis for Example 6 in Section 11-3. Identify each of the quantities computed in the example with the corresponding quantity in the MINITAB output.

Assignment   Perform regression analysis and answer all the questions in Exercise 14 in Section 11-3.
  We now type STOP and conclude the MINITAB session using the logging off procedure.

MINITAB output for the fertilizer-yield of grain problem.

The Regression Equation Is Yield = 31.6 + 0.377 Fert.

Predictor	Coef		Stdev		t-ratio	
Constant	31.590	a	3.259	s_a	9.69	a/s_a
fert	0.37714	b	0.05659	s_b	6.66	b/s_b

$s = 2.367$ $\leftarrow s_e$ R-sq = 91.7%   R-sq (adj) = 89.7%

Analysis of Variance

Source	DF	SS	MS
Regression	1	248.91	248.91
Error	4	22.42	5.60
Total	5	271.33	

$$y_i - \hat{y}_i$$
$$\downarrow$$

Obs.	Fert	Yield	Fit	Stdev. Fit	Residual	St. Resid
1	30.0	43.000	42.905	1.713	0.095	0.06
2	40.0	45.000	46.676	1.286	−1.676	−0.84
3	50.0	54.000	50.448	1.007	3.552	1.66
4	60.0	53.000	54.219	1.007	−1.219	−0.57
5	70.0	56.000	57.990	1.286	−1.990	−1.00
6	80.0	63.000	61.762	1.713	1.238	0.76

$$s_{\hat{y}_i} = s_e \sqrt{\frac{1}{n} + \frac{(x_i - \bar{x})^2}{\sum (x_i - \bar{x})^2}}$$

x_i y_i

$(y_i - \hat{y}_i)/s_{res}^{*}$

$$\hat{y}_i = 31.59 + 0.377 x_i$$

* The following is the formula for s_{res}.

$$s_{res} = s_e \sqrt{\left[\left(1 - \frac{1}{n}\right) - \frac{(x_i - \bar{x})^2}{\sum (x_i - \bar{x})^2}\right]}$$

Commands you have learned in this session:

```
CORRELATION
REGRESS
REGRESS;
PREDICT.
```

Analysis of Variance

12

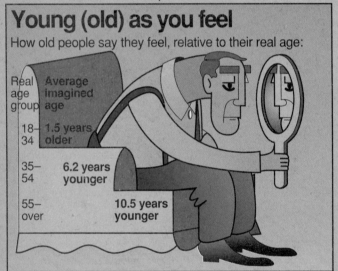

USA SNAPSHOTS®

A look at statistics that shape our lives

Young (old) as you feel

How old people say they feel, relative to their real age:

Real age group | Average imagined age

18–34 | 1.5 years older

35–54 | 6.2 years younger

55–over | 10.5 years younger

▲ Source: Graphic "Young (Old) As You Feel" by John Riley and Marcia Staimer *USA Today*. Data from Roper Organization for Replens. Reprinted by permission of *USA Today*, October 25, 1993, Section D, p. I.

INTRODUCTION

....►

In Chapter 9 we described a test for comparing two population means by using a *t* test when the population variances are not known but are assumed equal. We will now generalize this to more than two populations. For example, we might wish to compare several fertilizers, manufacturing processes, teaching methods, or advertising techniques. As usual, we will take samples from these populations. More than likely the sample means obtained from them will be different. The intent of our investigation is to find out if the observed difference in the sample means can be ascribed to chance. Or, are these differences too large to be attributed to chance, and due instead to the fact that the samples are from populations that are not alike?

In testing a hypothesis regarding more than two population means, we would prefer not to compare them two at a time, running a battery of *t* tests. First of all, this would be cumbersome, because the number of *t* tests involved would be very large, even for comparing a moderate number of populations. For example, to compare eight populations we would have to carry out twenty-eight *t* tests. (A student familiar with counting techniques can determine that there are twenty-eight possible combinations, that is $\binom{8}{2}$, of the eight populations taken two at a time.) Secondly, there are other complications that arise, especially by way of increasing the Type 1 error to a prohibitive level. We need a different test, one that, at the very least, is an extension of the *t* test used for testing two means. Such an extension is provided by the **analysis of variance (ANOVA)** method, which gives a single test for comparing several means. True to its name, it analyzes variances while testing for equality of means in the process.

The analysis of variance method owes its beginning to Sir Ronald A. Fisher (1890–1962), and received its early impetus in agricultural research applications. It is such a powerful statistical tool that it has since found applications in nearly every branch of scientific research, including economics, psychology, marketing research, and industrial engineering, to mention just a few. We consider here only the simplest case, that of **one-way classification,** where the observations are classified into groups on the basis of a single criterion of classification.

12-1   ONE-WAY CLASSIFICATION—EQUAL SAMPLE SIZES

THE TEST CRITERION AND THE RATIONALE FOR IT

Consider the following example, in which a car rental company is interested in comparing three brands of gasoline designated Brand 1, Brand 2, and Brand 3. Suppose it is agreed to make four runs with each brand, obtaining the mileage per gallon for each run as the response variable. Certainly we do not expect the runs to be made for different brands with different-sized engines. As a matter of fact, it might be desirable to make all the runs using one car, with Brand 1 tested at random, say, on runs 2, 5, 9, and 11; Brand

2 on runs 1, 6, 8, and 12; and Brand 3 on runs 3, 4, 7, and 10. We randomize in this way to minimize any bias that might result as the tune-up of the car deteriorates. Also, it is to be hoped that, as far as possible, the runs will be made on similar terrains. Such considerations as these form the essence of a well-designed experiment. You will recall from Section 6-3 that the design of experiment adopted here is referred to as the *completely randomized design*. As much as possible, we reduce the influence of the factors that might impair the results of the experiment. All other factors that might contribute to variability and cannot be eliminated are then lumped together and designated as the **experimental error.**

Suppose the results of the twelve runs are as presented in Table 12-1.

TABLE 12-1 Miles per Gallon on Twelve Runs, Four with Each Brand

	Brand 1	Brand 2	Brand 3
	25	29	32
	29	27	27
	24	26	29
	26	30	28
Mean mileage per gallon	26	28	29

From the table we see that individual readings vary, as do the means of the brands. Are these observed differences in the brand means due to chance fluctuations or are they due to the fact that the brands are basically different? In other words, if μ_1, μ_2, and μ_3 represent the true mean mileage per gallon for Brands 1, 2, and 3, respectively, the null hypothesis that we are interested in testing is

$$H_0: \mu_1 = \mu_2 = \mu_3.$$

In what follows we denote this common value of the population means as μ. The null hypothesis will be tested against the alternative hypothesis

H_A: Not all the brands have the same mean mileage per gallon.

Thus, the alternative hypothesis entertains possibilities in which all the population means are different from each other, or two of the means are equal but different from the third mean.

At this point we make some assumptions that are basic to applying the analysis of variance technique. These assumptions are the following.

1 The *response variable* (miles per gallon in our illustration) is normally distributed for each brand.

2 The variance of the distribution is the same for each brand. This assumption implies that although different brands of gasoline may have different mean mileages per gallon, the spread of the distribution for each of them is the same.

3 The samples are drawn in such a way that the results of one sample do not influence and are not influenced by those of others. In other words, the samples are independent.

As you will recall, these assumptions are the same as those made in Chapter 9 for testing equality of two means using the t test.

Our common sense would indicate that if the sample means differed substantially from each other, we would be inclined to doubt H_0 and regard the brands as being different. On the other hand, if the sample means differed very little from each other, we would be inclined to support the validity of H_0. However, we cannot rely entirely on this kind of argument. This will be made clear in the following discussion, in which we consider the two sets of data in Tables 12-2 and 12-3, where the means for respective brands in the two tables are equal.

TABLE 12-2 Miles per Gallon Where the Response for Each Brand Is Erratic

	Brand 1	Brand 2	Brand 3
	23	24	29
	25	29	25
	29	31	34
	27	28	28
Mean mileage per gallon	26	28	29

TABLE 12-3 Miles per Gallon Where the Response for the Brands Is Not Greatly Dispersed

	Brand 1	Brand 2	Brand 3
	25.2	28.5	29.6
	26.1	27.5	29.3
	26.8	28.2	28.4
	25.9	27.8	28.7
Mean mileage per gallon	26.0	28.0	29.0

The data in Table 12-2 would support the contention under H_0 that the population means are equal. The performance of the brands is so erratic and the overlap is so complete that we might regard the three samples corresponding to the three brands as one large sample drawn from one population, as shown in Figure 12-1.

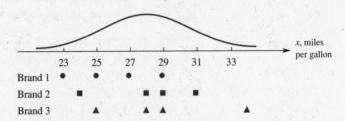

Figure 12-1

The data in Table 12-2, which can be conceived of as coming from one single population

The observed differences among the brand means could be ascribed to experimental or chance variations. In this case, it would stand to reason to regard the observed values of 26, 28, and 29 for brand means simply as three point estimates of the common mean μ.

On the other hand, a different picture emerges from the data in Table 12-3. The variability within brands is so small that the situation shown in Figure 12-2 seems more plausible. It would support the contention under the alternative hypothesis that the differences in the sample means are due to real differences among the brands. (Later, in Example 1, we will actually carry out the test of significance.)

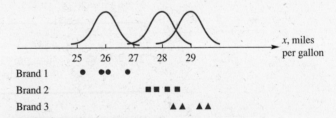

Figure 12-2

The data in Table 12-3, which can be conceived of as coming from three distinct populations

From the data in Tables 12-2 and 12-3, even though the observed sample means are exactly the same for respective brands, we are led to different conclusions. This situation arises because in the decision procedure it is appropriate to take into account not only the extent to which the sample means differ from each other, but also variability within the samples. This is why we call this procedure "analysis of variance."

Having determined the importance of the variability between the sample means together with the variability within the samples, exactly what criterion do we use that will take into account both these variabilities? Also, how do we provide a rationale for it? To accomplish this, we will describe a procedure that involves obtaining two unbiased estimates of σ^2 when H_0 is true and comparing them.

Consider a general situation where we have k populations, which we will refer to as *treatments,* and where, for the sake of simplicity, we assume that each sample consists of the same number n of observations (we generalize this later in Section 12-2). We might present the results as in Table 12-4, where, using double subscripts, we represent the outcomes as $x_{i,\,j}$, with the first subscript i designating the treatment and the second subscript j denoting the observation number within that treatment. For example, $x_{2,4}$ represents the fourth outcome in the second treatment. Thus, with reference to Table 12–1, we would have $x_{2,4} = 30$, $x_{3,1} = 32$, $x_{1,4} = 26$, and so on.

TABLE 12-4   General Response with n Observations per Treatment

	Treatment 1	Treatment 2	$\cdots$	Treatment k
	$x_{1,1}$	$x_{2,1}$		$x_{k,1}$
	$x_{1,2}$	$x_{2,2}$		$x_{k,2}$
	$\vdots$	$\vdots$		$\vdots$
	$x_{1,n}$	$x_{2,n}$		$x_{k,n}$
Mean	$\bar{x}_1 = \dfrac{\sum x_{1,j}}{n}$	$\bar{x}_2 = \dfrac{\sum x_{2,j}}{n}$	$\cdots$	$\bar{x}_k = \dfrac{\sum x_{k,j}}{n}$

The **grand mean,** designated $\bar{\bar{x}}$ and read as *x double bar*, is obtained by adding observations in all the samples and dividing the total by kn, the total number of observations. Because we are assuming that the sample sizes are equal, the grand mean is equal to the mean of the sample means. That is,

$$\bar{\bar{x}} = \frac{\bar{x}_1 + \bar{x}_2 + \cdots + \bar{x}_k}{k}$$

The Within-Samples Sum of Squares

Basic to our approach are the assumptions that the parent populations have the same variance σ^2 and that they are normally distributed. (If for some reason these assumptions are not satisfied, one should not apply the ANOVA technique for testing equality of means.*) Because $x_{1,1}, x_{1,2}, \ldots, x_{1,n}$ is a sample,

$$s_1^2 = \frac{\sum (x_{1,j} - \bar{x}_1)^2}{n - 1}$$

is an unbiased estimate of the variance of the population from which it was picked. Also,

$$s_2^2 = \frac{\sum (x_{2,j} - \bar{x}_2)^2}{n - 1}, \ldots, s_k^2 = \frac{\sum (x_{k,j} - \bar{x}_k)^2}{n - 1}$$

are unbiased estimates of the variances of their respective parent populations. Now, because all the populations are assumed to have the same variance σ^2, we can justifiably obtain an improved estimate of σ^2 by pooling whatever information $s_1^2, s_2^2, \ldots, s_k^2$ can contribute. Such a pooled unbiased estimate is given as

$$\frac{\sum (x_{1,j} - \bar{x}_1)^2 + \sum (x_{2,j} - \bar{x}_2)^2 + \cdots + \sum (x_{k,j} - \bar{x}_k)^2}{(n - 1) + (n - 1) + \cdots + (n - 1)}$$

(This is analogous to what we did in Section 9-4 when comparing means of two populations using the t test.) Using double summation, this pooled estimate s_p^2 can be written more compactly as

$$s_p^2 = \frac{\sum\sum (x_{i,j} - \bar{x}_i)^2}{k(n - 1)}.$$

The double summation might look intimidating. Only keep in mind that the inside summation on the right is carried out within each sample. When you carry out this process for each of the k samples you will get k values. The outside summation suggests adding these k values.

In statistical literature the term appearing in the numerator of s_p^2 is called the **error sum of squares,** and is commonly denoted as SSE. It is also called the **within-samples sum of squares,** as it is obtained by pooling variability *within* the samples. The number of degrees of freedom associated with SSE is equal to the sum of the degrees of freedom associated with $s_1^2, s_2^2, \ldots, s_k^2$, and is equal to $k(n - 1)$. Thus,

$$SSE = \sum\sum (x_{i,j} - \bar{x}_i)^2.$$

The quantity $SSE/[k(n - 1)]$, which is an estimate of σ^2, is called the *mean error sum of squares* and is denoted by MSE. *This estimate is an unbiased estimate of* σ^2 *irrespective of whether* H_0 *is valid or not.*

---

*In some situations transformations exist to bring the data into line with the assumptions.

For the data in Table 12-1 we see that

Brand 1: $\sum (x_{1,j} - \bar{x}_1)^2 = (25 - 26)^2 + (29 - 26)^2$
$$+ (24 - 26)^2 + (26 - 26)^2 = 14$$

Brand 2: $\sum (x_{2,j} - \bar{x}_2)^2 = (29 - 28)^2 + (27 - 28)^2$
$$+ (26 - 28)^2 + (30 - 28)^2 = 10$$

and

Brand 3: $\sum (x_{3,j} - \bar{x}_3)^2 = (32 - 29)^2 + (27 - 29)^2$
$$+ (29 - 29)^2 + (28 - 29)^2 = 14.$$

Consequently, the within sum of squares obtained by pooling variability within samples is

$$SSE = \sum\sum (x_{i,j} - \bar{x}_i)^2 = 14 + 10 + 14 = 38.$$

The pooled estimate of σ^2 is now obtained as

$$SSE/[k(n - 1)] = 38/[3(4 - 1)] = 4.222.$$

The Between Means Sum of Squares

The null hypothesis H_0 that we are interested in testing is

$$H_0: \mu_1 = \mu_2 = \cdots = \mu_k.$$

If the null hypothesis is true, which implies that all the populations have the same mean μ, we are in a position to obtain another estimate of σ^2. To H_0 being true, add our basic asumption that the populations have the same variance. The means $\bar{x}_1, \bar{x}_2, \ldots, \bar{x}_k$ then represent a sample of k means drawn from one single population with mean μ and variance σ^2. Using the formula for s^2 in Chapter 2, if we compute the quantity

$$\frac{(\bar{x}_1 - \bar{\bar{x}})^2 + (\bar{x}_2 - \bar{\bar{x}})^2 + \cdots + (\bar{x}_k - \bar{\bar{x}})^2}{k - 1} = \frac{\sum (\bar{x}_i - \bar{\bar{x}})^2}{k - 1}$$

and denote it as $s_{\bar{x}}^2$, we realize that $s_{\bar{x}}^2$ is simply an unbiased estimate of the variance of the distribution of the sample means. But the variance of the distribution of $\bar{X}$ is σ^2/n, as will be recalled from Chapter 6. It follows then that $s_{\bar{x}}^2$ is an unbiased estimate of σ^2/n so that $ns_{\bar{x}}^2$ is an unbiased estimate of σ^2. (We accept this statement intuitively.) That is,

$$\frac{n \sum (\bar{x}_i - \bar{\bar{x}})^2}{k - 1}$$

is another unbiased estimate of σ^2 if the null hypothesis is true. The term in the numerator, $n \sum (\bar{x}_i - \bar{\bar{x}})^2$, is called the **between means sum of squares,** as it is based on the variation between the means of the samples. The between means sum of squares is also referred to as the *sum of squares for treatments.*

It is commonly denoted as SSB and has $k - 1$ degrees of freedom. (We will have $3 - 1$, or 2, degrees of freedom in our illustration.) Thus,

$$SSB = n \sum (\bar{x}_i - \bar{\bar{x}})^2.$$

The quantity $SSB/(k - 1)$ is a new estimate of σ^2. It is called the *mean between sum of squares* and is denoted as *MSB*.

For the data in Table 12-1, the grand mean is given by

$$\bar{\bar{x}} = \frac{26 + 28 + 29}{3}$$

$$= 27.67.$$

Therefore,

$$SSB = n[(\bar{x}_1 - \bar{\bar{x}})^2 + (\bar{x}_2 - \bar{\bar{x}})^2 + (\bar{x}_3 - \bar{\bar{x}})^2]$$
$$= 4[(26 - 27.67)^2 + (28 - 27.67)^2$$
$$+ (29 - 27.67)^2]$$
$$= 18.667.$$

Thus, $SSB/(k - 1) = 18.667/2 = 9.333$ is an estimate of σ^2 based on the assumption that H_0 is true, that is, based on the assumption that the three population means are equal.

The Test of Hypothesis

We have obtained two estimates of σ^2, 4.222 and 9.333. The estimate 4.222 is obtained assuming equal population variances. It should be emphasized that the second estimate 9.333 is contingent on the assumption that the population variances are equal and, more importantly, on the truth of the null hypothesis that the population means are all equal. If the null hypothesis is true, then the two estimates $SSE/[k(n - 1)]$ and $SSB/(k - 1)$ are valid unbiased estimates of σ^2 and should be of the same order of magnitude. For the purpose of comparison, we use the following ratio, called the **variance ratio.**

$$f = \frac{ns_{\bar{x}}^2}{s_p^2} = \frac{SSB/(k - 1)}{SSE/[k(n - 1)]}$$

We might ascribe a slight departure of the above ratio from 1 to chance fluctuations. However, if the ratio is substantially larger than 1, it will cast doubts on the validity of H_0. For, if H_0 is not true, then $ns_{\bar{x}}^2$ tends to overestimate σ^2, because it now incorporates not only the experimental error but also real differences in the population means. The question that we are faced with now is "How much larger than 1 must the variance ratio be to cause us to reject H_0?" This will, of course, depend on the distribution of the variance ratio. The corresponding statistic, whcih we denote as F, has a distribution called F **distribution,** which we now consider.

F DISTRIBUTION

As a broad theoretical statement, a statistic F, the ratio of two independent chi-square variables, each divided by the corresponding degrees of freedom, has an F distribution. If the chi-square in the numerator has ν_1 degrees of freedom and the one in the denominator has ν_2 degrees of freedom, then the F distribution is said to have ν_1 and ν_2 degrees of freedom. It is important that the number of degrees of freedom be given in this specific order.

Properties of the F Distribution

1. It is a continuous distribution.
2. The curve describing the distribution is located to the right of the vertical axis at zero because F is a ratio of two positive terms (see Figure 12-3).

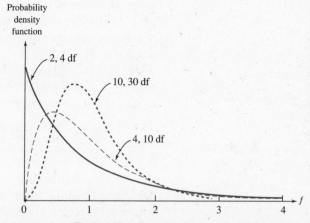

Figure 12-3 Probability Density Curves for F Distributions with Different Pairs of Degrees of Freedom

3. It depends on two parameters, the degrees of freedom associated with the numerator and those associated with the denominator. Thus, it constitutes what is called a **two parameter family of distributions.**
4. The distribution is skewed to the right, but the skewness disappears rapidly as the numbers of degrees of freedom increase (see Figure 12-3).

Appendix III, Table A-6, gives certain probabilities and corresponding f values for different F distributions specified by the parameters ν_1 and ν_2. Since an F distribution depends on two parameters, we are restricted even more than in the cases of the t distribution and the chi-square distribution in providing complete tables. In Table A-6, the top row in each table is assigned to the set of degrees of freedom ν_1, the number associated with the numerator. The left-hand column is assigned to the other set of degrees of freedom, ν_2. Each table then presents an f value (in the body of the table) for specific

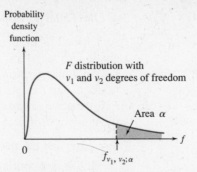

Figure 12-4 An f Value Leaving an Area α in the Right Tail

probability α contained in the right tail of the distribution. An f value for an F distribution with ν_1 and ν_2 degrees of freedom leaving an area α in the right tail of the distribution is commonly denoted as $f_{\nu_1, \nu_2; \alpha}$, and is called an upper percentage value of the F distribution. This is shown in Figure 12-4. There will be as many tables of f values as there are α levels for which one might wish to provide tables. For most statistical purposes, the α levels of interest are $\alpha = 0.1, 0.05, 0.025,$ and 0.01. For $\nu_1 = 4$ and $\nu_2 = 10$, the following are the upper percentage points $f_{4, 10; \alpha}$ for these α values.

Probability in the right tail	0.1	0.05	0.025	0.01
$f_{4, 10; \alpha}$	2.61	3.48	4.47	5.99

These are obtained from Table A-6 in Appendix III and are the colored entries in the tables. The locations of the preceding f values along the horizontal axis and the corresponding probabilities are illustrated in Figure 12-5.

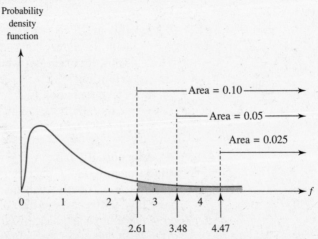

Figure 12-5 Probability Density Curve for F Distribution with 4, 10 df.
Area to the right of 2.61 is 0.10, etc.

Returning to the testing aspect, we have agreed that if the variance ratio is substantially *larger than* 1, it will cast doubts on the validity of H_0. Now it can be shown that the statistic giving the variance ratio on page 690 has an F distribution with $(k - 1)$ and $k(n - 1)$ degrees of freedom. Consequently, *the critical region of the test consists of values in the right tail of the F distribution with $(k - 1)$ and $k(n - 1)$ degrees of freedom.* If the computed value of the ratio exceeds the table value of the F distribution for a given α, we reject the null hypothesis at that level of significance and conclude that at least two of the population means are different from each other.

In our illustration, because $SSB = 18.667$ and $SSE = 38$, we get

$$f = \frac{18.667/2}{38/9} = 2.21.$$

Because $k = 3$ and $n = 4$ in our example, the corresponding F distribution has 2 (or $3 - 1$) and 9 (or $3(4 - 1)$) degrees of freedom. The table value at the 5 percent level of significance is $f_{2, 9; 0.05} = 4.26$. Because the computed value 2.21 does not exceed the table value 4.26, we do not reject the null hypothesis. There is no evidence indicating that the brands of gasoline are different.

THE ANOVA TABLE

We have obtained two estimates of σ^2. Actually, if the null hypothesis is true, it is possible to obtain one more estimate as

$$\frac{\sum\sum (x_{i,j} - \bar{\bar{x}})^2}{N - 1}$$

where N is the total number of observations. N is equal to kn for the data in Table 12-4 and equal to 3×4, or 12, for the data in Table 12-1. The numerator of this expression is called the **total sum of squares** and is usually denoted as SST. To compute it, we subtract each observation from the grand mean $\bar{\bar{x}}$, square these differences, and then total them. The denominator $N - 1$ is the number of degrees of freedom associated with SST. Thus,

$$SST = \sum\sum (x_{i,j} - \bar{\bar{x}})^2.$$

For the data in Table 12-1, since $\bar{\bar{x}} = 27.67$, we see that

$$
\begin{aligned}
SST = {} & (25 - 27.67)^2 + (29 - 27.67)^2 + (24 - 27.67)^2 \\
& + (26 - 27.67)^2 + (29 - 27.67)^2 + (27 - 27.67)^2 \\
& + (26 - 27.67)^2 + (30 - 27.67)^2 + (32 - 27.67)^2 \\
& + (27 - 27.67)^2 + (29 - 27.67)^2 + (28 - 27.67)^2 \\
= {} & 56.667.
\end{aligned}
$$

In summary, when we consider the three sums of squares that we have computed, we see that $SST = 56.667$, $SSB = 18.667$, $SSE = 38$, and, interestingly enough,

$$SST = SSB + SSE.$$

The partition of the total sum of squares in this way as a sum of the two sums of squares, *SSB* and *SSE*, always holds. Therefore, in practice we would usually compute *SST* and *SSB* and then obtain *SSE* by subtracting as $SSE = SST - SSB$. Also notice that *the degrees of freedom associated with SST are equal to the sum of the degrees of freedom associated with SSB and SSE*. This relation also is always true. These relations enable us to present the results of our procedure in a table called the analysis of variance table (abbreviated as the ANOVA table), as in Table 12-5. The last column of the ANOVA table gives the variance ratio.

TABLE 12-5 Analysis of Variance Table for the Data in Table 12-1

Source of Variation	Degrees of Freedom	Sum of Squares	Mean Sum of Squares	Computed Variance Ratio
Between	2	18.667	9.334	2.21
Within (error)	9	38.0	4.222	
Total	11	56.667		

The statistical test procedure for comparing means of *k* populations is summarized as follows.

COMPARISON OF *k* POPULATION MEANS

1. H_0: $\mu_1 = \mu_2 = \cdots = \mu_k$
2. H_A: Not all the population means are equal.
3. Test statistic: The variance ratio which has *F* distribution with $k - 1$ and $N - k$ degrees of freedom. To compute its value prepare the ANOVA table.
4. At the level of significance α, the decision rule is: *Reject H_0 if the computed value of the test statistic is greater than $f_{k-1, N-k; \alpha}$.* *Note:* The rejection region is in the right tail of the distribution.

EXAMPLE Analyze the data in Table 12-3, which is reproduced here.

	Brand 1	Brand 2	Brand 3
	25.2	28.5	29.6
	26.1	27.5	29.3
	26.8	28.2	28.4
	25.9	27.8	28.7
Mean mileage per gallon	26.0	28.0	29.0

SOLUTION

1. $H_0: \mu_1 = \mu_2 = \mu_3$

2. H_A: Not all the brand means are equal.

3. The test criterion is provided by the variance ratio, which we will compute next. Because the brand means are the same here as for the data in Table 12-1, *SSB* will be the same as before. We have

$$SSB = 18.667$$

$$\begin{aligned}
SST = &(25.2 - 27.67)^2 + (26.1 - 27.67)^2 \\
&+ (26.8 - 27.67)^2 + (25.9 - 27.67)^2 \\
&+ (28.5 - 27.67)^2 + (27.5 - 27.67)^2 \\
&+ (28.2 - 27.67)^2 + (27.8 - 27.67)^2 \\
&+ (29.6 - 27.67)^2 + (29.3 - 27.67)^2 \\
&+ (28.4 - 27.67)^2 + (28.7 - 27.67)^2 \\
= &\ 21.447.
\end{aligned}$$

Therefore,

$$\begin{aligned}
SSE &= 21.447 - 18.667 \\
&= 2.78.
\end{aligned}$$

Table 12-6 is the ANOVA table summarizing these results and other computations.

TABLE 12-6 ANOVA Table for the Data in Table 12-3

Source of Variation	Degrees of Freedom	Sum of Squares	Mean Sum of Squares	Computed Variance Ratio
Between	2	18.667	9.334	30.11
Within	9	2.780	0.310	
Total	11	21.447		

4. Using $\alpha = 0.05$, the decison rule is: *Reject H_0 if the computed value of the variance ratio is greater than 4.26.* (Here 4.26 is the table value of the *F* distribution with 2 and 9 degrees of freedom for $\alpha = 0.05$.)

5. The computed value is 30.11, as can be seen from the last column of the ANOVA table. Because it is greater than 4.26, we reject H_0 and conclude that the brand means are different. In our earlier discussion we had suspected that this was the case. ●

Section 12-1 Exercises

1. Five populations are being compared for their means and samples of size 7 are drawn from each population. How many df does each of the following have?
   (a) *SSB*   (b) *SSE*   (c) *SST*

2. In an analysis of variance problem having 5 categories, when samples of size 8 from each category were picked, the within sum of squares was found to be 85.75. Find an estimate of σ^2.

3. In Exercise 2 suppose the between sum of squares was given as 13.40. Assuming that the population means of the 5 categories are equal, obtain another estimate of σ^2.

4. The daily dissolved oxygen concentration (in milligrams per liter) of three streams was observed on 6 days for each stream. The observations are given in the following table.

Stream 1	Stream 2	Stream 3
4.02	2.88	2.76
3.98	3.06	3.50
3.85	3.70	2.82
4.32	2.98	3.22
4.16	3.88	3.15
3.92	3.28	3.80

   It may be assumed that the dissolved oxygen concentrations of the streams are normally distributed with the same variance, and are statistically independent. At the 5 percent level, are the true mean concentrations of the streams significantly different?

5. When samples of reinforcing bars supplied by four companies were tested for their yield strength, the following data were obtained.

Company A	Company B	Company C	Company D
22	18	25	24
19	28	20	18
15	26	25	25
18	24	24	19
24	20	26	21
19	21	18	22

   Assume that the yield strength of the bars is normally distributed for each company with the same variance σ^2. At the 5 percent level of significance, are the true mean yield strengths different for the companies?

12-2 ONE-WAY CLASSIFICATION—ARBITRARY SAMPLE SIZES

The ANOVA technique for one-way cassification is not limited to equal sample sizes. We assumed equal sample sizes of n in order to simplify matters in our previous discussion. With slight modifications in the formulas, we can apply the technique to any number of samples and any number of observations within each sample. In the following discussion we assume that there are k samples 1, 2, . . . , k with n_1, n_2, . . . , n_k observations, respectively. A general representation of the response collected is given in Table 12-7, where $x_i,\bullet$ is the total for Sample i.

TABLE 12-7 A General Representation of the Response with Unequal Sample Sizes

	Sample 1	Sample 2	$\cdots$	Sample k
	$x_{1,1}$	$x_{2,1}$	$\cdots$	$x_{k,1}$
	$x_{1,2}$	$x_{2,2}$	$\cdots$	$x_{k,2}$
	$\vdots$	$\vdots$		$\vdots$
	x_{1,n_1}	x_{2,n_2}	$\cdots$	x_{k,n_k}
Sample size	n_1	n_2	$\cdots$	n_k
Sample total	$x_1,\bullet$	$x_2,\bullet$	$\cdots$	$x_k,\bullet$
Sample mean	$\bar{x}_1$	$\bar{x}_2$	$\cdots$	$\bar{x}_k$

The total number of observations N is given by

$$N = n_1 + n_2 + \cdots + n_k$$

and the grand total, the sum of the entire set of observations, is given by

$$\text{Grand total} = x_1,\bullet + x_2,\bullet + \cdots + x_k,\bullet.$$

Shortcut Routine for Computations Leading to the ANOVA Table

The computations involved in calculating the sums of squares in an ANOVA problem can be quite tedious. It is helpful to use the following shortcut routine.

1 Obtain what is called the **correction term** (abbreviated **C.T.**) as

$$C.T. = \frac{(\text{grand total})^2}{N}.$$

2
$$SSB = \frac{x_1^2,\bullet}{n_1} + \frac{x_2^2,\bullet}{n_2} + \cdots + \frac{x_k^2,\bullet}{n_k} - C.T.$$

where $x_i^2,\bullet$ is the square of the total of the ith sample.

3 $SST = \sum\sum x_{ij}^2 - C.T.$

That is, square each of the observations and find the sum of these squares. From the total thus obtained, subtract the correction term.

4 SSE is found by subtracting SSB from SST.

The results can be presented now in an ANOVA table, Table 12-8.

TABLE 12-8   Analysis of Variance Table When the Sample Sizes Are Arbitrary

Source of Variation	Degrees of Freedom	Sum of Squares	Mean Sum of Squares	Computed Ratio
Between	$k - 1$	$\sum \dfrac{x_{i,\bullet}^2}{n_i} - C.T.$	$SSB/(k - 1)$	$\dfrac{SSB/(k - 1)}{SSE/(N - k)}$
Within (error)	$N - k$	$SST - SSB$	$SSE/(N - k)$	
Total	$N - 1$	$\sum\sum x_{ij}^2 - C.T.$		

We will illustrate the procedure using the following data.

	Sample 1	Sample 2	Sample 3	Sample 4
	3	6	11	3
	7	8	7	10
	8	12	16	
	15		9	
			15	
Sample size	4	3	5	2
Sample total	33	26	58	13

Here $n_1 = 4, n_2 = 3, n_3 = 5, n_4 = 2$. Thus $N = 4 + 3 + 5 + 2 = 14$. Also, the grand total is $33 + 26 + 58 + 13 = 130$. We now carry out the following steps.

1. $C.T. = \dfrac{(130)^2}{14}$

$= 1207.143$

2. $SSB = \dfrac{33^2}{4} + \dfrac{26^2}{3} + \dfrac{58^2}{5} + \dfrac{13^2}{2} - 1207.143$

$= 1254.883 - 1207.143$

$= 47.740$

3. $SST = 3^2 + 7^2 + 8^2 + 15^2 + 6^2 + 8^2 + 12^2 + 11^2$
$\qquad + 7^2 + 16^2 + 9^2 + 15^2 + 3^2 + 10^2 - 1207.143$
$\qquad = 1432.0 - 1207.143$
$\qquad = 224.857$
4. $SSE = 224.857 - 47.740$
$\qquad = 177.117$

The computed quantities are presented in the following ANOVA table.

Source of Variation	Degrees of Freedom	Sum of Squares	Mean Sum of Squares	Computed Ratio
Between	3	47.740	15.913	0.898
Within	10	177.117	17.712	
Total	13	224.857		

EXAMPLE 1

In order to test whether Yuri's productivity is the same on the five weekdays of a week, unknown to Yuri, his employer kept records on eighteen randomly picked days, as the data in Table 12-9 show. (Productivity is measured in terms of market value, in dollars, of the items produced by Yuri.) Carry out the test using $\alpha = 0.05$.

TABLE 12-9 Productivity on Eighteen Randomly Picked Days Classified according to the Day of the Week

	Monday	Tuesday	Wednesday	Thursday	Friday
	143	162	160	138	110
	128	136	132	168	130
	110	144	180	120	135
		158	160		
			138		
Sample size	3	4	5	3	3
Sample total	381	600	770	426	375
Sample mean	127	150	154	142	125

SOLUTION We assume that Yuri's productivity on any weekday is normally distributed with the same variance and that his productivity on one day does not influence his productivity on another day. Our approach now consists of the following steps.

1. $H_0: \mu_1 = \mu_2 = \mu_3 = \mu_4 = \mu_5$
2. H_A: At least two of the means are different. That is, the mean productivity on at least two days is different.

3. $C.T. = \dfrac{(2552)^2}{18} = 361{,}817$

$SST = 143^2 + 128^2 + \cdots + 130^2 + 135^2 - 361{,}817$

$\qquad = 368{,}334 - 361{,}817$

$\qquad = 6517$

$SSB = \dfrac{381^2}{3} + \dfrac{600^2}{4} + \dfrac{770^2}{5} + \dfrac{426^2}{3} + \dfrac{375^2}{3} - 361{,}817$

$\qquad = 364{,}334 - 361{,}817$

$\qquad = 2517$

$SSE = 6517 - 2517$

$\qquad = 4000$

We get the following ANOVA table as shown in Table 12-10.

TABLE 12-10   ANOVA Table for Data in Table 12-9

Source of Variation	Degrees of Freedom	Sum of Squares	Mean Sum of Squares	Computed Ratio
Between	4	2517	629.3	2.05
Within	13	4000	307.7	
Total	17	6517		

4. The F distribution has 4 and 13 degrees of freedom. From the table of F distribution, $f_{4,\,13;\,0.05} = 3.18$. The decision rule is: *Reject H_0 if the computed value of the variance ratio exceeds 3.18.*
5. From the ANOVA table, the computed value is 2.05, and so we do not reject H_0. The evidence does not substantiate that the mean productivity of Yuri is significantly different on five weekdays. ●

EXAMPLE 2   Table 12-11 gives the hourly wages (in dollars) of fifteen randomly picked workers classified according to their occupations. At the 1 percent level of significance, are the true mean wages different for the four occupations?

SOLUTION   We take for granted the basic assumptions—that the hourly wages for each occupation are normally distributed with the same variance, and the hourly wages of one occupation are not influenced by those of another occupation.

1. H_0: $\mu_1 = \mu_2 = \mu_3 = \mu_4$
2. H_A: At least two of the population means are different.

TABLE 12-11 Hourly Wages of Workers Classified according to Their Occupations

	Plumber	Electrician	Painter	Carpenter
	17.6	18.5	16.8	17.4
	18.3	18.7	16.7	16.5
	17.6	17.7	16.6	16.8
		18.3	16.4	
		17.8		
Sample size	3	5	4	3
Sample total	53.5	91.0	66.5	50.7
Sample mean	17.83	18.20	16.63	16.90

3. The grand total $= 53.5 + 91.0 + 66.5 + 50.7$

$$= 261.7$$

$$C.T. = \frac{(261.7)^2}{15}$$

$$= 4{,}565.793$$

$$SST = 17.6^2 + 18.3^2 + \cdots + 16.5^2$$
$$+ 16.8^2 - 4{,}565.793$$
$$= 4{,}574.270 - 4{,}565.793 = 8.477$$

$$SSB = \frac{53.5^2}{3} + \frac{91.0^2}{5} + \frac{66.5^2}{4} + \frac{50.7^2}{3} - 4{,}565.793$$
$$= 4{,}572.675 - 4{,}565.793$$
$$= 6.882$$

$$SSE = 8.477 - 6.882$$
$$= 1.595$$

These and other results are summarized in Table 12-12.

TABLE 12-12 ANOVA Table for Data in Table 12-11

Source of Variation	Degrees of Freedom	Sum of Squares	Mean Sum of Squares	Computed Ratio
Between	3	6.882	2.294	15.82
Within	11	1.595	0.145	
Total	14	8.477		

4. The F distribution has 3 and 11 degrees of freedom. Since $f_{3,11;0.01} = 6.22$, the decision rule is: *Reject H_0 if the computed value of the variance ratio is greater than 6.22.*

5. Because the computed value is 15.82, we reject H_0 at the 1 percent level of significance. There is rather strong evidence that the true mean wages differ significantly according to occupations. ⬤

Confidence Intervals

Our discussion in Section 1 showed that the mean within sum of squares provides a pooled estimate of the common variance σ^2. Following the methods in Chapters 7 and 8, we can now use this estimate for constructing confidence intervals for individual population means and differences in population means as follows.

CONFIDENCE INTERVAL FOR μ_i AND $\mu_i - \mu_j$

A $(1 - \alpha)$ 100 percent confidence interval for μ_i is given by

$$\bar{x}_i - t_{N-k,\alpha/2} \frac{s_p}{\sqrt{n_i}} < \mu_i < \bar{x}_i + t_{N-k,\alpha/2} \frac{s_p}{\sqrt{n_i}}$$

A $(1 - \alpha)$ 100 percent confidence interval for $\mu_i - \mu_j$ is given by

$$(\bar{x}_i - \bar{x}_j) \pm t_{N-k,\alpha/2} s_p \sqrt{\frac{1}{n_i} + \frac{1}{n_j}}$$

Here s_p is the positive square root of the mean within sum of squares given by

$$s_p = \sqrt{\frac{SSE}{N - k}}$$

and the t distribution has $N - k$ degrees of freedom, the number associated with the within sum of squares.

EXAMPLE 3

Consider the data in Example 2. Compute a 95 percent confidence interval for:

(a) the mean hourly wages of an electrician

(b) the difference in mean hourly wages of an electrician and a carpenter.

SOLUTION In the analysis of variance of the data in Table 12-12 we see that $s_p = \sqrt{0.145} = 0.38$. Also, since the within sum of squares has 11 df, we find $t_{11,0.025} = 2.201$. Thus we get the confidence intervals in the two cases as follows.

(a) The sample mean for electricians is 18.2 based on a sample size of 5. Therefore, a 95 percent confidence interval for the mean wages of an electrician is

$$\left(18.2 - 2.201 \frac{(0.38)}{\sqrt{5}}, \ 18.2 + 2.201 \frac{(0.38)}{\sqrt{5}}\right)$$

that is (17.83, 18.57).

(b) The sample mean for electricians based on a sample of 5 is 18.2, and that for carpenters based on a sample of 3 is 16.9. Therefore, a 95 percent confidence interval for the difference is given by

$$(18.2 - 16.9) \pm (2.201)(0.38) \sqrt{\frac{1}{5} + \frac{1}{3}}$$

that is, (0.69, 1.91).

Section 12-2 Exercises

1. Complete the following analysis of variance table.

Source of Variation	Degrees of Freedom	Sum of Squares	Mean Sum of Squares	Computed Ratio
Between	7			
Within		187.2		
Total	25	397.9		

2. In an analysis of variance problem having five categories with 40 observations altogether, the within-samples sum of squares was found to be 147. Obtain an estimate of σ^2.

In each of the following exercises, assume that the sampled populations are normally distributed with the same variance and that the samples are independent.

3. A golfer wants to compare a new brand of golf ball on the market with her old brand. The figures below give the distances (in feet) of six drives with the old brand and seven drives with the new brand.

Old Brand	New Brand
138	165
162	126
140	195
175	203
190	180
151	150
	185

At the 5 percent level, is there significant difference in the true mean distances on a drive for the two brands? Carry out the test using:

(a) the method for comparison of two means developed in Section 9-4

(b) The ANOVA technique.

4. A farm was divided into eighteen plots, and three concentrations of a fertilizer were allotted to them at random, with each concentration tried on six plots. The following figures give the yield of grain (in bushels) for each plot.

Concentration 1	Concentration 2	Concentration 3
15	32	28
24	21	16
18	16	25
28	35	28
22	26	22
20	18	29

At the 5 percent level of significance, do the data provide evidence that the three concentrations are significantly different in their effect on the yield of grain?

5. The figures in the following table give the tar content (in milligrams) in a pack of cigarettes for four brands, when five packs of each brand were tested.

Brand A	Brand B	Brand C	Brand D
340	358	335	358
323	320	318	345
319	340	330	330
330	360	320	350
335	338	338	340

Is there significant difference in the true mean tar content in a pack for the four brands? Use $\alpha = 0.1$.

6. The figures in the following table give the lengths (in inches) of trout caught in three lakes, Blue Lake, Clear Lake, and Fresh Lake.

Blue Lake	Clear Lake	Fresh Lake
11.9	10.6	11.2
10.8	10.4	9.6
9.8	10.3	11.2
10.6	10.2	10.5
11.2		10.3
9.7		

Is there a significant difference in the true mean lengths of trout found in the three lakes? Use $\alpha = 0.025$.

7. A laboratory experiment was conducted to compare three brands of tires. The following figures give the number of miles when five tires of each brand were tested.

Brand A	Brand B	Brand C
32,000	34,000	37,000
35,000	27,000	29,000
41,000	36,000	41,000
31,000	30,000	37,000
36,000	27,000	38,000

Do the data lend support to the contention that there are significant differences between the three brands? Use $\alpha = 0.025$.

8. Three horses, Prima Donna, Princess Jane, and Flying Princess, were timed running along a course. The following figures give times (in minutes) in five runs along the course.

Prima Donna	Princess Jane	Flying Princess
5.9	5.6	6.2
6.3	6.3	6.3
5.8	5.9	6.4
6.4	6.2	6.1
6.1	5.9	6.2

At the 5 percent level, are there significant differences between the true mean times along the course for three horses?

9. In order to compare three different types of insulations, a building contractor built fifteen rooms, five with each type of insulation. The following figures refer to the drop in temperature in each room during a 4-hour period on a night.

Insulation A	Insulation B	Insulation C
16	10	15
19	12	16
18	20	20
10	12	16
17	14	13

Do the data indicate significant differences among the three types of insulation? Use a suitable α level.

10. The manager of a bank decided to compare the speed on the job of four tellers working in the bank. The following data give the amount of time (in minutes) that they spend serving their customers, picked at random.

Teller 1	Teller 2	Teller 3	Teller 4
0.8	8.7	9.7	0.7
1.6	4.2	2.0	3.2
8.8	0.2	5.9	4.2
7.7	9.0	2.3	6.7
3.3		5.8	7.5
			1.4

Is there evidence indicating that there are significant differences among the true mean serving times for the four tellers? Use $\alpha = 0.025$.

11. An experimental agriculture station is interested in comparing three corn hybrids. Hybrids A and B are each planted on six plots and Hybrid C is planted on five plots. The yields per acre (in bushels) are as follows.

Hybrid A	Hybrid B	Hybrid C
86	83	79
94	75	75
99	62	53
61	86	55
70	77	72
89	79	

At the 10 percent level of significance, is the true mean yield per acre different for the three hybrids?

12. A pharmacology lab conducted an experiment to compare four pain-relieving drugs. Twenty-four subjects were used, with six allotted at random to Drug A, seven to Drug B, five to Drug C, and six to Drug D. The following figures give the number of hours of pain relief provided subsequent to administering a drug.

Drug A	Drug B	Drug C	Drug D
8	9	2	2
7	3	3	9
3	1	5	8
8	8	9	2
1	1	6	7
4	9		5
	8		

At the 5 percent level of significance, are the drugs different in the true mean number of hours of relief provided?

13. An experiment was carried out with sixty brown hens to examine what effect starch in the diet has on the water intake of the hens. Five levels of starch in the meal were considered. The hens given any diet were allotted to that diet at random. Ten hens were given Diet 1 with no starch, twelve hens Diet 2 with 100 g/kg starch, eight hens Diet 3 with 200 g/kg starch, fifteen hens Diet 4 with 300 g/kg starch, and fifteen hens Diet 5 with 400 g/kg starch. The data presented are the amount of water consumed over the duration of the experiment (*Journ. of the Sci. of Food & Agric.* 1984; *35:* pp. 36–40).

WATER CONSUMPTION (in milliliters)				
Diet 1	Diet 2	Diet 3	Diet 4	Diet 5
214	218	242	224	288
215	195	213	230	316
220	218	171	239	311
241	250	185	308	310
226	206	165	269	267
201	187	174	286	297
202	184	213	281	251
221	176	252	251	274
232	164		223	252
230	211		267	294
	193		309	279
	189		312	294
			273	266
			252	286
			238	263

Does the true mean water consumption of the hens differ with the level of starch in the diet? Carry out the test at the 1 percent level of significance.

14. Seven broiler chickens approximately 8 weeks of age, ten hens approximately 62 weeks of age, and eight turkeys approximately 17 weeks of age were slaughtered and the cytochrome c content in the breast meat measured (μg/g of tissue) for each bird (*Journ. of the Sci. of Food & Agric.* 1986; *37:* pp. 1236–1240).

Broiler Chicken	Hens	Turkeys
11.57	13.85	13.54
10.91	13.57	13.69
11.37	13.91	13.46
11.16	14.43	12.78
11.25	14.11	13.11
10.43	12.30	12.37
12.28	15.45	13.51
	13.81	12.73
	14.67	
	13.67	

(a) On the basis of the data, is there a significant difference among the true mean cytochrome c contents in the breast meats of the three types of poultry? Use $\alpha = 0.025$.

(b) Compute a 95 percent confidence interval for the true mean cytochrome c content in the breast meat of hens.

(c) Estimate the difference between the true mean cytochrome c contents of broiler chickens and turkeys using a 95 percent confidence interval.

15. Fertilization of fish-rearing ponds is necessary to improve the nutrition of planktonic algae and bacteria. The following three combinations were used to fertilize nursery ponds: (1) fodder yeast, (2) a combination of fodder yeast and mineral

fertilizers (superphosphate and ammonia nitrate), and (3) a combination of organic and mineral fertilizers (superphosphate ammonia nitrate, ordinary and green manure). Of the fifteen ponds considered for the experiment, five were treated with Combination 1, six with Combination 2, and four with Combination 3. The fish production was measured (kg/hectare) when the fish matured. The data are presented in the following table (*Hydrobiological Journal* 1982; *18*: pp. 58–61).

Combination 1	Combination 2	Combination 3
1703.7	1791.0	1528.3
1689.5	1845.9	1556.7
1638.0	1807.0	1528.1
1650.8	1854.6	1556.9
1669.8	1806.3	
	1817.7	

(a) At the 5 percent level, is there a significant difference in the true mean fish production among the three fertilizer combinations?

(b) Find a 95 percent confidence interval for the difference between the true mean amounts of fish produced using Combinations 1 and 2.

16. Sphagnum peat moss from bogs located in the province of Newfoundland, Canada, was used as substrate to cultivate oyster mushrooms (*Pleurotus ostreatus*). Acid hydrolyzed peat and nonhydrolyzed peat were used in preparing the spawn and in growing the mushroom fruiting bodies. For comparison, the mushroom was also cultivated on paper. (Paper contains 70 percent cellulose.) After harvesting, the crude protein content was determined by multiplying the percentage nitrogen by 6.25 (*Journ. of the Sci. of Food & Agric.* 1986; *37*: pp. 833–838). The following data refer to the amount of crude protein.

(a) Based on the data presented, does the true mean crude protein content differ depending on the substrate used? Carry out the test at the 5 percent level of significance.

(b) Estimate the difference between the true mean crude protein contents for hydrolyzed peat and nonhydrolyzed peat using a 98 percent confidence interval.

Hydrolyzed Peat	Nonhydrolyzed Peat	Paper
41.14	42.94	27.11
36.41	39.79	17.80
36.07	34.66	23.00
32.28	40.61	17.28
36.97	29.66	20.19
39.38	36.39	24.51
34.95	34.03	23.56
34.58	35.42	26.92
34.94	29.52	
	28.58	

CHAPTER **12 Summary**

✓ *CHECKLIST: KEY TERMS AND EXPRESSIONS*

❏ analysis of variance (ANOVA), page 683
❏ one-way classification, page 683
❏ experimental error, page 684
❏ grand mean, page 687

❏ error sum of squares (*SSE*), page 688
❏ within-samples sum of squares (*SSE*), page 688
❏ between means sum of squares (*SSB*), page 689

❏ variance ratio, page 690
❏ *F* distribution, page 690
❏ two parameter family of distributions, page 691
❏ total sum of squares (*SST*), page 693
❏ correction term (C.T.), page 697

KEY FORMULAS

sample total $x_{i},\bullet$ = sample total for sample *i*.

grand total   grand total = total of all the samples

correction term   **C.T. = (grand total)^2/N, where *N* is the total number of observations**

between means sum of squares

$$SSB = \frac{x_{1,\bullet}^2}{n_1} + \frac{x_{2,\bullet}^2}{n_2} + \cdots + \frac{x_{k,\bullet}^2}{n_k} - C.T.$$

total sum of squares

$$SST = \sum\sum x_{ij}^2 - C.T.$$

within-samples sum of squares $SSE = SST - SSB$

CHAPTER **12 Concept and Discussion Questions**

1. Analysis of variance deals with
   (a) quantitative measurement of a variable
   (b) counts of the numbers falling in different cells
   (c) frequency data of cases falling in different columns in a table

2. When comparing several populations, why is it not a sound approach to perform a battery of *t* tests by considering the populations in pairs?

3. What assumptions are made in carrying out analysis of variance?

4. What hypothesis do we test using analysis of variance?

5. For providing a rationale for carrying out the test of hypothesis we obtained two estimates of σ^2. Under what assumptions are these estimates obtained?

1. What assumptions are made in the analysis of variance for a one-way classification problem?

2. Decide whether the following statement is true or false.
   The mean total sum of squares
   = the mean between means sum of squares
   + the mean within samples sum of squares

3. In an analysis of variance problem with 4 categories and 30 observations altogether, the within sum of squares was obtained as 32.24. Find an estimate of σ^2.

4. In a laboratory experiment in which a certain strain of fruit flies was classified according to their genotypes, *AA*, *Aa*, and *aa*, some flies were picked at random from each genotype and the total wing length (in microunits) of each fly was measured. The data are presented in the following table.

AA	Aa	aa
10	8	10
17	16	9
19	13	16
14	10	9
	12	12

   Do the data provide sufficient evidence to suggest that the true mean wing lengths for the three genotypes are different? Use a suitable level of α.

5. Three fish-rearing ponds were fertilized with fodder yeast and mineral fertilizer containing superphosphate and ammonium nitrate. Five determinations of dissolved oxygen concentration (mg/l) were made in two of the ponds and four determinations were made in the third pond (*Hydrobiological Journal* 1982; *18:* pp. 58–61).

   DISSOLVED OXYGEN (mg/l)

Pond 1	Pond 2	Pond 3
5.5	5.3	5.8
5.8	5.2	5.9
5.6	5.4	5.7
5.5	5.1	5.6
5.4	5.3	

   This problem was considered in Chapter 9, where three comparisons of pairs were made (see Exercise 24, Section 9-4, page 553). Carry out the analysis of variance to test the null hypothesis that the true means of dissolved oxygen in the three ponds are not different. Test using $\alpha = 0.05$.

6. Rats weighing 45–65 grams were used in 4-week feeding trials. Hulls from Tower canola (TCH), Regent canola (RCH), and soybeans (SBH) were included

in the diets at 10 percent or 20 percent (*Journal of the Science of Food and Agriculture* 1984; *35*: pp. 625–631). At the end of a 4-week feeding period, the weight gain (g) for each rat was recorded. The data are presented in the following table.

			WEIGHT GAIN (g)			
Control	TCH 10%	TCH 20%	RCH 10%	RCH 20%	SBH 10%	SBH 20%
179	200	164	180	178	171	176
180	184	171	187	180	181	180
189	181	167	184	169	174	184
170	179	182	180	171	179	182
182	181	170	191	171	172	194
	189		179	176		180
				172		

Based on the data, are there significant differences among the diets with regard to the true mean gain in weight? Test using $\alpha = 0.10$.

7. The following data were obtained when five brands of sanders were tested for their trouble-free service (in hours).

Brand A	Brand B	Brand C	Brand D	Brand E
680	900	520	835	1020
735	580	680	450	630
540	600	720	700	700
865	780	925	730	550

If all the assumptions necessary for applying the ANOVA method hold, at the 5 percent level, are the five brands different?

8. Forty 5- to 6-week-old male rats of approximately 100 grams body weight were used in an experiment. Ten rats were allocated to each of four diets A, B, C, D containing casein treated with methanol (HCHO) solution with, respectively, 40, 70, 100, and 130 grams HCHO/liter. The figures for the dry matter digestibility of the rats are shown in the following table.

Diet A	Diet B	Diet C	Diet D
0.894	0.885	0.881	0.843
0.899	0.889	0.881	0.845
0.897	0.888	0.882	0.843
0.897	0.887	0.884	0.847
0.902	0.892	0.886	0.847
0.896	0.887	0.889	0.844
0.899	0.887	0.881	0.845
0.896	0.890	0.878	0.847
0.898	0.889	0.883	0.848
0.900	0.886	0.885	0.848

Is there a significant difference in the dry matter digestibility of the rats with changes in the level of methanol treatment of casein? Carry out the test with $\alpha = 0.05$.

9. The consistency of pig fatty tissue is important in meat processing and consumer acceptability. The main problem is backfat which is too soft caused, at least in part, by high proportions of linoleic acid in the lipid. The concentration of linoleic acid in the backfat is related to its concentration in the diet and to the level of fat in the diet. Linoleic acid, however, is an essential fatty acid and so must be present in the diet.

Three groups of fifteen female pigs were fed starter diets containing 0.8 percent, 1.1 percent, or 1.8 percent (percentage of dry feed) linoleic acid from 20 kilograms to 35 kilograms followed by a finisher diet consisting of 1.0 percent, 1.2 percent, or 1.4 percent linoleic acid, respectively, from 35 kilograms to slaughter at 85 kilograms live weight. Backfat thickness was measured 6.5 cm from the midline at the last rib, 45 minutes after slaughter. The data in the following table give the fat thickness (mm) measured on each cold carcass in the three diet regimens characterized: low (0.8 percent, 1.0 percent), medium (1.1 percent, 1.2 percent), and high (1.8 percent, 1.4 percent).

FAT THICKNESS (mm)		
Low Diet	**Medium Diet**	**High Diet**
15	17	13
16	17	14
16	16	14
16	16	13
15	17	13
16	16	14
16	16	14
14	15	14
16	16	15
16	15	15
17	16	14
15	15	14
14	17	14
15	16	14
16	15	14

Are there significant differences in the true mean fat thickness among the three diet groups? Use $\alpha = 0.10$.

▣ GRAPHING CALCULATOR INVESTIGATIONS

Read the section of the Appendix entitled "Using the TI-82 as a Data Analysis Tool," on obtaining summary analysis of variance results. Then redo several of the exercises from Chapter 12 of the text using the ANOVA1 program from the TI-82 Programs Appendix and your TI-82. In particular, try:

Section 12.2: Problems 4–15.
Review Exercises: Problems 4–9.

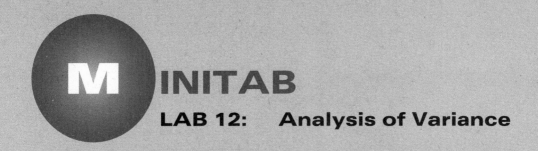

MINITAB

LAB 12:   Analysis of Variance

As you have seen in this chapter, in a one-way analysis of variance we are interested in comparing means of several populations. In this context the common variance σ^2 plays a very important role. So one of the things that we will do in this lab is simulate samples from normal populations with different means and study how the magnitude of σ^2 plays a role in discerning differences among population means. The normal procedure for carrying out analysis of variance involves reading the data for each level or treatment in separate columns and entering the command AOVONEWAY. (The first three letters AOV of the command stand for *Analysis of Variance*.) A slight variation of this, which is particularly useful when there are unequal sample sizes, will be explained while working one of the problems in this lab.

1 We consider four populations with different means, each one having $\sigma = 6$. We store the commands in the file named 'ANAVAR1' so that we can execute it repeatedly.

```
MTB >NOTE PLACE SAMPLES IN COLUMNS C1-C4

MTB >NOTE POPULATIONS NORMAL WITH SIGMA 6

MTB >STORE 'ANAVAR1'

STOR>NOECHO

STOR>RANDOM 80 OBSERVATIONS IN C1;

STOR>NORMAL MEAN 62 SIGMA 6.

STOR>RANDOM 75 OBSERVATIONS IN C2;

STOR>NORMAL MEAN 60 SIGMA 6.

STOR>RANDOM 50 OBSERVATIONS IN C3;

STOR>NORMAL MEAN 65 SIGMA 6.

STOR>RANDOM 45 OBSERVATIONS IN C4;

STOR>NORMAL MEAN 57 SIGMA 6.

STOR>AOVONEWAY ON DATA IN C1-C4
```

```
STOR>END

MTB >EXECUTE 'ANAVAR1'
```

The command AOVONEWAY performs the one-way analysis of variance. The printout gives an analysis of variance table similar to Table 12-5 in the text except that the between sum of squares is called FACTOR sum of squares. As you know, the within mean sum of squares provides an unbiased estimate of σ^2, so that its square root provides an estimate of σ. This square root is presented in the printout as POOLED STDEV. Because we simulated from populations with $\sigma = 6$, check whether the value printed for the estimate is near 6.

To test the null hypothesis H_0: $\mu_1 = \cdots = \mu_4$ compare the F-ratio from the analysis of variance table (the computed variance ratio) with the value from the F-table with 3 and 246 degrees of freedom. At the 5 percent level of significance, you can take $F_{3,246;\,0.05}$ as 2.60. It is highly likely that your conclusion will be to reject the null hypothesis of equality of the population means.

Assignment   Execute the program ANAVAR1 four more times with EXECUTE 'ANAVAR1' 4 TIMES. Each time see if the comments above apply.

2   In the following we simulate samples from normal populations with the same set of means that we had previously but with a different value for σ. We take $\sigma = 16$, a value much larger than the previous value 6.

```
MTB >NOTE POPULATION NORMAL WITH SIGMA 16

MTB >STORE 'ANAVAR2'

STOR>NOECHO

STOR>RANDOM 80 OBSERVATIONS IN C1;

STOR>NORMAL MEAN 62 SIGMA 16.

STOR>RANDOM 75 OBSERVATIONS IN C2;

STOR>NORMAL MEAN 60 SIGMA 16.

STOR>RANDOM 50 OBSERVATIONS IN C3;

STOR>NORMAL MEAN 65 SIGMA 16.

STOR>RANDOM 45 OBSERVATIONS IN C4;

STOR>NORMAL MEAN 57 SIGMA 16.

STOR>AOVONEWAY ON DATA IN C1-C4

STOR>END

MTB >EXECUTE 'ANAVAR2'
```

Upon execution of the program the pooled estimate of σ should be near 16. Check whether this is the case with your data. Also, the F-ratio in the

ANOVA table is likely to be reduced considerably compared with the case when σ was 6. You are less prone to reject the null hypothesis of equality of means. The explanation is that because of the large value of σ the variability between is completely masked by the variability within, and the large overlap of the populations makes it seem that the four samples have come from just one population.

Assignment   Execute the program 'ANAVAR2' six times and each time take note of the points raised above.

Assignment   Repeat the simulation process above by changing σ to 25.

3   We will now work out the Example in Section 12-1. In this example all the sample sizes are equal, so we use the command READ to enter the data in three columns.

```
MTB >NOTE EXAMPLE IN SECTION 12-1

MTB >READ DATA IN C1-C3

DATA>25.2 28.5 29.6

DATA>26.1 27.5 29.3

DATA>26.8 28.2 28.4

DATA>25.9 27.8 28.7

DATA>END OF DATA

MTB >PRINT C1-C3

MTB >AOVONEWAY FOR DATA IN C1-C3
```

The results presented in the analysis of variance should be the same as those in Table 12-6 on page 695. Besides giving the analysis of variance table, MINITAB also prints out a summary of the sample sizes, and the sample mean and standard deviation for each of the treatments. Furthermore, using the pooled estimate of σ, it presents a plot of the 95 percent confidence intervals for each of the treatment means.

Assignment   Work out Exercise 9 in Review Exercises.

4   Finally, consider Example 1 in Section 12-2. In this case the sample sizes are not equal, and we cannot read in the data using the command READ. We could, of course, enter the data in columns with five separate SET commands, and perform the analysis of variance with the command AOVONEWAY.

There is an alternate way to enter the data and a different command that can be used to perform the analysis. First we assign numerical codes to the various treatments. The entire data are then entered into two columns with the command READ, so that one column contains the numerical codes for the treatments and the other column contains the corresponding responses. The MINITAB command to carry out the analysis is ONEWAY AOV, properly specifying the two column where the response and the treatment codes are located. The data for our example are read as follows after identifying Monday with the code 1, Tuesday with 2, and so on.

```
MTB >NOTE TREATMENT CODES IN COLUMN C1 RESPONSE IN C2
MTB >NAME C1 'CODE', C2 'RESPONSE'
MB  >READ DATA IN C1 C2
DATA>1 143
DATA>1 128
DATA>1 110
DATA>2 162
DATA>2 136
DATA>2 144
DATA>2 158
DATA>3 160
DATA>3 132
DATA>3 180
DATA>3 160
DATA>3 138
DATA>4 138
DATA>4 168
DATA>4 120
DATA>5 110
DATA>5 130
DATA>5 135
DATA>END OF DATA
MTB >PRINT C1 C2
MTB >ONEWAY AOV RESPONSE IN C2, NUMERIC CODES IN C1
```

Assignment   Work out Exercise 14 in Section 12-2.

Assignment   Work out Exercise 13 in Section 12-2.

We now type STOP and conclude the MINITAB session using the logging off procedure.

Commands you have learned in this session:

```
AOVONEWAY
```

```
ONEWAY
```

CAREER *PROFILE*

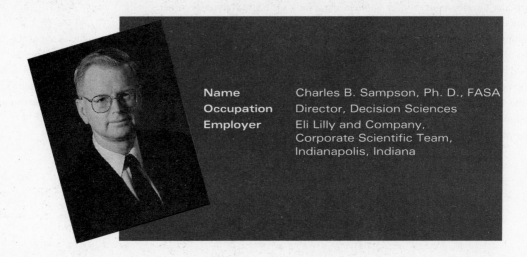

Name   Charles B. Sampson, Ph. D., FASA
Occupation Director, Decision Sciences
Employer   Eli Lilly and Company,
            Corporate Scientific Team,
            Indianapolis, Indiana

As I planned for my college career, almost everyone encouraged me to study engineering. This was because I seemed to have an aptitude for mathematics and science, and many of the high school advisors of that era (Sputnik) believed that the most important discipline for overtaking Russia's lead in the space program was engineering. However, I soon discovered that engineering was only one of my interests and that my most exciting courses during my first year of college were those in mathematics and physics. I changed my majors to mathematics and physics and continued my education with little knowledge of the job market or of other areas of study that were compatible with my interests in science and mathematics. If there were counselors who were aware of the science and the profession of statistics, I certainly did not find any. I entered graduate school in mathematics with the goal of becoming a college professor or working in industry as some type of applied mathematician. However, at the University of Iowa, two professors of statistics had fantastic reputations as teachers and I chose an elective in probability and statistics. This exposure to statistics was a major event in my life although I didn't realize it for some time.

As I completed my MS degree in mathematics, I decided that it made sense for me to obtain industrial experience before continuing my education. The placement office advised me that my best chances for employment were as a computer programmer or as a statistician. I did not know for sure what a statistician did (the courses I had in probability and statistics were theory courses), but I pursued the statistics option anyway and began my career as a statistician in quality engineering for IBM in upstate New York. I soon learned that practicing statistics was an exciting profession and that I did not have enough education to be effective as a consulting statistician, so I attended Iowa State University and received a PhD in statistics.

I am currently Director of Decision Sciences at Eli Lilly and Company after serving as Director of Statistical and Mathematical Sciences for many years. Eli Lilly is a research intensive pharmaceutical firm. As a consulting statistician, I had the opportunity to work with scientists, physicians, engineers, psychologists, and business persons. I helped design experiments, and subsequently analyzed and reported the results of these experiments to scientists and managers in basic research, clinical trials, process and product development, quality engineering, marketing, manufacturing, and the human resources component. Consulting in these varied areas of application is most exciting and provides for a stimulating and rewarding career. The field of statistics is increasingly being recognized as one that should play a larger role in decision- and policy-making in the face of uncertainty and variability. The job market continues to be excellent for MS and PhD statisticians. Those who choose to practice statistics as a profession can expect an extremely rewarding and interesting (and well paid) career.

For those of you interested in statistics, I suggest getting an excellent background in mathematics by taking as many statistics courses as possible in undergraduate school. Then go to graduate school in statistics and seek experience by using summer employment and/or internship opportunities in statistics.

Nonparametric Methods

USA SNAPSHOTS®

A look at statistics that shape the nation

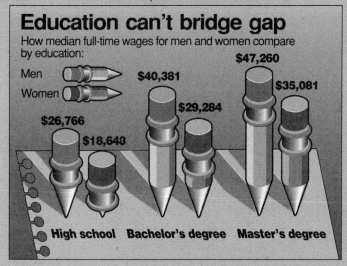

Education can't bridge gap

How median full-time wages for men and women compare by education:

Men
Women

$47,260
$40,381
$35,081
$29,284
$26,766
$18,643

High school   **Bachelor's degree**   **Master's degree**

▲ Source: Graphic "Education Can't Bridge Gap," by Marty Baumann, *USA Today.* Data from Census Bureau (1992 median incomes of people 25 and older). Reprinted by permission of *USA Today,* November 5–7, 1993, Section A, p. 1.

INTRODUCTION

Except in Chapter 10, where we tested independence of two attributes, the statistical test procedures with which we have been concerned have involved tests regarding population parameters that required making assumptions about some aspects of the distribution of the sampled population, for example, that it was normally distributed. In short, we have carried out tests of hypotheses about parameters of known population types. Such tests are commonly known as classical, standard, or **parametric tests.**

We can, of course, envision situations where the assumptions required in a classical test are not justified. For example, in comparing means of two populations using the *t* test we assumed that the populations were normally distributed and had equal variances. What if these assumptions are not justified? Statistical tests that are not based on assumptions about the distribution or the parameters of the sample population have been devised and they are referred to as **nonparametric** or **distribution-free tests.** Thus, the tests developed in Chapter 10 could have been included here.

Nonparametric techniques are appealing because they are intuitively easy to understand, and, of course, there is no need to make assumptions about the distribution of the population under investigation. Also, many nonparametric tests can be applied to situations where classical tests cannot be applied because of the nature of the response variable. This is especially true when the data cannot be measured quantitatively except by assigning ranks, as is the case in judging a contest.

The picture is not completely rosy. There are also some disadvantages in using nonparametric methods. These methods very often ignore the actual numerical value of an observation and take into account only its relative standing (ranking) in relation to other observations in the sample. As a result, acquired information is wasted. Furthermore, if indeed the assumptions under which a classical test can be applied hold true, then a nonparametric test is less efficient. This is understandable, because a nonparametric test assumes very little about the population distributions, and thus cannot be expected to be as proficient as tests that make use of known properties of the population. For a given level of significance α, a nonparametric test tends to have a higher probability of accepting H_0 *erroneously* as compared with the classical test (when the assumptions of the latter test are met).

This chapter presents some of the commonly applied nonparametric methods. These are (1) the *two-sample sign test*; (2) the rank-sum test—also called the Mann-Whitney U test; (3) the runs test, a test of randomness; and (4) Spearman's rank-correlation method.

13-1   THE TWO-SAMPLE SIGN TEST FOR MATCHED PAIRS

The **two-sample sign test** is one of the simplest nonparametric tests. It was originally developed to test for the median of a population with continuous distribution. However, it can also be used in treating data consisting of paired observations when the two samples are not independent and the differences of the matched pairs are not normally distributed. Recall that if the differences are normally distributed, or approximately so, then we would test for the equality of means using the paired *t*-test developed in Chapter 9. In this section we discuss the sign test in the context of comparison of two populations using paired data. The test is appropriately named *sign test*, because rather than using numerical values from the two samples as data, it uses *plus* signs ($+$) and *minus* signs ($-$) as the information on which to base the test. Consider the following situations.

a. To test whether a particular diet is effective in reducing weight, the weight of each individual involved is obtained before and after the diet. Certainly we feel that the weights of each individual before and after the diet are related. But, at the same time, we also feel that weight changes for different individuals are independent.

b. Ten judges grade two vintages of wine, grading each vintage on a scale of 1 to 5 points. The response of each judge as regards the two vintages is certainly not independent. However, we do feel justified in assuming that the responses are independent from judge to judge.

We replace each pair of observations by a $+$ sign if the first component of the pair is greater than the second, and by a $-$ sign if the first component is smaller than the second. Under the sign test, no ties are allowed. Those pairs for which ties do occur are omitted from the sample. The effective sample size n is then given as the original number of pairs reduced by the number of ties. To see the general approach, consider the following illustration.

Suppose an airline company wants to compare two models of planes, one manufactured by an English company and another by a French company. It is decided to fly fifteen planes of each model. Fifteen pilots of proven ability are each assigned to fly an English and a French model and are asked to rate them on a scale from 1 to 10. The results of the tests are given in Table 13-1.

The null hypothesis we wish to test is that the two types of planes are identical, that is, the population distributions of the two models are the same. There are various alternative hypotheses against which we could test the null hypothesis: namely, (1) the two models are not identical, (2) the French model is superior, or (3) the English model is superior.

TABLE 13-1   Scores Given to English and French Models

Pilot	English Model	French Model	Sign of Difference
1	7	4	+
2	3	5	−
3	5	7	−
4	6	3	+
5	8	6	+
6	4	4	← Tie
7	6	8	−
8	7	2	+
9	5	6	−
10	4	7	−
11	7	9	−
12	8	7	+
13	4	6	−
14	3	5	−
15	7	9	−

Suppose we wish to test

H_0: The two plane models are identical

against

H_A: The French model is superior

at the 5 percent level of significance. As a first step we obtain Column 4 in Table 13-1 giving the sign of the difference for each pair, the score of a pilot for an English plane minus his or her score for a French plane. There is one tie. Therefore, we take the effective sample size n as $15 - 1$, or 14. There are 5 plus signs and 9 minus signs.

If there are too many plus signs our inclination will be to judge in favor of the English model. By the same token, too few plus signs (that is, a preponderance of minus signs) will lead us to lean in favor of the French model. Thus, an appropriate test statistic for carrying out the test is

$$X = \textbf{number of plus signs.}$$

If the two models are not different, then a pilot is just as likely to decide in favor of an English model as in favor of the French model. That is, the probability of getting a $+$ sign for a pair is the same as getting a $-$ sign. Let p be the probability of getting a $+$ sign, that is, p is the probability of a pilot saying that the English model is superior. The null hypothesis that the two models are identical in performance is then stated equivalently as

$$H_0: p = \frac{1}{2}.$$

If the French model is superior, then the probability of deciding in favor of the English model is less than that of deciding against it. In other words, the probability of getting a $+$ sign for a pair is less than $\frac{1}{2}$. Thus, the alternative hypothesis, the French model is superior, can be reformulated as

$$H_A: p < \frac{1}{2}.$$

In view of the alternative hypothesis, we would prefer to reject H_0 for small values of X. Thus, the critical region consists of the left-tail of the distribution of X. Now, assuming that the responses of the pilots are independent, the distribution of X is binomial with n trials (the effective sample size after excluding the ties) and probability of success p. Under the null hypothesis, X has a binomial distribution with probability of success $\frac{1}{2}$. This distribution can be obtained from Appendix III, Table A-1 for appropriate n (14 in our example) and $p = \frac{1}{2}$. The distribution is reproduced here.

Successes x	$P(X = x)$		Successes x	$P(X = x)$
0	0.000		8	0.183
1	0.001		9	0.122
2	0.006		10	0.061
3	0.022		11	0.022
4	0.061		12	0.006
5	0.122		13	0.001
6	0.183		14	0.000
7	0.209			

$$\left(\begin{array}{c}\text{Total} \\ \text{probability}\end{array}\right) = 0.029 \qquad \left(\begin{array}{c}\text{Total} \\ \text{probability}\end{array}\right) = 0.09$$

From the distribution, $P(X \le 4) = 0.09$. This probability is greater than 0.05 and so 4 is too large to be the critical value. On the other hand, $P(X \le 3) = 0.029$, which is in between the commonly used significance levels of 0.05 and 0.01. Thus, we agree to take the value 3 as the critical value and arrive at the following decision rule.

Reject H_0 if the observed number of $+$ signs is less than or equal to 3.

In our data, there are 5 plus signs. Because this number is greater than 3 we do not reject H_0. There is no evidence to show that the French model is superior.

An alternative method: Because we are not in a position to provide a critical value corresponding to the exact level of significance α of 0.05, a better way to carry out the test is by providing the P-value. From our data, the observed number of plus signs x is 5, and

$$P(X \le 5) = 0.001 + 0.006 + 0.022 + 0.061 + 0.122$$
$$= 0.212.$$

The P-value is 0.212, which is higher than $\alpha = 0.05$. Because the P-value is greater than the level of significance, we cannot reject H_0 at the 5 percent level.

EXAMPLE

In Table 13-2 are given the volumes of monthly sales (in thousands of dollars) of seventeen salespersons before and after a three-month course in marketing. Is the course beneficial in promoting sales?

TABLE 13-2   Sales, before and after a Course in Marketing (in Thousands of Dollars)

Salesperson	Sales after the Course	Sales before the Course	Sign of Difference
1	6	3	+
2	9	8	+
3	12	7	+
4	8	10	−
5	8	6	+
6	7	7	← Tie
7	6	12	−
8	8	7	+
9	12	6	+
10	10	8	+
11	6	9	−
12	11	7	+
13	10	10	← Tie
14	10	7	+
15	9	4	+
16	8	5	+
17	11	7	+

SOLUTION   The null hypothesis H_0 and the alternative hypothesis H_A can be stated as follows.

H_0: The course in marketing does not affect sales.

H_A: The course improves sales.

The signs assigned to the differences in sales (sales after minus sales before) are given in Column 4 of Table 13-2. There are two ties. Thus, the effective sample size is $17 - 2$, or 15. There are 12 plus and 3 minus signs.

If p represents the probability of increased sales for a salesperson after the training, then we are really interested in testing the null hypothesis

$$H_0: p = \frac{1}{2}$$

against the alternative hypothesis

$$H_A: p > \frac{1}{2}.$$

We will opt in favor of the training program and reject the null hypothesis if there is a preponderance of $+$ signs. In other words, the critical region consists of values in the right tail of the distribution of X, the number of $+$ signs. Assuming independence, the distribution of X is binomial with 15 trials and, under the null hypothesis, $p = \frac{1}{2}$. We reproduce below its distribution from Table A-1, Appendix III, corresponding to $n = 15$ (effective sample size) and $p = \frac{1}{2}$.

Successes	$P(X = x)$	Successes	$P(X = x)$
0	0.000	8	0.196
1	0.000	9	0.153
2	0.003	10	0.092
3	0.014	11	0.042
4	0.042	12	0.014
5	0.092	13	0.003
6	0.153	14	0.000
7	0.196	15	0.000

Total probability = 0.017 (corresponding to Successes 12–15)

To test the hypothesis we first compute the P-value. Since the critical region consists of large values of X, and the observed number of plus signs is 12, the P-value is equal to

$$P(X \geq 12) = 0.014 + 0.003 = 0.017.$$

Because the P-value is less than 0.05, we reject the null hypothesis at the 5 percent level of significance. Thus, at the 5 percent level of significance we conclude that the course is beneficial in promoting sales. ●

If n is large, even as low as 15, then a normal distribution with mean $n/2$ (or np with $p = \frac{1}{2}$) and variance $n/4$ (or $np(1 - p)$ with $p = \frac{1}{2}$) provides a reasonably good approximation for the distribution of X under H_0. Consequently, the test can be carried out using the statistic

$$\frac{X - \dfrac{n}{2}}{\sqrt{\dfrac{n}{4}}} = \frac{2X - n}{\sqrt{n}}.$$

It is appropriate to use the continuity correction term in this context. The critical region is determined as usual on the basis of the level of significance and the nature of the alternative hypothesis.

We will now apply this approach in the preceding Example, where $n = 15$, which is large enough for a normal approximation. Since $n = 15$ and $x = 12$, and we want $P(X \geq 12)$, the computed value of the test statistic is

$$\frac{2(x - 0.5) - n}{\sqrt{n}} = \frac{2(11.5) - 15}{\sqrt{15}} = 2.07.$$

Because the computed value is greater than 1.645 ($z_{0.05}$), we reject the null hypothesis at the 5 percent level of significance. We arrive at the same conclusion as before.

In summary, use the binomial tables for n less than 15 and use the normal approximation for n greater than or equal to 15.

Section 13-1 Exercises

1. In a controlled experiment, a dietician tested a diet on fifteen men. The following figures give their weights before and after the diet.

Weight before	138	158	160	172	160	210	185	245
Weight after	143	149	159	180	160	192	170	205
Weight before	182	172	132	250	180	129	176	
Weight after	180	163	137	247	176	133	167	

Use the sign test to test the null hypothesis that the diet is not effective in reducing a man's weight against the alternative hypothesis that it is effective in reducing weight. Use the 5 percent level of significance.

2. A student can take a certain course in statistics either as a lecture course or as a personalized system of instruction course (PSI), where the student sets his or her own pace. Fifteen instructors who have taught both the lecture course and the PSI course were asked to give their evaluation of the effectiveness of the two methods by assigning scores from 0 to 5. The following figures give the instructors' ratings.

Instructor	PSI	Lecture		Instructor	PSI	Lecture
A	4	1		I	4	0
B	2	5		J	3	3
C	4	3		K	1	0
D	4	2		L	4	2
E	3	4		M	5	3
F	3	1		N	5	0
G	0	4		O	4	1
H	2	4				

Using the sign test, at the 5 percent level of significance, test the null hypothesis that there is no preference for one type of course over the other.

3. Under a special assistance program, fifty students were provided special assistance in their studies. At the end of the program, thirty-six showed a marked

improvement, four did not show any change, and ten showed a deterioration. Using the sign test, at the 2 percent level of significance, test the null hypothesis that the program is not beneficial.

4. Two presidential candidates, Candidate *A* and Candidate *B*, were rated by eighteen randomly picked voters on a scale of 0 to 5, as follows.

Candidate A	0	1	5	3	5	4	1	4	3	4	1	4	0	5	4	3	4	5
Candidate B	4	4	3	0	0	2	3	2	2	0	2	1	0	2	2	1	2	1

Use the sign test to test the null hypothesis that there is no significant preference for one candidate over the other.

5. To study the effect of temperature on skiing, sixteen skiers were clocked on a downhill course on two different days, once when the temperature was subzero and once when it was considerably warmer. The figures are given in the following table.

Skier	Time (Subzero) (in minutes)	Time (Warm) (in minutes)
A	8.20	8.35
B	8.01	8.10
C	9.98	10.01
D	8.85	8.90
E	7.88	7.91
F	8.26	8.20
G	9.30	9.10
H	8.50	8.38
I	9.66	9.69
J	8.76	8.70
K	7.98	8.78
L	8.98	9.21
M	9.22	9.86
N	8.62	8.60
O	9.00	9.00
P	8.44	8.48

At the 5 percent level of significance, test the null hypothesis that temperature has no effect on skiing skill against the alternative hypothesis that warm weather affects it adversely.

6. The following figures give the systolic blood pressure for fourteen subjects before and after yogic meditation. At the 5 percent level of significance, does meditation help lower systolic blood pressure?

Subject	Before	After
A	138	139
B	124	124
C	175	172
D	158	155
E	153	147
F	160	157

(continued)

Subject	Before	After
G	148	140
H	150	144
I	160	152
J	145	147
K	156	158
L	150	148
M	148	146
N	166	155

7. The following data represent two-hour blood glucose levels (in milligrams per milliliter) of seventeen patients before and after a treatment.

Patient Number	BLOOD GLUCOSE LEVEL		Patient Number	BLOOD GLUCOSE LEVEL	
	Before	After		Before	After
1	174	168	9	194	180
2	157	159	10	144	135
3	135	130	11	108	110
4	102	105	12	224	198
5	144	140	13	192	180
6	132	134	14	187	189
7	131	122	15	137	139
8	112	112	16	181	173
			17	159	148

At the 5 percent level of significance, is it true that the treatment results in reducing blood glucose level?

8. The following figures give uric acid levels (in milligrams per 100 milliliters) of twelve subjects before and after a special diet for one week. Do the data indicate that the diet reduces the uric acid level of a subject?

Subject	URIC ACID	
	Before	After
A	5.2	5.2
B	6.3	6.2
C	6.4	6.3
D	5.5	5.6
E	5.8	5.4
F	5.7	5.6
G	5.9	5.5
H	6.0	6.1
I	6.1	6.1
J	5.7	5.8
K	5.9	5.6
L	6.3	6.4

9. The productivity of twelve workers was observed during one week when they were under strict supervision and during another week when they were not. The following scores provide a measure of their productivity. Do the data indicate that strict supervision improves productivity? Use $\alpha = 0.05$.

Worker	With Supervision	With No Supervision
A	81	72
B	59	52
C	78	83
D	85	88
E	60	58
F	50	30
G	77	82
H	79	83
I	53	40
J	68	66
K	89	76
L	67	63

10. Twenty pairs of identical twins were considered in an experiment to test the effectiveness of a special training program. One member of each pair was given special training, with the other member of the pair acting as a control. At the end of the program it was found that fourteen of the twins given special training performed better than their controls, and six fared worse. Does the special training result in improved performance? Use $\alpha = 0.025$.

13-2   THE RANK-SUM TEST

The *rank-sum test* developed in this section is a procedure for comparing two populations when independent samples are drawn from them. Recall that for comparing the means of two populations by using the *t*-test, we assumed that the populations were normally distributed, both of them having the same variance. In many cases, it may happen that the two populations under investigation have distributions which, even at best, are not approximately normal, thereby invalidating the use of the standard *t*-test. Two nonparametric tests based on "rank-sums" were proposed in the 1940s for comparing two populations when there are indications that conditions necessary for applying the classical *t*-test are not met. One test, proposed by F. Wilcoxon in 1945, is called the **Wilcoxon rank-sum test,** and the other, proposed by Mann and Whitney in 1947, is commonly referred to as the **Mann-Whitney U test.** Since these two tests are equivalent and lead to identical conclusions, we explain only the Mann-Whitney U test in what follows.

TABLE 13-3 Contributions to Charities (in Dollars)

Community A	Community B
120	42
155	190
88	320
420	680
360	82
60	130
650	500
82	32
180	160
98	82
	890
	1020

To show how this test is applied, consider the data in Table 13-3, which refer to the charitable contributions (in dollars) made during a year by two samples of households picked at random from two communities in the same economic bracket.

We are interested in testing whether the two communities are similar with regard to helping charitable causes. We might consider comparing the means by using the t-test, as in Chapter 9. But a quick glance will point out that the data in Community B show considerably more variability than the data in Community A, thus making the assumption of equality of variances in the populations highly questionable. Besides, it seems that even the assumption of normality in the populations is open to question.

Under the rank-sum test, the data in the two samples are combined and then the observations are arranged in order of their size as though they represent one sample. That is, they are arranged by rank. The smallest observation is assigned the rank 1, the next highest the rank 2, and so on, until all observations are ranked. This ranking is shown in Table 13-4. Whenever

TABLE 13-4 The Contributors in the Two Communities Combined and Ranked according to Their Contributions

Contributions (in Dollars)	Community	Rank	
32	B	1	
42	B	2	
60	A	3	
82	B	5	⎫ Tied observations
82	B	5	⎬ all assigned the
82	A	5	⎭ rank 5
88	A	7	
98	A	8	
120	A	9	
130	B	10	
155	A	11	
160	B	12	
180	A	13	
190	B	14	
320	B	15	
360	A	16	
420	A	17	
500	B	18	
650	A	19	
680	B	20	
890	B	21	
1020	B	22	

there are tied observations, we take the mean rank for these observations and assign this value to each of the tied observations. For example, in our data the fourth, fifth, and sixth observations are all 82. Therefore, we assign the rank $\frac{4 + 5 + 6}{3} = 5$ to the three observations. As a result of the ranking, the households in Community A occupy ranks 3, 5, 7, 8, 9, 11, 13, 16, 17, and 19, while those in Community B occupy ranks 1, 2, 5, 5, 10, 12, 14, 15, 18, 20, 21, and 22.

The **rank sum** for each community is now found. The sum of the ranks occupied by the households in Community A is

$$3 + 5 + 7 + 8 + 9 + 11 + 13 + 16 + 17 + 19 = 108,$$

and that in Community B is

$$1 + 2 + 5 + 5 + 10 + 12 + 14 + 15 + 18 \\ + 20 + 21 + 22 = 145.$$

Suppose m represents the size of one of the samples and R_1 the sum of the ranks assigned to its members. Also, let n be the size of the other sample and R_2 the sum of the ranks assigned to its members. It can be established mathematically that the sum of the ranks R_1 and R_2 is always equal to $(m + n)(m + n + 1)/2$. Thus, we always have

$$R_1 + R_2 = \frac{(m + n)(m + n + 1)}{2}.$$

Notice that, in our illustration, assuming that R_1 is the sum of the ranks assigned to the members of Community A, $m = 10, n = 12, R_1 = 108$, and $R_2 = 145$. The sum of the ranks is

$$R_1 + R_2 = 108 + 145 = 253$$

and, indeed, $(10 + 12)(10 + 12 + 1)/2$ is also equal to 253.

The null hypothesis under test is

H_0: The two communities have identical distributions of contributions to charities.

against the alternative hypothesis

H_A: The distributions are different in that the distribution for Community A is shifted to the left or right of Community B.

In view of the fact that, for given m and n, $R_1 + R_2$ is always fixed, a small value of R_1 will result in large value of R_2, and conversely. Thus, in the case when both samples are of the same size, the smaller the value of R_1, the more inclined we will be to say that the underlying populations are different. In fact, we would strongly feel that the distribution for Community B is shifted to the right of the distribution for Community A. On the other hand, the larger the value of R_1, the stronger the evidence that the distribution for Community A is shifted to the right. The test is based on the statistic

$$U = mn + \frac{m(m + 1)}{2} - R_1.$$

If the null hypothesis that the two populations are identical is valid, it can be shown that the distribution of U has mean μ_U and standard deviation σ_U.

MEAN AND STANDARD DEVIATION OF U UNDER H_o

The mean and the standard deviation of the statistic U where

$$U = mn + \frac{m(m + 1)}{2} - R_1$$

are given by

$$\mu_U = \frac{mn}{2}$$

and

$$\sigma_U = \sqrt{\frac{mn(m + n + 1)}{12}}$$

In our case, since $m = 10$, $n = 12$, and $R_1 = 108$, we get

$$u = (10)(12) + \frac{10(10 + 1)}{2} - 108 = 67$$

$$\mu_U = \frac{(10)(12)}{2} = 60$$

and

$$\sigma_U = \sqrt{\frac{(10)(12)(10 + 12 + 1)}{12}} = 15.166.$$

If, additionally, both m and n are at least 10, the distribution of U is approximately normal. Changing to the z scale, we can then use the test statistic

$$\frac{U - \dfrac{mn}{2}}{\sqrt{\dfrac{mn(m + n + 1)}{12}}}.$$

Suppose we agree to carry out the test at the 5 percent level. In our example, the alternative hypothesis is two-sided. Therefore, using a two-tail test, we will reject the null hypothesis if the computed value is less than $-z_{\alpha/2}$ or greater than $z_{\alpha/2}$. Because $u = 67$, $\mu_U = 60$, and $\sigma_U = 15.166$, we get

$$\frac{u - \dfrac{mn}{2}}{\sqrt{\dfrac{mn(m + n + 1)}{12}}} = \frac{67 - 60}{15.166} = 0.46.$$

We do not reject the null hypothesis at the 5 percent level of significance. In other words, there is no evidence to indicate that the communities are dissimilar with regard to their contributions to charity.

EXAMPLE

A class of twenty students was divided into two groups, Group I of ten students and Group II of ten students. Group I was given special training using audiovisual facilities, while in Group II no such facilities were used. A commnon test was given to the groups and their scores are as recorded in Table 13-5. Would it be reasonable to say that the two instructional methods produce different results? (Use $\alpha = 0.05$.)

SOLUTION The null hypothesis is

H_0: The two methods produce identical results.

We will test this against the alternative hypothesis that the two methods do not produce the same results at the 5 percent level of significance. The combined data of the two samples are arranged according to magnitude in Table 13-6. R_1, the sum of the ranks occupied by Group I, is

$$3 + 6 + 7 + 13 + 14 + 15 + 17 + 18 + 19 + 20$$
$$= 132.$$

Now $m = 10$ and $n = 10$. Therefore,

$$u = (10)(10) + \frac{(10)(11)}{2} - 132 = 23.$$

$$\mu_U = \frac{(10)(10)}{2} = 50$$

TABLE 13-5 Scores of Two Groups of Students

Group I	Group II
95	49
66	39
42	80
82	65
71	30
68	62
52	53
86	49
50	55
88	60

TABLE 13-6 Students of the Two Groups Combined and Ranked according to Their Scores

Score	30	39	42	49	49	50	52	53	55	60	62	65	66	68	71	80	82	86	88	95
Group	II	II	I	II	II	I	I	II	II	II	II	II	I	I	I	II	I	I	I	I
Rank	1	2	3	4.5	4.5	6	7	8	9	10	11	12	13	14	15	16	17	18	19	20

Tied
Observations

and

$$\sigma_U = \sqrt{\frac{(10)(10)(10 + 10 + 1)}{12}} = 13.229.$$

Therefore,

$$\frac{u - \mu_U}{\sigma_U} = \frac{23 - 50}{13.229} = -2.04.$$

At the 5 percent level of significance, we reject the null hypothesis, because the computed value is less than -1.96. We conclude that the two instructional methods produce different results and, as a matter of fact, that students will benefit from the use of audiovisual facilities.   ●

Section 13-2 Exercises

1. Eleven voters were picked at random from those who voted in favor of a certain proposition, and thirteen from those who voted against it. The following figures give their ages.

In favor	28	33	27	31	29	25	58	30	25	41	68	
Against	31	43	45	37	40	41	48	35	33	39	42	36	75

   Using the rank-sum test, test the null hypothesis that the age distributions of those voting in favor and those voting against the proposition were identical. Use a 5 percent level of significance.

2. The following figures relate to the number of hours needed by two groups of workers to learn some skills.

Group I	7.2	6.6	7.7	6.9	7.4	8.0	6.5	8.1	7.5	7.1	6.9	7.4
Group II	7.4	7.9	6.2	8.2	7.8	8.6	7.6	7.1	7.0	6.7	8.3	8.8

   Use the rank-sum test to test the null hypothesis that the two samples come from identical populations. Use a 0.02 level of significance.

3. Twelve students who took a course under a lecture session program and fifteen students who took it under a self-pace program were given a common final test. The following figures give the scores of the students.

Lecture	46	89	85	61	87	35	57	87	76	79	49	92			
Self-pace	92	41	36	72	34	67	99	72	83	82	76	40	84	67	24

   Use the rank-sum test to test the null hypothesis that the two methods of instruction do not produce significantly different results.

4. In the first grade in a school, the teacher divided the class into two groups: Group A, consisting of ten students, and Group B, consisting of fourteen students. Students in each group were asked to memorize, in a 1-hour period, spellings of forty words from a given list. Students in Group A, however, were secretly promised an incentive that any student in their group would get as many pieces of candy as the number of words spelled correctly. No similar promise was made to students in Group B. The following figures refer to the number of words spelled correctly.

Group A	27	22	31	16	39	36	25	24	32	19				
Group B	28	11	32	35	6	24	14	16	39	19	26	20	10	40

Using the rank-sum test, at the 0.05 level of significance, test the null hypothesis that the incentive is ineffective.

5. Horses of two breeds, Breed A and Breed B, were entered in a race. They finished in the following order.

Standings	1	2	3	4	5	6	7	8	9	10
Breed	A	B	B	A	A	A	A	B	A	A

Standings	11	12	13	14	15	16	17	18	19	20	21	22
Breed	A	B	A	B	B	A	B	B	B	B	B	A

Using the rank-sum test, test the null hypothesis that there is no difference between the two breeds of racing horses. Use the 0.05 level of significance.

6. The following data were aimed at comparing two marine stations, one on the coastal waters and the other at the beginning of the open sea off the Israeli coast. At each station, chlorophyll a was determined by filtration of a 100-milliliter water sample on a 0.22 μm membrane filter. The pigments were extracted with 90 percent acetone, and their concentrations were evaluated using a fluorometer (A. Azov. "Seasonal patterns of phytoplankton productivity . . . " *Journal of Plankton Research,* Vol. 8, 1–3, Jan–Jun 1986).

CHLOROPHYLL a (mg/m³)	
Coastal Station	**Open Sea Station**
0.311	0.213
0.942	0.522
0.297	0.057
0.252	0.078
0.284	0.060
0.133	0.087
0.226	0.029
0.197	0.125
0.263	0.085
0.215	0.120
0.458	0.083
0.293	0.203
0.206	0.091
0.494	0.094
0.352	0.070
0.503	0.198
0.365	0.265
0.165	0.112
0.137	0.128
0.128	0.186
0.261	0.073
0.915	

At the 5 percent level, do the two stations differ significantly with regard to chlorophyll a content? Apply the rank-sum test.

13-3   THE RUNS TEST

To get a good cross section of the population, we require that the sample drawn be a random sample. We have relied heavily on this premise and have often initiated our discussion by saying, "Suppose we have a random sample . . . ." Intuitively, the concept of drawing a random sample seems rather easy, and as a result, we have not given much consideration to it. We will now address the important issue of testing the randomness of a sample.

In recent years, several tests for randomness have been developed. The one presented here is based on the order in which the observations are obtained and is called a **runs test.** We begin with the definition of a run.

If there is a sequence of symbols of two kinds, then a **run** is defined as a maximum-length unbroken subsequence of identical symbols.

For example, consider the following sequence of letters d and r, where we have drawn slashes to separate each block of consecutive identical letters.

$$d\ d\ d\ d/r\ r\ r/d/r\ r/d\ d\ d/r\ r$$

There are 6 runs, with 3 runs of letter r and 3 of letter d.

If in a sample drawn from a population containing letters d and r we obtained any of the following sequences

$$d\ d\ d\ .\ .\ .\ d, \text{ or}$$
$$r\ r\ r\ .\ .\ .\ r, \text{ or}$$
$$d\ d\ .\ .\ .\ d\ r\ r\ .\ .\ .\ r, \text{ or}$$
$$d\ r\ d\ r\ .\ .\ .\ d\ r\ d\ r$$

we would immediately question their randomness. Thus, too few runs or too many of them would clearly be indicative of a lack of randomness. We now provide a test when the case is not as clear-cut as above.

As a specific example, suppose twenty-five people are interviewed as to whether they are Democrats (recorded d) or Republicans (recorded r). Their responses (in the order in which they were interviewed) are as follows.

$$d\ d\ d/r\ r/d/r\ r/d\ d\ d/r\ r\ r\ r/d\ d\ d/r/d\ d\ d/r/d\ d$$

Counting the different blocks, we see that there are 11 runs. Also, there are 15 letters d and 10 letters r. We want to test the hypothesis

H_0: The sequence is random

against the alternative hypothesis

H_A: The sequence is not random

using, for example, $\alpha = 0.05$. When twenty-five people are interviewed and their responses are recorded sequentially as they are interviewed, the minimum number of runs possible is 1, which happens when all the individuals are of the same party. The maximum number of runs possible is 25, when the responses alternate as Republicans and Democrats (or, as Democrats and Republicans). These situations clearly indicate a pattern and exhibit no randomness. Actually, any arrangement having number of runs close to 1 or close to 25 will make the hypothesis of randomness highly suspect in our eyes. Since too few or too many runs indicate lack of randomness, *the runs test for randomness is always a two-tail test*.

The test of hypothesis regarding randomness is based on the distribution of runs. Suppose R denotes the total number of runs, m the number of symbols of one kind, and n the number of the other kind. It can be shown that the mean μ_R and the standard deviation σ_R of R are as follows.

MEAN AND STANDARD DEVIATION OF R

$$\mu_R = \frac{2mn}{m + n} + 1$$

$$\sigma_R = \sqrt{\frac{2mn(2mn - m - n)}{(m + n)^2(m + n - 1)}}$$

In our illustration, $m = 15$ and $n = 10$. (Because the formulas for μ_R and σ_R are symmetric in m and n, it would not matter if we took $m = 10$ and $n = 15$.) Thus,

$$\mu_R = \frac{2(15)(10)}{15 + 10} + 1 = 13$$

and

$$\sigma_R = \sqrt{\frac{2(15)(10)[2(15)(10) - 15 - 10]}{(15 + 10)^2(15 + 10 - 1)}}$$
$$= \sqrt{5.5}$$
$$= 2.345.$$

If m and n are large (both at least 10), the distribution of R can be approximated by a normal distribution with mean μ_R and standard deviation σ_R. The decision criterion is then provided by computing the value of

$$\frac{R - \mu_R}{\sigma_R}.$$

where the distribution of the statistic is, approximately, the standard normal distribution.

In our example, $R = 11$, $\mu_R = 13$, and $\sigma_R = 2.345$. Therefore, the computed value of $(R - \mu_R)/\sigma_R$ is

$$\frac{11 - 13}{2.345} = -0.85.$$

The alternative hypothesis is that there is a lack of randomness. Thus, with a two-tailed test, we reject the null hypothesis if the computed value is less than $-z_{\alpha/2}$ or greater than $z_{\alpha/2}$. Taking $\alpha = 0.05$, because $z_{0.025} = 1.96$, we see that there is no evidence showing a lack of randomness.

EXAMPLE 1

Maya believes that her statistics instructor's lectures have a pattern to their coherence. She records whether the lecture is understandable on twenty-five consecutive class days, recording g if she understands the lecture and b if she does not. The following sequence gives the results.

$$g\ g\ g/b\ b/g/b/g\ g\ g/b/g/b\ b/g\ g/b/g\ g\ g/b\ b\ b/g/b$$

Using $\alpha = 0.05$, test for randomness.

SOLUTION We have the following.

H_0: There is randomness.

H_A: There is lack of randomness.

Separating runs by slashes, we see that there are 14 runs. There are 14 letters g and 11 letters b. Thus $m = 14$, $n = 11$, and $R = 14$. As a result,

$$\mu_R = \frac{2(14)(11)}{14 + 11} + 1$$

$$= 13.32$$

and

$$\sigma_R = \sqrt{\frac{2(14)(11)[2(14)(11) - 14 - 11]}{(14 + 11)^2(14 + 11 - 1)}}$$

$$= \sqrt{5.811}$$

$$= 2.41.$$

Therefore, the computed value of $(R - \mu_R)/\sigma_R$ is

$$\frac{14 - 13.32}{2.41} = 0.282.$$

There is no evidence to show that the coherence of the lectures has any pattern, as the computed value is between -1.96 and 1.96. ●

The runs test can also be applied to check randomness of a sequence of numerical data. This can be accomplished by converting the original data into a sequence of symbols by, for example, writing the letter *a* if a value is above the median and the letter *b* if it is below the median. If a value is equal to the median, it is deleted from consideration. This done, we proceed as before to check the randomness of the sequence of letters *a* and *b*.

EXAMPLE 2   Suppose the annual rainfall (in inches) over 24 consecutive years is as follows.

38	29	56	41	43	37	57	46
62	38	54	59	60	36	62	43
57	40	46	35	62	46	52	55

Test for randomness at the 5 percent level of significance.

SOLUTION   We state H_0 and H_A as

H_0: There is randomness in the sequence.

H_A: There is no randomness.

Arranging the data in order of magnitude as

29	35	36	37	38	38	40	41
43	43	46	46	46	52	54	55
56	57	57	59	60	62	62	62

it can be seen that the median is 46. In the original data (taken in the order), if the value is above the median, we replace it by the letter *a*, and if it is below the median, we replace it by the letter *b*. The three observations equal to 46 are deleted, since 46 is the median. The following sequence of letters is obtained as a result.

b b/a/b b b/a a/b/a a a/b/a/b/a/b b/a a a

There are 10 *b*s, 11 *a*s, and 12 runs; that is, $m = 10$, $n = 11$, and $R = 12$. Thus,

$$\mu_R = \frac{2(10)(11)}{10 + 11} + 1$$

$$= 11.48$$

$$\sigma_R = \sqrt{\frac{2(10)(11)[2(10)(11) - 10 - 11]}{(10 + 11)^2(10 + 11 - 1)}}$$

$$= \sqrt{4.96}$$

$$= 2.23$$

and the computed value of $(R - \mu_R)/\sigma_R$ is

$$\frac{12 - 11.48}{2.23} = 0.23.$$

Because the computed value is so close to zero, there is little evidence that would cause us to question randomness.

Section 13-3 Exercises

1. A true-false test consisting of forty questions had the following sequence of answers (recorded T for true and F for false).

F	T	T	F	T	F	T	F	F	T
T	T	F	T	F	F	T	F	T	F
T	T	F	T	F	T	F	F	T	F
T	F	T	T	T	F	F	T	F	F

Test the hypothesis that the questions are arranged such that the sequence of answers $(T–F)$ is random. Use $\alpha = 0.05$.

2. A customer who regularly visits a particular restaurant orders either a bowl of soup or a tossed salad. The waitress recorded the orders (B for bowl of soup and T for tossed salad) over 30 consecutive days.

T	T	B	T	T	T	T	B	B	T	T	T	B	T	B
B	T	T	T	B	T	T	B	T	B	T	T	T	B	B

At the 5 percent level of significance, test the null hypothesis that there is randomness.

3. At the ticket counter in a movie house, the following standing order in a line was observed (recorded M for male, F for female).

M	F	M	M	F	M	M	M	F	F	M	F	F
M	M	F	F	F	M	F	M	F	M	M	M	

At the 5 percent level, test whether the members of the two sexes were standing in random order.

4. In a quality-control inspection, forty successive tubes produced by a machine are inspected as to whether they are defective (d) or nondefective (g). The following data were obtained.

g	g	g	g	g	d	d	g	g	g
g	g	g	g	g	d	g	d	g	g
g	g	d	g	g	d	d	g	g	g
d	g	g	g	g	g	g	d	d	g

Test the hypothesis that there was no particular trend in which the defective tubes occurred. Use $\alpha = 0.025$.

5. A copper wire coated with enamel was inspected, and the following figures give the number of defects on thirty consecutive pieces, each of length 1000 yards.

9	1	11	8	9	3	12	9	14	8
10	12	15	7	15	10	13	6	10	8
12	9	15	6	14	16	12	4	10	13

At the 5 percent level of significance, test for randomness of the number of defects.

6. A volunteer helping in a hospital spent the following number of hours on 30 consecutive days.

4	6	5	4	6	2	1	5	4	6	3	4	3	2	7
4	6	4	5	1	3	5	4	3	6	8	3	2	6	5

Test for randomness of the number of hours spent, at the 5 percent level of significance.

7. Recording D if there was a drought during a year and N if the precipitation was normal, the following arrangement indicates the nature of precipitation during 50 consecutive years from 1936 to 1985.

$$
\begin{array}{cccccccccc}
N & N & D & N & D & N & N & D & N & D \\
N & N & N & D & N & N & D & N & D & N \\
D & N & N & N & D & N & N & D & D & N \\
D & N & N & D & N & N & D & N & D & N \\
D & N & N & D & N & D & N & N & N & D
\end{array}
$$

At the 5 percent level of significance, test the null hypothesis that the occurrence of drought years and normal years is random.

8. The following sentence, *"Then, his voice thick with emotion, he echoed one of his most familiar campaign themes,"* was obtained from a composition. At the 5 percent level of significance, test whether consonants and vowels occur at random.

9. Pick a column from the random numbers table (Table A-7, Appendix III), and select from it fifty successive digits one below the other. Test for randomness of the digits using $\alpha = 0.05$.

10. The following figures represent the number of migratory birds observed in a certain marshland over 25 consecutive years.

248	510	410	380	310
540	395	480	360	330
501	489	380	398	440
370	488	362	620	405
378	260	440	524	320

At the 5 percent level of significance, test for randomness of the number of birds.

13-4   RANK CORRELATION

TABLE 13-7 Rankings in a Beauty Contest by Two Judges

Judge 1 x	Judge 2 y
2	4
7	6
1	1
3	3
5	5
6	2
4	7

We have already discussed the concept of coefficient of correlation. It will be recalled that to carry out tests regarding the population coefficient of correlation, one of the assumptions is that the two variables involved have a joint normal distribution. When this assumption is not met, we can often use the rank correlation coefficient, a statistic developed by C. Spearman in 1904, to test if there is an association between two variables. This nonparametric method can be used to test an association between any two numerical variables x and y measured on each item by ranking the x values and y values in order of their magnitude (see Example 4 in this section). However, it is most useful to analyze the relationship between two variables that cannot be expressed by exact measurement but from which ranked data can be obtained. The ranking procedure might be based on some qualitative factor such as taste, appearance, or some other criterion. For example, suppose two judges are asked to judge a beauty contest in which seven contestants have entered. Each judge ranks the contestants from 1 to 7. Consider the rankings as given in Table 13-7.

The **rank correlation coefficient,** which we denote as r_{rank}, is so named because it is a correlation coefficient between two variables where, it turns out, the variables are expressed as ranks. We could compute it by employing the formula in Section 11-2. However, a quicker method is to use the following formula.

RANK CORRELATION COEFFICIENT

$$r_{rank} = 1 - \frac{6 \sum d_i^2}{n(n^2 - 1)}$$

where
   d_i = difference between paired ranks
   n = number of paired items in the sample

The computations for finding r_{rank} for the data in Table 13-7 are presented in Table 13-8. Because $n = 7$ and $\sum d_i^2 = 30$, substituting in the formula, we get

$$r_{rank} = 1 - \frac{6(30)}{7(7^2 - 1)}$$

$$= 0.464.$$

TABLE 13-8   Computations for Finding r_{rank} for Table 13-7

Judge 1 x	Judge 2 y	d $x - y$	d^2
2	4	−2	4
7	6	1	1
1	1	0	0
3	3	0	0
5	5	0	0
6	2	4	16
4	7	−3	9
Total			30

Next, suppose there is a perfect agreement between the rankings of the two judges, as shown in Table 13-9. In this case $\Sigma\, d_i^2 = 0$ and we get

$$r_{rank} = 1 - \frac{6(0)}{7(7^2 - 1)} = 1.$$

At the other extreme, assume that the ranks assigned by the judges are exactly inverse, as in Table 13-10. An individual who gets the highest score from one judge gets the lowest score from the other, and so on. Substituting in the formula,

$$r_{rank} = 1 - \frac{6(112)}{7(7^2 - 1)}$$
$$= -1.$$

TABLE 13-9 Perfect Agreement in the Rankings of the Two Judges

Judge 1 x	Judge 2 y	d x − y
2	2	0
7	7	0
1	1	0
3	3	0
5	5	0
6	6	0
4	4	0

TABLE 13-10 Assignments of Ranks by the Two Judges Are Inverse

Judge 1 x	Judge 2 y	x − y	d²
2	6	−4	16
7	1	6	36
1	7	−6	36
3	5	−2	4
5	3	2	4
6	2	4	16
4	4	0	0
Total			112

In summary, the rank correlation coefficient, as would be expected, ranges between −1 and 1, attaining the value of 1 when there is a perfect agreement and a value of −1 when there is an inverse agreement. Thus,

$$-1 \le r_{rank} \le 1.$$

EXAMPLE 1   Ten applicants applying for professor's position at a university are ranked by the chairperson of the department and the dean. The results are shown in Table 13-11 (along with other computations). Find the rank correlation coefficient.

SOLUTION   From Table 13-11, $\Sigma\, d_i^2 = 96$. Because $n = 10$, we get

$$r_{rank} = 1 - \frac{6(96)}{10(10^2 - 1)}$$
$$= 0.42.$$

TABLE 13-11 Rankings of Applicants by the Chairperson and the Dean, Along with Other Computations

Applicant	Ranking by the Chairperson x	Ranking by the Dean y	d	d^2
A	4	9	−5	25
B	3	5	−2	4
C	6	10	−4	16
D	7	6	1	1
E	10	8	2	4
F	5	1	4	16
G	1	3	−2	4
H	8	7	1	1
I	9	4	5	25
J	2	2	0	0
Total				96

If the sample size is moderately large (even as low as 10), it can be shown that the sampling distribution of r_{rank} is approximately normal, with a mean equal to the population rank correlation coefficient, ρ_{rank}, and the standard deviation $\sigma_{r_{rank}}$ given as follows.

STANDARD DEVIATION OF r_{rank}

$$\sigma_{r_{rank}} = \frac{1}{\sqrt{n-1}}$$

Thus, the test statistic employed is

$$\frac{r_{rank} - \rho_{rank}}{\sigma_{r_{rank}}}$$

and its distribution is approximately standard normal. Upon simplification, this test statistic reduces to

$$\sqrt{n-1}(r_{rank} - \rho_{rank}).$$

The procedure for testing a hypothesis is given in the following example.

EXAMPLE 2

Using the data given in Example 1, test the hypothesis that there is no relation between the rankings given ten applicants by the chairperson of the department and the dean.

SOLUTION   We want to test the null hypothesis that the rankings of the dean and the chairperson are not related, that is, $\rho_{rank} = 0$. Actually, the statement is that the rankings are randomly matched. The alternative hypothesis is that $\rho_{rank} \neq 0$. Thus, we have

H_0: Rankings of the dean and the chairperson are not related.

H_A: Rankings are related.

We therefore compute

$$\sqrt{n-1}(r_{rank} - 0) = \sqrt{n-1}(r_{rank}).$$

In our example, $r_{rank} = 0.42$ and $\sqrt{n-1} = \sqrt{9} = 3$. Therefore,

$$\sqrt{n-1}(r_{rank}) = 3(0.42) = 1.26.$$

Because the computed value is between -1.96 and 1.96, we do not reject the null hypothesis at the 5 percent level. There is no evidence to indicate that the rankings of the chairperson and the dean are related.   ●

EXAMPLE 3

Two wire services, International and Global, ranked twelve football teams as shown in Table 13-12. Test the hypothesis that $\rho_{rank} = 0$, at the 5 percent level of significance.

SOLUTION   The necessary computations are shown in Table 13-12. Since $n = 12$ and $\Sigma d_i^2 = 28$, we get

$$r_{rank} = 1 - \frac{6(28)}{12(12^2 - 1)}$$

$$= 0.9.$$

TABLE 13-12   Rankings of Twelve Football Teams by Two Wire Services, and Other Computations

Team	Global x	International y	d	d^2
A	7	8	-1	1
B	1	3	-2	4
C	3	2	1	1
D	8	7	1	1
E	6	5	1	1
F	5	4	1	1
G	4	6	-2	4
H	2	1	1	1
I	10	9	1	1
J	9	12	-3	9
K	11	11	0	0
L	12	10	2	4
Total				28

We want to test

$$H_0: \rho_{rank} = 0 \quad \text{against} \quad H_A: \rho_{rank} \neq 0.$$

We therefore compute

$$\sqrt{n-1} \, (r_{rank}) = \sqrt{12-1}(0.9) = 2.98.$$

Because the computed value is greater than 1.96, we reject the null hypothesis at the 5 percent level and conclude that the rankings of the two wire services are related.   ●

In the next example we compute Spearman's rank correlation when the data consist of actual numerical values x and y.

EXAMPLE 4

A physical education department has obtained the following data on heights and weights of twelve athletes, as given in Table 13-13. Find the rank correlation and test whether the height and weight of an athlete are related.

TABLE 13-13   Heights and Weights of Twelve Athletes

Athlete	Height (in Inches) x	Weight (in Pounds) y
A	68	152
B	64	162
C	72	182
D	67	197
E	71	149
F	73	224
G	70	184
H	69	159
I	66	167
J	74	230
K	76	218
L	65	146

SOLUTION   As a first step, we replace the x values by their ranks assigned in order of increasing magnitude, the smallest value receiving a rank of 1 and the largest a rank of 12. The same is done with the y values. The resultant rankings and pairings are shown in Table 13-14. Because $n = 12$, we get

$$r_{rank} = 1 - \frac{6(106)}{12(12^2 - 1)}$$

$$= 0.63.$$

TABLE 13-14   Heights and Weights Replaced by Ranks

Athlete	Ranks Assigned to Heights	Ranks Assigned to Weights	d	d^2
A	5	3	2	4
B	1	5	−4	16
C	9	7	2	4
D	4	9	−5	25
E	8	2	6	36
F	10	11	−1	1
G	7	8	−1	1
H	6	4	2	4
I	3	6	−3	9
J	11	12	−1	1
K	12	10	2	4
L	2	1	1	1
Total				106

The null hypothesis is that the height and weight of an athlete are uncorrelated in the population. Now

$$\sqrt{n-1}(r_{rank}) = \sqrt{11}(0.63) = 2.09.$$

At the 5 percent level, the value is significant. We reject the null hypothesis, concluding that height and weight are correlated.   ●

Section 13-4 Exercises

1. In a 1000-meter race in which eleven mothers and a teenage daughter of each participated, the following standings were observed among mothers and daughters. Find the rank correlation and test the null hypothesis that it is zero in the population. Use $\alpha = 0.05$.

Family Name	Mother's Rank	Daughter's Rank
Tucker	2	4
Moore	5	7
Jackson	7	6
Hagopian	3	5
Alvarez	1	1
Fletcher	8	10
Harper	4	3
Chen	6	2
Price	11	8
Salerno	10	9
Mason	9	11

2. The following table gives the standings of twelve tennis pros at the end of six
   tournaments in the United States and at the end of as many tournaments in
   Europe.

Player	Standing in Europe	Standing in U.S.
Rashid	3	1
Jimmy	2	3
Ken	1	9
Raul	4	5
Ilie	5	2
Manuel	7	8
Vijay	10	6
Tony	12	7
John	6	10
Sandy	9	11
Mark	11	12
Arthur	8	4

Find the rank correlation coefficient. Using $\alpha = 0.05$, test the null hypothesis
that standings in Europe and the United States are not related.

3. Ten vice-presidents in a company were ranked according to their affability and
   level of competence. Find the rank correlation coefficient and test the null
   hypothesis that affability and competence are not related. Use $\alpha = 0.025$.

Name	Affability Rank	Competence Rank
Maroney	5	3
Jorgenson	8	1
Kozlowski	2	2
Ozawa	1	7
Meyer	6	4
Keating	3	10
Purcell	7	5
Ramirez	4	6
Simmons	10	8
Chao	9	9

4. The scores of eleven students on the midterm exam and the final exam were as
   follows.

Student	Midterm Score	Final Exam Score
Rita	82	94
Miguel	81	93
Hank	68	74
Mary	78	81
Jamo	92	96
John	76	67
Dominic	54	53
Ron	86	89
Tiffany	90	92
Jan	62	45
Ricardo	52	61

Compute the rank correlation coefficient and test the null hypothesis that it is zero in the population. Use $\alpha = 0.1$.

5. The following table lists fourteen comparable cars rated by the EPA according to their gasoline mileage and versatility.

Car Make	Mileage	Versatility
A	20.0	8
B	23.0	5
C	28.0	7
D	27.0	3
E	24.0	10
F	25.5	9
G	30.0	4
H	22.5	6
I	26.0	1
J	31.0	2
K	29.0	14
L	26.5	11
M	28.5	13
N	23.5	12

Compute the rank correlation coefficient, and at the 5 percent level, test the null hypothesis that it is zero in the population against the alternate hypothesis that it is not zero. (*Hint:* Assign lowest mileage Rank 1.)

6. The following table gives the IQs of father and son for twelve families. Calculate the rank-correlation coefficient. At the 5 percent level of significance, is there an association between father's IQ and son's IQ?

Family	Father	Son
Jones	110	130
McKee	140	120
Lutnetsky	138	122
Wong	166	150
Turnbull	120	140
Howard	128	116
Patel	142	156
Salazar	118	94
Tatum	95	125
Adler	150	133
Kim	108	100
Toth	133	110

CHAPTER 13 Summary

✓ *CHECKLIST: KEY TERMS AND EXPRESSIONS*

❏ parametric tests, page 720
❏ nonparametric tests, page 720
❏ distribution-free tests, page 720
❏ two-sample sign test, page 721

❏ Wilcoxon rank-sum test, page 729
❏ Mann-Whitney U test, page 729
❏ rank sum, page 731

❏ runs test, page 736
❏ run, page 736
❏ rank correlation coefficient, page 742

KEY FORMULAS

sign test

X = number of plus signs

Under H_0, X has binomial distribution with $p = \frac{1}{2}$. If $n \geq 15$, use the test statistic $\dfrac{2X - n}{\sqrt{n}}$, whose distribution is approximately standard normal.

Mann-Whitney U test

$$U = mn + \frac{m(m + 1)}{2} - R_1$$

where m is the size of one of the samples, R_1 is the sum of the ranks assigned to its members, and n is the size of the other sample.

If m and n are both greater than 9, then the distribution of the following statistic is approximately standard normal.

$$\frac{U - \dfrac{mn}{2}}{\sqrt{\dfrac{mn(m + n + 1)}{12}}}$$

runs test

If R is the number of runs, m is the number of symbols of one kind, n the number of the other kind, and if m and n are both greater than 10, then the distribution of

$$\left[R - \left(\frac{2mn}{m + n} + 1 \right) \right] \Big/ \sqrt{\frac{2mn(2mn - m - n)}{(m + n)^2(m + n - 1)}}$$

is approximately standard normal.

rank correlation

$$r_{rank} = 1 - \frac{6 \sum d_i^2}{n(n^2 - 1)}$$

where
  d_i = difference between paired ranks
  n = number of paired items.
If n is greater than 10, the statistic $\sqrt{n - 1}(r_{rank} - \rho_{rank})$ has a distribution that is approximately standard normal.

CHAPTER 13 Concept and Discussion Questions

1. What nonparametric test did you use and in what context did you use it, prior to discussing nonparametric tests in this chapter?

2. What are the parametric counterparts of the following nonparametric tests?
   (a) sign test   **(b)** rank-sum test
   When would you use these nonparametric tests in preference to their parametric counterparts?

3. What is the runs test supposed to determine?

4. Explain why the runs test for randomness is always a two-tail test.

5. Toss a coin 30 times and record sequentially heads (H) and tails (T) as they occur. Determine whether the occurrence of heads and tails is random or exhibits a pattern.

6. Suppose that pairs of values (x, y) are obtained on each unit in the sample and you wish to find out whether the variables are related.
   (a) How would you investigate this if the sample is from a normally distributed population?
   (b) What procedure would you recommend if the normality assumption is not valid?

CHAPTER **13   Review Exercises**

1. Discuss the advantages and disadvantages of nonparametric tests over parametric tests.

2. When would you use the rank-sum test rather than the classical t test?

3. What would you conclude about the randomness of a sequence in the following cases?
   (a) There were too many runs.
   (b) There were too few runs.

4. Suppose we have 20 letters a and 15 letters b and they are arranged in a random sequence. If R represents the number of runs, find the mean number of runs and the variance of the number. What is the approximate distribution of R?

5. A construction company wants to compare two mixes used in manufacturing structural beams. The following table represents the strengths of twenty-two experimental beams (in pounds per square inch).

Mix 1	Mix 2
996	1135
1035	860
1002	990
986	1200
895	870
1085	976
1110	1009
985	1182
1030	886
960	1090
	1230
	940

At the 5 percent level of significance, does one mix give stronger beams than the other? Use the rank-sum test.

6. A laboratory experiment was conducted to test whether a diuretic agent was effective in inducing more urine in dogs. The table below gives a 1-day collection of urine (in cubic centimeters) from thirteen dogs of a certain breed before and after the diuretic agent was administered.

Dog Number	Urine Before	Urine After
1	980	1200
2	1600	1480
3	850	850
4	1700	1500
5	960	1040
6	1010	980
7	860	1010
8	1850	1900
9	980	1060
10	1030	990
11	1400	1400
12	750	880
13	1130	1040

Use the sign test at the 5 percent level of significance to test the null hypothesis that the diuretic agent is not effective.

7. The written part of a driver's test consists of thirty questions, each with two choices. The following data give the correct choice (sequentially) for each of the questions (recorded 1 if the first is correct, 2 if the second is correct).

1	2	1	2	2	2	1	2	2	1	2	1	2	1	2
1	2	2	2	1	2	1	2	1	2	2	1	1	2	1

At the 5 percent level of significance, test the null hypothesis that the choices are arranged at random.

8. The following table gives the rankings of ten players based on their ability in tennis and Ping-Pong. Calculate the rank correlation coefficient for the data. At $\alpha = 0.01$, test the null hypothesis of no correlation.

Player	Rank in Tennis	Rank in Ping-Pong
1	4	7
2	2	4
3	3	2
4	1	8
5	8	3
6	9	5
7	7	6
8	6	9
9	5	1
10	10	10

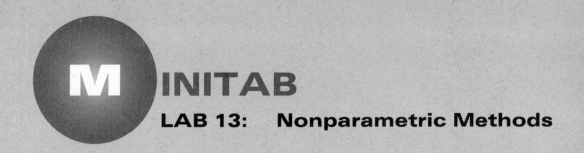

MINITAB

LAB 13:   Nonparametric Methods

Nonparametric tests are known for their simplicity. However, some of the features associated with these tests, such as ordering and assigning ranks to the data, can be very tedious, especially for large data sets. MINITAB does this work for us, and has built-in commands supporting the sign test with the command STEST, the Mann-Whitney test with the command MANNWHITNEY, and the runs test with the command RUNS. As for the Spearman rank correlation, once the data are ranked using the MINITAB command RANKS, the command CORRELATION, which was introduced in Lab 11, computes the correlation between the ranks.

1 In the following illustration of the sign test for paired data we consider the data in Table 13-1 in the text. The alternative hypothesis is left-sided, so we have added the subcommand ALTERNATIVE IS -1.

```
MTB >NOTE SIGN TEST FOR PAIRED DATA

MTB >NOTE DATA FROM TABLE 13-1

MTB >NAME C1 'ENGLISH' C2 'FRENCH'

MTB >READ DATA IN C1 C2

DATA>7    4

DATA>3    5

DATA>5    7

DATA>6    3

DATA>8    6

DATA>4    4

DATA>6    8

DATA>7    2

DATA>5    6

DATA>4    7
```

```
DATA>7    9

DATA>8    7

DATA>4    6

DATA>3    5

DATA>7    9

DATA>END OF DATA

MTB >PRINT C1 C2

MTB >NAME C3 'DIFFERENCE'

MTB >LET C3 = C1 - C2

MTB >PRINT C3

MTB >STEST OF DIFFERENCE 0 FOR DATA IN C3;

SUBC>ALTERNATIVE = -1.
```

The printout will give the number of $-$ signs (BELOW), ties (EQUAL), and $+$ signs (ABOVE). It will also give the *P*-value for the test of significance, in this case 0.212, which agrees with the answer in the text.

Assignment   Work out Exercises 1 and 5 in Section 13-1.

2   The Mann-Whitney test, which we now consider, is easily performed with the command MANNWHITNEY. Under the command we must specify the columns where the data from the two samples are located. This is illustrated in the following, using the data from Table 13-3 in the text.

```
MTB >NOTE DATA FROM TABLE 13-3

MTB >NAME C1 'COMM A', C2 'COMM B'

MTB >SET DATA IN C1

DATA>120 155 88 420 360 60 650 82 180 98

DATA>END OF DATA

MTB >PRINT C1

MTB >SET DATA IN C2

DATA>42 190 320 680 82 130 500 32 160 82 890

DATA>1020

DATA>END OF DATA

MTB >PRINT C2

MTB >MANNWHITNEY FOR DATA IN C1 C2
```

The output will display the medians for the two sets of data, a point estimate for the difference of the medians, and the *P*-value testing for equality of medians.

Assignment   Work out Exercises 1 and 6 in Section 13-2.

3   We now consider the runs test. When the data are given in a binary code, as in Example 1 in Section 13-3, where letters *g* and *b* are used, the runs test for randomness can be performed with the command RUNS. The first step in this process is to assign a numerical value to each of the two symbols involved. To show the procedure, in Example 1 we use 0 to identify the letter *g* and 1 to identify *b*.

```
MTB >NOTE EXAMPLE 1 IN SECTION 13-3

MTB >SET DATA IN C1

DATA>0 0 0 1 1 0 1 0 0 0 1 0 1

DATA>1 0 0 1 0 0 0 1 1 1 0 1

DATA> END OF DATA

MTB >RUNS ABOVE AND BELOW 0 FOR DATA IN C1
```

The numeral 0 mentioned under the command RUNS is the smaller of the two numerals in the assignment of the numerical codes. The MINITAB output includes the observed number of runs and the *P*-value.

Assignment   Work out Exercises 1 and 7 in Section 13-3.

4   Finally, we discuss the Spearman correlation. As mentioned earlier, to compute it we apply the command CORRELATION to ranked data. So, if the data are not ranked, we must first rank the values for each of the two variables. We now solve Example 4 in Section 13-4.

```
MTB >NOTE EXAMPLE 4 IN SECTION 13-4

MTB >NAME C1 'HEIGHT' C2 'WEIGHT'

MTB >NAME C3 'HT-RANK' C4 'WT-RANK'

MTB >READ DATA IN C1 C2

DATA>68  152

DATA>64  162

DATA>72  182

DATA>67  197

DATA>71  149
```

```
DATA>73  224

DATA>70  184

DATA>69  159

DATA>66  167

DATA>74  230

DATA>76  218

DATA>65  146

DATA>END OF DATA

MTB >PRINT C1 C2

MTB >RANKS OF C1 PUT IN C3

MTB >RANKS OF C2 PUT IN C4

MTB >PRINT C3 C4

MTB >NOTE SPEARMAN CORRELATION

MTB >CORRELATION OF C3 AND C4
```

There is no command to perform the analysis and to test for significance. You will have to do it yourself.

Assignment   Work out Exercises 3 and 4 in Section 13-4.

We now type STOP and conclude the MINITAB session using the logging off procedure.

Commands you have learned in this session:

```
STEST

MANNWHITNEY

RUNS

RANKS
```

APPENDIX

Summation Notation

Suppose we are recording values of a quantitative variable, for instance, the weights of a group of people. Instead of writing, in a cumbersome way, that the weight of the first person is 150 pounds, the weight of the second person is 175 pounds, and so on, we can symbolically write $x_1 = 150$ lbs, $x_2 = 175$ lbs, and so on, where the letter x stands for the weight variable and the subscript for the number of each person in the group. With this understanding, x_{50}, for instance, will represent the weight of the fiftieth person.

With six people in the group, we could represent their weights as x_1, x_2, x_3, x_4, x_5, x_6. Ordinarily, we would write the sum of these weight values as

$$x_1 + x_2 + x_3 + x_4 + x_5 + x_6.$$

But a mathematical shorthand that abbreviates the expression for the sum is

$$\sum_{i=1}^{i=6} x_i$$

where the symbol Σ, uppercase Greek sigma, is called the sigma notation, or **summation notation.** This notation tells us to carry out the summation; $i = 1$ at the bottom of Σ indicates that the subscript of the first term in the summation is 1, and the 6 at the top indicates that the subscript of the last term in the summation is 6.

In general, if there are n terms, or observations, $x_1, x_2, \ldots, x_n$, then $\sum_{i=1}^{n} x_i$ is a shorthand expression for their sum.

$$\sum_{i=1}^{n} x_i = x_1 + x_2 + \cdots + x_n$$

EXAMPLE 1 Suppose $x_1 = 10$, $x_2 = 5$, $x_3 = 7$, $x_4 = 9$, $x_5 = 16$, $x_6 = 10$, and $x_7 = 11$. Find the following.

(a) $\displaystyle\sum_{i=1}^{7} x_i$ (b) $\displaystyle\sum_{i=1}^{5} (x_i + 1)$

SOLUTION

(a) To find $\sum_{i=1}^{7} x_i$ we add every term from the first to the seventh.

$$\sum_{i=1}^{7} x_i = x_1 + x_2 + x_3 + x_4 + x_5 + x_6 + x_7$$
$$= 10 + 5 + 7 + 9 + 16 + 10 + 11 = 68$$

(b) Here we obtain

$$\sum_{i=1}^{5}(x_i + 1) = (x_1 + 1) + (x_2 + 1) + (x_3 + 1)$$
$$+ (x_4 + 1) + (x_5 + 1)$$
$$= (10 + 1) + (5 + 1) + (7 + 1) + (9 + 1)$$
$$+ (16 + 1)$$
$$= 52.$$

EXAMPLE 2 Suppose $x_1 = 13$, $x_2 = 9$, $x_3 = 10$, $x_4 = 20$, $f_1 = 3$, $f_2 = 5$, $f_3 = 2$, and $f_4 = 7$. Find the following.

(a) $\displaystyle\sum_{i=1}^{4} f_i x_i$ (b) $\displaystyle\sum_{i=1}^{4} f_i x_i^2$ (c) $\displaystyle\sum_{i=1}^{4} f_i(x_i - 10)^2$

SOLUTION

(a) The sigma notation tells us that we are adding four terms starting with $i = 1$ and ending with $i = 4$. The general term is given as $f_i x_i$, which is $f_1 x_1$ when $i = 1$, $f_2 x_2$ when $i = 2$, $f_3 x_3$ when $i = 3$, and $f_4 x_4$ when $i = 4$. Thus we have

$$\sum_{i=1}^{4} f_i x_i = f_1 x_1 + f_2 x_2 + f_3 x_3 + f_4 x_4$$
$$= 3(13) + 5(9) + 2(10) + 7(20)$$
$$= 244.$$

(b) Arguing as in Part a, we have

$$\sum_{i=1}^{4} f_i x_i^2 = f_1 x_1^2 + f_2 x_2^2 + f_3 x_3^2 + f_4 x_4^2$$
$$= 3(13)^2 + 5(9)^2 + 2(10)^2 + 7(20)^2$$
$$= 3912.$$

(c) Here we have

$$\sum_{i=1}^{4} f_i(x_i - 10)^2 = f_1(x_1 - 10)^2 + f_2(x_2 - 10)^2$$
$$+ f_3(x_3 - 10)^2 + f_4(x_4 - 10)^2$$
$$= 3(13 - 10)^2 + 5(9 - 10)^2$$
$$+ 2(10 - 10)^2 + 7(20 - 10)^2$$
$$= 732.$$

Appendix I Exercises

1. Write out the following sums explicitly. Do not compute the sums.

(a) $\displaystyle\sum_{i=1}^{6} 3^i$ (b) $\displaystyle\sum_{k=1}^{4} \frac{2}{k+5}$

(c) $\displaystyle\sum_{j=1}^{4} (2^j + j^2)$ (d) $\displaystyle\sum_{j=1}^{5} j(j-1)$

(e) $\displaystyle\sum_{k=1}^{4} (k^2 + 2k - 1)$ (f) $\displaystyle\sum_{j=1}^{5} \frac{j}{j+3}$

2. Compute $\displaystyle\sum_{i=1}^{5} x_i$ if $x_1 = 2.2$, $x_2 = 3.8$, $x_3 = 4.9$, $x_4 = 2.5$, and $x_5 = 4.9$.

3. Compute $\displaystyle\sum_{i=1}^{6} (x_i - 4)$ if $x_1 = 5$, $x_2 = 7$, $x_3 = -2$, $x_4 = 3$, $x_5 = 0$, and $x_6 = 4$.

4. Suppose $x_1 = 3$, $x_2 = 5$, $x_3 = -2$, $x_4 = 5$, $x_5 = -8$, and $x_6 = 4$. Compute the following.

(a) $\displaystyle\sum_{i=1}^{4} x_i$ (b) $\displaystyle\sum_{i=1}^{6} x_i + 1$ (c) $\displaystyle\sum_{i=1}^{6} (1 + x_i)$

(d) $\displaystyle\sum_{i=1}^{4} (x_i + 1)^2$ (e) $\displaystyle\sum_{i=1}^{4} x_i^2 + 1$ (f) $\displaystyle\sum_{i=1}^{6} x_i(x_i - 1)$

5. Suppose $x_1, x_2, \ldots, x_n$ and $y_1, y_2, \ldots, y_n$ are given values. Write the following expressions using summation notation.
(a) $3x_1 + 3x_2 + \cdots + 3x_n + 2y_1 + 2y_2 + \cdots + 2y_n$
(b) $2x_1^2 + 2x_2^2 + \cdots + 2x_n^2 + 2y_1^2 + 2y_2^2 + \cdots + 2y_n^2$
(c) $2(x_1 + y_1)^2 + 2(x_2 + y_2)^2 + \cdots + 2(x_n + y_n)^2$
(d) $x_1^2 y_1 + x_2^2 y_2 + \cdots + x_n^2 y_n$

6. (a) Suppose x_1, x_2, x_3, x_4, and x_5 are given numbers. Write the sum $\displaystyle\sum_{i=1}^{5} (3x_i - 2)^2$ explicitly without sigma notation.

(b) Compute $\displaystyle\sum_{i=1}^{5} (3x_i - 2)^2$ if $x_1 = 2$, $x_2 = 3$, $x_3 = -1$, $x_4 = 0$, and $x_5 = 6$.

7. If $x_1 = 3$, $x_2 = -4$, $x_3 = 5$, $x_4 = -2$, and $f_1 = 2$, $f_2 = 3$, $f_3 = 10$, $f_4 = 1$, compute the following.

(a) $\displaystyle\sum_{i=1}^{4} f_i x_i$ (b) $\displaystyle\sum_{i=1}^{4} f_i x_i^2$ (c) $\left(\displaystyle\sum_{i=1}^{4} f_i x_i \right)^2$ (d) $\displaystyle\sum_{i=1}^{4} (f_i x_i)^2$

8. If $x_1 = 3$, $x_2 = -4$, $x_3 = 5$, $x_4 = -2$, and $y_1 = -2$, $y_2 = -3$, $y_3 = 2$, $y_4 = 4$, compute the following.

(a) $\displaystyle\sum_{i=1}^{4} x_i y_i$ (b) $\displaystyle\sum_{i=1}^{4} (x_i - 1)(y_i - 3)$ (c) $\left(\displaystyle\sum_{i=1}^{4} x_i \right)\left(\displaystyle\sum_{i=1}^{4} y_i \right)$

(d) $\displaystyle\sum_{i=1}^{4} x_i^2 y_i^2$

9. If $\displaystyle\sum_{i=1}^{10} x_i = 238.6$, $x_1 = 12.6$, and $x_{10} = -28.0$, find the following.

(a) $\displaystyle\sum_{i=2}^{10} x_i$ **(b)** $\displaystyle\sum_{i=2}^{9} x_i$

10. If $\dfrac{\displaystyle\sum_{i=1}^{10} x_i}{10} = 22.6$ and $\displaystyle\sum_{i=1}^{9} x_i = 188.2$, find x_{10}.

11. If $\displaystyle\sum_{i=1}^{10} x_i = 150$ and $\displaystyle\sum_{i=1}^{10} x_i^2 = 2350$, compute $\displaystyle\sum_{i=1}^{10} x_i^2 - \dfrac{\left(\displaystyle\sum_{i=1}^{10} x_i\right)^2}{10}$.

12. If $x_1 = 1, x_2 = 2, x_3 = 4$, and $y_1 = -1, y_2 = 0, y_3 = 2$, compute the following.

$$\sum_{i=1}^{3} x_i y_i - \frac{\left(\displaystyle\sum_{i=1}^{3} x_i\right)\left(\displaystyle\sum_{i=1}^{3} y_i\right)}{3}$$

Poisson Distribution

APPENDIX II

INTRODUCTION

The Poisson distribution provides a probabilistic model for a wide class of phenomena. Typical examples are the number of telephone calls during a given period of time, the number of particles emitted from a radioactive source in a given period of time, and the number of bacteria on a plate of a given size.

We say that a random variable X has a Poisson distribution with parameter λ (lambda) if its probability function is given by

$$p(x) = P(X = x) = \frac{e^{-\lambda}\lambda^x}{x!}$$

where x can assume any value 0, 1, 2, . . . (the three dots mean *ad infinitum*).

Thus, a Poisson random variable is a discrete random variable and any nonnegative integer is in its range. Here e is a universal constant (base of the natural logarithms) and has the value 2.7182 . . . . In Table 1 in this appendix, we give the probabilities $p(x)$ for certain values of λ. For example, if $\lambda = 3.2$, then $p(3) = 0.2226$, $p(5) = 0.1140$, and so on. Most scientific pocket calculators are equipped to find $e^{-\lambda}$ for any value of λ, and the reader should experience no difficulty in finding $p(x)$ for any λ.

It turns out that the parameter λ is actually the mean (expected value) of the random variable. The variance of the random variable is also λ. Thus, the mean and the variance for a Poisson distribution coincide, both being equal to λ.

EXAMPLE 1

In an experiment conducted by Rutherford and Geiger, the mean number of α-particles emitted by a radioactive source during 7.5 seconds was found to be 3.9. Assuming that the number of α-particles emitted has a Poisson distribution, what is the probability that during 7.5 seconds the source will emit:

(a) exactly 2 α-particles

(b) at most 2 α-particles

(c) at least 2 α-particles.

SOLUTION The mean number of particles is 3.9. That is, $\lambda = 3.9$. Thus, referring to Table 1 with $\lambda = 3.9$, we get:

(a) $P(\text{exactly 2 } \alpha\text{-particles}) = p(2) = 0.1539$

(b) $P(\text{at most 2 } \alpha\text{-particles}) = p(0) + p(1) + p(2)$
$$= 0.0202 + 0.0789 + 0.1539$$
$$= 0.2530.$$

(c) "At least 2 α-particles" means *not* 0 or 1 α-particle. Therefore,
$$P(\text{at least 2 } \alpha\text{-particles}) = 1 - P(0 \text{ or } 1 \text{ } \alpha\text{-particle})$$
$$= 1 - [p(0) + p(1)]$$
$$= 1 - (0.0202 + 0.0789)$$
$$= 0.9009.$$

POISSON APPROXIMATION TO THE BINOMIAL PROBABILITIES

In Section 5-3, we saw that if p is not very near to 0 or 1, then for large n the binomial distribution can be approximated by a normal distribution. If p is near 0 or 1, the normal approximation is unsatisfactory. It can be shown, however, that if p is small and n is large, the Poisson distribution provides a satisfactory approximation to the binomial distribution. The statement of this result is as follows.

POISSON APPROXIMATION TO THE BINOMIAL

If n is large and p is small, the binomial probabilities $b(x; n, p)$ can be approximated by the Poisson probabilities with parameter np. Specifically

$$b(x; n, p) \approx \frac{e^{-np}(np)^x}{x!}, x = 0, 1, 2, \ldots .$$

If p is near 1, then of course, the probability of failure q will be small and, consequently, one can approximate the binomial probabilities of the number of failures by Poisson probabilities with parameter nq.

EXAMPLE 2 A computer has 1000 electronic components. The probability that a component will fail during one year of operation is 0.0022. Assuming the components function independently, find the approximate probability that during one year of operation:

(a) 3 components fail

(b) between 1 and 3 components, inclusive, fail.

SOLUTION

(a) Let X represent the number of components failing. We are given that $n = 1000$ and $p = 0.0022$. The exact answer to the probability that 3 components fail is

$$b(3; 1000, 0.0022) = \binom{1000}{3}(0.0022)^3(0.9978)^{997}.$$

Because p is small and n is large, we can use the Poisson approximation with parameter $np = 1000(0.0022) = 2.2$, and obtain an approximate answer as

$$b(3; 1000, 0.0022) \approx \frac{e^{-2.2}(2.2)^3}{3!} = 0.1966$$

referring to Table 1 with $\lambda = 2.2$ and $x = 3$.

(b) Here we get

$$P(1 \le X \le 3) = b(1; 1000, 0.0022) + b(2; 1000, 0.0022) + b(3; 1000, 0.0022)$$

$$\approx \frac{e^{-2.2}(2.2)^1}{1!} + \frac{e^{-2.2}(2.2)^2}{2!} + \frac{e^{-2.2}(2.2)^3}{3!}$$

$$= 0.2438 + 0.2681 + 0.1966$$

$$= 0.7085. \qquad \bullet$$

Appendix II Exercises

1. The number of elementary particles, recorded by a device in a space vehicle during a one-day flight, has a Poisson distribution with a mean of 3.5 particles. Find the probability that during a one-day flight there will be:
   (a) no particle recorded   (b) at least 4 particles recorded.

2. Suppose the number of telephone calls arriving in an office during a 15-minute period has a Poisson distribution with mean number of calls equal to 2.5. Find the probability that during a 15-minute period there will be:
   (a) no calls   (b) between 2 and 6 calls, inclusive.

3. Suppose a biological cell contains 400 genes. When treated radioactively, the probability that a gene will change into a mutant gene is 0.006 and is independent of the condition of the other genes. What is the approximate probability that there are at most 4 mutant genes after the treatment?

4. A book has 350 pages. The probability that a page has misprints (one or more) is 0.004. Find the approximate probability that the book has no page with misprints.

5. A salesman of heavy appliances has determined that the probability that a customer inquiring about a product will acutally buy it is 0.01. What is the approximate probability that when 120 people make inquiries (independently), 2 or less will buy?

TABLE 1 Poisson Probabilities*

x	0.1	0.2	0.3	0.4	λ 0.5	0.6	0.7	0.8	0.9	1.0
0	0.9048	0.8187	0.7408	0.6703	0.6065	0.5488	0.4966	0.4493	0.4066	0.3679
1	0.0905	0.1637	0.2222	0.2681	0.3033	0.3293	0.3476	0.3595	0.3659	0.3679
2	0.0045	0.0164	0.0333	0.0536	0.0758	0.0988	0.1216	0.1438	0.1647	0.1839
3	0.0002	0.0011	0.0033	0.0072	0.0126	0.0198	0.0284	0.0383	0.0494	0.0613
4		0.0001	0.0002	0.0007	0.0016	0.0030	0.0050	0.0077	0.0111	0.0153
5				0.0001	0.0002	0.0003	0.0007	0.0012	0.0020	0.0031
6							0.0001	0.0002	0.0003	0.0005
7										0.0001

x	1.1	1.2	1.3	1.4	λ 1.5	1.6	1.7	1.8	1.9	2.0
0	0.3329	0.3012	0.2725	0.2466	0.2231	0.2019	0.1827	0.1653	0.1496	0.1353
1	0.3662	0.3614	0.3543	0.3452	0.3347	0.3230	0.3106	0.2975	0.2841	0.2707
2	0.2014	0.2169	0.2303	0.2417	0.2510	0.2584	0.2640	0.2678	0.2700	0.2707
3	0.0738	0.0867	0.0998	0.1127	0.1255	0.1378	0.1496	0.1607	0.1710	0.1804
4	0.0203	0.0260	0.0324	0.0395	0.0471	0.0551	0.0636	0.0723	0.0812	0.0902
5	0.0045	0.0062	0.0084	0.0111	0.0141	0.0176	0.0215	0.0260	0.0309	0.0361
6	0.0008	0.0012	0.0018	0.0026	0.0035	0.0047	0.0061	0.0078	0.0098	0.0120
7	0.0001	0.0002	0.0003	0.0005	0.0008	0.0011	0.0015	0.0020	0.0027	0.0034
8			0.0001	0.0001	0.0001	0.0002	0.0003	0.0005	0.0006	0.0009
9							0.0001	0.0001	0.0001	0.0002

x	2.1	2.2	2.3	2.4.	λ 2.5	2.6	2.7	2.8	2.9	3.0
0	0.1225	0.1108	0.1003	0.0907	0.0821	0.0743	0.0672	0.0608	0.0550	0.0498
1	0.2572	0.2438	0.2306	0.2176	0.2052	0.1931	0.1815	0.1703	0.1596	0.1494
2	0.2700	0.2681	0.2652	0.2613	0.2565	0.2510	0.2450	0.2384	0.2314	0.2240
3	0.1890	0.1966	0.2033	0.2090	0.2138	0.2176	0.2205	0.2225	0.2237	0.2240
4	0.0992	0.1082	0.1169	0.1254	0.1336	0.1414	0.1488	0.1557	0.1622	0.1680
5	0.0416	0.0476	0.0538	0.0602	0.0668	0.0735	0.0804	0.0872	0.0940	0.1008
6	0.0146	0.0174	0.0206	0.0241	0.0278	0.0319	0.0362	0.0407	0.0455	0.0504
7	0.0044	0.0055	0.0068	0.0083	0.0099	0.0118	0.0139	0.0163	0.0188	0.0216
8	0.0011	0.0015	0.0019	0.0025	0.0031	0.0038	0.0046	0.0057	0.0068	0.0081
9	0.0003	0.0004	0.0005	0.0007	0.0009	0.0011	0.0014	0.0018	0.0022	0.0027
10	0.0001	0.0001	0.0001	0.0002	0.0002	0.0003	0.0004	0.0005	0.0006	0.0008
11						0.0001	0.0001	0.0001	0.0002	0.0002
12										0.0001

*The entries in the table give the Poisson probabilities $p(x) = \dfrac{e^{-\lambda}\lambda^x}{x!}$ for certain values of λ. For example, for $\lambda = 3.2$, $x = 3$, $p(3) = 0.2226$.

TABLE 1   (continued)

x	3.1	3.2	3.3	3.4	λ 3.5	3.6	3.7	3.8	3.9	4.0
0	0.0450	0.0408	0.0369	0.0334	0.0302	0.0273	0.0247	0.0224	0.0202	0.0183
1	0.1397	0.1304	0.1217	0.1135	0.1057	0.0983	0.0915	0.0850	0.0789	0.0733
2	0.2165	0.2087	0.2008	0.1929	0.1850	0.1771	0.1692	0.1615	0.1539	0.1465
3	0.2236	0.2226	0.2209	0.2186	0.2158	0.2125	0.2087	0.2046	0.2001	0.1954
4	0.1734	0.1781	0.1823	0.1858	0.1887	0.1912	0.1931	0.1944	0.1951	0.1954
5	0.1075	0.1140	0.1203	0.1264	0.1322	0.1377	0.1429	0.1477	0.1522	0.1563
6	0.0555	0.0607	0.0662	0.0716	0.0771	0.0826	0.0881	0.0936	0.0989	0.1042
7	0.0246	0.0278	0.0312	0.0348	0.0385	0.0425	0.0466	0.0508	0.0551	0.0595
8	0.0095	0.0111	0.0128	0.0148	0.0169	0.0191	0.0215	0.0241	0.0269	0.0298
9	0.0033	0.0040	0.0047	0.0056	0.0066	0.0076	0.0089	0.0102	0.0116	0.0132
10	0.0010	0.0013	0.0016	0.0018	0.0023	0.0028	0.0033	0.0039	0.0045	0.0053
11	0.0003	0.0004	0.0005	0.0006	0.0007	0.0009	0.0011	0.0013	0.0016	0.0019
12	0.0001	0.0001	0.0001	0.0002	0.0002	0.0003	0.0003	0.0004	0.0005	0.0006
13					0.0001	0.0001	0.0001	0.0001	0.0002	0.0002
14										0.0001

Tables

APPENDIX

p of X successes in n trials
p - prob. of success on any trial

TABLE A-1 Binomial Probabilities, Denoted $b(x; n, p)$

							p					
n	x	0.05	0.1	0.2	0.3	0.4	0.5	0.6	0.7	0.8	0.9	0.95
2	0	0.902	0.810	0.640	0.490	0.360	0.250	0.160	0.090	0.040	0.010	0.002
	1	0.095	0.180	0.320	0.420	0.480	0.500	0.480	0.420	0.320	0.180	0.095
	2	0.002	0.010	0.040	0.090	0.160	0.250	0.360	0.490	0.640	0.810	0.902
3	0	0.857	0.729	0.512	0.343	0.216	0.125	0.064	0.027	0.008	0.001	
	1	0.135	0.243	0.384	0.441	0.432	0.375	0.288	0.189	0.096	0.027	0.007
	2	0.007	0.027	0.096	0.189	0.288	0.375	0.432	0.441	0.384	0.243	0.135
	3		0.001	0.008	0.027	0.064	0.125	0.216	0.343	0.512	0.729	0.857
4	0	0.815	0.656	0.410	0.240	0.130	0.062	0.026	0.008	0.002		
	1	0.171	0.292	0.410	0.412	0.346	0.250	0.154	0.076	0.026	0.004	
	2	0.014	0.049	0.154	0.265	0.346	0.375	0.346	0.265	0.154	0.049	0.014
	3		0.004	0.026	0.076	0.154	0.250	0.346	0.412	0.410	0.292	0.171
	4			0.002	0.008	0.026	0.062	0.130	0.240	0.410	0.656	0.815
5	0	0.774	0.590	0.328	0.168	0.078	0.031	0.010	0.002			
	1	0.204	0.328	0.410	0.360	0.259	0.156	0.077	0.028	0.006		
	2	0.021	0.073	0.205	0.309	0.346	0.312	0.230	0.132	0.051	0.008	0.001
	3	0.001	0.008	0.051	0.132	0.230	0.312	0.346	0.309	0.205	0.073	0.021
	4			0.006	0.028	0.077	0.156	0.259	0.360	0.410	0.328	0.204
	5				0.002	0.010	0.031	0.078	0.168	0.328	0.590	0.774
6	0	0.735	0.531	0.262	0.118	0.047	0.016	0.004	0.001			
	1	0.232	0.354	0.393	0.303	0.187	0.094	0.037	0.010	0.002		
	2	0.031	0.098	0.246	0.324	0.311	0.234	0.138	0.060	0.015	0.001	
	3	0.002	0.015	0.082	0.185	0.276	0.312	0.276	0.185	0.082	0.015	0.002
	4		0.001	0.015	0.060	0.138	0.234	0.311	0.324	0.246	0.098	0.031
	5			0.002	0.010	0.037	0.094	0.187	0.303	0.393	0.354	0.232
	6				0.001	0.004	0.016	0.047	0.118	0.262	0.531	0.735

.049   .105
.154

TABLE A-1   (continued)

n	x	0.05	0.1	0.2	0.3	0.4	0.5	0.6	0.7	0.8	0.9	0.95
7	0	0.698	0.478	0.210	0.082	0.028	0.008	0.002				
	1	0.257	0.372	0.367	0.247	0.131	0.055	0.017	0.004			
	2	0.041	0.124	0.275	0.318	0.261	0.164	0.077	0.025	0.004		
	3	0.004	0.023	0.115	0.227	0.290	0.273	0.194	0.097	0.029	0.003	
	4		0.003	0.029	0.097	0.194	0.273	0.290	0.227	0.115	0.023	0.004
	5			0.004	0.025	0.077	0.164	0.261	0.318	0.275	0.124	0.041
	6				0.004	0.017	0.055	0.131	0.247	0.367	0.372	0.257
	7					0.002	0.008	0.028	0.082	0.210	0.478	0.698
8	0	0.663	0.430	0.168	0.058	0.017	0.004	0.001				
	1	0.279	0.383	0.336	0.198	0.090	0.031	0.008	0.001			
	2	0.051	0.149	0.294	0.296	0.209	0.109	0.041	0.010	0.001		
	3	0.005	0.033	0.147	0.254	0.279	0.219	0.124	0.047	0.009		
	4		0.005	0.046	0.136	0.232	0.273	0.232	0.136	0.046	0.005	
	5			0.009	0.047	0.124	0.219	0.279	0.254	0.147	0.033	0.005
	6			0.001	0.010	0.041	0.109	0.209	0.296	0.294	0.149	0.051
	7				0.001	0.008	0.031	0.090	0.198	0.336	0.383	0.279
	8					0.001	0.004	0.017	0.058	0.168	0.430	0.663
9	0	0.630	0.387	0.134	0.040	0.010	0.02					
	1	0.299	0.387	0.302	0.156	0.060	0.018	0.004				
	2	0.063	0.172	0.302	0.267	0.161	0.070	0.021	0.004			
	3	0.008	0.045	0.176	0.267	0.251	0.164	0.074	0.021	0.003		
	4	0.001	0.007	0.066	0.172	0.251	0.246	0.167	0.074	0.017	0.001	
	5		0.001	0.017	0.074	0.167	0.246	0.251	0.172	0.066	0.007	0.001
	6			0.003	0.021	0.074	0.164	0.251	0.267	0.176	0.045	0.008
	7				0.004	0.021	0.070	0.161	0.267	0.302	0.172	0.063
	8					0.004	0.018	0.060	0.156	0.302	0.387	0.299
	9						0.002	0.010	0.040	0.134	0.387	0.630
10	0	0.599	0.349	0.107	0.028	0.006	0.001					
	1	0.315	0.387	0.268	0.121	0.040	0.010	0.002				
	2	0.075	0.194	0.302	0.233	0.121	0.044	0.011	0.001			
	3	0.010	0.057	0.201	0.267	0.215	0.117	0.042	0.009	0.001		
	4	0.001	0.011	0.088	0.200	0.251	0.205	0.111	0.037	0.006		
	5		0.001	0.026	0.103	0.201	0.246	0.201	0.103	0.026	0.001	
	6			0.006	0.037	0.111	0.205	0.251	0.200	0.088	0.011	0.001
	7			0.001	0.009	0.042	0.117	0.215	0.267	0.201	0.057	0.010
	8				0.001	0.011	0.044	0.121	0.233	0.302	0.194	0.075
	9					0.002	0.010	0.040	0.121	0.268	0.387	0.315
	10						0.001	0.006	0.028	0.107	0.349	0.599

(continued)

TABLE A-1 (continued)

n	x	\(p\) 0.05	0.1	0.2	0.3	0.4	0.5	0.6	0.7	0.8	0.9	0.95
11	0	0.569	0.314	0.086	0.020	0.004						
	1	0.329	0.384	0.236	0.093	0.027	0.005	0.001				
	2	0.087	0.213	0.295	0.200	0.089	0.027	0.005	0.001			
	3	0.014	0.071	0.221	0.257	0.177	0.081	0.023	0.004			
	4	0.001	0.016	0.111	0.220	0.236	0.161	0.070	0.017	0.002		
	5		0.002	0.039	0.132	0.221	0.226	0.147	0.057	0.010		
	6			0.010	0.057	0.147	0.226	0.221	0.132	0.039	0.002	
	7			0.002	0.017	0.070	0.161	0.236	0.220	0.111	0.016	0.001
	8				0.004	0.023	0.081	0.177	0.257	0.221	0.071	0.014
	9				0.001	0.005	0.027	0.089	0.200	0.295	0.213	0.087
	10					0.001	0.005	0.027	0.093	0.236	0.384	0.329
	11							0.004	0.020	0.086	0.314	0.569
12	0	0.540	0.282	0.069	0.014	0.002						
	1	0.341	0.377	0.206	0.071	0.017	0.003					
	2	0.099	0.230	0.283	0.168	0.064	0.016	0.002				
	3	0.017	0.085	0.236	0.240	0.142	0.054	0.012	0.001			
	4	0.002	0.021	0.133	0.231	0.213	0.121	0.042	0.008	0.001		
	5		0.004	0.053	0.158	0.227	0.193	0.101	0.029	0.003		
	6			0.016	0.079	0.177	0.226	0.177	0.079	0.016		
	7			0.003	0.029	0.101	0.193	0.227	0.158	0.053	0.004	
	8			0.001	0.008	0.042	0.121	0.213	0.231	0.133	0.021	0.002
	9				0.001	0.012	0.054	0.142	0.240	0.236	0.085	0.017
	10					0.002	0.016	0.064	0.168	0.283	0.230	0.099
	11						0.003	0.017	0.071	0.206	0.377	0.341
	12							0.002	0.014	0.069	0.282	0.540
13	0	0.513	0.254	0.055	0.010	0.001						
	1	0.351	0.367	0.179	0.054	0.011	0.002					
	2	0.111	0.245	0.268	0.139	0.045	0.010	0.001				
	3	0.021	0.100	0.246	0.218	0.111	0.035	0.006	0.001			
	4	0.003	0.028	0.154	0.234	0.184	0.087	0.024	0.003			
	5		0.006	0.069	0.180	0.221	0.157	0.066	0.014	0.001		
	6		0.001	0.023	0.103	0.197	0.209	0.131	0.044	0.006		
	7			0.006	0.044	0.131	0.209	0.197	0.103	0.023	0.001	
	8			0.001	0.014	0.066	0.157	0.221	0.180	0.069	0.006	
	9				0.003	0.024	0.087	0.184	0.234	0.154	0.028	0.003
	10				0.001	0.006	0.035	0.111	0.218	0.246	0.100	0.021
	11					0.001	0.010	0.045	0.139	0.268	0.245	0.111
	12						0.002	0.011	0.054	0.179	0.367	0.351
	13							0.001	0.010	0.055	0.254	0.513

TABLE A-1   (continued)

							p					
n	x	0.05	0.1	0.2	0.3	0.4	0.5	0.6	0.7	0.8	0.9	0.95
14	0	0.488	0.229	0.044	0.007	0.001						
	1	0.359	0.356	0.154	0.041	0.007	0.001					
	2	0.123	0.257	0.250	0.113	0.032	0.006	0.001				
	3	0.026	0.114	0.250	0.194	0.085	0.022	0.003				
	4	0.004	0.035	0.172	0.229	0.155	0.061	0.014	0.001			
	5		0.008	0.086	0.196	0.207	0.122	0.041	0.007			
	6		0.001	0.032	0.126	0.207	0.183	0.092	0.023	0.002		
	7			0.009	0.062	0.157	0.209	0.157	0.062	0.009		
	8			0.002	0.023	0.092	0.183	0.207	0.126	0.032	0.001	
	9				0.007	0.041	0.122	0.207	0.196	0.086	0.008	
	10				0.001	0.014	0.061	0.155	0.229	0.172	0.035	0.004
	11					0.003	0.022	0.085	0.194	0.250	0.114	0.026
	12					0.001	0.006	0.032	0.113	0.250	0.257	0.123
	13						0.001	0.007	0.041	0.154	0.356	0.359
	14							0.001	0.007	0.044	0.229	0.488
15	0	0.463	0.206	0.035	0.005							
	1	0.366	0.343	0.132	0.031	0.005						
	2	0.135	0.267	0.231	0.092	0.022	0.003					
	3	0.031	0.129	0.250	0.170	0.063	0.014	0.002				
	4	0.005	0.043	0.188	0.219	0.127	0.042	0.007	0.001			
	5	0.001	0.010	0.103	0.206	0.186	0.092	0.024	0.003			
	6		0.002	0.043	0.147	0.207	0.153	0.061	0.012	0.001		
	7			0.014	0.081	0.177	0.196	0.118	0.035	0.003		
	8			0.003	0.035	0.118	0.196	0.177	0.081	0.014		
	9			0.001	0.012	0.061	0.153	0.207	0.147	0.043	0.002	
	10				0.003	0.024	0.092	0.186	0.206	0.103	0.010	0.001
	11				0.001	0.007	0.042	0.127	0.219	0.188	0.043	0.005
	12					0.002	0.014	0.063	0.170	0.250	0.129	0.031
	13						0.003	0.022	0.092	0.231	0.267	0.135
	14							0.005	0.031	0.132	0.343	0.366
	15								0.005	0.035	0.206	0.463

Note: Due to rounding error, in some cases the column total may be slightly different from 1 for a given n and p.

TABLE A-2 The Standard Normal Distribution (Area from 0 to *z*, *z* Positive)

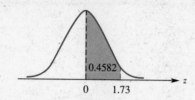

z	0.00	0.01	0.02	0.03	0.04	0.05	0.06	0.07	0.08	0.09
0.0	0.0000	0.0040	0.0080	0.0120	0.0160	0.0199	0.0239	0.0279	0.0319	0.0359
0.1	0.0398	0.0438	0.0478	0.0517	0.0557	0.0596	0.0636	0.0675	0.0714	0.0753
0.2	0.0793	0.0832	0.0871	0.0910	0.0948	0.0987	0.1026	0.1064	0.1103	0.1141
0.3	0.1179	0.1217	0.1255	0.1293	0.1331	0.1368	0.1406	0.1443	0.1480	0.1517
0.4	0.1554	0.1591	0.1628	0.1664	0.1700	0.1736	0.1772	0.1808	0.1844	0.1879
0.5	0.1915	0.1950	0.1985	0.2019	0.2054	0.2088	0.2123	0.2157	0.2190	0.2224
0.6	0.2257	0.2291	0.2324	0.2357	0.2389	0.2422	0.2454	0.2486	0.2517	0.2549
0.7	0.2580	0.2611	0.2642	0.2673	0.2704	0.2734	0.2764	0.2794	0.2823	0.2852
0.8	0.2881	0.2910	0.2939	0.2967	0.2995	0.3023	0.3051	0.3078	0.3106	0.3133
0.9	0.3159	0.3186	0.3212	0.3238	0.3264	0.3289	0.3315	0.3340	0.3365	0.3389
1.0	0.3413	0.3438	0.3461	0.3485	0.3508	0.3531	0.3554	0.3577	0.3599	0.3621
1.1	0.3643	0.3665	0.3686	0.3708	0.3729	0.3749	0.3770	0.3790	0.3810	0.3830
1.2	0.3849	0.3869	0.3888	0.3907	0.3925	0.3944	0.3962	0.3980	0.3997	0.4015
1.3	0.4032	0.4049	0.4066	0.4082	0.4099	0.4115	0.4131	0.4147	0.4162	0.4177
1.4	0.4192	0.4207	0.4222	0.4236	0.4251	0.4265	0.4279	0.4292	0.4306	0.4319
1.5	0.4332	0.4345	0.4357	0.4370	0.4382	0.4394	0.4406	0.4418	0.4429	0.4441
1.6	0.4452	0.4463	0.4474	0.4484	0.4495	0.4505	0.4515	0.4525	0.4535	0.4545
1.7	0.4554	0.4564	0.4573	0.4582	0.4591	0.4599	0.4608	0.4616	0.4625	0.4633
1.8	0.4641	0.4649	0.4656	0.4664	0.4671	0.4678	0.4686	0.4692	0.4699	0.4706
1.9	0.4713	0.4719	0.4726	0.4732	0.4738	0.4744	0.4750	0.4756	0.4761	0.4767
2.0	0.4772	0.4778	0.4783	0.4788	0.4793	0.4798	0.4803	0.4808	0.4812	0.4817
2.1	0.4821	0.4826	0.4830	0.4834	0.4838	0.4842	0.4846	0.4850	0.4854	0.4857
2.2	0.4861	0.4864	0.4868	0.4871	0.4875	0.4878	0.4881	0.4884	0.4887	0.4890
2.3	0.4893	0.4896	0.4898	0.4901	0.4904	0.4906	0.4909	0.4911	0.4913	0.4916
2.4	0.4918	0.4920	0.4922	0.4925	0.4927	0.4929	0.4931	0.4932	0.4934	0.4936
2.5	0.4938	0.4940	0.4941	0.4943	0.4945	0.4946	0.4948	0.4949	0.4951	0.4952
2.6	0.4953	0.4955	0.4956	0.4957	0.4959	0.4960	0.4961	0.4962	0.4963	0.4964
2.7	0.4965	0.4966	0.4967	0.4968	0.4969	0.4970	0.4971	0.4972	0.4973	0.4974
2.8	0.4974	0.4975	0.4976	0.4977	0.4977	0.4978	0.4979	0.4979	0.4980	0.4981
2.9	0.4981	0.4982	0.4982	0.4983	0.4984	0.4984	0.4985	0.4985	0.4986	0.4986
3.0	0.4987	0.4987	0.4987	0.4988	0.4988	0.4989	0.4989	0.4989	0.4990	0.4990

TABLE A-3   The Standard Normal Distribution (Areas in the Right Tail)

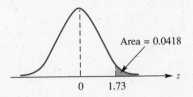

z	0.00	0.01	0.02	0.03	0.04	0.05	0.06	0.07	0.08	0.09
0.0	0.5000	0.4960	0.4920	0.4880	0.4840	0.4801	0.4761	0.4721	0.4681	0.4641
0.1	0.4602	0.4562	0.4522	0.4483	0.4443	0.4404	0.4364	0.4325	0.4286	0.4247
0.2	0.4207	0.4168	0.4129	0.4090	0.4052	0.4013	0.3974	0.3936	0.3897	0.3859
0.3	0.3821	0.3783	0.3745	0.3707	0.3669	0.3632	0.3594	0.3557	0.3520	0.3483
0.4	0.3446	0.3409	0.3372	0.3336	0.3300	0.3264	0.3228	0.3192	0.3156	0.3121
0.5	0.3085	0.3050	0.3015	0.2981	0.2946	0.2912	0.2877	0.2843	0.2810	0.2776
0.6	0.2743	0.2709	0.2676	0.2643	0.2611	0.2578	0.2546	0.2514	0.2483	0.2451
0.7	0.2420	0.2389	0.2358	0.2327	0.2296	0.2266	0.2236	0.2206	0.2177	0.2148
0.8	0.2119	0.2090	0.2061	0.2033	0.2005	0.1977	0.1949	0.1922	0.1894	0.1867
0.9	0.1841	0.1814	0.1788	0.1762	0.1736	0.1711	0.1685	0.1660	0.1635	0.1611
1.0	0.1587	0.1562	0.1539	0.1515	0.1492	0.1469	0.1446	0.1423	0.1401	0.1379
1.1	0.1357	0.1335	0.1314	0.1292	0.1271	0.1251	0.1230	0.1210	0.1190	0.1170
1.2	0.1151	0.1131	0.1112	0.1093	0.1075	0.1056	0.1038	0.1020	0.1003	0.0985
1.3	0.0968	0.0951	0.0934	0.0918	0.0901	0.0885	0.0869	0.0853	0.0838	0.0823
1.4	0.0808	0.0793	0.0778	0.0764	0.0749	0.0735	0.0721	0.0708	0.0694	0.0681
1.5	0.0668	0.0655	0.0643	0.0630	0.0618	0.0606	0.0594	0.0582	0.0571	0.0559
1.6	0.0548	0.0537	0.0526	0.0516	0.0505	0.0495	0.0485	0.0475	0.0465	0.0455
1.7	0.0446	0.0436	0.0427	0.0418	0.0409	0.0401	0.0392	0.0384	0.0375	0.0367
1.8	0.0359	0.0351	0.0344	0.0336	0.0329	0.0322	0.0314	0.0307	0.0301	0.0294
1.9	0.0287	0.0281	0.0274	0.0268	0.0262	0.0256	0.0250	0.0244	0.0239	0.0233
2.0	0.0228	0.0222	0.0217	0.0212	0.0207	0.0202	0.0197	0.0192	0.0188	0.0183
2.1	0.0179	0.0174	0.0170	0.0166	0.0162	0.0158	0.0154	0.0150	0.0146	0.0143
2.2	0.0139	0.0136	0.0132	0.0129	0.0125	0.0122	0.0119	0.0116	0.0113	0.0110
2.3	0.0107	0.0104	0.0102	0.0099	0.0096	0.0094	0.0091	0.0089	0.0087	0.0084
2.4	0.0082	0.0080	0.0078	0.0075	0.0073	0.0071	0.0069	0.0068	0.0066	0.0064
2.5	0.0062	0.0060	0.0059	0.0057	0.0055	0.0054	0.0052	0.0051	0.0049	0.0048
2.6	0.0047	0.0045	0.0044	0.0043	0.0041	0.0040	0.0039	0.0038	0.0037	0.0036
2.7	0.0035	0.0034	0.0033	0.0032	0.0031	0.0030	0.0029	0.0028	0.0027	0.0026
2.8	0.0026	0.0025	0.0024	0.0023	0.0023	0.0022	0.0021	0.0021	0.0020	0.0019
2.9	0.0019	0.0018	0.0018	0.0017	0.0016	0.0016	0.0015	0.0015	0.0014	0.0014
3.0	0.0013	0.0013	0.0013	0.0012	0.0012	0.0011	0.0011	0.0011	0.0010	0.0010

TABLE A-4 Values of *t* for Given Probability Levels

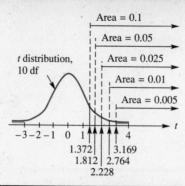

t distribution, 10 df

Area = 0.1
Area = 0.05
Area = 0.025
Area = 0.01
Area = 0.005

-3 -2 -1 0 1   4   *t*
1.372   3.169
1.812   2.764
2.228

Degrees of Freedom ν	\multicolumn{5}{c}{Area in the Right Tail under the Curve}
	0.1	0.05	0.025	0.01	0.005
	\multicolumn{5}{c}{Critical Values of t}
	$t_{\nu,\,0.1}$	$t_{\nu,\,0.05}$	$t_{\nu,\,0.025}$	$t_{\nu,\,0.01}$	$t_{\nu,\,0.005}$
1	3.078	6.314	12.706	31.821	63.657
2	1.886	2.920	4.303	6.965	9.925
3	1.638	2.353	3.182	4.541	5.841
4	1.533	2.132	2.776	3.747	4.604
5	1.476	2.015	2.571	3.365	4.032
6	1.440	1.943	2.447	3.143	3.707
7	1.415	1.895	2.365	2.998	3.499
8	1.397	1.860	2.306	2.896	3.355
9	1.383	1.833	2.262	2.821	3.250
10	1.372	1.812	2.228	2.764	3.169
11	1.363	1.796	2.201	2.718	3.106
12	1.356	1.782	2.179	2.681	3.055
13	1.350	1.771	2.160	2.650	3.012
14	1.345	1.761	2.145	2.624	2.977
15	1.341	1.753	2.131	2.602	2.947
16	1.337	1.746	2.120	2.583	2.921
17	1.333	1.740	2.110	2.567	2.898
18	1.330	1.734	2.101	2.552	2.878
19	1.328	1.729	2.093	2.539	2.861
20	1.325	1.725	2.086	2.528	2.845
21	1.323	1.721	2.080	2.518	2.831
22	1.321	1.717	2.074	2.508	2.819
23	1.319	1.714	2.069	2.500	2.807
24	1.318	1.711	2.064	2.492	2.797
25	1.316	1.708	2.060	2.485	2.787
26	1.315	1.706	2.056	2.479	2.779
27	1.314	1.703	2.052	2.473	2.771
28	1.313	1.701	2.048	2.467	2.763
29	1.311	1.699	2.045	2.462	2.756
30	1.310	1.697	2.042	2.457	2.750
⋮	⋮	⋮	⋮	⋮	⋮
∞	1.282	1.645	1.960	2.326	2.576

Table A-4 is taken from Table III of Fisher and Yates: *Statistical Tables for Biological, Agricultural and Medical Research,* published by Longman Group Ltd., London (previously published by Oliver & Boyd, Edinburgh), and by permission of the authors and publishers.

TABLE A-5 Values of the Chi-Square for Given Probability Levels

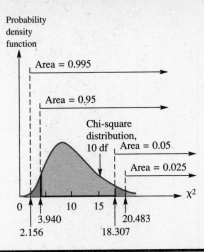

Area in the Right Tail under the Curve

Degrees of Freedom ν	0.995 $\chi^2_{\nu,\,0.995}$	0.99 $\chi^2_{\nu,\,0.99}$	0.975 $\chi^2_{\nu,\,0.975}$	0.95 $\chi^2_{\nu,\,0.95}$	0.05 $\chi^2_{\nu,\,0.05}$	0.025 $\chi^2_{\nu,\,0.025}$	0.01 $\chi^2_{\nu,\,0.01}$	0.005 $\chi^2_{\nu,\,0.005}$
1	0.0^4393	0.0^3157	0.0^3982	0.0^2393	3.841	5.024	6.635	7.879
2	0.0100	0.0201	0.0506	0.103	5.991	7.378	9.210	10.597
3	0.0717	0.115	0.216	0.352	7.815	9.348	11.345	12.838
4	0.207	0.297	0.484	0.711	9.488	11.143	13.277	14.860
5	0.412	0.554	0.831	1.145	11.070	12.832	15.086	16.750
6	0.676	0.872	1.237	1.635	12.592	14.449	16.812	18.548
7	0.989	1.239	1.690	2.167	14.067	16.013	18.475	20.278
8	1.344	1.646	2.180	2.733	15.507	17.535	20.090	21.955
9	1.735	2.088	2.700	3.325	16.919	19.023	21.666	23.589
10	2.156	2.558	3.247	3.940	18.307	20.483	23.209	25.188
11	2.603	3.053	3.816	4.575	19.675	21.920	24.725	26.757
12	3.074	3.571	4.404	5.226	21.026	23.337	26.217	28.300
13	3.565	4.107	5.009	5.892	22.362	24.736	27.688	29.819
14	4.075	4.660	5.629	6.571	23.685	26.119	29.141	31.319
15	4.601	5.229	6.262	7.261	24.996	27.488	30.578	32.801
16	5.142	5.812	6.908	7.962	26.296	28.845	32.000	34.267
17	5.697	6.408	7.564	8.672	27.587	30.191	33.409	35.718
18	6.265	7.015	8.231	9.390	28.869	31.526	34.805	37.156
19	6.844	7.633	8.907	10.117	30.144	32.852	36.191	38.582
20	7.434	8.260	9.591	10.851	31.410	34.170	37.566	39.997
21	8.034	8.897	10.283	11.591	32.671	35.479	38.932	41.401
22	8.643	9.542	10.982	12.338	33.924	36.781	40.289	42.796
23	9.260	10.196	11.689	13.091	35.172	38.076	41.638	44.181
24	9.886	10.856	12.401	13.848	36.415	39.364	42.980	45.558
25	10.520	11.524	13.120	14.611	37.652	40.646	44.314	46.928
26	11.160	12.198	13.844	15.379	38.885	41.923	45.652	48.290
27	11.808	12.879	14.573	16.151	40.113	43.194	46.963	49.645
28	12.461	13.565	15.308	16.928	41.337	44.461	48.278	50.993
29	13.121	14.256	16.047	17.708	42.557	45.722	49.588	52.336
30	13.787	14.953	16.791	18.493	43.773	46.979	50.892	53.672

Abridged from Table 8 of *Biometrika Tables for Statisticians*, Vol. 1, by permission of the Biometrika Trustees.

TABLE A-6 Percentage Points of the F Distribution

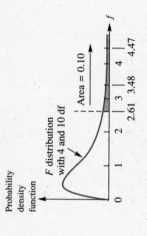

F distribution with 4 and 10 df

Area = 0.10

2.61 3.48   4.47

(a) Upper 10 percent points

UPPER 10 PERCENT POINTS

v_1, Degrees of Freedom of the Numerator

v_2, Degrees of freedom of the denominator

v_2	1	2	3	4	5	6	7	8	9	10	12	15	20	24	30	40	60	120	∞
1	39.86	49.50	53.59	55.83	57.24	58.20	58.91	59.44	59.86	60.19	60.71	61.22	61.74	62.00	62.26	62.53	62.79	63.06	63.33
2	8.53	9.00	9.16	9.24	9.29	9.33	9.35	9.37	9.38	9.39	9.41	9.42	9.44	9.45	9.46	9.47	9.47	9.48	9.49
3	5.54	5.46	5.39	5.34	5.31	5.28	5.27	5.25	5.24	5.23	5.22	5.20	5.18	5.18	5.17	5.16	5.15	5.14	5.13
4	4.54	4.32	4.19	4.11	4.05	4.01	3.98	3.95	3.94	3.92	3.90	3.87	3.84	3.83	3.82	3.80	3.79	3.78	3.76
5	4.06	3.78	3.62	3.52	3.45	3.40	3.37	3.34	3.32	3.30	3.27	3.24	3.21	3.19	3.17	3.16	3.14	3.12	3.10
6	3.78	3.46	3.29	3.18	3.11	3.05	3.01	2.98	2.96	2.94	2.90	2.87	2.84	2.82	2.80	2.78	2.76	2.74	2.72
7	3.59	3.26	3.07	2.96	2.88	2.83	2.78	2.75	2.72	2.70	2.67	2.63	2.59	2.58	2.56	2.54	2.51	2.49	2.47
8	3.46	3.11	2.92	2.81	2.73	2.67	2.62	2.59	2.56	2.54	2.50	2.46	2.42	2.40	2.38	2.36	2.34	2.32	2.29
9	3.36	3.01	2.81	2.69	2.61	2.55	2.51	2.47	2.44	2.42	2.38	2.34	2.30	2.28	2.25	2.23	2.21	2.18	2.16
10	3.29	2.92	2.73	2.61	2.52	2.46	2.41	2.38	2.35	2.32	2.28	2.24	2.20	2.18	2.16	2.13	2.11	2.08	2.06
11	3.23	2.86	2.66	2.54	2.45	2.39	2.34	2.30	2.27	2.25	2.21	2.17	2.12	2.10	2.08	2.05	2.03	2.00	1.97
12	3.18	2.81	2.61	2.48	2.39	2.33	2.28	2.24	2.21	2.19	2.15	2.10	2.06	2.04	2.01	1.99	1.96	1.93	1.90
13	3.14	2.76	2.56	2.43	2.35	2.28	2.23	2.20	2.16	2.14	2.10	2.05	2.01	1.98	1.96	1.93	1.90	1.88	1.85
14	3.10	2.73	2.52	2.39	2.31	2.24	2.19	2.15	2.12	2.10	2.05	2.01	1.96	1.94	1.91	1.89	1.86	1.83	1.80
15	3.07	2.70	2.49	2.36	2.27	2.21	2.16	2.12	2.09	2.06	2.02	1.97	1.92	1.90	1.87	1.85	1.82	1.79	1.76
16	3.05	2.67	2.46	2.33	2.24	2.18	2.13	2.09	2.06	2.03	1.99	1.94	1.89	1.87	1.84	1.81	1.78	1.75	1.72
17	3.03	2.64	2.44	2.31	2.22	2.15	2.10	2.06	2.03	2.00	1.96	1.91	1.86	1.84	1.81	1.78	1.75	1.72	1.69
18	3.01	2.62	2.42	2.29	2.20	2.13	2.08	2.04	2.00	1.98	1.93	1.89	1.84	1.81	1.78	1.75	1.72	1.69	1.66
19	2.99	2.61	2.40	2.27	2.18	2.11	2.06	2.02	1.98	1.96	1.91	1.86	1.81	1.79	1.76	1.73	1.70	1.67	1.63

TABLE A-6 (continued)

UPPER 10 PERCENT POINTS

ν_1, Degrees of Freedom of the Numerator

ν_2	1	2	3	4	5	6	7	8	9	10	12	15	20	24	30	40	60	120	∞
20	2.97	2.59	2.38	2.25	2.16	2.09	2.04	2.00	1.96	1.94	1.89	1.84	1.79	1.77	1.74	1.71	1.68	1.64	1.61
21	2.96	2.57	2.36	2.23	2.14	2.08	2.02	1.98	1.95	1.92	1.87	1.83	1.78	1.75	1.72	1.69	1.66	1.62	1.59
22	2.95	2.56	2.35	2.22	2.13	2.06	2.01	1.97	1.93	1.90	1.86	1.81	1.76	1.73	1.70	1.67	1.64	1.60	1.57
23	2.94	2.55	2.34	2.21	2.11	2.05	1.99	1.95	1.92	1.89	1.84	1.80	1.74	1.72	1.69	1.66	1.62	1.59	1.55
24	2.93	2.54	2.33	2.19	2.10	2.04	1.98	1.94	1.91	1.88	1.83	1.78	1.73	1.70	1.67	1.64	1.61	1.57	1.53
25	2.92	2.53	2.32	2.18	2.09	2.02	1.97	1.93	1.89	1.87	1.82	1.77	1.72	1.69	1.66	1.63	1.59	1.56	1.52
26	2.91	2.52	2.31	2.17	2.08	2.01	1.96	1.92	1.88	1.86	1.81	1.76	1.71	1.68	1.65	1.61	1.58	1.54	1.50
27	2.90	2.51	2.30	2.17	2.07	2.00	1.95	1.91	1.87	1.85	1.80	1.75	1.70	1.67	1.64	1.60	1.57	1.53	1.49
28	2.89	2.50	2.29	2.16	2.06	2.00	1.94	1.90	1.87	1.84	1.79	1.74	1.69	1.66	1.63	1.59	1.56	1.52	1.48
29	2.89	2.50	2.28	2.15	2.06	1.99	1.93	1.89	1.86	1.83	1.78	1.73	1.68	1.65	1.62	1.58	1.55	1.51	1.47
30	2.88	2.49	2.28	2.14	2.05	1.98	1.93	1.88	1.85	1.82	1.77	1.72	1.67	1.64	1.61	1.57	1.54	1.50	1.46
40	2.84	2.44	2.23	2.09	2.00	1.93	1.87	1.83	1.79	1.76	1.71	1.66	1.61	1.57	1.54	1.51	1.47	1.42	1.38
60	2.79	2.39	2.18	2.04	1.95	1.87	1.82	1.77	1.74	1.71	1.66	1.60	1.54	1.51	1.48	1.44	1.40	1.35	1.29
120	2.75	2.35	2.13	1.99	1.90	1.82	1.77	1.72	1.68	1.65	1.60	1.55	1.48	1.45	1.41	1.37	1.32	1.26	1.19
∞	2.71	2.30	2.08	1.94	1.85	1.77	1.72	1.67	1.63	1.60	1.55	1.49	1.42	1.38	1.34	1.30	1.24	1.17	1.00

ν_2, Degrees of freedom of the denominator

(continued)

TABLE A-6 (continued)

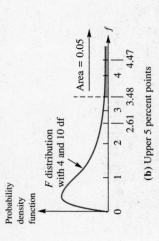

Probability density function

F distribution with 4 and 10 df

Area = 0.05

0   1   2   3   4   f
2.61 3.48 4.47

(b) Upper 5 percent points

UPPER 5 PERCENT POINTS

ν_1, Degrees of Freedom of the Numerator

ν_2	1	2	3	4	5	6	7	8	9	10	12	15	20	24	30	40	60	120	∞
1	161.4	199.5	215.7	224.6	230.2	234.0	236.8	238.9	240.5	241.9	243.9	245.9	248.0	249.1	250.1	251.1	252.2	253.3	254.3
2	18.51	19.00	19.16	19.25	19.30	19.33	19.35	19.37	19.38	19.40	19.41	19.43	19.45	19.45	19.46	19.47	19.48	19.49	19.50
3	10.13	9.55	9.28	9.12	9.01	8.94	8.89	8.85	8.81	8.79	8.74	8.70	8.66	8.64	8.62	8.59	8.57	8.55	8.53
4	7.71	6.94	6.59	6.39	6.26	6.16	6.09	6.04	6.00	5.96	5.91	5.86	5.80	5.77	5.75	5.72	5.69	5.66	5.63
5	6.61	5.79	5.41	5.19	5.05	4.95	4.88	4.82	4.77	4.74	4.68	4.62	4.56	4.53	4.50	4.46	4.43	4.40	4.36
6	5.99	5.14	4.76	4.53	4.39	4.28	4.21	4.15	4.10	4.06	4.00	3.94	3.87	3.84	3.81	3.77	3.74	3.70	3.67
7	5.59	4.74	4.35	4.12	3.97	3.87	3.79	3.73	3.68	3.64	3.57	3.51	3.44	3.41	3.38	3.34	3.30	3.27	3.23
8	5.32	4.46	4.07	3.84	3.69	3.58	3.50	3.44	3.39	3.35	3.28	3.22	3.15	3.12	3.08	3.04	3.01	2.97	2.93
9	5.12	4.26	3.86	3.63	3.48	3.37	3.29	3.23	3.18	3.14	3.07	3.01	2.94	2.90	2.86	2.83	2.79	2.75	2.71
10	4.96	4.10	3.71	3.48	3.33	3.22	3.14	3.07	3.02	2.98	2.91	2.85	2.77	2.74	2.70	2.66	2.62	2.58	2.54
11	4.84	3.98	3.59	3.36	3.20	3.09	3.01	2.95	2.90	2.85	2.79	2.72	2.65	2.61	2.57	2.53	2.49	2.45	2.40
12	4.75	3.89	3.49	3.26	3.11	3.00	2.91	2.85	2.80	2.75	2.69	2.62	2.54	2.51	2.47	2.43	2.38	2.34	2.30
13	4.67	3.81	3.41	3.18	3.03	2.92	2.83	2.77	2.71	2.67	2.60	2.53	2.46	2.42	2.38	2.34	2.30	2.25	2.21
14	4.60	3.74	3.34	3.11	2.96	2.85	2.76	2.70	2.65	2.60	2.53	2.46	2.39	2.35	2.31	2.27	2.22	2.18	2.13
15	4.54	3.68	3.29	3.06	2.90	2.79	2.71	2.64	2.59	2.54	2.48	2.40	2.33	2.29	2.25	2.20	2.16	2.11	2.07
16	4.49	3.63	3.24	3.01	2.85	2.74	2.66	2.59	2.54	2.49	2.42	2.35	2.28	2.24	2.19	2.15	2.11	2.06	2.01
17	4.45	3.59	3.20	2.96	2.81	2.70	2.61	2.55	2.49	2.45	2.38	2.31	2.23	2.19	2.15	2.10	2.06	2.01	1.96
18	4.41	3.55	3.16	2.93	2.77	2.66	2.58	2.51	2.46	2.41	2.34	2.27	2.19	2.15	2.11	2.06	2.02	1.97	1.92
19	4.38	3.52	3.13	2.90	2.74	2.63	2.54	2.48	2.42	2.38	2.31	2.23	2.16	2.11	2.07	2.03	1.98	1.93	1.88

ν_2, Degrees of freedom of the denominator

UPPER 5 PERCENT POINTS

	ν_1, Degrees of Freedom of the Numerator																		
	1	2	3	4	5	6	7	8	9	10	12	15	20	24	30	40	60	120	∞
20	4.35	3.49	3.10	2.87	2.71	2.60	2.51	2.45	2.39	2.35	2.28	2.20	2.12	2.08	2.04	1.99	1.95	1.90	1.84
21	4.32	3.47	3.07	2.84	2.68	2.57	2.49	2.42	2.37	2.32	2.25	2.18	2.10	2.05	2.01	1.96	1.92	1.87	1.81
22	4.30	3.44	3.05	2.82	2.66	2.55	2.46	2.40	2.34	2.30	2.23	2.15	2.07	2.03	1.98	1.94	1.89	1.84	1.78
23	4.28	3.42	3.03	2.80	2.64	2.53	2.44	2.37	2.32	2.27	2.20	2.13	2.05	2.01	1.96	1.91	1.86	1.81	1.76
24	4.26	3.40	3.01	2.78	2.62	2.51	2.42	2.36	2.30	2.25	2.18	2.11	2.03	1.98	1.94	1.89	1.84	1.79	1.73
25	4.24	3.99	2.99	2.76	2.60	2.49	2.40	2.34	2.28	2.24	2.16	2.09	2.01	1.96	1.92	1.87	1.82	1.77	1.71
26	4.23	3.37	2.98	2.74	2.59	2.47	2.39	2.32	2.27	2.22	2.15	2.07	1.99	1.95	1.90	1.85	1.80	1.75	1.69
27	4.21	3.35	2.96	2.73	2.57	2.46	2.37	2.31	2.25	2.20	2.13	2.06	1.97	1.93	1.88	1.84	1.79	1.73	1.67
28	4.20	3.34	2.95	2.71	2.56	2.45	2.36	2.29	2.24	2.19	2.12	2.04	1.96	1.91	1.87	1.82	1.77	1.71	1.65
29	4.18	3.33	2.93	2.70	2.55	2.43	2.35	2.28	2.22	2.18	2.10	2.03	1.94	1.90	1.85	1.81	1.75	1.70	1.64
30	4.17	3.32	2.92	2.69	2.53	2.42	2.33	2.27	2.21	2.16	2.09	2.01	1.93	1.89	1.84	1.79	1.74	1.68	1.62
40	4.08	3.23	2.84	2.61	2.45	2.34	2.25	2.18	2.12	2.08	2.00	1.92	1.84	1.79	1.74	1.69	1.64	1.58	1.51
60	4.00	3.15	2.76	2.53	2.37	2.25	2.17	2.10	2.04	1.99	1.92	1.84	1.75	1.70	1.65	1.59	1.53	1.47	1.39
120	3.92	3.07	2.68	2.45	2.29	2.17	2.09	2.02	1.96	1.91	1.83	1.75	1.66	1.61	1.55	1.50	1.43	1.35	1.25
∞	3.84	3.00	2.60	2.37	2.21	2.10	2.01	1.94	1.88	1.83	1.75	1.67	1.57	1.52	1.46	1.39	1.32	1.22	1.00

ν_2, Degrees of freedom of the denominator

(continued)

TABLE A-6 (continued)

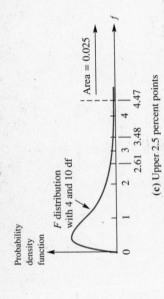

Probability
density
function

F distribution
with 4 and 10 df

Area = 0.025

2.61 3.48   4.47

(c) Upper 2.5 percent points

UPPER 2.5 PERCENT POINTS

ν_2	ν_1, Degrees of Freedom of the Numerator																		
	1	2	3	4	5	6	7	8	9	10	12	15	20	24	30	40	60	120	∞
1	647.8	799.5	864.2	899.6	921.8	937.1	948.2	956.7	963.3	968.6	976.7	984.9	993.1	997.2	1001.	1006.	1010.	1014.	1018.
2	38.51	39.00	39.17	39.25	39.30	39.33	39.36	39.37	39.39	39.40	39.41	39.43	39.45	39.46	39.46	39.47	39.48	39.49	30.50
3	17.44	16.04	15.44	15.10	14.88	14.73	14.62	14.54	14.47	14.42	14.34	14.25	14.17	14.12	14.08	14.04	13.99	13.95	13.90
4	12.22	10.65	9.98	9.60	9.36	9.20	9.07	8.98	8.90	8.84	8.75	8.66	8.56	8.51	8.46	8.41	8.36	8.31	8.26
5	10.01	8.43	7.76	7.39	7.15	6.98	6.85	6.76	6.68	6.62	6.52	6.43	6.33	6.28	6.23	6.18	6.12	6.07	6.02
6	8.81	7.26	6.60	6.23	5.99	5.82	5.70	5.60	5.52	5.46	5.37	5.27	5.17	5.12	5.07	5.01	4.96	4.90	4.85
7	8.07	6.54	5.89	5.52	5.29	5.12	4.99	4.90	4.82	4.76	4.67	4.57	4.47	4.42	4.36	4.31	4.25	4.20	4.14
8	7.57	6.06	5.42	5.05	4.82	4.65	4.53	4.43	4.36	4.30	4.20	4.10	4.00	3.95	3.89	3.84	3.78	3.73	3.67
9	7.21	5.71	5.08	4.72	4.48	4.32	4.20	4.10	4.03	3.96	3.87	3.77	3.67	3.61	3.56	3.51	3.45	3.39	3.33
10	6.94	5.46	4.83	4.47	4.24	4.07	3.95	3.85	3.78	3.72	3.62	3.52	3.42	3.37	3.31	3.26	3.20	3.14	3.08
11	6.72	5.26	4.63	4.28	4.04	3.88	3.76	3.66	3.59	3.53	3.43	3.33	3.23	3.17	3.12	3.06	3.00	2.94	2.88
12	6.55	5.10	4.47	4.12	3.89	3.73	3.61	3.51	3.44	3.37	3.28	3.18	3.07	3.02	2.96	2.91	2.85	2.79	2.72
13	6.41	4.97	4.35	4.00	3.77	3.60	3.48	3.39	3.31	3.25	3.15	3.05	2.95	2.89	2.84	2.78	2.72	2.66	2.60
14	6.30	4.86	4.24	3.89	3.66	3.50	3.38	3.29	3.21	3.15	3.05	2.95	2.84	2.79	2.73	2.67	2.61	2.55	2.49
15	6.20	4.77	4.15	3.80	3.58	3.41	3.29	3.20	3.12	3.06	2.96	2.86	2.76	2.70	2.64	2.59	2.52	2.46	2.40
16	6.12	4.69	4.08	3.73	3.50	3.34	3.22	3.12	3.05	2.99	2.89	2.79	2.68	2.63	2.57	2.51	2.45	2.38	2.32
17	6.04	4.62	4.01	3.66	3.44	3.28	3.16	3.06	2.98	2.92	2.82	2.72	2.62	2.56	2.50	2.44	2.38	2.32	2.25
18	5.98	4.56	3.95	3.61	3.38	3.22	3.10	3.01	2.93	2.87	2.77	2.67	2.56	2.50	2.44	2.38	2.32	2.26	2.19
19	5.92	4.51	3.90	3.56	3.33	3.17	3.05	2.96	2.88	2.82	2.72	2.62	2.51	2.45	2.39	2.33	2.27	2.20	2.13

ν_2, *Degrees of freedom of the denominator*

UPPER 2.5 PERCENT POINTS

	ν_1, Degrees of Freedom of the Numerator																		
	1	2	3	4	5	6	7	8	9	10	12	15	20	24	30	40	60	120	∞
20	5.87	4.46	3.86	3.51	3.29	3.13	3.01	2.91	2.84	2.77	2.68	2.57	2.46	2.41	2.35	2.29	2.22	2.16	2.09
21	5.83	4.42	3.82	3.48	3.25	3.09	2.97	2.87	2.80	2.73	2.64	2.53	2.42	2.37	2.31	2.25	2.18	2.11	2.04
22	5.79	4.38	3.78	3.44	3.22	3.05	2.93	2.84	2.76	2.70	2.60	2.50	2.39	2.33	2.27	2.21	2.14	2.08	2.00
23	5.75	4.35	3.75	3.41	3.18	3.02	2.90	2.81	2.73	2.67	2.57	2.47	2.36	2.30	2.24	2.18	2.11	2.04	1.97
24	5.72	4.32	3.72	3.38	3.15	2.99	2.87	2.78	2.70	2.64	2.54	2.44	2.33	2.27	2.21	2.15	2.08	2.01	1.94
25	5.69	4.29	3.69	3.35	3.13	2.97	2.85	2.75	2.68	2.61	2.51	2.41	2.30	2.24	2.18	2.12	2.05	1.98	1.91
26	5.66	4.27	3.67	3.33	3.10	2.94	2.82	2.73	2.65	2.59	2.49	2.39	2.28	2.22	2.16	2.09	2.03	1.95	1.88
27	5.63	4.24	3.65	3.31	3.08	2.92	2.80	2.71	2.63	2.57	2.47	2.36	2.25	2.19	2.13	2.07	2.00	1.93	1.85
28	5.61	4.22	3.63	3.29	3.06	2.90	2.78	2.69	2.61	2.55	2.45	2.34	2.23	2.17	2.11	2.05	1.98	1.91	1.83
29	5.59	4.20	3.61	3.27	3.04	2.88	2.76	2.67	2.59	2.53	2.43	2.32	2.21	2.15	2.09	2.03	1.96	1.89	1.81
30	5.57	4.18	3.59	3.25	3.03	2.87	2.75	2.65	2.57	2.51	2.41	2.31	2.20	2.14	2.07	2.01	1.94	1.87	1.79
40	5.42	4.05	3.46	3.13	2.90	2.74	2.62	2.53	2.45	2.39	2.29	2.18	2.07	2.01	1.94	1.88	1.80	1.72	1.64
60	5.29	3.93	3.34	3.01	2.79	2.63	2.51	2.41	2.33	2.27	2.17	2.06	1.94	1.88	1.82	1.74	1.67	1.58	1.48
120	5.15	3.80	3.23	2.89	2.67	2.52	2.39	2.30	2.22	2.16	2.05	1.94	1.82	1.76	1.69	1.61	1.53	1.43	1.31
∞	5.02	3.69	3.12	2.79	2.57	2.41	2.29	2.19	2.11	2.05	1.94	1.83	1.71	1.64	1.57	1.48	1.39	1.27	1.00

ν_2, Degrees of freedom of the denominator

(continued)

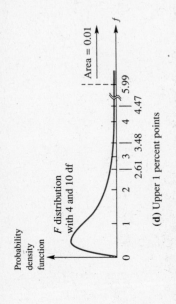

(d) Upper 1 percent points

F distribution with 4 and 10 df — Area = 0.01

UPPER 1 PERCENT POINTS

ν_1, Degrees of Freedom of the Numerator

ν_2	1	2	3	4	5	6	7	8	9	10	12	15	20	24	30	40	60	120	∞
1	4052	4999.5	5403	5625	5764	5859	5928	5981	6022	6056	6106	6157	6209	6235	6261	6287	6313	6339	6366
2	98.50	99.00	99.17	99.25	99.30	99.33	99.36	99.37	99.39	99.40	99.42	99.43	99.45	99.46	99.47	99.47	99.48	99.49	99.50
3	34.12	30.82	29.46	28.71	28.24	27.91	27.67	27.49	27.35	27.23	27.05	26.87	26.69	26.60	26.50	26.41	26.32	26.22	26.13
4	21.20	18.00	16.69	15.98	15.52	15.21	14.98	14.80	14.66	14.55	14.37	14.20	14.02	13.93	13.84	13.75	13.65	13.56	13.46
5	16.26	13.27	12.06	11.39	10.97	10.67	10.46	10.29	10.16	10.05	9.89	9.72	9.55	9.47	9.38	9.29	9.20	9.11	9.02
6	13.75	10.92	9.78	9.15	8.75	8.47	8.26	8.10	7.98	7.87	7.72	7.56	7.40	7.31	7.23	7.14	7.06	6.97	6.88
7	12.25	9.55	8.45	7.85	7.46	7.19	6.99	6.84	6.72	6.62	6.47	6.31	6.16	6.07	5.99	5.91	5.82	5.74	5.65
8	11.26	8.65	7.59	7.01	6.63	6.37	6.18	6.03	5.91	5.81	5.67	5.52	5.36	5.28	5.20	5.12	5.03	4.95	4.86
9	10.56	8.02	6.99	6.42	6.06	5.80	5.61	5.47	5.35	5.26	5.11	4.96	4.81	4.73	4.65	4.57	4.48	4.40	4.31
10	10.04	7.56	6.55	5.99	5.64	5.39	5.20	5.06	4.94	4.85	4.71	4.56	4.41	4.33	4.25	4.17	4.08	4.00	3.91
11	9.65	7.21	6.22	5.67	5.32	5.07	4.89	4.74	4.63	4.54	4.40	4.25	4.10	4.02	3.94	3.86	3.78	3.69	3.60
12	9.33	6.93	5.95	5.41	5.06	4.82	4.64	4.50	4.39	4.30	4.16	4.01	3.86	3.78	3.70	3.62	3.54	3.45	3.36
13	9.07	6.70	5.74	5.21	4.86	4.62	4.41	4.30	4.19	4.10	3.96	3.82	3.66	3.59	3.51	3.43	3.34	3.25	3.17
14	8.86	6.51	5.56	5.04	4.69	4.46	4.28	4.14	4.03	3.94	3.80	3.66	3.51	3.43	3.35	3.27	3.18	3.09	3.00
15	8.68	6.36	5.42	4.89	4.56	4.32	4.14	4.00	3.89	3.80	3.67	3.52	3.37	3.29	3.21	3.13	3.05	2.96	2.87
16	8.53	6.23	5.29	4.77	4.44	4.20	4.03	3.89	3.78	3.69	3.55	3.41	3.26	3.18	3.10	3.02	2.93	2.84	2.75
17	8.40	6.11	5.18	4.67	4.34	4.10	3.93	3.79	3.68	3.59	3.46	3.31	3.16	3.08	3.00	2.92	2.83	2.75	2.65
18	8.29	6.01	5.09	4.58	4.25	4.01	3.84	3.71	3.60	3.51	3.37	3.23	3.08	3.00	2.92	2.84	2.75	2.66	2.57
19	8.18	5.93	5.01	4.50	4.17	3.94	3.77	3.63	3.52	3.43	3.30	3.15	3.00	2.92	2.84	2.76	2.67	2.58	2.49

ν_2, Degrees of freedom of the denominator

TABLE A-6 (continued)

UPPER 1 PERCENT POINTS

ν_1, Degrees of Freedom of the Numerator

ν_2	1	2	3	4	5	6	7	8	9	10	12	15	20	24	30	40	60	120	∞
20	8.10	5.85	4.94	4.43	4.10	3.87	3.70	3.56	3.46	3.37	3.23	3.09	2.94	2.86	2.78	2.69	2.61	2.52	2.42
21	8.02	5.78	4.87	4.37	4.04	3.81	3.64	3.51	3.40	3.31	3.17	3.03	2.88	2.80	2.72	2.64	2.55	2.46	2.36
22	7.95	5.72	4.82	4.31	3.99	3.76	3.59	3.45	3.35	3.26	3.12	2.98	2.83	2.75	2.67	2.58	2.50	2.40	2.31
23	7.88	5.66	4.76	4.26	3.94	3.71	3.54	3.41	3.30	3.21	3.07	2.93	2.78	2.70	2.62	2.54	2.45	2.35	2.26
24	7.82	5.61	4.72	4.22	3.90	3.67	3.50	3.36	3.26	3.17	3.03	2.89	2.74	2.66	2.58	2.49	2.40	2.31	2.21
25	7.77	5.57	4.68	4.18	3.85	3.63	3.46	3.32	3.22	3.13	2.99	2.85	2.70	2.62	2.54	2.45	2.36	2.27	2.17
26	7.72	5.53	4.64	4.14	3.82	3.59	3.42	3.29	3.18	3.09	2.96	2.81	2.66	2.58	2.50	2.42	2.33	2.23	2.13
27	7.68	5.49	4.60	4.11	3.78	3.56	3.39	3.26	3.15	3.06	2.93	2.78	2.63	2.55	2.47	2.38	2.29	2.20	2.10
28	7.64	5.45	4.57	4.07	3.75	3.53	3.36	3.23	3.12	3.03	2.90	2.75	2.60	2.52	2.44	2.35	2.26	2.17	2.06
29	7.60	5.42	4.54	4.04	3.73	3.50	3.33	3.20	3.09	3.00	2.87	2.73	2.57	2.49	2.41	2.33	2.23	2.14	2.03
30	7.56	5.39	4.51	4.02	3.70	3.47	3.30	3.17	3.07	2.98	2.84	2.70	2.55	2.47	2.39	2.30	2.21	2.11	2.01
40	7.31	5.18	4.31	3.83	3.51	3.29	3.12	2.99	2.89	2.80	2.66	2.52	2.37	2.29	2.20	2.11	2.02	1.92	1.80
60	7.08	4.98	4.13	3.65	3.34	3.12	2.95	2.82	2.72	2.63	2.50	2.35	2.20	2.12	2.03	1.94	1.84	1.73	1.60
120	6.85	4.79	3.95	3.48	3.17	2.96	2.79	2.66	2.56	2.47	2.34	2.19	2.03	1.95	1.86	1.76	1.66	1.53	1.38
∞	6.63	4.61	3.78	3.32	3.02	2.80	2.64	2.51	2.41	2.32	2.18	2.04	1.88	1.79	1.70	1.59	1.47	1.32	1.00

ν_2, Degrees of freedom of the denominator

This table is reproduced from Table 18 of the *Biometrika Tables for Statisticians*, Vol. 1, by permission of the Biometrika Trustees.

TABLE A-7 Table of Random Numbers

				Random Digits					
12159	66144	05091	13446	45653	13684	66024	91410	51351	22772
30156	90519	95785	47544	66735	35754	11088	67310	19720	08379
59069	01722	53338	41942	65118	71236	01932	70343	25812	62275
54107	58081	82470	59407	13475	95872	16268	78436	39251	64247
99681	81295	06315	28212	45029	57701	96327	85436	33614	29070
27252	37875	53679	01889	35714	63534	63791	76342	47717	73684
93259	74585	11863	78985	03881	46567	93696	93521	54970	37607
84068	43759	75814	32261	12728	09636	22336	75629	01017	45503
68582	97054	28251	63787	57285	18854	35006	16343	51867	67979
60646	11298	19680	10087	66391	70853	24423	73007	74958	29020
97437	52922	80739	59178	50628	61017	51652	40915	94696	67843
58009	20681	98823	50979	01237	70152	13711	73916	87902	84759
77211	70110	93803	60135	22881	13423	30999	07104	27400	25414
54256	84591	65302	99257	92970	28924	36632	54044	91798	78018
36493	69330	94069	39544	14050	03476	25804	49350	92525	87941
87569	22661	55970	52623	35419	76660	42394	63210	62626	00581
22896	62237	39635	63725	10463	87944	92075	90914	30599	35671
02697	33230	64527	97210	41359	79399	13941	88378	68503	33609
20080	15652	37216	00679	02088	34138	13953	68939	05630	27653
20550	95151	60557	57449	77115	87372	02574	07851	22428	39189
72771	11672	67492	42904	64647	94354	45994	42538	54885	15983
38472	43379	76295	69406	96510	16529	83500	28590	49787	29822
24511	56510	72654	13277	45031	42235	96502	25567	23653	36707
01054	06674	58283	82831	97048	42983	06471	12350	49990	04809
94437	94907	95274	26487	60496	78222	43032	04276	70800	17378
97842	69095	25982	03484	25173	05982	14624	31653	17170	92785
53047	13486	69712	33567	82313	87631	03197	02438	12374	40329
40770	47013	63306	48154	80970	87976	04939	21233	20572	31013
52733	66251	69661	58387	72096	21355	51659	19003	75556	33095
41749	46502	18378	83141	63920	85516	75743	66317	45428	45940
10271	85184	46468	38860	24039	80949	51211	35411	40470	16070
98791	48848	68129	51024	53044	55039	71290	26484	70682	56255
30196	09295	47685	56768	29285	06272	98789	47188	35063	24158
99373	64343	92433	06388	65713	35386	43370	19254	55014	98621
27768	27552	42156	23239	46823	91077	06306	17756	84459	92513
67791	35910	56921	51976	78475	15336	92544	82601	17996	72268
64018	44004	08136	56129	77024	82650	18163	29158	33935	94262
79715	33859	10835	94936	02857	87486	70613	41909	80667	52176
20190	40737	82688	07099	65255	52767	65930	45861	32575	93731
82421	01208	49762	66360	00231	87540	88302	62686	38456	25872
00083	81269	35320	72064	10472	92080	80447	15259	62654	70882
56558	09762	20813	48719	35530	96437	96343	21212	32567	34305
41183	20460	08608	75273	43401	25888	73405	35639	92114	48006
39977	10603	35052	53751	64219	36235	84687	42091	42587	16996
29310	84031	03052	51356	44747	19678	14619	03600	08066	93899

TABLE A-7   (continued)

47360	03571	95657	85065	80919	14890	97623	57375	77855	15735
48481	98262	50414	41929	05977	78903	47602	52154	47901	84523
48097	56362	16342	75261	27751	28715	21871	37943	17850	90999
20648	30751	96515	51581	43877	94494	80164	02115	09738	51938
60704	10107	59220	64220	23944	34684	83696	82344	19020	84834
25906	55812	91669	89718	23250	04815	17631	71270	15212	79339
71700	66325	76311	17034	21766	07705	48877	42317	04105	42030
03016	85673	47567	12916	68175	33340	96422	80878	24803	47406
51869	68055	17763	50758	06197	20462	50212	61152	09784	36499
32800	33619	44111	31961	55915	32361	73344	41828	77001	73221
19033	15965	24010	82435	61754	71963	47249	65025	72642	51584
68532	20248	51858	92816	90176	78898	15669	78373	87683	53987
32624	64167	43811	47066	86723	91338	21399	57289	64512	76225
06896	13417	33832	99204	01869	22261	02591	65699	28016	36314
15643	72081	43395	27506	45992	63462	69685	59743	10789	83838
02655	14695	35834	28596	82124	43435	86245	03798	83691	57642
28169	30715	75768	32547	28968	76950	63945	47760	46249	40666
44827	78135	42579	65416	44144	07656	03949	26107	62280	40485
35245	53285	43338	22138	34026	00753	81526	47610	59944	02252
97720	31595	84314	78326	15899	54686	23842	38401	79888	64313
96119	74549	17179	37592	80940	53257	49241	92202	44984	98852
88216	11170	60618	35715	76336	12700	91283	71581	47406	13439
29428	72252	96633	65837	28356	10246	52263	38764	78422	90000
57020	50087	62210	76160	99240	14258	99000	77387	61801	93913
46518	21876	20660	60898	35424	96365	35586	84191	23140	98133
96912	98556	81929	73875	99719	04683	52514	50332	94262	68185
18365	85411	16407	85145	40903	22339	59446	25583	59193	88923
79863	03951	18832	89268	66313	68361	28818	07181	83525	42084
95044	01570	85590	69782	95322	42212	37103	47333	55931	76976
14954	64265	08799	65488	79740	70215	37568	39848	56313	86346
49227	10283	49155	20201	52220	60997	78868	90255	59693	42359
86673	39383	86702	45646	00020	31072	25273	43667	49225	35578
30442	30784	12122	64652	87058	00246	66809	90213	85167	47616
00585	86972	72633	57337	76142	75035	93392	83142	56102	07470
58299	63144	61881	92204	68931	39479	06376	69292	74701	94930
59842	85732	60448	74306	56895	97138	09514	14733	32763	41907
63578	50517	45400	05435	19713	23480	77878	47612	30154	05205
04638	41365	37885	36303	86611	03942	52933	48006	78687	39614
92139	74190	82902	28458	75624	24817	60091	18320	00991	95534
50496	12184	93603	23585	21626	30149	29708	72907	26905	03942
79485	31771	86195	01580	95535	29550	85860	55704	93395	68575
24104	67713	33278	45818	04408	50762	96800	59926	87225	94079
88542	84112	39205	49448	93668	11231	32445	69237	78294	65571
39954	27322	28657	46837	80829	15769	51005	91973	59555	10699
51764	20132	26744	59451	24426	93136	97392	76192	03512	76835

(continued)

TABLE A-7 (continued)

29085	80646	94619	09620	07512	66945	87580	48101	05973	76955
63166	14260	28167	61932	19123	43138	18608	73115	40175	05768
01103	68710	40632	59878	91195	72530	91708	50102	86958	60608
18031	90380	49055	55267	31208	81041	83200	25310	68162	91569
82124	16104	14324	10556	65863	02737	43070	00293	78355	92542
83443	15963	24036	15700	70067	00306	02331	53632	10004	72661
13126	67011	23482	61148	29214	11958	27569	44972	38537	69399
31844	91106	05210	30675	95843	20081	65857	72739	63990	10627
28138	68226	65633	33619	74139	57886	74823	94054	45760	40250
26292	36728	39389	84023	85120	06707	89701	12968	53511	49679
66677	57037	75632	09737	19172	18524	19920	79653	67386	48610
33568	61248	93534	00096	67637	21500	85447	63971	08600	55290
70551	71480	92572	27863	00367	77983	49226	69089	89523	57676
10832	50993	90483	66691	87680	10618	14705	32507	77765	74713
28856	35125	59697	29078	29165	50225	78650	33304	17317	33497
79872	66425	01553	55736	67129	66077	76136	36858	92126	34143
19480	53710	86437	71272	14237	17374	34525	70864	37316	15849
75368	39736	15713	84441	40463	97703	44277	24260	09284	98421
76223	93540	59992	91093	77951	04668	77613	10655	14759	94607
38993	82617	85490	56227	46984	89975	12727	34354	34498	58699
38475	68290	66451	15683	89570	24799	62775	61476	85338	94374
45712	72135	99597	22317	14929	83304	58805	45952	54166	06511
47613	55353	52248	01038	34471	41854	73294	94662	09926	44544
55512	45546	00352	89543	63076	62056	09210	36946	32114	50815
65697	03207	57833	27002	95131	05194	91931	33872	17684	77560
70741	03589	54335	50715	62424	20945	49530	88690	52815	34342
33004	17068	51494	53038	38255	18158	63751	44279	91954	78787
04122	77272	80245	51166	76277	13895	18823	05736	20492	22357
32115	01951	67968	51498	82075	96771	96527	52954	11649	49712
87189	31344	97705	26424	79585	15829	92914	64283	44940	01806
09546	76235	00582	65658	55783	53592	46602	57663	14667	71212
21765	29910	50763	77505	57949	85202	39491	46077	69583	84572
16934	59911	22790	93603	13805	42794	27915	91231	51858	48399
18776	22383	52940	98907	79405	73795	03552	04272	06388	51853
41488	90158	39267	61639	88854	58318	82398	94890	73580	87792
84867	74055	63182	09192	47756	13791	73837	41443	83747	81256
46779	15579	97278	02677	36922	39826	00932	07150	51089	34142
76145	23408	14038	73888	29423	60738	55806	31998	12821	68379
05483	94743	88796	40173	24085	27398	40523	08570	12091	47730
69255	72428	33431	06186	08386	90139	36214	56712	59704	44895
24212	55567	11831	74160	55868	41657	21755	55426	28591	78206
71350	01169	62429	67192	89912	02402	32078	23227	99553	81219
29020	47291	75514	86384	70145	10491	98777	02533	77676	70458
97530	44651	49644	65718	58660	32793	57059	73484	84470	02112
75596	45727	02374	38086	85321	88521	18992	12940	30189	84026

TABLE A-7 (continued)

32235	70425	47161	99015	78308	68746	18932	81745	13714	09601
88590	81787	29137	06616	23506	93413	93199	79503	82394	54764
23395	96060	57098	48890	06131	59647	10580	57042	20211	86994
95790	90688	64878	37752	90466	22768	01843	99499	35842	32184
10910	18277	52633	37227	95731	56639	71329	17690	56293	94392
75288	48328	88973	77899	26111	15014	30216	22236	72901	49626
73024	59219	80867	44331	80990	30296	47530	96640	79462	50845
26586	63772	11278	08598	96406	87996	90243	70530	98040	66436
29077	82702	01762	12078	60659	21436	50670	64601	66717	57684
40107	12987	36466	81268	29108	66157	27443	82875	04289	93856
17788	08023	14266	81079	96293	59991	34847	51054	26524	37035
86032	60360	84837	00912	26527	63684	60704	85236	33467	01641
29125	40001	99700	78836	79937	55608	26941	92279	29113	00396
63463	64969	37360	81225	15569	37200	93398	94244	27103	11242
30349	16191	49713	15246	15640	77334	90879	57999	48409	33489
88293	45401	12350	19040	81561	01155	85253	49479	66144	36486
74400	78899	06127	24365	88646	03944	87215	27085	16372	53548
87891	01263	68595	82315	46193	33306	66011	32972	92802	15708
03275	22982	83272	43570	29817	17323	45466	20498	08228	69682
95126	94417	09943	03316	64978	79651	97371	17634	62956	17714
40601	87085	51394	58140	80641	11547	57397	79825	62665	78796
78981	34943	92315	98737	85007	17558	77808	03537	46872	63504
54784	95754	48786	94402	62005	16589	08267	61878	69562	52089
24596	32468	84259	39563	77353	36180	34350	53331	18863	25424
45777	74383	58012	32923	44106	52121	30948	98812	39109	30748
94305	58672	25440	61911	16594	96747	73515	02587	72237	03004
74423	24195	01270	62733	08841	92999	82907	48049	05309	81854
87243	87076	04321	45628	31593	19783	24674	39222	37639	97573
95279	96674	37550	92395	17821	47986	56822	50499	37384	59050
10669	10427	09299	84045	39184	37102	94060	37288	30669	51740
13983	63696	08604	96115	34877	93673	16767	37258	32137	88081
25520	13548	78676	39472	41957	58478	45877	84216	79001	11734
81754	89975	07744	82864	63632	81103	66748	28539	81244	71407
99249	06587	10989	74629	17264	09220	16293	21958	99275	33075
25692	49122	53820	61407	66479	50095	43462	66380	52481	17731
71062	25537	03787	00508	70346	72869	66697	66426	92430	03241
98835	64436	73086	73216	37983	82543	88681	78385	26327	23654
31631	10493	96276	59088	17480	13347	60948	92191	33074	18170
48222	82827	64181	51377	39636	28938	83204	89441	06125	58756
97700	56405	25858	42033	11729	91962	62762	16535	61465	59891
02050	26216	39216	73287	56887	26522	20127	64522	58368	00710
75973	65876	36777	52742	08272	03969	14125	68998	93575	95479
01373	65250	08317	65902	05034	12660	92903	11015	52893	11155
83866	83199	85406	39590	17954	62519	87775	56581	74495	67094
12597	05140	36748	15697	43206	03194	47216	45552	19465	09413

(continued)

TABLE A-7 (continued)

33387	64696	72102	44081	67644	91011	54737	95279	96971	81654
76084	35282	10085	53580	28924	87971	56631	72301	11510	75086
73769	66032	11119	88801	30169	73856	09346	25188	54381	24477
04307	32351	86338	86420	10259	55707	52711	02482	41349	78143
29933	72064	94327	94859	18890	84470	58420	55774	74052	88976
83787	23267	80357	20523	58215	19706	45552	81944	90820	48073
00437	77777	57832	01058	25654	59456	36924	17398	72197	19795
87816	27435	01573	42907	98043	21332	21732	42079	91177	01928
82756	67233	82534	40832	48525	53269	26225	89933	32494	84807
70426	94290	17064	68483	64842	47695	13623	55646	29305	51719
42711	61143	40516	12203	14367	95095	44703	64297	13381	40965
76510	88343	65246	28697	10606	83368	24310	90199	84181	33045
57261	17829	07486	29959	71893	99581	39680	13235	11465	41203
99713	22576	71336	09523	09491	18354	69516	29568	16788	95639
51808	19367	76265	97323	96197	72764	60674	51051	76627	51398
00180	75937	95135	68397	27720	23011	58415	18328	39360	64450
24387	70467	99776	54982	87625	30810	34591	20667	09254	04537
44516	44771	71969	11540	51710	40042	19607	72235	29191	98230
23436	83139	43405	03055	14175	05963	91920	06619	99717	46565
30311	13461	76003	22280	12694	95205	12666	86677	85569	76320

From *A Million Random Digits with 100,000 Normal Deviates*, 1955, by permission of the Rand Corporation.

TABLE A-8 Expected Locations of Ordered Data from a Standard Normal Population

$n \to$ i	5 Prob.	5 Nscore	6 Prob.	6 Nscore	7 Prob.	7 Nscore	8 Prob.	8 Nscore
1	0.1190	−1.180	0.1000	−1.282	0.0862	−1.365	0.0758	−1.434
2	0.3095	−0.497	0.2600	−0.643	0.2241	−0.758	0.1970	−0.852
3	0.5000	0.000	0.4200	−0.202	0.3621	−0.353	0.3182	−0.473
4	0.6905	0.497	0.5800	0.202	0.5000	0.000	0.4394	−0.152
5	0.8810	1.180	0.7400	0.643	0.6379	0.353	0.5606	0.152
6			0.9000	1.282	0.7759	0.758	0.6818	0.473
7					0.9138	1.365	0.8030	0.852
8							0.9242	1.434

$n \to$ i	9 Prob.	9 Nscore	10 Prob.	10 Nscore	11 Prob.	11 Nscore	12 Prob.	12 Nscore
1	0.0676	−1.494	0.0610	−1.546	0.0555	−1.594	0.0510	−1.635
2	0.1757	−0.932	0.1585	−1.001	0.1414	−1.074	0.1327	−1.114
3	0.2838	−0.572	0.2561	−0.655	0.2333	−0.728	0.2143	−0.792
4	0.3919	−0.274	0.3537	−0.375	0.3222	−0.462	0.2959	−0.536
5	0.5000	0.000	0.4512	−0.123	0.4111	−0.225	0.3776	−0.312
6	0.6081	0.274	0.5488	0.123	0.5000	0.000	0.4592	−0.102
7	0.7162	0.572	0.6463	0.375	0.5889	0.225	0.5408	0.102
8	0.8243	0.932	0.7439	0.655	0.6778	0.462	0.6224	0.312
9	0.9324	1.494	0.8415	1.001	0.7667	0.728	0.7041	0.536
10			0.9390	1.546	0.8556	1.074	0.7857	0.792
11					0.9444	1.594	0.8673	1.114
12							0.9490	1.635

$n \to$ i	13 Prob.	13 Nscore	14 Prob.	14 Nscore	15 Prob.	15 Nscore	16 Prob.	16 Nscore
1	0.0472	−1.673	0.0439	−1.707	0.0410	−1.739	0.0385	−1.768
2	0.1226	−1.162	0.1140	−1.206	0.1066	−1.245	0.1000	−1.282
3	0.1981	−0.848	0.1842	−0.899	0.1721	−0.946	0.1615	−0.988
4	0.2736	−0.602	0.2544	−0.661	0.1377	−0.714	0.2231	−0.762
5	0.3491	−0.388	0.3246	−0.455	0.3033	−0.515	0.2846	−0.569
6	0.4245	−0.190	0.3947	−0.267	0.3689	−0.335	0.3462	−0.396
7	0.5000	0.000	0.4649	−0.088	0.4344	−0.165	0.4077	−0.233
8	0.5755	0.190	0.5351	0.088	0.5000	0.000	0.4692	−0.077
9	0.6509	0.388	0.6053	0.267	0.5656	0.165	0.5308	0.077
10	0.7264	0.602	0.6754	0.455	0.6311	0.335	0.5923	0.233
11	0.8019	0.848	0.7456	0.661	0.6967	0.515	0.6538	0.396
12	0.8774	1.162	0.8158	0.899	0.7623	0.714	0.7154	0.569
13	0.9528	1.673	0.8860	1.206	0.8279	0.946	0.7769	0.762
14			0.9561	1.707	0.8934	1.245	0.8385	0.988
15					0.9590	1.739	0.9000	1.282
16							0.9615	1.768

Large Data Sets

TABLE B-1 Sales Data on Houses Sold in South Bay and
North Bay Areas in Humboldt County in
California from 1989 to 1994

Row	SP	DOM	BEDS	BATHS	LVG-AREA*
1	36000	19	1	1.00	700
2	37500	153	1	1.00	542
3	47500	141	2	1.00	700
4	48000	176	2	1.00	1028
5	36000	53	2	1.00	750
6	41000	225	1	1.00	560
7	42500	63	2	1.00	816
8	50000	93	2	1.00	900
9	55000	155	2	2.00	1080
10	59000	170	2	1.00	850
11	61000	128	2	1.00	1080
12	55000	21	1	1.00	825
13	56000	89	2	1.00	960
14	59900	74	2	1.00	1050
15	59900	57	3	1.00	1008
16	57500	70	2	1.00	865
17	60000	6	2	1.00	840
18	62500	182	2	1.50	1020
19	60500	198	3	1.00	1000
20	60000	180	3	1.50	1100
21	63000	110	3	1.75	1200
22	65000	42	1	1.00	741
23	57500	235	2	1.25	1400

*SP: Sale price　　　　　　　　　　　　　　　(continued)
DOM: Days on market
BEDS: Number of bedrooms
BATHS: Number of baths
LVG-AREA: Living area

TABLE B-1   (continued)

Row	SP	DOM	BEDS	BATHS	LVG-AREA*
24	70000	132	2	1.00	1200
25	62000	91	3	1.00	1400
26	68842	84	3	1.00	960
27	66000	72	2	1.00	920
28	68500	12	2	1.00	810
29	65000	39	2	2.00	1456
30	61900	63	2	1.00	806
31	67500	166	2	1.00	890
32	68950	16	2	1.00	1100
33	66000	196	1	1.00	850
34	63500	117	2	1.00	780
35	74200	21	2	1.00	980
36	72000	126	3	1.00	1325
37	70500	19	2	1.00	870
38	72500	32	2	1.00	1500
39	71500	91	2	1.50	1020
40	71800	210	2	1.00	1050
41	72900	13	3	2.00	1184
42	65000	204	2	1.00	1350
43	73000	85	3	2.00	1650
44	67000	27	2	1.00	1040
45	74500	28	2	1.00	1000
46	74900	15	2	1.00	1027
47	74250	34	2	1.00	700
48	73900	116	3	2.00	1600
49	75750	5	2	1.00	1000
50	75000	6	3	1.25	1550
51	70000	28	2	1.00	800
52	72000	251	2	1.00	1100
53	76000	34	2	1.00	1028
54	76500	73	3	1.00	1180
55	76000	25	2	1.00	950
56	78900	40	2	1.00	950
57	76000	70	2	1.00	910
58	76000	45	3	1.00	1060
59	74000	225	2	1.00	1200
60	73000	45	2	1.00	1320
61	70000	96	3	1.00	950
62	127000	56	3	2.00	1310
63	129900	94	4	2.00	1600
64	127900	27	3	2.00	1600
65	125000	43	3	1.75	1600
66	134000	13	3	2.00	1600
67	134000	238	3	2.00	1600
68	138500	80	3	2.00	1550
69	137500	53	2	2.00	1500
70	127000	164	3	2.00	1650
71	140000	70	3	2.00	1762
72	135000	257	6	3.00	2786
73	145000	8	3	2.00	1620
74	145000	18	3	2.00	1580

(continued)

TABLE B-1   (continued)

Row	SP	DOM	BEDS	BATHS	LVG-AREA*
75	150000	248	4	2.50	2170
76	150000	52	4	3.00	2000
77	150000	190	4	2.00	1530
78	155000	20	3	2.50	1850
79	155000	61	3	1.50	1300
80	169000	99	3	3.50	1964
81	179000	47	3	2.00	1550
82	179000	72	3	2.00	2000
83	189000	64	3	2.50	1800
84	185000	24	4	2.50	2420
85	198500	23	3	2.00	2000
86	79500	84	3	1.00	1050
87	79900	34	3	2.00	1184
88	78900	74	3	1.00	1220
89	75000	9	2	1.00	950
90	75000	89	2	1.00	740
91	80000	6	3	2.00	1200
92	80000	70	3	2.00	1000
93	81000	16	3	1.00	900
94	79000	155	3	1.00	1350
95	79000	9	2	2.00	1025
96	80000	15	3	1.50	1200
97	75000	29	2	1.00	925
98	76050	137	2	1.00	1300
99	81000	11	4	2.00	1350
100	81500	6	3	1.50	1200
101	79900	283	2	1.00	1550
102	70000	125	4	2.00	2558
103	82500	117	3	1.00	1200
104	82900	10	2	1.00	1125
105	83000	106	3	1.00	1180
106	76000	180	3	1.00	970
107	83000	71	3	1.50	1225
108	85500	112	2	1.00	1050
109	84900	9	3	1.50	1100
110	83500	52	3	1.00	1000
111	82000	135	3	2.00	1820
112	84000	187	4	1.50	1750
113	85000	141	3	1.75	1550
114	88500	18	3	1.00	1590
115	80000	355	2	1.00	900
116	85000	8	2	1.00	910
117	71500	136	1	1.00	850
118	83000	107	3	2.00	1400
119	80000	92	2	1.00	1320
120	85500	5	2	1.00	1000
121	84000	101	3	1.00	1200
122	86900	39	3	2.00	1150
123	85000	34	3	1.75	1058
124	88000	137	4	2.00	1500
125	87000	23	2	1.50	1124

TABLE B-1   (continued)

Row	SP	DOM	BEDS	BATHS	LVG-AREA*
126	87900	91	3	1.50	1250
127	85000	32	2	1.00	1147
128	89000	4	2	1.00	900
129	88500	4	2	1.00	900
130	80000	83	2	1.00	1150
131	86000	7	3	1.00	1008
132	79000	50	4	2.00	2100
133	92000	50	3	2.00	1400
134	87500	5	3	1.00	1050
135	85000	42	3	1.00	1215
136	84000	0	2	1.00	981
137	88500	277	3	1.50	1640
138	85000	228	3	1.00	1260
139	85000	84	3	1.50	1150
140	86000	23	3	1.00	1200
141	86605	19	3	1.00	1125
142	87750	30	2	1.00	1060
143	85000	16	4	2.00	1700
144	78100	74	3	2.00	1377
145	90000	235	5	2.00	2740
146	86000	93	2	1.00	1150
147	92500	26	3	2.00	1800
148	92500	26	4	2.00	1800
149	88000	76	3	1.00	1360
150	90500	22	4	2.00	1560
151	89500	67	3	1.00	1300
152	93000	175	3	1.50	1300
153	93000	7	2	1.00	850
154	91000	51	3	1.00	1090
155	93000	63	2	1.00	1450
156	97000	261	3	2.00	1135
157	94500	64	3	1.00	1000
158	95000	337	3	3.00	2900
159	90000	163	3	2.00	1614
160	96000	24	3	1.00	1550
161	94000	235	2	1.00	1250
162	97500	154	3	1.50	1100
163	92500	88	2	1.00	1046
164	95000	10	3	1.00	1100
165	94500	28	3	1.00	1200
166	92000	128	3	1.00	1900
167	96000	86	3	2.00	1775
168	98500	41	5	1.75	1800
169	98500	15	3	1.50	1400
170	98000	9	2	1.00	974
171	87500	154	2	1.00	1200
172	99000	31	4	1.00	1416
173	97500	7	3	2.00	1200
174	90000	64	3	2.00	1150
175	98000	20	3	2.00	1200
176	96000	197	3	1.00	1040

(continued)

TABLE B-1   (continued)

Row	SP	DOM	BEDS	BATHS	LVG-AREA*
177	98000	69	2	2.00	1800
178	98500	97	3	1.00	1150
179	109500	34	3	2.00	1850
180	102500	25	3	1.00	1370
181	100000	40	2	2.00	1102
182	90000	179	4	2.00	1300
183	104900	101	3	2.00	1300
184	104500	69	3	2.00	1225
185	102000	184	3	2.00	1230
186	92250	104	4	2.00	1800
187	105000	9	3	1.00	1100
188	101000	0	4	2.00	1300
189	104000	141	3	2.00	1460
190	107000	298	3	2.00	1275
191	108000	76	3	2.00	1250
192	105000	53	3	2.00	1480
193	105000	87	3	2.00	1242
194	109900	89	3	2.00	1450
195	112500	48	5	1.00	1764
196	102000	214	4	2.50	1900
197	109700	91	2	2.00	1206
198	112000	48	3	1.75	1400
199	112000	306	3	2.00	1425
200	108500	102	3	2.00	1275
201	112000	36	3	1.00	1225
202	110000	30	3	2.00	1390
203	112000	258	3	2.00	1375
204	106000	374	3	1.50	1500
205	115000	358	3	2.00	1650
206	115500	10	3	2.00	1440
207	113500	84	3	2.00	1380
208	116000	152	3	1.50	1388
209	111500	118	3	2.00	1440
210	110000	122	3	2.00	1400
211	113000	99	3	2.00	1500
212	119800	296	3	2.00	1400
213	115000	28	3	2.00	1260
214	116000	175	3	2.00	1325
215	109000	277	2	1.00	1203
216	118000	353	3	2.00	1300
217	117000	69	3	2.00	1775
218	115000	88	3	2.00	1800
219	119000	41	3	2.00	1200
220	118000	167	3	2.00	1480
221	118000	70	3	2.50	1384
222	112000	97	3	2.00	1320
223	121500	155	4	1.50	1750
224	114500	119	3	2.00	1350
225	120000	56	3	2.00	1600
226	121500	67	3	2.00	1350

TABLE B-1　(continued)

Row	SP	DOM	BEDS	BATHS	LVG-AREA*
227	117500	63	3	2.00	1250
228	115000	68	3	2.00	1607
229	122500	44	3	2.00	1600
230	125000	389	3	2.00	1500
231	125500	87	4	2.00	1675
232	123000	56	3	2.00	1500
233	125000	87	3	2.00	1400
234	122000	204	3	2.00	1900
235	129500	170	3	2.50	1525
236	125000	181	3	2.00	1300
237	127500	72	3	2.50	1650
238	129900	52	3	2.00	1485
239	135000	9	3	2.50	1600
240	115000	30	2	1.00	1164
241	132450	75	2	1.50	1558
242	131000	39	3	2.50	1900
243	134900	67	3	2.00	1245
244	134000	25	3	2.00	1950
245	125000	44	3	2.00	1400
246	130000	80	3	2.00	1800
247	133500	173	4	1.50	1849
248	133000	52	3	2.00	1404
249	134000	9	3	2.00	1575
250	133000	42	4	2.00	1940
251	132000	153	3	3.00	1650
252	127500	75	3	2.00	1546
253	134500	28	3	2.00	1450
254	133500	54	3	2.00	2050
255	132900	139	3	2.00	1550
256	139900	18	4	2.00	1650
257	129900	253	3	2.50	2400
258	138000	36	4	2.00	2000
259	141700	77	3	2.00	1500
260	144000	0	5	2.00	2530
261	135000	170	3	2.50	2050
262	137500	52	3	2.00	1550
263	143000	86	4	1.50	2180
264	145000	75	3	2.00	1400
265	147000	36	2	2.00	1450
266	147000	36	3	2.00	1450
267	149500	1	3	2.00	1700
268	146100	60	4	2.00	1700
269	145000	203	3	2.00	1800
270	150000	77	3	2.00	2050
271	142500	52	3	2.00	2600
272	135000	214	3	1.50	1425
273	145000	111	3	2.00	1950
274	147500	165	3	2.00	2000
275	155000	0	3	2.00	2000
276	156500	59	3	2.00	1580

(continued)

TABLE B-1 (continued)

Row	SP	DOM	BEDS	BATHS	LVG-AREA*
277	155900	195	3	2.50	1780
278	155000	64	3	2.00	1700
279	140000	180	3	2.00	1566
280	157000	87	4	2.00	2300
281	157900	191	3	2.00	1800
282	162500	0	3	2.00	1356
283	170000	489	3	2.00	1888
284	155000	235	2	2.00	1300
285	175000	96	3	2.00	1950
286	175000	164	3	2.50	1840
287	160875	99	3	2.00	1900
288	173000	384	3	2.00	1600
289	180000	100	2	2.00	1260
290	185000	55	3	2.50	1900
291	185000	208	3	2.00	1660
292	185000	86	3	2.50	1722
293	189500	319	3	2.50	2350
294	180000	449	3	2.50	1900
295	182500	50	3	2.50	2600
296	175000	1	3	1.00	1200
297	182000	260	4	2.50	2590
298	189000	113	3	2.00	2100
299	190000	52	5	3.00	2400
300	187500	147	3	2.50	2500
301	50000	351	1	1.00	736
302	50000	185	2	1.50	850
303	50000	44	2	1.50	750
304	54000	373	3	1.50	1000
305	55000	54	2	1.50	850
306	57500	193	2	1.50	847
307	53500	210	2	1.00	900
308	56000	109	2	2.00	735
309	61000	14	2	1.50	620
310	75000	19	2	1.00	1065
311	68000	139	2	1.00	1088
312	65000	207	3	2.00	1536
313	75000	123	3	1.00	1064
314	73000	161	2	1.00	800
315	75750	121	3	1.50	1000
316	76900	116	3	1.00	1076
317	70000	167	2	1.00	960
318	82000	22	2	1.00	1260
319	81000	246	4	2.00	1460
320	85500	113	3	2.00	1120
321	83000	147	3	1.00	1200
322	83000	222	2	1.00	850
323	91400	29	3	1.00	1000
324	88500	142	3	1.50	1050
325	87000	146	3	2.00	1320
326	89900	60	3	1.00	900

TABLE B-1 (continued)

Row	SP	DOM	BEDS	BATHS	LVG-AREA*
327	89000	178	4	2.00	1300
328	85000	681	2	2.00	1000
329	90000	38	3	2.00	1000
330	97500	79	3	2.00	1200
331	100000	164	2	1.00	1250
332	98000	11	3	2.00	1200
333	100800	56	3	1.00	955
334	101500	211	3	2.00	1500
335	106000	127	3	2.00	1100
336	107000	38	4	2.00	2160
337	104000	147	3	1.00	1250
338	125000	14	3	1.00	1000
339	115000	61	3	2.00	1450
340	115000	120	4	2.00	1300
341	115000	41	4	2.00	1350
342	117900	19	3	2.00	1730
343	113500	67	3	2.00	1370
344	121750	58	3	2.50	1600
345	120000	30	3	2.00	1224
346	126000	56	3	1.50	1250
347	120000	20	3	2.00	1770

TABLE B-2   SAT Scores of a Sample of Freshman Students Entering
Humboldt State University, California, in the Fall of 1994

Row	Sex	Age	City	MATH	VERB	HSGPA*
1	F	17	Fresno	550	500	3.04
2	F	17	Claremont	530	420	3.40
3	F	17	Paso Robles	280	310	3.33
4	F	17	Wilmington	540	390	3.59
5	F	17	San Diego	360	400	2.75
6	F	17	Fallbrook	480	580	2.81
7	F	17	Stockton	480	400	3.48
8	F	17	Canyon Country	500	520	3.35
9	F	17	San Diego	490	550	3.62
10	F	17	Placentia	510	410	2.67
11	F	17	Arcata	700	640	3.81
12	F	17	Concord	570	520	2.90
13	F	17	Imperial Beach	550	520	4.00
14	F	17	Santa Cruz	550	410	*
15	F	17	Arcata	460	440	2.84
16	F	17	Arcata	570	610	3.08
17	F	18	McKinleyville	710	560	4.00
18	F	18	Covelo	460	390	3.82
19	F	18	Valencia	480	540	3.82
20	F	18	Eureka	390	400	2.56
21	F	18	San Diego	380	380	2.76
22	F	18	Walnut Creek	500	420	3.47
23	F	18	Templeton	430	390	3.08
24	F	18	Willow Creek	480	550	3.85
25	F	18	Simi Valley	330	490	3.40
26	F	18	Lodi	420	420	3.20
27	F	18	Oakland	550	450	2.92
28	F	18	Whitethorn	480	470	3.29
29	F	18	Garberville	380	320	2.56
30	F	18	Lancaster	440	460	3.00
31	F	18	Eureka	480	460	2.88
32	F	18	Oakland	250	250	2.68
33	F	18	Rancho Palos Verdes	560	570	2.52
34	F	18	Granada Hills	270	240	3.08
35	F	18	Hoopa	520	470	3.12
36	F	18	Santa Cruz	520	490	3.76
37	F	18	Fresno	440	460	3.55
38	F	18	Lafayette	520	430	3.65
39	F	18	Clayton	540	480	3.50
40	F	18	Bayside	530	500	*
41	F	18	Fairfield	590	450	3.70
42	F	18	Hacienda Heights	520	410	3.10
43	F	18	Huntington Beach	560	430	*
44	F	18	Davis	520	480	3.85

* MATH: Math scores
  VERB: Verbal scores
  HSGPA: High school GPA

TABLE B-2   (continued)

Row	Sex	Age	City	MATH	VERB	HSGPA*
45	F	18	Arcata	460	480	3.66
46	F	18	Covelo	470	470	3.39
47	F	18	Huntington Beach	530	450	3.08
48	F	18	Concord	480	450	5.00
49	F	18	McKinleyville	480	450	3.75
50	F	18	Grenada	550	540	3.42
51	F	18	Santa Barbara	670	520	4.30
52	F	18	Chula Vista	500	440	3.18
53	F	18	Thousand Oaks	540	470	2.96
54	F	18	San Jose	450	410	3.18
55	F	18	North Hollywood	440	500	3.40
56	F	18	San Bernardino	460	450	2.37
57	F	18	McKinleyville	540	590	3.50
58	F	18	Susanville	680	620	4.00
59	F	18	Arcata	420	520	3.23
60	F	18	Chico	460	300	3.00
61	F	18	Chico	530	470	3.41
62	F	18	Pomona	450	520	3.30
63	F	18	Pleasanton	540	480	3.08
64	F	18	Loomis	450	500	3.20
65	F	18	Clovis	420	410	3.20
66	F	18	Covina	410	430	3.60
67	F	18	Fallbrook	510	360	2.75
68	F	18	Coronado	370	460	3.42
69	F	18	Encinitas	730	700	3.98
70	F	18	Hemet	480	480	3.96
71	F	18	Hesperia	570	560	3.33
72	F	18	Helendale	480	420	4.00
73	F	18	Fallbrook	510	630	2.64
74	F	18	Jamul	540	520	2.68
75	F	18	El Cajon	510	480	3.10
76	F	18	San Diego	440	390	2.85
77	F	18	Vista	620	500	3.56
78	F	18	San Diego	350	400	2.85
79	F	18	Glendale	470	470	3.20
80	F	18	Laguna Niguel	450	480	*
81	F	18	Irvine	510	430	3.84
82	F	18	Sonoma	510	500	2.59
83	F	18	Atascadero	490	430	3.00
84	F	18	Yorba Linda	620	480	3.43
85	F	18	Arcata	590	480	2.54
86	F	18	Redding	530	420	3.68
87	F	18	Chico	570	390	3.36
88	F	18	Anaheim	570	400	3.30
89	F	18	Poway	460	450	3.56
90	F	18	Arcata	450	330	3.30
91	F	18	Arbuckle	590	530	3.36
92	F	18	San Ramon	640	560	3.59
93	F	18	Quincy	410	420	3.19

(continued)

TABLE B-2 (continued)

Row	Sex	Age	City	MATH	VERB	HSGPA*
94	F	18	Fort Bliss	560	540	2.78
95	F	18	Rancho Cordova	490	400	2.68
96	F	18	San Diego	580	560	2.86
97	F	18	Palo Alto	560	480	2.87
98	F	18	Blue Lake	600	470	3.89
99	F	18	Cupertino	690	500	2.86
100	F	18	Ferndale	360	380	3.26
101	F	18	El Toro	510	560	2.57
102	F	18	Santa Cruz	560	450	2.82
103	F	18	Arcata	400	420	3.57
104	F	18	Pomona	480	550	3.10
105	F	18	Arcata	330	480	3.60
106	F	18	Oceanside	510	630	3.90
107	F	18	Alameda	640	570	3.09
108	F	18	Arcata	450	380	3.10
109	F	18	Victorville	420	450	4.00
110	F	18	Montara	470	430	3.04
111	F	18	Santa Barbara	560	500	2.68
112	F	18	Simi Valley	710	620	4.00
113	F	18	Santa Cruz	480	570	3.50
114	F	18	Los Angeles	330	520	4.00
115	F	18	Arcata	310	420	3.56
116	F	18	Modesto	600	540	3.83
117	F	18	Folsom	470	500	3.00
118	F	18	Agoura Hills	540	500	3.27
119	F	18	Trinidad	470	400	3.30
120	F	18	Grass Valley	610	560	4.00
121	F	18	Arcata	470	470	3.78
122	F	18	Ferndale	550	430	3.92
123	F	18	Chula Vista	580	530	3.71
124	F	18	Sacramento	400	450	3.45
125	F	18	San Jose	580	400	3.25
126	F	18	Sacramento	530	430	3.10
127	F	18	Fullerton	440	440	3.40
128	F	18	Valley Springs	460	510	3.17
129	F	18	San Anselmo	560	550	3.95
130	F	18	McKinleyville	450	370	3.57
131	F	18	Beaumont	580	520	3.38
132	F	18	Arcata	640	490	4.00
133	F	18	Arcata	360	550	3.11
134	F	18	Vacaville	510	440	2.99
135	F	18	Valley Center	380	330	2.74
136	F	18	Redding	450	420	2.89
137	F	18	Santa Ana	460	510	2.91
138	F	18	La Mirada	570	500	3.14
139	F	18	Crafton	500	480	3.80
140	F	18	Anaheim	610	410	3.67
141	F	18	Alta Loma	510	480	4.00
142	F	18	Hollywood	440	530	3.50
143	F	18	Clements	410	490	3.32

TABLE B-2   (continued)

Row	Sex	Age	City	MATH	VERB	HSGPA*
144	F	18	Mission Viejo	480	380	3.10
145	F	18	Redwood City	510	530	3.50
146	F	18	Concord	510	480	3.28
147	F	18	San Jose	450	390	*
148	F	18	Chula Vista	570	530	3.28
149	F	18	Arcata	420	490	3.82
150	F	18	Alta Loma	490	470	3.88
151	F	18	Eureka	630	510	3.37
152	F	18	Eureka	570	330	3.46
153	F	18	McKinleyville	480	530	4.00
154	F	18	McKinleyville	540	370	3.90
155	F	18	Arcata	400	430	3.00
156	F	18	Visalia	460	390	3.77
157	F	18	Guerneville	410	360	3.00
158	F	18	Arcata	460	390	3.42
159	F	18	Ridgecrest	590	540	3.50
160	F	18	Arcata	450	400	*
161	F	18	San Diego	420	360	3.00
162	F	18	Trinidad	500	570	2.78
163	F	18	SanJose	400	500	*
164	F	18	Fresno	460	490	2.97
165	F	18	Mililani	540	550	2.98
166	F	18	Riverside	500	460	2.84
167	F	18	Big Creek	590	470	3.09
168	F	18	Gustavus	330	490	3.62
169	F	18	Covina	620	730	3.93
170	F	18	McFarland	490	430	3.98
171	F	18	Simi Valley	400	320	2.60
172	F	18	Portland	490	420	3.20
173	F	18	Sierra Madre	370	420	3.00
174	F	18	Thousand Oaks	550	490	3.40
175	F	18	Fairfield	440	430	2.50
176	F	18	Long Beach	430	430	3.10
177	F	18	Essex	430	600	3.40
178	F	18	Sherman Oaks	450	350	2.78
179	F	18	Loomis	560	420	2.79
180	F	18	Oakland	580	480	3.24
181	F	18	Ventura	450	510	*
182	F	18	San Jose	390	370	2.70
183	F	18	Whittier	660	570	3.00
184	F	18	La Puente	560	420	3.20
185	F	18	Stockton	440	380	3.13
186	F	18	Eureka	610	560	3.12
187	F	18	Santa Rosa	560	480	3.56
188	F	18	Placerville	450	440	2.59
189	F	18	Rancho Cordova	610	560	3.71
190	F	18	Auburn	470	510	3.21
191	F	18	South Pasadena	440	460	3.14
192	F	18	San Francisco	230	390	3.00
193	F	18	San Andreas	450	430	4.00

(continued)

TABLE B-2   (continued)

Row	Sex	Age	City	MATH	VERB	HSGPA*
194	F	18	Santa Rosa	570	670	3.90
195	F	18	Arcata	550	520	3.00
196	F	18	Arcata	360	360	2.96
197	F	18	Pacifica	460	480	3.20
198	F	18	Arcadia	570	410	3.26
199	F	18	Placerville	660	510	3.00
200	F	18	Los Angeles	530	540	2.90
201	F	18	San Diego	460	340	*
202	F	18	El Sobrante	570	480	3.13
203	F	18	Sacramento	570	500	3.33
204	F	18	Galt	340	450	3.55
205	F	18	San Jose	330	360	3.28
206	F	18	Roseville	540	470	3.64
207	F	18	Eureka	420	440	3.16
208	F	18	Bolinas	470	530	2.43
209	F	18	Davis	480	410	2.52
210	F	18	Escondido	460	430	3.23
211	F	18	San Diego	510	530	3.04
212	F	18	Fullerton	470	490	3.00
213	F	18	Leona Valley	380	370	3.38
214	F	19	Arcata	340	380	2.61
215	F	19	Hollister	650	530	4.00
216	F	19	Danville	610	520	2.96
217	F	19	Orinda	690	480	3.25
218	F	19	Arcata	500	410	3.52
219	F	19	San Ramon	410	330	2.64
220	F	19	Napa	470	350	3.17
221	F	19	San Diego	570	390	3.18
222	F	19	Los Gatos	410	310	3.00
223	F	19	Arcata	620	630	3.50
224	F	19	Salyer	380	330	3.08
225	F	19	Walnut Creek	410	390	2.44
226	F	19	La Mesa	510	480	2.75
227	F	19	Danville	360	400	2.85
228	F	19	Arcata	620	600	3.88
229	F	19	Holland	500	530	*
230	F	19	Arcata	550	400	3.41
231	F	19	Woodland	510	470	3.57
232	F	19	South San Francisco	370	440	3.10
233	F	19	Woodland	400	410	3.12
234	F	19	Arcata	450	530	3.25
235	F	19	Merced	510	390	3.60
236	F	20	Arcata	450	440	3.20
237	M	17	Moraga	750	730	2.58
238	M	17	Miranda	410	510	2.85
239	M	17	Loleta	580	440	2.34
240	M	17	Bellflower	490	470	2.83
241	M	18	Grass Vallley	590	520	2.92
242	M	18	Hacienda Heights	620	540	3.25
243	M	18	Bayside	660	360	2.62

TABLE B-2   (continued)

Row	Sex	Age	City	MATH	VERB	HSGPA*
244	M	18	Villa Park	530	470	2.95
245	M	18	Ben Lomond	520	410	3.14
246	M	18	Hydesville	490	470	3.82
247	M	18	San Diego	670	540	3.91
248	M	18	Reseda	460	530	3.30
249	M	18	San Diego	560	410	2.89
250	M	18	Running Springs	460	330	3.10
251	M	18	Orange	580	390	2.09
252	M	18	Coronado	610	460	3.81
253	M	18	San Bruno	600	560	2.78
254	M	18	San Diego	410	440	2.42
255	M	18	San Francisco	700	530	2.57
256	M	18	Cupertino	560	490	2.83
257	M	18	Kula	610	490	3.10
258	M	18	Trinidad	540	450	3.47
259	M	18	Watsonville	500	410	2.57
260	M	18	Fortuna	400	440	2.50
261	M	18	Eureka	610	450	3.30
262	M	18	McKinleyville	700	590	2.92
263	M	18	Cypress	580	450	2.83
264	M	18	Arcata	560	420	3.00
265	M	18	Carlsbad	570	540	3.21
266	M	18	Fountain Valley	410	460	2.46
267	M	18	Alderpoint	570	570	3.46
268	M	18	Tollhouse	430	430	2.69
269	M	18	Culver	420	370	2.70
270	M	18	Tiburon	580	590	2.55
271	M	18	Malibu	380	450	3.00
272	M	18	Murphys	420	450	3.35
273	M	18	Registerstown	570	310	2.58
274	M	18	Palos Verdes	540	580	2.55
275	M	18	Los Angeles	370	410	3.66
276	M	18	Los Angeles	440	480	2.77
277	M	18	Bayside	630	510	3.10
278	M	18	Castroville	540	540	3.04
279	M	18	Upland	690	630	3.77
280	M	18	Arcadia	560	480	2.80
281	M	18	Cerritos	520	500	3.74
282	M	18	Arcata	490	450	2.56
283	M	18	La Habra	470	490	2.38
284	M	18	Calimesa	520	520	3.95
285	M	18	Orick	410	460	3.36
286	M	18	Valencia	570	530	2.96
287	M	18	Fort Bragg	610	500	3.95
288	M	18	West Covina	480	470	3.22
289	M	18	Merced	470	370	3.35
290	M	18	Arcata	510	350	3.36
291	M	18	Walnut Creek	610	460	2.48
292	M	18	Manhattan Beach	490	390	3.45
293	M	18	Mission Viejo	330	350	2.55

(continued)

TABLE B-2   (continued)

Row	Sex	Age	City	MATH	VERB	HSGPA*
294	M	18	Fair Oaks	670	580	2.60
295	M	18	Fremont	450	420	2.69
296	M	18	San Miguel	490	470	3.76
297	M	18	S San Francisco	500	530	2.50
298	M	18	Santa Rosa	490	500	3.13
299	M	18	Little Norway	570	400	*
300	M	18	Carpinteria	720	580	2.58
301	M	18	Arcata	560	550	3.38
302	M	18	Tracy	500	580	3.36
303	M	18	Moss Beach	540	410	*
304	M	18	San Diego	460	440	2.50
305	M	18	Santa Cruz	480	500	3.36
306	M	18	Oxnard	440	420	3.55
307	M	18	Orange	630	530	3.00
308	M	18	Blue Lake	460	450	3.86
309	M	18	San Jose	570	460	2.88
310	M	18	Concord	560	410	2.95
311	M	18	Pleasanton	630	450	3.14
312	M	18	San Diego	490	460	3.13
313	M	18	Palmdale	530	390	3.72
314	M	18	Oakland	590	470	3.32
315	M	18	San Juan Capistrano	510	450	3.18
316	M	18	Carlsbad	610	460	*
317	M	18	Berkeley	560	360	2.40
318	M	18	Monrovia	640	490	2.58
319	M	18	Santa Maria	570	470	2.57
320	M	18	Long Beach	670	420	2.83
321	M	18	Carmel	590	510	2.97
322	M	18	Danville	610	470	3.03
323	M	18	Bayside	630	550	3.41
324	M	18	Thousand Oaks	580	360	2.54
325	M	18	Claremont	560	470	2.80
326	M	18	Campbell	470	450	3.45
327	M	18	Novato	520	540	2.29
328	M	18	Santa Rosa	680	580	4.00
329	M	18	Anaheim	570	490	3.80
330	M	18	Edwards AFB	410	450	2.60
331	M	18	Fairfax	560	450	3.10
332	M	18	Pinole	570	340	3.61
333	M	18	Piedmont	550	390	2.71
334	M	18	San Diego	530	580	2.20
335	M	18	Berkeley	520	480	2.60
336	M	18	Berkeley	560	560	2.81
337	M	18	Stockton	570	610	3.32
338	M	18	Santa Rosa	740	500	3.28
339	M	18	Ojai	620	470	3.42
340	M	18	Arcata	500	500	2.72
341	M	18	San Jose	600	570	3.35
342	M	18	Tiburon	630	530	3.08
343	M	18	Sebastopol	540	350	3.69
344	M	18	Eureka	550	440	2.90
345	M	18	Anderson	570	550	2.53

Row	Sex	Age	City	MATH	VERB	HSGPA*
346	M	18	Placerville	420	500	2.91
347	M	18	Van Nuys	540	510	3.07
348	M	18	Walnut Creek	500	560	2.79
349	M	18	Milpitas	510	430	2.41
350	M	18	Tahoe City	480	450	3.00
351	M	18	Belmont	480	470	2.50
352	M	18	Cutler	480	370	3.21
353	M	18	Manhattan Beach	660	490	3.35
354	M	18	Berkeley	510	400	3.06
355	M	18	Angels Camp	450	450	3.40
356	M	18	Glendale	550	510	3.20
357	M	18	San Diego	590	450	3.35
358	M	18	San Dimas	510	490	3.83
359	M	18	Burney	690	570	3.61
360	M	18	Carmel	460	460	3.49
361	M	18	McKinleyville	550	500	3.70
362	M	18	Oroville	570	490	3.94
363	M	18	Arcata	630	450	4.00
364	M	19	Thousand Oaks	490	420	4.00
365	M	19	Nevada City	530	440	3.20
366	M	19	Santa Cruz	700	520	3.37
367	M	19	Pleasanton	540	420	3.00
368	M	19	Santa Cruz	590	540	3.20
369	M	19	Walnut Creek	440	400	2.77
370	M	19	Santa Cruz	570	480	3.25
371	M	19	Alamo	490	450	2.36
372	M	19	Arcata	480	390	2.85
373	M	19	Kentfield	450	420	3.21
374	M	19	Eureka	480	430	3.15
375	M	19	Walnut Creek	440	440	3.27
376	M	19	Santa Rosa	390	350	3.06
377	M	19	Danville	390	480	2.70
378	M	19	Suisun	380	430	2.86
379	M	19	McKinleyville	440	310	3.43
380	M	19	Grass Valley	490	400	3.50
381	M	19	Walnut Creek	480	650	*
382	M	19	Solano Beach	460	350	2.48
383	M	19	Fort Dick	540	520	4.00
384	M	19	Lafayette	570	520	2.96
385	M	19	Roseville	550	410	3.10
386	M	19	Long Beach	510	400	3.35
387	M	19	Petaluma	590	500	3.67
388	M	19	Newport Beach	520	480	*
389	M	19	Bakersfield	680	380	3.22
390	M	19	Sonoma	510	600	3.83
391	M	19	Cabazon	320	330	2.77
392	M	19	Petrolia	590	600	3.20
393	M	19	Yorba Linda	420	490	2.50
394	M	19	Oceanside	410	400	2.60
395	M	20	Arcata	450	440	2.55
396	M	20	Nevada City	460	520	3.28
397	M	20	Poway	500	420	3.15
398	M	20	Livermore	590	440	2.54
399	M	20	San Jose	630	590	3.18
400	M	23	Lakeport	490	500	2.60

TABLE B-3 A Sample of Medical Data from a Diabetes Study Data Base

Row	Sex	Race	Age	HTCM	WTKG	SYSBP	DIASBP	FBSGLU	CHOLES*
1	0	1	86	165.1	125.2	122	78	112	191
2	0	0	86	161.3	108.4	130	90	*†	195
3	0	0	43	175.3	195.5	130	80	126	203
4	0	0	86	167.6	★	208	102	★	149
5	0	0	79	158.8	107.0	120	84	104	180
6	0	0	35	165.1	★	134	88	★	189
7	0	1	82	160.0	★	128	92	★	★
8	0	0	83	174.0	110.9	★	★	96	21
9	0	0	41	172.7	108.0	★	★	93	177
10	0	0	37	170.2	108.9	★	★	97	232
11	0	0	35	162.6	154.2	120	80	105	235
12	0	1	81	157.5	106.6	144	98	102	235
13	0	0	41	165.1	108.9	130	68	121	294
14	0	0	49	160.7	103.8	138	88	95	246
15	1	0	32	172.7	197.3	130	88	116	187
16	1	0	83	172.7	★	152	98	★	★
17	0	0	35	161.3	★	124	86	★	203
18	0	0	81	166.4	★	120	82	★	193
19	0	0	26	163.8	154.2	138	100	114	217
20	1	0	83	184.2	163.3	120	78	145	212
21	0	0	51	165.1	★	150	88	★	217
22	0	0	13	170.2	188.2	140	80	114	203
23	0	0	46	167.6	★	★	★	★	259
24	0	1	83	165.1	108.9	130	96	113	159
25	1	0	48	188.0	167.4	124	80	111	227
26	0	0	43	157.5	100.2	160	100	★	★
27	0	0	45	163.8	120.2	112	72	144	203
28	0	1	82	167.6	108.9	136	64	124	214
29	0	0	82	163.0	★	100	80	★	140
30	0	0	29	160.0	99.1	134	98	102	180
31	0	0	36	167.6	174.7	★	★	134	194
32	0	0	33	167.6	★	120	84	92	154
33	0	0	37	167.6	★	132	84	★	★
34	0	0	40	164.5	118.2	★	★	92	215
35	0	0	36	153.7	105.0	★	★	87	165
36	0	0	31	168.9	125.7	★	★	92	205
37	0	0	63	158.8	118.4	170	76	134	306

*Note: One or more variables are undefined.

* Sex: 1 male 0 female

Race: 1 white 0 black

HTCM: Height in centimeters

WTKG: Weight in kilograms

SYSBP: Systolic blood pressure

DIASBP: Diastolic blood pressure

FBSGLU: Fasting blood sugar glucose

CHOLES: Cholesterol

† A star★ indicates a missing value, called *nonresponse* in statistical jargon. This is one of the aspects of data collection in the field with which an investigator must contend.

TABLE B-3 (continued)

Row	Sex	Race	Age	HTCM	WTKG	SYSBP	DIASBP	FBSGLU	CHOLES*
38	0	1	39	167.6	109.1	120	70	195	262
39	0	0	50	164.5	⋆	130	80	⋆	⋆
40	0	0	46	165.1	130.0	150	100	182	209
41	0	1	35	172.7	115.0	150	86	97	200
42	0	0	37	161.3	⋆	130	80	⋆	⋆
43	0	0	46	162.0	116.3	150	92	268	226
44	1	0	21	177.0	167.3	138	84	100	230
45	0	0	37	172.7	120.5	120	80	97	165
46	1	0	32	193.0	239.0	120	80	91	189
47	0	0	36	167.6	⋆	132	90	⋆	⋆
48	0	0	40	160.0	114.5	115	75	184	220
49	1	0	21	174.0	⋆	140	84	⋆	⋆
50	1	0	34	177.8	148.2	148	84	92	190
51	1	0	54	176.5	146.4	130	90	397	230
52	0	0	55	165.1	132.3	160	90	176	223
53	0	0	37	154.9	117.4	⋆	⋆	113	170
54	0	0	23	157.5	106.8	⋆	⋆	104	172
55	0	0	29	168.9	⋆	140	98	⋆	⋆
56	1	1	28	191.8	186.8	⋆	⋆	100	228
57	0	0	53	166.4	100.7	160	90	145	227
58	0	0	41	165.1	104.5	⋆	⋆	119	205
59	1	1	34	165.0	172.7	⋆	⋆	119	264
60	0	0	41	166.4	⋆	140	100	⋆	⋆
61	0	0	59	165.1	104.3	170	80	186	281
62	0	0	38	156.2	101.2	120	72	166	230
63	0	0	51	163.0	⋆	⋆	⋆	287	270
64	0	0	81	167.6	93.2	⋆	⋆	107	176
65	0	0	24	174.0	112.5	112	70	99	196
66	1	0	36	177.8	228.6	160	100	109	179
67	0	1	32	166.0	142.9	⋆	⋆	⋆	159
68	0	0	76	161.0	⋆	122	80	⋆	⋆
69	0	0	74	154.9	⋆	142	90	⋆	⋆
70	1	0	25	187.0	172.0	130	90	134	175
71	0	0	51	167.0	126.5	⋆	⋆	111	231
72	0	1	25	152.0	⋆	140	102	178	189
73	1	0	36	172.7	⋆	158	92	⋆	⋆
74	0	0	34	165.1	115.5	110	76	96	185
75	0	1	34	162.6	137.7	160	70	100	241
76	0	0	40	166.4	⋆	120	80	⋆	⋆
77	1	0	41	184.2	235.0	⋆	s⋆	123	195
78	1	1	37	181.6	167.3	135	73	331	200
79	0	0	22	168.9	111.6	⋆	⋆	94	330
80	1	0	38	182.9	147.3	⋆	⋆	106	160
81	0	0	36	162.6	⋆	112	72	⋆	247
82	1	0	36	182.0	157.4	132	80	110	219
83	1	0	30	177.8	⋆	150	90	⋆	⋆
84	0	0	30	171.5	⋆	138	70	⋆	⋆
85	0	0	23	165.1	⋆	126	90	⋆	⋆
86	1	0	36	170.2	⋆	160	100	⋆	⋆

(continued)

TABLE B-3 (continued)

Row	Sex	Race	Age	HTCM	WTKG	SYSBP	DIASBP	FBSGLU	CHOLES*
87	0	0	41	167.6	124.3	176	90	261	220
88	0	0	28	160.0	122.7	140	80	99	202
89	0	0	34	165.1	118.2	128	80	94	224
90	1	0	32	171.5	145.9	146	80	113	175
91	0	1	22	160.0	130.2	120	78	99	162
92	0	1	43	166.4	*	130	80	*	*
93	0	0	23	177.8	109.1	110	64	90	170
94	0	1	37	172.0	113.6	110	84	99	195
95	0	0	36	165.1	126.6	*	*	104	161
96	0	0	58	162.6	110.7	130	80	275	313
97	0	0	31	170.2	147.7	140	80	366	155
98	0	0	23	165.1	115.9	158	78	101	224
99	0	0	40	167.6	108.2	124	80	101	194
100	1	1	46	174.0	119.0	*	*	93	235
101	1	0	44	174.0	*	134	78	*	*
102	0	0	43	167.6	*	144	92	*	*
103	0	0	38	156.2	*	122	74	*	*
104	0	0	21	167.6	122.7	110	78	93	184
105	1	0	43	172.7	135.2	110	70	298	232
106	0	0	50	172.7	*	146	90	*	198
107	0	1	30	165.1	*	132	74	*	*
108	0	0	31	162.6	*	146	90	*	*
109	0	0	39	167.6	*	152	96	*	191
110	0	0	33	177.8	127.0	130	70	110	222
111	1	1	27	189.2	189.5	*	*	103	266
112	0	1	30	160.0	*	140	86	*	*
113	0	0	32	177.8	*	172	98	*	*
114	0	0	29	168.9	*	150	94	*	*
115	0	1	33	157.5	123.8	150	102	*	184
116	0	0	41	160.0	108.9	150	102	164	227
117	0	1	29	157.5	158.8	150	110	100	*
118	0	0	34	152.4	97.3	132	68	103	214
119	0	0	42	161.3	117.5	140	80	79	214
120	0	0	48	165.1	93.9	116	84	220	322
121	0	0	41	162.6	108.2	140	90	85	208
122	0	0	23	171.5	130.0	130	80	103	165
123	1	1	42	176.0	*	135	82	*	*
124	0	0	21	168.9	*	128	70	*	*
125	0	0	17	162.6	140.5	*	*	82	155
126	0	0	40	175.3	*	110	78	*	*
127	0	1	27	162.6	*	148	80	*	*
128	0	0	49	158.8	115.2	150	76	142	124
129	0	0	32	167.6	*	138	90	*	*
130	0	0	44	167.6	106.1	142	102	99	311
131	0	1	27	168.9	*	130	66	*	*
132	0	1	27	149.9	122.5	120	64	91	183
133	0	1	36	167.6	*	160	108	*	*
134	0	0	31	166.4	*	138	80	*	*
135	0	0	42	166.4	*	130	82	*	*

TABLE B-3   (continued)

Row	Sex	Race	Age	HTCM	WTKG	SYSBP	DIASBP	FBSGLU	CHOLES*
136	0	1	33	160.0	★	124	82	★	178
137	0	0	39	158.8	★	130	80	★	★
138	0	0	52	168.9	★	140	88	★	★
139	0	0	28	163.8	115.7	126	82	94	147
140	0	0	37	157.5	111.1	156	96	103	224
141	0	0	46	176.5	★	104	72	★	★
142	0	0	37	172.7	★	122	100	★	★
143	0	0	25	156.2	★	126	76	★	★
144	1	1	38	175.3	191.9	156	84	★	★
145	0	1	34	160.0	102.7	128	100	90	245
146	0	0	36	162.6	104.6	★	★	95	190
147	1	0	32	180.3	★	170	96	★	★
148	0	0	29	147.0	93.0	148	76	99	202
149	0	0	41	170.2	★	130	80	★	★
150	1	0	41	179.1	★	160	100	★	★
151	0	0	32	162.6	113.4	142	90	101	252
152	0	0	46	172.7	110.7	★	★	103	205
153	0	0	25	163.8	158.8	156	98	123	★
154	0	0	35	160.0	106.1	112	62	119	190
155	0	0	39	167.6	★	140	92	★	★
156	0	0	30	166.4	147.4	134	100	110	167
157	0	0	35	163.8	★	112	70	★	★
158	0	1	28	167.6	165.6	★	★	109	190
159	0	0	48	168.9	★	144	86	★	343
160	0	0	29	165.1	117.9	122	84	107	170
161	0	1	41	161.3	137.4	150	98	★	★
162	0	1	37	165.1	★	184	112	★	★
163	0	0	42	167.6	★	118	72	★	★
164	0	1	43	154.9	119.8	130	100	116	267
165	0	0	40	156.2	127.0	150	82	97	154
166	0	0	51	161.3	99.8	110	70	116	187
167	0	1	38	160.0	★	140	100	★	★
168	0	0	50	167.6	★	140	88	★	432
169	0	1	35	165.1	173.9	136	78	122	242
170	0	1	41	162.6	137.4	132	82	111	205
171	0	0	34	171.5	★	120	88	★	★
172	0	0	48	162.6	★	110	68	★	★
173	0	0	34	160.0	108.0	★	★	113	204
174	0	1	53	152.4	133.4	176	84	239	365
175	0	0	26	174.0	★	126	86	★	164
176	0	0	52	156.2	104.3	140	84	122	179
177	1	0	18	182.9	★	122	88	★	★
178	1	0	38	179.1	★	132	72	★	★
179	0	0	39	169.5	★	130	80	★	★
180	0	0	50	158.8	★	140	60	★	★
181	0	0	39	168.9	★	142	94	★	★
182	0	0	52	162.6	★	★	★	92	210
183	0	1	51	151.1	113.9	128	80	81	308
184	0	0	35	170.2	136.1	130	80	88	143

(continued)

TABLE B-3 (continued)

Row	Sex	Race	Age	HTCM	WTKG	SYSBP	DIASBP	FBSGLU	CHOLES*
185	0	0	45	160.0	122.7	130	94	303	198
186	0	0	46	157.5	⋆	120	68	⋆	⋆
187	0	0	36	160.0	114.3	⋆	⋆	94	180
188	0	0	42	168.3	132.0	140	60	⋆	273
189	0	0	47	161.3	111.8	⋆	⋆	88	177
190	0	1	34	161.3	158.8	120	76	84	180
191	0	0	36	171.5	149.3	142	98	112	288
192	0	0	30	170.2	127.7	140	90	93	194
193	1	0	25	161.3	148.8	⋆	⋆	94	246
194	0	0	51	161.3	108.4	130	90	97	195
195	1	0	35	171.5	⋆	140	88	⋆	⋆
196	0	1	29	166.4	102.1	⋆	⋆	93	183
197	0	0	39	165.1	111.1	⋆	⋆	⋆	248
198	0	0	50	157.5	97.7	⋆	⋆	108	221
199	1	0	40	188.0	⋆	140	90	⋆	⋆
200	0	0	32	170.2	105.2	⋆	⋆	104	199
201	1	0	46	177.8	169.2	180	90	114	182
202	0	0	49	157.5	102.1	⋆	⋆	121	201
203	0	0	36	168.9	133.8	⋆	⋆	92	200
204	0	1	31	170.2	142.9	110	80	⋆	205
205	1	0	35	185.4	146.1	144	100	107	237
206	0	0	57	162.6	128.8	144	90	298	249
207	1	0	43	182.9	⋆	140	90	156	140
208	0	0	25	161.3	122.5	⋆	⋆	80	190
209	0	0	35	167.6	⋆	148	104	99	185
210	0	0	32	162.6	149.7	140	88	112	251
211	0	0	34	158.8	131.1	120	68	137	191
212	1	0	43	167.6	⋆	⋆	⋆	⋆	240
213	0	0	24	177.8	147.3	140	98	105	170
214	0	1	37	170.8	⋆	140	90	⋆	⋆
215	1	0	35	185.4	⋆	150	80	⋆	⋆
216	0	0	50	160.0	⋆	170	100	⋆	⋆
217	0	1	47	161.3	100.2	182	98	198	474
218	0	0	17	163.8	⋆	138	78	⋆	⋆
219	0	0	41	161.3	151.0	140	90	89	197
220	0	0	45	158.8	121.6	130	70	115	236
221	0	0	37	161.3	152.0	128	86	169	160
222	1	0	32	185.4	⋆	110	62	⋆	158
223	0	0	42	167.6	⋆	120	80	⋆	209
224	0	0	34	174.0	⋆	138	80	⋆	163
225	1	1	44	182.9	⋆	120	80	⋆	425
226	0	0	45	154.9	157.9	140	98	249	242
227	1	0	42	188.0	131.5	180	114	127	245
228	1	0	55	170.9	147.4	128	80	147	182
229	0	0	52	162.6	125.2	120	98	⋆	⋆
230	0	0	31	165.1	136.1	150	80	92	147
231	0	0	58	157.5	153.0	170	80	⋆	352
232	0	0	31	157.5	⋆	150	80	⋆	⋆
233	1	0	37	185.4	145.2	160	116	103	162

TABLE B-3 (continued)

Row	Sex	Race	Age	HTCM	WTKG	SYSBP	DIASBP	FBSGLU	CHOLES*
234	0	0	30	161.9	111.4	★	★	100	180
235	0	0	35	172.7	113.9	★	★	87	237
236	0	0	49	160.0	★	★	★	164	289
237	0	0	30	154.9	102.1	132	78	110	171
238	1	0	17	182.9	★	190	108	★	★
239	0	1	24	160.0	★	112	76	★	★
240	0	0	33	166.4	127.5	148	88	129	204
241	0	0	16	164.5	★	160	106	★	237
242	0	0	37	167.6	★	150	80	★	★
243	1	0	35	178.4	215.5	130	80	115	126
244	0	0	39	163.8	111.1	118	82	85	169
245	0	1	46	162.6	121.6	116	78	188	170
246	0	0	34	162.6	124.2	150	90	101	226
247	0	0	52	157.5	105.5	★	★	82	300
248	0	0	43	167.6	144.2	146	82	★	330
249	0	0	40	165.1	★	130	80	★	★
250	0	0	34	160.0	112.0	★	★	105	136
251	0	0	37	158.1	122.5	130	100	★	★
252	0	0	30	167.6	127.9	130	70	85	185
253	1	1	31	181.6	162.2	110	58	★	★
254	0	0	38	166.4	167.4	130	80	261	227
255	0	0	40	170.2	★	★	★	★	146
256	0	0	35	165.1	120.4	★	★	97	176
257	0	0	31	166.4	145.4	122	90	95	187
258	0	0	25	154.9	104.3	156	98	115	183
259	1	1	29	181.6	161.5	160	130	★	★
260	0	1	57	160.0	89.1	178	100	★	★
261	1	0	36	170.2	★	130	94	120	204
262	0	0	41	167.6	123.8	170	110	86	★
263	1	0	34	175.3	77.3	170	98	★	★
264	1	0	44	180.3	163.3	148	88	135	152
265	0	0	33	159.4	48.1	102	54	88	★
266	1	0	28	200.7	89.1	125	68	89	★
267	0	0	27	165.1	46.3	98	70	81	★
268	0	0	24	162.6	55.3	90	60	80	★
269	0	0	60	172.7	123.8	180	100	117	269
270	0	0	30	160.0	120.2	128	88	104	188
271	0	0	35	155.0	★	★	★	★	257
272	1	0	27	188.0	84.0	110	64	86	★
273	0	0	19	152.4	44.9	110	70	84	★
274	1	0	30	167.6	64.6	98	60	88	★
275	0	0	25	167.6	63.5	104	80	83	★
276	0	0	24	172.7	56.7	118	60	76	★
277	0	0	28	160.0	47.6	90	50	82	★
278	1	0	23	193.0	74.8	90	70	86	★
279	0	0	33	152.4	127.3	110	70	121	★
280	0	0	21	167.6	51.3	100	40	79	★
281	0	0	29	162.6	49.0	86	60	82	★
282	1	0	25	182.9	74.8	108	74	88	★

(continued)

TABLE B-3 (continued)

Row	Sex	Race	Age	HTCM	WTKG	SYSBP	DIASBP	FBSGLU	CHOLES*
283	0	0	34	170.2	58.5	90	60	79	⋆
284	1	0	38	182.9	80.7	116	68	89	⋆
285	1	0	31	190.5	69.4	104	60	74	⋆
286	1	2	27	170.2	56.2	120	60	72	⋆
287	0	0	24	162.6	59.0	76	48	77	⋆
288	0	2	38	153.7	49.0	70	50	79	⋆
289	0	1	26	177.8	69.9	80	48	77	⋆
290	0	0	28	156.2	54.4	98	70	83	⋆
291	0	0	31	154.9	68.9	116	60	88	⋆
292	0	0	44	157.5	53.5	104	64	71	⋆
293	0	1	39	166.4	54.0	100	60	76	⋆
294	1	0	39	167.6	61.2	124	68	91	⋆
295	0	0	25	162.6	56.7	110	64	83	⋆
296	0	0	35	170.2	53.5	100	68	85	⋆
297	1	0	34	170.2	68.0	118	60	84	⋆
298	0	2	34	170.2	64.9	102	70	97	⋆
299	1	0	27	179.1	69.9	110	72	83	⋆
300	0	0	38	182.9	62.1	108	64	73	⋆
301	1	0	24	180.3	86.2	118	70	90	⋆
302	1	0	38	182.9	72.6	115	60	84	⋆
303	1	0	31	172.7	72.6	110	68	83	⋆
304	1	0	41	182.9	68.0	95	60	97	⋆
305	0	0	27	165.1	59.0	108	64	86	⋆
306	0	0	27	160.0	55.1	100	70	72	⋆
307	1	1	60	170.2	80.5	90	60	⋆	⋆
308	0	1	43	158.8	93.0	130	82	⋆	⋆
309	0	0	53	171.5	⋆	⋆	⋆	⋆	192
310	0	0	42	170.2	⋆	156	100	⋆	⋆
311	0	1	48	152.4	141.5	140	74	⋆	⋆
312	1	0	28	185.4	98.4	152	90	⋆	⋆
313	0	0	52	152.4	107.0	150	82	89	⋆
314	0	1	19	167.6	269.4	135	95	92	⋆
315	0	1	27	167.6	199.1	150	120	⋆	⋆

APPENDIX

Using the TI-82 as a Data Analysis Tool

DESCRIPTIVE STATISTICS

The following example will serve to illustrate how the TI-82 can be used to create graphs and generate summary descriptive statistics.

EXAMPLE   A personnel director is interested in determining if there are any differences between the male and female employees at his company with respect to the number of sick days taken each year. The following are data for 27 randomly selected female and 33 randomly selected male employees.

NUMBER OF SICK DAYS TAKEN THIS YEAR

Females			Males		
15	6	10	2	3	1
18	5	8	4	5	3
14	7	9	4	1	1
7	8	12	2	2	0
12	10	5	3	0	2
6	6	7	4	2	5
16	9	8	0	3	2
15	11	6	1	4	3
12	14	11	3	6	2
			5	2	0
			2	1	5

DATA ENTRY

The easiest way to enter data on the TI-82 is to press **STAT** and select **EDIT.**
This brings up a screen that allows you to enter your data in columns or lists.
Enter the female data in list L_1 and the male data in list L_2.

HISTOGRAMS

To display a histogram for the female data press **STAT PLOT** and then, with
Plot 1 selected, highlight **ON**, the **histogram icon Xlist: L_1** and **Freq: 1.**
Next, turn **Plots 2** and **3 OFF** using the **STAT PLOT** menu. Press the
WINDOW key to select the range for both the data (x) and the frequency
scale (y).

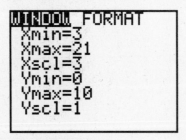

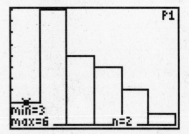

Calculator tip   If you prefer you can let the TI-82 do the window settings
for you. Press **ZOOM,** then select **ZoomStat.** If you now want to alter the
display, you can press **WINDOW** to view the settings and make changes.

The **TRACE** button in conjunction with the arrow keys allows you to read off the class boundaries and corresponding frequencies. This is useful in establishing a frequency table for your data. Using the **TRACE** option with the histogram generated on our display, we could construct the following frequency table for the female data:

Number of Days Out	Frequency
3 to 6	2
6 to 9	10
9 to 12	6
12 to 15	5
15 to 18	3
18 to 21	10

L 3	L 4	L 5
4.5	2	▬▬▬
7.5	10	
10.5	6	
13.5	5	
16.5	3	
19.5	1	
------	------	

L 5 (1)=

An examination of the data reveals that the upper endpoints of the class boundaries are not included, whereas the lower endpoints are included. So the class interval "3 to 6" includes all cases from 3 up to, but not including, 6. As a result it covers all female employees who were out 3, 4, or 5 days.

If data have been previously summarized into a frequency table, you can enter these grouped data on the TI-82 by storing the class midpoints in one list and the corresponding frequencies in the adjacent list. The display shows the data input for the frequency table above where list L_3 was used to enter the class marks and list L_4 was used to enter the frequencies.

Calculator tip   The TI-82 allows you to enter a maximum of 99 values in a given list. For larger data sets a statistical computer program is recommended.

STEM-AND-LEAF PLOTS

The TI-82 does not have a built-in routine to create stem-and-leaf plots; however, you can option **2:SortA(** under the **EDIT** menu to sort lists in ascending order. This makes the task of creating your own stem-and-leaf graph much easier.

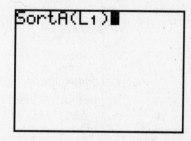

SUMMARY STATISTICS

To display the summary statistics for the female data select **CALC** from the **STAT** menu, then select **1-VarStats** from the **CALC** menu. Following the prompt **1-VarStats** enter L_1 since the female data were stored in list L_1. Similarly, repeat this sequence using L_2 to display the summary statistics for the male data.

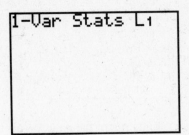

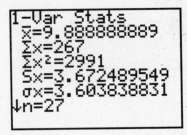

To display the boxplots for both female and male data sets press **STAT PLOT** and then, with **Plot 1** selected, highlight **ON**, the **boxplot** icon, and **XList:L_1** and **Freq:1**. Next, turn **Plot 2 ON**, highlight the **boxplot** icon and **Xlist:L_2** and **Freq:1**. Press **WINDOW** to set up the horizontal axes with a minimum of zero and a maximum of 20, since the data ranged from 0 to 18. You can adjust the window settings to vary your boxplot displays. The **TRACE** button can be used in conjunction with the arrow keys to mark the quartiles as well as the min and max values.

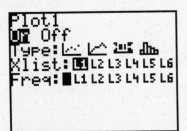

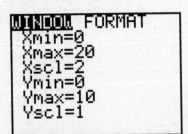

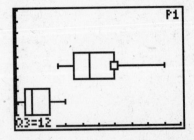

BINOMIAL PROBABILITIES

Your TI-82 comes with a number of useful programs to compute certain summary statistics and produce several useful displays. You can also program your TI-82 to perform additional calculations and to display other statistical graphs. At the end of this appendix is a program called BINOMIAL. This program displays a histogram to help you visualize the probability distribution for a binomial experiment and computes the binomial probabilities for the number of successes. You can use this output to answer a variety of questions related to a binomial variable.

EXAMPLE

Suppose that 80 percent of all families own a TV set. If ten families are selected at random, find the probability that

(a)   seven families own a TV set.
(b)   at least seven families own a TV set.
(c)   at most three families own a TV set.
(d)   Create a graph of the probability distribution for $x = \#$ of TV sets.
(e)   Find the mean and standard deviation of the distribution.   ●

Make sure the program **BINOMIAL** has been entered on your calculator. Then press the **PRGM** key, highlight **BINOMIAL,** and press **ENTER**. Following the **prgmBINOMIAL** prompt, press **ENTER** again. Next, type a value for **N,** the number of trials, followed by a value for **P,** the number of successes. In a few seconds you will be asked if you would like a graph depicting the shape of the probability distribution for x. If you respond **1** for Yes, a graph will be displayed. Press **ENTER** again and you will get a table of values. The column labeled Y1 gives the binomial probabilities corresponding to X successes. So we can determine from the output that the probability that exactly 7 families own a TV set is 0.20133.

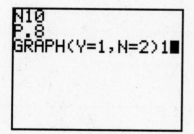
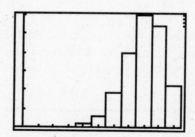

The binomial probabilities are also stored in list L_1. You may want to investigate the three lists that the program **BINOMIAL** created. Press the **STAT** key and select **EDIT.** Notice that list L_1 contains the binomial probabilities, list L_2 contains the x values, and list L_3 displays the binomial probabilities as percents rounded to the closest whole number. Hence, the chance that exactly 7 out of 10 households will have a TV is .20133 (from list L_1) or 20% (from list L_3). To answer Parts (b) and (c) of the example, we just have to add the appropriate probabilities in list L_1. Be careful, however, since $L_1(1)$ corresponds to $x = 0$ and $L_1(2)$ corresponds to $x = 1, \ldots$, and $L_1(11)$ corresponds to $x = 10$. Hence, to answer Part (b) of the example we should type:

$$L_1(8) + L_1(9) + L_1(10) + L_1(11)$$

Your display indicates that this probability is 0.879. Similarly, to answer Part (c) type:

$$L_1(1) + L_1(2) + L_1(3) + L_1(4)$$

Your display indicates that this probability is 0.000864.

To get a feeling for the concept of the mean and standard deviation of a probability distribution, we could calculate these values using the x values in list L_2 and the rounded percents in list L_3. So, press **STAT**, highlight **CALC**, Highlight **1-Var Stats**, and, following the prompt 1-Var Stats, enter L_2, L_3. This indicates that list L_2 contains our X values and list L_3 contains our $p(x)$ values. The output gives the mean as 7.98, the standard deviation as 1.297, and the sample size as 101. These can be compared to the theoretical mean of 8.0 and the theoretical standard deviation of 1.265, gotten by using the formulas $u = np$ and $\sigma = \sqrt{npq}$. The discrepancies occur as a result of rounding the figures in list L_3.

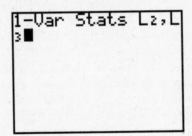

Calculator tip   To exit the EDIT mode of the STAT program press QUIT. Also, if the number of trials exceeds 46, a histogram cannot be displayed, so the program does not ask if you want a graph when $n > 46$.

NORMAL PROBABILITIES

Another useful program that is supplied at the end of this Appendix is the program **ZPROG**. This program allows you to determine areas under the standard normal curve. First, make sure the program has been entered on your calculator. Then press the **PRGM** key. The program requires you to input two z-scores, a lower bound A and an upper bound B.

EXAMPLE   If *z* is the standard normal variable, find the following probabilities:

(a) $P(-1.2 < z < 2.8)$
(b) $P(z > 1.4)$ ●

To use the program **ZPROG** to answer Part (a), press **PRGM**, highlight **ZPROG**, and press **ENTER** following the **prgmZPROG** prompt. Following the **LOWER Z** prompt type -1.2 and following the **UPPER Z** prompt type 2.8. If you would like a graph of the standard normal curve with the area between $-.12$ and 2.8 shaded, respond 1 to the Graph inquiry and press **ENTER**. The screen after the graph indicates that the probability that *z* is between -1.2 and 2.8 is .88237.

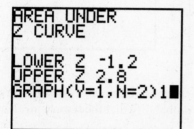

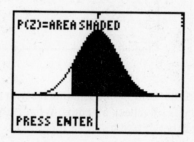

To answer Part (b), follow the same sequence of instructions using $z = 10$ as your upper bound *B*. This works since the probability that *z* falls above 10 is essentially zero. The display gives the probability for *z* greater than 1.4 as .0807566.

NONSTANDARD NORMAL PROBABILITIES
CENTRAL LIMIT THEOREM
NORMAL APPROXIMATION TO THE BINOMIAL

The program **NSNORMAL** can be used to quickly evaluate probabilities for normally distributed variables, central limit theorem problems, and normal approximation to the binomial problems. Make sure the program **NSNORMAL** has been entered on your calculator. Then press the **PRGM** key, highlight **NSNORMAL,** and press **ENTER.** Following the **prgmNSNORMAL** prompt, press **ENTER** again. The next three examples will illustrate how to use this program.

EXAMPLE   An automobile manufacturer claims the new model car gets 28 mpg on average, with a standard deviation of 3.5 mpg. Assume the mpg are normally distributed. What percent of these autos get more than 32 mpg?   ●

To solve this problem select **1: NORMAL PROB** from the **CHOICES** menu. Do this by highlighting **1** and pressing **ENTER.** Then select **1:P(X > A)** from the NORMAL PROB menu. Do this by entering the number **1** following the CHOICE prompt. Enter the values for the population mean *M*, standard deviation, *S*, and *X* value *A*. The display indicates that .1265 or 12.65% of the autos get more than 32 mpg.

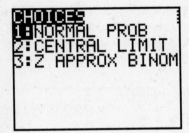
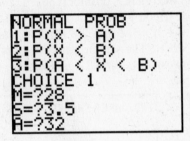
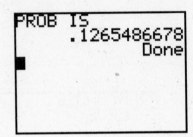

EXAMPLE

For the automobile manufacturer in the previous example, a consumer group decides to check 50 autos. What is the probability that the sample mean mpg for these 50 autos is 28 mpg or less? ●

To solve this problem select **2:CENTRAL LIMIT** from the CHOICES menu. Then select **5:P(MEAN < B)** from the CENTRAL LIMIT menu. Then enter the values for the population mean *M*, standard deviation *S*, sample size *N*, and the *x* value *B*. The display indicates that the probability of obtaining a mean of 27 mpg or less based on a sample of 50 autos is .0217.

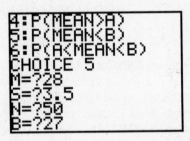
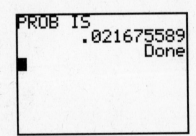

EXAMPLE

According to the *1990 Statistical Abstract of the United States*, 38.2 percent of all households own at least one dog. In a random sample of 1000 households, what is the probability that more than 400 will own at least one dog? ●

This time select **3:Z APPROX BINOM** from the CHOICES menu, then **7:P(X > A)** from the **Z APPROX BINOM** menu. Enter the values for *N, P,* and *A*. The display indicates that the probability that 400 or fewer households out of 1,000 will own at least one dog is 0.1274.

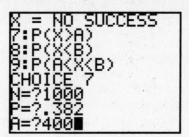
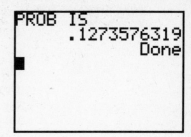

CONFIDENCE INTERVALS HYPOTHESIS TESTING

There are three programs at the end of this appendix that can assist you in evaluating confidence intervals and testing hypotheses. They are: **MEANONE, MEANTWO,** and **PROPTWO.** First, make sure these programs have been entered into your TI-82, then try the following three examples to see how these programs work.

EXAMPLE

Refer to the data at the beginning of this Appendix concerning the sick days for male and female employees. Recall that sample values for the females were $\bar{x} = 9.8889$, $s = 3.6725$, and $n = 27$.   ●

Suppose we wanted to use these data to test the null hypothesis that the mean number of sick days for female employees is 9 days. To test this hypothesis at the 0.05 level of significance, select **MEANONE** from the **PRGM** menu and enter the MEAN, ST DEV, N, and T values as requested. Note the T value of 2.056 is the tabled value of T for 26 degrees of freedom using a two-tailed alternative. The display gives a 95% confidence interval for the mean and indicates that the calculated t-value for the null hypothesis is $t = 1.2577$. Hence, on average females take between 8.43 and 11.34 sick days ($p < 0.05$).

```
MEAN 9.8889
ST DEV 3.6725
N 27
Z OR T 2.056
```

```
CONFIDENCE INT
LOWER
        8.435774805
UPPER
        11.34202519
PRESS ENTER
```

```
TEST AN HYPOTH
(1=Y,2=N) 1
HYPOTH MEAN= 9
TEST STAT FOR
H0 IS
        1.257688193
        Done
```

EXAMPLE   If we now consider the sick day data for the male employees, you will recall that $\bar{x} = 2.5152$, $s = 1.6417$, and $n = 33$. ●

Suppose we wanted to use these data to test the null hypothesis of no difference in the mean number of sick days taken by male and female employees at the 0.05 level of significance. To do this, select **MEANTWO** from the **PRGM** menu, enter the female MEAN, ST DEV, and N, followed by the same information for the males. We will enter $Z = 1.96$ since the degrees of freedom here are $(33 + 27 - 2) = 58$. The display estimates the true difference in the mean number of sick days to be between 5.98 and 8.77 days ($p < 0.05$) and gives the value of the test statistic for the null hypothesis of no difference in mean number of days out as $t = 10.35$. Hence, on average females take between 5.98 and 8.77 more sick days than males ($p < 0.05$).

```
X1 MEAN 9.8889
S1 3.6725
N1 27
X2 MEAN 2.5152
S2 1.6417
N2 33
INPUT T1.96■
```

```
SAMPLE DIF IS
              7.3737
LOWER BND
        5.977726993
UPPER BND
        8.769673007
PRESS ENTER
```

```
DO YOU WANT TO
TEST H0: M1=M2
(1=Y, 2=N)1
TEST STAT FOR
H0 IS
         10.3529595
             Done
```

EXAMPLE   The following data on smoking patterns are taken from the *1990 Statistical Abstract of the United States*. ●

% Smoker	1965	1970	1974	1976	1977	1978	1979	1980	1983	1985	1987
Male	50.2	44.3	43.4	42.1	40.9	39.0	38.4	38.5	35.5	33.2	31.5
Female	31.9	30.8	31.4	31.3	31.4	29.6	29.2	29	28.7	28	26.2

Enter the male data in one list and the female data in another list on your TI-82. Since these are paired data, we need to create a third list consisting of the male values minus the female values. To accomplish this, quit STAT and type : $L_1 - L_2$ **STO** L_3 on the home screen. This puts the paired difference of the male and female data in list L_3. Now, from the STAT menu select CALC, **1:1-Var Stats** L_3 to obtain the mean, standard deviation, and sample size for the difference data. The display indicates the mean difference in the percent of male and female smokers is 9.9545 with a standard deviation of 3.7766 and sample size 11. Next, input these values in the program **MEANONE** using a t value of 2.228 for 95% confidence. The display indicates that on average there are between 7.4% and 12.5% more male smokers in the population than females ($p < 0.05$) and the test that the mean difference is zero yields a t value of 8.7421. Hence, over the years spanning 1965 to 1987 there were, on average, a significantly higher percentage of male smokers than female smokers ($p < 0.05$).

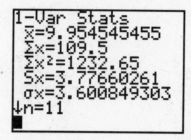

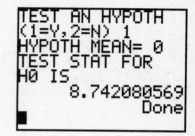

EXAMPLE
Suppose a sample of 55 men and 60 women age 18 or older indicated that 25.2% of the males and 23.6% of the females never eat breakfast. Establish a 95% confidence interval for the true difference in the proportion of men and women who never eat breakfast and test the null hypothesis that there is no difference in the male and female proportions. ●

Select **PROPTWO** from the program menu and enter the information as indicated on the displays. The test statistic for the null hypothesis of no difference in proportions is 0.1197 and hence not significant at the 0.05 level. We conclude that these data give no evidence of a significant difference in the proportion of men and women who skip breakfast.

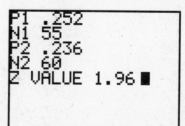

TESTING INDEPENDENCE: CONTINGENCY TABLES

Make sure the program **CHISQ** is loaded in your TI-82. This program allows you to test for the independence of two attributes when data are arranged in a contingency table. We illustrate its operation using the following example:

EXAMPLE

The most recent study of criminal "victimization" was conducted by the Census in 1979. It reports the number of offenses people say happened to them that year, and whether they notified the police that the crime occurred. The data are in units of 1,000 incidents (Source: *A Statistical Portrait of the United States*).

	Number of Claimed Incidents Not Reported to Authorities	Number Claimed Reported to Authorities
Rapes and attempted rapes	95	97
Robberies	497	619
Aggravated assaults	862	907
Residential burglaries	11,465	5,850
Motor vehicle thefts or attempted thefts	443	950

To analyze the data select **CHISQ** from the **PRGM** menu and respond as indicated in the display boxes. Be careful—you must enter the frequencies row by row, not column by column. Also, do not enter commas. The critical value of chi-square with 4 degrees of freedom at 0.005 level of significance is 14.86. So, the calculated chi-square for these data of 94.89 indicates there is a statistically significant relationship between the type of victimization and whether it is reported to the authorities. To view the observed values press **MATRIX**, highlight [**A**], and press ENTER twice, then use the cursor keys to scroll the observed value matrix. To view the expected values, press **MATRIX**, highlight [**E**], and press ENTER twice. Again scroll to view the entire

```
ROW                1
X=?95
X=?97
```

```
CHISQ
         94.89261
DF
              4
          Done
```

```
[E]
...117.76    74.24...
...684.51    431.4...
...1085.03   683.9...
...10620.29  6694....
...854.41    538.5...
...13362     8423  ...
```

matrix. The row and column totals are included in both matrix [A] and [E]. An examination of these matrices reveals that victims are more likely to report rapes, robberies, and aggravated assaults, but less likely to report residential burglaries and motor vehicle theft.

SIMPLE LINEAR REGRESSION ANALYSIS

Make sure the program **REG** is loaded in your TI-82. The following example will illustrate how the TI-82 can aid in the analysis of simple linear regression.

EXAMPLE

Refer to the data on the percent of male smokers for the years 1965 to 1987 given in the section on confidence intervals and hypothesis testing. Enter the year in list L_1 and the percent of males who smoke in list L_2. Then go to the **CALC** menu and select **5:LinReg(ax + b)**. Then, following the **linReg(ax + b)** prompt, enter L_1, L_2. This sets the x-values to those stored in L_1 and the y-values to those stored in L_2. Press ENTER to see the regression coefficients and the correlation coefficient. The regression equation is $y = -.8193x + 1659.9684$.   ●

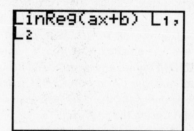

For additional summary statistics select **REG** from the **PRGM** menu. To estimate the percent male smokers for the year 1990 enter 1990 for X. Since there are 11 pairs of values we will use a T with 9 degrees of freedom. So, to construct 95% confidence intervals we will use $T = 2.262$. The display gives the 95% confidence interval for the slope, followed by the 95% confidence interval for the predicted percent of male smokers in 1990, followed by the test statistic for the null hypothesis that the population correlation coefficient is zero. The confidence interval for the slope indicates that we would expect the percent of male smokers to decrease from between .82 to .91 percent per year ($p < 0.05$). The confidence interval for Y indicates that we would expect on average between 27.3% and 31.8% of the males to be smokers in 1990. The test statistic $T = -20.16$ indicates that there is a significant linear correlation between the studied year and the percent of male smokers. As time goes on, the percentage of male smokers is decreasing.

There are two main steps to creating a scatter plot with the regression line superimposed on the graph. First, to create the scatter plot, press **2nd STAT PLOT**. Highlight **Plot 1** and press **ENTER.** Then make the selections as indicated on the display. If your X lists and Y lists are not L_1 and L_2, then enter the appropriate lists for these variables. Finally, make sure Plots 2 and 3 are turned OFF.

Next, press the **Y =** button. Enter the regression equation as Y_1. You may have to delete an already existing Y_1 equation before you do this. Finally, press the **ZOOM** button and select **9:ZoomStat** and press ENTER. Once the graph is displayed you can use the TRACE button together with the cursor keys to trace the data values or along the line. Using the up arrow key you can toggle between tracing the points and the line. The display indicates that in the "year" 1968.1362 the predicted percent of male smokers is 47.47.

Calculator tip   If you run the program REG before you attempt to create a scatter plot, you do not have to enter the equation Y_1. The REG program sets up Y_1 as "$aX + b$" and also turns Plots 2 and 3 off. After running the REG program all you need to do to obtain a scatter diagram is to press **2nd STAT PLOT** and select the appropriate X list and Y list, then press **ZOOM** and select **9:ZoomStat.**

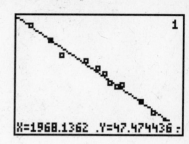

ANALYSIS OF VARIANCE

Make sure the program ANOVA1 has been entered on your TI-82. We will use the data from Example 1 of Chapter 12 to illustrate the program. The productivity data for the five days of the week will be stored in lists L_1 to L_5, so make sure these lists are cleared, then enter the data using the STAT menu. Next, select ANOVA1 from the PRGM menu. The only required input for this program is the number of groups, so enter 5. The program then looks for the data in the *first* 5 lists. On successive screens the program gives the sums of squares, degrees of freedom, mean squares, and the computed F ratio.

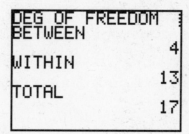

TI-82 Programs

BINOMIAL PROGRAM

```
ClrHome
FnOff
ClrList L₁,L₂,L₃
PlotsOff
Input "N",N
Input "P",P
"N nCr X*(P)^X*(1-P)^(N-X)"→Y₁
Y₁(0)→L₁(1)
For(K,1,N)
Y₁(K)→L₁(K+1)
End
round(100*L₁,0)→L₃
If N>46:Then:Goto 2:End
Input "GRAPH(Y=1,N=2)",R
If R=1:Then:Goto 1:Else:Goto 2:End
Lbl 1
N→XMax
-.5→Xmin
```

```
(N+.5)→Xmax
1→Xscl
0→Ymin
Max(L₃)→Ymax
5→Yscl
0→L₂(1)
For(I,1,N)
I→L₂(I+1)
End
Plot 1(Histogram, L₂,L₃)
DispGraph
Pause
Lbl 2
0→TblMin
1→ΔTBl
DispTable
ClrHome
Stop
```

ZPROG PROGRAM

```
:ClrHome
:Disp "AREA UNDER"
:Disp "Z CURVE"
:Disp "  "
:Input "LOWER Z",A
:Input "UPPER Z",B
:Input "GRAPH(Y=1,N=2)",N
:If N=1:Then:Goto 1:Else:Goto 2:End
:Lbl 1
```

```
:PlotsOff
:"1/√(2π)*e^(-X²/2)"→Y₁
:-4→Xmin
:4→Xmax
:-.2→Ymin
:.5→Ymax
:1→Xscl
:.1→Yscl
:Shade (0, Y₁,1, A,B)
:Text(3,2,"P(Z)=AREA SHADED")
:Text (55,2,"PRESS  ENTER")
:DispGraph
:Pause
:ClrHome
:Lbl 2
:"1/√(2π)*e^(-X²/2)"→Y₁
:fnInt(Y₁,X,A,B)→W
:Disp "AREA IS",W
```

NSNORMAL PROGRAM

```
:ClrHome
:Menu("CHOICES","NORMAL PROB", 1,"CENTRAL LIMIT",2,
"Z APPROX BINOMIAL",3)
:Lbl 1
:Disp "NORMAL PROB"
:Disp "1:P(X > A)"
:Disp "2:P(X < B)"
:Disp "3:P(A < X < b)"
:Input "CHOICE",C
```

```
:Goto 4
:Lbl 2
:ClrHome
:Disp "CTRL LIMIT THM"
:Disp "4:P(MEAN>A)"
:Disp "5:P(MEAN<B)"
:Disp "6:P(A<MEAN<B)"
:Input "CHOICE",C
:Goto 4
:Lbl 3
:ClrHome
:Disp "Z APPROX BINOM"
:Disp "X = NO SUCCESS"
:Disp "7:P(X>A)"
:Disp "8:P(X<B)"
:Disp "9:P(A<X<B)"
:Input "CHOICE",C
:Lbl 4
:If C≤6
:Then
:Prompt M:Prompt S:End
:If C≤6 and C≥4
:Then
:Prompt N
:S/√N→S
:End
:If C=1 or C=4
:Then
:Prompt A:(A−M)/S→L:abs (L)+5→U
```

```
:End
:If C=2 or C=5
:Then
:Prompt B:(B−M)/S→U:-abs (U)−5→L
:End
:If C=3 or C=6
:Then
:Prompt A:Prompt B:(A−M)/S→L:(B−M)/S→U
:End
:If C>6
:Then
:Prompt N
:Prompt P
:End
:If C=7:Then:Prompt A
:(A−.5−NP)/√(NP(1−P))→L
:abs (L)+5→U
:End
:If C=8:Then:Prompt B
:(B+.5−NP)/√(NP(1−P))→U
:-abs (U)−5→L
:End
:If C=9:Then:Prompt A:Prompt B
:(A−.5−NP)/√(NP(1−P))→L
:(B+.5−NP)/√(NP(1−P))→U
:End
:If L<−5
:Then
:L=−6:End
:If U<−5
```

```
:Then
:U=−5:End
:If U>5
:Then
:U=6:End
:If L>5
:Then
:L=5:End
:ClrHome
:Disp L:Disp U
:"1/√(2π)*e^(−X²/2)"→Y₁
:fnInt (Y₁,X,L,U)→W
:Disp "PROB IS",W
ZPROG                 •Program
:ClrHome
:Disp "AREA UNDER"
:Disp "Z CURVE"
:Disp "   "
:Input "LOWER Z",A
:Input "UPPER Z",B
:Input "GRAPH(Y=1, N=2)",N
:If N=1:Then:Goto 1:Else:Goto 2:End
:Lbl 1
:PlotsOff
:"1/√(2π)*e^(−X²/2)"→Y₁
:−4→Xmin
:4→Xmax
:−.2→Ymin
:.5→Ymax
:1→Xscl
```

```
:.1→Yscl
:Shade(0,Y₁,1,A,B)
:Text(3,2,"P(Z)=AREA SHADED")
:Text(55,2,"PRESS  ENTER")
:DispGraph
:Pause
:ClrHome
:Lbl 2
:"1/√(2π)*e^(−X²/2)"→Y₁
:fnInt(Y₁,X,A,B)→W
:Disp "AREA IS",W
```

MEANONE PROGRAM

```
Input "MEAN",M
Input "ST DEV",S
Input "N",N
Input "Z OR T",Z
ClrHome
Disp "CONFIDENCE INT"
Disp "LOWER",M−ZS/√N
Disp "UPPER",M+ZS/√N
Disp "PRESS ENTER":Pause :ClrHome
Disp "DO YOU WANT TO"
Disp "TEST AN HYPOTH"
Input "(1=Y,2=N)",R
If R=1:Then:Goto 3:Else:Goto 4:End
Lbl 3
Input "HYPOTH MEAN=",C
Disp "TEST STAT FOR
Disp "H0 IS",(M−C/(S/√N)
Lbl 4
```

MEANTWO PROGRAM

```
ClrHome
Input "X1 MEAN",X
Input "S1",S
Input "N1",N
Input "X2 MEAN",Y
Input "S2",T
Input "N2",M
If M<30 or N<30
Then
Input "INPUT T",Z
√(((N−1)S²+(M−1)T²)/(N+M−2))→P
abs (X−Y) − ZP√(N⁻¹+M⁻¹)→L
abs (X−Y)+ZP√(N⁻¹+M⁻¹)→U
End
If M ≥ 30 and N ≥ 30
Then
Input "INPUT Z",Z
abs (X−Y)−Z√(S²/N+T²/M)→L
abs (X−Y)+Z√(S²/N+T²/M)→U
End
ClrHome
Disp "CONFIDENCE INT"
Disp "SAMPLE DIF IS",abs(X−Y)
Disp "LOWER BND",L
Disp "UPPER BND",U
Disp "PRESS ENTER":Pause :ClrHome
Disp "DO YOU WANT TO"
Disp "TEST H0: M1=M2"
Input "(1=Y, 2=N)",R
If R=1:Then:Goto 3:Else:Goto 4:End
```

```
Lbl 3
If M≥30 and N≥30
Then
Disp "TEST STAT FOR"
Disp "H0 IS",abs (X−Y)/√(S²/N+T²/M)
End
If M<30 or N<30
Then
Disp "TEST STAT FOR"
Disp "H0 IS",(X−Y)/P√(M⁻¹+N⁻¹)
End
Lbl 4
```

PROPTWO PROGRAM

```
ClrHome
Input "P1",P
(1−P)→Q
Input "N1",N
Input "P2",X
Input "N2",M
(1−X)→Y
Input "Z VALUE",Z
ClrHome
Disp "CONFIDENCE INT"
Disp "SAMPLE DIF",abs (P−X)
(PQ/N+XY/M)→D
Z√D→E
Disp "LOWER",(abs (P−X)−E)
```

```
Disp "UPPER",(abs (P−X)+E)
Disp "PRESS ENTER":Pause :ClrHome
Disp "DO YOU WANT TO"
Disp "TEST H0:P1=P2"
Input "(1=Y,2=N)",R
If R=1:Then:Goto 3:Else:Goto 4:End
Lbl 3
ClrHome
(NP+MX)/(N+M)→Y
Disp "TEST STAT FOR"
Disp "H0 IS",abs (P − X)√(Y(1 − Y)/N+Y(1−Y)/M)
Lbl 4
```

CHISQ PROGRAM

```
ClrHome
Input "NO ROWS",R
Input "NO COLMS",C
{R+1,C+1}→dim [A]
{R+1,C+1}→dim [E]
{R,1}→dim [B]
{1,C}→dim [C]
{R,C}→dim [D]
For(1,1,R)
For(J,1,C)
0→[A](I,J)
End
End
For(I,1,R)
ClrHome
```

```
Disp "ROW",I
For(J,1,C)
Prompt X
X→[A](I,J)
End
End
For(I,1,R)
[A](I,1)→[B](I,1)
End
For(I,1,R)
For(J,2,C)
[B](I,1)+[A](I,J)→[B](I,1)
End
End
For(J,1,C)
[A](1,J)→[C](1,J)
End
For(J,1,C)
For(I,2,R)
[C](1,J)+[A](I,J)→[C](1,J)
End
End
0→T
For(I,1,R)
T+[B](I,1)→T
End
For(I,1,R)
For(J,1,C)
```

```
[B](I,1)*[C](1,J)/T→[E](I,J)
int ([E](I,J)*100+.5)/100→[E](I,J)
End
End
For(I,1,R)
 [B](I,1)→[E](I,C+1)
 [B](I,1)→[A](I,C+1)
End
For(J,1,C)
 [C](1,J)→[E](R+1,J)
 [C](1,J)→[A](R+1,J)
End
T→[E](R+1,C+1)
T→[A](R+1,C+1)
For(I,1,R)
For(J,1,C)
([A](I,J)−[E](I,J))²/[E](I,J)→[D](I,J)
End
End
0→S
For(I,1,R)
For(J,1,C)
S+[D](I,J)→S
End
End
ClrHome
Disp "CHISQ",int (S*10000+.5)/100000
Disp "DF",(R−1)*(C−1)
```

REG PROGRAM

```
ClrHome
Prompt X
Prompt T
LinReg(ax+b)
aX+b→Z
Σx²−(Σx)²/n→L
Σy²−(Σy)²/n→M
Σxy−(Σx)(Σy)/n→N
M−aN→P
√(P/(n−2))→S
Disp "PRESS ENTER"2−Var Stats
Pause
ClrHome
Disp "SLOPE A",a
Disp "LOWER A",a−TS/√L
Disp "UPPER A",a+TS/√L
Disp "PRESS ENTER"
Pause
ClrHome
Disp "PRED Y",Z
TS√(1+N⁻¹+(X−x̄)²/L)→E
Disp "LOWER Y",Z−E
Disp "UPPER Y",Z+E
r√(n−2)/√(1−r²)→R
Disp "PRESS ENTER"
Pause
ClrHome
```

```
Disp "SAMPLE R",r
Disp "TEST R=0",R
FnOff
"aX+b"→Y₁
PlotsOff 2,3
PlotsOn 1
Plot1(Scatter,L₁,L₂,■)
```

ANOVA1 PROGRAM

```
Disp "NUMBER OF GROUPS"
Input G
dim L₁+dim L₂→D
sum L₁+sum L₂→W
(sum L₁)²/dim L₁+(sum L₂)²/dim L₂→B
sum L₁²+sum L₂²→T
If G=2
Then
Goto 7
Else
D+dim L₃→D
W+sum L₃→W
B+(sum L₃)²/dim L₃→B
T+sum L₃²→T:End
If G=3
Then
Goto 7
Else
D+dim L₄→D
```

```
W+sum L₄→W

B+(sum L₄)²/dim L₄→B

T+sum L₄²→T: End

If G=4

Then

Goto 7

Else

D+dim L₅→D

W+sum L₅→W

B+(sum L₅)²/dim L₅→B

T+sum L₅²→T:End

If G=5

Then

Goto 7

Else

D+dim L₆→D

W+sum L₆→W

B+(sum L₆)²/dim L₆→B

T+sum 6²→T:End

Lbl 7

B−W²/D→B

T−W²/D→T

T−B→E

Disp "B",B

Disp "E",E

Disp "T",T
```

Answers to Odd-Numbered Exercises

CHAPTER 1

Section 1-1

1. a. quantitative   c. qualitative   e. quantitative
   g. quantitative   i. quantitative   k. qualitative

3. a. continuous   c. continuous   e. discrete
   g. continuous   i. continuous   k. discrete
   m. discrete

5. The totality of all the tires conceivably manufactured by the plant.

7. a. The results of all the tosses that could conceivably be carried out.
   b. The results of the twenty tosses of the coin.

9. i. Distance of the earthquake from Eureka; continuous.
   ii. Strength of the earthquake; quantitative when Richter scale reading is given; qualitative, when characterized as mild-strong.
   iii. Number of individuals who reported feeling it to the USGS.

11. i. Color of a rat; qualitative   ii. Age of a rat; quantitative; continuous
    iii. Sex of a rat; qualitative   iv Weight of a rat; quantitative; continuous
    v. Side effects of space travel; qualitative

13. i. The type of company; qualitative (employing fewer than 100, between 100 and 500, etc.)
    ii. Net new jobs added to the U.S. economy; quantitative; discrete
    iii. Number of American workers who earn paychecks; quantitative; discrete

15. 166

17. a. 43   b. 12

Section 1-2

1.

Weight	Frequency
15.6	6
15.8	4
15.9	5
16.0	6
16.2	9

3.

Class	Frequency
5.00–9.99	7
10.00–14.99	8
15.00–19.99	7
20.00–24.99	3
25.00–29.99	2
30.00–34.99	1

5. With six classes, we could form the following frequency distribution.

Class	Frequency
8.70–9.29	5
9.30–9.89	8
9.90–10.49	7
10.50–11.09	4
11.10–11.69	5
11.70–12.29	1

7. a. 10   f. 23 (In Parts b., c., d., and e. no answers possible.)

9. a. 8, 13, 18, 23, 28, 33   b. 5.5, 10.5, 15.5, 20.5, 25.5, 30.5, 35.5   c. 5

11.

Class	Relative Frequency	Percentage Frequency
1400–1499	0.08	8
1500–1599	0.20	20
1600–1699	0.12	12
1700–1799	0.35	35
1800–1899	0.18	18
1900–1999	0.07	7

13.

Class	Relative Frequency	Percentage Frequency
12.1–16.0	0.089	8.9
16.1–20.0	0.156	15.6
20.1–28.0	0.222	22.2
28.1–32.0	0.256	25.6
32.1–40.0	0.200	20.0
40.1–48.0	0.078	7.8

15. a. During most months (50 months) the earnings were between $900.00 and $2299.00; only during 13 months the earnings were between the wide range $2300.00 and $3899.00
    b. 999.5, 1199.5, 1399.5, 1699.5, 2099.5, 2649.5, 3449.5
    c. 899.5, 1099.5, 1299.5, 1499.5, 1899.5, 2299.5, 2999.5, 3899.5
    d. The widest class is 3000–3899 with length 900.
    e. 1300–1499

17. a. 25
    b. 387.5, 412.5, 437.5, 462.5, 487.5, 512.5, 537.5
    c. 388–412, 413–437, 438–462, 463–487, 488–512, 513–537.

Section 1-3

1.

Relative Frequency	Central Angle (in Degrees)
0.2742	98.7
0.5032	181.2
0.2226	80.1

3.

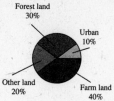

Forest land 30%
Urban 10%
Other land 20%
Farm land 40%

5.

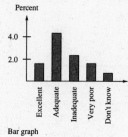

Fair 15%
Good 30%
Very good 35%
Excellent 20%

7. a.

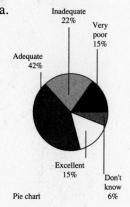

Inadequate 22%
Very poor 15%
Adequate 42%
Excellent 15%
Don't know 6%

Pie chart

b.

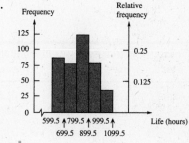

Percent

4.0
2.0

Excellent
Adequate
Inadequate
Very poor
Don't know

Bar graph

9.

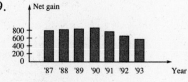

Net gain

800
600
400
200
0

'87 '88 '89 '90 '91 '92 '93   Year

Population increased during the
period 1987 to 1990 and then
began to decline.

11.

Frequency

125
100
75
50
25
0

599.5   799.5   999.5
   699.5   899.5   1099.5

Relative frequency

0.25
0.125

Life (hours)

15. a. 49   b. 46   c. 94   d. 48

17.

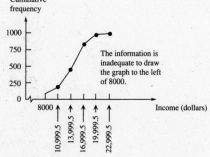

Cumulative frequency

1000
750
500
250
0

The information is
inadequate to draw
the graph to the left
of 8000.

8000
10,999.5
13,999.5
16,999.5
19,999.5
22,999.5

Income (dollars)

19. a.

Speed	Cumulative Frequency
less than 47.5	0
less than 50.5	12
less than 53.5	44
less than 56.5	94
less than 59.5	179
less than 62.5	194
less than 65.5	200

b.

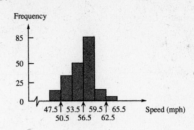

c. Cumulative frequency

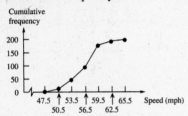

21. a., b.

Class	Class Mark	Relative Frequency	Percent Relative Frequency
20–29	24.5	0.125	12.5
30–39	34.5	0.300	30.0
40–49	44.5	0.225	22.5
50–59	54.5	0.150	15.0
60–69	64.5	0.100	10.0
70–79	74.5	0.062	6.3
80–89	84.5	0.038	3.8

c.

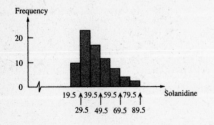

d.

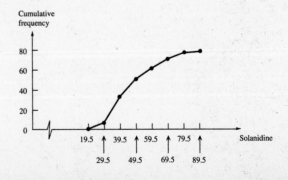

23.

Stem	Leaves
18	5, 6
19	5, 0, 7, 8
20	3, 5
21	0, 4, 5, 6
22	5, 4
23	0, 8, 7, 3
24	0
25	4

25.

Stem	Leaves
93	8
94	1, 9
95	4, 6
96	5, 7
97	1, 4, 4, 6, 9
98	0, 2, 4, 4, 5, 6, 7, 8
99	0, 5, 8, 8
100	0, 2
101	0, 3, 5, 8
102	1
103	
104	7

27. a. No

c. Percentages

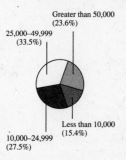

b. 70.2

d. Usually (373); always (224); rarely/never (82); sometimes (156)

Chapter 1 Review Exercises

*2. a. quantitative; discrete
    e. quantitative; continuous
    i. quantitative; discrete
    m. qualitative
    q. quantitative; discrete

    c. quantitative; discrete
    g. quantitative; discrete
    k. quantitative; discrete
    o. quantitative; continuous
    s. quantitative; continuous

3.

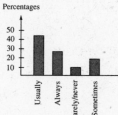

Greater than 50,000 (23.6%)
25,000–49,999 (33.5%)
Less than 10,000 (15.4%)
10,000–24,999 (27.5%)

5.

Stem	Leaves
2.3	1, 0, 8, 9
2.4	0, 6, 4, 6, 3, 3, 8, 6
2.5	8, 9, 2, 9, 4
2.6	5, 8, 4, 4, 8, 6, 0
2.7	1, 3, 3, 1, 3
2.8	0

---

*In some instances the author has chosen to give answers to even exercises.

7. a.

Amount (in Dollars)	Cumulative Frequency
less than −0.5	0
less than 199.5	120
less than 399.5	205
less than 599.5	279
less than 799.5	348
less than 999.5	354

b.

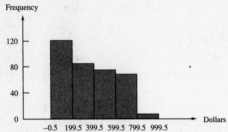

c.

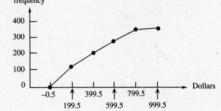

9. a.

Stem	Leaves
2	0, 1, 4
2	5, 6, 7, 8
3	0, 0, 0, 0, 1, 2, 3, 3, 3, 3, 4, 4, 4, 4
3	5, 5, 5, 5, 6, 6, 7, 7, 8
4	0, 1, 2, 2
4	6

11. a.

Classes	Frequency
20.0–22.9	2
23.0–25.9	4
26.0–28.9	7
29.0–31.9	3
32.0–34.9	8
35.0–37.9	3
38.0–40.9	0
41.0–43.9	3

b.

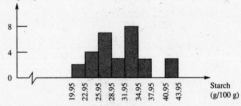

c. Stem-and-leaf diagram with 5 lines per stem.

Stem	Leaves
2	1
2	2, 3, 3
2	4, 4
2	6, 7, 7, 7
2	8, 8, 8
3	0, 0, 1
3	2, 2, 2, 3, 3, 3, 3
3	4, 5
3	6, 7
3	
4	1
4	2, 2

13. a.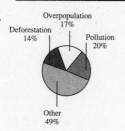

b. pollution (380); overpopulation (323); deforestation (266); other (931)

CHAPTER 2

Section 2-1

1. 67

3. 2.9

5. a. False b. True c. True

7. $850,000

9. 35.56

11. a. 34 b. 230
Conclusion: The number of observations is large, and even a fluke observation (an outlier) like 230 can change the mean only slightly.

13. a. 7 b. 3(7) = 21 c. 7(7) = 49

15. a. 7.333 hours b. $62.33 c. same answer

17. 387

19. 2.1 inches

21. a. Mean = 2.465; median = 2.55   b. Mean = 1.285; median = 1.16

23. a. 4037.3   b. 4003.5

27. a. $7 + \frac{1}{4}$; Q_1 lies between the 7th and 8th largest observations, one-fourth of the way above the 7th.
    c. $10 + \frac{3}{4}$; Q_1 lies between the 10th and 11th largest observations, three-fourths of the way above the 10th.
    e. $12 + \frac{1}{4}$; Q_1 lies between the 12th and 13th largest observations, one-fourth of the way above the 12th.
    g. $17 + \frac{3}{4}$; Q_1 lies between the 17th and 18th largest observations, three-fourths of the way above the 17th.

28. a. $21 + \frac{3}{4}$; Q_3 lies between the 21st and 22nd largest observations, three-fourths of the way above the 21st.
    c. $32 + \frac{1}{4}$; Q_3 lies between the 32nd and 33rd largest observations, one-fourth of the way above the 32nd.
    e. $36 + \frac{3}{4}$; Q_3 lies between the 36th and 37th largest observations, three-fourths of the way above the 36th.
    g. $53 + \frac{1}{4}$; Q_3 lies between the 53rd and 54th largest observations, one-fourth of the way above the 53rd.

29. $Q_1 = 0.39$ $Q_2 = 0.47$ $Q_3 = 0.62$

31. a. i. 0.85   ii. 0.56   iii. $Q_1 = 0.27$ $Q_3 = 1.10$

    b.

    c. Data are skewed to the right. Also, median is not in the center of the data indicating that the middle 50 percent is also skewed to the right.

33. 14   35. 17

37. −6.8; −4.0; bimodal   39. −7; −5; −4; trimodal

41. 3.5

Section 2-2

1. No; consider −7, −7, 0, 0, 0, 0, 4, 4, 3, 3 which has 0 as the mean, median, and mode but is not symmetric.

3. The median

5. Mean = 20; Median = 20

7. Mean = 27.5; Median = 14; Mode = 12; mean is the least satisfactory.

9. Median; 51

11. New mean = $49,800; new median = $47,800

13. a. Median   b. 6–8 years

Section 2-3

1. 43 56 69 75 82 87; range $= 44$
3. 10,900
5. $98.25
7. a. 4.75
   b. 1.25, 1.75, 2.75, 5.25, 2.75, 0.25, 0.75, 1.25
   c. 2
9. 0.92
11. 2
13. 3 weeks
15. Variance $= 101.67$; standard deviation $= 10.1$
17. Variance $= 36.5$; standard deviation $= 6$
19. All the values are identical.
21. Mean $= 2.49$; standard deviation $= 0.102$
23. Mean $= 150$; variance $= 11,111.11$
25. 0.036
27. Mean $= 78$; variance $= 144$; standard deviation $= 12$
29. a. Variance $= 0.049$; standard deviation $= 0.22$
    b. All the observations are in (2.496, 3.378).
31. a.

Calcium

Stem	Leaves
12	0, 3, 8
13	0, 5, 8
14	0, 2, 7
15	1, 1, 2, 6, 7
16	0, 2, 3, 4, 8
17	2, 3
18	
19	6

Oxalate

Stem	Leaves
2	0, 3, 4, 4
2	7, 8
3	2, 3, 3, 3
3	5, 6, 7, 7, 7, 8
4	1, 2, 3
4	7
5	1
5	
6	
6	6

   b. Median for calcium $= 15.15$; median for oxalate $= 3.55$
   d. Calcium: mean $= 15.13$, standard deviation $= 1.85$. Oxalate: mean $= 3.58$, standard deviation $= 1.04$
   f. Mean $= 4.60$; standard deviation $= 1.53$
33. a. Mean of Pond 1 $= 5.56$; mean of Pond 2 $= 5.26$; mean of Pond 3 $= 5.75$
    b. 5.5071
    c. 5.523; not the same. However, $[5(5.56) + 5(5.26) + 4(5.75)]/14 = 5.5071$.
    d. s for Pond 1 $= 0.15$; s for Pond 2 $= 0.11$; s for Pond 3 $= 0.13$
    e. 0.24

35. a.   Mean = 74.58; median = 73.00
     c. $s = 24.05$; the interval is (26.5, 122.7); 31 observations are in it, that is, 93.9 percent.

37. 5 places

39. a.   70
     b.   Variance = 45.756; standard deviation = 6.76
     c.   (56.48, 83.52); number of observations = 24; 96 percent.
     d.   (49.72, 90.28); number of observations = 25; 100 percent.
     e.   yes

41. a.   32.96   b.   5.65   c.   29 cows
     d. $Q_1 = 30.05$; $Q_3 = 36.00$; interquartile range = 5.95; the range of the middle 50 percent of the
          data is 5.95
     e.   16 cows, that is, 53.3 percent

43. At least 45 data points in the interval (72, 120); at least 53 in (60, 132).

Section 2-4

1.   a.   166.25   b.   16.0

3.   Mean = 153.5; variance = 60.60; standard deviation = 7.78

5. $\bar{x} = 3.656$; $s = 0.065$

7. $729.15

9. $s^2 = 43.97$

11. $\bar{x} = 166.83$; $s = 14.19$

13. a.   Mean = 369.5; standard deviation = 124.3
     b.   At least 16 books have number of pages in the interval (220.3, 518.7).

15. a.   32.2   b.   5.9

17.

Time	Frequency
0.375	12
0.625	11
0.875	12
1.125	7
1.375	10
1.625	11
1.875	13
2.125	8
2.375	7
2.625	8
2.875	4
3.125	4
3.375	2

mean length = 1.57 seconds; standard deviation = 0.8315

Chapter 2 Review Exercises

5. 8

7. a. 39.35   b. 38.8

9. a. 6.03   b. Variance = 0.016; standard deviation = 0.125   c. 9, that is, 90 percent

11. a. 4.78   b. 4.75   c. 0.18

13. a. 10.95   b. Modal class 8.0–11.9; mode = 9.95

15. a. 20.00%, 17.24%, 22.40%, 16.92%, 15.09%, 15.48%
    b. 17.855

17. s^2 as (ohms)^2; s as ohms

19. Variance = 4.42; standard deviation = 2.1

21. Mean = 20.66; range = 13.5; standard deviation = 3.28

23. a. 0.588   b. 2.125   c. 0.575   d. 2.15
    e. $Q_1 = 0.5675$ $Q_3 = 0.6125$
    f. $Q_1 = 2.0750$ $Q_3 = 2.2025$
    g. 0.045   h. 0.1275

25. According to Chebyshev's rule at least 15 students should receive a grade of D, C, or B. The assignment is in agreement because, from Exercise 24, there are actually 18 students.

27. $\bar{x} = 2001$; $s^2 = 40.67$; $s = 6.4$.

29. b. Mean = 0.408; median = 0.41
    c. Range = 0.2; standard deviation = 0.0505
    d. 34 out of 35, that is, 97.1% in $(\bar{x} - 2s, \bar{x} + 2s)$;
       35 out of 35, that is, 100% in $(\bar{x} - 3s, \bar{x} + 3s)$;
       The percentages are close to those given by the Empirical rule.

CHAPTER 3

Section 3-2

1. $S = \{sss, ssd, sds, dss, dds, dsd, sdd, ddd\}$

3. a. G_1G_2, G_1G_3, G_2D_3, G_4D_1, D_1D_2, D_2D_3
   b. G_1D_1, G_1D_2, G_1D_3, G_3D_2, G_3D_3, G_4D_1

5. Acceptable assignment in c. and d.

7. yes

9. a. 4/10   b. 6/10

11. a. $\{1, 2, 4, 6\}$ b. 4/6

13. a. {(1, 1), (1, 2), (1, 3), (1, 4), (1, 5), (1, 6)
       (2, 1), (2, 2), (2, 3), (2, 4), (2, 5), (2, 6)
       (3, 1), (3, 2), (3, 3), (3, 4), (3, 5), (3, 6)
       (4, 1), (4, 2), (4, 3), (4, 4), (4, 5), (4, 6)
       (5, 1), (5, 2), (5, 3), (5, 4), (5, 5), (5, 6)
       (6, 1), (6, 2), (6, 3), (6, 4), (6, 5), (6, 6)}
       Probability = 1
    b. {(1, 1), (2, 2), (3, 3), (4, 4), (5, 5), (6, 6)}; Probability = 6/36 = 1/6
    c. {(1, 6), (2, 5), (3, 4), (4, 3), (5, 2), (6, 1)}; Probability = 6/36 = 1/6
    d. {(1, 1), (1, 3), (1, 5), (2, 2), (2, 4), (2, 6), (3, 1), (3, 3), (3, 5), (4, 2), (4, 4), (4, 6), (5, 1),
       (5, 3), (5, 5), (6, 2), (6, 4), (6, 6)}; Probability = 18/36 = 1/2
    e. {(1, 1), (1, 2), (1, 3), (1, 4), (2, 1), (2, 2), (2, 3), (3, 1), (3, 2), (4, 1)};
       Probability = 10/36 = 5/18
    f. {(5, 1), (5, 2), (5, 3), (5, 4), (5, 5), (5, 6), (6, 1), (6, 2), (6, 3), (6, 4), (6, 5), (6, 6)};
       Probability = 12/36 = 1/3
    g. {(2, 2, (4, 4), (6, 6)}; Probability = 3/36 = 1/12
15. a. $P(\{e_1\}) = 3/11$; $P(\{e_2\}) = 1/11$; $P(\{e_3\}) = 3/11$; $P(\{e_4\}) = 4/11$
    b. 10/11
17. a. 0.05   b. 0.8   c. 0.35
19. a. $S = \{v_1, v_2, v_3, v_4, v_5\}$ b. i. 1/5   ii. 3/5
21. a. S = {(TV 825, ref 650, sofa 750), (TV 825, ref 750, sofa 650), (TV 650, ref 825, sofa 750),
       (TV 650, ref 750, sofa 825), (TV 750, ref 650, sofa 825), (TV 750, ref 825, sofa 650)}
    b. i. 1/6   ii. 3/6   iii. 0   iv. 2/6   v. 5/6
23. a. {conservative, libertarian, populist, liberal, undesignated}
    b. 0.34   c. 0.53   d. 0.39
25. a. {(urban, more strict), (urban, kept as are), (urban, no opinion), (suburban, more strict), (suburban,
       kept as are), (suburban, no opinion), (rural, more strict), (rural, kept as are), (rural, no opinion)}
    b. i. 0.71   ii. 0.29   iii. 0.06

Section 3-3

1. 0.65   3. 0.8   5. 0.08   7. 0.35
9. a. 4/7   b. 1/3   c. 100/101   d. 6/14
11. 1 to 6   13. 0.0005   15. 0.992   17. 0.7
19. 0.6
21. a. *P*(*A* and *B*) cannot be greater than *P*(*A*)
    b. *P*(*A*) cannot be greater than *P*(*A* or *B*)
23. a. 0.35   b. 0.25
25. a. 0.89   b. 0.11
27. a. 0.4   b. 0.5
29. a. 0.5   b. 0.1

31. a. 0.6   b. 0.8   c. 0.3
33. 37 to 63
35. 13 to 37

Section 3-4

1. a. 20   b. 6,720   c. 720   d. 720
   e. 24   f. 380   g. 28   h. 15
   i. 364   j. 1   k. 1   l. 125
3. 30
5. 5184; (1, 2, 3, 4, H, T), (1, 1, 3, 2, T, T), (3, 3, 6, 6, T, H)
7. 60 ways
9. $\binom{10}{2} = 45$
11. a. 210   b. 90
13. 8,008
15. 970,200
17. 1/18,009,460
19. a. 0.00022   b. 0.097
21. a. 64 possible RNA codons
    b. 3   c. 4   d. 864
23. a. 2/7   b. 5/7   c. 1/42
25. a. 1/24   b. 1/4   c. 1/12
27. 3! (3! 2! 3!)/8!
29. $\binom{6}{2}\binom{6}{3} \Big/ \binom{12}{5}$
31. a. 10/51   b. 7/238   c. 231/238
33. 3/14

Section 3-5

1. a. 1/3   b. 1/3   c. 5/8
3. a. 0.92   b. 0.96   c. 0.84   d. 0.78
5. 6 percent
7. 0.06
9. a. 0.12   b. 0.63   c. 0.75
11. a. 0.2   b. 0.12   c. 0.56   d. 0.28
13. a. 0.3   b. 0.86
15. a. no   b. yes

17.   a.   0.42   b.   0.28

19.   a.   0.000064   b.   0.984   c.   0.016

21.   a.   0.16   b.   0.3   c.   0.36

     d.   0.24   e.   0.7

23.   a.   0.048   b.   0.952

25.   a.   0.25   b.   0.06   c.   0.56

27.   0.000024

29.   a.   0.28   b.   0.18   c.   0.42   d.   0.54

31.   0.72

33.   0.98

35. $1 - \left(\dfrac{499}{500}\right)^{346} \approx 0.5$

Section 3-6

1.   a.   7/24   b.   2/7

3.   a.   31/48   b.   25/31

5.   a.   17/30   b.   5/17

7.   0.65

Chapter 3 Review Exercises

3.   64 mixes; 320 tests

5.   a.   0.8   b.   0.74   c.   0.52   d.   0.76

7.   0

9.   0.675

11.   a.   1/120   b.   1/20   c.   1/4   d.   1/6

13.   1/3

15.   28,800/749,398

17.   a.   0.12   b.   0.18   c.   0.46   d.   0.42

CHAPTER 4

Section 4-1

1.   The random variable assumes the values as follows:

     a.   1, 2, 3, 4, 5, 6

     b.   0, 1, 2, 3, 4, 5

     c.   −3, 4

3. continuous
7. continuous
11. discrete
15. discrete with values 0, 1, 2, . . . , 18080

5. discrete
9. continuous
13. discrete with values 0, 1, 2, 3, 4, 5, 6

Section 4-2

1. Values assumed 0, 1, 2, and 3 with respective probabilities 0.05, 0.45, 0.45, 0.05.

3. a. There are two Republicans on the committee; $\binom{10}{2}\binom{15}{4}\Big/\binom{25}{6}$

    b. There are 2, or 3, or 4 Republicans on the committee;
    $$\left[\binom{10}{2}\binom{15}{4} + \binom{10}{3}\binom{15}{3} + \binom{10}{4}\binom{15}{2}\right]\Big/\binom{25}{6}$$

    c. $4 \le X \le 6$; $\left[\binom{10}{4}\binom{15}{2} + \binom{10}{5}\binom{15}{1} + \binom{10}{6}\binom{15}{0}\right]\Big/\binom{25}{6}$

5. a. 2.1   b. Variance = 1.39; standard deviation = 1.18

7. -20¢; if he played the game several times he would lose an average of 20¢ per play

13. a. 0.4   b. 0.4   c. 0.55

15. $P(X = 2) = 0.2$; $P(X = 4) = 0.5$

17. a. $21   b. $24
    From Exercise 16, the expected profit when 4 cakes are baked is $15.50. Therefore, it is most profitable to bake 2 cakes.

19. a. 0.75   b. 0.8874

21. a.

x	118	73	43	−2
$p(x)$	0.0002	0.0098	0.0198	0.9702

    b. −0.35 dollars   c. $9.766

23. a. 0.99971042   b. 0.035
    In the long run you will win, on the average, 3.5 cents; not worthwhile to spend 32 cents for postage.

25. The expected payout is $133. So the expected net profit is $29. The company can expect to make $435,000.

Section 4-3

1. a. $(0.4)^3(0.6)^4$ b. $\binom{7}{3}(0.4)^3(0.6)^4$

3. 1/64

5. a. 0.230   b. 0.233   c. 0.158
    d. 0.273

7. 0.251

9. a. 0   b. 0.046   c. 0.168

11. a. 0.002   b. 0.177

13. Expected number of defective items = 10; variance = 9

15. a. 0.207   b. 8.4   c. 3.36

17. 6.6

19. a. 34.6   b. 52.6   c. 13
     d. 34.6

21. a. 0.037   b. 0.094   c. 0
     d. 0.041   e. 0.344

23. a. $(0.1)^8$ b. 0.005

25. a. The probability of getting 7 correct answers is 0.001; the probability of passing the exam is 0.001.
     b. $b(2; 5, 0.2) + b(3; 5, 0.2) + b(4; 5, 0.2) + b(5; 5, 0.2) = 0.262$

27. a. $\dbinom{4}{3}\left(\dfrac{9}{19}\right)^3\left(\dfrac{10}{19}\right)$

     b. i. $\dbinom{4}{3}\left(\dfrac{9}{19}\right)^3\left(\dfrac{10}{19}\right)$ ii. $\dbinom{4}{2}\left(\dfrac{9}{19}\right)^2\left(\dfrac{10}{19}\right)^2$ iii. 0

29. a. expected value 1.8, most probable number 2
     b. expected value 14.25, most probable number 15
     c. expected value 4.4, most probable number 4
     d. expected value 0.65, most probable number 0
     The two values are close in each case. The mathematical statement is that the most probable number is the *integer part* of $np + p$.

Section 4-4

1. a. $\dbinom{30000}{4}\dbinom{70000}{6}\Big/\dbinom{100000}{10} = 0.20013$ b. 0.02824
     c. 0.00014   d. 0.99841

3. a.

x	0	1	2	3	4	5	6
$p(x)$	0.317	0.423	0.207	0.047	0.00513	0.000252	0

     b. Mean = $0.9998 \approx 1$; standard deviation = 0.873

5. a. 1/30   b. 0.7

7. 0.074

9. 0.0088

Chapter 4 Review Exercises

3. a. i. 0.1   ii. 0.1   iii. 0.4
      iv. 0   v. 0.9   vi. 0.3
     b. $121.60   c. 11,325.44 (dollars)^2

5. $2.10

7. Expected number = 127.5; variance = 318.75

9. a. 0.84   b. 1.6496   c. 0.978

11. a. 1/64   b. 27/64   c. 27/64

13. 0.346

15. Expected number = 3(3/4) = 2.25

     a. False   b. False   c. False   d. False

     e. False   f. True   g. True

CHAPTER 5

Section 5-1

9. a.   b.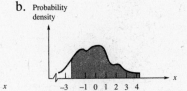

     c.

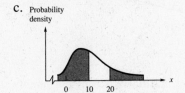

11. a. 0.2   b. 0.4   c. 0.4

13. The probability density curve in Exercise 12.

15. a. 0.2   b. 0.12   c. 0.52

     d. 0.32   e. 0.3   f. 0.7

17. a. 0.22   b. 0.36   c. 0.1

     d. 0.58   e. 0.18   f. 0.68

Section 5-2

1. a. 0.3849   c. 0.8849   e. 0.1525

     g. 0.8306   i. 0.2450   k. 0.7517

2. a. 0.48   c. −1.26   e. 2.16

     g. 0.71   i. −2.26   k. −0.42

3. a. 1.33   c. −1.97   e. 0.32

4. a. 0.0122   c. 0.0268   e. 0.0818
5. a. 0.3345   b. 0.7734   c. 0.7066
    d. 0.3085   e. 0.6915   f. 0.0122
7. a. 0.4772   b. 0.9332   c. 0.927
    d. 0.9332   e. 0.1525   f. 0.1525
9. 0.0021; approximately 21 in 10,000 cows
11. 1.906 inches
13. 1.632
15. a. 0.6554   b. approximately 655 trees
17. a. 0.8413   b. 5384 pounds
18. 0.383
19. a. $\binom{6}{2}(0.383)^2(0.617)^4$ b. $(0.383)^6$
21. a. 0.0139   b. 0.3462
23. $(0.0475)^2$ 25. 0.9772   27. 5.35 years
29. 575 points
31. 1.74 percent
33. a.

c.

e.

In Parts a, c, and e, normality seems reasonably valid.

Section 5-3

1. a. Binomial; $n = 200$; $p = 0.7$
    b. Mean = 140; standard deviation = 6.481
    c. n is large; $np = 140$ and $n(1 - p) = 60$, both greater than 5
    d. 0.0638
    e. 0.7887
3. a. 0.0028   b. 0.9945   c. 0.3594

5. 0.0985
7. a. 0.0344   b. 0.0436
9. 0.0023
11. a. 0.0472   b. 0.1020
13. 0.281
15. 0.0384
17. a. 0.0334; no   b. 0.1112   c. no
19. a. i. 16.8   ii. 0.9852
    b. 0; the quoted rate of 10 percent is highly suspect
    c. 0.0423; with 10 percent rate of cancer, 22 women getting it is not all that improbable
21. 0.6746

Chapter 5 Review Exercises

1. a.   b.

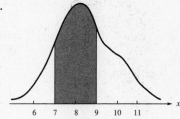   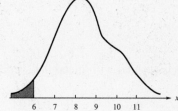

    c.

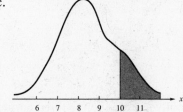

3. a. 0.18   b. 0.08   c. 0.26
5. a. 0.32   b. 0.68   c. 0.82
7. a. 0.1012   c. 0.0642   e. 0.673
8. a. 1.64   c. 0.42   e. −1.52
9. a. 0.6368   c. 0.0861
10. a. 21.75   c. 2.30
11. approximately 4 percent
13. 20.47 percent
15. 0.0562
17. a. 0.1788   b. 0.0212

CHAPTER 6

Section 6-1

1. a. 112.8   b. 112.8
3. Mean = 60; standard deviation = 1.598
5. a. 401.48   b. 437.5
7. a. 8.5   b. 2.63
9. a. Mean = 2.3; standard deviation = 4.67
   b. Mean = 2.3; standard deviation = 0.467
   c. approximately normal with mean 2.3, standard deviation 0.467
   d. i. 0.8185   ii. 0.0228   iii. 0.0228
11. 0.0228
13. a. 0.0062   b. 0.0668   c. 0.9544
15. 0.7971
17. 0.0062
19. 0.8413
21. a. 0.1   b. 0.0009
   c. approximately normal with mean 0.1, variance 0.0009
   d. 0.0228
23. a. 0.0019   b. 0.0104
25. 0.2709
27. a. 0.0125   b. 0.6172
29. $P(\overline{X} > 0) = P(Z > 16.67) = 0$

Section 6-3

3. a. completely randomized design
7. two blocks in each age group; each block will have 3 cows randomly assigned to the three hormones; altogether, there will be six blocks
9. weight; each block will involve 4 volunteers; two blocks with excessively overweight volunteers, one block with moderately overweight, and one with slightly overweight to normal

Chapter 6 Review Exercises

1. False
3. a. True   b. False   c. True
   d. False   e. True   f. False
5. a. False   b. True
7. 0.0207
9. a. 0.003   b. 0.9505
11. a. 0.0037   b. 0.0125   c. 0.9945
13. a. 0.0823   b. 0.7842

CHAPTER 7

Section 7-1

1. statements b and c are inferential
3. a. x/n b. $\bar{x}$ c. p
   d. p e. μ
5. $\bar{x} = 2.3$; $s^2 = 0.0686$. Estimate of μ is 2.3, of σ^2 is 0.0686, and of σ is 0.262.
9. 0.205
11. 0.68
15. a. 1.645   c. -1.73 e. 2.16
17. 0.0182
19. 0.5438
21. (0.751, 0.849)
23. (43.533, 48.468)
25. (139.6, 160.4)
27. a. (5.037, 6.651)
29. a. 1.76   b. 13.32
31. (b), to have as narrow an interval as possible with a high level of confidence
33. a. increase   b. decrease
35. 299
37. 174
39. 95.44 percent
41. 94.3 percent

Section 7-2

1. a. 1.734   b. 2.16   c. 3.143
   d. 2.262
2. a. 2.447   c. 2.624   e. 2.921
3. (12.663, 17.737)
5. (143.132, 161.468)
7. (21.718, 25.882)
9. (4.805, 6.195)
11. a.   b. (11.977, 13.689)

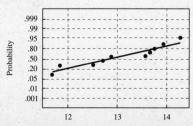

13. (11.392, 12.284)
15. (100.27, 117.85)
17. (3.938, 5.062)
19. (66.631, 73.369)
21. a. (647.0956, 716.9044)   b. (161,815,961, 179,272,699)
23. (4.675, 4.885)
25. a. (0.554, 0.622)   b. (2.037, 2.213)
27. (15.66, 21.54)
29. (148.034, 149.766)
31. a. (224.517, 235.483)   b. (224.368, 235.632)
33. (9711.7, 10108.3)

Section 7-3

1. a. 14.067   c. 6.571   e. 9.542
2. a. 11.070   c. 18.475
3. a. 7.962   c. 2.733
4. a. 26.119   c. 1.646
5. a. $\chi^2_{5,0.08}$ c. $\chi^2_{20,0.92}$ e. $\chi^2_{12,0.03}$
7. (0.618, 6.416)
9. (8.301, 22.824)
11. (0.167, 0.530)
13. a. (8.672, 54.708)   b. (2.945, 7.396)
15. a. i. (9.53, 12.47)   ii. (1.41, 3.75)
17. a. (3367.64, 3395.26)   b. (28.365, 48.962)
19. (0.0087, 0.0187)
21. (0.540, 2.678)

Section 7-4

1. a. 0.02   b. 0.0219   c. 0.0173
3. (0.184, 0.416)
5. (0.545, 0.695)
7. (0.468, 0.778)
9. (0.311, 0.489)
11. (0.828, 0.874)
13. a. (0.399, 0.461)   b. (0.214, 0.266)
15. a. 96   b. 2401   c. 960,400
As the error that we are prepared to tolerate decreases, the sample size increases.

17. 6670

19. a. (0.443, 0.517)   b. (0.133, 0.187)

23. point estimate = 0.297; margin of error = 0.038; conservative margin of error = 0.042

27. wider

Chapter 7 Review Exercises

1. a. 1.62   b. 0.0589   c. 0.243
3. a. 3.52   b. 0.347   c. 0.589
5. 57.3 percent
7. (9.916, 17.684)
9. (−4.951, −1.849); no assumptions are necessary because the sample is large
11. (6.228, 8.972); normal distribution
13. (47.835, 50.165)
15. a. (5.088, 5.512)   b. (0.152, 0.502)
17. (153.249, 172.151)
19. (0.384, 0.434)
21. (0.694, 0.858)
23. a. (0.0074, 0.0522)   b. (0.165, 0.372)
25. b. (0.152, 0.208)   c. (0.159, 0.201)

CHAPTER 8

Section 8-1

1. a. 3.05   b. 0.5
3. −0.66
5. (−1.268, 2.868)
7. (−1.922, 0.292)
9. a. (−0.489, 3.689)   b. (−3.689, 0.489)
11. (2.011, 2.789)
13. (23.686, 26.114)
15. b. (0.145, 0.515)   c. 0.37
17. a. (−18.886, −11.114)
19. a. (3398.92, 4601.08) (confidence interval for $\mu_W - \mu_E$)

Section 8-2

1. 1.2758
3. 0.0238
5. 0.5685
7. a. 1.7370   b. 20   c. 1.7
    d. 1.283   e. (0.417, 2.983)
9. a. 1.489 (Model 2 *minus* Model 1)   b. 2.127
    c. (−0.638, 3.616)
11. a. (0.14, 0.52)   b. 0.38; 2.63 percent shorter
13. (−211.106, 31.106) (confidence interval for $\mu_A - \mu_B$)
15. b. (−2.663, −0.087)
17. (0.217, 0.495)
19. (−1363.22, 343.22)
21. (0.011, 0.109)

Section 8-3

1. a. 0.154   b. (−0.189, 0.119)
3. a. 0.108   b. (−0.099, 0.118)
5. a. (−0.175, 0.079)
7. a. (−0.013, 0.142)
9. a. (−0.009, 0.213)

Chapter 8 Review Exercises

1. (−0.225, 4.825)
3. (−0.236, 0.126)
5. (2.367, 4.033)
7. a. (−2.833, −0.807)
9. (−0.0423, 0.0873)
11. (−5.713, 0.313)
13. (0.013, 0.085)
15. (0.981, 2.219)

CHAPTER 9

Section 9-1

1. $H_0: p \geq 0.001$
   $H_A: p < 0.001$

3. $H_0: \mu \leq 3$
   $H_A: \mu > 3$

5. $H_0: p \leq 0.6$
   $H_A: p > 0.6$

7. a. One possible error is not to market the serum when it is effective. The other possible error is to decide to market the serum when it is not effective.
   b. not to market the serum when it is indeed effective
   c. to market the serum when it is not effective

9. H_0: The serum is not effective.
   H_A: The serum is effective.

11. $H_0: \mu$, the mean nicotine content of the new brand, is ≥ 1 mg.
    $H_A: \mu$, the mean nicotine content of the new brand, is < 1 mg.

12. a. False   c. False   e. True   g. True

15. Type I error; probability $= \alpha \approx 0.1$

17. H_0: The body is not that of Joseph Mengele.
    H_A: The body is that of Joseph Mengele.
    The level of significance is extremely small.

Section 9-2

1. a. not acceptable; $\bar{x}$ is not a parameter   c. not acceptable; there is overlap
   e. acceptable   g. acceptable   i. not acceptable; there is overlap

3. a. 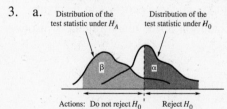   b.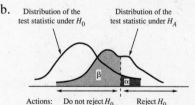

   c.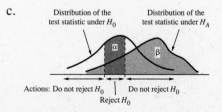

5. $\alpha = 1/5; \beta = 0$

7. $\alpha = 0.0228; \beta = 0.0918$

9. Reject H_0

11. a. Do not reject H_0 b. 0.0436

13. a. 0.0274 c. 0.0054

15. a. H_0: The stockbroker is guessing, that is, $p = 0.5$.
    H_A: The stockbroker is not guessing, that is, $p = 0.8$.
    b. Type I error: Conclusion that the stockbroker is not guessing when she is in fact guessing.
    Type II error: Conclusion that the stockbroker is guessing when she is not guessing.

17. No

Section 9-3

1. Test statistic: $\dfrac{\overline{X} - 10.2}{1.3/\sqrt{16}}$
    Decision rule: Reject H_0 if the computed value is greater than 1.645.

3. Test statistic: $\dfrac{\overline{X} - 32.8}{6.8/\sqrt{14}}$
    Decision rule: Reject H_0 if the computed value is less than -2.33.

5. Test statistic: $\dfrac{\overline{X} + 3.3}{4.8/\sqrt{20}}$
    Decision rule: Reject H_0 if the computed value is less than -2.33 or greater than 2.33.

7. Computed value $= 1.83$
    a. Do not reject H_0 because $z_{0.025} = 1.96$
    b. Reject H_0 because $z_{0.05} = 1.645$.

9. H_0: $\mu = 50$; H_A: $\mu > 50$; computed value $= 2.88$; reject H_0; the company is correct in its claim.

11. H_0: $\mu = 5$; H_A: $\mu > 5$; computed value $= 2$; reject H_0; the contention of the management is justified.

13. Test statistic: $\dfrac{\overline{X} - 110}{S/\sqrt{25}}$
    Decision rule: Reject H_0 if the computed value is less than -2.064.

15. Test statistic: $\dfrac{\overline{X} - 17.8}{S/\sqrt{20}}$
    Decision rule: Reject H_0 if the computed value is greater than 2.093.

17. Test statistic: $\dfrac{\overline{X} - 11.6}{S/\sqrt{12}}$
    Decision rule: Reject H_0 if the computed value is less than -1.796 or greater than 1.796.

19. H_0: $\mu = 13$; H_A: $\mu < 13$; computed value $= -1.845$; do not reject H_0 because $-t_{9,0.025} = -2.262$; the true mean time is not less than 13 minutes.

21. H_0: $\mu = 36{,}000$; H_A: $\mu > 36{,}000$; computed value $= 1.415$; $t_{15,0.05} = 1.753$; do not reject H_0; the true mean mileage is not significantly greater than 36,000 miles at the 5 percent level.

23. H_0: $\mu = 60$; H_A: $\mu < 60$; computed value $= -2.676$; reject H_0; at the 1 percent level the true mean battery life is less than 60 hours.

25. H_0: $\mu = 12$; H_A: $\mu > 12$; computed value $= 3.023$; reject H_0; recommend to buy the brand.

27. H_0: $\mu = 12$; H_A: $\mu \neq 12$; computed value $= -0.645$; do not reject H_0; the true mean photometer reading is not different from 12.

29. a. (4.063, 4.929)
     b. H_0: $\mu = 4$; H_A: $\mu > 4$; computed value $= 2.345$; reject H_0 because $t_{28,0.05} = 1.701$; the true mean cardiac output is greater than 4.

31. H_0: $\mu = 85$; H_A: $\mu \neq 85$; computed value $= -0.295$; do not reject H_0; the true mean daily selenium intake is not different from 85.

33. H_0: $\mu = 10.2$; H_A: $\mu \neq 10.2$; computed value $= 2.09$; do not reject H_0; the true mean soluble sugar is not different from 10.2 g/100 ml.

35. Test statistic: $\dfrac{\overline{X} - 82}{S/\sqrt{64}}$

    Decision rule: Reject H_0 if the computed value is less than -2.33 or greater than 2.33, because $z_{0.01} = 2.33$.

37. Test statistic: $\dfrac{\overline{X}}{S/\sqrt{50}}$

    Decision rule: Reject H_0 if the computed value is less than -1.645 or greater than 1.645.

39. H_0: $\mu = 250$; H_A: $\mu < 250$; computed value $= -1.5$; do not reject H_0 because $-z_{0.05} = -1.645$; the true mean cholesterol is not less than 250 mg.

41. H_0: $\mu = 4$; H_A: $\mu > 4$; computed value $= 3.05$; reject H_0 at the 5 percent level and at the 1 percent level; P-value $= 0.0011$; the true mean stay is longer than 4 days.

43. Test statistic: $6S^2/0.36$
    Decision rule: Reject H_0 if the computed value is less than $\chi^2_{6,0.975} = 1.237$.

45. Test statistic: $17S^2/10$
    Decision rule: Reject H_0 if the computed value is less than 8.672 or greater than 27.587.

47. Test statistic: $10S^2/9$
    Decision rule: Reject H_0 if the computed value is greater than 18.307.

49. H_0: $\sigma = 0.5$; H_A: $\sigma > 0.5$; computed value $= 24.2$; do not reject H_0; the true standard deviation is not significantly greater than 0.5, at the 2.5 percent level.

51. H_0: $\sigma = 5$; H_A: $\sigma \neq 5$; computed value $= 7.11$; do not reject H_0 at the 5 percent level; the population standard deviation is not different from 5 pounds.

53. a. H_0: $\mu = 6$; H_A: $\mu \neq 6$; computed value $= -1.53$; do not reject H_0; the true mean is not significantly different from 6.
     b. H_0: $\sigma = 0.7$; H_A: $\sigma < 0.7$; computed value $= 5.93$; reject H_0 at the 5 percent level; the population standard deviation is less than 0.7.

55. a. $H_0: \mu = 35$; $H_A: \mu < 35$; computed value $= -1.974$; reject H_0 because $-t_{29,0.05} = -1.699$; the true mean is less than 35.
   b. $H_0: \sigma = 5$; $H_A: \sigma \neq 5$; computed value $= 37.035$; do not reject H_0; the true standard deviation is not different from 5.

57. Test statistic: $\dfrac{(X/60) - 0.4}{\sqrt{\dfrac{(0.4)(0.6)}{60}}}$
   Decision rule: Reject H_0 if the computed value is less than -1.96 or greater than 1.96.

59. Test statistic: $\dfrac{(X/36) - 0.7}{\sqrt{\dfrac{(0.7)(0.3)}{36}}}$
   Decision rule: Reject H_0 if the computed value is greater than 1.645.

61. Test statistic: $\dfrac{(X/120) - 0.8}{\sqrt{\dfrac{(0.8)(0.2)}{120}}}$
   Decision rule: Reject H_0 if the computed value is less than -2.33.

63. $H_0: p = 0.8$; $H_A: p > 0.8$; computed value $= 2.0$; reject H_0 because $z_{0.03} = 1.88$; the claim of the drug company is valid.

65. Computed value $= 2.421$; reject H_0 because $z_{0.025} = 1.96$; the true proportion is greater than 0.35.

67. $H_0: p = 0.2$; $H_A: p < 0.2$; computed value $= -1.046$; do not reject H_0; the credibility is not less than 20 percent.

69. $H_0: p = 0.5$; $H_A: p > 0.5$; computed value $= 4.19$; reject H_0; majority of female executives are harassed.

71. Let p be the probability that an individual selected at random feels that the federal government has become a lot more liberal under President Clinton.
   $H_0: p = 0.2$; $H_A: p > 0.2$; computed value $= 3.167$; reject H_0; it is valid that at least one in five feels that the government has become a lot more liberal.

Section 9-4

1. Computed value $= 1.789$
   a. Do not reject H_0 because $z_{0.025} = 1.96$, and 1.789 is in the interval $(-1.96, 1.96)$
   b. Reject H_0 because $1.789 > z_{0.05} = 1.645$

3. Computed value $= -2.902$
   a. Reject H_0 because $z_{0.05} = 1.645$.
   b. Reject H_0 because $-z_{0.1} = -1.28$.

5. $H_0: \mu_T = \mu_F$; $H_A: \mu_T \neq \mu_F$; computed value $= -4.80$; reject H_0 because $z_{0.01} = 2.33$; the two models are different.

7. $H_0: \mu_A = \mu_B$; $H_A: \mu_A < \mu_B$; computed value $= -3.84$; reject H_0; Feed B is better than Feed A.

9. a. One has to take into consideration sample sizes and standard deviations.
   b. $H_0: \mu_{T_1} = \mu_{T_2}$; $H_A: \mu_{T_1} < \mu_{T_2}$; computed value $= -2.7$; reject H_0 because $z_{0.05} = 1.645$; Teller 2 is slower than Teller 1.

11. Computed value $= 2.285$
    a. Reject H_0 because $t_{20,0.025} = 2.086$.
    b. Reject H_0 because $t_{20,0.05} = 1.725$.

13. H_0; $\mu_A = \mu_B$; H_A: $\mu_A \neq \mu_B$; computed value $= -1.55$; do not reject H_0 because $t_{20,0.025} = 2.086$; there is no significant difference between the two companies.

15. H_0: $\mu_w = \mu_p$; H_A: $\mu_w < \mu_p$; computed value $= -0.59$; do not reject H_0; children from the well-to-do neighborhood are not more adept than those in the poor neighborhood.

17. H_0: $\mu_1 = \mu_2$; H_A: $\mu_1 < \mu_2$; computed value $= -1.797$; reject H_0 because $t_{17,0.05} = 1.74$; Model 1 is superior to Model 2 for acceleration.

19. H_0: $\mu_A = \mu_B$; H_A: $\mu_A \neq \mu_B$; computed value $= 1.554$; do not reject H_0 because $t_{10,0.025} = 2.228$; the two varieties are not significantly different.

21. H_0: $\mu_p = \mu_w$; H_A: $\mu_p \neq \mu_w$; computed value $= -1.182$; do not reject H_0 because $t_{25,0.01} = 2.485$; the two cooking methods are not different.

23. H_0: $\mu_1 = \mu_2$; H_A: $\mu_1 \neq \mu_2$; computed value $= 3.426$; reject H_0; the true mean urea concentrations are different for the two diets.

25. a. $(1.543, 6.217)$
    b. H_0: $\mu_1 = \mu_2$; H_A: $\mu_1 > \mu_2$; computed value $= 3.869$; reject H_0; the true mean body protein at 20% level is less than that at 0% level.

27. H_0: $\mu_{CAH} = \mu_{TPH}$; H_A: $\mu_{CAH} < \mu_{TPH}$; computed value $= -4.024$; reject H_0; the TPH method is superior to the CAH method.

29. H_0: $\mu_P = \mu_W$; H_A: $\mu_P \neq \mu_W$; computed value $= -4.012$; reject H_0; the true mean lipid content is different when beef is pressure cooked and when it is water cooked.

31. Computed value $= -2.967$
    a. Reject H_0 because $t_{14,0.025} = 2.145$.
    b. Reject H_0 because $-t_{14,0.05} = -1.761$.

33. H_0: $\mu_D = 0$; H_A: $\mu_D < 0$; $n = 9$; $\bar{d} = -6.78$; $s_d = 13.526$; computed value $= -1.504$; do not reject H_0 because $-t_{8,0.05} = -1.86$; there is no evidence that the special coaching benefits the students.

35. H_0: $\mu_D = 0$; H_A: $\mu_D < 0$; computed value $= -2.048$; reject H_0 because $-t_{9,0.05} = -1.833$; the true mean glutamic acid in the core tissue is greater than that in the pericarp tissue.

37. $(-2.609, 0.859)$

39. $(-2.390, 0.036)$

43. H_0: $p_D = p_R$; H_A: $p_D < p_R$; computed value $= -1.127$; do not reject H_0 because $-z_{0.01} = -2.33$; the true proportion among Republicans is not greater than that among Democrats.

45. H_0: $p_V = p_{NV}$; H_A: $p_V < p_{NV}$ (Here p is the probability of catching cold); computed value $= -1.273$; do not reject H_0 because $-z_{0.02} = -2.05$; the vaccine is not effective.

47. H_0: $p_{PD} = p_{LLA}$; H_A: $p_{PD} \neq p_{LLA}$; computed value $= -2.266$; reject H_0; the true proportions differ significantly at the 5 percent level.

49. a. H_0: $p_c = p_s$; H_A: $p_c \neq p_s$; computed value $= 0.35$; do not reject H_0; the true proportions for city dwellers and suburban dwellers are not different.
    b. H_0: $p_c = p_r$; H_A: $p_c > p_r$; computed value $= 2.90$; reject H_0 because $z_{0.05} = 1.645$; the true proportion among city dwellers is significantly higher than that among rural dwellers.

Chapter 9 Review Exercises

3. Type I error: To assert that Vitamin C is not effective in controlling colds when, in fact, it is.
   Type II error: To assert that Vitamin C is effective in controlling colds when, in fact, it is not.

5. a. The distribution is approximately normal with mean 0.4 and standard deviation $= \sqrt{\dfrac{(0.4)(0.6)}{100}}$,
   or 0.049.

   b. The distribution is approximately normal with mean 0.3 and standard deviation $= \sqrt{\dfrac{(0.3)(0.7)}{100}}$,
   or 0.046.

7. a. $H_0: \mu = 16$; $H_A: \mu \neq 16$; computed value $= 0.626$; do not reject H_0; the true mean is not different from 16 ounces.

   b. $H_0: \sigma^2 = 0.16$; $H_A: \sigma^2 \neq 0.16$; computed value $= 14.374$; do not reject H_0; the true variance is not different from 0.16 ounces squared.

9. $H_0: \sigma = 1$; $H_A: \sigma \neq 1$; computed value $= 18.786$; do not reject H_0; the true standard deviation is not different from 1.

11. $H_0; \mu = 9.2$; $H_A: \mu > 9.2$; computed value $= 1.874$; reject H_0 because $t_{11,0.025} = 2.201$; do not reject H_0; the claim is not valid.

13. $H_0: \mu_S = \mu_U$; $H_A: \mu_S \neq \mu_U$; computed value $= -1.915$; do not reject H_0 because $t_{28,0.01} = 2.467$; the two ways do not yield different true means for saponin.

15. $H_0: \mu_F = \mu_C$; $H_A: \mu_F \neq \mu_C$; computed value $= 4.412$; reject H_0; there is a significant difference in the true mean magnesium content of the two topsoils.

17. $H_0: \mu_C = \mu_M$; $H_A: \mu_C \neq \mu_M$; computed value $= 18.76$; reject H_0; the true mean carbohydrate content is different in the two types of fish.

19. $H_0: \mu_S = \mu_W$; $H_A: \mu_S \neq \mu_W$; computed value $= 5.357$; reject H_0; there is a difference in the true mean casein content under the two diets.

21. $H_0: \mu_L = \mu_G$; $H_A: \mu_L \neq \mu_G$; computed value $= 1.53$; do not reject H_0; the true mean height growths for the two classifications do not differ at the 5 percent level.

23. $H_0: \mu_A = \mu_B$; $H_A: \mu_A \neq \mu_B$; computed value $= -1.574$; do not reject H_0; the two brands are not significantly different at the 5 percent level.

25. Let $d =$ protein before *minus* protein after; $H_0: \mu_D = 0$; $H_A: \mu_D > 0$; computed value $= 2.447$; reject H_0 because $t_{7,0.025} = 2.365$; there is significant decrease in the true mean sarcoplasmic protein after irradiation.

27. Let $p =$ probability child is infected with aids when mother is treated with AZT; $H_0: p = 0.25$; $H_A: p < 0.25$; computed value $= -3.4$; reject H_0; data indicate that AZT is effective in reducing the risk.

CHAPTER 10

Section 10-1

1. $H_0: p_{Green} = 3/4$; $p_{Yellow} = 1/4$

a. Computed χ^2 value = 0.775; because $\chi^2_{1,0.05}$ = 3.841, do not reject H_0; the segregation ratio is 3:1.

b. Computed value = $\dfrac{\dfrac{x}{n} - 0.75}{\sqrt{\dfrac{(0.75)(0.25)}{172}}}$ = 0.88, because x = 134 and n = 172; do not reject H_0. We

get the same conclusion in both cases. A worthwhile observation is that, $(0.88)^2 \approx 0.775$, the computed χ^2 value in Part a.

3. Computed value = 9.587; $\chi^2_{5,0.05}$ = 11.07; do not reject H_0; the distribution is the same as in 1960.

5. H_0: p_{RY} = 9/16, p_{RG} = 3/16, p_{WY} = 3/16, p_{WG} = 1/16; H_A: The probabilities are not as specified under H_0. Computed value = 6.634; $\chi^2_{3,0.05}$ = 7.815; do not reject H_0; the segregation ratio is 9 : 3 : 3 : 1.

7. a. Using the binomial formula with n = 4 and p = 1/2, the probabilities of 0, 1, 2, 3, 4 sons are, respectively, 1/16, 4/16, 6/16, 4/16, 1/16.

b. 10, 40, 60, 40, 10

c. Computed value = 5,842; $\chi^2_{4,0.05}$ = 9.488; do not reject H_0; the probability that a child is a son in four-children families is 1/2.

9. Computed value = 11.4; $\chi^2_{4,0.05}$ = 9.488; reject H_0; the ratio is not 30 : 15 : 10 : 5 : 20.

11. H_0: p_1 = p_2 = p_3 = p_4 = p_5 = 1/5. H_A: The probabilities are not as specified under H_0. Computed value = 66.411; $\chi^2_{4,0.05}$ = 9.488; reject H_0; the child occupant is more likely to be in younger age group of 1 to 2 years.

Section 10-2

1. Computed value = 17.017; $\chi^2_{1,0.025}$ = 5.024; reject H_0; smoking and drinking are related.

3. Computed value = 16.837; $\chi^2_{2,0.01}$ = 9.21; reject H_0; the two conditions are related.

5. Computed value = 35.096; $\chi^2_{9,0.05}$ = 16.919; reject H_0; the grade point average and evaluation of the instructor are related.

7. Computed value = 12.746; $\chi^2_{4,0.025}$ = 11.143; reject H_0; weight and blood pressure are related.

9. Computed value = 20.679; $\chi^2_{6,0.025}$ = 14.449; reject H_0; fabric preference and the age group of women are related.

Section 10-3

1. a. Computed χ^2 value is 1.924; $\chi^2_{1,0.05}$ = 3.841; do not reject H_0; there is no difference in the true proportions of curly wings among the two crosses.

b. x = 63; m = 180 and y = 40; n = 144.

Computed value = $\dfrac{0.35 - 0.2778}{\sqrt{(0.318)(0.682)\left(\dfrac{1}{180} + \dfrac{1}{144}\right)}}$ = 1.387;

do not reject H_0; the proportions of the curly wings in the two crosses are the same. Notice, incidentally, that $(1.387)^2$ = 1.924, the value computed in Part a.

3. Computed value = 1.443; $\chi^2_{2,0.05}$ = 5.991; do not reject H_0; there is no significant difference between the pads at the 5 percent level.

5. Computed value = 5.093; $\chi^2_{3,0.025}$ = 9.348; do not reject H_0; the sentiment is not different in the four regions.

7. Computed value = 6.945; $\chi^2_{2,0.025}$ = 7.378; do not reject H_0; the two methods are not different in their effectiveness.

9. Computed value = 11.356; $\chi^2_{4,0.05}$ = 9.488; reject H_0; the three pain relievers are not the same in their effectiveness.

11. Computed value = 6.298; $\chi^2_{2,0.01}$ = 9.21; do not reject H_0; the true proportions of patients getting relief are not significantly different for the three drugs, at the 1 percent level.

13. Computed value = 26.927; $\chi^2_{4,0.025}$ = 11.143; reject H_0; there is a significant difference in the blood alcohol levels of the two sexes, at the 2.5 percent level.

Chapter 10 Review Exercises

1. H_0: The data are in conformity with a hypothesized distribution. That is, the probability that an item belongs to Category c_1 is p_1, that it belongs to Category c_2 is p_2, and so on.
H_A: The data are not in conformity.

3. Computed value = 6.10; $\chi^2_{3,0.05}$ = 7.815; do not reject H_0; there is no evidence that the numbers are not picked at random.

5. H_0: p_A = 0.4; H_A: $P_A \ne 0.4$; computed value = 3.50; there are 2df; $\chi^2_{2,0.05}$ = 5.991; therefore P-value is greater than 0.05; do not reject H_0 at the 5 percent level; the proportion of A alleles in the population is not different from 0.4 at the 5 percent level.

CHAPTER 11

Section 11-1

1.

x	3	0	-2
y	-4.5	1.5	5.5

3. a.
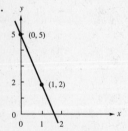

    b.

c.

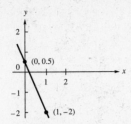

5. a. *y*-intercept = −10; slope = −15   b. *y*-intercept = −5; slope = 4
    c. *y*-intercept = 5.8; slope = 4.9   d. *y*-intercept = 15.2; slope = −24.9

7. −7.5

9. 0.75

Section 11-2

1. a. 492; 1254; 34,628; 227,948; 88,420   b. 0.712

3. a. 1   b. −1

5. a.   b. 0.032

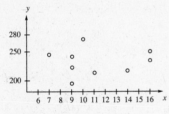

7. a.   b. 0.321

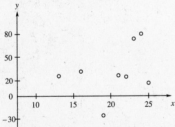

    c. The magnitude of *r* does not indicate a *strong* linear relation; but the relation is positive.

9. a.   b. 0.946

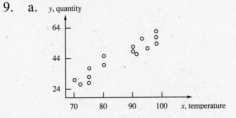

11. 0.901; coefficient of correlation is not affected.

General comment: If the variables are multipled by constants, the coefficient of correlation is affected by a factor of ± 1; by $+1$ if the constants have the same sign as in this exercise; by -1 if the constants have opposite signs.

13. a. $b = 1.952$; $a = -5.377$
    b. $\hat{y} = -5.377 + 1.952x$

15. $\hat{y} = 99.467 - 6.457x$

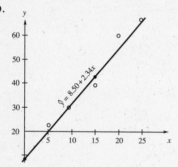

The line provides a satisfactory representation for the data.

17. a. b.

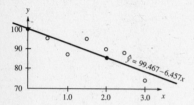

   c. yes

19. a. $\hat{y} = 11.83 - 8.53x$
    b. -35.085

21. a. $\hat{y} = -3.239 + 3.544x$
    b. 22.278

23. a. $\hat{y} = 14.33 + 0.826x$
    b. $\hat{y} = 111.56 - 0.584x$

c.
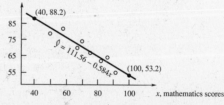

25. a.

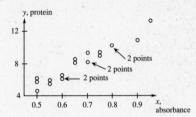

b. $\hat{y} = -3.101 + 16.79x$
c. 0.959

27. a.

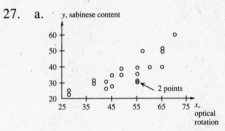

b. $\hat{y} = 1.351 + 0.7x$
c. 0.849

29. $\hat{y} = -45.466 + 1.058x$; 44,464 gallons.

31. a. b.

y, mileage

c. 55 miles per hour.

Section 11-3

1. a. $n = 8$; $\Sigma\, x_i = 194$; $\Sigma\, y_i = 32.5$; $\Sigma\, x_i^2 = 5,300$; $\Sigma\, y_i^2 = 146.75$; $\Sigma\, x_i y_i = 880$

   b. $a = 0.32116$; $b = 0.15428$; $\hat{y} = 0.3212 + 0.1543x$

   c.

x	y	$\hat{y}$
12	2.0	2.1728
16	3.0	2.7900
16	2.5	2.7900
24	4.0	4.0244
24	4.5	4.0244
30	5.0	4.9502
36	6.0	5.8760
36	5.5	5.8760

   $$\sum_{i=1}^{8} (y_i - \hat{y}_i)^2 = 0.5441$$

   d. $\dfrac{1}{6}(0.5441) = 0.091$

   e. $\dfrac{1}{6}[146.75 - (0.32116)(32.5) - (0.15428)(880)] = 0.091$

3.   3.323

5.   a. $a = 40.285714; b = 4.2857143$ b.   7.578
     c.   10.43   d.   0.573

7.   a.   0.000153   b.   (0.130, 0.178)

9.   a.   11   b. $s_a^2 = 0.3483; s_b^2 = 0.11$

11. a. $a = 39.9333; b = -1.9771$
          $\hat{y} = 39.933 - 1.977x$
     b. $s_e = 4.72; s_a = 4.394; s_b = 0.2256$ c.   (27.735, 52.131)
     d.   (-2.603, -1.351)   e.   (-1.011, 9.705)

13. a.   b. $\hat{y} = -89.986 + 5.705x$

y, stopping distance

     c.   195.264   d.   155.574
     e. $s_a^2 = 160.705; s_b^2 = 0.0733$ f.   Compute value = 2.605; reject H_0 because
                                                    $t_{8,0.025} = 2.306$.
     g.   (185.542, 204.986)

15. a. $\hat{y} = -18.571 + 0.099x$ b.   11.129
     c.   0.4211   d. $s_a = 1.78; s_b = 0.00634; s_{\hat{y}_0} = 0.298$
     e.   confidence interval for A: (-23.523, -13.619)
          confidence interval for B: (0.082, 0.116)
          confidence interval for μ_{Y1300}: (10.307, 11.951)
     f.   (9.146, 13.111)   g.   Computed value = -0.159; do not reject
                                                    H_0.

17. a. $\hat{y} = -2.7 + 2.607x$ b.   (2.468, 2.746)
     c.   Compute value = 4.14; reject H_0. d.   (10.057, 10.613)
     e.   (9.61, 11.06)

19. a. $\hat{y} = -27.095 + 2.382x$ b.   (1.942, 2.822)
     c.   (17.373, 23.717)   d.   (13.698, 27.392)

Section 11-4

1.   a.   0.66538   b.   63.5279
     c.   1040.9   d.   812.2
     e.

Component	Sum of Squares	Explained/Total
Explained (regression)	228.7	0.22
Unexplained (error)	812.2	
Total	1040.9	

     f.   0.22

3. ± 0.7

5. a. 0.7128
   b. 0.5081
   c. 50.81 percent of the total variation in annual wheat yield is explained by the linear regression relation.

7. Computed value of $r \sqrt{\dfrac{n-2}{1-r^2}} = -4.619$; reject H_0 because $t_{64, 0.025} \approx z_{0.025} = 1.96$.

9. Computed value $= -1.732$; $t_{9, 0.025} = 2.262$; do not reject H_0.

11. Computed value $= \pm 5.881$; reject H_0.

13. a.

Component	Sum of Squares	Explained/Total
Explained	1673.312	0.507
Unexplained	1629.545	
Total	3302.857	

   b. 0.712
   c. Computed value $= 2.267$; $t_{5, 0.025} = 2.571$; do not reject H_0.

15. a. 81 percent
   b. ± 0.9; however, from our experience we suspect that the coefficient of correlation is $+0.9$ rather than -0.9.
   c. Computed value $= 16.52$; reject H_0 at the 5 percent level because the P-value is less than 0.05.

17. a. $\hat{y} = 0.364 + 0.0156x$ (x is temperature, y is liquid water)
   b. $r = 0.96$
   c. $H_0: \rho = 0$, $H_A: \rho > 0$; computed value 8.4; reject H_0 since $t_{6, 0.025} = 2.447$.

Chapter 11 Review Exercises

5. The regression coefficient b and the correlation coefficient r must have the same sign.

7. a. $\hat{y} = 2.01 + 0.0797x$
   b. 0.934

9. $r = 0.943$; computed value $= 8.0$; $t_{8, 0.025} = 2.306$; reject H_0.

11. a. $\hat{y} = 468.709 + 6.008x$
   b. (5.657, 6.359)
   c. Computed value $= 2.81$; $t_{6, 0.05} = 1.943$; reject H_0.
   d. (2793.2, 2950.6)
   e. (2639.8, 3104.0)

13. a.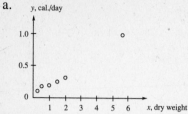

b. $\hat{y} = 0.0541 + 0.152x$

c. 0.993

d. H_0: $\rho = 0$; H_A: $\rho > 0$; computed value $= 16.81$; $t_{4,0.05} = 2.132$; reject H_0.

15. $r < -0.4683$

17. H_0: $\rho = 0$; H_A: $\rho \neq 0$; because $n = 12$, based on the conclusion of Exercise 16, we would reject H_0 if $r < -0.576$ or $r > 0.576$; from the data $r = 0.762$; reject H_0.

CHAPTER 12

Section 12-1

1. a. 4   b. 30   c. 34

3. 3.35

5.

Source	df	Sum of Squares	Mean Sum of Squares	Computed Ratio
Between	3	47.120	15.705	1.483
Within	20	211.838	10.592	
Total	23	258.958		

Computed ratio $= 1.483$; $f_{3,20;0.05} = 3.10$; do not reject H_0; the true mean yield strengths are not different.

Section 12-2

1.

Source	df	Sum of Squares	Mean Sum of Squares	Computed Ratio
Between	7	210.7	30.1	2.89
Within	18	187.2	10.4	
Total	25	397.9		

3. $H_0: \mu_0 = \mu_N: H_A: \mu_0 \neq \mu_N.$

a. Computed value $= \dfrac{\bar{x} - \bar{y}}{s_p\sqrt{\dfrac{1}{6} + \dfrac{1}{7}}} = -0.939$; do not reject H_0; the two brands are not different.

b.

Source	df	Sum of Squares	Mean Sum of Squares	Computed Ratio
Between	1	518.359	518.359	0.882
Within	11	6463.333	587.576	
Total	12	6981.692		

Computed ratio $= 0.88$; $f_{1,11;0.05} = 4.84$; do not reject H_0; the two brands are not significantly different. (Incidentally, notice that $0.882 = (-0.939)^2$, the square of the computed value in Part a.)

5.

Source	df	Sum of Squares	Mean Sum of Squares	Computed Ratio
Between	3	1148.55	382.85	2.88
Within	16	2130.00	133.12	
Total	19	3278.55		

Computed ratio $= 2.88$; $f_{3,16;0.1} = 2.46$; reject H_0; the true mean tar contents are different for the four brands.

7.

Source	df	Sum of Squares	Mean Sum of Squares	Computed Ratio
Between	2	84,933,330	42,466,665	2.45
Within	12	208,000,000	17,333,333	
Total	14	292,933,330		

Computed ratio $= 2.45$: $f_{2,12;0.025} = 5.10$; do not reject H_0; there is no significant difference between the brands.

9.

Source	df	Sum of Squares	Mean Sum of Squares	Computed Ratio
Between	2	19.2	9.60	0.85
Within	12	135.2	11.27	
Total	14	154.4		

Computed ratio $= 0.85$; $f_{2,12;0.05} = 3.89$; do not reject H_0; the three insulations are not significantly different.

11.

Source	df	Sum of Squares	Mean Sum of Squares	Computed ratio
Between	2	736.8	368.4	2.58
Within	14	1997.7	142.7	
Total	16	2734.5		

Computed ratio = 2.58; $f_{2,14;0.1}$ = 2.73; do not reject H_0; the true mean yields are not significantly different for the three hybrids, at the 10 percent level.

13.

Source	df	Sum of Squares	Mean sum of Squares	Computed Ratio
Between	4	72,200	18,050	28.93
Within	55	34,307	624	
Total	59	106,507		

Computed ratio 28.93; $f_{4,55;0.01} \approx$ 3.70; reject H_0; the true mean water consumption differs with the level of starch in the diet.

15. a.

Source	df	Sum of Squares	Mean Sum of Squares	Computed Ratio
Between	2	190,412	95,206	168.11
Within	12	6,796	566.33	
Total	14	197,208		

Computed value = 168.11; reject H_0; the combinations are significantly different.
b. (−181.4, −118.6)

Chapter 12 Review Exercises

3. 1.24

5.

Source	df	Sum of Squares	Mean Sum of Squares	Computed Ratio
Between	2	0.5553	0.2776	15.77
Within	11	0.1940	0.0176	
Total	13	0.7493		

Reject H_0 because the computed ratio is greater than $f_{2,11;0.05}$ = 3.98; the true mean dissolved oxygen is different for the three ponds.

7.

Source	df	Sum of Squares	Mean Sum of Squares	Computed Ratio
Between	4	4,832	1,208	0.04
Within	15	414,688	27,646	
Total	19	419,520		

Do not reject H_0; the five brands are not significantly different.

9.

Source	df	Sum of Squares	Mean Sum of Squares	Computed Ratio
Between	2	35.244	17.622	32.63
Within	42	22.667	0.540	
Total	44	57.911		

Computed ratio $= 32.63$; $f_{2,42;0.1} \approx 2.44$; reject H_0; there is a significant difference in the true mean fat thickness in the three diet groups.

CHAPTER 13

Section 13-1

1. Taking differences, before *minus* after, H_0; $p = 1/2$; H_A: $p > 1/2$; P-value $= 0.09$; do not reject H_0 at the 5 percent level; the diet is not effective in reducing weight.

3. H_0: $p = 1/2$; H_A: $p > 1/2$; computed value $= \dfrac{2(35.5) - 46}{\sqrt{46}} = 3.69$; $z_{0.02} = 2.055$; reject H_0; the special assistance program is beneficial.

5. Taking differences, subzero *minus* warm, H_0: $p = 1/2$; H_A: $p < 1/2$; P-value $= 0.151$; warm weather does not affect adversely, at the 5 percent level.

7. Taking differences, before *minus* after, H_0: $p = 1/2$; H_A: $p > 1/2$; computed value $\dfrac{2(9.5) - 16}{\sqrt{16}} = 0.75$; do not reject H_0 at the 5 percent level because P-value $= 0.227$; the treatment does not reduce blood glucose level.

9. Taking differences, with *minus* without, H_0: $p = 1/2$; H_A: $p > 1/2$; P-value $= 0.194$; do not reject H_0 at the 5 percent level; strict supervision does not improve productivity.

Section 13-2

1. $m = 11$; $n = 13$; $R_1 = 99.5$; $u = 109.5$; $\mu_U = 71.5$; $\sigma_U = 17.26$; $\dfrac{u - \mu_U}{\sigma_U} = 2.20$; reject H_0; the age distributions are not identical.

3. $m = 12$; $n = 15$; $R_1 = 185$; $u = 73$; $\mu_U = 90$; $\sigma_U = 20.494$; $\dfrac{u - \mu_U}{\sigma_U} = -0.83$; do not reject H_0; the two methods of instruction do not produce different results.

5. $m = 11$; $n = 11$; $R_1 = 104$; $u = 83$; $\mu_U = 60.5$; $\sigma_U = 15.229$; $\dfrac{u - \mu_U}{\sigma_U} = 1.48$; do not reject H_0; the two breeds are not significantly different at the 5 percent level.

Section 13-3

1. $m = 20$; $n = 20$; $R = 29$; $\mu_R = 21$; $\sigma_R = 3.121$; $\dfrac{R - \mu_R}{\sigma_R} = 2.56$; reject H_0; the sequence of answers is not random.

3. $m = 14$; $n = 11$; $R = 15$; $\mu_R = 13.32$; $\sigma_R = 2.41$; $\dfrac{R - \mu_R}{\sigma_R} = 0.70$; do not reject H_0 that there was randomness.

5. $m = 13$; $n = 13$; $R = 18$; $\mu_R = 14$; $\sigma_R = 2.498$; $\dfrac{R - \mu_R}{\sigma_R} = 1.6$; do not reject H_0; the number of defects are randomly distributed along the length of the wire.

7. $m = 31$; $n = 19$; $R = 36$; $\mu_R = 24.56$; $\sigma_R = 3.29$; $\dfrac{R - \mu_R}{\sigma_R} = 3.48$; reject H_0; the occurrences of drought years and normal years cannot be regarded as random.

Section 13-4

1. $r_{rank} = 0.782$; $\sqrt{n - 1}\ (r_{rank}) = 2.47$; reject H_0; the rankings are related.
3. $r_{rank} = 0.067$; $\sqrt{n - 1}\ (r_{rank}) = 0.2$; do not reject H_0; the rankings are not related.
5. $r_{rank} = -0.077$; $\sqrt{n - 1}\ (r_{rank}) = -0.28$; do not reject H_0; the rankings are not related.

Chapter 13 Review Exercises

5. $m = 10$; $n = 12$; $R_1 = 113$; $u = 62$; $\mu_U = 60$; $\sigma_U = 15.166$; $\dfrac{u - \mu_U}{\sigma_U} = 0.13$; do not reject H_0; the distributions are not dissimilar.

7. $m = 13$; $n = 17$; $R = 23$; $\mu_R = 15.733$; $\sigma_R = 2.641$; $\dfrac{R - \mu_R}{\sigma_R} = 2.75$; reject H_0; the choices are not arranged at random.

Index